Multivariate Statistical Inference and Applications

WILEY SERIES IN PROBABILITY AND STATISTICS
TEXTS AND REFERENCES SECTION

Established by WALTER A. SHEWHART and SAMUEL S. WILKS

A complete list of the titles in this series appears at the end of this volume.

Multivariate Statistical Inference and Applications

ALVIN C. RENCHER

Department of Statistics
Brigham Young University

A Wiley-Interscience Publication

JOHN WILEY & SONS, INC.

New York • Chichester • Weinheim • Brisbane • Singapore • Toronto

This text is printed on acid-free paper. ♾

Library of Congress Cataloging-in-Publication Data:

Rencher, Alvin C., 1934–
Multivariate statistical inference and applications / Alvin C. Rencher.
p. cm. — (Wiley series in probability and statistics. Texts and references section)
"A Wiley-Interscience publication."
Includes bibliographical references and index.
ISBN 0-471-57151-2 (cloth : alk. paper)
1. Multivariate analysis. I. Title. II. Series.
QA278.R453 1998
519.5′35—dc21 97-5255
CIP

Printed in the United States of America

10 9 8 7 6 5 4 3 2 1

Contents

Preface

Scientists and clinicians who measure more than one variable on each product or subject have a vital need for multivariate methodology that permits analyzing these variables simultaneously. A univariate analysis of each variable ignoring the other variables provides only partial information about an intertwined complex system. A simultaneous analysis of all variables reveals the key features of the entire system and greatly enhances statistical inference. Any univariate analyses should be interpreted in light of the multivariate analysis.

Multivariate analysis is at once challenging and fascinating. We study the effect of several variables on a test statistic or on a descriptive statistic. The effect of each variable in the presence of the other variables depends on two things: (1) the correlation of the variable with the other variables and (2) the influence the other variables have on the variable in question as it affects the test statistic or the descriptive statistic. I have long been intrigued by this interplay of variables, especially the second aspect, the impact of each variable in the presence of the other variables. I have provided an explicit formulation of this effect for several techniques in this book.

In the preparation of this book, my objectives were the following:

1. To provide a theoretical foundation for multivariate analysis at an accessible level. I have searched for or devised proofs that are easy to follow for students with a minimum of theoretical background in statistics. For some readers, these proofs are an essential part of understanding the material and furnish insights obtainable in no other way. Others may wish to skip the proofs on a first reading.
2. To cover a wide range of areas of application. The book is driven by applications in its choice of topics and its choice of literature to cite. I have carefully selected new developments that can be applied, not those of theoretical interest only.
3. To provide a comprehensive survey of the literature in multivariate analysis. The bibliography in this book has over 900 entries and is fairly comprehensive (I have listed only those references cited in the text). In addition to providing original sources for fundamental techniques, I have combed the multivariate literature for the past 20 years and have included most developments that appear to have potential for application. In the classroom setting, these recent

citations will provide sources for miniprojects. As a reference book, the statistical researcher will find definitive original sources, and the practitioner will gain access to methodological details and algorithms for applying the latest useful techniques.

4. Above all else, to provide clarity of exposition. I hope that students, instructors, researchers, and practitioners alike will find this book more comfortable than most.

The mathematical prerequisites for this book are calculus and matrix algebra. Statistical prerequisites include some exposure to statistical theory, including distributions of random variables, expected values, moment-generating functions, and an introduction to estimation and testing hypotheses. One or two statistical methods courses would also be helpful, with coverage of basic procedures such as t-tests, regression analysis, and analysis of variance.

The book provides a substantial number of theoretical problems and a smaller number of applied problems using real data sets. The problems, along with the answers in Appendix C, extend the book in two significant ways: (1) the theoretical problems and answers fill in nearly all gaps in derivations and proofs and also extend the coverage of material in the text and (2) the numerical problems and answers extend the examples illustrating the theory. As an instructor, I find that supplying answers saves a great deal of class time and enables me to cover more material and cover it better. The students, of course, find answers to be very helpful. Answers are even more useful to a reader who is learning the material outside the formal classroom setting.

Many multivariate techniques are extensions of analogous univariate methods. In many cases, I have provided reviews of the univariate procedures. I hope that these reviews will help the reader feel more comfortable with the new material. Some of these univariate reviews may even provide insights that the student missed in previous courses. For examples of such reviews, see Section 4.5.1 on contrasts, Sections 4.8.2 and 4.8.4 on unbalanced models in analysis of variance, Sections 4.10.1–4.10.4 on analysis of covariance, Sections 7.2.1–7.2.6 on multiple regression with fixed x's, and Sections 7.3.1–7.3.4 on multiple regression with random x's.

I have made an effort to use standard notation and to be consistent in notational usage throughout the book. I have largely refrained from the use of abbreviations and mnemonic devices. I find these to be annoying when I use a book as a reference.

Equations, tables, and figures are numbered sequentially throughout each chapter. Theorems are numbered sequentially within a section, for example, Theorems 2.3A, 2.3B, and so on. Examples are not numbered sequentially; an example is identified by the same number as the section in which it appears and is placed at the end of the section.

Standard format has been used in citing references in the text. For a journal article, the year alone suffices, for example, Box and Cox (1964). For books, I have in most cases included a page number or section, as in Hogg and Craig (1995, Section 7.4).

My selection of topics for this volume reflects years of teaching and consulting with researchers in many disciplines. Chapters 1 and 2 provide an introduction to sampling from multivariate populations. Chapters 3, 4, 7, and 8 present multivariate extensions of standard univariate procedures such as t-tests, analysis of variance, multiple regression, and multiple correlation. Chapters 5, 6, 9, and 10 cover additional multivariate techniques such as methods for separating groups of multivariate observations, procedures for allocating observations to groups, and techniques for reducing dimension. Appendix A provides a review of the requisite matrix theory and also serves as reference for matrix manipulations in the text. Appendix B supplies tables for many multivariate distributions and tests. These enable the reader to carry out an exact test in many cases for which the software packages provide only approximate tests. Appendix C gives hints or answers for most theoretical problems and provides complete answers for all numerical problems.

The diskette included with the text contains (1) all the data sets, including those referred to but not shown in the text, (2) SAS command files for all examples in the text, and (3) SAS command files for all numerical problems. The contents and usage of the diskette are described in Appendix D. I believe these data sets and command files make the book more valuable as a reference as well as a text.

The reader may have a question about how this book relates to my previous book, *Methods of Multivariate Analysis* (Wiley, 1995). The 1995 volume is directed toward methodological applications and is written at a lower mathematical level. The emphasis in the present volume is shifted more toward theory, at a level somewhat below that of Anderson (1984). The two volumes could be used in a two-semester course, but they are primarily designed to be read independently.

With the two volumes, I hope to (1) provide broader coverage of useful techniques than is possible in a single volume and (2) offer the instructor a choice between an intuitive approach and a more theoretical approach. The present volume provides derivations of many results in the 1995 volume and also covers many topics not discussed in the 1995 volume (for example, power and sample size, unbalanced data analysis in MANOVA, multivariate analysis of covariance, and multivariate quality control). Likewise, there are topics in the 1995 volume that are not covered in the second volume, and for some of these I have referred the reader to Rencher (1995). Some examples of coverage unique to the 1995 volume are the following: graphical approaches to plotting multivariate observation vectors, methods for assessing multivariate normality, repeated measures, growth curves, and the relationship of canonical correlation analysis to discriminant analysis and MANOVA.

To summarize, this book would serve well as a text in an applied course with a theoretical orientation or in a theory course with an applied orientation. It also functions as an unmatched reference book with its extensive up-to-date coverage of the literature.

I will be very grateful to readers who take the time to notify me of errors or of other suggestions they may have for improvements.

I am indebted to many individuals in the preparation of this manuscript. I thank the following for reading the manuscript and making many valuable suggestions: Mike Speed, Bruce Brown, Don Norton, Kim Greenburg, James Gartside, Ellen Burns, Susan Wolfe, David Smith, and Douglas Rosenquist. Gale Bryce and Del Scott read Chapter 4 and made several beneficial suggestions. I am grateful to the following students at BYU who helped with computations and typing: Greg Jones, Russell Earl, Christopher Bodily, William Christensen, Rachel Jones, Karla Wasden, Scott G. Curtis, and Julie Thomas. My wife LaRue provided vital support by patiently assuming many household, yard, and family duties that would have been mine during the several years of writing and revising the manuscript.

ALVIN C. RENCHER

Acknowledgments

I thank the authors and publishers of tables and data for permission to use the following materials:

- Figure 6.1, Anderson (1982), Reprinted by permission of Elsevier North-Holland Publishing Co.
- Table 1.2, Reaven and Miller (1979), Reprinted by permission of *Diabetologia* and Springer-Verlag.
- Table 1.3, Timm, *Multivariate Analysis: With applications in Education and Psychology* (1975), Reprinted by permission of Brooks/Cole and the author.
- Table 1.4, O'Sullivan and Mahan (1966), Reprinted by permission of Springer-Verlag.
- Table 3.1, Kramer and Jensen (1969a), Reprinted by permission of *The Journal of Quality Technology.*
- Table 3.3, Simes (1986), Reprinted by permission of the Biometrika Trustees.
- Tables 3.4 and 3.5, Everitt (1979), Reprinted by permission of the *Journal of the American Statistical Association.*
- Table 3.6, Cox and Martin (1937), Reprinted by permission of the Iowa Agricultural Experiment Station, Iowa State University, Ames, Iowa.
- Table 3.7, Stevens (1980), Copyright © 1980 by the American Psychological Association. Reprinted by permission.
- Table 3.8, Lubischew (1962), Reprinted by permission of *Biometrics.*
- Table 4.5, Allison et al. (1962), Reproduced from *Journal of Experimental Medicine* by copyright permission of The Rockefeller University Press.
- Table 4.6, Woodward (©1931, *The Journal of the American Dental Association,* Volume 18, pp. 419–442), Reprinted by permission of ADA Publishing Co., Inc.
- Table 4.10, Ratkowsky and Martin (1974), Reprinted by permission of *Australian Journal of Agricultural Research.*
- Table 4.11, Smith, Gnanadesikan, and Hughes (1962), reprinted by permission of *Biometrics.*

- Table 5.2, Rencher and Larson (1980), Reprinted by permission of the American Statistical Association.
- Table 5.3, Pollock, Jackson and Pate (1980), Reprinted by permission of *Research Quarterly*.
- Tables 6.2, 6.3, 6.4, 6.5, 6.7, 6.8, Espahbodi (1991), Reprinted by permission of *Journal of Banking and Finance*.
- Table 6.6, Rencher (1992a), Reprinted by permission of *Communications in Statistics—Part B, Simulation and Computation*.
- Table 7.1, Pickard and Berk (1990), Reprinted by permission of *The American Statistician*.
- Table 9.1, Longley (1967), Reprinted by permission of the *Journal of the American Statistical Association*.
- Table B1, Kramer and Jensen (1969), Reprinted by permission of the *Journal of Quality Technology* and the American Society for Quality Control.
- Table B2, Bailey (1977) and Table B3, Tiku (1967), Reprinted by permission of the *Journal of the American Statistical Association*.
- Table B4, Wall (1967), Reprinted by permission of the author, Albuquerque, NM.
- Table B6, Schuurman et al. (1975), Reprinted by permission of the *Journal of Statistical Computation and Simulation*.
- Table B7, Davis (1970a, 1970b, 1980a), Reprinted by permission of the Biometrika Trustees, the *Annals of the Institute of Statistical Mathematics*, and *Communications in Statistics—Simulation and Computation*.
- Table B8, Lee et al. (1977), Reprinted by permission of Elsevier North-Holland Publishing Company.
- A 3×3 correlation matrix in Example 9.7.1a, Pearce (1965), Reprinted by permission of *Biometrie-Praximetrie*.
- Two 3×3 correlation matrices in Example 9.7.1b, Constable and Mardia (1992), Reprinted by permission of Carfax Publishing Co., P.O. Box 25, Abington, Oxfordshire OX 14 34E, United Kingdom.
- Table Ex3-10a.dat on the diskette, Frets (1921), Reprinted by permission of *Genetica*.
- Table Ex3-5-3a.dat on the diskette, Beall (1945), Reprinted by permission of *Psychometrika*.
- Table Ex3-5-3b.dat on the diskette, Andrews and Herzberg, *Data*, (1985, pp. 223–8), Reprinted with permission of Springer-Verlag.
- Table 8-6-3.dat on the diskette, Timm (1975), Reprinted by permission of the *Journal of Quality Technology* and the author.

Multivariate Statistical
Inference and Applications

CHAPTER 1

Some Properties of Random Vectors and Matrices

1.1 INTRODUCTION

The development of multivariate analysis had its germination in the late 1920s with the work of Wishart, Hotelling, Wilks, Fisher, and Mahalanobis. Rao (1983) gives a brief but excellent historical review.

For many years, applications were hindered by the lack of computing power. However, with the availability of modern computers and software packages, almost any desired analysis can be readily carried out.

In multivariate analysis, each observation consists of a vector,

$$\mathbf{y} = \begin{pmatrix} y_1 \\ y_2 \\ \vdots \\ y_p \end{pmatrix}.$$

The p variables $y_1, y_2, \ldots, y_p$ represent measurements on a single subject or object. Since the variables arise from the same sampling unit, they are typically intercorrelated. Multivariate procedures make allowance for these correlations among the variables.

In some cases the variables are measured on the same scale, such as several exam scores expressed as percentages. Ordinarily, however, the scales differ, as, for example, in height, weight, percent body fat, and resting heart rate. In most cases, multivariate techniques do not require that the variables be commensurate (similar in scale of measurement), and many procedures are not affected by a change of scale.

Most of the multivariate techniques we will consider can be categorized as either *descriptive* or *inferential*. Descriptive procedures may characterize the correlation structure within the observation vectors or show how the variables contribute to the grouping patterns among the observations. Other descriptive procedures attempt to disentangle the overlapping information in the correlated variables by constructing a small number of uncorrelated variables (linear combinations of the original variables) that reveal the essential dimensionality of the system. Interpretation of these new dimensions is of interest.

Multivariate inferential procedures include hypothesis tests that allow for any correlation structure among the variables. The tests also provide control of experimentwise error rates, no matter how many variables are tested. Many multivariate inferential techniques are extensions of univariate procedures such as t-tests or F-tests. If a multivariate test rejects the hypothesis, we are interested in determining the relative contribution of the variables to rejection.

For most procedures, we will consider only *continuous* random variables. In many cases, however, multivariate techniques yield good results when applied to discrete ordinal data. Categorical or discrete data can be analyzed by methods such as *log linear models* [see, for example, Imrey, Koch, and Stokes (1981, 1982)], *generalized linear models* [see, for example, McCullagh and Nelder (1989)], or *correspondence analysis* [see, for example, Jambu (1991) or Rijckevorsel and Leeuw (1988)].

We review some properties of univariate and bivariate random variables in Section 1.2 and then extend to vectors of higher dimension in the remainder of Chapter 1.

1.2 UNIVARIATE AND BIVARIATE RANDOM VARIABLES

1.2.1 Univariate Random Variables

For reasons that will be made clear in Section 1.3, we do not distinguish notationally between a random variable y and an observed value of y. If the *density* $f(y)$ is known, the *population mean* or *expected value* of y is defined as

$$E(y) = \mu = \int_{-\infty}^{\infty} y f(y)\, dy. \tag{1.1}$$

The expectation of a function of y can be found in a similar manner:

$$E[g(y)] = \int_{-\infty}^{\infty} g(y) f(y)\, dy. \tag{1.2}$$

From (1.2), it follows that for a constant a and a function $h(y)$

$$E(ay) = aE(y), \tag{1.3}$$

$$E[g(y) + h(y)] = E[g(y)] + E[h(y)]. \tag{1.4}$$

For a sample of n observations, $y_1, y_2, \ldots, y_n$, the *sample mean* is given by

$$\overline{y} = \frac{1}{n} \sum_{i=1}^{n} y_i. \tag{1.5}$$

If $y_1, y_2, \ldots, y_n$ is a random sample from a population with mean μ and variance σ^2, then the sample mean $\overline{y}$ is an unbiased estimator for the population mean μ; that is,

$$E(\overline{y}) = \mu. \tag{1.6}$$

The sample mean has a property analogous to (1.3); that is, if $z_i = ay_i$, for $i = 1, 2, \ldots, n$, then $\bar{z} = a\bar{y}$:

$$\bar{z} = \frac{1}{n}\sum_{i=1}^{n} z_i = \frac{1}{n}\sum_i ay_i = \frac{1}{n}a\sum_i y_i$$
$$= a\left(\frac{1}{n}\sum_i y_i\right) = a\bar{y}. \tag{1.7}$$

The *population variance* of y is defined as

$$\text{var}(y) = \sigma^2 = E(y - \mu)^2. \tag{1.8}$$

By (1.3) and (1.4), the variance can be expressed as

$$\sigma^2 = E(y^2) - \mu^2. \tag{1.9}$$

For a sample of n observations, the *sample variance* is defined as

$$s^2 = \frac{\sum_{i=1}^{n}(y_i - \bar{y})^2}{n-1}, \tag{1.10}$$

which can be written in the form

$$s^2 = \frac{\left(\sum_{i=1}^{n} y_i^2\right) - n\bar{y}^2}{n-1}. \tag{1.11}$$

If $y_1, y_2, \ldots, y_n$ is a random sample from a population with mean μ and variance σ^2, then the sample variance s^2 is an unbiased estimator for the population variance σ^2; that is,

$$E(s^2) = \sigma^2. \tag{1.12}$$

For a constant a, the population variance of ay is given by

$$\text{var}(ay) = a^2\sigma^2. \tag{1.13}$$

Similarly, the sample variance of $z_i = ay_i$, $i = 1, 2, \ldots, n$, is given by $s_z^2 = a^2 s^2$:

$$s_z^2 = \frac{\sum_{i=1}^{n}(z_i - \bar{z})^2}{n-1} = \frac{\sum_i (ay_i - a\bar{y})^2}{n-1}$$
$$= \frac{\sum_i \left[a(y_i - \bar{y})\right]^2}{n-1} = \frac{a^2\sum_i (y_i - \bar{y})^2}{n-1}$$
$$= a^2 s^2. \tag{1.14}$$

Table 1.1. Height and Weight for a Sample of 20 College-Age Males

Person Number	Height x	Weight y	Person Number	Height x	Weight y
1	69	153	11	72	140
2	74	175	12	79	265
3	68	155	13	74	185
4	70	135	14	67	112
5	72	172	15	66	140
6	67	150	16	71	150
7	66	115	17	74	165
8	70	137	18	75	185
9	76	200	19	75	210
10	68	130	20	76	220

1.2.2 Bivariate Random Variables

A *bivariate random variable* (x, y) arises when two variables x and y are measured on each object or subject. Typically, x and y will be correlated. To illustrate, heights (inches) and weights (pounds) for a sample of 20 college-age males are given in Table 1.1 and plotted in Figure 1.1, where we see a tendency for y to increase when x increases. We now discuss measures of the strength of the relationship between x and y.

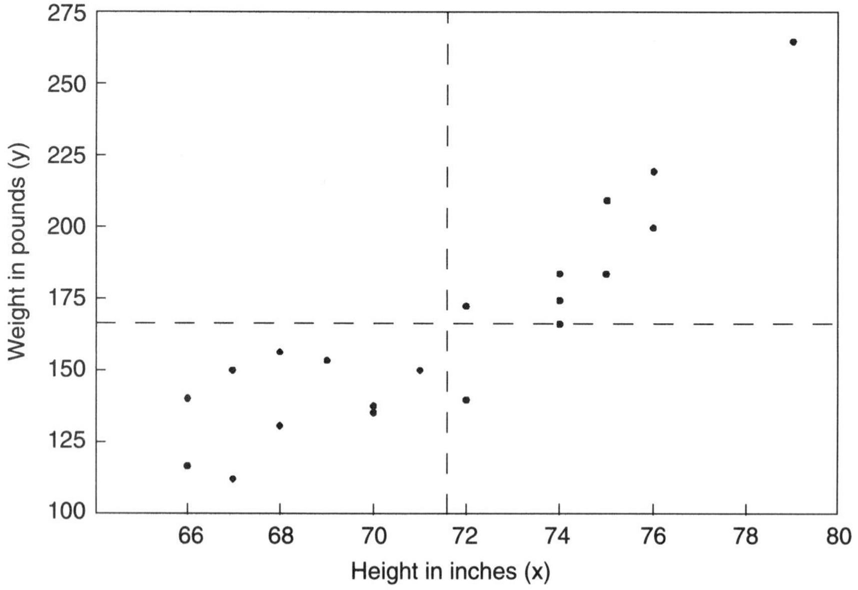

Figure 1.1. Bivariate scatter plot of data in Table 1.1.

The *population covariance* of x and y is defined as

$$\text{cov}(x, y) = \sigma_{xy} = E\left[(x - \mu_x)(y - \mu_y)\right], \tag{1.15}$$

where μ_x and μ_y are the means of x and y. By (1.3) and (1.4), this can be written as

$$\sigma_{xy} = E(xy) - \mu_x\mu_y. \tag{1.16}$$

The two random variables x and y in a bivariate random variable (x, y) are said to be *independent* if their joint density factors into the product of their marginal densities: $f(x, y) = g(x)h(y)$. With this definition, we obtain the following properties for expectations involving bivariate random variables x and y:

1. $E(x + y) = E(x) + E(y)$. (1.17)
2. $E(xy) = E(x)E(y)$ if x and y are independent. (1.18)
3. $\sigma_{xy} = 0$ if x and y are independent. (1.19)

The first property is true in general, but as noted, the second and third properties hold only when x and y are independently distributed. The third property is obtainable from the second. (See Problem 1.9 at the end of this chapter.)

The converse of property 3 (and of property 2) is not true; that is, $\sigma_{xy} = 0$ does not imply independence. It is easy to construct examples of bivariate x and y that are dependent (in a nonlinear pattern) but that have zero covariance, as illustrated in Figure 1.2. However, if x and y have a bivariate normal distribution, then $\sigma_{xy} = 0$

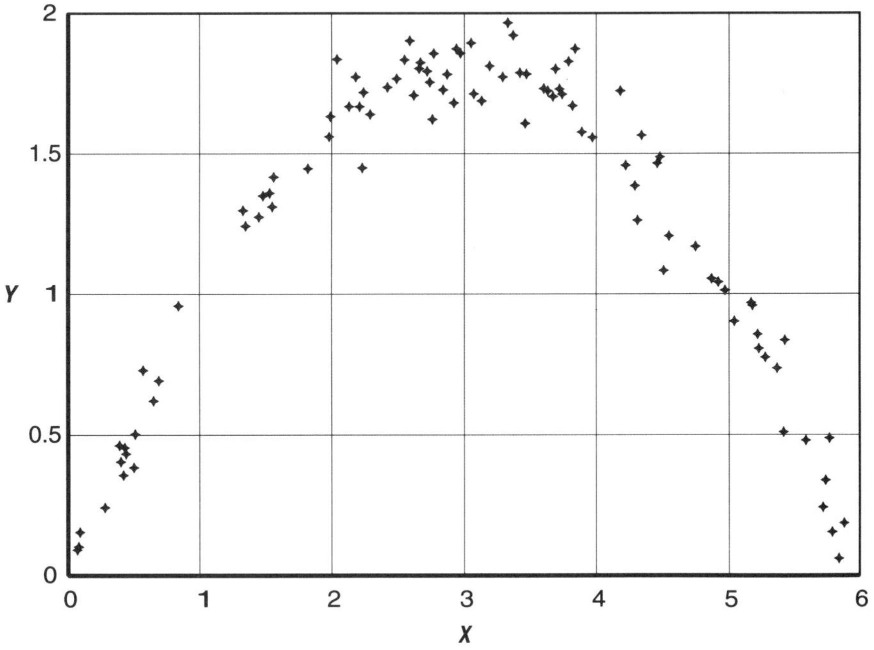

Figure 1.2. A sample from a population in which x and y have zero covariance and are dependent.

implies independence of x and y (Theorem 2.2D), because in the bivariate normal case, $E(y|x)$ and $E(x|y)$ are linear (Theorem 2.2E).

For a sample $(x_1, y_1), (x_2, y_2), \ldots, (x_n, y_n)$, the *sample covariance* is defined as

$$s_{xy} = \frac{\sum_{i=1}^{n}(x_i - \bar{x})(y_i - \bar{y})}{n-1}, \tag{1.20}$$

which can readily be shown to equal

$$s_{xy} = \frac{(\sum_{i=1}^{n} x_i y_i) - n\bar{x}\,\bar{y}}{n-1}. \tag{1.21}$$

The sample covariance s_{xy} is an unbiased estimator for σ_{xy}; that is,

$$E(s_{xy}) = \sigma_{xy}. \tag{1.22}$$

The covariance σ_{xy} depends on the scale of measurement of both x and y. To standardize σ_{xy}, we divide it by the standard deviations of x and y to obtain the *population correlation*:

$$\rho_{xy} = \operatorname{corr}(x, y) = \frac{\sigma_{xy}}{\sigma_x \sigma_y} = \frac{E(x - \mu_x)(y - \mu_y)}{\sqrt{E(x - \mu_x)^2}\sqrt{E(y - \mu_y)^2}}. \tag{1.23}$$

The corresponding *sample correlation* is given by

$$r_{xy} = \frac{s_{xy}}{s_x s_y} = \frac{\sum_{i=1}^{n}(x_i - \bar{x})(y_i - \bar{y})}{\sqrt{\sum_{i=1}^{n}(x_i - \bar{x})^2 \sum_{i=1}^{n}(y_i - \bar{y})^2}}. \tag{1.24}$$

Neither of the correlations defined in (1.23) and (1.24) depends on the units of measurement; both correlations range between -1 and 1.

The sample correlation r_{xy} is a biased estimator of the population correlation ρ_{xy}, except when $\rho_{xy} = 0$. Olkin and Pratt (1958) proposed an adjustment to r_{xy} to make it more nearly unbiased,

$$r_{xy}^* = r_{xy}\left[1 + \frac{1 - r_{xy}^2}{2(n-3)}\right], \tag{1.25}$$

and suggested that this bias correction is accurate to within .01 for $n \geq 8$ and within .001 for $n \geq 18$. Anderson (1984, p. 119) gave a series representation of an unbiased estimator of ρ; the first two terms of the series are similar to (1.25).

1.3 MEAN VECTORS AND COVARIANCE MATRICES FOR RANDOM VECTORS

In many texts, an uppercase letter is used to represent or name a random variable, and the corresponding lowercase letter denotes a realization of the random variable. This notation is useful in univariate analysis, but it may be confusing in a multivariate context in which we need to distinguish between vectors and matrices. Consequently, uppercase boldface letters represent matrices of random variables or constants, lowercase boldface letters denote vectors of random variables or constants, and nonbolded (usually lowercase) letters are used for univariate random variables or constants.

Suppose we have a random sample of n *observation vectors*, $\mathbf{y}_1, \mathbf{y}_2, \ldots, \mathbf{y}_n$. Two vectors, say $\mathbf{y}_1$ and $\mathbf{y}_2$, are *independent* if each variable y_{1j} in $\mathbf{y}_1$ is independent of every variable y_{2j} in $\mathbf{y}_2$. Since $\mathbf{y}_1, \mathbf{y}_2, \ldots, \mathbf{y}_n$ constitute a random sample, these n vectors are independent.

For reviews of graphical presentations of multivariate data vectors $\mathbf{y}_1, \mathbf{y}_2, \ldots, \mathbf{y}_n$, see Rencher (1995, Section 3.4); Buja and Tukey (1991); Mihalisin, Schwegler, and Timlin (1992); and Everitt (1994).

The n observation vectors are transposed and listed as rows in the *data matrix* $\mathbf{Y}$:

$$\mathbf{Y} = \begin{pmatrix} \mathbf{y}_1' \\ \mathbf{y}_2' \\ \vdots \\ \mathbf{y}_i' \\ \vdots \\ \mathbf{y}_n' \end{pmatrix} = (\text{units}) \begin{matrix} \\ 1 \\ 2 \\ \\ i \\ \\ n \end{matrix} \overset{\text{(variables)}}{\begin{pmatrix} \overset{1}{y_{11}} & \overset{2}{y_{12}} & \cdots & \overset{j}{y_{1j}} & \cdots & \overset{p}{y_{1p}} \\ y_{21} & y_{22} & \cdots & y_{2j} & \cdots & y_{2p} \\ \vdots & \vdots & & \vdots & & \vdots \\ y_{i1} & y_{i2} & \cdots & y_{ij} & \cdots & y_{ip} \\ \vdots & \vdots & & \vdots & & \vdots \\ y_{n1} & y_{n2} & \cdots & y_{nj} & \cdots & y_{np} \end{pmatrix}}. \tag{1.26}$$

In the matrix $\mathbf{Y}$, the first subscript represents units (subjects or objects), and the second subscript corresponds to variables. Typically, $n > p$, so this array is more convenient for tabular listing or computer entry than defining $\mathbf{Y}$ to be the $p \times n$ matrix $(\mathbf{y}_1, \mathbf{y}_2, \ldots, \mathbf{y}_n)$ with the n observation vectors as columns.

If we wish to discuss both the columns and rows of $\mathbf{Y}$, the columns will be denoted as follows:

$$\mathbf{Y} = (\mathbf{y}_{(1)}, \mathbf{y}_{(2)}, \ldots, \mathbf{y}_{(p)}). \tag{1.27}$$

Thus, for example, $\mathbf{y}_2$ is the p-vector of variables measured on the second sampling unit, while $\mathbf{y}_{(2)}$ is the n-vector of observations on the second variable.

The *sample mean vector* is defined as

$$\bar{\mathbf{y}} = \frac{1}{n}\sum_{i=1}^{n} \mathbf{y}_i = \begin{pmatrix} \bar{y}_1 \\ \bar{y}_2 \\ \vdots \\ \bar{y}_p \end{pmatrix}. \tag{1.28}$$

Thus the average of the n vectors yields the average on each variable.

Using (A.1.7) and (A.2.15) in Appendix A, we can calculate $\bar{\mathbf{y}}$ directly from $\mathbf{Y}$:

$$\bar{\mathbf{y}} = \frac{1}{n}\mathbf{Y}'\mathbf{j}, \tag{1.29}$$

where $\mathbf{j}$ is an $n \times 1$ vector of 1s.

The *population mean* or *expected value* of the random vector $\mathbf{y}$ is defined as the vector of expected values of the p variables,

$$E(\mathbf{y}) = E\begin{pmatrix} y_1 \\ y_2 \\ \vdots \\ y_p \end{pmatrix} = \begin{pmatrix} E(y_1) \\ E(y_2) \\ \vdots \\ E(y_p) \end{pmatrix} = \begin{pmatrix} \mu_1 \\ \mu_2 \\ \vdots \\ \mu_p \end{pmatrix} = \boldsymbol{\mu}, \tag{1.30}$$

where $E(y_j) = \mu_j$ is obtained as $\int y_j f_j(y_j)\, dy_j$ using $f_j(y_j)$, the marginal density of y_j. Since $E(\overline{y}_j) = \mu_j$, it follows that

$$E(\overline{\mathbf{y}}) = \begin{pmatrix} \mu_1 \\ \mu_2 \\ \vdots \\ \mu_p \end{pmatrix} = \boldsymbol{\mu}, \tag{1.31}$$

from which $\overline{\mathbf{y}}$ is an unbiased estimator of $\boldsymbol{\mu}$.

The symmetric matrix of sample variances and covariances is called the *sample covariance matrix*:

$$\mathbf{S} = (s_{jk}) = \begin{pmatrix} s_{11} & s_{12} & \cdots & s_{1p} \\ s_{21} & s_{22} & \cdots & s_{2p} \\ \vdots & \vdots & & \vdots \\ s_{p1} & s_{p2} & \cdots & s_{pp} \end{pmatrix}. \tag{1.32}$$

The jth diagonal element s_{jj} (also denoted by s_j^2, the square of the standard deviation s_j) is the sample variance of the jth variable:

$$s_{jj} = s_j^2 = \frac{1}{n-1}\sum_{i=1}^{n}(y_{ij} - \overline{y}_j)^2 \tag{1.33}$$

$$= \frac{1}{n-1}\left(\sum_{i=1}^{n} y_{ij}^2 - n\overline{y}_j^2\right). \tag{1.34}$$

The typical off-diagonal element s_{jk} in (1.32) is the sample covariance of the jth and kth variables:

$$s_{jk} = \frac{1}{n-1}\sum_{i=1}^{n}(y_{ij} - \overline{y}_j)(y_{ik} - \overline{y}_k) \tag{1.35}$$

$$= \frac{1}{n-1}\left(\sum_{i=1}^{n} y_{ij}y_{ik} - n\overline{y}_j\overline{y}_k\right). \tag{1.36}$$

A measure of overall variability can be based on $|\mathbf{S}|$ or $\text{tr}(\mathbf{S})$, known as the *generalized sample variance* and *total sample variance*, respectively. For minimization properties of $|\mathbf{S}|$ and $\mathrm{tr}(\mathbf{S})$ see Problem 1.20. For properties of $|\mathbf{S}|$ and $\mathrm{tr}(\mathbf{S})$ as measures of overall variability, see Rencher (1995, Section 3.10).

The sample covariance matrix $\mathbf{S}$ can be computed three ways. The first approach uses (1.34) and (1.36). The second method uses the observation vectors:

$$\mathbf{S} = \frac{1}{n-1}\sum_{i=1}^{n}(\mathbf{y}_i - \bar{\mathbf{y}})(\mathbf{y}_i - \bar{\mathbf{y}})' \tag{1.37}$$

$$= \frac{1}{n-1}\left(\sum_{i=1}^{n}\mathbf{y}_i\mathbf{y}_i' - n\bar{\mathbf{y}}\bar{\mathbf{y}}'\right). \tag{1.38}$$

The third approach for obtaining $\mathbf{S}$ uses the data matrix $\mathbf{Y}$. From (1.29), we have $\bar{\mathbf{y}} = \mathbf{Y}'\mathbf{j}/n$ and $n\bar{\mathbf{y}} = \mathbf{Y}'\mathbf{j}$. By (A.3.8), we can write $\sum_i \mathbf{y}_i\mathbf{y}_i'$ in the form

$$\sum_{i=1}^{n}\mathbf{y}_i\mathbf{y}_i' = \mathbf{Y}'\mathbf{Y}. \tag{1.39}$$

Then (1.38) becomes

$$\mathbf{S} = \frac{1}{n-1}\left(\mathbf{Y}'\mathbf{Y} - \frac{1}{n}\mathbf{Y}'\mathbf{j}\mathbf{j}'\mathbf{Y}\right)$$

$$= \frac{1}{n-1}\left[\mathbf{Y}'\mathbf{Y} - \mathbf{Y}'\left(\frac{1}{n}\mathbf{J}\right)\mathbf{Y}\right] \qquad \text{[by (A.1.8)]}$$

$$= \frac{1}{n-1}\mathbf{Y}'\left(\mathbf{I} - \frac{1}{n}\mathbf{J}\right)\mathbf{Y}, \tag{1.40}$$

where $\mathbf{I}$ and $\mathbf{J}$ are $n \times n$ ($\mathbf{J}$ is a square matrix of 1s). Expression (1.40) is a convenient computational device, provided n is not too large.

We can use (1.40) to show that $\mathbf{S}$ is at least positive semidefinite (see Section A.6). It can easily be shown by direct multiplication that

$$\left(\mathbf{I} - \frac{1}{n}\mathbf{J}\right)'\left(\mathbf{I} - \frac{1}{n}\mathbf{J}\right) = \mathbf{I} - \frac{1}{n}\mathbf{J}, \tag{1.41}$$

so that (1.40) can be written

$$\mathbf{S} = \frac{1}{n-1}\mathbf{Y}'\left(\mathbf{I} - \frac{1}{n}\mathbf{J}\right)\mathbf{Y} = \frac{1}{n-1}\mathbf{Y}'\left(\mathbf{I} - \frac{1}{n}\mathbf{J}\right)'\left(\mathbf{I} - \frac{1}{n}\mathbf{J}\right)\mathbf{Y}$$

$$= \frac{1}{n-1}\mathbf{Y}_c'\mathbf{Y}_c, \tag{1.42}$$

where $\mathbf{Y}_c = \left(\mathbf{I} - \frac{1}{n}\mathbf{J}\right)\mathbf{Y}$ is the *centered* form of the data matrix $\mathbf{Y}$. Since $\mathbf{S}$ is proportional to $\mathbf{Y}_c'\mathbf{Y}_c$, it is at least positive semidefinite by (A.6.1) and a remark following (A.6.1). If the variables are continuous and not linearly related and if $n - 1 > p$, then the probability is 1 that $\mathbf{S}$ is positive definite (Siotani, Hayakawa, and Fujikoshi 1985, p. 60).

We now show that the *centering matrix* $\mathbf{I} - \frac{1}{n}\mathbf{J}$ centers $\mathbf{Y}$:

$$\begin{aligned}\mathbf{Y}_c &= \left(\mathbf{I} - \frac{1}{n}\mathbf{J}\right)\mathbf{Y} = \mathbf{Y} - \frac{1}{n}\mathbf{J}\mathbf{Y} \\ &= \mathbf{Y} - \frac{1}{n}\mathbf{j}\mathbf{j}'\mathbf{Y} = \mathbf{Y} - \mathbf{j}\overline{\mathbf{y}}' \qquad \text{[by (1.29)]} \\ &= \begin{pmatrix} y_{11} & y_{12} & \cdots & y_{1p} \\ y_{21} & y_{22} & \cdots & y_{2p} \\ \vdots & \vdots & & \vdots \\ y_{n1} & y_{n2} & \cdots & y_{np} \end{pmatrix} - \begin{pmatrix} 1 \\ 1 \\ \vdots \\ 1 \end{pmatrix} (\overline{y}_1 \quad \overline{y}_2 \quad \cdots \quad \overline{y}_p) \\ &= \begin{pmatrix} y_{11} & y_{12} & \cdots & y_{1p} \\ y_{21} & y_{22} & \cdots & y_{2p} \\ \vdots & \vdots & & \vdots \\ y_{n1} & y_{n2} & \cdots & y_{np} \end{pmatrix} - \begin{pmatrix} \overline{y}_1 & \overline{y}_2 & \cdots & \overline{y}_p \\ \overline{y}_1 & \overline{y}_2 & \cdots & \overline{y}_p \\ \vdots & \vdots & & \vdots \\ \overline{y}_1 & \overline{y}_2 & \cdots & \overline{y}_p \end{pmatrix} \\ &= \begin{pmatrix} y_{11} - \overline{y}_1 & y_{12} - \overline{y}_2 & \cdots & y_{1p} - \overline{y}_p \\ y_{21} - \overline{y}_1 & y_{22} - \overline{y}_2 & \cdots & y_{2p} - \overline{y}_p \\ \vdots & \vdots & & \vdots \\ y_{n1} - \overline{y}_1 & y_{n2} - \overline{y}_2 & \cdots & y_{np} - \overline{y}_p \end{pmatrix}.\end{aligned} \tag{1.43}$$

Note that $\mathbf{J}$ is $n \times n$, $\mathbf{j}$ is $n \times 1$, and $\mathbf{j}\overline{\mathbf{y}}'$ is $n \times p$.

The *population covariance matrix* is defined as

$$\boldsymbol{\Sigma} = \text{cov}(\mathbf{y}) = E\left[(\mathbf{y} - \boldsymbol{\mu})(\mathbf{y} - \boldsymbol{\mu})'\right]. \tag{1.44}$$

Note the distinction in notation between $\boldsymbol{\Sigma}$ used as the covariance matrix and $\sum$ used as the summation symbol.

In (1.44), we have introduced the expected value of a random matrix. This expected value is defined as the matrix of expected values of the corresponding elements, which is analogous to the expected value of a vector in (1.30). The jkth element of $(\mathbf{y} - \boldsymbol{\mu})(\mathbf{y} - \boldsymbol{\mu})'$ is $(y_j - \mu_j)(y_k - \mu_k)$. Thus, by (1.15), the jkth element of $E[(\mathbf{y} - \boldsymbol{\mu})(\mathbf{y} - \boldsymbol{\mu})']$ is $E[(y_j - \mu_j)(y_k - \mu_k)] = \sigma_{jk}$. Hence

$$E\left[(\mathbf{y} - \boldsymbol{\mu})(\mathbf{y} - \boldsymbol{\mu})'\right] = \begin{pmatrix} \sigma_{11} & \sigma_{12} & \cdots & \sigma_{1p} \\ \sigma_{21} & \sigma_{22} & \cdots & \sigma_{2p} \\ \vdots & \vdots & & \vdots \\ \sigma_{p1} & \sigma_{p2} & \cdots & \sigma_{pp} \end{pmatrix} = \boldsymbol{\Sigma}, \tag{1.45}$$

a matrix that contains population variances $\sigma_{jj} = \sigma_j^2$ on the diagonal and population covariances off the diagonal. The covariance matrix $\boldsymbol{\Sigma}$ is symmetric because $\sigma_{jk} = \sigma_{kj}$, and it is positive definite if the y's are continuous random variables and if there are no linear relationships among the y's.

We can write (1.44) in a form analogous to (1.38):

$$\boldsymbol{\Sigma} = E(\mathbf{y}\mathbf{y}') - \boldsymbol{\mu}\boldsymbol{\mu}'. \tag{1.46}$$

By (1.12) and (1.22), $E(s_{jk}) = \sigma_{jk}$ for all j, k, and using (1.32), we have

$$E(\mathbf{S}) = \boldsymbol{\Sigma}. \tag{1.47}$$

1.4 CORRELATION MATRICES

By (1.24), the sample correlation between the jth and kth variables is given by

$$r_{jk} = \frac{s_{jk}}{\sqrt{s_{jj}s_{kk}}} = \frac{s_{jk}}{s_j s_k}. \tag{1.48}$$

The *sample correlation matrix* is defined as

$$\mathbf{R} = (r_{jk}) = \begin{pmatrix} 1 & r_{12} & \cdots & r_{1p} \\ r_{21} & 1 & \cdots & r_{2p} \\ \vdots & \vdots & & \vdots \\ r_{p1} & r_{p2} & \cdots & 1 \end{pmatrix}, \tag{1.49}$$

which is symmetric, since $r_{jk} = r_{kj}$.

To relate **R** to **S**, we define a diagonal matrix [see (A.1.4), (A.1.5), and a remark following (A.1.5)]:

$$\mathbf{D}_s = [\text{diag}(\mathbf{S})]^{1/2} = \text{diag}(s_1, s_2, \ldots, s_p). \tag{1.50}$$

Then, by (A.2.27) and (A.2.28),

$$\mathbf{R} = \mathbf{D}_s^{-1}\mathbf{S}\mathbf{D}_s^{-1}, \tag{1.51}$$

$$\mathbf{S} = \mathbf{D}_s\mathbf{R}\mathbf{D}_s. \tag{1.52}$$

If the data matrix $\mathbf{Y} = (y_{ij})$ is standardized to $\mathbf{Z} = (z_{ij})$, where $z_{ij} = (y_{ij} - \bar{y}_j)/s_j$, then the covariance matrix for the z's is equal to the correlation matrix for the y's:

$$\mathbf{S}_z = \frac{1}{n-1}\mathbf{Z}'\mathbf{Z} = \mathbf{R}. \tag{1.53}$$

Note that $\mathbf{S}_z = \mathbf{Z}'\mathbf{Z}/(n-1)$ is analogous to $\mathbf{S} = \mathbf{Y}_c'\mathbf{Y}_c/(n-1)$ in (1.42).

The *population correlation matrix* is defined as

$$\mathbf{P}_\rho = (\rho_{jk}) = \begin{pmatrix} 1 & \rho_{12} & \cdots & \rho_{1p} \\ \rho_{21} & 1 & \cdots & \rho_{2p} \\ \vdots & \vdots & & \vdots \\ \rho_{p1} & \rho_{p2} & \cdots & 1 \end{pmatrix}, \tag{1.54}$$

where $\rho_{jk} = \sigma_{jk}/\sigma_j\sigma_k$, as in (1.23). The subscript ρ in $\mathbf{P}_\rho$ is used as a reminder that $\mathbf{P}$ is the uppercase version of ρ. If we define $\mathbf{D}_\sigma = \text{diag}(\sigma_1, \sigma_2, \ldots, \sigma_p)$ to be a diagonal matrix of population standard deviations analogous to $\mathbf{D}_s$ in (1.50), then

$$\mathbf{P}_\rho = \mathbf{D}_\sigma^{-1}\boldsymbol{\Sigma}\mathbf{D}_\sigma^{-1}, \tag{1.55}$$

$$\boldsymbol{\Sigma} = \mathbf{D}_\sigma\mathbf{P}_\rho\mathbf{D}_\sigma. \tag{1.56}$$

While $\overline{\mathbf{y}}$ and $\mathbf{S}$ are unbiased estimators of $\boldsymbol{\mu}$ and $\boldsymbol{\Sigma}$, such is not the case with $\mathbf{R}$. The sample correlation matrix $\mathbf{R}$ is a biased estimator of the population correlation matrix $\mathbf{P}_\rho$, since r_{jk} is a biased estimator of ρ_{jk} [see (1.25) for a bias correction].

1.5 PARTITIONED MEAN VECTORS AND COVARIANCE MATRICES

In some cases, the variables measured on a sampling unit have a natural partitioning into two subsets. For convenience, we denote the two subvectors by $\mathbf{y}$ and $\mathbf{x}$, so that the n observation vectors in a sample have the form $\begin{pmatrix}\mathbf{y}_1\\ \mathbf{x}_1\end{pmatrix}, \begin{pmatrix}\mathbf{y}_2\\ \mathbf{x}_2\end{pmatrix}, \ldots, \begin{pmatrix}\mathbf{y}_n\\ \mathbf{x}_n\end{pmatrix}$. There are p measurements in each $\mathbf{y}_i$ and q measurements in each $\mathbf{x}_i$:

$$\begin{pmatrix}\mathbf{y}_i\\ \mathbf{x}_i\end{pmatrix} = \begin{pmatrix} y_{i1} \\ \vdots \\ y_{ip} \\ x_{i1} \\ \vdots \\ x_{iq} \end{pmatrix}, \qquad i = 1, 2, \ldots, n. \tag{1.57}$$

The sample mean vector and sample covariance matrix are given by

$$\begin{pmatrix}\overline{\mathbf{y}}\\ \overline{\mathbf{x}}\end{pmatrix} = \begin{pmatrix} \overline{y}_1 \\ \vdots \\ \overline{y}_p \\ \overline{x}_1 \\ \vdots \\ \overline{x}_q \end{pmatrix}, \tag{1.58}$$

$$\mathbf{S} = \begin{pmatrix} \mathbf{S}_{yy} & \mathbf{S}_{yx} \\ \mathbf{S}_{xy} & \mathbf{S}_{xx} \end{pmatrix}, \tag{1.59}$$

where $\overline{\mathbf{y}}$ is the $p \times 1$ mean vector of the y's; $\overline{\mathbf{x}}$ is the $q \times 1$ mean vector of the x's; $\mathbf{S}_{yy}$ is the $p \times p$ covariance matrix of the y's; $\mathbf{S}_{xx}$ is the $q \times q$ covariance matrix of the x's; and $\mathbf{S}_{yx}$ is the $p \times q$ matrix of covariances between the y's and the x's. Note that $\mathbf{S}_{xy}$ is $q \times p$ and that

$$\mathbf{S}_{xy} = \mathbf{S}'_{yx}. \tag{1.60}$$

We illustrate the structure of $\mathbf{S}_{yx}$ and $\mathbf{S}_{xy}$ with $p = 2$ and $q = 3$:

$$\mathbf{S}_{yx} = \begin{pmatrix} s_{y_1x_1} & s_{y_1x_2} & s_{y_1x_3} \\ s_{y_2x_1} & s_{y_2x_2} & s_{y_2x_3} \end{pmatrix},$$

$$\mathbf{S}_{xy} = \begin{pmatrix} s_{x_1y_1} & s_{x_1y_2} \\ s_{x_2y_1} & s_{x_2y_2} \\ s_{x_3y_1} & s_{x_3y_2} \end{pmatrix}.$$

The population mean vector and covariance matrix for a partitioned random vector are

$$E\begin{pmatrix} \mathbf{y} \\ \mathbf{x} \end{pmatrix} = \begin{pmatrix} E(\mathbf{y}) \\ E(\mathbf{x}) \end{pmatrix} = \begin{pmatrix} \boldsymbol{\mu}_y \\ \boldsymbol{\mu}_x \end{pmatrix}, \qquad (1.61)$$

$$\mathrm{cov}\begin{pmatrix} \mathbf{y} \\ \mathbf{x} \end{pmatrix} = \boldsymbol{\Sigma} = \begin{pmatrix} \boldsymbol{\Sigma}_{yy} & \boldsymbol{\Sigma}_{yx} \\ \boldsymbol{\Sigma}_{xy} & \boldsymbol{\Sigma}_{xx} \end{pmatrix}. \qquad (1.62)$$

The covariance matrix $\boldsymbol{\Sigma}_{yx}$ can be denoted by $\mathrm{cov}(\mathbf{y}, \mathbf{x})$:

$$\mathrm{cov}(\mathbf{y}, \mathbf{x}) = \boldsymbol{\Sigma}_{yx}. \qquad (1.63)$$

Note that $\boldsymbol{\Sigma}_{yx} = \boldsymbol{\Sigma}'_{xy}$, or

$$\mathrm{cov}(\mathbf{y}, \mathbf{x}) = [\mathrm{cov}(\mathbf{x}, \mathbf{y})]'. \qquad (1.64)$$

By definition,

$$\mathrm{cov}(\mathbf{y}, \mathbf{x}) = E\left[(\mathbf{y} - \boldsymbol{\mu}_y)(\mathbf{x} - \boldsymbol{\mu}_x)'\right], \qquad (1.65)$$

$$\mathrm{cov}(\mathbf{x}, \mathbf{y}) = E\left[(\mathbf{x} - \boldsymbol{\mu}_x)(\mathbf{y} - \boldsymbol{\mu}_y)'\right], \qquad (1.66)$$

and these likewise confirm that $\mathrm{cov}(\mathbf{y}, \mathbf{x}) = [\mathrm{cov}(\mathbf{x}, \mathbf{y})]'$, since $[(\mathbf{x} - \boldsymbol{\mu}_x)(\mathbf{y} - \boldsymbol{\mu}_y)']' = (\mathbf{y} - \boldsymbol{\mu}_y)(\mathbf{x} - \boldsymbol{\mu}_x)'$.

There is an important notational distinction between $\mathrm{cov}\binom{\mathbf{y}}{\mathbf{x}}$ in (1.62) and $\mathrm{cov}(\mathbf{y}, \mathbf{x})$ in (1.63) or (1.65). In fact, we have used the notation "cov" in three ways:

1. the covariance of two variables, as in (1.15),

$$\mathrm{cov}(x, y) = \sigma_{xy} = E\left[(x - \mu_x)(y - \mu_y)\right],$$

which is a scalar;

2. the covariance matrix for a vector $\mathbf{y}$, as in (1.44),

$$\mathrm{cov}(\mathbf{y}) = \boldsymbol{\Sigma} = E\left[(\mathbf{y} - \boldsymbol{\mu})(\mathbf{y} - \boldsymbol{\mu})'\right],$$

which is a square matrix containing variances and covariances [this is the sense in which we use $\mathrm{cov}\binom{\mathbf{y}}{\mathbf{x}}$];

3. the covariance matrix for two vectors, as in (1.63) and (1.65),

$$\text{cov}(\mathbf{y}, \mathbf{x}) = \mathbf{\Sigma}_{yx} = E\left[(\mathbf{y} - \boldsymbol{\mu}_y)(\mathbf{x} - \boldsymbol{\mu}_x)'\right],$$

which is a rectangular matrix containing the covariance of each y_i with each x_j (if $p = q$, $\mathbf{\Sigma}_{yx}$ is square).

If $p = q$, it is possible to find the sum of $\mathbf{x}$ and $\mathbf{y}$. In this case, the mean vector and covariance matrix of the sum are given by

$$E(\mathbf{x} + \mathbf{y}) = E(\mathbf{x}) + E(\mathbf{y}) = \boldsymbol{\mu}_x + \boldsymbol{\mu}_y, \tag{1.67}$$

$$\begin{aligned} \text{cov}(\mathbf{x} + \mathbf{y}) &= \text{cov}(\mathbf{x}) + \text{cov}(\mathbf{y}) + \text{cov}(\mathbf{x}, \mathbf{y}) + \text{cov}(\mathbf{y}, \mathbf{x}) \\ &= \mathbf{\Sigma}_{xx} + \mathbf{\Sigma}_{yy} + \mathbf{\Sigma}_{xy} + \mathbf{\Sigma}_{yx}. \end{aligned} \tag{1.68}$$

If $p = q$ and $\mathbf{x}$ and $\mathbf{y}$ are independent, then $\text{cov}(x_i, y_j) = 0$ for all i and j, and $\mathbf{\Sigma}_{xy} = \mathbf{\Sigma}_{yx} = \mathbf{O}$. In this case, (1.68) reduces to

$$\text{cov}(\mathbf{x} + \mathbf{y}) = \mathbf{\Sigma}_{xx} + \mathbf{\Sigma}_{yy} \qquad \text{(if } \mathbf{x} \text{ and } \mathbf{y} \text{ are independent).} \tag{1.69}$$

There are sample results that correspond to (1.67), (1.68), and (1.69).

1.6 LINEAR FUNCTIONS OF RANDOM VARIABLES

In multivariate analysis, we often use linear combinations of the variables $y_1, y_2, \ldots, y_p$. We discuss the sample means, sample variances, and sample covariances of linear combinations in Section 1.6.1 and their population counterparts in Section 1.6.2.

1.6.1 Sample Means, Variances, and Covariances

A linear combination of the variables in the random vector $\mathbf{y}$ can be written as

$$z = a_1 y_1 + a_2 y_2 + \cdots + a_p y_p = \mathbf{a}'\mathbf{y}, \tag{1.70}$$

where $\mathbf{a}' = (a_1, a_2, \ldots, a_p)$ is a vector of constants. Taking this linear combination of each $\mathbf{y}_i$ in a sample yields

$$z_i = a_1 y_{i1} + a_2 y_{i2} + \cdots + a_p y_{ip} = \mathbf{a}'\mathbf{y}_i, \qquad i = 1, 2, \ldots, n. \tag{1.71}$$

The sample mean of z is given by

$$\overline{z} = \frac{1}{n} \sum_{i=1}^{n} z_i = \mathbf{a}'\overline{\mathbf{y}}, \tag{1.72}$$

which can be demonstrated as follows:

$$\overline{z} = \frac{1}{n} \sum_{i=1}^{n} z_i = \frac{1}{n} \sum_{i=1}^{n} \mathbf{a}'\mathbf{y}_i = \frac{1}{n}(\mathbf{a}'\mathbf{y}_1 + \mathbf{a}'\mathbf{y}_2 + \cdots + \mathbf{a}'\mathbf{y}_n)$$

$$= \frac{1}{n}\mathbf{a}'(\mathbf{y}_1 + \mathbf{y}_2 + \cdots + \mathbf{y}_n) \qquad \text{[by (A.2.17)]}$$

$$= \mathbf{a}'\left(\frac{1}{n}\sum_{i=1}^{n}\mathbf{y}_i\right) = \mathbf{a}'\overline{\mathbf{y}}.$$

The sample variance of $z = a_1y_1 + a_2y_2 + \cdots + a_py_p$ is not equal to $a_1^2s_1^2 + a_2^2s_2^2 + \cdots + a_p^2s_p^2$, because in general the y_j's are not independent. We must also take the covariances into account:

$$s_z^2 = \sum_{j=1}^{p} a_j^2 s_j^2 + \sum_{j \neq k} a_j a_k s_{jk}$$

$$= \mathbf{a}'\mathbf{S}\mathbf{a} \qquad \text{[by (A.2.29)].} \tag{1.73}$$

The result in (1.73) is easily demonstrated using the basic definition of s_z^2 as the sample variance of $z_1, z_2, \ldots, z_n$:

$$s_z^2 = \frac{1}{n-1}\sum_{i=1}^{n}(z_i - \overline{z})^2 = \frac{1}{n-1}\sum_{i=1}^{n}(\mathbf{a}'\mathbf{y}_i - \mathbf{a}'\overline{\mathbf{y}})^2$$

$$= \frac{1}{n-1}\sum_{i}(\mathbf{a}'\mathbf{y}_i - \mathbf{a}'\overline{\mathbf{y}})(\mathbf{y}_i'\mathbf{a} - \overline{\mathbf{y}}'\mathbf{a})$$

$$= \frac{1}{n-1}\sum_{i}\mathbf{a}'(\mathbf{y}_i - \overline{\mathbf{y}})(\mathbf{y}_i - \overline{\mathbf{y}})'\mathbf{a}$$

$$= \mathbf{a}'\left[\frac{1}{n-1}\sum_{i}(\mathbf{y}_i - \overline{\mathbf{y}})(\mathbf{y}_i - \overline{\mathbf{y}})'\right]\mathbf{a} \qquad \text{[by (A.2.19)]}$$

$$= \mathbf{a}'\mathbf{S}\mathbf{a}.$$

Since $s_z^2 \geq 0$, we have $\mathbf{a}'\mathbf{S}\mathbf{a} \geq 0$ for all possible $\mathbf{a}$, and therefore $\mathbf{S}$ is at least positive semidefinite. This was also shown by (1.42) and a remark following (1.42).

For two linear combinations $z = \mathbf{a}'\mathbf{y}$ and $w = \mathbf{b}'\mathbf{y} = b_1y_1 + b_2y_2 + \cdots + b_py_p$, the sample covariance and correlation are given by

$$s_{zw} = \mathbf{a}'\mathbf{S}\mathbf{b}, \tag{1.74}$$

$$r_{zw} = \frac{s_{zw}}{\sqrt{s_z^2 s_w^2}} = \frac{\mathbf{a}'\mathbf{S}\mathbf{b}}{\sqrt{(\mathbf{a}'\mathbf{S}\mathbf{a})(\mathbf{b}'\mathbf{S}\mathbf{b})}}. \tag{1.75}$$

Suppose we have k linear transformations:

$$z_1 = a_{11}y_1 + a_{12}y_2 + \cdots + a_{1p}y_p$$

$$z_2 = a_{21}y_1 + a_{22}y_2 + \cdots + a_{2p}y_p$$

$$\vdots$$

$$z_k = a_{k1}y_1 + a_{k2}y_2 + \cdots + a_{kp}y_p.$$

These linear functions can be written in matrix notation as

$$\mathbf{z} = \mathbf{A}\mathbf{y},$$

where $\mathbf{z}$ is $k \times 1$, $\mathbf{A}$ is $k \times p$, and $\mathbf{y}$ is $p \times 1$. We thus map $(y_1, y_2, \ldots, y_p)$ into $(z_1, z_2, \ldots, z_k)$, where k could be greater than p, but we would ordinarily be interested only in $k \leq p$. (See principal components in Chapter 9, for instance, where we transform the p y's to fewer than p dimensions that capture most of the information in the y's.)

If a sample $\mathbf{y}_1, \mathbf{y}_2, \ldots, \mathbf{y}_n$ is transformed to $\mathbf{z}_1, \mathbf{z}_2, \ldots, \mathbf{z}_n$, where $\mathbf{z}_i = \mathbf{A}\mathbf{y}_i$, then by (A.2.18) and (A.2.20), the $k \times 1$ sample mean vector and $k \times k$ sample covariance matrix of the $\mathbf{z}$'s are given by

$$\overline{\mathbf{z}} = \mathbf{A}\overline{\mathbf{y}}, \tag{1.76}$$

$$\mathbf{S}_z = \mathbf{A}\mathbf{S}\mathbf{A}'. \tag{1.77}$$

If two sets of linear combinations are of interest, say $\mathbf{z} = \mathbf{A}\mathbf{y}$ and $\mathbf{w} = \mathbf{B}\mathbf{y}$, where $\mathbf{A}$ is $k \times p$ and $\mathbf{B}$ is $m \times p$, then the $k \times m$ matrix of sample covariances between the z's and the w's is given by

$$\mathbf{S}_{zw} = \widehat{\text{cov}}(\mathbf{z}, \mathbf{w}) = \widehat{\text{cov}}(\mathbf{A}\mathbf{y}, \mathbf{B}\mathbf{y}) = \mathbf{A}\mathbf{S}\mathbf{B}'. \tag{1.78}$$

The notation $\widehat{\text{cov}}(\mathbf{A}\mathbf{y}, \mathbf{B}\mathbf{y})$ represents the sample covariance matrix of $\mathbf{A}\mathbf{y}$ and $\mathbf{B}\mathbf{y}$. If $\mathbf{B}$ consists of a single row $\mathbf{b}'$, then (1.78) becomes

$$\widehat{\text{cov}}(\mathbf{A}\mathbf{y}, \mathbf{b}'\mathbf{y}) = \mathbf{A}\mathbf{S}\mathbf{b}. \tag{1.79}$$

Suppose $\mathbf{y}$ and $\mathbf{x}$ are subvectors measured on the same sampling unit, as in Section 1.5, and let $u = \mathbf{a}'\mathbf{y}$ and $v = \mathbf{b}'\mathbf{x}$ be linear combinations of these, where $\mathbf{a}$ and $\mathbf{y}$ are $p \times 1$ and $\mathbf{b}$ and $\mathbf{x}$ are $q \times 1$. Then the sample covariance of u and v is given by

$$s_{uv} = \widehat{\text{cov}}(\mathbf{a}'\mathbf{y}, \mathbf{b}'\mathbf{x}) = \mathbf{a}'\mathbf{S}_{yx}\mathbf{b}, \tag{1.80}$$

where $\mathbf{S}_{yx}$ is the $p \times q$ sample covariance matrix for $\mathbf{y}$ and $\mathbf{x}$ defined in (1.59). To find the sample correlation between u and v, we first obtain

$$s_u^2 = \mathbf{a}'\mathbf{S}_{yy}\mathbf{a}, \tag{1.81}$$

$$s_v^2 = \mathbf{b}'\mathbf{S}_{xx}\mathbf{b}, \tag{1.82}$$

where $\mathbf{S}_{yy}$ and $\mathbf{S}_{xx}$ are defined in the partitioned matrix $\mathbf{S}$ in (1.59). Then by (1.24),

$$r_{uv} = \frac{s_{uv}}{\sqrt{s_u^2 s_v^2}} = \frac{\mathbf{a}'\mathbf{S}_{yx}\mathbf{b}}{\sqrt{(\mathbf{a}'\mathbf{S}_{yy}\mathbf{a})(\mathbf{b}'\mathbf{S}_{xx}\mathbf{b})}}. \tag{1.83}$$

If $\mathbf{u} = \mathbf{A}\mathbf{y}$ represents several linear combinations of $\mathbf{y}$ and $\mathbf{v} = \mathbf{B}\mathbf{x}$ represents linear combinations of $\mathbf{x}$, then the sample covariance matrix of $\mathbf{u}$ and $\mathbf{v}$ is given by

$$\widehat{\text{cov}}(\mathbf{u}, \mathbf{v}) = \widehat{\text{cov}}(\mathbf{A}\mathbf{y}, \mathbf{B}\mathbf{x}) = \mathbf{A}\mathbf{S}_{yx}\mathbf{B}'. \tag{1.84}$$

If $p = q$, we can consider the special case of (1.84) in which $\mathbf{A} = \mathbf{B}$. Then (1.84) becomes

$$\widehat{\text{cov}}(\mathbf{Ay}, \mathbf{Ax}) = \mathbf{AS}_{yx}\mathbf{A}'. \tag{1.85}$$

Example 1.6.1. Reaven and Miller (1979; see also Andrews and Herzberg 1985, pp. 215–219) recorded five variables for normal patients and diabetics. Partial results for normal patients only are given in Table 1.2. The five variables are

$$y_1 = \text{relative weight},$$
$$y_2 = \text{fasting plasma glucose},$$
$$x_1 = \text{glucose intolerance},$$
$$x_2 = \text{insulin response to oral glucose},$$
$$x_3 = \text{insulin resistance}.$$

Table 1.2. Relative Weight, Blood Glucose, and Insulin Levels

Patient Number	y_1	y_2	x_1	x_2	x_3
1	.81	80	356	124	55
2	.95	97	289	117	76
3	.94	105	319	143	105
4	1.04	90	356	199	108
5	1.00	90	323	240	143
6	.76	86	381	157	165
7	.91	100	350	221	119
8	1.10	85	301	186	105
9	.99	97	379	142	98
10	.78	97	296	131	94
11	.90	91	353	221	53
12	.73	87	306	178	66
13	.96	78	290	136	142
14	.84	91	371	200	93
15	.74	86	312	208	68
16	.98	80	393	202	102
17	1.10	90	364	152	76
18	.85	99	359	185	37
19	.83	85	296	116	60
20	.93	90	345	123	50
21	.95	91	378	136	47
22	.74	88	304	134	50
23	.95	95	347	184	91
24	.97	91	327	192	124
25	.72	92	386	279	74

The covariance matrix for the two y's with the three x's is

$$\mathbf{S}_{yx} = \begin{pmatrix} .2740 & -.1997 & 1.0688 \\ 8.9733 & 17.4317 & -16.81 \end{pmatrix}.$$

This covariance matrix can be computed in two ways: (1) by obtaining the overall $\mathbf{S}$ for the y's and the x's and then partitioning it as in (1.59) or (2) by using the result of Problem 1.29(b), $\mathbf{S}_{yx} = \mathbf{Y}'(\mathbf{I} - \frac{1}{n}\mathbf{J})\mathbf{X}/(n-1)$.

To illustrate the covariance and correlation formulas (1.80) and (1.83), we define $u = 2y_1 - y_2 = \mathbf{a}'\mathbf{y}$ and $v = 2x_1 - 3x_2 + x_3 = \mathbf{b}'\mathbf{x}$. Then, by (1.80), the covariance of u and v is given by

$$s_{uv} = (2, -1)\begin{pmatrix} .2740 & -.1997 & 1.0688 \\ 8.9733 & 17.4317 & -16.81 \end{pmatrix}\begin{pmatrix} 2 \\ -3 \\ 1 \end{pmatrix} = 55.5903.$$

To convert the covariance $s_{uv} = 55.5903$ to a correlation, we first use (1.81) and (1.82) to obtain

$$s_u^2 = \mathbf{a}'\mathbf{S}_{yy}\mathbf{a} = 42.3725, \qquad s_v^2 = \mathbf{b}'\mathbf{S}_{xx}\mathbf{b} = 14,248.25.$$

Then by (1.83),

$$r_{uv} = \frac{s_{uv}}{\sqrt{s_u^2 s_v^2}} = \frac{55.5903}{\sqrt{(42.3725)(14,248.25)}} = .0715.$$

To illustrate (1.84), we define

$$u_1 = y_1 + y_2, \qquad u_2 = y_1 - y_2,$$
$$v_1 = x_1 + x_2 + x_3, \qquad v_2 = x_1 - 2x_2 + 2x_3.$$

These can be written in matrix form as

$$\mathbf{u} = \begin{pmatrix} u_1 \\ u_2 \end{pmatrix} = \begin{pmatrix} 1 & 1 \\ 1 & -1 \end{pmatrix}\begin{pmatrix} y_1 \\ y_2 \end{pmatrix} = \mathbf{Ay},$$

$$\mathbf{v} = \begin{pmatrix} v_1 \\ v_2 \end{pmatrix} = \begin{pmatrix} 1 & 1 & 1 \\ 1 & -2 & 2 \end{pmatrix}\begin{pmatrix} x_1 \\ x_2 \\ x_3 \end{pmatrix} = \mathbf{Bx}.$$

Then by (1.84), we have

$$\widehat{\text{cov}}(\mathbf{Ay}, \mathbf{Bx}) = \begin{pmatrix} 1 & 1 \\ 1 & -1 \end{pmatrix}\begin{pmatrix} .2740 & -.1997 & 1.0688 \\ 8.9733 & 17.4317 & -16.81 \end{pmatrix}\begin{pmatrix} 1 & 1 \\ 1 & -2 \\ 1 & 2 \end{pmatrix}$$

$$= \begin{pmatrix} 10.738 & -56.699 \\ -8.452 & 62.321 \end{pmatrix}.$$

Notice that $\widehat{\text{cov}}(\mathbf{Ay}, \mathbf{Bx})$ is asymmetric, as we would expect.

1.6.2 Population Means, Variances, and Covariances

If $z = \mathbf{a}'\mathbf{y}$, where $\mathbf{a}$ is a vector of constants, then the *population mean* and *variance* of z are given by

$$E(z) = E(\mathbf{a}'\mathbf{y}) = \mathbf{a}'E(\mathbf{y}) = \mathbf{a}'\boldsymbol{\mu}, \tag{1.86}$$

$$\sigma_z^2 = \text{var}(\mathbf{a}'\mathbf{y}) = \mathbf{a}'\boldsymbol{\Sigma}\mathbf{a}. \tag{1.87}$$

The result in (1.86) can be proved as follows:

$$\begin{aligned} E(\mathbf{a}'\mathbf{y}) &= E(a_1y_1 + a_2y_2 + \cdots + a_py_p) \\ &= E(a_1y_1) + E(a_2y_2) + \cdots + E(a_py_p) \\ &= a_1E(y_1) + a_2E(y_2) + \cdots + a_pE(y_p) \\ &= (a_1, a_2, \ldots, a_p)\begin{pmatrix} E(y_1) \\ E(y_2) \\ \vdots \\ E(y_p) \end{pmatrix} = \mathbf{a}'E(\mathbf{y}). \end{aligned}$$

We now present three additional properties of expected values of random vectors and matrices. These properties can be used to obtain the covariance and correlation results below in (1.91)–(1.99). Let $\mathbf{y}$ be a random vector, $\mathbf{X}$ be a random matrix, $\mathbf{a}$ and $\mathbf{b}$ be vectors of constants, and $\mathbf{A}$ and $\mathbf{B}$ be matrices of constants. Then assuming the sizes of the matrices and vectors are conformable,

$$E(\mathbf{Ay}) = \mathbf{A}E(\mathbf{y}), \tag{1.88}$$

$$E(\mathbf{a}'\mathbf{Xb}) = \mathbf{a}'E(\mathbf{X})\mathbf{b}, \tag{1.89}$$

$$E(\mathbf{AXB}) = \mathbf{A}E(\mathbf{X})\mathbf{B}. \tag{1.90}$$

For two linear combinations $z = \mathbf{a}'\mathbf{y}$ and $w = \mathbf{b}'\mathbf{y}$, the *population covariance* and *correlation* are given by

$$\text{cov}(z, w) = \sigma_{zw} = \mathbf{a}'\boldsymbol{\Sigma}\mathbf{b}, \tag{1.91}$$

$$\rho_{zw} = \text{corr}(\mathbf{a}'\mathbf{y}, \mathbf{b}'\mathbf{y}) = \frac{\sigma_{zw}}{\sigma_z\sigma_w} = \frac{\mathbf{a}'\boldsymbol{\Sigma}\mathbf{b}}{\sqrt{(\mathbf{a}'\boldsymbol{\Sigma}\mathbf{a})(\mathbf{b}'\boldsymbol{\Sigma}\mathbf{b})}}. \tag{1.92}$$

For several linear combinations $\mathbf{Ay}$, where $\mathbf{A}$ is $k \times p$, the *population mean vector* and *covariance matrix* are given by

$$E(\mathbf{Ay}) = \mathbf{A}E(\mathbf{y}), \tag{1.93}$$

$$\text{cov}(\mathbf{Ay}) = \mathbf{A}\boldsymbol{\Sigma}\mathbf{A}'. \tag{1.94}$$

If the rows of $\mathbf{B}$ represent coefficients for another set of linear combinations different from $\mathbf{Ay}$, then

$$\text{cov}(\mathbf{Ay}, \mathbf{By}) = \mathbf{A\Sigma B}'. \tag{1.95}$$

If $\mathbf{y}$ and $\mathbf{x}$ are subvectors measured on the same sampling unit, where $\mathbf{y}$ is $p \times 1$ and $\mathbf{x}$ is $q \times 1$, we have the following population results analogous to (1.80), (1.83), and (1.84):

$$\text{cov}(\mathbf{a}'\mathbf{y}, \mathbf{b}'\mathbf{x}) = \mathbf{a}'\mathbf{\Sigma}_{yx}\mathbf{b}, \tag{1.96}$$

$$\text{corr}(\mathbf{a}'\mathbf{y}, \mathbf{b}'\mathbf{x}) = \frac{\mathbf{a}'\mathbf{\Sigma}_{yx}\mathbf{b}}{\sqrt{(\mathbf{a}'\mathbf{\Sigma}_{yy}\mathbf{a})(\mathbf{b}'\mathbf{\Sigma}_{xx}\mathbf{b})}}, \tag{1.97}$$

$$\text{cov}(\mathbf{Ay}, \mathbf{Bx}) = \mathbf{A\Sigma}_{yx}\mathbf{B}', \tag{1.98}$$

where $\mathbf{\Sigma}_{yx}$, $\mathbf{\Sigma}_{yy}$, and $\mathbf{\Sigma}_{xx}$ are defined in (1.62) and (1.65). If $p = q$ and $\mathbf{B} = \mathbf{A}$, (1.98) becomes

$$\text{cov}(\mathbf{Ay}, \mathbf{Ax}) = \mathbf{A\Sigma}_{yx}\mathbf{A}'. \tag{1.99}$$

1.7 MEASURING INTERCORRELATION

It is sometimes desirable to obtain a single measure of the overall amount of intercorrelation among a set of variables $y_1, y_2, \ldots, y_p$, that is, to represent the correlations in $\mathbf{R}$ by a single number. If one variable in the set were of primary interest, we could use the multiple correlation of this variable with the other $p - 1$ variables (see Sections 7.2.5 and 7.3.3). In a similar vein, we can define the (canonical) correlation between a subset of variables and the remaining variables (see Chapter 8).

Various suggestions have been made for an index of intercorrelation among a set of variables. Many of these use the eigenvalues of the sample correlation matrix $\mathbf{R}$. Silvey (1969) proposed the use of the smallest eigenvalue of $\mathbf{R}$, while Peele and Ryan (1979) used the smallest and largest eigenvalues of $\mathbf{R}$. Mason, Gunst, and Webster (1975) introduced the *condition number* λ_1/λ_p, the ratio of the largest to the smallest eigenvalue of $\mathbf{R}$. Hoerl and Kennard (1970b) used $\sum_{j=1}^p (1/\lambda_j)$, where $\lambda_1, \lambda_2, \ldots, \lambda_p$ are all the eigenvalues of $\mathbf{R}$. All the above indexes vary with p, the number of variables. To reduce the effect of p, Rencher and Pun (1980) used $\sum_{j=1}^p (1/\lambda_j)/p$.

Farrar and Glauber (1967) and Haitovsky (1969) worked directly with $|\mathbf{R}|$, which equals 1 if the variables are independent and 0 if there is an exact linear dependence among the variables. Gleason and Staelin (1975) used

$$g = \sqrt{\frac{(\sum_{j=1}^p \lambda_j^2) - p}{p(p-1)}}, \tag{1.100}$$

which ranges from 0 when the variables are independent ($\mathbf{R} = \mathbf{I}$) to 1 when all the variables are perfectly correlated with each other ($\mathbf{R} = \mathbf{J}$).

Chatterjee and Price (1977) discussed the *variance inflation factor* for each variable,

$$\text{VIF}_j = r^{jj} = \frac{1}{1 - R_j^2}, \qquad j = 1, 2, \ldots, p, \tag{1.101}$$

where r^{jj} is the jth diagonal element of $\mathbf{R}^{-1}$ and R_j^2 is the squared multiple correlation of y_j regressed on the $p - 1$ remaining y's. It can be shown that

$$R_j^2 = 1 - \frac{1}{r^{jj}}, \tag{1.102}$$

from which the second equality in (1.101) follows.

Heo (1987) compared six different measures of intercorrelation based on the above suggestions:

$$q_1 = \left(1 - \frac{\text{smallest } \lambda_j}{\text{largest } \lambda_j}\right)^{p+2}, \qquad q_2 = 1 - \frac{p}{\sum_{j=1}^{p} (1/\lambda_j)},$$

$$q_3 = 1 - \sqrt{|\mathbf{R}|}, \qquad q_4 = \left(\frac{\text{largest } \lambda_j}{p}\right)^{3/2},$$

$$q_5 = \left(1 - \frac{\text{smallest } \lambda_j}{p}\right)^5, \qquad q_6 = \sum_{j=1}^{p} \frac{1 - 1/r^{jj}}{p} = \text{average } R^2.$$

The six indexes have been adjusted to a range of 0 to 1. In making the comparisons, the maximum population multiple correlation between any variable and the other $p - 1$ variables was used as a baseline indication of population intercorrelation. The most consistent performer over a wide range of values of p, n, and the amount of intercorrelation was q_2, the average reciprocal eigenvalue converted to a scale of 0 to 1.

Example 1.7.1. Timm (1975, p. 233; 1980, p. 47) reported the responses of 11 subjects to "probe words" at five positions in a sentence. The data are given in Table 1.3, in which the variables are y_i = response time for the ith probe word, $i = 1, 2, \ldots, 5$.

We illustrate some measures of intercorrelation. The correlation matrix is given by

$$\mathbf{R} = \begin{pmatrix} 1.0000 & .6144 & .7572 & .5751 & .4131 \\ .6144 & 1.0000 & .5474 & .7498 & .5477 \\ .7572 & .5474 & 1.0000 & .6053 & .6919 \\ .5751 & .7498 & .6053 & 1.0000 & .5239 \\ .4131 & .5477 & .6919 & .5239 & 1.0000. \end{pmatrix}.$$

The determinant and eigenvalues of $\mathbf{R}$ are

$$|\mathbf{R}| = .0409186,$$

$$\text{Eigenvalues} = (3.4165, .6144, .5723, .2712, .1256).$$

Table 1.3. Response Time for Five Probe Word Positions

Subject Number	y_1	y_2	y_3	y_4	y_5
1	51	36	50	35	42
2	27	20	26	17	27
3	37	22	41	37	30
4	42	36	32	34	27
5	27	18	33	14	29
6	43	32	43	35	40
7	41	22	36	25	38
8	38	21	31	20	16
9	36	23	27	25	28
10	26	31	31	32	36
11	29	20	25	26	25

The Gleason and Staelin index in (1.100) is

$$g = \sqrt{\frac{\sum_j \lambda_j^2 - p}{p(p-1)}} = \sqrt{\frac{12.46675 - 5}{5(4)}} = .611.$$

The indexes $q_1, q_2, \ldots, q_6$ are as follows:

$$q_1 = \left(1 - \frac{\text{smallest } \lambda_j}{\text{largest } \lambda_j}\right)^{p+2} = \left(1 - \frac{.1256}{3.4165}\right)^7 = .769,$$

$$q_2 = 1 - \frac{5}{\sum_{j=1}^5 (1/\lambda_j)} = 1 - \frac{5}{15.3173} = .674,$$

$$q_3 = 1 - \sqrt{|\mathbf{R}|} = 1 - \sqrt{.0409186} = .798,$$

$$q_4 = .565, \qquad q_5 = .881, \qquad q_6 = .661.$$

The indexes, which are in fairly general agreement, indicate a moderate amount of intercorrelation among the five variables.

1.8 MAHALANOBIS DISTANCE

To obtain a useful multivariate measure of distance between two vectors $\mathbf{y}_1$ and $\mathbf{y}_2$, we must take into account the covariances of the variables as well as their variances. The simple Euclidean distance $(\mathbf{y}_1 - \mathbf{y}_2)'(\mathbf{y}_1 - \mathbf{y}_2)$ does not do this. We standardize this distance with the inverse of the covariance matrix:

$$d^2 = (\mathbf{y}_1 - \mathbf{y}_2)'\mathbf{S}^{-1}(\mathbf{y}_1 - \mathbf{y}_2). \tag{1.103}$$

Some standardized distances we will encounter in Chapter 3 are

$$D^2 = (\overline{\mathbf{y}} - \boldsymbol{\mu})'\mathbf{S}^{-1}(\overline{\mathbf{y}} - \boldsymbol{\mu}), \tag{1.104}$$

$$\Delta^2 = (\boldsymbol{\mu} - \boldsymbol{\mu}_0)'\boldsymbol{\Sigma}^{-1}(\boldsymbol{\mu} - \boldsymbol{\mu}_0), \tag{1.105}$$

$$\Delta^2 = (\boldsymbol{\mu}_1 - \boldsymbol{\mu}_2)'\boldsymbol{\Sigma}^{-1}(\boldsymbol{\mu}_1 - \boldsymbol{\mu}_2). \tag{1.106}$$

Distances of this type are commonly called *Mahalanobis distances* (Mahalanobis 1936). The inverse covariance matrix in a Mahalanobis distance transforms the variables so that they are uncorrelated and have the same variance.

1.9 MISSING DATA

We often find missing values for one or more variables in some of the observation vectors. (This type of "missingness" differs from the case in which entire observation vectors are missing from a balanced design; see Section 4.8.) A small number of missing entries in the data matrix **Y** in (1.26) can easily be managed by simply deleting an entire row if it has a missing value. However, for a large number of missing values, this approach would lead to deletion of most of the data.

We present several methods for handling missing data. Ideally, to use these methods, we assume missing values that are scattered randomly throughout a data matrix rather than missing values that depend to some extent on the values of the variables. Rao (1983) argued that maximum likelihood methods for estimating missing values (to be discussed below) should not be used because of an assumption of randomness in the missing data. He asserted that in practical problems missing values usually occur in a nonrandom manner.

Rubin (1976) distinguished between *missing at random* (MAR) and *missing completely at random* (MCAR). The pattern of missing values is MCAR if missingness does not depend on either the observed y's or the missing y's. The pattern is MAR if missingness does not depend on the missing y's but may depend on the observed y's. For further discussion of the differences between MAR and MCAR, see Heitjan and Basu (1996).

We consider four basic methods that have been proposed for dealing with the problem of missing values:

1. The first approach, sometimes referred to as *listwise deletion*, uses only those observation vectors that are complete. This is a good solution if only a few vectors are incomplete but will not be as satisfactory if a larger proportion is affected. Note that if the missing values do not occur at random, then exclusion of their data vectors will leave a nonrandom sample that is no longer representative of the population.

A variation of this procedure is to delete either variables or observation vectors, that is, either columns or rows of the data matrix **Y** defined in (1.26). Hemel et al. (1987) proposed a stepwise technique to accomplish this deletion in a manner that retains as much of the data as possible.

2. Another option, often called *pairwise deletion*, is to use all available observations when calculating $\bar{\mathbf{y}}$ and all available pairs of values in the calculation of **S** or **R**. To illustrate, consider the following data matrix:

$$\begin{pmatrix} y_{11} & y_{12} & y_{13} \\ y_{21} & & y_{23} \\ y_{31} & y_{32} & y_{33} \\ y_{41} & y_{42} & \\ y_{51} & y_{52} & y_{53} \end{pmatrix}.$$

To compute $\bar{y}_1$, we have five observations; for $\bar{y}_2$ and $\bar{y}_3$, we have four available observations. For s_{12} and s_{13}, there are four pairs of observations; for s_{23}, only three pairs are available. At first glance this approach is appealing because it uses all available information, but the procedure is not generally recommended. The covariance matrix **S** (or correlation matrix **R**) that is obtained this way is usually not positive definite or even positive semidefinite, which severely limits applications. For example, when this method was tried on a data set with 85 variables, 10 of the 85 eigenvalues of **R** turned out to be negative. Such a result invalidates many of the ordinary multivariate techniques that use **R** or **S**. Heiberger (1977) found by simulation that the use of this approach for missing observations in regression was often less efficient than regression that was limited to the complete observation vectors.

3. The third method (Schwertman and Allen 1973, 1979) is a "smoothing" procedure that finds the positive semidefinite matrix **T** that is "closest" to the matrix **S** produced in method 2 above. The term *closest* is used in the least squares sense; the expression

$$\sum_{j=1}^{p} \sum_{k=1}^{p} (s_{jk} - t_{jk})^2 \tag{1.107}$$

is minimized, where $\mathbf{S} = (s_{jk})$ and $\mathbf{T} = (t_{jk})$. The matrix **T** that minimizes (1.107) is

$$\mathbf{T} = \sum_{\lambda_j > 0} \lambda_j \mathbf{a}_j \mathbf{a}_j', \tag{1.108}$$

where the sum is over all the positive eigenvalues of **S** and their corresponding eigenvectors $\mathbf{a}_j$, scaled so that $\mathbf{a}_j'\mathbf{a}_j = 1$. Since some negative eigenvalues will be deleted in most cases, the matrix **T** will typically be singular and cannot be used in applications that require an inverse. However, this method is a viable option in applications such as principal components (see Chapter 9) that do not involve an inverse.

4. The fourth alternative is to estimate the missing values, that is, to "fill the holes" in the data matrix. The remainder of this section will be devoted to a discussion of various schemes for estimation of the missing values. The use of these procedures is questionable unless the missing values occur at random. If the selection process for missing values depends somewhat on the variable-values, then substantial bias may be introduced in estimating the missing responses.

Wilks (1932b) proposed replacing each missing value by the mean of the available data in the corresponding column of the data matrix. The use of means reduces the sample variances as well as the absolute value of the covariances; therefore, the imputation of means for missing values leads to a sample covariance matrix **S** that is positive definite but biased.

Buck (1960) advocated a regression approach for estimating missing values. The data matrix is partitioned into a submatrix containing all the rows with missing entries and another submatrix with all the complete rows. To estimate a missing value y_{ij} in the ith row, we use the data in the submatrix with complete rows and regress the jth column on the other $p - 1$ columns to obtain $\hat{y}_j = b_0 + b_1 y_1 + \cdots + b_{j-1} y_{j-1} + b_{j+1} y_{j+1} + \cdots + b_p y_p$. Then, based on the nonmissing entries $y_{i1}, \ldots, y_{i,j-1}, y_{i,j+1}, \ldots, y_{ip}$ in the ith row, we obtain the predicted value, $\hat{y}_{ij}$.

Buck's regression method can be iterated as follows: Estimate all missing entries in the data matrix **Y** by using predicted values from regression, as above. Then calculate the covariance matrix **S** (or correlation matrix **R**). Use **S** to obtain new regression equations that will produce new predicted values to replace the previous ones. (Regression estimates based on **S** and **R** are given in Sections 7.2.3 and 7.3.2.) Using the revised data matrix, recompute **S**. Continue this process until the predicted values stabilize.

If most of the observation vectors have missing values, it may not be possible to find sufficient data to calculate the initial prediction equations. In such a situation, we could begin by using means, as in Wilks' approach above, and then use regression in subsequent iterations.

In most cases, the regression approach is preferable to Wilks' method of imputing means. However, if a variable to be predicted is not very highly correlated with the other variables, the regression technique offers little improvement over Wilks' approach. The regression technique underestimates the variances and covariances, but the bias is less than that for Wilks' method.

Several writers have suggested a maximum likelihood approach (see Section 2.3.1). This is ordinarily based on the assumption of a multivariate normal distribution (more generally, the exponential family of distributions), in contrast to the pragmatic approaches above that are not assumption based. Orchard and Woodbury (1972) proposed a general iterative approach to computing maximum likelihood estimates of parameters such as means, variances, and covariances. Dempster, Laird, and Rubin (1977) called this approach the EM (expectation–maximization) algorithm. The expectation step estimates a missing value by means of a conditional expectation, given the observed values in the vector and the current parameter estimates. The maximization step uses the estimated missing values and computes the parameter estimates by means of maximum likelihood. The parameter estimates are then used to find better estimates of the missing values, and these are used to obtain new parameter estimates. This iterative process continues until the estimated parameter values converge. In the case of the multivariate normal, the EM algorithm turns out to be essentially the same as the iterative regression approach above (Little 1992). For an extended treatment of the missing-data problem, with emphasis on the maximum likelihood approach, see Little and Rubin (1987). Reviews of missing data in multivariate analysis have been given by Murty and Federer (1991) and Little and Schenker (1992).

Dempster et al. (1977) discussed the EM algorithm for other families of distributions in addition to the multivariate normal. Szatrowski (1983) applied the EM algorithm to certain patterned mean vectors and covariance matrices. Srivastava (1985) used maximum likelihood estimators based on the EM algorithm to construct likelihood ratio tests for one-sample mean vectors (Section 3.3), two-sample mean vectors (Section 3.5), multivariate regression (Sections 7.4–7.8), and growth curves (Rencher 1995, Section 6.10). Little (1992) discussed regression with missing x's and applied many of the above procedures to this problem. Rubin (1978, 1987) recommended *multiple imputation*, in which several estimates are found for each missing value. The analysis is repeated for each of these multiple values, and final parameter estimates are obtained as averages. Rubin and Schenker (1987) applied multiple imputation to obtain improved confidence intervals for agriculture industry codes. Li, Raghunathan, and Rubin (1991), Li et al. (1991), and Meng and Rubin (1992) noted that multiple imputation yields better significance levels than single imputation.

Little and Smith (1987) and Little (1988) discussed robust estimation of the mean vector and covariance matrix. Todeschini (1990) proposed a k-nearest-neighbor method for estimating missing values. If the jth variable value is missing in an observation vector, it is estimated as the weighted (based on distance) average value of the jth variable in the k nearest observation vectors (whose jth variable is not missing). Rubin (1991) compared the EM algorithm to the following related techniques: data augmentation, stochastic relaxation (Gibbs sampler), and sampling importance resampling. Liu and Rubin (1994) considered extensions of the EM algorithm with improved convergence properties.

The last method we will discuss was proposed by Gleason and Staelin (1975). Their approach is based on the singular value decomposition of $\mathbf{Z}$, the standardized (by columns) data matrix. By (1.53), $\mathbf{Z}'\mathbf{Z}/(n-1) = \mathbf{R}$, the correlation matrix, and by (A.10.4), the normalized eigenvectors of $\mathbf{R}$ and $\mathbf{Z}'\mathbf{Z}$ are the same. Assuming $\mathbf{Z}$ is $n \times p$ of rank $p < n$, the *singular value decomposition* of $\mathbf{Z}$ is defined as

$$\mathbf{Z} = \mathbf{UDV}' = \sqrt{\lambda_1}\mathbf{u}_1\mathbf{v}_1' + \sqrt{\lambda_2}\mathbf{u}_2\mathbf{v}_2' + \cdots + \sqrt{\lambda_p}\mathbf{u}_p\mathbf{v}_p', \tag{1.109}$$

where $\lambda_1, \lambda_2, \ldots, \lambda_p$ are the eigenvalues of $\mathbf{Z}'\mathbf{Z} = (n-1)\mathbf{R}$, $\mathbf{D} = \text{diag}(\sqrt{\lambda_1}, \sqrt{\lambda_2}, \ldots, \sqrt{\lambda_p})$, $\mathbf{V}$ is the $p \times p$ matrix of (normalized) eigenvectors of $\mathbf{R}$, and $\mathbf{U}$ is the $n \times p$ matrix of the first p (normalized) eigenvectors (corresponding to the first p eigenvalues) of $\mathbf{ZZ}'$, or alternatively,

$$\mathbf{U} = \mathbf{ZVD}^{-1}. \tag{1.110}$$

Note that by (A.10.9), the eigenvalues $\lambda_1, \lambda_2, \ldots, \lambda_p$ of $\mathbf{Z}'\mathbf{Z}$ are also the nonzero eigenvalues of $\mathbf{ZZ}'$. The singular value decomposition of a full rank rectangular matrix is an extension of the spectral decomposition of a symmetric matrix given in

(A.11.4) and (A.11.5). It can be further extended to (nonsymmetric) square matrices or rectangular matrices not of full rank, in which case $\mathbf{D}$ is augmented by rows or columns of zeros.

For structural economy, we retain only the r largest eigenvalues $\lambda_1, \lambda_2, \ldots, \lambda_r$, where $\sum_{j=1}^{r} \lambda_j / \sum_{j=1}^{p} \lambda_j$ is a desired proportion. We denote the $r \times r$ submatrix of $\mathbf{D}$ that contains these r eigenvalues by $\hat{\mathbf{D}}$ and denote the corresponding $p \times r$ submatrix of $\mathbf{V}$ by $\hat{\mathbf{V}}$. We use (1.110) to obtain $\hat{\mathbf{U}} = \mathbf{Z}\hat{\mathbf{V}}\hat{\mathbf{D}}^{-1}$ and use (1.109) to obtain an approximation to $\mathbf{Z}$:

$$\begin{aligned} \hat{\mathbf{Z}} &= \hat{\mathbf{U}}\hat{\mathbf{D}}\hat{\mathbf{V}}' = (\mathbf{Z}\hat{\mathbf{V}}\hat{\mathbf{D}}^{-1})\hat{\mathbf{D}}\hat{\mathbf{V}}' \\ &= \mathbf{Z}\hat{\mathbf{V}}\hat{\mathbf{V}}'. \end{aligned} \tag{1.111}$$

The expression in (1.111) approximates $\mathbf{Z}$ in the least squares sense, as in (1.107) and (1.108); that is, $\sum_{jk}(z_{jk} - \hat{z}_{jk})^2$ is a minimum (Gabriel 1978, Householder and Young 1938, Jolliffe 1986, pp. 38–39). Now partition $\mathbf{Z}$ into $\mathbf{Z} = (\mathbf{Z}_1, \mathbf{Z}_2)$, where $\mathbf{Z}_1$ represents the columns that contain missing values, with a corresponding partitioning on $\mathbf{V}$ so that

$$\mathbf{V} = \begin{pmatrix} \mathbf{V}_1 \\ \mathbf{V}_2 \end{pmatrix}.$$

Then (1.111) becomes

$$(\hat{\mathbf{Z}}_1, \hat{\mathbf{Z}}_2) = (\mathbf{Z}_1, \mathbf{Z}_2)\hat{\mathbf{V}}\hat{\mathbf{V}}' = (\mathbf{Z}_1, \mathbf{Z}_2)\begin{pmatrix} \hat{\mathbf{V}}_1\hat{\mathbf{V}}_1' & \hat{\mathbf{V}}_1\hat{\mathbf{V}}_2' \\ \hat{\mathbf{V}}_2\hat{\mathbf{V}}_1' & \hat{\mathbf{V}}_2\hat{\mathbf{V}}_2' \end{pmatrix},$$

and by (A.3.1)

$$\hat{\mathbf{Z}}_1 = \mathbf{Z}_1\hat{\mathbf{V}}_1\hat{\mathbf{V}}_1' + \mathbf{Z}_2\hat{\mathbf{V}}_2\hat{\mathbf{V}}_1'. \tag{1.112}$$

At this point Gleason and Staelin replace $\mathbf{Z}_1$ on the right side of (1.112) with $\hat{\mathbf{Z}}_1$ in order to solve for $\hat{\mathbf{Z}}_1$ in terms of $\mathbf{Z}_2$:

$$\hat{\mathbf{Z}}_1 = \mathbf{Z}_2\hat{\mathbf{V}}_2\hat{\mathbf{V}}_1'(\mathbf{I} - \hat{\mathbf{V}}_1\hat{\mathbf{V}}_1')^{-1}. \tag{1.113}$$

Thus predicted values for the missing entries in $\mathbf{Z}_1$ are easily obtained using eigenvectors of $\mathbf{R}$. The process can then be iterated. An initial value of $\mathbf{R}$ must be provided, perhaps by Wilks' method of filling in means for missing values.

In a Monte Carlo study, Gleason and Staelin showed that their method is slightly better than the regression approach and that both are vastly superior to Wilks' method. They recommended their method because, for the cases they considered, it was 20 times as fast computationally as the regression method.

1.10 ROBUST ESTIMATORS OF $\boldsymbol{\mu}$ AND $\boldsymbol{\Sigma}$

Many authors have claimed that researchers can typically expect up to 10% of observations to have errors in measurement or recording. Another common assertion is that data are seldom normally distributed but rather tend to be "thicker" in the tails. A few values from such long tail areas can radically alter the sample mean and severely inflate the sample variance. *Robust estimators* of the mean and variance are less sensitive to extreme observations than are $\overline{y}$ and s^2. A robust estimator with a smaller estimate of variance than s^2 for long-tailed distributions is said to be *efficient*. For example, Mosteller and Tukey (1977) compared the *mean deviation* $\sum_i |y_i - \overline{y}|/n$ to s for a normal distribution and a *contaminated normal* where 1% of the observations came from a distribution with a standard deviation three times as large. They used large samples and found that for normality, mean deviation is 88% as efficient as s, whereas for 1% contamination, mean deviation is 144% as efficient as s.

A good review of robust estimators was given by Hogg (1979). In a large study at Princeton, 65 estimators of location were compared (Andrews et al. 1972). When contamination was present, the sample mean performed poorly. The study encourages the practice of checking for outliers and routinely discarding extreme observations. Using a Monte Carlo approach, Relles and Rogers (1977) compared some robust estimators of location with estimates by several statisticians who used their best judgment to reject outliers. They found (page 111) that the statisticians "stave off the tremendous disasters that one can experience using least squares. Statisticians are not robust, however, when compared with some of the estimators that are currently being advanced in the literature. The outlier rejecting statistician is at least 20 percent less efficient than the best estimator in all the situations studied here. In complicated multivariate contexts, the identification of outliers can be quite difficult, and the relative performance of the statistician may be even worse." They concluded (page 111) that "a simple algorithm for statisticians to become more robust might be to trim off one or two observations more than they think they should."

Hogg (1979) laments the lack of general acceptance and application of robust techniques and concedes that this may be due to an overabundance of robust schemes. He advocates concentrating our attention on Huber's (1964) M-estimator, which involves finding a solution for $\tilde{\mu}$ in

$$\sum_{i=1}^{n} \psi\left(\frac{y_i - \tilde{\mu}}{s}\right) = 0. \qquad (1.114)$$

Various forms of the function ψ have been suggested, one of which is given in (1.116) below in a multivariate context.

Gnanadesikan (1977, pp. 127–128) lists four robust multivariate estimators of $\boldsymbol{\mu}$, all of which are vectors of univariate robust estimators for each variable:

1. $\mathbf{y}_M^*$, the vector of medians (Mood 1941);
2. $\mathbf{y}_{\text{HL}}^*$, the vector of Hodges-Lehmann estimators, each of which is the median of the averages of pairs of observations (Bickel 1964);
3. $\mathbf{y}_{T(\alpha)}^*$, the vector of α-trimmed means, each of which is the mean remaining after discarding a proportion α of the smallest and largest observations [see Gnanadesikan and Kettenring (1972), who recommended $\alpha = .1$]; and

4. $\mathbf{y}^*_{\text{SINE}}$, the vector of M-estimators from (1.114) for each variable, where each $\tilde{\mu}_j$ in $\tilde{\boldsymbol{\mu}} = (\tilde{\mu}_1, \tilde{\mu}_2, \ldots, \tilde{\mu}_p)'$ is the solution to

$$\sum_{i=1}^{n} \psi\left(\frac{y_{ij} - \tilde{\mu}_j}{s_j}\right) = 0, \tag{1.115}$$

with ψ defined as

$$\psi(z) = \begin{cases} \sin(z/2.1) & \text{for } |z| < 2.1\pi, \\ 0 & \text{otherwise.} \end{cases} \tag{1.116}$$

Devlin, Gnanadesikan, and Kettenring (1981) compared five robust estimators of the covariance matrix under various conditions of distribution and contamination. We will discuss only the two estimators they found to perform best, $\tilde{\mathbf{S}}$(MVT) and $\tilde{\mathbf{S}}$(MLT). The *multivariate trimming* (MVT) approach is iterative but rather simple to carry out. At each step the squared distances

$$d_i^2 = (\mathbf{y}_i - \tilde{\boldsymbol{\mu}})'\tilde{\mathbf{S}}^{-1}(\mathbf{y}_i - \tilde{\boldsymbol{\mu}}), \qquad i = 1, 2, \ldots, n, \tag{1.117}$$

are calculated for the n observation vectors $\mathbf{y}_i$, $i = 1, 2, \ldots, n$, where $\tilde{\boldsymbol{\mu}}$ and $\tilde{\mathbf{S}}$ are the current estimates. A specified percentage (Devlin et al. used 10%) of the $\mathbf{y}_i$'s with the largest d_i^2 are set aside in each step, and the remaining observations are used to compute $\tilde{\boldsymbol{\mu}}$ and $\tilde{\mathbf{S}}$ for the next step using the ordinary formulas for $\bar{\mathbf{y}}$ and $\mathbf{S}$. Devlin et al. recommended the use of $\bar{\mathbf{y}}$ and $\mathbf{S}$ for initial estimates and suggested termination after 25 iterations or when Fisher's z-transform of the $\tilde{r}_{jk}$'s does not change by more than .001. Fisher's z-transform of $\tilde{r}_{jk}$ is defined as

$$z_{jk} = \frac{1}{2} \ln\left(\frac{1 + \tilde{r}_{jk}}{1 - \tilde{r}_{jk}}\right) = \tanh^{-1}\tilde{r}_{jk}. \tag{1.118}$$

The $\tilde{r}_{jk}$'s can be obtained at each step by converting $\tilde{\mathbf{S}}$ to $\tilde{\mathbf{R}}$ by the usual rescaling method in (1.51), $\tilde{\mathbf{R}} = \mathbf{D}^{-1}\tilde{\mathbf{S}}\mathbf{D}^{-1}$, where $\mathbf{D}$ is a diagonal matrix of square roots of diagonal elements of $\tilde{\mathbf{S}}$. The final estimates are denoted by $\tilde{\mathbf{S}}$(MVT) and $\tilde{\mathbf{R}}$(MVT).

The MLT (maximum likelihood t) procedure, a special case of an M-estimator, is similar to MVT but uses *weights* based on distances instead of trimming based on distances. The convergence criterion for this iterative procedure is the same as for MVT. The estimates at each step are

$$\tilde{\boldsymbol{\mu}} = \frac{\sum_{i=1}^{n} w_i \mathbf{y}_i}{\sum_{i=1}^{n} w_i}, \tag{1.119}$$

$$\tilde{\mathbf{S}} = \frac{1}{n} \sum_{i=1}^{n} w_i (\mathbf{y}_i - \tilde{\boldsymbol{\mu}})(\mathbf{y}_i - \tilde{\boldsymbol{\mu}})', \tag{1.120}$$

where

$$w_i = \frac{p + 1}{d_i^2 + 1}, \tag{1.121}$$

with d_i^2 defined in (1.117). The resulting estimates are denoted by $\tilde{\mathbf{S}}$(MLT) and $\tilde{\mathbf{R}}$(MLT), where $\tilde{\mathbf{R}} = \mathbf{D}^{-1}\tilde{\mathbf{S}}\mathbf{D}^{-1}$ as above.

Both of the above methods are *affine equivariant*, that is, invariant to linear transformations: If $\mathbf{z}_i = \mathbf{A}\mathbf{y}_i + \mathbf{b}$, then $\tilde{\boldsymbol{\mu}}_z = \mathbf{A}\tilde{\boldsymbol{\mu}}_y + \mathbf{b}$ and $\tilde{\mathbf{S}}_z = \mathbf{A}\tilde{\mathbf{S}}_y\mathbf{A}'$. In their Monte Carlo study, Devlin et al. (1981) found that $\tilde{\mathbf{R}}$(MLT) was least affected by the outlier types considered and provided the best robust estimates of eigenvalues and eigenvectors. However, when p is large, $\tilde{\mathbf{R}}$(MVT) is preferable because it has a higher *breakdown point*, which is defined as the percentage of outliers the procedure can handle. The breakdown point for MLT is inversely proportional to p, while that for MVT does not decrease with p but remains the same as the percentage trimmed.

Devlin et al. (1981) recommended that $\mathbf{R}$ and $\tilde{\mathbf{R}}$ be compared. Major differences in at least one element would indicate an outlier problem. In the presence of outliers, $\tilde{\mathbf{R}}$ obtained by either of the above approaches will be superior to $\mathbf{R}$.

Additional multivariate M-estimators were given by Maronna (1976) and extended by Huber (1981). These estimators are generalizations of maximum likelihood estimators (Section 2.3.1) of parameters of elliptical distributions (Section 2.4). The estimators $\tilde{\boldsymbol{\mu}}$ and $\tilde{\mathbf{S}}$ are defined as solutions of the simultaneous equations

$$\frac{1}{n}\sum_{i=1}^{n} v_1(d_i^2)(\mathbf{y}_i - \tilde{\boldsymbol{\mu}}) = \mathbf{0}, \tag{1.122}$$

$$\frac{1}{n}\sum_{i=1}^{n} \left[v_2(d_i^2)(\mathbf{y}_i - \tilde{\boldsymbol{\mu}})(\mathbf{y}_i - \tilde{\boldsymbol{\mu}})' - v_3(d_i^2)\tilde{\mathbf{S}}\right] = \mathbf{O}, \tag{1.123}$$

where v_1, v_2, and v_3 are real-valued functions chosen to achieve certain properties and where $d_i^2 = (\mathbf{y}_i - \tilde{\boldsymbol{\mu}})'\tilde{\mathbf{S}}^{-1}(\mathbf{y}_i - \tilde{\boldsymbol{\mu}})$ as in (1.117). Equations (1.122) and (1.123) must be solved iteratively.

Rousseeuw and Yohai (1984) proposed multivariate S-estimators of $\boldsymbol{\mu}$ and $\boldsymbol{\Sigma}$, denoted by $\tilde{\boldsymbol{\mu}}$ and $\tilde{\mathbf{S}}$, that minimize $|\tilde{\mathbf{S}}|$ subject to

$$\frac{1}{n}\sum_{i=1}^{n} \rho\left[(\mathbf{y}_i - \tilde{\boldsymbol{\mu}})'\tilde{\mathbf{S}}^{-1}(\mathbf{y}_i - \tilde{\boldsymbol{\mu}})\right] = b_0, \tag{1.124}$$

where $b_0 > 0$ and $\rho(d_i^2)$ are chosen to achieve certain properties. Rocke (1992) and Lopuhaä (1989) showed that S-estimators are a special case of M-estimators.

Rousseeuw (1984) introduced MVE estimators. The MVE estimator $\tilde{\boldsymbol{\mu}}$ is the center of the minimum-volume ellipsoid that covers at least $\lfloor n/2 \rfloor + 1$ points in the sample, where $\lfloor n/2 \rfloor$ is the greatest integer $\leq n/2$. The MVE estimator of $\boldsymbol{\Sigma}$ is the matrix $\tilde{\mathbf{S}}$ associated with the equation $(\mathbf{y} - \tilde{\boldsymbol{\mu}})'\tilde{\mathbf{S}}^{-1}(\mathbf{y} - \tilde{\boldsymbol{\mu}}) = 1$, which characterizes the minimum volume ellipsoid. For computational details of the M, S, and MVE robust estimators, see Rocke (1992), who notes that computational problems arise with these estimators when either p or n is moderately large.

The S and MVE estimators were introduced to improve on the *breakdown point* (the fraction of outliers or contaminated data that has a large effect on the estimator) of the M-estimator. The M-estimators were believed to have had a breakdown point of $1/(p+1)$, where p is the number of variables (Maronna 1976, Huber 1981). However, with appropriate choices of input functions, the breakdown point for M is the same as that of the S and MVE estimators, which approaches $\frac{1}{2}$ (Tyler 1991, Rocke 1992). Rocke and Woodruff (1993) and Woodruff and Rocke (1994) discuss a compound estimator using an MVE estimator as a starting point for an S-estimator.

Mehrotra (1995) proposed a robust estimator of the covariance matrix based on robust estimators of each variance and covariance separately. The method is noniterative, but it will sometimes produce an estimated covariance matrix that is not positive definite.

PROBLEMS

1.1 Show that $E(\overline{y}) = \mu$ as in (1.6).

1.2 Show that $E(y-\mu)^2 = E(y^2) - \mu^2$ as in (1.9).

1.3 Show that $\sum_{i=1}^n (y_i - \overline{y})^2 = \sum_{i=1}^n y_i^2 - n\overline{y}^2$ as in (1.11).

1.4 Show that the expected value of a quadratic form is given by

$$E(\mathbf{y}'\mathbf{A}\mathbf{y}) = \text{tr}(\mathbf{A}\boldsymbol{\Sigma}) + \boldsymbol{\mu}'\mathbf{A}\boldsymbol{\mu}, \tag{1.125}$$

where

$$E(\mathbf{y}) = \boldsymbol{\mu} \quad \text{and} \quad \text{cov}(\mathbf{y}) = \boldsymbol{\Sigma}.$$

1.5 Use (1.125) from Problem 1.4 to show that $E(s^2) = \sigma^2$ as in (1.12).

1.6 Show that $\text{var}(ay) = a^2\sigma^2$ as in (1.13).

1.7 Show that $E[(x-\mu_x)(y-\mu_y)] = E(xy) - \mu_x\mu_y$ as in (1.16).

1.8 Show that if x and y are independent, then $E(xy) = E(x)E(y)$ as in (1.18).

1.9 Show that $\sigma_{xy} = 0$ if x and y are independent, as in (1.19).

1.10 Show that $\sum_{i=1}^n (x_i - \overline{x})(y_i - \overline{y}) = \sum_{i=1}^n x_i y_i - n\overline{x}\,\overline{y}$, thus verifying (1.21).

1.11 Show that

$$E(\mathbf{x}'\mathbf{A}\mathbf{y}) = \text{tr}(\mathbf{A}\boldsymbol{\Sigma}_{yx}) + \boldsymbol{\mu}_x'\mathbf{A}\boldsymbol{\mu}_y, \tag{1.126}$$

where $\mathbf{x}' = (x_1, x_2, \ldots, x_n)$, $\mathbf{y}' = (y_1, y_2, \ldots, y_n)$, $\mathbf{A}$ is an $n \times n$ matrix of constants, $\boldsymbol{\Sigma}_{yx} = \text{cov}(\mathbf{y}, \mathbf{x})$ is the square matrix of covariances of the y's with the x's, $\boldsymbol{\mu}_x = E(\mathbf{x})$, and $\boldsymbol{\mu}_y = E(\mathbf{y})$. (This result holds also for the more general case where $\mathbf{x}$ and $\mathbf{y}$ are not the same size, in which case $\mathbf{A}$ and $\boldsymbol{\Sigma}_{yx}$ become rectangular.)

1.12 Use (1.126) from Problem 1.11 to prove that $E(s_{xy}) = \sigma_{xy}$ as in (1.22).

1.13 Evaluate the correction factor $1 + (1 - r_{xy}^2)/2(n-3)$ in (1.25) for the nine combinations of $r_{xy} = .3, .6, .9$ and $n = 10, 30, 50$.

1.14 Show that $\sum_{i=1}^n \mathbf{y}_i/n = (\overline{y}_1, \overline{y}_2, \ldots, \overline{y}_p)'$ as in (1.28).

1.15 Show that $\mathbf{Y}'\mathbf{j} = \sum_i \mathbf{y}_i$ as in (1.29).

1.16 Show that (1.37) is equivalent to (1.32); that is,

$$\frac{1}{n-1}\sum_{i=1}^{n}(\mathbf{y}_i - \overline{\mathbf{y}})(\mathbf{y}_i - \overline{\mathbf{y}})' = \begin{pmatrix} s_{11} & s_{12} & \cdots & s_{1p} \\ s_{21} & s_{22} & \cdots & s_{2p} \\ \vdots & \vdots & & \vdots \\ s_{p1} & s_{p2} & \cdots & s_{pp} \end{pmatrix}.$$

1.17 Show that $\sum_{i=1}^{n}(\mathbf{y}_i - \overline{\mathbf{y}})(\mathbf{y}_i - \overline{\mathbf{y}})' = \sum_i \mathbf{y}_i\mathbf{y}_i' - n\overline{\mathbf{y}}\,\overline{\mathbf{y}}'$ as in (1.38).

1.18 Show that $[\mathbf{I} - (1/n)\,\mathbf{J}]'[\mathbf{I} - (1/n)\,\mathbf{J}] = \mathbf{I} - (1/n)\,\mathbf{J}$ as in (1.41).

1.19 Show that we must have $n - 1 > p$ for $\mathbf{S}$ to be positive definite, as noted following (1.42).

1.20 Replace $\overline{\mathbf{y}}$ by $\mathbf{a}$ in (1.37) to obtain $\mathbf{S}_a = \sum_{i=1}^{n}(\mathbf{y}_i - \mathbf{a})(\mathbf{y}_i - \mathbf{a})'/(n-1)$.

(a) Show that $\mathbf{S}_a = \mathbf{S} + n(\overline{\mathbf{y}} - \mathbf{a})(\overline{\mathbf{y}} - \mathbf{a})'/(n-1)$.

(b) Show that $|\mathbf{S}_a| = |\mathbf{S}|[1 + n(\overline{\mathbf{y}} - \mathbf{a})'\mathbf{S}^{-1}(\overline{\mathbf{y}} - \mathbf{a})/(n-1)]$ and that $\min_{\mathbf{a}} |\mathbf{S}_a| = |\mathbf{S}|$.

(c) Show that $\min_{\mathbf{a}} \mathrm{tr}(\mathbf{S}_a) = \mathrm{tr}(\mathbf{S})$.

1.21 Show that $E[(\mathbf{y} - \boldsymbol{\mu})(\mathbf{y} - \boldsymbol{\mu})'] = E(\mathbf{y}\mathbf{y}') - \boldsymbol{\mu}\boldsymbol{\mu}'$ as in (1.46).

1.22 Show that $E(\mathbf{S}) = \boldsymbol{\Sigma}$, as in (1.47), using the following two approaches:

(a) Use (1.38).

(b) Show that the numerator of (1.37) can be expressed as

$$\sum_{i=1}^{n}(\mathbf{y}_i - \overline{\mathbf{y}})(\mathbf{y}_i - \overline{\mathbf{y}})' = \sum_{i=1}^{n}(\mathbf{y}_i - \boldsymbol{\mu})(\mathbf{y}_i - \boldsymbol{\mu})' - n(\overline{\mathbf{y}} - \boldsymbol{\mu})(\overline{\mathbf{y}} - \boldsymbol{\mu})'. \quad (1.127)$$

1.23 Define a generalized quadratic form $\mathbf{Y}'\mathbf{A}\mathbf{Y}$, where $\mathbf{Y} = (\mathbf{y}_1, \mathbf{y}_2, \ldots, \mathbf{y}_n)'$ as in (1.26) and $\mathbf{A}$ is a constant $n \times n$ symmetric matrix [see, for example, (1.40)]. If $\mathbf{y}_1, \mathbf{y}_2, \ldots, \mathbf{y}_n$ are independent (see Section 1.3) with $E(\mathbf{y}_i) = \boldsymbol{\mu}_i$ and $\mathrm{cov}(\mathbf{y}_i) = \boldsymbol{\Sigma}$, show that

$$E(\mathbf{Y}'\mathbf{A}\mathbf{Y}) = (\mathrm{tr}\,\mathbf{A})\boldsymbol{\Sigma} + E(\mathbf{Y}')\mathbf{A}E(\mathbf{Y}). \quad (1.128)$$

1.24 Use (1.128) from Problem 1.23 to obtain an alternative proof that $E(\mathbf{S}) = \boldsymbol{\Sigma}$ as in (1.47).

1.25 For $p = 3$, show that $\mathbf{D}_s^{-1}\mathbf{S}\mathbf{D}_s^{-1} = \mathbf{R}$, which illustrates (1.51).

1.26 Prove $\mathbf{S}_z = \mathbf{R}$, as in (1.53), using the following steps:

(a) Transform y_{ij} to $z_{ij} = (y_{ij} - \overline{y}_j)/s_j$. Then the ith row of $\mathbf{Z}$ becomes $\mathbf{z}_i' = (\mathbf{y}_i - \overline{\mathbf{y}})'\mathbf{D}_s^{-1}$, where $\mathbf{D}_s$ is defined by (1.50). Show that $\overline{\mathbf{z}}' = \mathbf{0}'$.

(b) Using (1.38) with $\overline{\mathbf{z}} = \mathbf{0}$, show that $\mathbf{S}_z = \sum_{i=1}^{n} \mathbf{z}_i\mathbf{z}_i'/(n-1) = \mathbf{D}_s^{-1}\mathbf{S}\mathbf{D}_s^{-1} = \mathbf{R}$.

1.27 Show that $\mathbf{Z}'\mathbf{Z}/(n-1) = \mathbf{R}$, as in (1.53), where $\mathbf{Z}$ is the standardized (by columns) form of the data matrix $\mathbf{Y}$.

1.28 Define the standardization $\mathbf{z}_i = \mathbf{S}^{-1/2}(\mathbf{y}_i - \bar{\mathbf{y}})$, $i = 1, 2, \ldots, n$, where $\mathbf{S}^{-1/2}$ is the inverse of the square root matrix $\mathbf{S}^{1/2}$ defined in (A.11.7). Show that $\bar{\mathbf{z}} = \mathbf{0}$ and $\mathbf{S}_z = \mathbf{I}$.

1.29 Show that $\mathbf{S}_{yx}$ in (1.59) can be expressed in the following two forms:

(a) $\mathbf{S}_{yx} = \sum_{i=1}^{n}(\mathbf{y}_i - \bar{\mathbf{y}})(\mathbf{x}_i - \bar{\mathbf{x}})'/(n-1)$.

(b) $\mathbf{S}_{yx} = \mathbf{Y}'[\mathbf{I} - (1/n)\mathbf{J}]\mathbf{X}/(n-1)$, where $\mathbf{Y}$ and $\mathbf{X}$ are the data matrices for $\mathbf{y}_1, \ldots, \mathbf{y}_n$ and $\mathbf{x}_1, \ldots, \mathbf{x}_n$. Thus $\mathbf{Y}$ is $n \times p$, $\mathbf{X}$ is $n \times q$, $\mathbf{I}$ is $n \times n$, and $\mathbf{J}$ is $n \times n$.

1.30 Let $p = 2$ and $q = 3$, so that $\mathbf{y} = (y_1, y_2)'$ and $\mathbf{x} = (x_1, x_2, x_3)'$ in Section 1.5. Write out (1.65) and (1.66) and show that, for this illustration, $\text{cov}(\mathbf{y}, \mathbf{x}) = \text{cov}(\mathbf{x}, \mathbf{y})'$.

1.31 Explain why $\text{cov}(\mathbf{x} + \mathbf{y})$ in (1.68) does not have the form $\mathbf{\Sigma}_{xx} + \mathbf{\Sigma}_{yy} + 2\mathbf{\Sigma}_{xy}$.

1.32 Show that the sample covariance of $z = \mathbf{a}'\mathbf{y}$ and $w = \mathbf{b}'\mathbf{y}$ is given by $s_{zw} = \mathbf{a}'\mathbf{S}\mathbf{b}$ as in (1.74).

1.33 Derive (1.76) and (1.77); that is, show that if $\mathbf{z}_i = \mathbf{A}\mathbf{y}_i$, $i = 1, 2, \ldots, n$, then $\bar{\mathbf{z}} = \mathbf{A}\bar{\mathbf{y}}$ and $\mathbf{S}_z = \mathbf{A}\mathbf{S}\mathbf{A}'$.

1.34 Show that if $\mathbf{z}_i = \mathbf{A}\mathbf{y}_i + \mathbf{b}$, $i = 1, 2, \ldots, n$, then $\bar{\mathbf{z}} = \mathbf{A}\bar{\mathbf{y}} + \mathbf{b}$ and $\mathbf{S}_z = \mathbf{A}\mathbf{S}\mathbf{A}'$.

1.35 Show that for $\mathbf{z} = \mathbf{A}\mathbf{y}$ and $\mathbf{w} = \mathbf{B}\mathbf{y}$ the sample covariance matrix is $\widehat{\text{cov}}(\mathbf{A}\mathbf{y}, \mathbf{B}\mathbf{y}) = \mathbf{A}\mathbf{S}\mathbf{B}'$ as in (1.78).

1.36 Show that (1.79) follows from (1.78); that is, obtain $\widehat{\text{cov}}(\mathbf{A}\mathbf{y}, \mathbf{b}'\mathbf{y}) = \mathbf{A}\mathbf{S}\mathbf{b}$ from $\widehat{\text{cov}}(\mathbf{A}\mathbf{y}, \mathbf{B}\mathbf{y}) = \mathbf{A}\mathbf{S}\mathbf{B}'$.

1.37 Derive (1.80), the sample covariance of $u = \mathbf{a}'\mathbf{y}$ and $v = \mathbf{b}'\mathbf{x}$.

1.38 Show that $\widehat{\text{cov}}(\mathbf{A}\mathbf{y}, \mathbf{B}\mathbf{x}) = \mathbf{A}\mathbf{S}_{yx}\mathbf{B}'$ as in (1.84).

1.39 Prove that $\text{var}(\mathbf{a}'\mathbf{y}) = \mathbf{a}'\mathbf{\Sigma}\mathbf{a}$ as in (1.87).

1.40 Prove that $\text{cov}(\mathbf{a}'\mathbf{y}, \mathbf{b}'\mathbf{y}) = \mathbf{a}'\mathbf{\Sigma}\mathbf{b}$ as in (1.91).

1.41 Prove that $E(\mathbf{A}\mathbf{y}) = \mathbf{A}E(\mathbf{y})$ as in (1.93).

1.42 Using (1.44) and (1.90), prove that $\text{cov}(\mathbf{A}\mathbf{y}) = \mathbf{A}\mathbf{\Sigma}\mathbf{A}'$ as in (1.94).

1.43 Show that $E(\mathbf{A}\mathbf{y} + \mathbf{b}) = \mathbf{A}\boldsymbol{\mu} + \mathbf{b}$ and that $\text{cov}(\mathbf{A}\mathbf{y} + \mathbf{b}) = \mathbf{A}\mathbf{\Sigma}\mathbf{A}'$, where $\mathbf{A}$ is a constant $k \times p$ matrix and $\mathbf{b}$ is a constant $k \times 1$ vector.

1.44 Using (1.65) and (1.90), show that $\text{cov}(\mathbf{A}\mathbf{y}, \mathbf{B}\mathbf{y}) = \mathbf{A}\mathbf{\Sigma}\mathbf{B}'$ as in (1.95).

1.45 Prove that $\text{cov}(\mathbf{a}'\mathbf{y}, \mathbf{b}'\mathbf{x}) = \mathbf{a}'\mathbf{\Sigma}_{yx}\mathbf{b}$ as in (1.96).

1.46 Prove that $\text{cov}(\mathbf{A}\mathbf{y}, \mathbf{B}\mathbf{x}) = \mathbf{A}\mathbf{\Sigma}_{yx}\mathbf{B}'$ as in (1.98).

1.47 For the Gleason–Staelin index of intercorrelation g in (1.100), show that $g = 0$ when the variables are independent ($\mathbf{R} = \mathbf{I}$) and $g = 1$ when the variables are perfectly correlated ($\mathbf{R} = \mathbf{J}$).

1.48 Verify (1.102), $R_j^2 = 1 - 1/r^{jj}$, for $j = 1$.

1.49 Verify (1.112), $\hat{\mathbf{Z}}_1 = \mathbf{Z}_1\hat{\mathbf{V}}_1\hat{\mathbf{V}}_1' + \mathbf{Z}_2\hat{\mathbf{V}}_2\hat{\mathbf{V}}_1'$.

1.50 Verify (1.113), $\hat{\mathbf{Z}}_1 = \mathbf{Z}_2\hat{\mathbf{V}}_2\hat{\mathbf{V}}_1'(\mathbf{I} - \hat{\mathbf{V}}_1\hat{\mathbf{V}}_1')^{-1}$.

1.51 Verify the statement in Section 1.10 that a Mahalanobis distance transforms the variables so that they are uncorrelated and have the same variance.

1.52 Calculate r_{xy} and r_{xy}^*, as given by (1.24) and (1.25), for the height and weight data in Table 1.1.

1.53 Calculate $\overline{\mathbf{y}}$ and $\mathbf{S}$ for the probe word data in Table 1.3.

1.54 For the probe word data in Table 1.3, define $z = 2y_1 - 3y_2 + y_3 - 2y_4 + 3y_5 = (2, -3, 1, -2, 3)\mathbf{y}$. Find $\overline{z}$ and s_z^2 in two ways:

(a) Evaluate z for each row of Table 1.3 and find $\overline{z}$ and s_z^2 directly from $z_1, z_2, \ldots, z_{11}$.

(b) Use $\overline{z} = \mathbf{a}'\overline{\mathbf{y}}$ and $s_z^2 = \mathbf{a}'\mathbf{S}\mathbf{a}$, as in (1.72) and (1.73).

1.55 For the probe word data in Table 1.3, define $w = -y_1 + 3y_2 + 2y_3 - 2y_4 - 3y_5$ and define z as in the previous problem.

(a) Find s_{zw} by (1.74).

(b) Evaluate z and w for each row of Table 1.3 and find r_{zw} from the 11 pairs (z_i, w_i), $i = 1, 2, \ldots, 11$, using (1.24).

(c) Find r_{zw} using (1.75).

1.56 For the probe word data in Table 1.3, find the correlation between $z = (y_1 + y_2)/2$ and $w = (y_3 + y_4 + y_5)/3$.

1.57 Define the following linear combinations for the probe word data in Table 1.3:

$$
\begin{aligned}
z_1 &= y_1 + y_2 + y_3 + y_4 + y_5 \\
z_2 &= 3y_1 - y_2 - 2y_3 + y_4 + 3y_5 \\
z_3 &= -2y_1 - 3y_2 + y_3 + 3y_4 - 2y_5.
\end{aligned}
$$

(a) Find $\overline{\mathbf{z}}$ using (1.76).

(b) Find $\mathbf{S}_z$ using (1.77).

1.58 Define the following linear combinations for the probe word data in Table 1.3:

$$
\begin{aligned}
w_1 &= 2y_1 + 3y_2 - y_3 - 2y_4 - 4y_5 \\
w_2 &= -y_1 + 4y_2 - 3y_3 + 2y_4 - y_5 .
\end{aligned}
$$

Using the three z's defined in the previous problem, find the covariance matrix between the z's and the w's, $\mathbf{S}_{zw}$, using (1.78).

1.59 For the diabetes data in Table 1.2, find the mean vector and covariance matrix for all five variables and partition them into $\binom{\mathbf{y}}{\mathbf{x}}$ as in (1.58), and

$$
\mathbf{S} = \begin{pmatrix} \mathbf{S}_{yy} & \mathbf{S}_{yx} \\ \mathbf{S}_{xy} & \mathbf{S}_{xx} \end{pmatrix}
$$

as in (1.59).

1.60 Use all five variables in the diabetes data in Table 1.2.

(a) Find **R** and the eigenvalues of **R**.

(b) Calculate the following measures of intercorrelation from Section 1.7: $g, q_1, q_2, q_3, q_4, q_5$, and q_6.

1.61 For the diabetes data in Table 1.2, define $u = 2y_1 - 3y_2$ and $v = x_1 - 2x_2 + x_3$. Find s_u^2, s_v^2, s_{uv}, and r_{uv}, using (1.81), (1.82), (1.80), and (1.83), respectively.

1.62 For the diabetes data in Table 1.2, define

$$u_1 = y_1 + y_2, \qquad u_2 = 2y_1 - y_2$$

and

$$v_1 = x_1 + x_2 - 2x_3, \qquad v_2 = x_1 - 2x_2 + x_3.$$

(a) Find $\mathbf{S}_u$ and $\mathbf{S}_v$ using (1.77).

(b) Find the matrix of sample covariances of the u's with the v's, $\widehat{\text{cov}}(\mathbf{u}, \mathbf{v})$, using (1.84).

1.63 Table 1.4 lists data from O'Sullivan and Mahan (1966, see also Andrews and Herzberg 1985, p. 214), where y_1, y_2, y_3 and x_1, x_2, x_3 represent measurements

Table 1.4. Glucose Measurements on Three Occasions

Fasting			One Hour After Sugar Intake		
y_1	y_2	y_3	x_1	x_2	x_3
60	69	62	97	69	98
56	53	84	103	78	107
80	69	76	66	99	130
55	80	90	80	85	114
62	75	68	116	130	91
74	64	70	109	101	103
64	71	66	77	102	130
73	70	64	115	110	109
68	67	75	76	85	119
69	82	74	72	133	127
60	67	61	130	134	121
70	74	78	150	158	100
66	74	78	150	131	142
83	70	74	99	98	105
68	66	90	119	85	109
78	63	75	164	98	138
77	68	74	144	71	153
66	77	68	77	82	89
70	70	72	114	93	122
75	65	71	77	70	109

Note: Measurements are in mg/100 ml.

of blood glucose levels on three occasions for 20 women. Find the mean vector and covariance matrix for all six variables and partition them into $\binom{\bar{\mathbf{y}}}{\bar{\mathbf{x}}}$ as in (1.58) and

$$\mathbf{S} = \begin{pmatrix} \mathbf{S}_{yy} & \mathbf{S}_{yx} \\ \mathbf{S}_{xy} & \mathbf{S}_{xx} \end{pmatrix}$$

as in (1.59).

1.64 For the glucose data in Table 1.4, define $u = 3y_1 - 2y_2 + y_3$ and $v = x_1 + 3x_2 - 2x_3$. Find s_u^2, s_v^2, s_{uv}, and r_{uv} using (1.81), (1.82), (1.80), and (1.83), respectively.

1.65 For the glucose data in Table 1.4, define

$$u_1 = y_1 + y_2 + y_3, \qquad u_2 = y_1 - 2y_2 + 2y_3$$

and

$$v_1 = 3x_1 + x_2 - 2x_3, \qquad v_2 = -x_1 - 2x_2 + x_3, \qquad v_3 = 2x_1 - 3x_2 + x_3.$$

(a) Find $\mathbf{S}_u$ and $\mathbf{S}_v$ using (1.77).

(b) Find the matrix of sample covariances of the u's with the v's, $\widehat{\text{cov}}(\mathbf{u}, \mathbf{v})$, using (1.84).

CHAPTER 2

The Multivariate Normal Distribution

The majority of multivariate inferential procedures are based on the multivariate normal distribution, which is a direct generalization of the univariate normal distribution. While real data may not often be exactly multivariate normal, the multivariate normal will frequently serve as a useful approximation to the true distribution. Fortunately, many of the procedures based on multivariate normality are robust to departures from normality.

2.1 UNIVARIATE AND MULTIVARIATE NORMAL DENSITY FUNCTIONS

We begin with a review of the univariate normal density in Section 2.1.1. In Section 2.1.2, we obtain the multivariate normal density directly from the univariate normal, and in Sections 2.1.3–2.1.5, we discuss some characteristics of the multivariate normal density.

2.1.1 Univariate Normal

A normally distributed random variable y with mean μ and variance σ^2 has the density function

$$g(y) = \frac{1}{\sqrt{2\pi}\sqrt{\sigma^2}} e^{-(y-\mu)^2/2\sigma^2}, \qquad -\infty < y < \infty. \tag{2.1}$$

We indicate that y has the density (2.1) by saying that y is distributed as $N(\mu, \sigma^2)$, or simply that y is $N(\mu, \sigma^2)$. The standardized variate $z = (y - \mu)/\sigma$ has mean 0 and variance 1. If z is normal, (2.1) becomes

$$f(z) = \frac{1}{\sqrt{2\pi}} e^{-z^2/2}. \tag{2.2}$$

We now discuss finding the density $g(y)$ in (2.1) from $f(z)$ in (2.2). The technique will be extended to the multivariate case in Section 2.1.2, in which we obtain the multivariate normal distribution by transforming independent univariate normal random variables into correlated variables and then finding their density.

The transformation from z to y is $y = \sigma z + \mu$, which is a monotone increasing function of z. To find the density $g(y)$, we work with the distribution function $G(y)$, defined as $G(y) = \int_{-\infty}^{y} g(t)\,dt$. By the fundamental theorem of calculus, $G'(y) = g(y)$. Likewise, the distribution function for z is $F(z)$, and we have $F'(z) = f(z)$. The correspondence $G(y) = F(z)$, where $z = (y - \mu)/\sigma$, can be established by noting that $G(y) = \int_{-\infty}^{y} g(t)\,dt = \int_{-\infty}^{(y-\mu)/\sigma} f(x)\,dx = F[(y - \mu)/\sigma] = F(z)$. Then by the chain rule,

$$g(y) = G'(y) = F'(z)\frac{dz}{dy} = f(z)\frac{dz}{dy}. \tag{2.3}$$

For other transformations, y may be a monotone decreasing function of z, in which case $G(y) = 1 - F(z)$. Then (2.3) becomes

$$g(y) = -f(z)\frac{dz}{dy}. \tag{2.4}$$

Note that $-dz/dy$ is positive since dz/dy is negative. Both (2.3) and (2.4) can be expressed using the absolute value of dz/dy:

$$g(y) = f(z)\left|\frac{dz}{dy}\right|. \tag{2.5}$$

To use (2.5) to find the density of y, it is clear that both z and dz/dy on the right side must be expressed in terms of y.

Let us apply (2.5) to $y = \sigma z + \mu$. The density $f(z)$ is given by (2.2), and for $z = (y - \mu)/\sigma$, $|dz/dy| = 1/\sigma$. Thus

$$\begin{aligned} g(y) = f(z)\left|\frac{dz}{dy}\right| &= f\left(\frac{y-\mu}{\sigma}\right)\frac{1}{\sigma} \\ &= \frac{1}{\sqrt{2\pi}\sigma} e^{-(y-\mu)^2/2\sigma^2}, \end{aligned}$$

which is the same as (2.1).

2.1.2 Multivariate Normal

We now obtain the multivariate normal density by transforming the random vector $\mathbf{z} = (z_1, z_2, \ldots, z_p)'$, where each z_i has the $N(0, 1)$ density in (2.2) and the z's are independent. Thus $E(\mathbf{z}) = \mathbf{0}$ and $\text{cov}(\mathbf{z}) = \mathbf{I}$. When random variables are independently distributed, their joint density is the product of their individual densities [see a remark following (1.16)]. Thus, using (2.2) and inserting subscripts to distinguish the univariate densities on the right side from the joint density on the left,

$$\begin{aligned} f(\mathbf{z}) &= f_1(z_1)f_2(z_2)\cdots f_p(z_p) \\ &= \frac{1}{\sqrt{2\pi}}e^{-z_1^2/2}\frac{1}{\sqrt{2\pi}}e^{-z_2^2/2}\cdots\frac{1}{\sqrt{2\pi}}e^{-z_p^2/2} \\ &= \frac{1}{(\sqrt{2\pi})^p}e^{-\sum_j z_j^2/2} \\ &= \frac{1}{(\sqrt{2\pi})^p}e^{-\mathbf{z}'\mathbf{z}/2}. \end{aligned} \tag{2.6}$$

We say that $\mathbf{z}$ has a multivariate normal density with mean vector $\mathbf{0}$ and covariance matrix $\mathbf{I}$, or simply that $\mathbf{z}$ is distributed as $N_p(\mathbf{0}, \mathbf{I})$.

We wish to obtain the multivariate normal density with arbitrary mean vector $\boldsymbol{\mu}$ and (positive definite) covariance matrix $\boldsymbol{\Sigma}$. By analogy with $y = \sigma z + \mu$ in Section 2.1.1, we define the transformation

$$\mathbf{y} = \boldsymbol{\Sigma}^{1/2}\mathbf{z} + \boldsymbol{\mu}, \tag{2.7}$$

where $\boldsymbol{\Sigma}^{1/2}$ is the (symmetric) square root matrix defined in (A.11.7). The mean vector and covariance matrix for the transformed random vector $\mathbf{y}$ are (see Problem 1.43)

$$\begin{aligned} E(\mathbf{y}) &= E(\boldsymbol{\Sigma}^{1/2}\mathbf{z} + \boldsymbol{\mu}) = \boldsymbol{\Sigma}^{1/2}E(\mathbf{z}) + E(\boldsymbol{\mu}) \\ &= \boldsymbol{\Sigma}^{1/2}\mathbf{0} + \boldsymbol{\mu} = \boldsymbol{\mu}, \\ \text{cov}(\mathbf{y}) &= \text{cov}(\boldsymbol{\Sigma}^{1/2}\mathbf{z} + \boldsymbol{\mu}) = \boldsymbol{\Sigma}^{1/2}\,\text{cov}(\mathbf{z})(\boldsymbol{\Sigma}^{1/2})' \\ &= \boldsymbol{\Sigma}^{1/2}\mathbf{I}\boldsymbol{\Sigma}^{1/2} = \boldsymbol{\Sigma}. \end{aligned}$$

Let us now find the density of $\mathbf{y} = \boldsymbol{\Sigma}^{1/2}\mathbf{z} + \boldsymbol{\mu}$ from the density of $\mathbf{z}$ in (2.6). By (2.5), the density of $y = \sigma z + \mu$ is $g(y) = f(z)|dz/dy| = f(z)|1/\sigma|$. The analogous expression for multivariate linear transformations such as $\mathbf{y} = \boldsymbol{\Sigma}^{1/2}\mathbf{z} + \boldsymbol{\mu}$ is

$$g(\mathbf{y}) = f(\mathbf{z})\text{abs}|\boldsymbol{\Sigma}^{-1/2}|, \tag{2.8}$$

where $\mathbf{z}$ must be written in terms of $\mathbf{y}$ and $\boldsymbol{\Sigma}^{-1/2}$ is defined as $(\boldsymbol{\Sigma}^{1/2})^{-1}$. Note that in (2.8) we use the vertical bars for the determinant of a matrix as opposed to the use of vertical bars to indicate the absolute value of a scalar quantity in (2.5). Thus the expression $\text{abs}|\boldsymbol{\Sigma}^{-1/2}|$ represents the absolute value of the determinant of $\boldsymbol{\Sigma}^{-1/2}$ and parallels the absolute-value expression $|dz/dy| = |1/\sigma|$ in the univariate illustration. (The determinant $|\boldsymbol{\Sigma}^{-1/2}|$ is the *Jacobian* of the transformation; see any advanced calculus text.) Just as $1/\sigma$ does not require an absolute-value designation, we can likewise dispense with the absolute value in (2.8), since $\boldsymbol{\Sigma}^{-1/2}$ is positive definite. Thus (2.8) can be written as

$$g(\mathbf{y}) = f(\mathbf{z})|\boldsymbol{\Sigma}^{-1/2}| \tag{2.9}$$

$$= f(\mathbf{z})|\boldsymbol{\Sigma}|^{-1/2} \qquad \text{[by (A.7.5) and (A.7.6)]}. \tag{2.10}$$

From (2.7), we have $\mathbf{z} = \boldsymbol{\Sigma}^{-1/2}(\mathbf{y} - \boldsymbol{\mu})$, and using (2.6) and (2.10), we can write the density of $\mathbf{y}$ as

$$
\begin{aligned}
g(\mathbf{y}) = f(\mathbf{z})|\boldsymbol{\Sigma}|^{-1/2} &= \frac{1}{(\sqrt{2\pi})^p|\boldsymbol{\Sigma}|^{1/2}} e^{-\mathbf{z}'\mathbf{z}/2} \\
&= \frac{1}{(\sqrt{2\pi})^p|\boldsymbol{\Sigma}|^{1/2}} e^{-[\boldsymbol{\Sigma}^{-1/2}(\mathbf{y}-\boldsymbol{\mu})]'[\boldsymbol{\Sigma}^{-1/2}(\mathbf{y}-\boldsymbol{\mu})]/2} \\
&= \frac{1}{(\sqrt{2\pi})^p|\boldsymbol{\Sigma}|^{1/2}} e^{-(\mathbf{y}-\boldsymbol{\mu})'(\boldsymbol{\Sigma}^{1/2}\boldsymbol{\Sigma}^{1/2})^{-1}(\mathbf{y}-\boldsymbol{\mu})/2} \\
&= \frac{1}{(\sqrt{2\pi})^p|\boldsymbol{\Sigma}|^{1/2}} e^{-(\mathbf{y}-\boldsymbol{\mu})'\boldsymbol{\Sigma}^{-1}(\mathbf{y}-\boldsymbol{\mu})/2}, \qquad (2.11)
\end{aligned}
$$

which is the multivariate normal density function with mean vector $\boldsymbol{\mu}$ and covariance matrix $\boldsymbol{\Sigma}$. When $\mathbf{y}$ has the density (2.11), we say that $\mathbf{y}$ is distributed as $N_p(\boldsymbol{\mu}, \boldsymbol{\Sigma})$, or simply that $\mathbf{y}$ is $N_p(\boldsymbol{\mu}, \boldsymbol{\Sigma})$.

A comparison of (2.11) and (2.1) shows the generalized (Mahalanobis) distance $(\mathbf{y} - \boldsymbol{\mu})'\boldsymbol{\Sigma}^{-1}(\mathbf{y} - \boldsymbol{\mu})$ in place of $(y - \mu)^2/\sigma^2$ in the exponent and the generalized variance $|\boldsymbol{\Sigma}|$ replacing σ^2 in the denominator. These functions serve analogous purposes in the two densities; $\mathbf{y}$ is less likely to be far from $\boldsymbol{\mu}$ than close, and a small value of $|\boldsymbol{\Sigma}|$ indicates that the $\mathbf{y}$'s are concentrated closer to $\boldsymbol{\mu}$ than is the case when $|\boldsymbol{\Sigma}|$ is larger. A small value of $|\boldsymbol{\Sigma}|$ may also indicate a high degree of multicollinearity among the variables, signifying that the $\mathbf{y}$'s tend to occupy a smaller dimensional subspace of the p dimensions.

2.1.3 Constant Density Ellipsoids

We illustrate the effect of the size of $|\boldsymbol{\Sigma}|$ for two bivariate normal distributions in Figure 2.1. The elliptical contours in Figure 2.1 can be found by setting the bivariate normal density function equal to a constant and solving for $\mathbf{y}$. Likewise, we find the equation of a constant density ellipsoid for a p-variate normal distribution by solving

$$
\begin{gathered}
f(\mathbf{y}) = c, \\
\frac{1}{(\sqrt{2\pi})^p|\boldsymbol{\Sigma}|^{1/2}} e^{-(\mathbf{y}-\boldsymbol{\mu})'\boldsymbol{\Sigma}^{-1}(\mathbf{y}-\boldsymbol{\mu})/2} = c.
\end{gathered}
$$

Taking the natural logarithm of both sides and rearranging give

$$
(\mathbf{y} - \boldsymbol{\mu})'\boldsymbol{\Sigma}^{-1}(\mathbf{y} - \boldsymbol{\mu}) = a. \qquad (2.12)
$$

For the value of a, see Problem 2.2.

To find the lengths and directions of the axes in (2.12), we note that $\boldsymbol{\Sigma}$ is symmetric and use (A.11.5) to write $\boldsymbol{\Sigma} = \mathbf{CDC}'$, where $\mathbf{D} = \text{diag}(\lambda_1, \lambda_2, \ldots, \lambda_p)$ is a diagonal matrix of eigenvalues of $\boldsymbol{\Sigma}$ and $\mathbf{C}$ is an orthogonal matrix with normalized

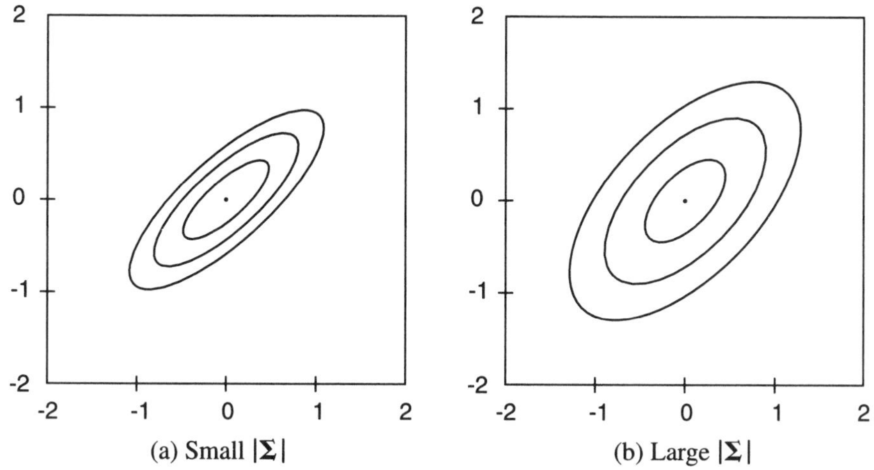

Figure 2.1. Contour plots for two bivariate normal distributions.

eigenvectors of $\boldsymbol{\Sigma}$ as columns. Then

$$\begin{aligned}\boldsymbol{\Sigma}^{-1} &= (\mathbf{CDC}')^{-1} = (\mathbf{C}')^{-1}\mathbf{D}^{-1}\mathbf{C}^{-1} \\ &= \mathbf{CD}^{-1}\mathbf{C}' \end{aligned} \tag{2.13}$$

since $\mathbf{C}$ is orthogonal. Using (2.13) in (2.12) gives

$$(\mathbf{y} - \boldsymbol{\mu})'\mathbf{CD}^{-1}\mathbf{C}'(\mathbf{y} - \boldsymbol{\mu}) = a.$$

Defining $\mathbf{z} = \mathbf{C}'(\mathbf{y} - \boldsymbol{\mu})$ yields

$$\mathbf{z}'\mathbf{D}^{-1}\mathbf{z} = a$$

or

$$\sum_{j=1}^{p} \frac{z_j^2}{\lambda_j} = a, \tag{2.14}$$

which is the equation of an ellipsoid with semimajor axes of length $\sqrt{a\lambda_j}$. To find the value of a that corresponds to a given proportion of observations contained in the ellipsoid, we can use the fact that $\Sigma_j z_j^2/\lambda_j$ has a chi-square distribution (see Theorem 2.2F); for example, $P(\sum_{j=1}^{p} z_j^2/\lambda_j \leq \chi^2_{.10,p}) = .90$. The directions of the axes are given by the eigenvectors of $\boldsymbol{\Sigma}$, since $\mathbf{z} = \mathbf{C}'(\mathbf{y} - \boldsymbol{\mu})$ is a rotation and $\mathbf{C}$ is orthogonal with eigenvectors of $\boldsymbol{\Sigma}$ as columns.

The constant density contour for the bivariate normal is illustrated in Figure 2.2. To construct the major axis, we use $\boldsymbol{\mu}$ for the origin and plot a point corresponding to the first eigenvector $\mathbf{c}_1$ or some multiple of $\mathbf{c}_1$. We then draw a line through $\boldsymbol{\mu}$ and $\mathbf{c}_1$. The other axis is similarly established using $\mathbf{c}_2$. The axes will be perpendicular because $\mathbf{c}_1$ and $\mathbf{c}_2$ are orthogonal.

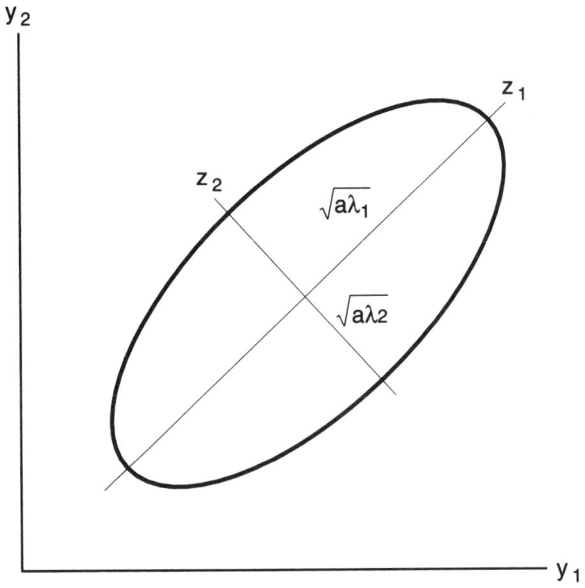

Figure 2.2. Constant density contour for bivariate normal.

2.1.4 Generating Multivariate Normal Data

The steps leading to (2.11) can be used to generate multivariate normal random vectors on the computer for simulation studies. If we desire to generate $\mathbf{y}$ from $N_p(\boldsymbol{\mu}, \boldsymbol{\Sigma})$ with given values of $\boldsymbol{\mu}$ and $\boldsymbol{\Sigma}$, we can use $\mathbf{y} = \boldsymbol{\Sigma}^{1/2}\mathbf{z} + \boldsymbol{\mu}$ as in (2.7), where $\mathbf{z}$ is $N_p(\mathbf{0}, \mathbf{I})$. (Note that $\mathbf{z}$ was defined differently in Section 2.1.3.) Alternatively, we can factor $\boldsymbol{\Sigma}$ into $\boldsymbol{\Sigma} = \mathbf{AA}'$ using the Cholesky procedure in Section A.6 and define $\mathbf{y} = \mathbf{Az} + \boldsymbol{\mu}$. The vector $\mathbf{z}$, consisting of independent standard normal variates, is easily obtained on the computer. If we wish to specify a population correlation matrix $\mathbf{P}_\rho$, we can use (1.56) to obtain $\boldsymbol{\Sigma} = \mathbf{D}_\sigma \mathbf{P}_\rho \mathbf{D}_\sigma$, where $\mathbf{D}_\sigma = \text{diag}(\sigma_1, \sigma_2, \ldots, \sigma_p)$ is a diagonal matrix containing the desired σ_i's for the variables. Bryce and Maynes (1979) have provided a method of generating random vectors with any desired eigenstructure so that, instead of specifying $\boldsymbol{\Sigma}$ or $\mathbf{P}_\rho$, one need only choose the eigenvalues of $\boldsymbol{\Sigma}$ or $\mathbf{P}_\rho$ (or of $\mathbf{S}$ or $\mathbf{R}$). Measures of overall intercorrelation are generally based on the eigenvalues (see Section 1.7).

2.1.5 Moments

The *rth moment* of a random variable y is defined as

$$E(y^r). \tag{2.15}$$

The *rth central moment* is defined similarly as

$$E(y - \mu)^r. \tag{2.16}$$

The distribution of a random variable can be at least partially characterized by the moments.

For the random vector $\mathbf{y} = (y_1, y_2, \ldots, y_p)'$, the first moment of each y_j is given by the corresponding element of the mean vector $\boldsymbol{\mu}$:

$$E(y_j) = \mu_j. \tag{2.17}$$

The elements of the covariance matrix $\boldsymbol{\Sigma}$ are the second central moments

$$E\big[(y_j - \mu_j)(y_k - \mu_k)\big] = \sigma_{jk}. \tag{2.18}$$

In the case of the multivariate normal distribution, all third central moments are zero; in fact, all central moments of odd order are zero. Fourth central moments of the multivariate normal are given by

$$E\big[(y_i - \mu_i)(y_j - \mu_j)(y_k - \mu_k)(y_l - \mu_l)\big] = \sigma_{ij}\sigma_{kl} + \sigma_{ik}\sigma_{jl} + \sigma_{il}\sigma_{jk}, \tag{2.19}$$

with special cases as follows:

$$E\big[(y_i - \mu_i)^2(y_j - \mu_j)(y_k - \mu_k)\big] = \sigma_{ii}\sigma_{jk} + 2\sigma_{ij}\sigma_{ik}, \tag{2.20}$$

$$E\big[(y_i - \mu_i)^2(y_j - \mu_j)^2\big] = \sigma_{ii}\sigma_{jj} + 2\sigma_{ij}^2, \tag{2.21}$$

$$E(y_i - \mu_i)^4 = 3\sigma_{ii}^2. \tag{2.22}$$

Since the multivariate normal distribution is fairly well characterized by its first four moments, these moments can be used in checking for normality. For a review of tests for multivariate normality and for detection of outliers, see Rencher (1995, Sections 4.4 and 4.5). Rocke and Woodruff (1996) discuss the nature of the difficulty of detecting multivariate outliers and why the difficulty increases with p, the number of variables. They propose improved techniques that detect a relatively high percentage of outliers in the data.

2.2 PROPERTIES OF MULTIVARIATE NORMAL RANDOM VECTORS

We begin with a review of moment generating functions, which can be used to obtain some of the properties of multivariate normal random variables.

The *moment generating function* for a univariate random variable y is defined as

$$M_y(t) = E(e^{ty}). \tag{2.23}$$

If y is distributed as the univariate normal $N(\mu, \sigma^2)$, the moment generating function is given by

$$M_y(t) = e^{t\mu + t^2\sigma^2/2}. \tag{2.24}$$

The moment generating function for a multivariate random vector $\mathbf{y}$ is defined as

$$M_{\mathbf{y}}(\mathbf{t}) = E\left(e^{t_1 y_1 + t_2 y_2 + \cdots + t_p y_p}\right) = E(e^{\mathbf{t}'\mathbf{y}}). \tag{2.25}$$

For a multivariate normal random vector $\mathbf{y}$, the moment generating function is given in the following theorem.

Theorem 2.2A. If $\mathbf{y}$ is distributed as $N_p(\boldsymbol{\mu}, \boldsymbol{\Sigma})$, its moment generating function is given by

$$M_{\mathbf{y}}(\mathbf{t}) = e^{\mathbf{t}'\boldsymbol{\mu} + \mathbf{t}'\boldsymbol{\Sigma}\mathbf{t}/2}. \tag{2.26}$$

Proof. By (2.25) and (2.11), the moment generating function is

$$M_{\mathbf{y}}(\mathbf{t}) = \int_{-\infty}^{\infty} \cdots \int_{-\infty}^{\infty} k e^{\mathbf{t}'\mathbf{y} - (\mathbf{y}-\boldsymbol{\mu})'\boldsymbol{\Sigma}^{-1}(\mathbf{y}-\boldsymbol{\mu})/2} \, d\mathbf{y}, \tag{2.27}$$

where $k = 1/(\sqrt{2\pi})^p |\boldsymbol{\Sigma}|^{1/2}$ and $d\mathbf{y} = dy_1 dy_2 \cdots dy_p$. By working with the exponent, we obtain

$$M_{\mathbf{y}}(\mathbf{t}) = \int_{-\infty}^{\infty} \cdots \int_{-\infty}^{\infty} k e^{\mathbf{t}'\boldsymbol{\mu} + \mathbf{t}'\boldsymbol{\Sigma}\mathbf{t}/2 - (\mathbf{y}-\boldsymbol{\mu}-\boldsymbol{\Sigma}\mathbf{t})'\boldsymbol{\Sigma}^{-1}(\mathbf{y}-\boldsymbol{\mu}-\boldsymbol{\Sigma}\mathbf{t})/2} \, d\mathbf{y} \tag{2.28}$$

$$= e^{\mathbf{t}'\boldsymbol{\mu} + \mathbf{t}'\boldsymbol{\Sigma}\mathbf{t}/2} \int_{-\infty}^{\infty} \cdots \int_{-\infty}^{\infty} k e^{-(\mathbf{y}-\boldsymbol{\mu}-\boldsymbol{\Sigma}\mathbf{t})'\boldsymbol{\Sigma}^{-1}(\mathbf{y}-\boldsymbol{\mu}-\boldsymbol{\Sigma}\mathbf{t})/2} \, d\mathbf{y} \tag{2.29}$$

$$= e^{\mathbf{t}'\boldsymbol{\mu} + \mathbf{t}'\boldsymbol{\Sigma}\mathbf{t}/2}.$$

The integral in (2.29) is equal to 1 because the multivariate normal density in (2.11) integrates to 1 for any value of the mean vector, including $\boldsymbol{\mu} + \boldsymbol{\Sigma}\mathbf{t}$. □

Note that the multivariate normal generating function in (2.26) is an extension of its univariate counterpart in (2.24) and reduces to it if $p = 1$.

Moment generating functions can be differentiated to obtain moments (see Problems 2.4 and 2.5). Two additional properties of moment generating functions are as follows:

1. If two random vectors have the same moment generating function, they have the same density.
2. Two random vectors are independent if and only if their joint moment generating function factors into the product of their two separate moment generating

functions. Thus, if $\mathbf{y}' = (\mathbf{y}_1', \mathbf{y}_2')$, then

$$M_{\mathbf{y}}(\mathbf{t}) = M_{\mathbf{y}_1}(\mathbf{t}_1)M_{\mathbf{y}_2}(\mathbf{t}_2) \tag{2.30}$$

if and only if $\mathbf{y}_1$ and $\mathbf{y}_2$ are independent.

The following theorem provides the distribution of linear transformations of multivariate normal random variables.

Theorem 2.2B. Let $\mathbf{y}$ be $N_p(\boldsymbol{\mu}, \boldsymbol{\Sigma})$ and define $\mathbf{a}$ to be a vector of constants and $\mathbf{A}$ to be a $q \times p$ matrix of constants with rank $q \leq p$. Then

(i) $z = \mathbf{a}'\mathbf{y}$ is $N(\mathbf{a}'\boldsymbol{\mu}, \mathbf{a}'\boldsymbol{\Sigma}\mathbf{a})$ and
(ii) $\mathbf{z} = \mathbf{A}\mathbf{y}$ is $N_q(\mathbf{A}\boldsymbol{\mu}, \mathbf{A}\boldsymbol{\Sigma}\mathbf{A}')$.

Proof. **(i)** The moment generating function for $z = \mathbf{a}'\mathbf{y}$ is given by

$$\begin{aligned} M_z(t) &= E(e^{tz}) = E(e^{t\mathbf{a}'\mathbf{y}}) = E(e^{(t\mathbf{a})'\mathbf{y}}) \\ &= e^{(t\mathbf{a})'\boldsymbol{\mu}+(t\mathbf{a})'\boldsymbol{\Sigma}(t\mathbf{a})/2} \qquad \text{[by (2.26)]} \\ &= e^{(\mathbf{a}'\boldsymbol{\mu})t+(\mathbf{a}'\boldsymbol{\Sigma}\mathbf{a})t^2/2}. \end{aligned} \tag{2.31}$$

On comparing (2.31) with (2.24), it is clear that $z = \mathbf{a}'\mathbf{y}$ is univariate normal with mean $\mathbf{a}'\boldsymbol{\mu}$ and variance $\mathbf{a}'\boldsymbol{\Sigma}\mathbf{a}$.

(ii) The moment generating function for $\mathbf{z} = \mathbf{A}\mathbf{y}$ is given by

$$M_{\mathbf{z}}(\mathbf{t}) = E(e^{\mathbf{t}'\mathbf{z}}) = E(e^{\mathbf{t}'\mathbf{A}\mathbf{y}}),$$

which becomes

$$M_{\mathbf{z}}(\mathbf{t}) = e^{\mathbf{t}'(\mathbf{A}\boldsymbol{\mu})+\mathbf{t}'(\mathbf{A}\boldsymbol{\Sigma}\mathbf{A}')\mathbf{t}/2} \tag{2.32}$$

(see Problem 2.10). Thus, by (2.26) and (2.32), $\mathbf{z} = \mathbf{A}\mathbf{y}$ is $N_q(\mathbf{A}\boldsymbol{\mu}, \mathbf{A}\boldsymbol{\Sigma}\mathbf{A}')$. The covariance matrix $\mathbf{A}\boldsymbol{\Sigma}\mathbf{A}'$ is positive definite because $\mathbf{A}$ is full rank. □

Corollary 1. Let $\mathbf{y}$ be $N_p(\boldsymbol{\mu}, \boldsymbol{\Sigma})$, let $\boldsymbol{\Sigma} = \mathbf{A}\mathbf{A}'$ be the (triangular) Cholesky factorization defined in Section A.6, and let $\boldsymbol{\Sigma}^{-1/2}$ be the inverse of the (symmetric) square root matrix defined in (A.11.7). Then a *standardized vector* $\mathbf{z}$ can be obtained in two ways:

$$\mathbf{z}_1 = \mathbf{A}^{-1}(\mathbf{y} - \boldsymbol{\mu}), \tag{2.33}$$

$$\mathbf{z}_2 = \boldsymbol{\Sigma}^{-1/2}(\mathbf{y} - \boldsymbol{\mu}). \tag{2.34}$$

In both cases,

$$\mathbf{z}_i \quad \text{is} \quad N_p(\mathbf{0}, \mathbf{I}), \quad i = 1, 2. \tag{2.35}$$

The converse of Theorem 2.2B(i) can be used as a definition of the multivariate normal:

Definition 2.2.1. If $\mathbf{a}'\mathbf{y}$ is (univariate) normal for all possible $\mathbf{a}$, then $\mathbf{y}$ is $N_p(\boldsymbol{\mu}, \boldsymbol{\Sigma})$.

The marginal distributions of multivariate normal variables are also normal, as established in the following theorem.

Theorem 2.2C. If $\mathbf{y}$ is $N_p(\boldsymbol{\mu}, \boldsymbol{\Sigma})$, then any subvector of $\mathbf{y}$ has a multivariate normal distribution with the means, variances, and covariances unchanged.

Proof. Without loss of generality, let the subvector of interest consist of the first r elements of $\mathbf{y}$, designated as $\mathbf{y}_1$, with the remaining variables denoted by $\mathbf{y}_2$, and let $\boldsymbol{\mu}$ and $\boldsymbol{\Sigma}$ be partitioned accordingly:

$$\mathbf{y} = \begin{pmatrix} \mathbf{y}_1 \\ \mathbf{y}_2 \end{pmatrix}, \quad \boldsymbol{\mu} = \begin{pmatrix} \boldsymbol{\mu}_1 \\ \boldsymbol{\mu}_2 \end{pmatrix}, \quad \boldsymbol{\Sigma} = \begin{pmatrix} \boldsymbol{\Sigma}_{11} & \boldsymbol{\Sigma}_{12} \\ \boldsymbol{\Sigma}_{21} & \boldsymbol{\Sigma}_{22} \end{pmatrix}.$$

Define $\mathbf{A} = (\mathbf{I}_r, \mathbf{O})$, where $\mathbf{I}_r$ is an $r \times r$ identity matrix and $\mathbf{O}$ is an $r \times (p - r)$ matrix of 0's. Then $\mathbf{A}\mathbf{y} = \mathbf{y}_1$, and $\mathbf{y}_1$ is $N_r(\boldsymbol{\mu}_1, \boldsymbol{\Sigma}_{11})$ by Theorem 2.2B(ii). □

For the next two theorems, the observation vector is partitioned into two subvectors denoted by $\mathbf{y}$ and $\mathbf{x}$, where $\mathbf{y}$ is $p \times 1$ and $\mathbf{x}$ is $q \times 1$, with a corresponding partitioning of $\boldsymbol{\mu}$ and $\boldsymbol{\Sigma}$ [see (1.61) and (1.62)]:

$$\boldsymbol{\mu} = E\begin{pmatrix} \mathbf{y} \\ \mathbf{x} \end{pmatrix} = \begin{pmatrix} \boldsymbol{\mu}_y \\ \boldsymbol{\mu}_x \end{pmatrix}, \quad \boldsymbol{\Sigma} = \operatorname{cov}\begin{pmatrix} \mathbf{y} \\ \mathbf{x} \end{pmatrix} = \begin{pmatrix} \boldsymbol{\Sigma}_{yy} & \boldsymbol{\Sigma}_{yx} \\ \boldsymbol{\Sigma}_{xy} & \boldsymbol{\Sigma}_{xx} \end{pmatrix}.$$

It was noted in a remark following (1.19) that if two random variables y_j and y_k are independent, then $\sigma_{jk} = 0$, but that the converse of that statement is not true in general. By extension, if two random vectors $\mathbf{y}$ and $\mathbf{x}$ are independent, then $\boldsymbol{\Sigma}_{yx} = \mathbf{O}$; that is, the covariance of each y_j with each x_k is 0. Likewise, the converse is not true in general, but it is true for multivariate normal random vectors. In other words, zero covariances imply independence for multivariate normal variables, as noted in the following theorem.

Theorem 2.2D. If $\mathbf{y}$ and $\mathbf{x}$ are jointly multivariate normal, and if $\boldsymbol{\Sigma}_{yx} = \mathbf{O}$, then $\mathbf{y}$ and $\mathbf{x}$ are independent.

Proof. If $\boldsymbol{\Sigma}_{yx} = \mathbf{O}$, then

$$\boldsymbol{\Sigma} = \begin{pmatrix} \boldsymbol{\Sigma}_{yy} & \mathbf{O} \\ \mathbf{O} & \boldsymbol{\Sigma}_{xx} \end{pmatrix},$$

and the exponent of the moment generating function in (2.26) becomes

$$\begin{aligned} \mathbf{t}'\boldsymbol{\mu} + \tfrac{1}{2}\mathbf{t}'\boldsymbol{\Sigma}\mathbf{t} &= (\mathbf{t}_y', \mathbf{t}_x')\begin{pmatrix} \boldsymbol{\mu}_y \\ \boldsymbol{\mu}_x \end{pmatrix} + \tfrac{1}{2}(\mathbf{t}_y', \mathbf{t}_x')\begin{pmatrix} \boldsymbol{\Sigma}_{yy} & \mathbf{O} \\ \mathbf{O} & \boldsymbol{\Sigma}_{xx} \end{pmatrix}\begin{pmatrix} \mathbf{t}_y \\ \mathbf{t}_x \end{pmatrix} \\ &= \mathbf{t}_y'\boldsymbol{\mu}_y + \mathbf{t}_x'\boldsymbol{\mu}_x + \tfrac{1}{2}\mathbf{t}_y'\boldsymbol{\Sigma}_{yy}\mathbf{t}_y + \tfrac{1}{2}\mathbf{t}_x'\boldsymbol{\Sigma}_{xx}\mathbf{t}_x. \end{aligned}$$

Then the moment generating function can be written as

$$e^{\mathbf{t}_y'\boldsymbol{\mu}_y+\mathbf{t}_y'\boldsymbol{\Sigma}_{yy}\mathbf{t}_y/2}e^{\mathbf{t}_x'\boldsymbol{\mu}_x+\mathbf{t}_x'\boldsymbol{\Sigma}_{xx}\mathbf{t}_x/2},$$

which is the product of the moment generating functions of $\mathbf{y}$ and $\mathbf{x}$. Hence, by (2.30), $\mathbf{y}$ and $\mathbf{x}$ are independent. □

Corollary 1. If $\mathbf{y}$ is $N_p(\boldsymbol{\mu}, \boldsymbol{\Sigma})$, then any two individual variables y_j and y_k are independent if $\sigma_{jk} = 0$.

Corollary 2. If $\mathbf{y}$ is $N_p(\boldsymbol{\mu}, \boldsymbol{\Sigma})$, then $\mathbf{Ay}$ and $\mathbf{By}$ are independent if $\mathbf{A\Sigma B}' = \mathbf{O}$.

Corollary 3. If the subvectors $\mathbf{y}$ and $\mathbf{x}$ are jointly multivariate normal with $\boldsymbol{\Sigma}_{yx} = \mathbf{O}$, and if $\mathbf{y}$ and $\mathbf{x}$ are the same size (say $p \times 1$), then they can be added or subtracted, and

$$\mathbf{y}+\mathbf{x} \quad \text{is} \quad N_p(\boldsymbol{\mu}_y + \boldsymbol{\mu}_x, \boldsymbol{\Sigma}_{yy} + \boldsymbol{\Sigma}_{xx}),$$
$$\mathbf{y}-\mathbf{x} \quad \text{is} \quad N_p(\boldsymbol{\mu}_y - \boldsymbol{\mu}_x, \boldsymbol{\Sigma}_{yy} + \boldsymbol{\Sigma}_{xx}).$$

The relationship of subvectors $\mathbf{y}$ and $\mathbf{x}$ when they are not independent is given in the following theorem.

Theorem 2.2E. If $\mathbf{y}$ and $\mathbf{x}$ are jointly multivariate normal with $\boldsymbol{\Sigma}_{yx} \neq \mathbf{O}$, then the conditional distribution of $\mathbf{y}$ given $\mathbf{x}$, $f(\mathbf{y}|\mathbf{x})$, is multivariate normal with mean vector and covariance matrix

$$E(\mathbf{y}|\mathbf{x}) = \boldsymbol{\mu}_y + \boldsymbol{\Sigma}_{yx}\boldsymbol{\Sigma}_{xx}^{-1}(\mathbf{x} - \boldsymbol{\mu}_x), \tag{2.36}$$

$$\text{cov}(\mathbf{y}|\mathbf{x}) = \boldsymbol{\Sigma}_{yy} - \boldsymbol{\Sigma}_{yx}\boldsymbol{\Sigma}_{xx}^{-1}\boldsymbol{\Sigma}_{xy}. \tag{2.37}$$

Proof. By definition, the conditional density is given by

$$f(\mathbf{y}|\mathbf{x}) = \frac{g(\mathbf{y}, \mathbf{x})}{h(\mathbf{x})}. \tag{2.38}$$

The ratio in (2.38) can be evaluated directly, but the following approach is more instructive. We write (2.38) in the form

$$g(\mathbf{y}, \mathbf{x}) = f(\mathbf{y}|\mathbf{x})h(\mathbf{x}). \tag{2.39}$$

Since $f(\mathbf{y}|\mathbf{x})h(\mathbf{x})$ is a product, we seek a function of $\mathbf{y}$ and $\mathbf{x}$ that is independent of $\mathbf{x}$. By Theorem 2.2B(ii), linear functions of $\binom{\mathbf{y}}{\mathbf{x}}$ are normal. We therefore consider $\mathbf{z} = \mathbf{y} - \mathbf{Bx}$, which gives

$$\text{cov}(\mathbf{z}, \mathbf{x}) = \boldsymbol{\Sigma}_{yx} - \mathbf{B}\boldsymbol{\Sigma}_{xx} \tag{2.40}$$

(see Problem 2.17). If we choose $\mathbf{B} = \boldsymbol{\Sigma}_{yx}\boldsymbol{\Sigma}_{xx}^{-1}$, then $\text{cov}(\mathbf{z}, \mathbf{x}) = \mathbf{O}$, and $\mathbf{z}$ is independent of $\mathbf{x}$ (Theorem 2.2D). We can therefore write (2.39) as

$$g(\mathbf{y}, \mathbf{x}) = f(\mathbf{y} - \boldsymbol{\Sigma}_{yx}\boldsymbol{\Sigma}_{xx}^{-1}\mathbf{x})h(\mathbf{x}),$$

from which,

$$f(\mathbf{y}|\mathbf{x}) = f(\mathbf{y} - \boldsymbol{\Sigma}_{yx}\boldsymbol{\Sigma}_{xx}^{-1}\mathbf{x}).$$

Now $f(\mathbf{y} - \boldsymbol{\Sigma}_{yx}\boldsymbol{\Sigma}_{xx}^{-1}\mathbf{x})$ is the multivariate normal density with

$$E(\mathbf{y} - \boldsymbol{\Sigma}_{yx}\boldsymbol{\Sigma}_{xx}^{-1}\mathbf{x}) = \boldsymbol{\mu}_y - \boldsymbol{\Sigma}_{yx}\boldsymbol{\Sigma}_{xx}^{-1}\boldsymbol{\mu}_x$$

and

$$\begin{aligned}
\text{cov}(\mathbf{y} - \boldsymbol{\Sigma}_{yx}\boldsymbol{\Sigma}_{xx}^{-1}\mathbf{x}) &= \text{cov}\left[(\mathbf{I}, -\boldsymbol{\Sigma}_{yx}\boldsymbol{\Sigma}_{xx}^{-1})\begin{pmatrix}\mathbf{y}\\ \mathbf{x}\end{pmatrix}\right] \\
&= (\mathbf{I}, -\boldsymbol{\Sigma}_{yx}\boldsymbol{\Sigma}_{xx}^{-1})\begin{pmatrix}\boldsymbol{\Sigma}_{yy} & \boldsymbol{\Sigma}_{yx}\\ \boldsymbol{\Sigma}_{xy} & \boldsymbol{\Sigma}_{xx}\end{pmatrix}\begin{pmatrix}\mathbf{I}\\ -\boldsymbol{\Sigma}_{xx}^{-1}\boldsymbol{\Sigma}_{xy}\end{pmatrix} \\
&= \boldsymbol{\Sigma}_{yy} - \boldsymbol{\Sigma}_{yx}\boldsymbol{\Sigma}_{xx}^{-1}\boldsymbol{\Sigma}_{xy}.
\end{aligned} \tag{2.41}$$

Thus $f(\mathbf{y} - \boldsymbol{\Sigma}_{yx}\boldsymbol{\Sigma}_{xx}^{-1}\mathbf{x})$ is of the form $N_p(\boldsymbol{\mu}_y - \boldsymbol{\Sigma}_{yx}\boldsymbol{\Sigma}_{xx}^{-1}\boldsymbol{\mu}_x, \boldsymbol{\Sigma}_{yy} - \boldsymbol{\Sigma}_{yx}\boldsymbol{\Sigma}_{xx}^{-1}\boldsymbol{\Sigma}_{xy})$, where p is the number of variables in $\mathbf{y}$. The exponent in this density contains the factor

$$\begin{aligned}
\mathbf{y} - \boldsymbol{\Sigma}_{yx}\boldsymbol{\Sigma}_{xx}^{-1}\mathbf{x} - (\boldsymbol{\mu}_y - \boldsymbol{\Sigma}_{yx}\boldsymbol{\Sigma}_{xx}^{-1}\boldsymbol{\mu}_x) &= \mathbf{y} - \boldsymbol{\mu}_y - \boldsymbol{\Sigma}_{yx}\boldsymbol{\Sigma}_{xx}^{-1}(\mathbf{x} - \boldsymbol{\mu}_x) \\
&= \mathbf{y} - \left[\boldsymbol{\mu}_y + \boldsymbol{\Sigma}_{yx}\boldsymbol{\Sigma}_{xx}^{-1}(\mathbf{x} - \boldsymbol{\mu}_x)\right].
\end{aligned}$$

Hence $\mathbf{y}|\mathbf{x}$ is multivariate normal with

$$E(\mathbf{y}|\mathbf{x}) = \boldsymbol{\mu}_y + \boldsymbol{\Sigma}_{yx}\boldsymbol{\Sigma}_{xx}^{-1}(\mathbf{x} - \boldsymbol{\mu}_x),$$

$$\text{cov}(\mathbf{y}|\mathbf{x}) = \boldsymbol{\Sigma}_{yy} - \boldsymbol{\Sigma}_{yx}\boldsymbol{\Sigma}_{xx}^{-1}\boldsymbol{\Sigma}_{xy}. \qquad \square$$

In Theorem 2.2E it is clear that $E(\mathbf{y}|\mathbf{x})$ is a linear function of $\mathbf{x}$, and therefore any pair of variables in a multivariate normal vector exhibits a linear trend. Hence the (population) covariance or correlation is a good measure of relationship between two normal variables. For example, $E(y_i|y_j) = \mu_i + (\sigma_{ij}/\sigma_{jj})(y_j - \mu_j)$, and the covariance σ_{ij} is related to the slope of the line. This is not the case with nonnormal variables that exhibit a curved trend.

By definition, the sum of squares of p independent standard normal random variables has a chi-square distribution. For multivariate normal random variables, there is a similar function that has a chi-square distribution.

Theorem 2.2F. If $\mathbf{y}$ is $N_p(\boldsymbol{\mu}, \boldsymbol{\Sigma})$, then the quadratic form $(\mathbf{y} - \boldsymbol{\mu})'\boldsymbol{\Sigma}^{-1}(\mathbf{y} - \boldsymbol{\mu})$ is distributed as $\chi^2(p)$.

Proof. Let $\mathbf{z}$ be a standardized vector as defined in (2.33) or (2.34), so that, by (2.35), $\mathbf{z}$ is $N_p(\mathbf{0}, \mathbf{I})$. Then $\mathbf{z}'\mathbf{z}$ has a chi-square distribution, since it is a sum of squares of p independent standard normal variables. For $\mathbf{z}$ defined by either (2.33) or (2.34), we have

$$\mathbf{z}'\mathbf{z} = (\mathbf{y} - \boldsymbol{\mu})'\boldsymbol{\Sigma}^{-1}(\mathbf{y} - \boldsymbol{\mu}). \tag{2.42}$$

□

2.3 ESTIMATION OF PARAMETERS IN THE MULTIVARIATE NORMAL DISTRIBUTION

We obtain maximum likelihood estimators of $\boldsymbol{\mu}$ and $\boldsymbol{\Sigma}$ in Section 2.3.1 and consider the properties of these estimators in Section 2.3.2.

2.3.1 Maximum Likelihood Method

If the density of a random vector $\mathbf{y}$ is assumed to be known, the method of *maximum likelihood* can often be used to find estimators of parameters. The joint density of the observation vectors $\mathbf{y}_1, \mathbf{y}_2, \ldots, \mathbf{y}_n$ is called the *likelihood function*. The *maximum likelihood estimators* are the values of the parameters (as functions of $\mathbf{y}_1, \mathbf{y}_2, \ldots, \mathbf{y}_n$) that maximize the likelihood function. These values can sometimes be found by differentiation. In the following theorem, we obtain the maximum likelihood estimators of $\boldsymbol{\mu}$ and $\boldsymbol{\Sigma}$ for a random sample from a multivariate normal distribution.

Theorem 2.3A. If $\mathbf{y}_1, \mathbf{y}_2, \ldots, \mathbf{y}_n$ is a random sample from $N_p(\boldsymbol{\mu}, \boldsymbol{\Sigma})$, then the maximum likelihood estimators of $\boldsymbol{\mu}$ and $\boldsymbol{\Sigma}$ are

$$\hat{\boldsymbol{\mu}} = \bar{\mathbf{y}}, \tag{2.43}$$

$$\begin{aligned}\hat{\boldsymbol{\Sigma}} &= \frac{1}{n}\sum_{i=1}^{n}(\mathbf{y}_i - \bar{\mathbf{y}})(\mathbf{y}_i - \bar{\mathbf{y}})' = \frac{1}{n}\mathbf{W} \\ &= \frac{n-1}{n}\mathbf{S},\end{aligned} \tag{2.44}$$

where $\mathbf{W} = \sum_{i=1}^{n}(\mathbf{y}_i - \bar{\mathbf{y}})(\mathbf{y}_i - \bar{\mathbf{y}})'$.

Proof. Since the $\mathbf{y}_i$'s are independent (because they arise from a random sample), the likelihood function (joint density) is the product of the densities of the $\mathbf{y}_i$'s:

$$\begin{aligned}L(\boldsymbol{\mu}, \boldsymbol{\Sigma}) &= \prod_{i=1}^{n} f(\mathbf{y}_i; \boldsymbol{\mu}, \boldsymbol{\Sigma}) \\ &= \prod_{i=1}^{n} \frac{1}{(\sqrt{2\pi})^p|\boldsymbol{\Sigma}|^{1/2}} e^{-(\mathbf{y}_i - \boldsymbol{\mu})'\boldsymbol{\Sigma}^{-1}(\mathbf{y}_i - \boldsymbol{\mu})/2} \\ &= \frac{1}{(\sqrt{2\pi})^{np}|\boldsymbol{\Sigma}|^{n/2}} e^{-\sum_{i=1}^{n}(\mathbf{y}_i - \boldsymbol{\mu})'\boldsymbol{\Sigma}^{-1}(\mathbf{y}_i - \boldsymbol{\mu})/2}.\end{aligned} \tag{2.45}$$

We use the notation $L(\boldsymbol{\mu}, \boldsymbol{\Sigma})$ because we consider $\mathbf{y}_1, \mathbf{y}_2, \ldots, \mathbf{y}_n$ to be known or available from a future sample. For the given values of $\mathbf{y}_1, \mathbf{y}_2, \ldots, \mathbf{y}_n$, we seek the values of $\boldsymbol{\mu}$ and $\boldsymbol{\Sigma}$ that maximize (2.45). We first express (2.45) in a form that will facilitate finding the maximum.

The scalar quantity $(\mathbf{y}_i - \boldsymbol{\mu})'\boldsymbol{\Sigma}^{-1}(\mathbf{y}_i - \boldsymbol{\mu})$ is equal to its trace. Hence, by (A.8.1) and (A.8.2), we have

$$\begin{aligned}\sum_{i=1}^{n}(\mathbf{y}_i - \boldsymbol{\mu})'\boldsymbol{\Sigma}^{-1}(\mathbf{y}_i - \boldsymbol{\mu}) &= \sum_{i=1}^{n} \operatorname{tr}(\mathbf{y}_i - \boldsymbol{\mu})'\boldsymbol{\Sigma}^{-1}(\mathbf{y}_i - \boldsymbol{\mu}) \\ &= \operatorname{tr}\left[\boldsymbol{\Sigma}^{-1}\sum_{i=1}^{n}(\mathbf{y}_i - \boldsymbol{\mu})(\mathbf{y}_i - \boldsymbol{\mu})'\right]. \qquad (2.46)\end{aligned}$$

Now by adding and subtracting $\overline{\mathbf{y}}$ in the sum in the right side of (2.46), we obtain

$$\begin{aligned}\sum_{i=1}^{n}(\mathbf{y}_i - \boldsymbol{\mu})(\mathbf{y}_i - \boldsymbol{\mu})' &= \sum_{i=1}^{n}(\mathbf{y}_i - \overline{\mathbf{y}} + \overline{\mathbf{y}} - \boldsymbol{\mu})(\mathbf{y}_i - \overline{\mathbf{y}} + \overline{\mathbf{y}} - \boldsymbol{\mu})' \\ &= \sum_{i=1}^{n}(\mathbf{y}_i - \overline{\mathbf{y}})(\mathbf{y}_i - \overline{\mathbf{y}})' + n(\overline{\mathbf{y}} - \boldsymbol{\mu})(\overline{\mathbf{y}} - \boldsymbol{\mu})' \\ &= \mathbf{W} + n(\overline{\mathbf{y}} - \boldsymbol{\mu})(\overline{\mathbf{y}} - \boldsymbol{\mu})'. \qquad (2.47)\end{aligned}$$

The other two terms in the expression preceding (2.47) vanish because $\sum_i(\mathbf{y}_i - \overline{\mathbf{y}}) = \mathbf{0}$. Using (2.46) and (2.47) in (2.45), we obtain

$$L(\boldsymbol{\mu}, \boldsymbol{\Sigma}) = \frac{1}{(\sqrt{2\pi})^{np}|\boldsymbol{\Sigma}|^{n/2}} e^{-\operatorname{tr}\boldsymbol{\Sigma}^{-1}[\mathbf{W}+n(\overline{\mathbf{y}}-\boldsymbol{\mu})(\overline{\mathbf{y}}-\boldsymbol{\mu})']/2}. \qquad (2.48)$$

Because the natural logarithm is an increasing function, the maximum of $\ln L$ will occur at the same point as the maximum of L. We prefer to work with $\ln L$ because of its simpler form for differentiation:

$$\begin{aligned}\ln L(\boldsymbol{\mu}, \boldsymbol{\Sigma}) &= -np \ln\sqrt{2\pi} - \tfrac{n}{2}\ln|\boldsymbol{\Sigma}| - \tfrac{1}{2}\operatorname{tr}\boldsymbol{\Sigma}^{-1}\left[\mathbf{W} + n(\overline{\mathbf{y}} - \boldsymbol{\mu})(\overline{\mathbf{y}} - \boldsymbol{\mu})'\right] \\ &= -np \ln\sqrt{2\pi} - \tfrac{n}{2}\ln|\boldsymbol{\Sigma}| - \tfrac{1}{2}\operatorname{tr}(\boldsymbol{\Sigma}^{-1}\mathbf{W}) - \tfrac{n}{2}(\overline{\mathbf{y}} - \boldsymbol{\mu})'\boldsymbol{\Sigma}^{-1}(\overline{\mathbf{y}} - \boldsymbol{\mu}). \\ &\qquad (2.49)\end{aligned}$$

To find the maximum likelihood estimator for $\boldsymbol{\mu}$, we differentiate $\ln L(\boldsymbol{\mu}, \boldsymbol{\Sigma})$ in (2.49) with respect to $\boldsymbol{\mu}$, using (A.13.2) and (A.13.3), and set the resulting expression equal to $\mathbf{0}$:

$$\frac{\partial \ln L(\boldsymbol{\mu}, \boldsymbol{\Sigma})}{\partial \boldsymbol{\mu}} = -\mathbf{0} - \mathbf{0} - \mathbf{0} + n(\boldsymbol{\Sigma}^{-1}\overline{\mathbf{y}} - \boldsymbol{\Sigma}^{-1}\boldsymbol{\mu}) = \mathbf{0}, \qquad (2.50)$$

which gives

$$\hat{\boldsymbol{\mu}} = \overline{\mathbf{y}}.$$

It is clear that $\hat{\boldsymbol{\mu}} = \overline{\mathbf{y}}$ maximizes $\ln L(\boldsymbol{\mu}, \boldsymbol{\Sigma})$ with respect to $\boldsymbol{\mu}$, because the last term of (2.49) is ≤ 0, and the term vanishes for $\hat{\boldsymbol{\mu}} = \overline{\mathbf{y}}$.

Before differentiating $\ln L(\boldsymbol{\mu}, \boldsymbol{\Sigma})$ to find $\hat{\boldsymbol{\Sigma}}$, we substitute $\hat{\boldsymbol{\mu}} = \overline{\mathbf{y}}$ in (2.49) and rewrite $\ln |\boldsymbol{\Sigma}|$ in terms of $\boldsymbol{\Sigma}^{-1}$ to obtain

$$\ln L(\hat{\boldsymbol{\mu}}, \boldsymbol{\Sigma}) = -np \ln \sqrt{2\pi} + \tfrac{n}{2} \ln |\boldsymbol{\Sigma}^{-1}| - \tfrac{1}{2} \operatorname{tr}(\boldsymbol{\Sigma}^{-1}\mathbf{W}). \tag{2.51}$$

We now differentiate (2.51) with respect to $\boldsymbol{\Sigma}^{-1}$ using (A.13.5) and (A.13.6):

$$\frac{\partial \ln L(\hat{\boldsymbol{\mu}}, \boldsymbol{\Sigma})}{\partial \boldsymbol{\Sigma}^{-1}} = -\mathbf{O} + n\boldsymbol{\Sigma} - \tfrac{n}{2} \operatorname{diag}(\boldsymbol{\Sigma}) - \mathbf{W} + \tfrac{1}{2} \operatorname{diag}(\mathbf{W}) = \mathbf{O}, \tag{2.52}$$

from which we have

$$\hat{\boldsymbol{\Sigma}} - \tfrac{1}{2} \operatorname{diag}(\hat{\boldsymbol{\Sigma}}) = \frac{1}{n} \left[\mathbf{W} - \tfrac{1}{2} \operatorname{diag}(\mathbf{W})\right] \tag{2.53}$$

or

$$\hat{\boldsymbol{\Sigma}} = \frac{1}{n}\mathbf{W}.$$

To see that $\hat{\boldsymbol{\Sigma}} = \mathbf{W}/n$ is the solution to (2.53), note that for the off-diagonal elements we have $\hat{\sigma}_{jk} = w_{jk}/n, j \neq k$, and for the diagonal elements, $\hat{\sigma}_{jj} - \hat{\sigma}_{jj}/2 = w_{jj}/n - w_{jj}/2n$ or $\hat{\sigma}_{jj}/2 = w_{jj}/2n$. □

Note that we solved (2.53) for $\boldsymbol{\Sigma}$ rather than $\boldsymbol{\Sigma}^{-1}$, even though we differentiated with respect to $\boldsymbol{\Sigma}^{-1}$. Otherwise we would have obtained $(\mathbf{W}/n)^{-1}$ as the maximum likelihood estimator for $\boldsymbol{\Sigma}^{-1}$. This illustrates the invariance property of maximum likelihood estimators in Theorem 2.3B below; that is, $[(\mathbf{W}/n)^{-1}]^{-1} = \mathbf{W}/n$ is the maximum likelihood estimator for $(\boldsymbol{\Sigma}^{-1})^{-1} = \boldsymbol{\Sigma}$. Since $\boldsymbol{\Sigma}$ is positive definite, $\mathbf{W}$ is positive definite with probability 1 [see a remark following (1.42)].

The maximum likelihood estimator $\hat{\boldsymbol{\Sigma}}$ has divisor n instead of $n-1$ and is therefore biased [see (1.47)]. We often use $\mathbf{S}$ in its place.

The *invariance property* of maximum likelihood estimators is given in the following theorem.

Theorem 2.3B. The maximum likelihood estimator of a function of one or more parameters is the same function of the maximum likelihood estimators of the parameters. Thus, if $\boldsymbol{\theta}$ represents a vector or matrix of parameters for which we have a maximum likelihood estimator $\hat{\boldsymbol{\theta}}$, then the maximum likelihood estimator of a function $g(\boldsymbol{\theta})$ is given by $g(\hat{\boldsymbol{\theta}})$. □

We illustrate the use of this invariance property by finding the maximum likelihood estimator of the population correlation matrix $\mathbf{P}_\rho$ when sampling from the multivariate normal distribution. By (1.55), the relationship between $\mathbf{P}_\rho$ and $\boldsymbol{\Sigma}$ is given by $\mathbf{P}_\rho =$

$\mathbf{D}_\sigma^{-1}\boldsymbol{\Sigma}\mathbf{D}_\sigma^{-1}$, where

$$\mathbf{D}_\sigma^{-1} = \text{diag}\left(\frac{1}{\sigma_1}, \frac{1}{\sigma_2}, \ldots, \frac{1}{\sigma_p}\right) = \text{diag}\left(\frac{1}{\sqrt{\sigma_{11}}}, \frac{1}{\sqrt{\sigma_{22}}}, \ldots, \frac{1}{\sqrt{\sigma_{pp}}}\right).$$

Now the maximum likelihood estimator of $1/\sqrt{\sigma_{jj}}$ is $1/\sqrt{\hat{\sigma}_{jj}}$, where $\hat{\sigma}_{jj} = \frac{1}{n}\sum_{i=1}^n (y_{ij} - \bar{y}_j)^2$. Thus $\hat{\mathbf{D}}_\sigma^{-1} = \text{diag}\left(1/\sqrt{\hat{\sigma}_{11}}, 1/\sqrt{\hat{\sigma}_{22}}, \ldots, 1/\sqrt{\hat{\sigma}_{pp}}\right)$, and we obtain

$$\begin{aligned}
\hat{\mathbf{P}}_\rho = \hat{\mathbf{D}}_\sigma^{-1}\hat{\boldsymbol{\Sigma}}\hat{\mathbf{D}}_\sigma^{-1} &= \left(\frac{\hat{\sigma}_{jk}}{\sqrt{\hat{\sigma}_{jj}}\sqrt{\hat{\sigma}_{kk}}}\right) \\
&= \left(\frac{\sum_i (y_{ij} - \bar{y}_j)(y_{ik} - \bar{y}_k)/n}{\sqrt{\sum_i (y_{ij} - \bar{y}_j)^2/n}\sqrt{\sum_i (y_{ik} - \bar{y}_k)^2/n}}\right) \\
&= \left(\frac{\sum_i (y_{ij} - \bar{y}_j)(y_{ik} - \bar{y}_k)}{\sqrt{\sum_i (y_{ij} - \bar{y}_j)^2}\sqrt{\sum_i (y_{ik} - \bar{y}_k)^2}}\right) \\
&= (r_{jk}) = \mathbf{R}.
\end{aligned}$$

2.3.2 Properties of $\bar{\mathbf{y}}$ and S

We now list some properties of $\bar{\mathbf{y}}$ and $\mathbf{S}$ based on a random sample $\mathbf{y}_1, \mathbf{y}_2, \ldots, \mathbf{y}_n$ from a multivariate normal distribution. Most of the properties are extensions of familiar univariate results.

By the Neyman factorization theorem, a vector statistic $\hat{\boldsymbol{\theta}}$ is *sufficient* for a vector parameter $\boldsymbol{\theta}$ if the likelihood function can be factored as

$$L(\mathbf{y}_1, \ldots, \mathbf{y}_n, \boldsymbol{\theta}) = g(\hat{\boldsymbol{\theta}}, \boldsymbol{\theta})h(\mathbf{y}_1, \ldots, \mathbf{y}_n), \tag{2.54}$$

where $h(\mathbf{y}_1, \ldots, \mathbf{y}_n)$ does not depend on $\boldsymbol{\theta}$. This implies that all the information in $\mathbf{y}_1, \mathbf{y}_2, \ldots, \mathbf{y}_n$ about $\boldsymbol{\theta}$ is captured by $\hat{\boldsymbol{\theta}}$. If $\hat{\boldsymbol{\theta}}$ is sufficient for $\boldsymbol{\theta}$ and if additional conditions are satisfied, then $\boldsymbol{\theta}$ is estimated by $\hat{\boldsymbol{\theta}}$ (or a function of $\hat{\boldsymbol{\theta}}$ that is unbiased) with "minimum" covariance matrix (among all unbiased estimators) (Blackwell 1947, Lehmann and Scheffé 1950). We now show that $\bar{\mathbf{y}}$ and $\mathbf{S}$ are sufficient for $\boldsymbol{\mu}$ and $\boldsymbol{\Sigma}$.

Theorem 2.3C. If $\bar{\mathbf{y}}$ and $\mathbf{S}$ are based on a random sample $\mathbf{y}_1, \mathbf{y}_2, \ldots, \mathbf{y}_n$ from $N_p(\boldsymbol{\mu}, \boldsymbol{\Sigma})$, then $\bar{\mathbf{y}}$ and $\mathbf{S}$ are jointly sufficient for $\boldsymbol{\mu}$ and $\boldsymbol{\Sigma}$.

Proof. By (2.45)–(2.49), the likelihood function can be written as

$$\begin{aligned}
L(\boldsymbol{\mu}, \boldsymbol{\Sigma}) &= \frac{1}{(\sqrt{2\pi})^{np}|\boldsymbol{\Sigma}|^{n/2}} e^{-[(n-1)\,\text{tr}\,\boldsymbol{\Sigma}^{-1}\mathbf{S} + n(\bar{\mathbf{y}}-\boldsymbol{\mu})'\boldsymbol{\Sigma}^{-1}(\bar{\mathbf{y}}-\boldsymbol{\mu})]/2} \qquad (2.55) \\
&= g(\bar{\mathbf{y}}, \mathbf{S}, \boldsymbol{\mu}, \boldsymbol{\Sigma})h(\mathbf{y}_1, \ldots, \mathbf{y}_n).
\end{aligned}$$

This has the same form as (2.54), with $h(\mathbf{y}_1, \ldots, \mathbf{y}_n) = 1$. □

Note that the right side of (2.55) is not of the form $g_1(\bar{\mathbf{y}}, \boldsymbol{\mu})g_2(\mathbf{S}, \boldsymbol{\Sigma})$. Thus $\bar{\mathbf{y}}$ and $\mathbf{S}$ are *jointly* sufficient for $\boldsymbol{\mu}$ and $\boldsymbol{\Sigma}$, not *independently* sufficient. However, $\bar{\mathbf{y}}$ and $\mathbf{S}$ are independently distributed.

Theorem 2.3D. If $\bar{\mathbf{y}}$ and $\mathbf{S}$ are based on a random sample $\mathbf{y}_1, \mathbf{y}_2, \ldots, \mathbf{y}_n$ from $N_p(\boldsymbol{\mu}, \boldsymbol{\Sigma})$, then $\bar{\mathbf{y}}$ and $\mathbf{S}$ are independent. □

The distribution of $\bar{\mathbf{y}}$ is given in the following two theorems.

Theorem 2.3E. If $\bar{\mathbf{y}}$ is the mean of a random sample $\mathbf{y}_1, \mathbf{y}_2, \ldots, \mathbf{y}_n$ from $N_p(\boldsymbol{\mu}, \boldsymbol{\Sigma})$, then $\bar{\mathbf{y}}$ is distributed as $N_p(\boldsymbol{\mu}, \frac{1}{n}\boldsymbol{\Sigma})$.

Proof. If $a_1, a_2, \ldots, a_n$ are constants, then

$$\sum_{i=1}^{n} a_i \mathbf{y}_i \text{ is distributed as } N_p\left(\sum_{i=1}^{n} a_i \boldsymbol{\mu}, \sum_{i=1}^{n} a_i^2 \boldsymbol{\Sigma}\right), \tag{2.56}$$

which is an immediate extension of Theorem 2.2B(ii); see also Problem 2.28. If we let $a_i = 1/n$, then $\bar{\mathbf{y}}$ can be expressed in the form $\sum_{i=1}^{n} a_i \mathbf{y}_i$:

$$\bar{\mathbf{y}} = \frac{1}{n}\sum_{i=1}^{n} \mathbf{y}_i = \frac{1}{n}\mathbf{y}_1 + \frac{1}{n}\mathbf{y}_2 + \cdots + \frac{1}{n}\mathbf{y}_n,$$

and we have

$$\sum_{i=1}^{n} a_i = \frac{1}{n} + \cdots + \frac{1}{n} = 1, \qquad \sum_{i=1}^{n} a_i^2 = \frac{1}{n^2} + \cdots + \frac{1}{n^2} = \frac{1}{n}.$$

Thus $\sum_i a_i \boldsymbol{\mu} = \boldsymbol{\mu}$, $\sum_i a_i^2 \boldsymbol{\Sigma} = \frac{1}{n}\boldsymbol{\Sigma}$, and it follows that

$$\bar{\mathbf{y}} \text{ is } N_p\left(\boldsymbol{\mu}, \frac{1}{n}\boldsymbol{\Sigma}\right). \tag{2.57}$$

□

Even if $\mathbf{y}$ is not normally distributed, the distribution of $\bar{\mathbf{y}}$ will still be approximately normal. This result, known as the *multivariate central limit theorem*, is given in Theorem 2.3F.

Theorem 2.3F. If $\bar{\mathbf{y}}$ is the mean vector of a random sample $\mathbf{y}_1, \mathbf{y}_2, \ldots, \mathbf{y}_n$ from a population with mean vector $\boldsymbol{\mu}$ and covariance matrix $\boldsymbol{\Sigma}$, then as $n \to \infty$, the distribution of $\sqrt{n}(\bar{\mathbf{y}} - \boldsymbol{\mu})$ approaches the multivariate normal $N_p(\mathbf{0}, \boldsymbol{\Sigma})$. Thus, for large n, $\bar{\mathbf{y}}$ is approximately $N_p(\boldsymbol{\mu}, \frac{1}{n}\boldsymbol{\Sigma})$. □

The distribution of $\mathbf{S}$ is given in the following section.

2.3.3 Wishart Distribution

We now introduce the *Wishart distribution*, which is the multivariate analogue of the χ^2-distribution, or more precisely of the $\sigma^2\chi^2$-distribution. By definition, the sum of squares of independent standard normal (univariate) random variables is distributed as χ^2:

$$\sum_{i=1}^{n} z_i^2 = \sum_{i=1}^{n} \frac{(y_i - \mu)^2}{\sigma^2} \text{ is } \chi^2(n). \tag{2.58}$$

Thus $\sum_{i=1}^{n}(y_i - \mu)^2$ is distributed as $\sigma^2\chi^2$. If $\bar{y}$ is substituted for μ, it can be shown that $\sum_{i=1}^{n}(y_i - \bar{y})^2/\sigma^2$ is $\chi^2(n-1)$.

In the multivariate case, we have a definition analogous to (2.58):

$$\sum_{i=1}^{n}(\mathbf{y}_i - \boldsymbol{\mu})(\mathbf{y}_i - \boldsymbol{\mu})' \text{ is } W_p(n, \boldsymbol{\Sigma}), \tag{2.59}$$

where $\mathbf{y}_1, \mathbf{y}_2, \ldots, \mathbf{y}_n$ are independently distributed as $N_p(\boldsymbol{\mu}, \boldsymbol{\Sigma})$. Thus the $p(p+1)/2$ distinct variables in $\sum_{i=1}^{n}(\mathbf{y}_i - \boldsymbol{\mu})(\mathbf{y}_i - \boldsymbol{\mu})'$ are jointly distributed as the Wishart distribution $W_p(n, \boldsymbol{\Sigma})$, where n is the degrees of freedom. If we define $\mathbf{z}_i = \mathbf{y}_i - \boldsymbol{\mu}$, then $\mathbf{z}_i$ is $N_p(\mathbf{0}, \boldsymbol{\Sigma})$, and (2.59) becomes

$$\sum_{i=1}^{n} \mathbf{z}_i\mathbf{z}_i' \text{ is } W_p(n, \boldsymbol{\Sigma}). \tag{2.60}$$

By (A.3.8), $\sum_{i=1}^{n} \mathbf{z}_i\mathbf{z}_i' = \mathbf{Z}'\mathbf{Z}$, where $\mathbf{z}_i'$ is the ith row of the $n \times p$ matrix $\mathbf{Z}$. Thus $\mathbf{Z}'\mathbf{Z}$ is $W_p(n, \boldsymbol{\Sigma})$, where

$$\mathbf{Z} = \begin{pmatrix} \mathbf{z}_1' \\ \mathbf{z}_2' \\ \vdots \\ \mathbf{z}_n' \end{pmatrix},$$

and the $\mathbf{z}_i$'s are independent and identically distributed as $N_p(\mathbf{0}, \boldsymbol{\Sigma})$.

A condition under which the generalized quadratic form $\mathbf{Z}'\mathbf{A}\mathbf{Z}$ is distributed as a Wishart is given in the following theorem.

Theorem 2.3G. Let $\mathbf{Z} = (\mathbf{z}_1, \mathbf{z}_2, \ldots, \mathbf{z}_n)'$, where $\mathbf{z}_1, \mathbf{z}_2, \ldots, \mathbf{z}_n$ are independent and each $\mathbf{z}_i$ is $N_p(\mathbf{0}, \boldsymbol{\Sigma})$. Then $\mathbf{Z}'\mathbf{A}\mathbf{Z}$ is distributed as $W_p(r, \boldsymbol{\Sigma})$ if $\mathbf{A}$ is an $n \times n$ constant idempotent matrix of rank r.

Proof. By (A.11.4), $\mathbf{A} = \sum_{i=1}^{n} \lambda_i\mathbf{x}_i\mathbf{x}_i'$, where λ_i is the ith eigenvalue of $\mathbf{A}$, and $\mathbf{x}_i$ is the corresponding $n \times 1$ eigenvector. By (A.12.1), r of the λ_i's equal 1 and the rest equal 0. Thus $\mathbf{A} = \sum_{i=1}^{r} \mathbf{x}_i\mathbf{x}_i'$. Now

$$\begin{aligned} \mathbf{Z}'\mathbf{A}\mathbf{Z} &= \mathbf{Z}'\left(\sum_{i=1}^{r} \mathbf{x}_i\mathbf{x}_i'\right)\mathbf{Z} = \sum_{i=1}^{r} \mathbf{Z}'\mathbf{x}_i\mathbf{x}_i'\mathbf{Z} \\ &= \sum_{i=1}^{r} (\mathbf{Z}'\mathbf{x}_i)(\mathbf{Z}'\mathbf{x}_i)'. \end{aligned}$$

It can be shown (see Problem 2.30) that each $\mathbf{Z}'\mathbf{x}_i$ is $N_p(\mathbf{0}, \boldsymbol{\Sigma})$ and that the $\mathbf{Z}'\mathbf{x}_i$'s are independent. Thus by (2.60)

$$\mathbf{Z}'\mathbf{A}\mathbf{Z} = \sum_{i=1}^{r} (\mathbf{Z}'\mathbf{x}_i)(\mathbf{Z}'\mathbf{x}_i)' \text{ is } W_p(r, \boldsymbol{\Sigma}). \qquad \square$$

A condition for the independence of two Wishart matrices $\mathbf{Z}'\mathbf{A}\mathbf{Z}$ and $\mathbf{Z}'\mathbf{B}\mathbf{Z}$ is given in the following theorem.

Theorem 2.3H. Let the $n \times p$ matrix $\mathbf{Z}$ be defined as in Theorem 2.3G and let $\mathbf{A}$ and $\mathbf{B}$ be $n \times n$ idempotent matrices of rank r and s, respectively. Then the Wishart matrices $\mathbf{Z}'\mathbf{A}\mathbf{Z}$ and $\mathbf{Z}'\mathbf{B}\mathbf{Z}$ are independent if $\mathbf{AB} = \mathbf{O}$.

Proof. As in the proof of Theorem 2.3G, we write

$$\mathbf{Z}'\mathbf{A}\mathbf{Z} = \sum_{i=1}^{r} (\mathbf{Z}'\mathbf{x}_i)(\mathbf{Z}'\mathbf{x}_i)', \qquad \mathbf{Z}'\mathbf{B}\mathbf{Z} = \sum_{i=1}^{s} (\mathbf{Z}'\mathbf{u}_i)(\mathbf{Z}'\mathbf{u}_i)',$$

where the $\mathbf{u}_i$'s are eigenvectors of $\mathbf{B}$. The two generalized quadratic forms $\mathbf{Z}'\mathbf{A}\mathbf{Z}$ and $\mathbf{Z}'\mathbf{B}\mathbf{Z}$ will be independent if $\text{cov}(\mathbf{Z}'\mathbf{x}_i, \mathbf{Z}\mathbf{u}_j) = \mathbf{O}$ for all i, j. This value of the covariance matrix follows from $\mathbf{AB} = \mathbf{O}$ (see Problem 2.31). $\square$

When $\overline{\mathbf{y}}$ is substituted for $\boldsymbol{\mu}$ in $\sum_{i=1}^{n}(\mathbf{y}_i - \boldsymbol{\mu})(\mathbf{y}_i - \boldsymbol{\mu})'$ in (2.59), the distribution remains Wishart with the degrees of freedom reduced by 1.

Theorem 2.3I. If $\mathbf{S}$ is based on a random sample from $N_p(\boldsymbol{\mu}, \boldsymbol{\Sigma})$, then

$$(n-1)\mathbf{S} = \sum_{i=1}^{n} (\mathbf{y}_i - \overline{\mathbf{y}})(\mathbf{y}_i - \overline{\mathbf{y}})' \text{ is } W_p(n-1, \boldsymbol{\Sigma}). \tag{2.61}$$

$\square$

The Wishart distribution has a reproductive property analogous to that of the χ^2-distribution.

Theorem 2.3J. If $\mathbf{W}_1$ is $W_p(\nu_1, \boldsymbol{\Sigma})$, $\mathbf{W}_2$ is $W_p(\nu_2, \boldsymbol{\Sigma})$, and $\mathbf{W}_1$ and $\mathbf{W}_2$ are independent, then

$$\mathbf{W}_1 + \mathbf{W}_2 \text{ is } W_p(\nu_1 + \nu_2, \boldsymbol{\Sigma}). \tag{2.62}$$

$\square$

Another property of the Wishart distribution is given in the following theorem.

Theorem 2.3K. If $\mathbf{W}$ is $W_p(\nu, \boldsymbol{\Sigma})$ and $\mathbf{C}$ is a constant $q \times p$ matrix of rank $q \leq p$, then

$$\mathbf{CWC}' \text{ is } W_q(\nu, \mathbf{C\Sigma C}'). \tag{2.63}$$

□

Note that because $\mathbf{CWC}'$ is $q \times q$, the dimension of the Wishart distribution is q. The degrees-of-freedom parameter ν remains the same as the degrees of freedom of $\mathbf{W}$.

2.4 ADDITIONAL TOPICS

There are distributions available that will model some data sets better than does the multivariate normal, but in most cases percentage points of these distributions have not been tabulated. Jensen (1985) gives a good review of multivariate distributions, including some symmetric distributions that are longer tailed than the normal and may be potentially useful for such data.

Elliptically contoured distributions (Section 3.12.1) have contours of equal density (Section 2.1.3) with elliptical shape similar to that of the multivariate normal. This class of distributions includes distributions that are longer tailed as well as shorter tailed than the normal. It has been shown that many properties of this class of distributions are very similar to analogous properties of the multivariate normal distribution. These properties include the following: estimation of $\boldsymbol{\mu}$ and $\boldsymbol{\Sigma}$ (Section 2.3.1), testing H_0: $\boldsymbol{\mu} = \boldsymbol{\mu}_0$ (Section 3.3), testing H_0: $\boldsymbol{\mu}_1 = \boldsymbol{\mu}_2$ (Section 3.5), and testing H_0: $\boldsymbol{\mu}_1 = \boldsymbol{\mu}_2 = \cdots = \boldsymbol{\mu}_k$ (Section 4.1). Fang and Zhang (1990) give a comprehensive review of multivariate procedures based on elliptically contoured distributions.

The *multivariate t-distribution* (Dunnett and Sobel 1954; Tong 1990, Chapter 9) is the joint distribution of $\mathbf{t} = (t_1, t_2, \ldots, t_p)'$, where each t_i has the t-distribution. As an example of this distribution, let $\mathbf{y}$ be $N_p(\boldsymbol{\mu}, \sigma^2\mathbf{P}_\rho)$, where $\mathbf{P}_\rho$ is a correlation matrix, and let s^2 be an unbiased estimator for σ^2 such that $\nu s^2/\sigma^2$ is distributed as χ^2_ν and s^2 is independent of $\mathbf{y}$. Then $\mathbf{t} = \mathbf{y}/s$ has a (central) multivariate t-distribution with ν degrees of freedom. The density is given by

$$f(\mathbf{t}; \mathbf{P}_\rho, \nu) = \frac{\Gamma\left[(\nu + p)/2\right]\left[1 + (1/\nu)\mathbf{t}'\mathbf{P}_\rho^{-1}\mathbf{t}\right]^{-(\nu+p)/2}}{\Gamma\left(\nu/2\right)|\nu\pi\mathbf{P}_\rho|^{1/2}}. \tag{2.64}$$

Liu and Rubin (1994) discuss estimation of the parameters of the multivariate t-distribution.

An alternative approach to nonnormal data is to transform to approximate normality. Andrews, Gnanadesikan, and Warner (1971) proposed a multivariate generalization of the power transformation of Box and Cox (1964). Each variable y_j,

$j = 1, 2, \ldots, p$, is transformed by

$$y_j^{(\lambda_j)} = \begin{cases} \frac{y_j^{\lambda_j} - 1}{\lambda_j} & \text{for } \lambda_j \neq 0, \\ \ln y_j & \text{for } \lambda_j = 0. \end{cases} \quad (2.65)$$

The λ_j's are estimated from the data by maximum likelihood so that

$$(y_1^{(\lambda_1)}, y_2^{(\lambda_2)}, \ldots, y_p^{(\lambda_p)})'$$

is approximately multivariate normal. For further discussion of this transformation, see Gnanadesikan (1977, Section 5.3), McLachlan (1992, Section 6.3), and Velilla (1993, 1994, 1995).

PROBLEMS

2.1 Show that (2.10) is equal to (2.9); that is, show that $|\mathbf{\Sigma}^{-1/2}| = |\mathbf{\Sigma}|^{-1/2}$.

2.2 Find a in (2.12) in terms of c, $(\sqrt{2\pi})^p$, and $|\mathbf{\Sigma}|^{1/2}$.

2.3 Show that $\mathbf{z}'\mathbf{D}^{-1}\mathbf{z} = \sum_{j=1}^{p} z_j^2/\lambda_j$ as in (2.14).

2.4 Show that for the multivariate normal distribution, all third central moments are zero, as noted in Section 2.1.5.

2.5 Derive the expression for the fourth central moment in (2.19).

2.6 Show that the special cases (2.20), (2.21), and (2.22) follow from (2.19).

2.7 Suppose $\mathbf{y}$ is $N_3(\boldsymbol{\mu}, \mathbf{\Sigma})$, where

$$\boldsymbol{\mu} = \begin{pmatrix} 3 \\ 1 \\ 4 \end{pmatrix}, \qquad \mathbf{\Sigma} = \begin{pmatrix} 6 & 1 & -2 \\ 1 & 13 & 4 \\ -2 & 4 & 4 \end{pmatrix}.$$

Find the following central moments:

(a) $E\left[(y_1 - \mu_1)(y_3 - \mu_3)\right]$

(b) $E(y_3 - \mu_3)^2$

(c) $E\left[(y_1 - \mu_1)(y_2 - \mu_2)(y_3 - \mu_3)\right]$

(d) $E\left[(y_1 - \mu_1)^2(y_3 - \mu_3)\right]$

(e) $E(y_2 - \mu_2)^3$

(f) $E\left[(y_1 - \mu_1)^2(y_2 - \mu_2)(y_3 - \mu_3)\right]$

(g) $E\left[(y_1 - \mu_1)^2(y_3 - \mu_3)^2\right]$

(h) $E(y_1 - \mu_1)^4$

2.8 Suppose $\mathbf{y}$ is $\mathrm{N}_4(\boldsymbol{\mu}, \mathbf{\Sigma})$, where

$$\boldsymbol{\mu} = \begin{pmatrix} -2 \\ 3 \\ -1 \\ 5 \end{pmatrix}, \qquad \mathbf{\Sigma} = \begin{pmatrix} 11 & -8 & 3 & 9 \\ -8 & 9 & -3 & -6 \\ 3 & -3 & 2 & 3 \\ 9 & -6 & 3 & 9 \end{pmatrix}.$$

Find the following central moments:

(a) $E\left[(y_2 - \mu_2)(y_4 - \mu_4)\right]$
(b) $E(y_3 - \mu_3)^2$
(c) $E\left[(y_1 - \mu_1)(y_2 - \mu_2)(y_4 - \mu_4)\right]$
(d) $E\left[(y_3 - \mu_3)(y_4 - \mu_4)^2\right]$
(e) $E(y_2 - \mu_2)^3$
(f) $E\left[(y_1 - \mu_1)(y_2 - \mu_2)(y_3 - \mu_3)(y_4 - \mu_4)\right]$
(g) $E\left[(y_1 - \mu_1)^2(y_3 - \mu_3)(y_4 - \mu_4)\right]$
(h) $E\left[(y_1 - \mu_1)(y_2 - \mu_2)^2(y_4 - \mu_4)\right]$
(i) $E\left[(y_1 - \mu_1)^2(y_3 - \mu_3)^2\right]$
(j) $E(y_2 - \mu_2)^4$

2.9 Show that the exponent in (2.27) can be expressed in the form given in (2.28); that is, show that $\mathbf{t}'\mathbf{y} - (\mathbf{y} - \boldsymbol{\mu})'\boldsymbol{\Sigma}^{-1}(\mathbf{y} - \boldsymbol{\mu})/2 = \mathbf{t}'\boldsymbol{\mu} + \mathbf{t}'\boldsymbol{\Sigma}\mathbf{t}/2 - (\mathbf{y} - \boldsymbol{\mu} - \boldsymbol{\Sigma}\mathbf{t})'\boldsymbol{\Sigma}^{-1}(\mathbf{y} - \boldsymbol{\mu} - \boldsymbol{\Sigma}\mathbf{t})/2$.

2.10 Show that $E\left(e^{\mathbf{t}'\mathbf{A}\mathbf{y}}\right) = e^{\mathbf{t}'(\mathbf{A}\boldsymbol{\mu}) + \mathbf{t}'(\mathbf{A}\boldsymbol{\Sigma}\mathbf{A}')\mathbf{t}/2}$ as in (2.32).

2.11 Show that if $\mathbf{A}$ is $q \times p$ of rank $q \leq p$, then $\mathbf{A}\boldsymbol{\Sigma}\mathbf{A}'$ is positive definite, as required in Theorem 2.2B(ii).

2.12 Prove Corollary 1 to Theorem 2.2B; that is, show that $\mathbf{z}$ is $N_p(\mathbf{0}, \mathbf{I})$ for $\mathbf{z}$ defined by (2.33) and (2.34).

2.13 If $\mathbf{y}$ is $N_p(\boldsymbol{\mu}, \sigma^2\mathbf{I})$ and $\mathbf{C}$ is an orthogonal matrix, show that $\mathbf{C}\mathbf{y}$ is $N_p(\mathbf{C}\boldsymbol{\mu}, \sigma^2\mathbf{I})$.

2.14 Let $\mathbf{A} = (\mathbf{I}_r, \mathbf{O})$, as defined in the proof of Theorem 2.2C. Show that $\mathbf{A}\mathbf{y} = \mathbf{y}_1$, $\mathbf{A}\boldsymbol{\mu} = \boldsymbol{\mu}_1$, and $\mathbf{A}\boldsymbol{\Sigma}\mathbf{A}' = \boldsymbol{\Sigma}_{11}$.

2.15 Prove Corollary 2 to Theorem 2.2D.

2.16 Prove Corollary 3 to Theorem 2.2D.

2.17 Show that $\text{cov}(\mathbf{z}, \mathbf{x}) = \boldsymbol{\Sigma}_{yx} - \mathbf{B}\boldsymbol{\Sigma}_{xx}$ as in (2.40), where $\mathbf{z} = \mathbf{y} - \mathbf{B}\mathbf{x}$.

2.18 Show that $\text{cov}(\mathbf{y} - \boldsymbol{\Sigma}_{yx}\boldsymbol{\Sigma}_{xx}^{-1}\mathbf{x}) = \boldsymbol{\Sigma}_{yy} - \boldsymbol{\Sigma}_{yx}\boldsymbol{\Sigma}_{xx}^{-1}\boldsymbol{\Sigma}_{xy}$ as in (2.41).

2.19 Prove Theorem 2.2E by direct evaluation of (2.38).

2.20 Show that $\mathbf{z}'\mathbf{z} = (\mathbf{y} - \boldsymbol{\mu})'\boldsymbol{\Sigma}^{-1}(\mathbf{y} - \boldsymbol{\mu})$, as in (2.42), where $\mathbf{z}$ is defined in (2.33) and (2.34).

2.21 Show that $\sum_{i=1}^n \text{tr}\,(\mathbf{y}_i - \boldsymbol{\mu})'\boldsymbol{\Sigma}^{-1}(\mathbf{y}_i - \boldsymbol{\mu}) = \text{tr}\left[\boldsymbol{\Sigma}^{-1}\sum_{i=1}^n(\mathbf{y}_i - \boldsymbol{\mu})(\mathbf{y}_i - \boldsymbol{\mu})'\right]$, thus verifying (2.46).

2.22 Show that $\sum_{i=1}^n(\mathbf{y}_i - \boldsymbol{\mu})(\mathbf{y}_i - \boldsymbol{\mu})' = \sum_{i=1}^n(\mathbf{y}_i - \bar{\mathbf{y}})(\mathbf{y}_i - \bar{\mathbf{y}})' + n(\bar{\mathbf{y}} - \boldsymbol{\mu})(\bar{\mathbf{y}} - \boldsymbol{\mu})'$, thus verifying (2.47).

2.23 Show that $\text{tr}\,\boldsymbol{\Sigma}^{-1}\left[\mathbf{W} + n(\bar{\mathbf{y}} - \boldsymbol{\mu})(\bar{\mathbf{y}} - \boldsymbol{\mu})'\right] = \text{tr}(\boldsymbol{\Sigma}^{-1}\mathbf{W}) + n(\bar{\mathbf{y}} - \boldsymbol{\mu})'\boldsymbol{\Sigma}^{-1}(\bar{\mathbf{y}} - \boldsymbol{\mu})$ as in (2.49).

2.24 Show that $\partial \ln L(\boldsymbol{\mu}, \boldsymbol{\Sigma})/\partial\boldsymbol{\mu} = n(\boldsymbol{\Sigma}^{-1}\bar{\mathbf{y}} - \boldsymbol{\Sigma}^{-1}\boldsymbol{\mu})$ as in (2.50).

2.25 Verify the result in (2.52) for $\partial \ln L(\hat{\boldsymbol{\mu}}, \boldsymbol{\Sigma})/\partial\boldsymbol{\Sigma}^{-1}$.

2.26 Show that the exponent in (2.55) follows from (2.46) and (2.47).

2.27 Prove Theorem 2.3D in the following two ways:

(a) By (1.29) and (1.42), $\bar{\mathbf{y}} = \frac{1}{n}\mathbf{Y}'\mathbf{j}$ and $\mathbf{S} = \mathbf{Y}'(\mathbf{I} - \frac{1}{n}\mathbf{J})'(\mathbf{I} - \frac{1}{n}\mathbf{J})\mathbf{Y}/(n-1)$. Let $\mathbf{A} = \mathbf{I} - \frac{1}{n}\mathbf{J}$. By (1.27), $\mathbf{j}'\mathbf{Y} = (\mathbf{j}'\mathbf{y}_{(1)}, \ldots, \mathbf{j}'\mathbf{y}_{(p)})$ and $\mathbf{AY} = (\mathbf{Ay}_{(1)}, \ldots, \mathbf{Ay}_{(p)})$. Show that $\text{cov}(\mathbf{j}'\mathbf{y}_{(i)}, \mathbf{Ay}_{(j)}) = \mathbf{0}'$ for all i, j. This implies independence since $\mathbf{y}$ is $N_p(\boldsymbol{\mu}, \boldsymbol{\Sigma})$.

(b) Show that $\text{cov}(\mathbf{y}_i - \bar{\mathbf{y}}, \bar{\mathbf{y}}) = \mathbf{O}$ for all $i = 1, 2, \ldots, n$, which implies that $\bar{\mathbf{y}}$ is independent of all $\mathbf{y}_i - \bar{\mathbf{y}}$, and hence $\bar{\mathbf{y}}$ is independent of $\mathbf{S} = \sum_{i=1}^n (\mathbf{y}_i - \bar{\mathbf{y}})(\mathbf{y}_i - \bar{\mathbf{y}})'/(n-1)$.

2.28 Show that (2.56) holds; that is, if $\mathbf{y}_1, \mathbf{y}_2, \ldots, \mathbf{y}_n$ is a random sample from $N_p(\boldsymbol{\mu}, \boldsymbol{\Sigma})$, then $\sum_i a_i \mathbf{y}_i$ is $N_p(\sum_i a_i \boldsymbol{\mu}, \sum_i a_i^2 \boldsymbol{\Sigma})$. Use the following two approaches:

(a) Define $\mathbf{y} = (\mathbf{y}_1', \mathbf{y}_2', \ldots, \mathbf{y}_n')'$ and $\mathbf{A} = (a_1\mathbf{I}, a_2\mathbf{I}, \ldots, a_n\mathbf{I})$, where each $\mathbf{I}$ is $p \times p$. Then by Theorem 2.2B, $\mathbf{Ay}$ is $N_p(\mathbf{A}\boldsymbol{\mu}_y, \mathbf{A}\boldsymbol{\Sigma}_y\mathbf{A}')$ for appropriately defined $\boldsymbol{\mu}_y$ and $\boldsymbol{\Sigma}_y$.

(b) Use the moment generating function in Theorem 2.2A.

2.29 Using the methods of Problem 2.28(a), show that $\text{cov}\left(\sum_{i=1}^n a_i \mathbf{y}_i, \sum_{i=1}^n b_i \mathbf{y}_i\right) = \sum_{i=1}^n a_i b_i \boldsymbol{\Sigma}$, which is equal to $\mathbf{O}$ if $\sum_{i=1}^n a_i b_i = 0$.

2.30 In the proof of Theorem 2.3G, show the following:

(a) $\mathbf{Z}'\mathbf{x}_i$ is $N_p(\mathbf{0}, \boldsymbol{\Sigma})$ for $i = 1, 2, \ldots, r$;

(b) $\text{cov}(\mathbf{Z}'\mathbf{x}_i, \mathbf{Z}'\mathbf{x}_j) = \mathbf{O}$ for $i \neq j$.

2.31 In the proof of Theorem 2.3H, show that $\text{cov}(\mathbf{Z}'\mathbf{x}_i, \mathbf{Zu}_j) = \mathbf{O}$.

2.32 Prove Theorem 2.3I.

2.33 Suppose $\mathbf{y}_1, \mathbf{y}_2, \ldots, \mathbf{y}_{25}$ is a random sample of size $n = 25$ from $N_8(\boldsymbol{\mu}, \boldsymbol{\Sigma})$. What is the distribution of each of the following?

(a) $\bar{\mathbf{y}}$,

(b) $24\mathbf{S}$, and

(c) $24\mathbf{CSC}'$, where $\mathbf{C}$ is a constant 3×8 matrix of rank 3.

2.34 Prove Theorem 2.3J.

2.35 Prove Theorem 2.3K.

2.36 If $\mathbf{y}$ is $N_p(\boldsymbol{\mu}, \boldsymbol{\Sigma})$, show that $E(\mathbf{y} - \boldsymbol{\mu})'\boldsymbol{\Sigma}^{-1}(\mathbf{y} - \boldsymbol{\mu}) = p$.

2.37 Consider a random sample $\mathbf{y}_1, \mathbf{y}_2, \ldots, \mathbf{y}_n$ with mean vector $\bar{\mathbf{y}}$ and covariance matrix $\mathbf{S}$, where each $\mathbf{y}_i$ is $p \times 1$. Define $u_{ii} = (\mathbf{y}_i - \bar{\mathbf{y}})'\mathbf{S}^{-1}(\mathbf{y}_i - \bar{\mathbf{y}})$ and $u_{ij} = (\mathbf{y}_i - \bar{\mathbf{y}})'\mathbf{S}^{-1}(\mathbf{y}_j - \bar{\mathbf{y}})$ for $i \neq j$.

(a) Show that $E(u_{ii}) = (n-1)p/n$.

(b) Show that $E(u_{ij}) = -p/n$ for $i \neq j$.

(c) Let $\mathbf{U} = (u_{ij})$. Show that $E(\mathbf{U}) = p(\mathbf{I} - \frac{1}{n}\mathbf{J})$.

2.38 Suppose $\mathbf{W}$ is distributed as $W_p(\nu, \boldsymbol{\Sigma})$. Show that $\boldsymbol{\Sigma}^{-1/2}\mathbf{W}\boldsymbol{\Sigma}^{-1/2}$ is $W_p(\nu, \mathbf{I})$.

2.39 Suppose $\mathbf{W}$ is $W_p(\nu, \boldsymbol{\Sigma})$. Show that $\mathbf{a}'\mathbf{Wa}/\mathbf{a}'\boldsymbol{\Sigma}\mathbf{a}$ is $\chi^2(\nu)$, where $\mathbf{a}$ is a constant vector.

2.40 If $\mathbf{W} = \sum_{i=1}^n (\mathbf{y}_i - \boldsymbol{\mu})(\mathbf{y}_i - \boldsymbol{\mu})'$ is $W_p(n, \boldsymbol{\Sigma})$ as in (2.59), show that $E(\mathbf{W}) = n\boldsymbol{\Sigma}$ and $\text{cov}(w_{ij}, w_{kl}) = n(\sigma_{ik}\sigma_{jl} + \sigma_{il}\sigma_{jk})$.

CHAPTER 3

Hotelling's T^2-Tests

3.1 INTRODUCTION

Multivariate hypothesis testing differs from univariate testing in many ways. For example, (1) the number of possible hypotheses is very large and (2) in many cases, there are several alternative test statistics from which to choose. The number of possible hypotheses is large because of the large number of parameters. For example, the p-variate normal distribution has $\frac{1}{2}p(p+3)$ parameters, for each of which we could specify a hypothesis, as well as for subsets or functions of the parameters. We will consider various approaches to test construction, including the likelihood ratio test, the union intersection method, and other techniques.

There are many advantages to testing p variables in a single multivariate test rather than in p separate univariate tests. These advantages include preserving the α-level, testing with greater power, and determining the contribution of each variable in the presence of the other variables. The multivariate tests make allowance for the intercorrelations among the variables, which accounts in part for the increase in power. For additional comparisons of multivariate and univariate tests, see Rencher (1995, Sections 5.1 and 5.2).

3.2 TEST FOR $H_0\colon \boldsymbol{\mu} = \boldsymbol{\mu}_0$ WITH $\boldsymbol{\Sigma}$ KNOWN

In order to illustrate the principles of multivariate testing, we begin with the single-sample case in which $\boldsymbol{\Sigma}$ is known. We wish to test $H_0\colon \boldsymbol{\mu} = \boldsymbol{\mu}_0$ versus $H_1\colon \boldsymbol{\mu} \neq \boldsymbol{\mu}_0$, where $\boldsymbol{\mu}_0$ is completely specified. The implication of the vector inequality in H_1 is that at least one $\mu_j \neq \mu_{0j}$.

We assume the availability of a random sample of n observation vectors $\mathbf{y}_1$, $\mathbf{y}_2, \ldots, \mathbf{y}_n$ from $N_p(\boldsymbol{\mu}, \boldsymbol{\Sigma})$, with $\boldsymbol{\Sigma}$ known, from which we calculate $\overline{\mathbf{y}} = \sum_{i=1}^{n} \mathbf{y}_i/n$. To compare $\overline{\mathbf{y}}$ with $\boldsymbol{\mu}_0$, we use

$$Z^2 = n(\overline{\mathbf{y}} - \boldsymbol{\mu}_0)'\boldsymbol{\Sigma}^{-1}(\overline{\mathbf{y}} - \boldsymbol{\mu}_0). \tag{3.1}$$

By Theorem 2.2F, Z^2 is distributed as χ^2_p if H_0 is true. We therefore reject H_0 if $Z^2 \geq \chi^2_{\alpha,p}$, where $\chi^2_{\alpha,p}$ is the upper α quantile of the χ^2_p-distribution, and α is the probability of a Type I error (rejecting H_0 when it is true). To compare $\overline{\mathbf{y}}$ with $\boldsymbol{\mu}_0$, we have simply chosen a convenient test statistic with a known distribution. However, it can be shown that the test using Z^2 in (3.1) is equivalent to the likelihood ratio test (see Sections 3.3.2 and 3.3.7 and Problem 3.1).

It is often suggested that if $\boldsymbol{\Sigma}$ is replaced by $\mathbf{S}$ in (3.1), then Z^2 has an approximate χ^2-distribution. However, for this to hold, n must be larger than in the analogous univariate situation ($p = 1$). The value of n required for Z^2 based on $\mathbf{S}$ to approach the χ^2-distribution increases with p [for additional comments, see Rencher (1995, Section 5.3.2)].

3.3 HOTELLING'S T^2-TEST FOR $H_0\colon \boldsymbol{\mu} = \boldsymbol{\mu}_0$ WITH $\boldsymbol{\Sigma}$ UNKNOWN

We begin with a review of the univariate t-test in order to introduce some test properties and the likelihood ratio method of test construction. We extend to the analogous multivariate test in Section 3.3.3.

3.3.1 Univariate t-Test for $H_0\colon \mu = \mu_0$ with σ^2 Unknown

To test $H_0\colon \mu = \mu_0$ versus $H_1\colon \mu \neq \mu_0$ when $p = 1$, we assume that a random sample of n observations $y_1, y_2, \ldots, y_n$ has been obtained from $N(\mu, \sigma^2)$ with σ^2 unknown. From the sample, we calculate $\overline{y}$ and s^2. If H_0 is true,

$$t = \frac{\overline{y} - \mu_0}{s/\sqrt{n}} = \frac{\sqrt{n}(\overline{y} - \mu_0)}{s} \tag{3.2}$$

is distributed as t_{n-1}. We reject H_0 if $|t| \geq t_{\alpha/2,n-1}$ or if $t^2 = n(\overline{y}-\mu_0)^2/s^2 \geq t^2_{\alpha/2,n-1}$, where $t^2_{\alpha/2,n-1}$ is the square of the upper critical value $t_{\alpha/2,n-1}$ from the t-table.

The *formal definition* of a t random variable with ν degrees of freedom is

$$t = \frac{z}{\sqrt{w/\nu}}, \tag{3.3}$$

where z is $N(0, 1)$, w is χ^2_ν, and z and w are independent. From this definition, the density of t can be derived and tables of critical values can be computed. To see that (3.2) can be expressed in the form of (3.3), we write (3.2) as follows:

$$\frac{\sqrt{n}(\overline{y} - \mu_0)}{s} = \frac{(\overline{y} - \mu_0)/(\sigma/\sqrt{n})}{\sqrt{\sum_{i=1}^n (y_i - \overline{y})^2/(n-1)\sigma^2}}. \tag{3.4}$$

Then, assuming that the y_i's are normal,

$$z = \frac{\overline{y} - \mu_0}{\sigma/\sqrt{n}} \text{ is } N(0, 1),$$

$$w = \frac{\sum_{i=1}^n (y_i - \overline{y})^2}{\sigma^2} \text{ is } \chi^2_{n-1},$$

and z and w are independent because of the independence of $\overline{y}$ and s^2 when sampling from a normal distribution.

The t-test has the following important properties:

1. The t-statistic in (3.2) is invariant to linear transformations on the y's; that is, if $z_i = ay_i + b$, $i = 1, 2, \ldots, n$, then

$$t_z^2 = t_y^2. \tag{3.5}$$

This property is easy to demonstrate:

$$\begin{aligned} t_z^2 &= \frac{n(\overline{z} - \mu_{0z})^2}{s_z^2} = \frac{n\left[a\overline{y} + b - (a\mu_0 + b)\right]^2}{a^2 s_y^2} \\ &= \frac{na^2(\overline{y} - \mu_0)^2}{a^2 s_y^2} = t_y^2. \end{aligned}$$

2. The t-test has optimal power among all unbiased tests. An *unbiased test* is one that has power greater than or equal to α for all values of μ. For example, the t-test with a one-tailed rejection region $t \geq t_{\alpha,n-1}$ corresponding to $H_1\colon \mu > \mu_0$ would not be unbiased, since $P(t \geq t_{\alpha,n-1})$ is less than α when $\mu < \mu_0$.

3.3.2 Likelihood Ratio Method of Test Construction

The t-test is intuitively appealing as a standardized distance between $\overline{y}$ and μ_0, and as noted in Section 3.3.1, it has other desirable properties. In addition, the t-test is the *likelihood ratio test.* We now describe the likelihood ratio approach to constructing tests. The likelihood function is the joint density of the observations $y_1, y_2, \ldots, y_n$ (Section 2.3.1). Because of the independence assumed in a random sample, the likelihood function is the product of the densities of the y_i's:

$$L(\mu, \sigma^2) = \prod_{i=1}^{n} f(y_i; \mu, \sigma^2). \tag{3.6}$$

The likelihood ratio method compares the maximum value of $L(\mu, \sigma^2)$ restricted by H_0 to the maximum of $L(\mu, \sigma^2)$ under H_1, which is essentially unrestricted for a two-sided H_1, such as $H_1\colon \mu \neq \mu_0$. Thus, if μ_0 is a plausible value of μ for the sample, $\max_{H_0} L(\mu, \sigma^2)$ should be close to $\max_{H_1} L(\mu, \sigma^2)$. If $\max_{H_0} L(\mu, \sigma^2)$ is not close to $\max_{H_1} L(\mu, \sigma^2)$, we would conclude that $y_1, y_2, \ldots, y_n$ apparently did not come from $N(\mu_0, \sigma^2)$ because the likelihood function is much greater when μ is not restricted to be μ_0.

To find the maximum value of $L(\mu, \sigma^2)$ under H_0, μ is set equal to μ_0 and σ^2 is determined by the data; that is, σ^2 is estimated to be the value that maximizes $L(\mu_0, \sigma^2)$. Under H_1, μ and σ^2 are both determined by the data to be the values that maximize $L(\mu, \sigma^2)$. In describing this unrestricted maximum as $\max_{H_1} L(\mu, \sigma^2)$, we are ignoring the restriction imposed by H_1 that $\mu \neq \mu_0$.

It is customary to describe the likelihood ratio method in terms of maximizing L subject to ω, the set of all values of μ and σ^2 satisfying H_0, and maximizing L subject to Ω, the set of all values of μ and σ^2 without restrictions (other than natural restrictions such as $\sigma^2 > 0$). However, to simplify notation in this case in which σ^2 is unspecified and H_1 includes all values of μ except μ_0, we refer to maximizing L under H_0 and H_1.

The restricted maximum under H_0 is compared with the unrestricted maximum under H_1 by the *likelihood ratio*

$$\text{LR} = \frac{\max_{H_0} L(\mu, \sigma^2)}{\max_{H_1} L(\mu, \sigma^2)} \tag{3.7}$$

$$= \frac{\max L(\mu_0, \sigma^2)}{\max L(\mu, \sigma^2)}.$$

It is clear that $0 \leq \text{LR} \leq 1$, because the maximum restricted to $\mu = \mu_0$ cannot be as large as the unrestricted maximum. Smaller values of LR would favor H_1, and larger values would favor H_0. We would thus reject H_0 if $\text{LR} \leq c$, where c is selected so that $P(\text{LR} \leq c) = \alpha$ if H_0 is true. Wald (1943) showed that, for large n,

$$-2 \ln \text{LR} \text{ is approximately } \chi^2_\nu, \tag{3.8}$$

where ν is equal to the number of parameters estimated under H_1 minus the number estimated under H_0. In the case of $H_0\colon \mu = \mu_0$ versus $H_1\colon \mu \neq \mu_0$, ν is 1 because μ and σ^2 are estimated under H_1, while only σ^2 is estimated under H_0. In some cases the χ^2-approximation is not needed because LR turns out to be a function of a familiar test statistic such as t or F whose exact distribution is available.

We now show that the t-test in (3.2) is the likelihood ratio test.

Theorem 3.3A. For a random sample $y_1, y_2, \ldots, y_n$ from $N(\mu, \sigma^2)$, the t-test in (3.2) for $H_0\colon \mu = \mu_0$ versus $H_1\colon \mu \neq \mu_1$ is the likelihood ratio test.

Proof. The likelihood function (3.6) can be written as

$$L(\mu, \sigma^2) = \prod_{i=1}^{n} f(y_i; \mu, \sigma^2) = \prod_{i=1}^{n} \frac{1}{\sqrt{2\pi\sigma^2}} e^{-(y_i - \mu)^2/2\sigma^2}$$

$$= \frac{1}{(\sqrt{2\pi})^n (\sigma^2)^{n/2}} e^{-\Sigma_i (y_i - \mu)^2/2\sigma^2}. \tag{3.9}$$

Under H_1, the values of μ and σ^2 that maximize L are the maximum likelihood estimators $\hat{\mu} = \bar{y}$ and $\hat{\sigma}^2 = \sum_{i=1}^{n} (y_i - \bar{y})^2/n$. When these are substituted for μ and

σ^2 in $L(\mu, \sigma^2)$, we have

$$\max_{H_1} L(\mu, \sigma^2) = \frac{1}{(\sqrt{2\pi})^n (\hat{\sigma}^2)^{n/2}} e^{-\Sigma_i (y_i - \hat{\mu})^2 / 2\hat{\sigma}^2}$$

$$= \frac{n^{n/2}}{(\sqrt{2\pi})^n \left[\sum_i (y_i - \bar{y})^2\right]^{n/2}} e^{-n\Sigma_i (y_i - \bar{y})^2 / 2\Sigma_i (y_i - \bar{y})^2}$$

$$= \frac{n^{n/2}}{(\sqrt{2\pi})^n \left[\sum_i (y_i - \bar{y})^2\right]^{n/2}} e^{-n/2}. \tag{3.10}$$

Under H_0, μ is set equal to μ_0 in (3.9), and the value of σ^2 that maximizes $L(\mu_0, \sigma^2)$ turns out to be $\hat{\sigma}_0^2 = \sum_{i=1}^n (y_i - \mu_0)^2 / n$. When μ_0 and $\hat{\sigma}_0^2$ are substituted for μ and σ^2 in $L(\mu, \sigma^2)$, we obtain

$$\max_{H_0} L(\mu, \sigma^2) = \frac{n^{n/2}}{(\sqrt{2\pi})^n \left[\sum_i (y_i - \mu_0)^2\right]^{n/2}} e^{-n/2}. \tag{3.11}$$

The likelihood ratio (3.7) is thus given by the ratio of (3.11) and (3.10):

$$\text{LR} = \frac{\max_{H_0} L}{\max_{H_1} L} = \left[\frac{\sum_i (y_i - \bar{y})^2}{\sum_i (y_i - \mu_0)^2}\right]^{n/2}. \tag{3.12}$$

To express LR in (3.12) as a function of the t-statistic in (3.2), we write $\sum_{i=1}^n (y_i - \mu_0)^2$ as

$$\sum_{i=1}^n (y_i - \mu_0)^2 = \sum_{i=1}^n (y_i - \bar{y} + \bar{y} - \mu_0)^2$$

$$= \sum_i (y_i - \bar{y})^2 + n(\bar{y} - \mu_0)^2 + 2(\bar{y} - \mu_0) \sum_i (y_i - \bar{y})$$

$$= \sum_i (y_i - \bar{y})^2 + n(\bar{y} - \mu_0)^2,$$

and

$$\text{LR} = \left[\frac{\sum_{i=1}^n (y_i - \bar{y})^2}{\sum_{i=1}^n (y_i - \bar{y})^2 + n(\bar{y} - \mu_0)^2}\right]^{n/2}$$

$$= \left[\frac{1}{1 + n(\bar{y} - \mu_0)^2 / \sum_{i=1}^n (y_i - \bar{y})^2}\right]^{n/2}$$

$$= \left[\frac{1}{1 + t^2/(n-1)}\right]^{n/2}. \tag{3.13}$$

Thus LR is small when t^2 is large, and rejection of H_0 for LR $\le c$ is equivalent to rejection of H_0 for $t^2 \ge t^2_{\alpha/2}$. The t-test in (3.2) is therefore a likelihood ratio test statistic. □

The likelihood ratio method of test construction usually leads to tests that are relatively powerful and sometimes produces tests with optimum power over a wide class of alternatives. Many multivariate tests are derived in this fashion.

3.3.3 One-Sample T^2-Test

We now consider the multivariate case in which the hypothesis $H_0\colon \boldsymbol{\mu} = \boldsymbol{\mu}_0$ is p-dimensional. In order to test $H_0\colon \boldsymbol{\mu} = \boldsymbol{\mu}_0$ versus $H_1\colon \boldsymbol{\mu} \ne \boldsymbol{\mu}_0$, we assume that a random sample $\mathbf{y}_1, \mathbf{y}_2, \ldots, \mathbf{y}_n$ is available from $N_p(\boldsymbol{\mu}, \boldsymbol{\Sigma})$, with $\boldsymbol{\Sigma}$ unknown. By analogy with the univariate t-statistic in (3.2) and the Z^2-statistic in (3.1), we use the sample mean vector $\overline{\mathbf{y}}$ and the sample covariance matrix $\mathbf{S}$ to construct the test statistic,

$$T^2 = n(\overline{\mathbf{y}} - \boldsymbol{\mu}_0)'\mathbf{S}^{-1}(\overline{\mathbf{y}} - \boldsymbol{\mu}_0), \tag{3.14}$$

which is the standardized distance from $\overline{\mathbf{y}}$ to $\boldsymbol{\mu}_0$. We show in Section 3.3.7 that this is the likelihood ratio test statistic.

If H_0 is true and if sampling is from $N_p(\boldsymbol{\mu}, \boldsymbol{\Sigma})$, then T^2 in (3.14) has Hotelling's (1931) T^2-distribution with dimension p and degrees of freedom $n - 1$. We reject H_0 if $T^2 \ge T^2_{\alpha,p,n-1}$ and accept H_0 otherwise. We use the terminology "accept H_0" for expositional convenience; it means only that we have failed to reject H_0. Critical values $T^2_{\alpha,p,n-1}$ of the T^2-distribution are given in Table B.1 in Appendix B (Kramer and Jensen 1969a). For an analysis of patterns in the T^2-table and the insights that these patterns provide about the T^2-test and multivariate testing in general, see Rencher (1995, Section 5.3.2).

For $p = 1$, the T^2-statistic in (3.14) reduces to the square of the univariate t in (3.2):

$$T^2 = n(\overline{y} - \mu_0)(s^2)^{-1}(\overline{y} - \mu_0) = \frac{n(\overline{y} - \mu_0)^2}{s^2} = t^2.$$

Another link between T^2 and the univariate t is that in cases where there is an analogous t-test, the degrees of freedom for the T^2-test will be the same as for the univariate t-test. Thus the one-sample T^2-test has $n - 1$ degrees of freedom, and the two-sample T^2-test (to be defined in Section 3.5.2) has $n_1 + n_2 - 2$ degrees of freedom.

To avoid inverting $\mathbf{S}$, an alternative formula for computing T^2 in (3.14) can be obtained using (A.7.10):

$$T^2 = \frac{|\mathbf{S} + n(\overline{\mathbf{y}} - \boldsymbol{\mu}_0)(\overline{\mathbf{y}} - \boldsymbol{\mu}_0)'|}{|\mathbf{S}|} - 1. \tag{3.15}$$

A key assumption in the T^2-distribution is the independence of $\bar{\mathbf{y}}$ and $\mathbf{S}$; this assumption holds when sampling from a multivariate normal population (Theorem 2.3D). Usually $\bar{\mathbf{y}}$ and $\mathbf{S}$ are obtained from the same sample, although this is not necessary. As long as $\mathbf{S}$ is an unbiased estimator of $\boldsymbol{\Sigma}$ and is independent of $\bar{\mathbf{y}}$, the estimator $\mathbf{S}$ could come wholly or partly from another sample. For example, suppose $\bar{\mathbf{y}}_1$ and $\mathbf{S}_1$ arise from a sample of size n_1 from $N_p(\boldsymbol{\mu}_1, \boldsymbol{\Sigma})$ and that a supplementary estimate $\mathbf{S}_2$ is available from a sample of size n_2 from $N_p(\boldsymbol{\mu}_2, \boldsymbol{\Sigma})$, where the two distributions have a common covariance matrix $\boldsymbol{\Sigma}$ and the two samples are independent. We estimate $\boldsymbol{\Sigma}$ by the pooled estimator $\mathbf{S}_{\text{pl}} = [(n_1 - 1)\mathbf{S}_1 + (n_2 - 1)\mathbf{S}_2]/(n_1 + n_2 - 2)$, which has $n_1 + n_2 - 2$ degrees of freedom. Then a T^2-statistic could be based on $\mathbf{S}_1, \mathbf{S}_2$, or $\mathbf{S}_{\text{pl}}$:

$$n_1(\bar{\mathbf{y}}_1 - \boldsymbol{\mu}_1)'\mathbf{S}_1^{-1}(\bar{\mathbf{y}}_1 - \boldsymbol{\mu}_1) \text{ is } T^2_{p,n_1-1},$$

$$n_1(\bar{\mathbf{y}}_1 - \boldsymbol{\mu}_1)'\mathbf{S}_2^{-1}(\bar{\mathbf{y}}_1 - \boldsymbol{\mu}_1) \text{ is } T^2_{p,n_2-1},$$

$$n_1(\bar{\mathbf{y}}_1 - \boldsymbol{\mu}_1)'\mathbf{S}_{\text{pl}}^{-1}(\bar{\mathbf{y}}_1 - \boldsymbol{\mu}_1) \text{ is } T^2_{p,n_1+n_2-2}.$$

The coefficient of the quadratic form in all three cases is n_1 because each of $\mathbf{S}_1/n_1, \mathbf{S}_2/n_1$, and $\mathbf{S}_{\text{pl}}/n_1$ estimates $\text{cov}(\bar{\mathbf{y}}_1) = \boldsymbol{\Sigma}/n_1$. Thus the leading coefficient in these T^2-statistics is the sample size for $\bar{\mathbf{y}}$, and the degrees of freedom is the denominator of the unbiased estimator of $\boldsymbol{\Sigma}$.

3.3.4 Formal Definition of T^2 and Relationship to F

The *formal definition* of a T^2 random variable is similar to the formal definition of the t random variable given in (3.3). Let $\mathbf{z}$ be distributed as the multivariate normal $N_p(\mathbf{0}, \boldsymbol{\Sigma})$ and $\mathbf{W}$ be distributed as the Wishart $W_p(\nu, \boldsymbol{\Sigma})$, with $\mathbf{z}$ and $\mathbf{W}$ independent. Then the T^2 random variable with dimension p and degrees of freedom ν is defined as

$$T^2 = \mathbf{z}'\left(\frac{\mathbf{W}}{\nu}\right)^{-1}\mathbf{z}. \tag{3.16}$$

The distribution of Hotelling's T^2 can be derived from this definition.

It is easy to show that the T^2-statistic (3.14) satisfies the formal definition (3.16). Define $\bar{\mathbf{v}} = \sqrt{n}(\bar{\mathbf{y}} - \boldsymbol{\mu}_0)$ and $\mathbf{W} = (n-1)\mathbf{S}$. Then $\bar{\mathbf{v}}$ is $N_p(\mathbf{0}, \boldsymbol{\Sigma})$ if $\boldsymbol{\mu} = \boldsymbol{\mu}_0$, $\mathbf{W}$ is $W_p(n-1, \boldsymbol{\Sigma})$, and $\bar{\mathbf{v}}$ and $\mathbf{W}$ are independent. Hence $T^2 = \bar{\mathbf{v}}'[\mathbf{W}/(n-1)]^{-1}\bar{\mathbf{v}}$ satisfies (3.16) and can be expressed as

$$\begin{aligned} T^2 &= \bar{\mathbf{v}}'\left(\frac{\mathbf{W}}{n-1}\right)^{-1}\bar{\mathbf{v}} \\ &= [\sqrt{n}(\bar{\mathbf{y}} - \boldsymbol{\mu}_0)]'\left[\frac{(n-1)\mathbf{S}}{n-1}\right]^{-1}[\sqrt{n}(\bar{\mathbf{y}} - \boldsymbol{\mu}_0)] \\ &= n(\bar{\mathbf{y}} - \boldsymbol{\mu}_0)'\mathbf{S}^{-1}(\bar{\mathbf{y}} - \boldsymbol{\mu}_0), \end{aligned}$$

which is (3.14).

The square of a univariate t has an F-distribution. In the multivariate case, a simple function of T^2 also has an F-distribution, as shown in the following theorem.

Theorem 3.3B. The T^2-statistic with ν degrees of freedom can be transformed to an F-statistic with p and $\nu - p + 1$ degrees of freedom:

$$\frac{\nu - p + 1}{\nu p} T^2_{p,\nu} = F_{p,\nu-p+1}. \tag{3.17}$$

Proof. An F random variable is defined as the ratio of two independent χ^2 random variables, each divided by its degrees of freedom. To express (3.16) in this form, multiply and divide by $\mathbf{z}'\boldsymbol{\Sigma}^{-1}\mathbf{z}$ to obtain

$$T^2 = \nu\mathbf{z}'\mathbf{W}^{-1}\mathbf{z} = \frac{\nu\mathbf{z}'\boldsymbol{\Sigma}^{-1}\mathbf{z}}{\mathbf{z}'\boldsymbol{\Sigma}^{-1}\mathbf{z}/\mathbf{z}'\mathbf{W}^{-1}\mathbf{z}}, \tag{3.18}$$

where $\mathbf{z}$ is $N_p(\mathbf{0}, \boldsymbol{\Sigma})$, $\mathbf{W}$ is $W_p(\nu, \boldsymbol{\Sigma})$, and $\mathbf{z}$ and $\mathbf{W}$ are independent. By Theorem 2.2F, the quadratic form $\mathbf{z}'\boldsymbol{\Sigma}^{-1}\mathbf{z}$ in the numerator is distributed as χ^2_p. It can be shown (Seber 1984, pp. 30–31; Styan 1989) that the denominator $\mathbf{z}'\boldsymbol{\Sigma}^{-1}\mathbf{z}/\mathbf{z}'\mathbf{W}^{-1}\mathbf{z}$ is distributed as $\chi^2_{\nu-p+1}$ and is independent of $\mathbf{z}'\boldsymbol{\Sigma}^{-1}\mathbf{z}$. If each of these two independent χ^2 random variables is divided by its degrees of freedom, the resulting ratio will have an F-distribution. Multiplying both sides of (3.18) by the ratio $(\nu - p + 1)/\nu p$, we obtain

$$\frac{\nu - p + 1}{\nu p} T^2 = \frac{\mathbf{z}'\boldsymbol{\Sigma}^{-1}\mathbf{z}/p}{(\mathbf{z}'\boldsymbol{\Sigma}^{-1}\mathbf{z}/\mathbf{z}'\mathbf{W}^{-1}\mathbf{z})/(\nu - p + 1)},$$

which, by definition, has an F-distribution with p and $\nu - p + 1$ degrees of freedom. □

3.3.5 Effect on T^2 of Adding a Variable

The addition of a variable to T^2 may either strengthen the evidence against the hypothesis or weaken it. For example, from Table B.1, we obtain, for 20 degrees of freedom,

$$T^2_{.05,6,20} - T^2_{.05,5,20} = 22.324 - 17.828 = 4.496.$$

Thus, if a sixth variable is added to the five already present, the critical value is increased by 4.496. If the new variable does not potentially increase the calculated T^2 by that amount, then T^2 is less likely to reject H_0. (A test of the significance of the increase in T^2 is given in Section 3.11.4.)

Rencher (1993) has given a breakdown of the factors that influence the increase in T^2 caused by an additional variable. The term "additional variable" is for convenience. In some cases, new variables may be available, but typically we are interested in the effect on T^2 of each of the present variables. Let x designate the variable of interest or additional variable to be added to $\mathbf{y} = (y_1, y_2, \ldots, y_p)'$. We denote the sample mean and variance of x by $\overline{x}$ and s_x^2, the vector of sample covariances of x with the y's by $\mathbf{s}_{xy}$, and the hypothesized mean of x by μ_{0x}. For consistency, we denote the sample covariance matrix of $\mathbf{y}$ by $\mathbf{S}_{yy}$ and the sample mean and hypothesized mean of $\mathbf{y}$ by $\overline{\mathbf{y}}$ and $\boldsymbol{\mu}_{0y}$, respectively. Then the effect of x on T^2 is given in the following theorem.

Theorem 3.3C. For the $p + 1$ variables $(y_1, y_2, \ldots, y_p, x) = (\mathbf{y}', x)$, T^2 can be expressed as

$$T^2_{y,x} = T^2_y + \frac{n[\hat{\boldsymbol{\beta}}'(\overline{\mathbf{y}} - \boldsymbol{\mu}_{0y}) - (\overline{x} - \mu_{0x})]^2}{s^2_x(1 - R^2)} \tag{3.19}$$

$$= T^2_y + \frac{(\hat{t}_x - t_x)^2}{1 - R^2}, \tag{3.20}$$

where $T^2_{y,x}$ is the value of T^2 based on the y's and x, T^2_y is the value of T^2 based on the y's alone, $\hat{\boldsymbol{\beta}} = \mathbf{S}^{-1}_{yy}\mathbf{s}_{xy}$ is the vector of regression coefficients of x with the y's [corrected for their means, see (7.26)], $R^2 = \mathbf{s}'_{xy}\mathbf{S}^{-1}_{yy}\mathbf{s}_{xy}/s^2_x$ is the squared multiple correlation of x regressed on the y's [see (7.66)], $t_x = \sqrt{n}(\overline{x} - \mu_{0x})/s_x$, and

$$\hat{t}_x = \frac{\hat{\boldsymbol{\beta}}'(\overline{\mathbf{y}} - \boldsymbol{\mu}_{0y})}{s_x/\sqrt{n}}. \qquad \square$$

Thus t_x is the ordinary t-statistic for x by itself, and $\hat{t}_x$ can be interpreted as a "predicted" value of t_x based on the information about $\overline{x} - \mu_{0x}$ already available in the y's. If t_x and $\hat{t}_x$ are of the same sign, there are three ways in which the contribution of x can be important: (a) t_x substantially larger in absolute value than $\hat{t}_x$, (b) $\hat{t}_x$ substantially larger in absolute value than t_x, and (c) R^2 large. Otherwise, if t_x is close to $\hat{t}_x$, so that most of the evidence $\overline{x}$ provides against the hypothesis is predictable from $\overline{\mathbf{y}}$, there is little reason to include x. If t_x and $\hat{t}_x$ are of opposite signs, their effect combines to increase T^2. Theorem 3.3C also demonstrates that if x were orthogonal to the y's ($\hat{\boldsymbol{\beta}} = \mathbf{0}$), the addition of x to T^2_y would reduce to t^2_x.

Note that (3.19) proves that the addition of a variable can only increase T^2. It may seem surprising that this increase in T^2 is inversely related to $1 - R^2$ rather than to R^2; that is, the larger the value of R^2, the larger the increase in T^2. Perhaps we can draw an analogy to simple linear regression, in which a given difference between y and $\hat{y}$ is more important if the squared correlation r^2 is larger.

The net effect of a variable on T^2 is given by the second term on the right side of (3.19) or (3.20). This effect can be either greater or less than what would be expected from its univariate contribution. It is intuitively obvious that overlap with other variables can render a variable partially redundant so that its multivariate contribution is less than its univariate effect, but heretofore it has not been easy to grasp how the contribution of a variable can be enhanced in the presence of the others. [For illustrations of such situations, see Flury (1989) and Hamilton (1987).] In Theorem 3.3C, the breakdown of the effect of each variable makes clear how this can happen. Note the linearity inherent in the effect of each variable, as manifested by the presence of $\hat{\boldsymbol{\beta}}$ and R^2.

Table 3.1. Calcium in Soil and Turnip Greens

Observation Number	y_1	y_2	y_3
1	35	3.5	2.80
2	35	4.9	2.70
3	40	30.0	4.38
4	10	2.8	3.21
5	6	2.7	2.73
6	20	2.8	2.81
7	35	4.6	2.88
8	35	10.9	2.90
9	35	8.0	3.28
10	30	1.6	3.20

Example 3.3.5. Table 3.1 lists observations of three types of calcium measurements in soil and turnip greens (Kramer and Jensen 1969a).

Target values of these three variables are 15.0, 6.0, and 2.85. Using $\boldsymbol{\mu}_0 = (15.0, 6.0, 2.85)'$ in (3.14) gives $T^2_{y,x} = 24.559$. We now examine the effect of each variable on T^2 by using (3.19) and (3.20). We first consider the effect of y_3 as it is added to T^2_y based on y_1 and y_2. With $x = y_3$ we have

$$\begin{pmatrix} \overline{\mathbf{y}} \\ \overline{x} \end{pmatrix} = \left(\begin{array}{c} 28.1 \\ 7.18 \\ \hline 3.089 \end{array}\right), \qquad \begin{pmatrix} \mathbf{S}_{yy} & \mathbf{s}_{xy} \\ \mathbf{s}'_{xy} & s_x^2 \end{pmatrix} = \left(\begin{array}{cc|c} 140.54 & 49.68 & 1.94 \\ 49.68 & 72.25 & 3.68 \\ \hline 1.94 & 3.68 & .25 \end{array}\right).$$

The value of T^2 based on $\mathbf{y} = (y_1, y_2)'$ is

$$\begin{aligned} T_y^2 &= n(\overline{\mathbf{y}} - \boldsymbol{\mu}_{0y})'\mathbf{S}_{yy}^{-1}(\overline{\mathbf{y}} - \boldsymbol{\mu}_{0y}) \\ &= 10\begin{pmatrix} 28.1 - 15.0 \\ 7.18 - 6.0 \end{pmatrix}'\begin{pmatrix} 140.54 & 49.68 \\ 49.68 & 72.25 \end{pmatrix}^{-1}\begin{pmatrix} 28.1 - 15.0 \\ 7.18 - 6.0 \end{pmatrix} \\ &= 14.388. \end{aligned}$$

Thus, without y_3, T^2 falls from $T^2_{y,x} = 24.559$ to $T^2_y = 14.388$, a reduction of 10.171. To see what factors contribute to this difference, we examine the elements of the second term on the right side of both (3.19) and (3.20):

$$\hat{\boldsymbol{\beta}} = \mathbf{S}_{yy}^{-1}\mathbf{s}_{xy} = \begin{pmatrix} 140.54 & 49.68 \\ 49.68 & 72.25 \end{pmatrix}^{-1}\begin{pmatrix} 1.94 \\ 3.68 \end{pmatrix} = \begin{pmatrix} -.00551 \\ .05467 \end{pmatrix},$$

$$\hat{t}_x = \frac{\hat{\boldsymbol{\beta}}'(\overline{\mathbf{y}} - \boldsymbol{\mu}_{0y})}{s_x/\sqrt{n}} = \frac{-.007720}{\sqrt{.2501/10}} = -.0488,$$

$$t_x = \frac{\overline{x} - \mu_{0x}}{s_x/\sqrt{n}} = \frac{3.089 - 2.85}{\sqrt{.2501/10}} = 1.511,$$

$$R^2 = \frac{\mathbf{s}'_{xy}\mathbf{S}_{yy}^{-1}\mathbf{s}_{xy}}{s_x^2} = .7607,$$

$$T_{y,x}^2 - T_y^2 = \frac{(\hat{t}_x - t_x)^2}{1 - R^2} = \frac{(-.0488 - 1.511)^2}{1 - .7607} = 10.171.$$

Thus the increase in T^2 due to $x = y_3$ is largely induced by $t_x = 1.511$ and the fairly high squared multiple correlation of y_3 with y_1 and y_2, $R^2 = .7607$. For $x = y_1$ and $x = y_2$, we have

x	$\hat{t}_x$	t_x	R^2	$T_{y,x}^2 - T_y^2$
y_2	2.009	.438	.797	12.198
y_1	−.227	3.494	.282	19.289

For y_2, the increase in T^2 is due to $\hat{t}_x$ and R^2. For y_1, the increase is due almost entirely to t_x.

3.3.6 Properties of the T^2-Test

Some important properties of the T^2-test of $H_0\colon \boldsymbol{\mu} = \boldsymbol{\mu}_0$ versus $H_1\colon \boldsymbol{\mu} \neq \boldsymbol{\mu}_0$ are as follows:

1. The T^2-statistic is invariant to transformations of the form $\mathbf{z}_i = \mathbf{A}\mathbf{y}_i + \mathbf{b}$, where $\mathbf{A}$ is a nonsingular matrix, that is, $T_z^2 = T_y^2$ (see problems 3.9 and 3.37). Invariance of this type, often referred to as *affine* invariance (or equivariance), includes changes in scale, as, for example, from inches to centimeters.
2. The T^2-test is the uniformly most powerful invariant test (Anderson 1984, Section 5.6.1).
3. The T^2-test is sensitive to certain departures from normality (see Section 3.7.2).
4. The T^2-test is the likelihood ratio test (see Section 3.3.7).
5. The T^2-test is the union–intersection test (see Section 3.3.8).

In the univariate case ($p = 1$), the t-statistic can be adapted to serve for a one-sided alternative hypothesis such as $H_1\colon \mu > \mu_0$. The resulting one-tailed t-test has some optimal properties. In the multivariate case, however, the T^2-test cannot be similarly adapted to have optimal properties for a one-sided alternative. For $p > 1$, the one-sided alternative $H_1\colon \boldsymbol{\mu} > \boldsymbol{\mu}_0$ can be defined to mean that $\mu_j > \mu_{0j}$ for all $j = 1, 2, \ldots, p$. Kariya and Cohen (1992) showed that for this case there is no scale invariant test statistic with satisfactory properties. The T^2-test is invariant but cannot be recommended for obvious reasons—it would reject H_0 when we want to accept it, namely, when some or all $\overline{y}_j$ are considerably less than the corresponding μ_{0j}. For additional discussion of the one-sided multivariate problem, see Perlman (1969), Marden (1982), and Troendle (1996).

3.3.7 Likelihood Ratio Test

The likelihood ratio method of test construction was introduced in Section 3.3.2 and illustrated for the univariate test of $H_0\colon \mu = \mu_0$ with σ^2 unknown. We now consider the analogous multivariate case.

Theorem 3.3D. If $\mathbf{y}_1, \mathbf{y}_2, \ldots, \mathbf{y}_n$ constitute a random sample from $N_p(\boldsymbol{\mu}, \boldsymbol{\Sigma})$ with $\boldsymbol{\Sigma}$ unknown, then for $H_0\colon \boldsymbol{\mu} = \boldsymbol{\mu}_0$ versus $H_1\colon \boldsymbol{\mu} \neq \boldsymbol{\mu}_0$, the likelihood ratio approach leads to the T^2 test statistic (3.14).

Proof. The likelihood function for the sample is given by (2.45) as

$$L(\boldsymbol{\mu}, \boldsymbol{\Sigma}) = \frac{1}{(\sqrt{2\pi})^{np}|\boldsymbol{\Sigma}|^{n/2}} e^{-\sum_{i=1}^{n}(\mathbf{y}_i - \boldsymbol{\mu})'\boldsymbol{\Sigma}^{-1}(\mathbf{y}_i - \boldsymbol{\mu})/2}. \tag{3.21}$$

By (2.48), $L(\boldsymbol{\mu}, \boldsymbol{\Sigma})$ in (3.21) can be written in the form

$$L(\boldsymbol{\mu}, \boldsymbol{\Sigma}) = \frac{1}{(\sqrt{2\pi})^{np}|\boldsymbol{\Sigma}|^{n/2}} e^{-\operatorname{tr}\boldsymbol{\Sigma}^{-1}[\mathbf{W} + n(\bar{\mathbf{y}} - \boldsymbol{\mu})(\bar{\mathbf{y}} - \boldsymbol{\mu})']/2},$$

where $\mathbf{W} = \sum_{i=1}^{n}(\mathbf{y}_i - \bar{\mathbf{y}})(\mathbf{y}_i - \bar{\mathbf{y}})'$. By Theorem 2.3A, the values of $\boldsymbol{\mu}$ and $\boldsymbol{\Sigma}$ that maximize $L(\boldsymbol{\mu}, \boldsymbol{\Sigma})$ under H_1 are the maximum likelihood estimators $\hat{\boldsymbol{\mu}} = \bar{\mathbf{y}}$ and $\hat{\boldsymbol{\Sigma}} = \mathbf{W}/n$, respectively. (We ignore the restriction $\boldsymbol{\mu} \neq \boldsymbol{\mu}_0$ and consider $\boldsymbol{\mu}$ to be unrestricted.) When these estimators are substituted for $\boldsymbol{\mu}$ and $\boldsymbol{\Sigma}$ in $L(\boldsymbol{\mu}, \boldsymbol{\Sigma})$, the result is

$$\begin{aligned}
\max_{H_1} L &= \frac{1}{(\sqrt{2\pi})^{np}|\hat{\boldsymbol{\Sigma}}|^{n/2}} e^{-\operatorname{tr}(\mathbf{W}/n)^{-1}\mathbf{W}/2} \\
&= \frac{n^{np/2}}{(\sqrt{2\pi})^{np}|\sum_i(\mathbf{y}_i - \bar{\mathbf{y}})(\mathbf{y}_i - \bar{\mathbf{y}})'|^{n/2}} e^{-\operatorname{tr}(n\mathbf{I})/2} \\
&= \frac{n^{np/2}}{(\sqrt{2\pi})^{np}|\sum_i(\mathbf{y}_i - \bar{\mathbf{y}})(\mathbf{y}_i - \bar{\mathbf{y}})'|^{n/2}} e^{-np/2}.
\end{aligned} \tag{3.22}$$

Under H_0, $\boldsymbol{\mu}$ is restricted to be equal to $\boldsymbol{\mu}_0$, and the value of $\boldsymbol{\Sigma}$ that maximizes $L(\boldsymbol{\mu}_0, \boldsymbol{\Sigma})$ is $\hat{\boldsymbol{\Sigma}}_0 = \sum_{i=1}^{n}(\mathbf{y}_i - \boldsymbol{\mu}_0)(\mathbf{y}_i - \boldsymbol{\mu}_0)'/n$. When these values are substituted for $\boldsymbol{\mu}$ and $\boldsymbol{\Sigma}$ in $L(\boldsymbol{\mu}, \boldsymbol{\Sigma})$, we obtain

$$\max_{H_0} L = \frac{1}{(\sqrt{2\pi})^{np}|\hat{\boldsymbol{\Sigma}}_0|^{n/2}} e^{-\sum_i(\mathbf{y}_i - \boldsymbol{\mu}_0)'\hat{\boldsymbol{\Sigma}}_0^{-1}(\mathbf{y}_i - \boldsymbol{\mu}_0)/2} \tag{3.23}$$

$$= \frac{n^{np/2}}{(\sqrt{2\pi})^{np}|\sum_i(\mathbf{y}_i - \boldsymbol{\mu}_0)(\mathbf{y}_i - \boldsymbol{\mu}_0)'|^{n/2}} e^{-np/2}. \tag{3.24}$$

The likelihood ratio is then given by the ratio of (3.24) and (3.22):

$$\begin{aligned}
\text{LR} &= \frac{\max_{H_0} L}{\max_{H_1} L} = \frac{|\sum_{i=1}^n (\mathbf{y}_i - \overline{\mathbf{y}})(\mathbf{y}_i - \overline{\mathbf{y}})'|^{n/2}}{|\sum_{i=1}^n (\mathbf{y}_i - \boldsymbol{\mu}_0)(\mathbf{y}_i - \boldsymbol{\mu}_0)'|^{n/2}} \\
&= \frac{|\sum_i (\mathbf{y}_i - \overline{\mathbf{y}})(\mathbf{y}_i - \overline{\mathbf{y}})'|^{n/2}}{|\sum_i (\mathbf{y}_i - \overline{\mathbf{y}})(\mathbf{y}_i - \overline{\mathbf{y}})' + n(\overline{\mathbf{y}} - \boldsymbol{\mu}_0)(\overline{\mathbf{y}} - \boldsymbol{\mu}_0)'|} \qquad \text{[by (2.47)]} \\
&= \frac{|\sum_i (\mathbf{y}_i - \overline{\mathbf{y}})(\mathbf{y}_i - \overline{\mathbf{y}})'/(n-1)|^{n/2}}{|\sum_i (\mathbf{y}_i - \overline{\mathbf{y}})(\mathbf{y}_i - \overline{\mathbf{y}})'/(n-1) + n(\overline{\mathbf{y}} - \boldsymbol{\mu}_0)(\overline{\mathbf{y}} - \boldsymbol{\mu}_0)'/(n-1)|^{n/2}} \\
&= \frac{|\mathbf{S}|^{n/2}}{|\mathbf{S} + n(\overline{\mathbf{y}} - \boldsymbol{\mu}_0)(\overline{\mathbf{y}} - \boldsymbol{\mu}_0)'/(n-1)|^{n/2}} \qquad (3.25) \\
&= \frac{|\mathbf{S}|^{n/2}}{|\mathbf{S}|^{n/2}\left[1 + n(\overline{\mathbf{y}} - \boldsymbol{\mu}_0)'\mathbf{S}^{-1}(\overline{\mathbf{y}} - \boldsymbol{\mu}_0)/(n-1)\right]^{n/2}} \qquad \text{[by (A.7.10)]} \; (3.26) \\
&= \left[\frac{1}{1 + T^2/(n-1)}\right]^{n/2}. \qquad (3.27)
\end{aligned}$$

From (3.27), we see that rejecting H_0 for LR $\leq c$ is equivalent to rejecting H_0 for $T^2 \geq T^2_\alpha$, where T^2 is given in (3.14). □

3.3.8 Union–Intersection Test

The *union–intersection* method of test construction was introduced by Roy (1957). We now demonstrate that for $H_0\colon \boldsymbol{\mu} = \boldsymbol{\mu}_0$ versus $H_1\colon \boldsymbol{\mu} \neq \boldsymbol{\mu}_0$, the union–intersection approach leads to the T^2 test statistic (3.14). The procedure involves linear functions of the parameters.

By (A.4.5), it is clear that H_0 is true if and only if $\mathbf{a}'\boldsymbol{\mu} = \mathbf{a}'\boldsymbol{\mu}_0$ for all possible $\mathbf{a}$ and that H_0 is false if $\mathbf{a}'\boldsymbol{\mu} \neq \mathbf{a}'\boldsymbol{\mu}_0$ for at least one $\mathbf{a}$. We thus reject $H_0\colon \boldsymbol{\mu} = \boldsymbol{\mu}_0$ if at least one vector $\mathbf{a}$ can be found such that the univariate hypothesis $\mathbf{a}'\boldsymbol{\mu} = \mathbf{a}'\boldsymbol{\mu}_0$ is rejected. The rejection region for H_0 is therefore the union over all $\mathbf{a}$ of the rejection regions for the univariate hypotheses. Similarly, we accept $H_0\colon \boldsymbol{\mu} = \boldsymbol{\mu}_0$ only if every univariate hypothesis $\mathbf{a}'\boldsymbol{\mu} = \mathbf{a}'\boldsymbol{\mu}_0$ is accepted. The acceptance region is therefore the intersection over all $\mathbf{a}$ of the acceptance regions of the univariate hypotheses.

Since $\overline{\mathbf{y}}$ is $N_p(\boldsymbol{\mu}, \frac{1}{n}\boldsymbol{\Sigma})$, the linear combination $\mathbf{a}'\overline{\mathbf{y}}$ is distributed as $N(\mathbf{a}'\boldsymbol{\mu}, \mathbf{a}'\boldsymbol{\Sigma}\mathbf{a}/n)$, and we can test the univariate hypothesis $\mathbf{a}'\boldsymbol{\mu} = \mathbf{a}'\boldsymbol{\mu}_0$ by the t-statistic

$$t(\mathbf{a}) = \frac{\mathbf{a}'\overline{\mathbf{y}} - \mathbf{a}'\boldsymbol{\mu}_0}{\sqrt{\mathbf{a}'\mathbf{S}\mathbf{a}/n}}, \qquad (3.28)$$

which, for a given $\mathbf{a}$, has acceptance region

$$|t(\mathbf{a})| < c,$$

where c is selected to provide the correct α for the test. The acceptance region for H_0: $\boldsymbol{\mu} = \boldsymbol{\mu}_0$ is therefore

$$\bigcap_{\mathbf{a}} \left[|t(\mathbf{a})| < c \right], \tag{3.29}$$

where the intersection over all $\mathbf{a}$ defines the smallest interval that contains all $t(\mathbf{a})$ and is bounded by $\pm c$. The rejection region for H_0: $\boldsymbol{\mu} = \boldsymbol{\mu}_0$ is

$$\bigcup_{\mathbf{a}} \left[|t(\mathbf{a})| \geq c \right], \tag{3.30}$$

and we will reject H_0 if any $|t(\mathbf{a})|$ is greater than or equal to c, or equivalently if $\max_{\mathbf{a}} |t(\mathbf{a})| \geq c$.

To find $\max_{\mathbf{a}} |t(\mathbf{a})|$, it is more convenient mathematically to work with $t^2(\mathbf{a})$, which we write in the form

$$t^2(\mathbf{a}) = \frac{n[\mathbf{a}'(\overline{\mathbf{y}} - \boldsymbol{\mu}_0)]^2}{\mathbf{a}'\mathbf{S}\mathbf{a}}. \tag{3.31}$$

The value of $\max_{\mathbf{a}} t^2(\mathbf{a})$ is given in the following theorem.

Theorem 3.3E. If $\mathbf{y}$ is $N_p(\boldsymbol{\mu}_0, \boldsymbol{\Sigma})$, then $\max_{\mathbf{a}} t^2(\mathbf{a}) = T^2$, where $t^2(\mathbf{a})$ is given by (3.31) and T^2 by (3.14).

Proof. The value of $\mathbf{a}$ that maximizes (3.31) can be found by differentiating $t^2(\mathbf{a})$ with respect to $\mathbf{a}$ and setting the result equal to $\mathbf{0}$. Using (A.13.2) and (A.13.3), we obtain

$$\frac{\partial t^2(\mathbf{a})}{\partial \mathbf{a}} = \frac{n\mathbf{a}'\mathbf{S}\mathbf{a}2[\mathbf{a}'(\overline{\mathbf{y}} - \boldsymbol{\mu}_0)](\overline{\mathbf{y}} - \boldsymbol{\mu}_0) - n[\mathbf{a}'(\overline{\mathbf{y}} - \boldsymbol{\mu}_0)]^2 2\mathbf{S}\mathbf{a}}{(\mathbf{a}'\mathbf{S}\mathbf{a})^2} = \mathbf{0},$$

which simplifies to

$$\mathbf{a}'\mathbf{S}\mathbf{a}(\overline{\mathbf{y}} - \boldsymbol{\mu}_0) - \mathbf{a}'(\overline{\mathbf{y}} - \boldsymbol{\mu}_0)\mathbf{S}\mathbf{a} = \mathbf{0},$$

from which

$$\mathbf{a} = \frac{\mathbf{a}'\mathbf{S}\mathbf{a}}{\mathbf{a}'(\overline{\mathbf{y}} - \boldsymbol{\mu}_0)} \mathbf{S}^{-1}(\overline{\mathbf{y}} - \boldsymbol{\mu}_0) = c\mathbf{S}^{-1}(\overline{\mathbf{y}} - \boldsymbol{\mu}_0). \tag{3.32}$$

Substituting (3.32) into (3.31), we obtain

$$\begin{aligned}
\max_{\mathbf{a}} t^2(\mathbf{a}) &= \max_{\mathbf{a}} \frac{n[\mathbf{a}'(\overline{\mathbf{y}} - \boldsymbol{\mu}_0)]^2}{\mathbf{a}'\mathbf{S}\mathbf{a}} \\
&= \frac{n[c(\overline{\mathbf{y}} - \boldsymbol{\mu}_0)'\mathbf{S}^{-1}(\overline{\mathbf{y}} - \boldsymbol{\mu}_0)]^2}{c(\overline{\mathbf{y}} - \boldsymbol{\mu}_0)'\mathbf{S}^{-1}\mathbf{S}c\mathbf{S}^{-1}(\overline{\mathbf{y}} - \boldsymbol{\mu}_0)} \\
&= n(\overline{\mathbf{y}} - \boldsymbol{\mu}_0)'\mathbf{S}^{-1}(\overline{\mathbf{y}} - \boldsymbol{\mu}_0) = T^2.
\end{aligned} \tag{3.33}$$

□

Hence, in the case of H_0: $\boldsymbol{\mu} = \boldsymbol{\mu}_0$, both the likelihood ratio and union–intersection approaches lead to the same test.

From (3.32), it is clear that any multiple of $\mathbf{a} = \mathbf{S}^{-1}(\bar{\mathbf{y}} - \boldsymbol{\mu}_0)$ maximizes $t^2(\mathbf{a})$, that is, maximally separates $\mathbf{a}'\bar{\mathbf{y}}$ from $\mathbf{a}'\boldsymbol{\mu}_0$. The linear function $z = \mathbf{a}'\mathbf{y}$, with coefficient vector $\mathbf{a} = \mathbf{S}^{-1}(\bar{\mathbf{y}} - \boldsymbol{\mu}_0)$, is called the *discriminant function*. We will discuss the discriminant function $z = \mathbf{a}'\mathbf{y}$ further in Sections 3.4.5, 3.5.5, 5.2, 5.4, 5.5.1, 5.7.2a, and 5.11.1.

3.4 CONFIDENCE INTERVALS AND TESTS FOR LINEAR FUNCTIONS OF $\boldsymbol{\mu}$

We now consider confidence intervals and tests for various linear combinations of $\boldsymbol{\mu}$, including the individual elements μ_j. We begin with a confidence region for the entire mean vector $\boldsymbol{\mu}$.

3.4.1 Confidence Region for $\boldsymbol{\mu}$

Since $n(\bar{\mathbf{y}} - \boldsymbol{\mu})'\mathbf{S}^{-1}(\bar{\mathbf{y}} - \boldsymbol{\mu})$ is distributed as T^2, we can make the probability statement $P[n(\bar{\mathbf{y}} - \boldsymbol{\mu})'\mathbf{S}^{-1}(\bar{\mathbf{y}} - \boldsymbol{\mu}) \leq T^2_{\alpha,p,n-1}] = 1 - \alpha$, from which a $100(1 - \alpha)\%$ *confidence region* for $\boldsymbol{\mu}$ is given by all vectors $\boldsymbol{\mu}$ that satisfy

$$n(\bar{\mathbf{y}} - \boldsymbol{\mu})'\mathbf{S}^{-1}(\bar{\mathbf{y}} - \boldsymbol{\mu}) \leq T^2_{\alpha,p,n-1}. \tag{3.34}$$

This hyperellipsoidal region for $\boldsymbol{\mu}$ is centered at $\boldsymbol{\mu} = \bar{\mathbf{y}}$. However, the values of $\boldsymbol{\mu}$ that satisfy (3.34) are not easy to visualize except in the case $p = 2$, where we can draw an ellipse. For $p > 2$, we can substitute various values of $\boldsymbol{\mu}$ into (3.34) to determine if they are inside the region. But this is equivalent to finding those values of $\boldsymbol{\mu}_0$ that would not be rejected by the T^2-test of H_0: $\boldsymbol{\mu} = \boldsymbol{\mu}_0$ in (3.14). Thus we are back to the hypothesis test, and (3.34) provides little additional insight into the possible position of $\boldsymbol{\mu}$.

3.4.2 Confidence Interval for a Single Linear Combination $\mathbf{a}'\boldsymbol{\mu}$

By (3.28), a $100(1 - \alpha)\%$ confidence interval for $\mathbf{a}'\boldsymbol{\mu}$ is given by

$$\mathbf{a}'\bar{\mathbf{y}} - t_{\alpha/2,n-1}\sqrt{\frac{\mathbf{a}'\mathbf{S}\mathbf{a}}{n}} \leq \mathbf{a}'\boldsymbol{\mu} \leq \mathbf{a}'\bar{\mathbf{y}} + t_{\alpha/2,n-1}\sqrt{\frac{\mathbf{a}'\mathbf{S}\mathbf{a}}{n}}. \tag{3.35}$$

If confidence intervals are desired for several linear combinations, see Sections 3.4.3 and 3.4.4.

3.4.3 Simultaneous Confidence Intervals for μ_j and $\mathbf{a}'\boldsymbol{\mu}$

Because the confidence region for $\boldsymbol{\mu}$ in (3.34) is unwieldy, we look for confidence intervals for μ_j or for $\mathbf{a}'\boldsymbol{\mu}$ for arbitrary $\mathbf{a}$. The linear combination $\mathbf{a}'\boldsymbol{\mu}$ allows for contrasts of the form $\mu_1 - \mu_2$ or $\mu_1 - 2\mu_2 + \mu_3$ and also yields each μ_j by choosing

$\mathbf{a}' = (0, \ldots, 0, 1, 0, \ldots, 0)$ with a 1 in the jth position. Confidence intervals for all possible $\mathbf{a}'\boldsymbol{\mu}$ are given in the following theorem.

Theorem 3.4A. If $\mathbf{y}_1, \mathbf{y}_2, \ldots, \mathbf{y}_n$ is a random sample from $N_p(\boldsymbol{\mu}, \boldsymbol{\Sigma})$ with mean $\overline{\mathbf{y}}$ and covariance matrix $\mathbf{S}$, then a set of *simultaneous confidence intervals* for $\mathbf{a}'\boldsymbol{\mu}$ with confidence coefficient $100(1-\alpha)\%$ is given by

$$\mathbf{a}'\overline{\mathbf{y}} - T_{\alpha,p,n-1}\sqrt{\frac{\mathbf{a}'\mathbf{S}\mathbf{a}}{n}} \leq \mathbf{a}'\boldsymbol{\mu} \leq \mathbf{a}'\overline{\mathbf{y}} + T_{\alpha,p,n-1}\sqrt{\frac{\mathbf{a}'\mathbf{S}\mathbf{a}}{n}}, \tag{3.36}$$

where $T_{\alpha,p,n-1}$ is the (positive) square root of $T^2_{\alpha,p,n-1}$ from Table B.1 in Appendix B. The probability that all such intervals generated by all possible choices of $\mathbf{a}$ simultaneously contain $\mathbf{a}'\boldsymbol{\mu}$ is $1-\alpha$.

Proof. By Theorem 3.3E, $\max_{\mathbf{a}} t^2(\mathbf{a}) = T^2$, where $t^2(\mathbf{a})$ is given by (3.31). Thus

$$P\left\{\max_{\mathbf{a}} t^2(\mathbf{a}) \leq T^2_{\alpha,p,n-1}\right\} = P\left\{T^2 \leq T^2_{\alpha,p,n-1}\right\} = 1-\alpha,$$

or

$$P\left\{\max_{\mathbf{a}}\left(\frac{n\left[\mathbf{a}'(\overline{\mathbf{y}}-\boldsymbol{\mu}_0)\right]^2}{\mathbf{a}'\mathbf{S}\mathbf{a}}\right) \leq T^2_{\alpha,p,n-1}\right\} = 1-\alpha.$$

Hence, for all $\mathbf{a}$ simultaneously,

$$P\left\{\left[\mathbf{a}'(\overline{\mathbf{y}}-\boldsymbol{\mu}_0)\right]^2 \leq T^2_{\alpha,p,n-1}\frac{\mathbf{a}'\mathbf{S}\mathbf{a}}{n}\right\} = 1-\alpha,$$

and for any particular $\mathbf{a}$,

$$P\left\{\left[\mathbf{a}'(\overline{\mathbf{y}}-\boldsymbol{\mu}_0)\right]^2 \leq T^2_{\alpha,p,n-1}\frac{\mathbf{a}'\mathbf{S}\mathbf{a}}{n}\right\} \geq 1-\alpha.$$

The inequality inside the braces is equivalent to

$$|\mathbf{a}'(\overline{\mathbf{y}}-\boldsymbol{\mu}_0)| \leq T_{\alpha,p,n-1}\sqrt{\frac{\mathbf{a}'\mathbf{S}\mathbf{a}}{n}},$$

which, with $\boldsymbol{\mu} = \boldsymbol{\mu}_0$, can be rewritten to obtain (3.36). □

Notice that the intervals in (3.36) are in the usual form, $\mathbf{a}'\overline{\mathbf{y}} \pm k_\alpha\sqrt{\widehat{\text{var}}(\mathbf{a}'\overline{\mathbf{y}})}$. However, in this case the coefficient $k_\alpha = T_{\alpha,p,n-1}$ is very large in order to allow for all possible $\mathbf{a}$, including those chosen after seeing the data. If only a few combinations $\mathbf{a}'\boldsymbol{\mu}$ are considered, the true confidence coefficient may well exceed $100(1-\alpha)\%$.

To obtain simultaneous confidence intervals for $\mu_1, \mu_2, \ldots, \mu_p$, we set $\mathbf{a}_1' = (1, 0, \ldots, 0)$, $\mathbf{a}_2' = (0, 1, 0, \ldots, 0), \ldots, \mathbf{a}_p' = (0, \ldots, 0, 1)$ in (3.36) to obtain

$$\begin{aligned}
\overline{y}_1 - T_{\alpha,p,n-1}\sqrt{\frac{s_{11}}{n}} &< \mu_1 < \overline{y}_1 + T_{\alpha,p,n-1}\sqrt{\frac{s_{11}}{n}},\\
\overline{y}_2 - T_{\alpha,p,n-1}\sqrt{\frac{s_{22}}{n}} &< \mu_2 < \overline{y}_2 + T_{\alpha,p,n-1}\sqrt{\frac{s_{22}}{n}},\\
&\vdots\\
\overline{y}_p - T_{\alpha,p,n-1}\sqrt{\frac{s_{pp}}{n}} &< \mu_p < \overline{y}_p + T_{\alpha,p,n-1}\sqrt{\frac{s_{pp}}{n}}.
\end{aligned} \tag{3.37}$$

However, because (3.36) allows for all possible $\mathbf{a}'\boldsymbol{\mu}$, the intervals in (3.37) are ordinarily much wider than necessary to provide an overall confidence level of $100(1-\alpha)\%$ for these p intervals. The coefficient $T_{\alpha,p,n-1}$ is considerably greater than $t_{\alpha/2,n-1}$ used in (3.35) (except for $p = 1$). For example, with $n = 25$ and $p = 10$, we have $T_{.05,10,24} = 6.380$ and $t_{.025,24} = 2.064$. But clearly the use of p intervals of the form $\overline{y}_j \pm t_{\alpha/2,n-1}\sqrt{s_{jj}/n}$, $j = 1, 2, \ldots, p$, would be inappropriate because the overall confidence level for all p intervals would be less than $100(1-\alpha)\%$. Some coefficient between $t_{\alpha/2,n-1}$ and $T_{\alpha,p,n-1}$ is needed in order for the p intervals to achieve an overall confidence level closer to the nominal $100(1-\alpha)\%$. We will discuss such a coefficient in Section 3.4.4.

Note that when $p = 1$, the vector $\overline{\mathbf{y}}$ has only the single element $\overline{y}$, and the intervals in (3.37) reduce to

$$\overline{y} - T_{\alpha,1,n-1}\sqrt{\frac{s^2}{n}} \leq \mu \leq \overline{y} + T_{\alpha,1,n-1}\sqrt{\frac{s^2}{n}}.$$

This is the usual univariate confidence interval, since $T_{\alpha,1,n-1} = t_{\alpha/2,n-1}$.

Example 3.4.3. We illustrate the computation of simultaneous confidence intervals for the calcium data in Table 3.1. The mean vector and covariance matrix are

$$\overline{\mathbf{y}} = \begin{pmatrix} 28.1 \\ 7.18 \\ 3.09 \end{pmatrix}, \qquad \mathbf{S} = \begin{pmatrix} 140.54 & 49.68 & 1.94 \\ 49.68 & 72.25 & 3.68 \\ 1.94 & 3.68 & .25 \end{pmatrix}.$$

With $p = 3$ and $n = 10$, we obtain $T_{.05,3,9} = \sqrt{16.776} = 4.0946$ from Table B.1 in Appendix B. From (3.37), the intervals are as follows:

$$\begin{aligned}
&\overline{y}_1 \pm T_{.05,3,9}\sqrt{\frac{s_{11}}{n}}, \quad 28.1 \pm 4.0946\sqrt{\frac{140.54}{10}}, \quad 28.1 \pm 15.35, \quad (12.75, 43.45);\\
&\overline{y}_2 \pm T_{.05,3,9}\sqrt{\frac{s_{22}}{n}}, \quad 7.18 \pm 4.0946\sqrt{\frac{72.25}{10}}, \quad 7.18 \pm 11.009, \quad (-3.829, 18.189);\\
&\overline{y}_3 \pm T_{.05,3,9}\sqrt{\frac{s_{33}}{n}}, \quad 3.09 \pm 4.0946\sqrt{\frac{.25}{10}}, \quad 3.09 \pm .648, \quad (2.442, 3.738).
\end{aligned}$$

These intervals are very wide due to the large coefficient, 4.0946.

3.4.4 Bonferroni Confidence Intervals for μ_j and $\mathbf{a}'\boldsymbol{\mu}$

The Bonferroni approach is developed from the best known of a family of probability inequalities due to Bonferroni (1936). If we specify a *few* linear combinations *before* we see the data, the Bonferroni method provides narrower intervals while still guaranteeing the overall confidence level.

Suppose we desire confidence intervals for k linear combinations $\mathbf{a}_1'\boldsymbol{\mu}, \mathbf{a}_2'\boldsymbol{\mu}, \ldots, \mathbf{a}_k'\boldsymbol{\mu}$. Let E_i denote the event that the ith interval contains $\mathbf{a}_i'\boldsymbol{\mu}$ and E_i^c the complementary event that the ith interval does not contain $\mathbf{a}_i'\boldsymbol{\mu}$, where $P(E_i^c) = \alpha_i$ and $P(E_i) = 1-\alpha_i$, $i = 1, 2, \ldots, k$. Now

$$\begin{aligned} P[\text{all } E_i] &= 1 - P(\text{not all } E_i) = 1 - P(\text{at least one } E_i^c) \\ &= 1 - P(E_1^c \cup E_2^c \cup \cdots \cup E_k^c) \\ &\geq 1 - \sum_{i=1}^{k} P(E_i^c) = 1 - (\alpha_1 + \alpha_2 + \cdots + \alpha_k). \end{aligned} \tag{3.38}$$

The inequality

$$1 - P(E_1^c \cup E_2^c \cup \cdots \cup E_k^c) \geq 1 - \sum_{i=1}^{k} P(E_i^c)$$

can be rewritten as

$$\sum_{i=1}^{k} P(E_i^c) \geq P(E_1^c \cup E_2^c \cup \cdots \cup E_k^c).$$

We illustrate this inequality for $k = 3$ by means of a Venn diagram in Figure 3.1. Because of overlap, it is clear that $P(E_1^c) + P(E_2^c) + P(E_3^c) \geq P(E_1^c \cup E_2^c \cup E_3^c)$.

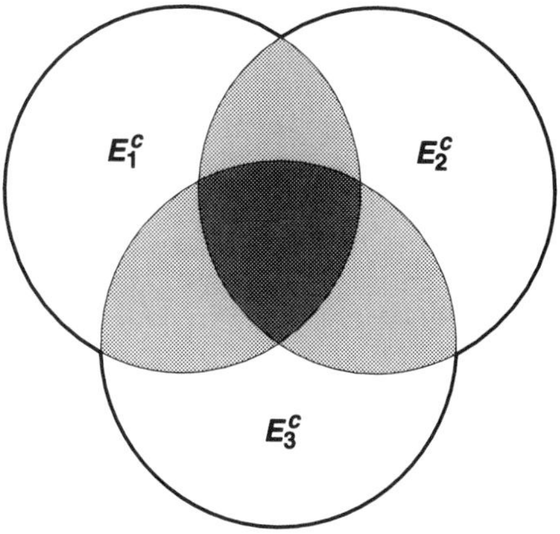

Figure 3.1. Venn diagram showing $\Sigma_{i=1}^{3} P(E_i^c) \geq P(E_1^c \cup E_2^c \cup E_3^c)$.

By means of (3.38), we can control the overall confidence level for the k intervals by using $t_{\alpha_i/2}$, $i = 1, 2, \ldots, k$, with $\sum_{i=1}^{k} \alpha_i = \alpha$. The probability statement in (3.38) holds for any correlation structure of the variables in $\mathbf{y}$. It is possible to use larger α_i's for the important intervals and smaller α_i's for the less important intervals. However, we often consider the k intervals to be equally important and use $t_{\alpha/2k}$ in place of each $t_{\alpha_i/2}$. For k linear combinations $\mathbf{a}_1'\boldsymbol{\mu}, \mathbf{a}_2'\boldsymbol{\mu}, \ldots, \mathbf{a}_k'\boldsymbol{\mu}$, the Bonferroni intervals are therefore given by

$$\mathbf{a}_i'\bar{\mathbf{y}} - t_{\alpha/2k,n-1}\sqrt{\frac{\mathbf{a}_i'\mathbf{S}\mathbf{a}_i}{n}} \le \mathbf{a}_i'\boldsymbol{\mu} \le \mathbf{a}_i'\bar{\mathbf{y}} + t_{\alpha/2k,n-1}\sqrt{\frac{\mathbf{a}_i'\mathbf{S}\mathbf{a}_i}{n}}, \qquad i = 1, 2, \ldots, k. \tag{3.39}$$

If k is not too large, $t_{\alpha/2k,n-1} < T_{\alpha,n-1}$, and the intervals in (3.39) will be narrower than those in (3.36). Table B.2 in Appendix B (Bailey 1977) contains Bonferroni t critical values $t_{\alpha/2r,\nu}$ for $r = 1, 2, \ldots, 19$.

We can use (3.39) to develop Bonferroni intervals for $\mu_1, \mu_2, \ldots, \mu_p$ in $\boldsymbol{\mu}$. Using $k = p$ and $\mathbf{a}_i' = (0, \ldots, 0, 1, 0, \ldots, 0)$ with a 1 in the ith position, we obtain

$$\begin{aligned}
\bar{y}_1 - t_{\alpha/2p,n-1}\sqrt{\frac{s_{11}}{n}} &< \mu_1 < \bar{y}_1 + t_{\alpha/2p,n-1}\sqrt{\frac{s_{11}}{n}}, \\
\bar{y}_2 - t_{\alpha/2p,n-1}\sqrt{\frac{s_{22}}{n}} &< \mu_2 < \bar{y}_2 + t_{\alpha/2p,n-1}\sqrt{\frac{s_{22}}{n}}, \\
&\vdots \\
\bar{y}_p - t_{\alpha/2p,n-1}\sqrt{\frac{s_{pp}}{n}} &< \mu_p < \bar{y}_p + t_{\alpha/2p,n-1}\sqrt{\frac{s_{pp}}{n}}.
\end{aligned} \tag{3.40}$$

By (3.38), these p statements hold with an overall confidence level greater than or equal to $100(1 - \alpha)\%$.

The Bonferroni intervals in (3.40) are conservative because the overall confidence level is $\ge 1 - \alpha$. However, they are less conservative than the simultaneous T_α intervals, since

$$t_{\alpha/2p,n-1} < T_{\alpha,p,n-1}.$$

We therefore recommend the Bonferroni intervals for most purposes because they are generally much narrower than the simultaneous intervals based on T_α in (3.37). The widths of the two types of intervals are $2T_{\alpha,p,n-1}\sqrt{s_{jj}/n}$ and $2t_{\alpha/2p,n-1}\sqrt{s_{jj}/n}$. These two widths are compared in Table 3.2 for selected values of $n - 1$ and p.

Table 3.2. Values of the Ratio $T_{\alpha,p,n-1}/t_{\alpha/2p,n-1}$ for $1 - \alpha = .95$

	p			
$n - 1$	3	6	10	20
10	1.36	2.35		
20	1.24	1.61	2.28	
50	1.19	1.43	1.71	2.38
100	1.18	1.38	1.61	2.08
∞	1.17	1.35	1.52	1.85

Note that the Bonferroni intervals are designed for linear functions of $\boldsymbol{\mu}$ of prior interest, that is, individual μ_j's, or some values of $\mathbf{a}_i'\boldsymbol{\mu}$ specified before seeing the data. Thus, $t_{\alpha/2k,n-1}$ in (3.39) allows for k intervals to be constructed but does not allow for linear combinations that emphasize large differences observed among $\bar{y}_1, \bar{y}_2, \ldots, \bar{y}_p$ in $\bar{\mathbf{y}}$. If we wish to choose linear functions of $\boldsymbol{\mu}$ on the basis of the data, we must use the simultaneous T_α intervals (3.36) in order to preserve the α-level.

We noted in Section 3.3.3 that $\bar{\mathbf{y}}$ and $\mathbf{S}$ need not be from the same sample. Suppose $\bar{\mathbf{y}}$ and $\mathbf{S}_1$ are based on n_1 observations from $N_p(\boldsymbol{\mu}, \boldsymbol{\Sigma})$, and suppose we have an independent estimate $\mathbf{S}_2$ based on n_2 observations from $N_p(\boldsymbol{\mu}_2, \boldsymbol{\Sigma})$. We can construct a pooled estimate of the common covariance matrix $\boldsymbol{\Sigma}$:

$$\mathbf{S}_{\text{pl}} = \frac{(n_1 - 1)\mathbf{S}_1 + (n_2 - 1)\mathbf{S}_2}{n_1 + n_2 - 2}. \tag{3.41}$$

Then (3.36), (3.37), and (3.40) become

$$\mathbf{a}'\bar{\mathbf{y}} \pm T_{\alpha,p,n_1+n_2-2}\sqrt{\frac{\mathbf{a}'\mathbf{S}_{\text{pl}}\mathbf{a}}{n_1}}, \tag{3.42}$$

$$\bar{y}_j \pm T_{\alpha,p,n_1+n_2-2}\sqrt{\frac{s_{\text{pl},jj}}{n_1}}, \tag{3.43}$$

$$\bar{y}_j \pm t_{\alpha/2p,n_1+n_2-2}\sqrt{\frac{s_{\text{pl},jj}}{n_1}}. \tag{3.44}$$

Thus when n_1 is small, we can take advantage of auxiliary information about $\boldsymbol{\Sigma}$, if available, to obtain a more precise estimator of $\boldsymbol{\mu}$.

Example 3.4.4. We illustrate the computation of the Bonferroni intervals (3.40) for μ_1, μ_2, and μ_3 for the calcium data in Table 3.1. Using the means and variances given in Example 3.4.3 and $t_{\alpha/2p,n-1} = t_{.025/3,9}$ from Table B.2, we obtain the following intervals:

$$\bar{y}_1 \pm t_{.0083,9}\sqrt{\frac{s_{11}}{n}},\quad 28.1 \pm 2.9333\sqrt{\frac{140.54}{10}},\quad 28.1 \pm 11.00,\ (17.1, 39.1) \qquad (\text{for } \mu_1);$$

$$7.18 \pm 7.884,\ (-.704, 15.064) \qquad (\text{for } \mu_2);$$

$$3.09 \pm .464,\ (2.626, 3.554) \qquad (\text{for } \mu_3).$$

Since $2.9333 < 4.0946$, these intervals are narrower than the simultaneous intervals obtained in Example 3.4.3.

3.4.5 Tests for $H_0\colon \mathbf{a}'\boldsymbol{\mu} = \mathbf{a}'\boldsymbol{\mu}_0$ and $H_0\colon \mu_j = \mu_{0j}$

To test $H_0\colon \mathbf{a}'\boldsymbol{\mu} = \mathbf{a}'\boldsymbol{\mu}_0$ for arbitrary values of $\mathbf{a}$, a test statistic can easily be obtained from (3.28) and Theorems 3.3E and 3.4A. We compute

$$t(\mathbf{a}) = \frac{\mathbf{a}'(\bar{\mathbf{y}} - \boldsymbol{\mu}_0)\sqrt{n}}{\sqrt{\mathbf{a}'\mathbf{S}\mathbf{a}}} \tag{3.45}$$

and reject H_0 if $|t(\mathbf{a})| \geq T_{\alpha,p,n-1}$. Any number of these tests can be made, even for values of $\mathbf{a}$ suggested by the data, with the assurance of overall protection at the α-level of significance.

To obtain tests on individual variables, we choose the coefficient vector $\mathbf{a}$ to have a 1 in the appropriate position and 0s elsewhere. To test $H_0\colon \mu_2 = \mu_{02}$, for example, we use $\mathbf{a}'\bar{\mathbf{y}} = (0, 1, 0, \ldots, 0)\bar{\mathbf{y}} = \bar{y}_2$ and $\mathbf{a}'\mathbf{S}\mathbf{a} = s_{22}$. In this case, $t(\mathbf{a}) = (\bar{y}_2 - \mu_{02})\sqrt{n}/\sqrt{s_{22}}$ is the same as the univariate t but the critical value $T_{\alpha,p,n-1}$ is larger than $t_{\alpha/2,n-1}$. In general, to test $H_{0j}\colon \mu_j = \mu_{0j},\ j = 1, 2, \ldots, p$, we calculate

$$t_j = \frac{(\bar{y}_j - \mu_{0j})\sqrt{n}}{\sqrt{s_{jj}}} \tag{3.46}$$

and reject H_{0j} if $|t_j| \geq T_{\alpha,p,n-1}$. Note that if $p = 1$, the test statistic (3.46) reduces to the usual $t = (\bar{y} - \mu_0)\sqrt{n}/\sqrt{s^2}$, and the critical value $T_{\alpha,1,n-1}$ becomes $t_{\alpha/2,n-1}$.

If only a few a priori hypotheses are to be tested, such as $H_{0i}\colon \mathbf{a}_i'\boldsymbol{\mu} = \mathbf{a}_i'\boldsymbol{\mu}_0$ for $i = 1, 2, \ldots, k$ or $H_{0j}\colon \mu_j = \mu_{0j}$ for $j = 1, 2, \ldots, p$, Bonferroni critical values $t_{\alpha/2k,n-1}$ or $t_{\alpha/2p,n-1}$ from Table B.2 can be used for (3.45) or (3.46). These critical values will be less conservative than $T_{\alpha,p,n-1}$ but will still offer protection at the overall α-level of significance for all k or p tests.

Bonferroni tests can also be carried out using p-values, which are routinely provided by most software programs. For an upper one-tailed test, the p-value is defined as the probability of obtaining a test statistic value at least as great as the observed value if H_0 is true. If we denote the p-value of such a test by P, then we would reject H_{0i} if $P \leq \alpha/k$, where k is the number of tests. In the case of two-sided tests, the p-values should be two-tailed; that is, reject H_{0i} if $P \leq \alpha/2k$.

Simes (1986) proposed a modification of the Bonferroni procedure that leads to slightly less conservative Type I error rates. If k hypotheses are tested, the overall hypothesis H_0 is that all k subhypotheses $H_{0i},\ i = 1, 2, \ldots, k$, are true. Let $P_{(1)} \leq P_{(2)} \leq \cdots \leq P_{(k)}$ be the ordered p-values based, for example, on the ordinary t-distribution. Simes suggested rejecting the overall hypothesis if any $P_{(i)} \leq i\alpha/k$ for $i = 1, 2, \ldots, k$. If all the tests are independent, Simes showed that the overall error rate for this procedure is α. For nonindependent tests, Simes investigated the error rates by simulation. He considered multivariate normal and chi-square distributions with common correlation ρ between all pairs of variables. His results are shown in Table 3.3 for a target α-value of .05. Useful improvement (error rate closer to the intended α of .05) is seen only for the higher values of ρ in these limited simulation results.

Hochberg (1988) extended Simes's procedure as follows: Reject H_0 if the ordered p-value $P_{(i)}$ satisfies the inequality

$$P_{(i)} \leq \frac{\alpha}{k - i + 1} \tag{3.47}$$

Table 3.3. Type I Error Rates[a] for Modified (M) and Classical (C) Bonferroni Test Procedures for k Tests

	Normal Distribution		χ_1^2-Distribution		χ_5^2-Distribution	
ρ	M	C	M	C	M	C
			$k = 5$			
0	.049	.048	.050	.049	.049	.048
.3	.049	.048	.045	.040	.044	.040
.6	.043	.039	.044	.029	.039	.033
.9	.033	.024	.048	.016	.041	.019
			$k = 10$			
0	.049	.048	.049	.048	.049	.048
.3	.047	.045	.043	.037	.042	.039
.6	.039	.034	.042	.026	.035	.029
.9	.028	.017	.047	.012	.039	.014

[a]Based on 100,000 simulations each; estimated standard error ≤ 0.007.

for any $i = 1, 2, \ldots, k$, where H_0 is an overall hypothesis such as $H_{0i}\colon \mathbf{a}_i'\boldsymbol{\mu} = \mathbf{a}_i'\boldsymbol{\mu}_0$ is true for all $i = 1, 2, \ldots, k$. If the inequality in (3.47) holds for some i, the corresponding individual hypothesis H_{0i} is also rejected. A similar statement can be made for $H_{0j}\colon \mu_j = \mu_{0j}$ for $j = 1, 2, \ldots, p$.

Another extension of Simes's procedure was given by Hommel (1988): Reject H_0 (and the corresponding individual hypothesis H_{0i}) if

$$P_{(i)} < \frac{\alpha}{j}$$

for some i, where j is defined as

$$j = \max\left\{ i \in [1, 2, \ldots, k] \text{ such that } P_{(k-i+m)} > \frac{k\alpha}{m} \text{ for } m = 1, 2, \ldots, i \right\}.$$

Additional procedures designed to improve the Bonferroni procedure have been suggested by Holm (1979), Shaffer (1986), Rom (1990), and Holland and Copenhaver (1987). Wright (1992) discussed these procedures and recommended wider use of such adjustments to p-values. Holland (1991) compared the procedures of Holm, Hommel, and Hochberg and concluded that Hommel's method is more powerful, though not uniformly so. Broadbent (1993) compared six procedures: traditional Bonferroni, Holm, Simes, Hochberg, Hommel, and Rom. She found the Hommel procedure to be slightly superior to the others in terms of both Type I error and power. However, there was not a large difference between the procedures of Hommel, Hochberg, and Rom.

Bristol (1993) presented approximate methods for adjusting p-values. Bristol's methods are less conservative than Bonferroni-based procedures. Troendle (1993) proposed a resampling method of testing hypotheses in the individual variables. His technique takes the covariance structure into account without making distributional assumptions. In a simulation study, Troendle showed that his method was slightly superior to that of Hochberg (1988) in some cases.

When several subtests are run, as described above, the overall α-level is known as the *experimentwise error rate*. If a group of hypotheses such as $H_{0j}\colon \mu_j = \mu_{0j}$, $j = 1, 2, \ldots, p$, is tested, the experimentwise error rate is the probability of rejecting one or more tests when all H_{0j} are true, that is, when $\boldsymbol{\mu} = \boldsymbol{\mu}_0$. Thus either of the critical values $T_{\alpha,p,n-1}$ or $t_{\alpha/2p,n-1}$ will preserve the desired experimentwise error rate α. However, the use of $t_{\alpha/2,n-1}$ for each of the p tests would not preserve the experimentwise error rate.

In the foregoing discussion of tests of hypotheses on individual variables, H_{0j} : $\mu_j = \mu_{0j}$, $j = 1, 2, \ldots, p$, we have assumed that the tests are carried out independently of (or ignoring) a T^2-test of $H_0\colon \boldsymbol{\mu} = \boldsymbol{\mu}_0$. The use of tests of $H_{0j}\colon \mu_j = \mu_{0j}$ following rejection of the T^2-test $\boldsymbol{\mu} = \boldsymbol{\mu}_0$ involves a different error structure (see the last paragraph of Section 3.6.4).

Rejection of $H_0\colon \boldsymbol{\mu} = \boldsymbol{\mu}_0$ by the T^2-test implies that $\boldsymbol{\mu} \neq \boldsymbol{\mu}_0$ and therefore that $\mu_j \neq \mu_{0j}$ for at least one j, but rejection of H_0 does not imply that one of the hypotheses $H_{0j}\colon \mu_j = \mu_{0j}$, $j = 1, 2, \ldots, p$, will be rejected, even if $t_{\alpha/2,n-1}$ is used as a critical value for (3.46). However, if the T^2-test leads to rejection of H_0, we can find at least one linear combination $\mathbf{a}'\overline{\mathbf{y}}$ for use in (3.45) such that $|t(\mathbf{a})| \geq T_{\alpha,p,n-1}$, since, by Theorem 3.3E, $\max_{\mathbf{a}} t^2(\mathbf{a}) = n(\overline{\mathbf{y}} - \boldsymbol{\mu}_0)'\mathbf{S}^{-1}(\overline{\mathbf{y}} - \boldsymbol{\mu}_0) = T^2$. By (3.32), the vector $\mathbf{a}$ that maximizes $t^2(\mathbf{a})$ is (any multiple of) the discriminant function coefficient vector $\mathbf{a} = \mathbf{S}^{-1}(\overline{\mathbf{y}} - \boldsymbol{\mu}_0)$. Thus if we reject $H_0\colon \boldsymbol{\mu} = \boldsymbol{\mu}_0$ due to $T^2 \geq T^2_\alpha$, then $\mathbf{a}'\overline{\mathbf{y}}$ with $\mathbf{a} = \mathbf{S}^{-1}(\overline{\mathbf{y}} - \boldsymbol{\mu}_0)$ will lead to rejection of $H_0\colon \mathbf{a}'\boldsymbol{\mu} = \mathbf{a}'\boldsymbol{\mu}_0$ using (3.45).

We note in passing that if an auxiliary estimate $\mathbf{S}_2$ of $\boldsymbol{\Sigma}$ is available (as illustrated in the last paragraph of Sections 3.3.3 and 3.4.4), we can take advantage of the increased precision this offers. Using the pooled sample covariance matrix $\mathbf{S}_{\text{pl}}$ as in (3.41), the test statistic in (3.45) becomes

$$t(\mathbf{a}) = \frac{\mathbf{a}'(\overline{\mathbf{y}} - \boldsymbol{\mu}_0)\sqrt{n}}{\sqrt{\mathbf{a}'\mathbf{S}_{\text{pl}}\mathbf{a}}}, \tag{3.48}$$

and we reject $H_0\colon \mathbf{a}'\boldsymbol{\mu} = \mathbf{a}'\boldsymbol{\mu}_0$ if $|t(\mathbf{a})| \geq T_{\alpha,p,n_1+n_2-2}$. Similarly, (3.46) becomes

$$t_j = \frac{(\overline{y}_j - \mu_{0j})\sqrt{n}}{\sqrt{s_{\text{pl},jj}}}, \qquad j = 1, 2, \ldots, p, \tag{3.49}$$

and we reject $H_{0j}\colon \mu_j = \mu_{0j}$ if $|t_j| \geq T_{\alpha,p,n_1+n_2-2}$ or if $|t_j| \geq t_{\alpha/2p,n_1+n_2-2}$.

Example 3.4.5. We use the calcium data in Table 3.1 to illustrate the t_j-test in (3.46) for each variable. In Example 3.3.5 we tested the multivariate hypothesis

$$H_0\colon \boldsymbol{\mu} = \begin{pmatrix} \mu_{01} \\ \mu_{02} \\ \mu_{03} \end{pmatrix} = \begin{pmatrix} 15.0 \\ 6.0 \\ 2.85 \end{pmatrix}.$$

We now test each variable separately using (3.46). The means and variances are found in Example 3.3.5:

$$y_1\colon t_1 = \frac{(\overline{y}_1 - \mu_{01})\sqrt{n}}{\sqrt{s_{11}}} = \frac{(28.1 - 15.0)\sqrt{10}}{\sqrt{140.54}} = 3.494,$$

$$y_2\colon t_2 = \frac{(7.18 - 6.0)\sqrt{10}}{\sqrt{72.25}} = .439,$$

$$y_3\colon t_3 = \frac{(3.09 - 2.85)\sqrt{10}}{\sqrt{.25}} = 1.518.$$

Using the simultaneous critical value $T_{.05,3,9} = 4.0946$, we would accept $H_{0j}\colon \mu_j = \mu_{0j}$ in all three cases. Using the Bonferroni critical value $t_{.025/3,9} = 2.9333$ from Table B.2, we would reject $H_{01}\colon \mu_1 = 15.0$, since $t_1 = 3.494 > 2.933$.

3.4.6 Tests for $H_0\colon \mathbf{C}\boldsymbol{\mu} = \mathbf{0}$

The hypothesis $H_0\colon \boldsymbol{\mu} = \boldsymbol{\mu}_0$ specifies each μ_j completely. Often a linear relationship among the μ_j's is of more interest. Several linear functions may be examined jointly in the expression $\mathbf{C}\boldsymbol{\mu} = \mathbf{0}$. In most of the remainder of this section, we illustrate the use of $H_0\colon \mathbf{C}\boldsymbol{\mu} = \mathbf{0}$ to express the hypothesis

$$H_0\colon \mu_1 = \mu_2 = \cdots = \mu_p. \tag{3.50}$$

At the end of the section, we consider the more general hypothesis $H_0\colon \mathbf{C}\boldsymbol{\mu} = \boldsymbol{\gamma}$ for a specified $\boldsymbol{\gamma}$ of interest.

If the p variables $y_1, y_2, \ldots, y_p$ in the random vector $\mathbf{y}$ are commensurate (measured in the same units and with comparable variances), the hypothesis in (3.50) may be of interest. Note that the hypothesis in (3.50) cannot be tested using an ordinary analysis-of-variance (ANOVA) approach, which assumes independence of all y_{ij}. However, in a multivariate framework, we can readily construct a test that allows for the correlations among the variables.

The hypothesis in (3.50) is equivalent to $H_{0j}\colon \mu_1 - \mu_j = 0$, $j = 2, 3, \ldots, p$, or, alternatively, $H_{0j}\colon \mu_j - \mu_{j+1} = 0$, $j = 1, 2, \ldots, p-1$. These two expressions can be rewritten as $H_0\colon \mathbf{C}_1\boldsymbol{\mu} = \mathbf{0}$ and $H_0\colon \mathbf{C}_2\boldsymbol{\mu} = \mathbf{0}$, where

$$\mathbf{C}_1 = \begin{pmatrix} 1 & -1 & 0 & \cdots & 0 \\ 1 & 0 & -1 & \cdots & 0 \\ \vdots & \vdots & \vdots & & \vdots \\ 1 & 0 & 0 & \cdots & -1 \end{pmatrix}, \qquad \mathbf{C}_2 = \begin{pmatrix} 1 & -1 & 0 & \cdots & 0 \\ 0 & 1 & -1 & \cdots & 0 \\ \vdots & \vdots & \vdots & & \vdots \\ 0 & 0 & 0 & \cdots & -1 \end{pmatrix}.$$

Thus $\mathbf{C}$ is not unique; in fact, if $\mathbf{C}$ is any full-rank $(p-1) \times p$ matrix such that $\mathbf{Cj} = \mathbf{0}$, then $H_0\colon \mathbf{C}\boldsymbol{\mu} = \mathbf{0}$ is equivalent to (3.50).

To test $H_0\colon \mathbf{C}\boldsymbol{\mu} = \mathbf{0}$ corresponding to (3.50), we transform to $\mathbf{z} = \mathbf{Cy}$, which, by (1.76) and (1.77), has sample mean vector and covariance matrix $\overline{\mathbf{z}} = \mathbf{C}\overline{\mathbf{y}}$ and $\mathbf{S}_z = \mathbf{CSC}'$. If $\mathbf{y}$ is $N_p(\boldsymbol{\mu}, \boldsymbol{\Sigma})$, then by Theorem 2.2B(ii), $\mathbf{z} = \mathbf{Cy}$ is $N_{p-1}(\mathbf{C}\boldsymbol{\mu}, \mathbf{C}\boldsymbol{\Sigma}\mathbf{C}')$.

Under H_0, $\mathbf{Cy}$ is $N_{p-1}(\mathbf{0}, \mathbf{C\Sigma C}')$; and, by (3.16) and Theorem 2.3K,

$$T^2 = n\bar{\mathbf{z}}'\mathbf{S}_z^{-1}\bar{\mathbf{z}} = n(\mathbf{C}\bar{\mathbf{y}})'(\mathbf{CSC}')^{-1}(\mathbf{C}\bar{\mathbf{y}}) \tag{3.51}$$

is distributed as $T^2_{p-1,n-1}$ under H_0. We reject H_0: $\mathbf{C}\boldsymbol{\mu} = \mathbf{0}$ if $T^2 \geq T^2_{\alpha,p-1,n-1}$. Note that the dimension $p-1$ corresponds to the number of rows of $\mathbf{C}$, which is the number of transformed variables in $\mathbf{z} = \mathbf{Cy}$.

The test statistic in (3.51) does not reduce to $T^2 = n\bar{\mathbf{y}}'\mathbf{S}^{-1}\bar{\mathbf{y}}$ because $\mathbf{C}$ is $(p-1)\times p$ and has no inverse. However, if we append the row $(1, 1, \ldots, 1)$ to $\mathbf{C}$ and call the augmented matrix $\mathbf{A}$, then $\mathbf{A}$ is $p \times p$ and full rank. With

$$\mathbf{A} = \begin{pmatrix} \mathbf{j}' \\ \mathbf{C} \end{pmatrix}$$

in place of $\mathbf{C}$, (3.51) would reduce to $T^2 = n\bar{\mathbf{y}}'\mathbf{S}^{-1}\bar{\mathbf{y}}$, which tests H_0: $\boldsymbol{\mu} = \mathbf{0}$. The first element in $\mathbf{A}\boldsymbol{\mu} = \mathbf{0}$ is $(1, 1, \ldots, 1)\boldsymbol{\mu} = 0$ or $\sum_{j=1}^{p} \mu_j = 0$. This, combined with $\mu_1 = \mu_2 = \cdots = \mu_p$, is equivalent to $\boldsymbol{\mu} = \mathbf{0}$. Note that

$$n(\mathbf{A}\bar{\mathbf{y}})'(\mathbf{ASA}')^{-1}(\mathbf{A}\bar{\mathbf{y}}) = n\bar{\mathbf{y}}'\mathbf{S}^{-1}\bar{\mathbf{y}} > n(\mathbf{C}\bar{\mathbf{y}})'(\mathbf{CSC}')^{-1}\mathbf{C}\bar{\mathbf{y}}. \tag{3.52}$$

As noted earlier, if $\mathbf{C}$ is any full-rank $(p-1) \times p$ matrix such that $\mathbf{Cj} = \mathbf{0}$, then $\mathbf{C}\boldsymbol{\mu} = \mathbf{0}$ is equivalent to $\mu_1 = \mu_2 = \cdots = \mu_p$. Since $\mathbf{Cj} = \mathbf{0}$, each row of $\mathbf{C}$ sums to 0 by (A.2.15), and $\mathbf{C}$ is sometimes called a *contrast matrix*. A linear combination $c_{i1}\mu_1 + c_{i2}\mu_2 + \cdots + c_{ip}\mu_p$ is a *contrast* in the μ's if $\sum_j c_{ij} = 0$. In order to express H_0: $\mu_1 = \mu_2 = \cdots = \mu_p$ in (3.50), $\mathbf{C}$ must be full (row) rank; that is, the rows (contrasts) must be linearly independent. Since $\mathbf{C}$ is not unique, H_0 can be represented in many ways. Linearly independent contrasts such as those illustrated in $\mathbf{C}_1$ or $\mathbf{C}_2$ can be used, as well as orthogonal contrasts or orthogonal polynomials in the μ_j's.

To show that T^2 in (3.51) is unique even though $\mathbf{C}$ is not unique, note that there exists a nonsingular matrix $\mathbf{K}$ that yields $\mathbf{C} = \mathbf{KC}_1$ (see, for example, Bronson 1975, Section 3.4). Hence, by the invariance of T^2 to nonsingular linear transformations (property 1, Section 3.3.6), a test based on $\mathbf{z}_i = \mathbf{Cy}_i = \mathbf{KC}_1\mathbf{y}_i$ gives the same value of T^2 as a test based on $\mathbf{w}_i = \mathbf{C}_1\mathbf{y}_i$.

To test hypotheses H_0: $\mathbf{C}\boldsymbol{\mu} = \mathbf{0}$ other than (3.50), note that if $\mathbf{C}$ is $k \times p$ of rank $k < p-1$, then T^2 in (3.51) is distributed as $T^2_{k,n-1}$ under H_0: $\mathbf{C}\boldsymbol{\mu} = \mathbf{0}$ (assuming $\mathbf{C}$ is full rank). To test the more general hypothesis H_0: $\mathbf{C}\boldsymbol{\mu} = \boldsymbol{\gamma}$, where $\mathbf{C}$ is $k \times p$ of rank k, we use

$$T^2 = n(\mathbf{C}\bar{\mathbf{y}} - \boldsymbol{\gamma})'(\mathbf{CSC}')^{-1}(\mathbf{C}\bar{\mathbf{y}} - \boldsymbol{\gamma}), \tag{3.53}$$

which is distributed as $T^2_{k,n-1}$ (under H_0: $\mathbf{C}\boldsymbol{\mu} = \boldsymbol{\gamma}$).

Example 3.4.6. We illustrate the test of H_0: $\mu_1 = \mu_2 = \mu_3$ for the calcium data in Table 3.1. In this case, the variables do not appear to be commensurate, and a direct comparison of means may not be of interest, but we will carry out the test for illustrative purposes. The hypothesis H_0: $\mu_1 = \mu_2 = \mu_3$ is equivalent to H_0: $\mathbf{C}\boldsymbol{\mu} = \mathbf{0}$, with

$$\mathbf{C} = \begin{pmatrix} 1 & -1 & 0 \\ 0 & 1 & -1 \end{pmatrix}.$$

Using $\overline{\mathbf{y}}$ and $\mathbf{S}$ from Example 3.3.5, we obtain

$$\mathbf{C}\overline{\mathbf{y}} = \begin{pmatrix} 20.92 \\ 4.09 \end{pmatrix}, \qquad \mathbf{CSC}' = \begin{pmatrix} 113.43 & -20.83 \\ -20.83 & 65.15 \end{pmatrix},$$

and by (3.51),

$$\begin{aligned} T^2 &= n(\mathbf{C}\overline{\mathbf{y}})'(\mathbf{CSC}')^{-1}(\mathbf{C}\overline{\mathbf{y}}) \\ &= (10)(20.92, 4.09)\begin{pmatrix} 113.43 & -20.83 \\ -20.83 & 65.15 \end{pmatrix}^{-1}\begin{pmatrix} 20.92 \\ 4.09 \end{pmatrix} \\ &= 48.846. \end{aligned}$$

We reject H_0, since $48.846 > T^2_{.01,2,9} = 19.460$.

3.5 TESTS OF $H_0\colon \boldsymbol{\mu}_1 = \boldsymbol{\mu}_2$ ASSUMING $\boldsymbol{\Sigma}_1 = \boldsymbol{\Sigma}_2$

We review the univariate likelihood ratio test in Section 3.5.1 and discuss the analogous T^2-test and its properties in Sections 3.5.2–3.5.5.

3.5.1 Review of Univariate Likelihood Ratio Test for $H_0\colon \mu_1 = \mu_2$ When $\sigma_1^2 = \sigma_2^2$

In the univariate case ($p = 1$), two independent random samples $y_{11}, y_{12}, \ldots, y_{1n_1}$ and $y_{21}, y_{22}, \ldots, y_{2n_2}$ are drawn from $N(\mu_1, \sigma_1^2)$ and $N(\mu_2, \sigma_2^2)$, respectively, where σ_1^2 and σ_2^2 are unknown but assumed to be equal, $\sigma_1^2 = \sigma_2^2 = \sigma^2$, say. The hypothesis of interest is

$$H_0\colon \mu_1 = \mu_2 \qquad \text{versus} \qquad H_1\colon \mu_1 \neq \mu_2.$$

Let $\text{SS}_1 = \sum_{i=1}^{n_1}(y_{1i} - \overline{y}_1)^2 = (n_1 - 1)s_1^2$ and $\text{SS}_2 = \sum_{i=1}^{n_2}(y_{2i} - \overline{y}_2)^2 = (n_2 - 1)s_2^2$. Then the pooled variance is

$$s_{\text{pl}}^2 = \frac{\text{SS}_1 + \text{SS}_2}{n_1 + n_2 - 2} = \frac{(n_1 - 1)s_1^2 + (n_2 - 1)s_2^2}{n_1 + n_2 - 2},$$

which is the minimum variance unbiased estimator for σ^2. Note that the denominator $n_1 + n_2 - 2$ is the sum of the weights $n_1 - 1$ and $n_2 - 1$ in the numerator.

For a test statistic, we use

$$t = \frac{\overline{y}_1 - \overline{y}_2}{s_{\text{pl}}\sqrt{1/n_1 + 1/n_2}}, \tag{3.54}$$

which has a t-distribution with $n_1 + n_2 - 2$ degrees of freedom, provided that H_0 holds. We therefore reject H_0 if $|t| \geq t_{\alpha/2, n_1+n_2-2}$. The t-statistic in (3.54) can be written in the form

$$t = \frac{\overline{y}_1 - \overline{y}_2}{s_{\text{pl}}\sqrt{1/n_1 + 1/n_2}} = \sqrt{\frac{n_1 n_2}{n_1 + n_2}} \left(\frac{\overline{y}_1 - \overline{y}_2}{s_{\text{pl}}} \right)$$

or

$$t^2 = \frac{n_1 n_2}{n_1 + n_2} \left[\frac{(\overline{y}_1 - \overline{y}_2)^2}{s_{\text{pl}}^2} \right]. \tag{3.55}$$

It can be shown that (3.54) also satisfies the *formal definition* of a t random variable as given by (3.3) and accompanying assumptions.

The test statistic (3.54) is obtainable from the likelihood ratio approach. The assumptions $\sigma_1^2 = \sigma_2^2$ and independence of the two samples are both necessary for the likelihood ratio to reduce to a function of t^2 in (3.55). If $\sigma_1^2 \neq \sigma_2^2$ and we use $s_1^2/n_1 + s_2^2/n_2$ as an estimate of $\text{var}(\overline{y}_1 - \overline{y}_2)$, the resulting statistic $(\overline{y}_1 - \overline{y}_2)/\sqrt{s_1^2/n_1 + s_2^2/n_2}$ does not have a t-distribution.

Theorem 3.5A. For two independent random samples $y_{11}, y_{12}, \ldots, y_{1n_1}$ and $y_{21}, y_{22}, \ldots, y_{2n_2}$ from $N(\mu_1, \sigma_1^2)$ and $N(\mu_2, \sigma_2^2)$, respectively, the t-test in (3.54) is the likelihood ratio test, provided that $\sigma_1^2 = \sigma_2^2$.

Proof. As in (3.6), the likelihood function is the product of the densities of $y_{11}, \ldots, y_{1n_1}$ and $y_{21}, \ldots, y_{2n_2}$:

$$\begin{aligned}
L(\mu_1, \sigma_1^2, \mu_2, \sigma_2^2) \\
&= \prod_{i=1}^{n_1} f(y_{1i}, \mu_1, \sigma_1^2) \prod_{i=1}^{n_2} f(y_{2i}, \mu_2, \sigma_2^2) \\
&= \prod_{i=1}^{n_1} \frac{1}{\sqrt{2\pi}\sigma_1} e^{-(y_{1i}-\mu_1)^2/2\sigma_1^2} \prod_{i=1}^{n_2} \frac{1}{\sqrt{2\pi}\sigma_2} e^{-(y_{2i}-\mu_2)^2/2\sigma_2{}^2} \\
&= \frac{1}{(\sqrt{2\pi})^{n_1}(\sigma_1^2)^{n_1/2}} e^{-\sum_{i=1}^{n_1}(y_{1i}-\mu_1)^2/2\sigma_1^2} \frac{1}{(\sqrt{2\pi})^{n_2}(\sigma_2^2)^{n_2/2}} e^{-\sum_{i=1}^{n_2}(y_{2i}-\mu_2)^2/2\sigma_2^2}.
\end{aligned}$$

If σ_1^2 is set equal to σ_2^2 and if parameters are estimated under H_0 and H_1, the resulting likelihood ratio is

$$\text{LR} = \left[\frac{1}{1 + t^2/(n_1 + n_2 - 2)} \right]^{(n_1+n_2)/2}, \tag{3.56}$$

which is a decreasing function of t^2 as given by (3.55). Rejecting $H_0\colon \mu_1 = \mu_2$ for $\text{LR} \leq c$ is equivalent to rejecting H_0 for $t^2 \geq t_{\alpha/2}^2$, and the t-test in (3.54) is a likelihood ratio test. □

3.5.2 Test for $H_0\colon \boldsymbol{\mu}_1 = \boldsymbol{\mu}_2$ When $\boldsymbol{\Sigma}_1 = \boldsymbol{\Sigma}_2$

In the multivariate case, we wish to compare the mean vectors from two populations. We assume that two independent random samples $\mathbf{y}_{11}, \mathbf{y}_{12}, \ldots, \mathbf{y}_{1n_1}$ and $\mathbf{y}_{21}, \mathbf{y}_{22}, \ldots, \mathbf{y}_{2n_2}$ are drawn from $N_p(\boldsymbol{\mu}_1, \boldsymbol{\Sigma}_1)$ and $N_p(\boldsymbol{\mu}_2, \boldsymbol{\Sigma}_2)$, respectively, where $\boldsymbol{\Sigma}_1$ and $\boldsymbol{\Sigma}_2$ are unknown. In order to obtain a T^2-test, we must assume that $\boldsymbol{\Sigma}_1 = \boldsymbol{\Sigma}_2 = \boldsymbol{\Sigma}$, say. (A test of $H_0\colon \boldsymbol{\Sigma}_1 = \boldsymbol{\Sigma}_2$ is given in Section 4.3.) From the two samples, we calculate $\bar{\mathbf{y}}_1 = \sum_{i=1}^{n_1} \mathbf{y}_{1i}/n_1$, $\bar{\mathbf{y}}_2 = \sum_{i=1}^{n_2} \mathbf{y}_{2i}/n_2$, $\mathbf{W}_1 = \sum_{i=1}^{n_1} (\mathbf{y}_{1i} - \bar{\mathbf{y}}_1)(\mathbf{y}_{1i} - \bar{\mathbf{y}}_1)' = (n_1 - 1)\mathbf{S}_1$, and $\mathbf{W}_2 = \sum_{i=1}^{n_2} (\mathbf{y}_{2i} - \bar{\mathbf{y}}_2)(\mathbf{y}_{2i} - \bar{\mathbf{y}}_2)' = (n_2 - 1)\mathbf{S}_2$. A pooled estimator of the covariance matrix is calculated as

$$\mathbf{S}_{\text{pl}} = \frac{\mathbf{W}_1 + \mathbf{W}_2}{n_1 + n_2 - 2} = \frac{(n_1 - 1)\mathbf{S}_1 + (n_2 - 1)\mathbf{S}_2}{n_1 + n_2 - 2},$$

for which $E(\mathbf{S}_{\text{pl}}) = \boldsymbol{\Sigma}$.

To test

$$H_0\colon \boldsymbol{\mu}_1 = \boldsymbol{\mu}_2 \quad \text{versus} \quad H_1\colon \boldsymbol{\mu}_1 \neq \boldsymbol{\mu}_2,$$

we use the test statistic

$$T^2 = \frac{n_1 n_2}{n_1 + n_2} (\bar{\mathbf{y}}_1 - \bar{\mathbf{y}}_2)' \mathbf{S}_{\text{pl}}^{-1} (\bar{\mathbf{y}}_1 - \bar{\mathbf{y}}_2), \tag{3.57}$$

which is distributed as T^2_{p,n_1+n_2-2} when H_0 is true. We reject H_0 if $T^2 \geq T^2_{\alpha,p,n_1+n_2-2}$. Critical values T^2_{α,p,n_1+n_2-2} are given in Table B.1.

We note again the assumptions that must hold in order for (3.57) to have the T^2_{p,n_1+n_2-2} distribution:

1. The two samples are drawn from $N_p(\boldsymbol{\mu}_1, \boldsymbol{\Sigma}_1)$ and $N_p(\boldsymbol{\mu}_2, \boldsymbol{\Sigma}_2)$, respectively.
2. $\boldsymbol{\Sigma}_1 = \boldsymbol{\Sigma}_2$.
3. The two samples are independent.

By Theorem 3.3B, the T^2-statistic (3.57) can be transformed to an F-statistic:

$$\frac{n_1 + n_2 - p - 1}{(n_1 + n_2 - 2)p} T^2 = F_{p,n_1+n_2-p-1}. \tag{3.58}$$

An alternative version of the T^2 formula can be obtained using (A.7.10):

$$T^2 = \frac{|\mathbf{S}_{\text{pl}} + n_1 n_2 (\bar{\mathbf{y}}_1 - \bar{\mathbf{y}}_2)(\bar{\mathbf{y}}_1 - \bar{\mathbf{y}}_2)'/(n_1 + n_2)|}{|\mathbf{S}_{\text{pl}}|} - 1. \tag{3.59}$$

3.5.3 Effect on T^2 of Adding a Variable

The change in the one-sample T^2 due to an additional variable was discussed in Section 3.3.5. In the two-sample case, if a variable x is added to each $\mathbf{y}$ in the two samples, expressions analogous to (3.19) and (3.20) for the increase in T^2 are given in the following theorem (Rencher 1993).

Theorem 3.5B. For the $p+1$ variables $(y_1, y_2, \ldots, y_p, x) = (\mathbf{y}', x)$, T^2 in (3.57) can be expressed as

$$T^2_{y,x} = \frac{n_1 n_2}{n_1+n_2}(\bar{\mathbf{y}}_1 - \bar{\mathbf{y}}_2)'\mathbf{S}_{\text{pl}}^{-1}(\bar{\mathbf{y}}_1 - \bar{\mathbf{y}}_2) + \frac{n_1 n_2}{n_1+n_2}\frac{\left[\mathbf{s}'_{xy}\mathbf{S}_{\text{pl}}^{-1}(\bar{\mathbf{y}}_1 - \bar{\mathbf{y}}_2) - (\bar{x}_1 - \bar{x}_2)\right]^2}{s_x^2 - \mathbf{s}'_{xy}\mathbf{S}_{\text{pl}}^{-1}\mathbf{s}_{xy}}$$

$$= T_y^2 + \frac{n_1 n_2}{n_1+n_2}\frac{\left[\hat{\boldsymbol{\beta}}'(\bar{\mathbf{y}}_1 - \bar{\mathbf{y}}_2) - (\bar{x}_1 - \bar{x}_2)\right]^2}{s_x^2(1-R^2)} \tag{3.60}$$

$$= T_y^2 + \frac{(\hat{t}_x - t_x)^2}{1-R^2}, \tag{3.61}$$

where $T^2_{y,x}$ is based on $(y_1, \ldots, y_p, x)$, T_y^2 is based on $(y_1, \ldots, y_p)$, $\mathbf{S}_{\text{pl}}$ is the pooled sample covariance matrix of the y's, $\mathbf{s}_{xy}$ is the vector of pooled covariances of x with the y's, s_x^2 is the pooled variance of x, $R^2 = \mathbf{s}'_{xy}\mathbf{S}_{\text{pl}}^{-1}\mathbf{s}_{xy}/s_x^2$ is the (within-sample) squared multiple correlation of x regressed on the y's [see (7.66)] $\hat{\boldsymbol{\beta}} = \mathbf{S}_{\text{pl}}^{-1}\mathbf{s}_{xy}$ is the regression coefficient vector [see (7.54)], and

$$\hat{t}_x = \frac{\hat{\boldsymbol{\beta}}'(\bar{\mathbf{y}}_1 - \bar{\mathbf{y}}_2)}{s_x[(n_1+n_2)/n_1 n_2]^{1/2}}, \qquad t_x = \frac{\bar{x}_1 - \bar{x}_2}{s_x[(n_1+n_2)/n_1 n_2]^{1/2}}. \qquad \square$$

In Theorem 3.5B, the contribution of x to T^2 depends on the same factors as in the one-sample case in Theorem 3.3C; t_x is the usual t-statistic for x by itself, and $\hat{t}_x$ is a "predicted" value of t_x based on the information about $\bar{x}_1 - \bar{x}_2$ already available in the y's. We can see in (3.61) how a variable can contribute more to group separation in the presence of other variables than it does by itself. This can happen in various ways: (1) $\hat{t}_x$, the predicted group separation based on other variables, may be much greater than t_x, the group separation for the variable by itself; (2) t_x and $\hat{t}_x$ may be of opposite sign; and (3) there may be a very high squared multiple correlation R^2 with the other variables. The first of these is illustrated by y_4 in the first example below [Example 3.5.3(a)].

It is also clear in (3.61) how a variable that is significant by itself (large t_x) can become redundant in the presence of other variables ($\hat{t}_x$ close to t_x). This is illustrated by y_1 and y_3 in the second example below [Example 3.5.3(b)]. With these two types of added information, the researcher has guidelines for designing or screening new variables that will be more effective in separating the two groups.

To illustrate, suppose a researcher is comparing subjects who have high blood cholesterol levels to subjects who have low cholesterol levels. The following variables are measured: level of dietary animal fat, level of dietary fiber, level of physical activity, and hereditary predisposition. If the researcher wishes to study the interdependence of these variables as they contribute jointly to group separation, this information is shown precisely in (3.60). We can examine $\hat{\boldsymbol{\beta}}$ for an indication of how much the variable in question is influenced by each of the other variables. The contribution of a variable is also affected by its overall relationship with the other variables, as measured by R^2.

Theorem 3.5B also demonstrates that T^2 is increased by the addition of x, because the second term on the right side of (3.60) or (3.61) is positive. However, the critical value also increases because the number of variables has been incremented from p to $p + 1$. Thus one must be careful about arbitrarily adding a variable. A formal test for the significance of the increase in T^2 when variables are added is given in Section 3.11.1.

Example 3.5.3(a). Four psychological tests were administered to 32 male subjects and 32 female subjects (Beall 1945, see diskette for data). The mean vectors and covariance matrices of the two samples are

$$\overline{\mathbf{y}}_1 = \begin{pmatrix} 15.97 \\ 15.91 \\ 27.19 \\ 22.75 \end{pmatrix}, \qquad \overline{\mathbf{y}}_2 = \begin{pmatrix} 12.34 \\ 13.91 \\ 16.66 \\ 21.94 \end{pmatrix},$$

$$\mathbf{S}_1 = \begin{pmatrix} 5.192 & 4.545 & 6.522 & 5.250 \\ 4.545 & 13.18 & 6.760 & 6.266 \\ 6.522 & 6.760 & 28.67 & 14.47 \\ 5.250 & 6.266 & 14.47 & 16.65 \end{pmatrix},$$

$$\mathbf{S}_2 = \begin{pmatrix} 9.136 & 7.549 & 4.864 & 4.151 \\ 7.549 & 18.60 & 10.22 & 5.446 \\ 4.864 & 10.22 & 30.04 & 13.49 \\ 4.151 & 5.446 & 13.49 & 28.00 \end{pmatrix}.$$

The pooled covariance matrix is

$$\mathbf{S}_{\text{pl}} = \frac{(32 - 1)\mathbf{S}_1 + (32 - 1)\mathbf{S}_2}{32 + 32 - 2}$$

$$= \begin{pmatrix} 7.164 & 6.047 & 5.693 & 4.701 \\ 6.047 & 15.89 & 8.492 & 5.856 \\ 5.693 & 8.492 & 29.36 & 13.98 \\ 4.701 & 5.856 & 13.98 & 22.32 \end{pmatrix}.$$

The two-sample T^2-statistic in (3.57) yields $T^2 = 97.6015$, which becomes $T^2_{y,x}$ in the present context. If any three of the variables constitute $\mathbf{y}$ and the other variable is x, then, by (3.61),

$$T^2_{y,x} - T^2_y = \frac{(\hat{t}_x - t_x)^2}{1 - R^2},$$

where

$$\hat{t}_x = \frac{\hat{\boldsymbol{\beta}}'(\overline{\mathbf{y}}_1 - \overline{\mathbf{y}}_2)}{s_x[(n_1 + n_2)/n_1 n_2]^{1/2}}, \qquad t_x = \frac{\overline{x}_1 - \overline{x}_2}{s_x[(n_1 + n_2)/n_1 n_2]^{1/2}},$$

and $R^2 = \mathbf{s}'_{xy}\mathbf{S}_{\text{pl}}^{-1}\mathbf{s}_{xy}/s_y^2$ is the squared multiple correlation of x regressed on $\mathbf{y}$. The values of $\hat{t}_x$, t_x, R^2, and $T^2_{y,x} - T^2_y$ are given below for each y_j as x:

x	$\hat{t}_x$	t_x	R^2	$T^2_{y,x} - T^2_y$
y_1	1.990	5.417	.373	18.728
y_2	4.094	2.007	.356	6.767
y_3	1.342	7.775	.363	64.976
y_4	4.620	.688	.328	23.009

Since the values of R^2 are small and approximately equal, the effect of each variable is due mostly to $\hat{t}_x$ or t_x. The variable that makes the greatest contribution to T^2 is y_3, whose effect is attributable to a large value of t_x as compared to $\hat{t}_x$. For y_4, the next greatest contributor, the increase in T^2 is due to a relatively large value of $\hat{t}_x$. By itself, y_4 does not significantly separate the two groups ($t_x = .688$), but in the presence of the other variables its contribution is greatly enhanced ($\hat{t}_x = 4.620$).

Example 3.5.3(b). Four enzymes were measured in an attempt to detect carriers of Duchenne muscular dystrophy (Andrews and Herzberg 1985, pp. 223–228, see diskette for data). The overall $T^2_{y,x}$ statistic that compared 39 carriers and 34 noncarriers on all four variables was 71.70 ($p = 8.9 \times 10^{-10}$). As in Example 3.5.3(a), the values of $T^2_{y,x} - T^2_y$, $\hat{t}_x$, t_x, and R^2 are given below for each variable (see Rencher 1993):

x	$\hat{t}_x$	t_x	R^2	$T^2_{y,x} - T^2_y$
y_1	−3.13	−4.37	.39	2.52
y_2	− .35	−5.40	.01	25.70
y_3	−3.94	−4.03	.47	.02
y_4	−2.43	−6.44	.25	21.59

The four variables have similar individual t_x-values, but their contribution to T^2 ranges from .02 to 25.70. Without the new information available in $\hat{t}_x$ and R^2, the researcher does not know why a variable makes a large addition or small addition to T^2.

Note that y_1 and y_3 increase T^2 by only 2.52 and .02, respectively, despite the significant individual t_x-values of these two variables, $t_x = -4.37$ and $t_x = -4.03$ ($p = .0004$ and $p = .017$, respectively). In both cases, $\hat{t}_x$ is so close to t_x that the effect of the variable on T^2 is effectively nullified and the variable becomes almost totally redundant. These two variables have the largest values of R^2, but they are not large enough to offset the closeness of $\hat{t}_x$ to t_x. In contrast to y_1 and y_3, the second and fourth variables, y_2 and y_4, make a large contribution to T^2 because their values of t_x are not as well predicted by the other variables.

The specific information about why y_3, for example, contributes almost nothing to T^2 in spite of its large t_x-value may be useful in a search for a variable to replace it. The researcher may wish to find variables that significantly boost the T^2-value so as to obtain a more sensitive screening procedure for the disease. A preliminary examination may reveal that certain types of variables have favorable patterns on $\hat{t}_x$ or R^2, with a consequent narrowing of the search.

3.5.4 Properties of the Two-Sample T^2-Statistic

The formal definition of a T^2 random variable is given in (3.16) as

$$T^2 = \mathbf{z}'\left(\frac{\mathbf{W}}{\nu}\right)^{-1}\mathbf{z},$$

where $\mathbf{z}$ is $N_p(\mathbf{0}, \boldsymbol{\Sigma})$, $\mathbf{W}$ is $W_p(\nu, \boldsymbol{\Sigma})$, and $\mathbf{z}$ and $\mathbf{W}$ are independent. It is easy to show that the two-sample T^2-statistic in (3.57) satisfies (3.16).

The properties of the two-sample T^2-test are analogous to those for the one-sample case given in Section 3.3.6, with the exception of property 3. The T^2-test or test statistic (3.57) is

1. invariant to nonsingular transformations $\mathbf{z}_i = \mathbf{A}\mathbf{y}_i + \mathbf{b}$,
2. uniformly most powerful among invariant tests (Siotani, Hayakawa, and Fujikoshi 1985, p. 208),
3. reasonably robust to departures from normality (see Section 3.7.2),
4. the likelihood ratio test (see Section 3.5.5), and
5. the union–intersection test (see Section 3.5.5).

3.5.5 Likelihood Ratio and Union–Intersection Tests

In Sections 3.3.7 and 3.3.8, we showed that the one-sample T^2-test of $H_0\colon \boldsymbol{\mu} = \boldsymbol{\mu}_0$ is both the likelihood ratio test and the union–intersection test. There are similar results for the two-sample case.

Theorem 3.5C. If $\mathbf{y}_{11}, \mathbf{y}_{12}, \ldots, \mathbf{y}_{1n_1}$ and $\mathbf{y}_{21}, \mathbf{y}_{22}, \ldots, \mathbf{y}_{2n_2}$ are two independent random samples from $N_p(\boldsymbol{\mu}_1, \boldsymbol{\Sigma}_1)$ and $N_p(\boldsymbol{\mu}_2, \boldsymbol{\Sigma}_2)$, respectively, where $\boldsymbol{\Sigma}_1 = \boldsymbol{\Sigma}_2$, then the likelihood ratio approach leads to the T^2-statistic (3.57).

Proof. The proof is similar to that of Theorem 3.3D. It is left as an exercise (Problem 3.24) to show that

$$\text{LR} = \left[\frac{1}{1 + T^2/(n_1 + n_2 - 2)}\right]^{(n_1+n_2)/2}, \tag{3.62}$$

where T^2 is given by (3.57). Thus rejection of $H_0\colon \boldsymbol{\mu}_1 = \boldsymbol{\mu}_2$ for small values of LR is equivalent to rejection of H_0 for large values of T^2. □

In the union–intersection approach, we work with linear combinations $\mathbf{a}'\mathbf{y}$ and reject $H_0\colon \boldsymbol{\mu}_1 = \boldsymbol{\mu}_2$ if $\max_{\mathbf{a}} t^2(\mathbf{a}) \geq c$, where

$$t^2(\mathbf{a}) = \frac{\left[\mathbf{a}'(\bar{\mathbf{y}}_1 - \bar{\mathbf{y}}_2)\right]^2}{[(n_1 + n_2)/n_1 n_2]\mathbf{a}'\mathbf{S}_{\text{pl}}\mathbf{a}}. \tag{3.63}$$

The value of $\max_{\mathbf{a}} t^2(\mathbf{a})$ is given in the following theorem.

Theorem 3.5D. Under the assumptions of Theorem 3.5C,

$$\max_{\mathbf{a}} t^2(\mathbf{a}) = T^2 = \frac{n_1 n_2}{n_1 + n_2}(\bar{\mathbf{y}}_1 - \bar{\mathbf{y}}_2)'\mathbf{S}_{\text{pl}}^{-1}(\bar{\mathbf{y}}_1 - \bar{\mathbf{y}}_2). \tag{3.64}$$

Proof. The proof is similar to that of Theorem 3.3E. We differentiate $t^2(\mathbf{a})$ in (3.63) with respect to $\mathbf{a}$, set the result equal to $\mathbf{0}$, solve for $\mathbf{a}$, and substitute into $t^2(\mathbf{a})$. □

The vector $\mathbf{a}$ that maximizes $t^2(\mathbf{a})$ in (3.63) is (any multiple of)

$$\mathbf{a} = \mathbf{S}_{\text{pl}}^{-1}(\bar{\mathbf{y}}_1 - \bar{\mathbf{y}}_2), \tag{3.65}$$

which is the coefficient vector in the discriminant function $z = \mathbf{a}'\mathbf{y}$.

The discriminant function provides a multivariate view of the effect of the individual variables on T^2. Suppose the T^2-test leads to rejection of $H_0\colon \boldsymbol{\mu}_1 = \boldsymbol{\mu}_2$. Then by Theorem 3.5D, $t^2(\mathbf{a})$ is maximized by $\mathbf{a}$ in (3.65), and the discriminant function $\mathbf{a}'\mathbf{y}$ with $\mathbf{a} = \mathbf{S}_{\text{pl}}^{-1}(\bar{\mathbf{y}}_1 - \bar{\mathbf{y}}_2)$ will therefore lead to rejection of $H_0\colon \mathbf{a}'\boldsymbol{\mu}_1 = \mathbf{a}'\boldsymbol{\mu}_2$ by $t^2(\mathbf{a})$. Thus we can examine the values of a_j, $j = 1, 2, \ldots, p$, to see which of the y_j's contribute most to rejection of $H_0\colon \boldsymbol{\mu}_1 = \boldsymbol{\mu}_2$. This multivariate comparison of the two samples is typically different from the univariate comparisons given by the tests on individual variables in Section 3.6.4 below. The effect of a variable on T^2 is often altered by the presence of other variables. Note that if the variables are not commensurate in scale or variance, the coefficients a_j in (3.65) should be standardized (see Section 5.4) before using them to rank the variables in order of their contribution to T^2. The discriminant function is discussed further in Chapters 4 and 5.

3.6 CONFIDENCE INTERVALS AND TESTS FOR LINEAR FUNCTIONS OF TWO MEAN VECTORS

We now consider confidence intervals and tests for linear combinations $\mathbf{a}'(\boldsymbol{\mu}_1 - \boldsymbol{\mu}_2)$, which include $\mu_{1j} - \mu_{2j}$, $j = 1, 2, \ldots, p$, for an appropriate choice of $\mathbf{a}$. We also discuss a test for several linear combinations $\mathbf{C}(\boldsymbol{\mu}_1 - \boldsymbol{\mu}_2)$. We begin with a confidence region for $\boldsymbol{\mu}_1 - \boldsymbol{\mu}_2$.

3.6.1 Confidence Region for $\boldsymbol{\mu}_1 - \boldsymbol{\mu}_2$

If $H_0\colon \boldsymbol{\mu}_1 = \boldsymbol{\mu}_2$ is not true, the statistic T^2 in Section 3.5.1 will still have Hotelling's T^2-distribution if we subtract $\boldsymbol{\delta} = \boldsymbol{\mu}_1 - \boldsymbol{\mu}_2$. Thus

$$P\left[\frac{n_1 n_2}{n_1 + n_2}(\overline{\mathbf{y}}_1 - \overline{\mathbf{y}}_2 - \boldsymbol{\delta})'\mathbf{S}_{\text{pl}}^{-1}(\overline{\mathbf{y}}_1 - \overline{\mathbf{y}}_2 - \boldsymbol{\delta}) \le T^2_{\alpha,p,n_1+n_2-2}\right] = 1 - \alpha,$$

where values of T^2_{α,p,n_1+n_2-2} are found in Table B.1 in Appendix B. A $100(1-\alpha)\%$ confidence region for $\boldsymbol{\delta} = \boldsymbol{\mu}_1 - \boldsymbol{\mu}_2$ is therefore given by all vectors $\boldsymbol{\delta}$ that satisfy

$$(\overline{\mathbf{y}}_1 - \overline{\mathbf{y}}_2 - \boldsymbol{\delta})'\mathbf{S}_{\text{pl}}^{-1}(\overline{\mathbf{y}}_1 - \overline{\mathbf{y}}_2 - \boldsymbol{\delta}) \le \frac{n_1 + n_2}{n_1 n_2} T^2_{\alpha,p,n_1+n_2-2}. \tag{3.66}$$

The values of $\boldsymbol{\delta}$ given by (3.66) are not easy to visualize except for $p = 2$, in which case an elliptical region in two dimensions can be plotted. For higher values of p, we can substitute various values of $\boldsymbol{\delta}$ in (3.66) to gain some information, but this is equivalent to testing $H_0\colon \boldsymbol{\delta} = \boldsymbol{\delta}_0$, which can be done with an obvious modification of (3.57). We therefore work with confidence intervals for individual means or linear combinations.

3.6.2 Simultaneous Confidence Intervals for $\mathbf{a}'(\boldsymbol{\mu}_1 - \boldsymbol{\mu}_2)$ and $\mu_{1j} - \mu_{2j}$

We begin with simultaneous confidence intervals for $\mathbf{a}'(\boldsymbol{\mu}_1 - \boldsymbol{\mu}_2)$ for arbitrary choices of $\mathbf{a}$.

Theorem 3.6A. If $\overline{\mathbf{y}}_1$, $\overline{\mathbf{y}}_2$, and $\mathbf{S}_{\text{pl}}$ are obtained from two independent samples from $N_p(\boldsymbol{\mu}_1, \boldsymbol{\Sigma})$ and $N_p(\boldsymbol{\mu}_2, \boldsymbol{\Sigma})$, respectively, where both distributions have the same covariance matrix, then $100(1-\alpha)\%$ *simultaneous confidence limits* for all linear combinations $\mathbf{a}'\boldsymbol{\delta} = \mathbf{a}'(\boldsymbol{\mu}_1 - \boldsymbol{\mu}_2)$ are given by

$$\mathbf{a}'(\overline{\mathbf{y}}_1 - \overline{\mathbf{y}}_2) \pm T_{\alpha,p,n_1+n_2-2}\sqrt{\frac{n_1 + n_2}{n_1 n_2}\mathbf{a}'\mathbf{S}_{\text{pl}}\mathbf{a}}, \tag{3.67}$$

where $T_{\alpha,p,n_1+n_2-2} = \sqrt{T^2_{\alpha,p,n_1+n_2-2}}$.

Proof. Since $\max_{\mathbf{a}} t^2(\mathbf{a}) = T^2$ by Theorem 3.3E, where $t^2(\mathbf{a})$ is given by (3.63), we have

$$P\left\{\max_{\mathbf{a}} \frac{\left[\mathbf{a}'(\overline{\mathbf{y}}_1 - \overline{\mathbf{y}}_2)\right]^2}{[(n_1 + n_2)/n_1 n_2]\mathbf{a}'\mathbf{S}_{\text{pl}}\mathbf{a}} \le T^2_{\alpha,p,n_1+n_2-2}\right\} = 1 - \alpha.$$

By an argument similar to that used in the proof of Theorem 3.4A, this probability statement leads to the intervals in (3.67). □

If $\mathbf{a}' = (0, \ldots, 0, 1, 0, \ldots, 0)$ is used in (3.67), we obtain simultaneous confidence intervals for individual means, $\mu_{1j} - \mu_{2j}$, $j = 1, 2, \ldots, p$,

$$\overline{y}_{1j} - \overline{y}_{2j} \pm T_{\alpha,p,n_1+n_2-2}\sqrt{\frac{n_1+n_2}{n_1 n_2} s_{\text{pl},jj}}, \tag{3.68}$$

where $s_{\text{pl},jj}$ is the jth diagonal element of $\mathbf{S}_{\text{pl}}$. These intervals are very conservative because (3.67) allows for all possible $\mathbf{a}$, including those chosen after seeing the data. The coefficient T_{α,p,n_1+n_2-2} is much greater than $t_{\alpha/2,n_1+n_2-2}$, but use of the latter would result in intervals that are too narrow to provide an overall confidence level of $100(1-\alpha)\%$.

3.6.3 Bonferroni Confidence Intervals for $\mathbf{a}'(\boldsymbol{\mu}_1 - \boldsymbol{\mu}_2)$ and $\mu_{1j} - \mu_{2j}$

For a small number of linear combinations specified *before* the data are examined, the *Bonferroni method* provides shorter intervals than those in (3.67) while still preserving the overall confidence level. As in Section 3.4.4, we merely replace T_{α,p,n_1+n_2-2} by a Bonferroni t-coefficient based on the number of intervals desired.

For k linear combinations $\mathbf{a}_i'(\boldsymbol{\mu}_1 - \boldsymbol{\mu}_2)$, $i = 1, 2, \ldots, k$, the Bonferroni intervals are given by

$$\mathbf{a}_i'(\overline{\mathbf{y}}_1 - \overline{\mathbf{y}}_2) \pm t_{\alpha/2k,n_1+n_2-2}\sqrt{\frac{n_1+n_2}{n_1 n_2}\mathbf{a}_i'\mathbf{S}_{\text{pl}}\mathbf{a}_i}. \tag{3.69}$$

Bonferroni critical values $t_{\alpha/2k}$ can be found in Table B.2 in Appendix B.

Bonferroni intervals for the differences between individual means, $\mu_{1j} - \mu_{2j}$, $j = 1, 2, \ldots, p$, are given by

$$\overline{y}_{1j} - \overline{y}_{2j} \pm t_{\alpha/2p,n_1+n_2-2}\sqrt{\frac{n_1+n_2}{n_1 n_2} s_{\text{pl},jj}}. \tag{3.70}$$

3.6.4 Tests for $H_0\colon \mathbf{a}'(\boldsymbol{\mu}_1 - \boldsymbol{\mu}_2) = \mathbf{a}'\boldsymbol{\delta}_0$ and $H_{0j}\colon \mu_{1j} - \mu_{2j} = 0$

If we wish to test $H_0\colon \mathbf{a}'\boldsymbol{\delta} = \mathbf{a}'\boldsymbol{\delta}_0$ for any and all arbitrary values of $\mathbf{a}$, where $\boldsymbol{\delta} = \boldsymbol{\mu}_1 - \boldsymbol{\mu}_2$, we can readily construct a test statistic by again using the result that $\max_{\mathbf{a}} t^2(\mathbf{a}) = T^2$, as in the proof of Theorem 3.6A. The test statistic is therefore the ordinary t-statistic using $\mathbf{a}'\overline{\mathbf{y}}_1 - \mathbf{a}'\overline{\mathbf{y}}_2$:

$$t(\mathbf{a}) = \frac{\mathbf{a}'(\overline{\mathbf{y}}_1 - \overline{\mathbf{y}}_2 - \boldsymbol{\delta}_0)}{\sqrt{[(n_1+n_2)/n_1 n_2]\mathbf{a}'\mathbf{S}_{\text{pl}}\mathbf{a}}}, \tag{3.71}$$

and we reject $H_0\colon \mathbf{a}'\boldsymbol{\delta} = \mathbf{a}'\boldsymbol{\delta}_0$ if $|t(\mathbf{a})| \geq T_{\alpha,p,n_1+n_2-2}$. Any number of these tests can be made without exceeding α, the overall (experimentwise) error rate. The advantage of (3.71) is that we may test hypotheses that are suggested by the data. We pay for this privilege by using the critical value T_α, which is much greater than $t_{\alpha/2}$.

If we wish to make k prespecified tests H_{0i}: $\mathbf{a}_i'\boldsymbol{\delta} = \mathbf{a}_i'\boldsymbol{\delta}_0$, $i = 1, 2, \ldots, k$, the Bonferroni critical value $t_{\alpha/2k,n_1+n_2-2}$ can be substituted for T_{α,p,n_1+n_2-2} so that we reject H_{0i} if $|t(\mathbf{a}_i)| \geq t_{\alpha/2k,n_1+n_2-2}$. If k is not too large, the Bonferroni tests will be less conservative and thereby more powerful than those using T_{α,p,n_1+n_2-2}.

If we use $\mathbf{a}' = (0, \ldots, 0, 1, 0, \ldots, 0)$ and $\boldsymbol{\delta}_0 = \mathbf{0}$ in (3.71), we obtain tests that compare the means for each variable; that is, to test H_{0j}: $\mu_{1j} - \mu_{2j} = 0$, $j = 1, 2, \ldots, p$, we use

$$t_j = \frac{\overline{y}_{1j} - \overline{y}_{2j}}{\sqrt{[(n_1 + n_2)/n_1 n_2] s_{jj}}} \tag{3.72}$$

and reject H_{0j} if $|t_j| \geq T_{\alpha,p,n_1+n_2-2}$ or if $|t_j| \geq t_{\alpha/2p,n_1+n_2-2}$, where s_{jj} is the jth diagonal element of $\mathbf{S}_{\text{pl}}$.

In (3.72), we test each variable (in every experiment) without regard to the result of any multivariate test; that is, we did not consider the overall T^2-test. If the p tests on individual variables in (3.72) are carried out only on the condition of rejection by the T^2-test of H_0: $\boldsymbol{\mu}_1 = \boldsymbol{\mu}_2$, then when H_0 is true, we will reject H_{0j} less often in the individual tests than if we ignore (or do not perform) the T^2-test. Hence, if used only after a rejection of $H_0 : \boldsymbol{\mu}_1 = \boldsymbol{\mu}_2$ by T^2, the t_j-tests in (3.72) using T_α or $t_{\alpha/2p}$ become even more conservative. Hummel and Sligo (1971) showed that if each t_j-test in (3.72) is carried out with a critical value $t_{\alpha/2}$ only after rejection of H_0: $\boldsymbol{\mu}_1 = \boldsymbol{\mu}_2$ by the overall T^2-test, then the overall (experimentwise) α-level for all p tests is close to the nominal α-value. Thus when these ordinary t-tests using $t_{\alpha/2}$ are "protected" by a multivariate T^2-test, the α-level is maintained (not inflated). In this case, the conservative critical values T_α or $t_{\alpha/2p}$ associated with (3.72) are not needed.

3.6.5 Test for H_0: $\mathbf{C}(\boldsymbol{\mu}_1 - \boldsymbol{\mu}_2) = \mathbf{0}$

Suppose we wish to compare certain contrasts among the components of $\boldsymbol{\mu}_1$ with the same contrasts in $\boldsymbol{\mu}_2$. This comparison can be expressed as H_0: $\mathbf{C}\boldsymbol{\mu}_1 = \mathbf{C}\boldsymbol{\mu}_2$. Let $\mathbf{y}_1$ be an observation vector from $N_p(\boldsymbol{\mu}_1, \boldsymbol{\Sigma})$ and $\mathbf{y}_2$ be an observation vector from $N_p(\boldsymbol{\mu}_2, \boldsymbol{\Sigma})$, where both groups have the same covariance matrix $\boldsymbol{\Sigma}$. If $\mathbf{C}$ is a $k \times p$ full-rank contrast matrix as in Section 3.4.6, then $\mathbf{C}\mathbf{y}_1$ and $\mathbf{C}\mathbf{y}_2$ are distributed as $N_k(\mathbf{C}\boldsymbol{\mu}_1, \mathbf{C}\boldsymbol{\Sigma}\mathbf{C}')$ and $N_k(\mathbf{C}\boldsymbol{\mu}_2, \mathbf{C}\boldsymbol{\Sigma}\mathbf{C}')$, respectively. Using $\overline{\mathbf{y}}_1, \overline{\mathbf{y}}_2$, and $\mathbf{S}_{\text{pl}}$ as estimators of $\boldsymbol{\mu}_1$, $\boldsymbol{\mu}_2$, and $\boldsymbol{\Sigma}$, the test statistic for H_0: $\mathbf{C}\boldsymbol{\mu}_1 = \mathbf{C}\boldsymbol{\mu}_2$ is

$$\begin{aligned} T^2 &= (\mathbf{C}\overline{\mathbf{y}}_1 - \mathbf{C}\overline{\mathbf{y}}_2)' \left[\left(\frac{1}{n_1} + \frac{1}{n_2} \right) \mathbf{C}\mathbf{S}_{\text{pl}}\mathbf{C}' \right]^{-1} (\mathbf{C}\overline{\mathbf{y}}_1 - \mathbf{C}\overline{\mathbf{y}}_2) \\ &= \frac{n_1 n_2}{n_1 + n_2} (\overline{\mathbf{y}}_1 - \overline{\mathbf{y}}_2)' \mathbf{C}' \left[\mathbf{C}\mathbf{S}_{\text{pl}}\mathbf{C}' \right]^{-1} \mathbf{C}(\overline{\mathbf{y}}_1 - \overline{\mathbf{y}}_2), \end{aligned} \tag{3.73}$$

which is distributed as T^2_{k,n_1+n_2-2} under H_0. Note that the dimension k is the number of rows of $\mathbf{C}$.

3.7 ROBUSTNESS OF THE T^2-TEST

The univariate two-sample t-test is remarkably robust to violation of the assumptions of equal variances and normality under the following conditions: (1) the ratio of the two sample variances is no greater than 20 to 1, (2) the ratio of sample sizes n_1 to n_2 is no greater than 4 to 1, and (3) the degrees of freedom for t is at least 10 (Harris 1985, p. 332).

The multivariate T^2-test is likewise somewhat robust to departures from the assumptions of $\mathbf{\Sigma}_1 = \mathbf{\Sigma}_2$ and multivariate normality. However, the robustness is not as universal as that of the t-test.

3.7.1 Robustness to $\mathbf{\Sigma}_1 \neq \mathbf{\Sigma}_2$

Ito and Schull (1964) claimed that the effect of $\mathbf{\Sigma}_1 \neq \mathbf{\Sigma}_2$ on the significance level and power of the two-sample T^2-test in (3.57) is slight if n_1 and n_2 are large and approximately equal and if $\mathbf{\Sigma}_1$ and $\mathbf{\Sigma}_2$ are not too disparate. When n_1 and n_2 differ widely, $\mathbf{\Sigma}_1 \neq \mathbf{\Sigma}_2$ has a sizable effect on both α and the power of the T^2-test.

Robustness to $\mathbf{\Sigma}_1 \neq \mathbf{\Sigma}_2$ was also investigated by Hakstian, Roed, and Lind (1979), Holloway and Dunn (1967), and Hopkins and Clay (1963). If we denote the actual Type I error rate as τ and the relationship between $\mathbf{\Sigma}_1$ and $\mathbf{\Sigma}_2$ as $\mathbf{\Sigma}_2 = d^2\mathbf{\Sigma}_1$, where $d > 1$, then all three of these studies were in agreement in finding $\tau > \alpha$ when $n_1 > n_2$ and $\tau < \alpha$ when $n_1 < n_2$, where α is the nominal level used in the critical value T^2_α. Based on these results, Hakstian et al. (1979) recommended that the T^2-test not be used in the case $d > 1$ and $n_1 > n_2$ (assuming $\mathbf{\Sigma}_2 = d^2\mathbf{\Sigma}_1$, with $d^2 > 1$).

In the above studies, the ratio of sample sizes (n_1/n_2 or n_2/n_1) ranged from 1.5 to 5. Algina and Oshima (1990) investigated the case of sample sizes ratios smaller than 1.5. Their results were consistent with previous studies: $\tau > \alpha$ when $n_1 > n_2$ and $\tau < \alpha$ when $n_1 < n_2$ (assuming $\mathbf{\Sigma}_2 = d^2\mathbf{\Sigma}_1$ with $d^2 > 1$). Sample sizes varied from 6 to 111 and the ratio of sample sizes varied from 1.1 to 1.25. In the case of $n_1 > n_2$ and $d^2 > 1$, the estimated value of τ for nominal $\alpha = .05$ varied from .045 to .202, with about half less than .1. In their simulation, Algina and Oshima used $\mathbf{\Sigma}_1 = \mathbf{\Sigma}_2 = \mathbf{I}$. It is unclear whether their results will hold for more general patterns in $\mathbf{\Sigma}_i$ and in the relationship $\mathbf{\Sigma}_2 = d^2\mathbf{\Sigma}_1$.

Christensen and Rencher (1997) studied the effect of $\mathbf{\Sigma}_1 \neq \mathbf{\Sigma}_2$ on T^2 when $n_1 = n_2$ as well as when $n_1 \neq n_2$. For $n_1 = n_2$, τ varied from .058 to .117. For $n_1 \neq n_2$, τ ranged from .01 to .44. Therefore, the use of T^2 is not recommended when $\mathbf{\Sigma}_1 \neq \mathbf{\Sigma}_2$. See Section 3.9.2 for solutions in this case.

3.7.2 Robustness to Nonnormality

Everitt (1979) used Monte Carlo methods to investigate the robustness to nonnormality of both the one-sample and the two-sample T^2-tests when $\mathbf{\Sigma}_1 = \mathbf{\Sigma}_2$. He compared the multivariate normal with the uniform, exponential, and lognormal. For the latter three distributions, he generated p variables independently, assuming $\mathbf{\Sigma} = \mathbf{I}$. This

Table 3.4. Percentage of One-Sample T^2-Values Exceeding 5% and 1% Points of the T^2-Distribution

			By Distributions Sampled							
			Multinormal		Uniform		Exponential		Lognormal	
n	p	N	5%	1%	5%	1%	5%	1%	5%	1%
5	2	5000	5.0	0.9	6.1	1.7	13.2	3.8	22.2	8.6
10	2	5000	5.2	1.1	5.5	1.6	13.1	5.9	22.8	12.2
	4	5000	5.4	1.2	5.2	1.6	14.0	5.1	27.7	13.7
	6	2000	4.7	0.9	5.9	1.3	13.9	4.3	26.5	9.6
15	2	2000	5.4	0.7	7.3	1.7	11.6	5.3	21.1	12.4
	4	2000	4.5	1.0	4.8	1.2	14.0	5.7	27.4	14.7
	6	2000	5.5	1.4	5.0	1.3	15.3	5.5	31.6	16.5
	8	2000	4.8	0.6	4.8	0.5	14.7	4.5	30.4	13.0
20	2	2000	4.3	1.0	5.4	1.1	10.7	4.9	16.7	8.9
	4	2000	4.7	0.9	5.3	1.0	13.3	5.4	26.1	14.9
	6	2000	4.0	0.9	5.0	0.9	14.3	5.2	30.9	15.9
	8	2000	4.8	0.9	7.1	1.8	13.1	4.2	32.0	16.5
	10	2000	4.8	0.4	6.1	0.9	14.5	4.2	33.0	16.3

Note: n = T^2-statistic sample size; p = number of variables; N = Monte Carlo sample size.

limitation on the covariance matrix may restrict the complete applicability of his results. The covariance matrix used for the multivariate normal was not given.

Everitt's results for the one-sample case are given in Table 3.4. We see that for the multivariate normal and uniform, the significance levels were acceptably close to the nominal 5% and 1% values. The samples from the skewed distributions (exponential and lognormal), on the other hand, produced unacceptably large increases in α-values. In most cases, the problem worsens as p increases.

In contrast to the one-sample case, the two-sample T^2 behaved very well, staying comfortably close to the nominal α-values for all four distributions, as seen in Table 3.5. We can see little effect due to changes in p or the relative sizes of the two samples, n_1 and n_2. The only effect of the skewed distributions is to decrease α slightly. This slight conservative tendency, if it holds up under more general conditions than those considered by Everitt, should make us less hesitant to use the two-sample T^2-test when we are unsure of normality.

Nachtsheim and Johnson (1988) noted that skewness and kurtosis were confounded in Everitt's (1979) study. They presented new distributions that depart from multivariate normality and, for $p = 2, 3$, performed simulations showing that Hotelling's T^2 is reasonably robust to general departures from multivariate normality.

3.8 PAIRED OBSERVATION TEST

In this section, we depart from the notation used in previous sections and denote the two samples as $\mathbf{y}_1, \mathbf{y}_2, \ldots, \mathbf{y}_n$ and $\mathbf{x}_1, \mathbf{x}_2, \ldots, \mathbf{x}_n$ to indicate that the samples are dependent because of a natural pairing between corresponding observation vectors

Table 3.5. Percentage of Two-Sample T^2-Values Exceeding 5% and 1% Points of the T^2-Distribution

				By Distributions Sampled							
				Multinormal		Uniform		Exponential		Lognormal	
p	N	n_1	n_2	5%	1%	5%	1%	5%	1%	5%	1%
2	5000	5	5	5.1	1.1	5.7	1.2	3.6	0.6	3.0	0.8
			10	4.8	1.0	5.2	1.3	4.0	1.2	3.8	0.7
			15	5.0	1.3	4.8	0.7	4.8	1.4	4.7	1.3
			20	5.2	0.9	4.8	1.3	5.6	1.5	5.2	1.6
		10	10	5.4	1.2	5.4	1.2	4.3	0.8	2.9	0.4
			15	4.8	1.0	5.2	1.1	4.5	0.7	3.2	0.5
			20	4.4	1.0	5.3	1.5	4.3	0.7	4.3	0.8
		15	15	5.2	0.9	5.0	1.2	4.4	0.8	3.4	0.4
			20	4.3	0.7	5.0	1.0	4.4	0.9	3.5	0.4
		20	20	5.3	1.0	5.0	1.0	4.2	0.8	3.8	0.3
4	5000	10	10	5.1	0.9	4.8	0.9	4.7	0.8	3.0	0.4
			15	4.9	1.0	5.0	1.1	4.5	0.9	3.4	0.4
			20	4.3	1.0	5.1	1.2	4.7	0.8	3.5	0.5
		15	15	5.1	1.0	4.9	1.1	4.6	0.7	3.0	0.4
			20	5.2	1.1	5.2	1.2	4.3	0.9	3.6	0.4
		20	20	4.6	1.1	5.2	1.0	4.0	0.6	3.2	0.5
6	2500	15	15	5.1	1.2	5.7	1.2	3.0	0.3	3.2	0.5
			20	4.1	0.8	4.6	1.0	3.8	0.4	3.7	0.6
		20	20	5.4	1.4	5.2	1.2	4.6	0.5	3.2	0.5
8	1000	20	20	5.5	1.1	5.3	1.5	4.4	0.8	3.5	0.2

$\mathbf{y}_i$ and $\mathbf{x}_i$. Such pairing may arise, for example, when a treatment is applied twice to the same individual or when subjects are matched according to some criterion such as IQ or family background. In such cases, the two samples are correlated, that is, $\text{cov}(\mathbf{y}_i, \mathbf{x}_i) \neq \mathbf{O}$, and we can work directly with the differences between the paired observations, $\mathbf{d}_i = \mathbf{y}_i - \mathbf{x}_i$.

The n pairs of observation vectors $\mathbf{y}_i$, $\mathbf{x}_i$ are reduced to a single sample of n differences, $\mathbf{d}_i$, $i = 1, 2, \ldots, n$. The hypothesis can be expressed as $H_0\colon \boldsymbol{\mu}_d = \mathbf{0}$, which is equivalent to $H_0\colon \boldsymbol{\mu}_y = \boldsymbol{\mu}_x$. To test H_0, we calculate

$$\bar{\mathbf{d}} = \frac{1}{n}\sum_{i=1}^{n} \mathbf{d}_i, \qquad \mathbf{S}_d = \frac{1}{n-1}\sum_{i=1}^{n}(\mathbf{d}_i - \bar{\mathbf{d}})(\mathbf{d}_i - \bar{\mathbf{d}})',$$

from which we obtain

$$T^2 = n\bar{\mathbf{d}}'\mathbf{S}_d^{-1}\bar{\mathbf{d}}. \tag{3.74}$$

If H_0 is true, T^2 in (3.74) is distributed as $T^2_{p,n-1}$. We reject H_0 if $T^2 \geq T^2_{\alpha,p,n-1}$.

Note that the assumption $\boldsymbol{\Sigma}_{yy} = \boldsymbol{\Sigma}_{xx}$ is not needed because $\mathbf{S}_d$ estimates $\text{cov}(\mathbf{y}_i - \mathbf{x}_i) = \boldsymbol{\Sigma}_{yy} - \boldsymbol{\Sigma}_{yx} - \boldsymbol{\Sigma}_{xy} + \boldsymbol{\Sigma}_{xx}$.

By analogy with the one-sample tests and confidence intervals in Sections 3.3 and 3.4, we have the discriminant function coefficient vector

$$\mathbf{a} = \mathbf{S}_d^{-1}\bar{\mathbf{d}}, \tag{3.75}$$

simultaneous confidence intervals

$$\bar{d}_j \pm T_{\alpha,p,n-1}\sqrt{\frac{s_{d,jj}}{n}}, \qquad j = 1, 2, \ldots, p, \tag{3.76}$$

Bonferroni intervals

$$\bar{d}_j \pm t_{\alpha/2p,n-1}\sqrt{\frac{s_{d,jj}}{n}}, \qquad j = 1, 2, \ldots, p, \tag{3.77}$$

and tests on individual variables

$$t_j = \frac{\bar{d}_j}{\sqrt{\frac{s_{d,jj}}{n}}}, \tag{3.78}$$

where we reject H_{0j}: $\mu_{d_j} = 0$, $j = 1, 2, \ldots, p$, if $|t_j| \geq T_{\alpha,p,n-1}$ or if $|t_j| \geq t_{\alpha/2p,n-1}$.

3.9 TESTING H_0: $\boldsymbol{\mu}_1 = \boldsymbol{\mu}_2$ WHEN $\boldsymbol{\Sigma}_1 \neq \boldsymbol{\Sigma}_2$

In the univariate case ($p = 1$), the problem of testing H_0: $\mu_1 = \mu_2$ when $\sigma_1^2 \neq \sigma_2^2$ (assuming two independent samples) is known as the Behrens–Fisher problem (Behrens 1929, Fisher 1939). The corresponding multivariate case is the problem of testing H_0: $\boldsymbol{\mu}_1 = \boldsymbol{\mu}_2$ when $\boldsymbol{\Sigma}_1 \neq \boldsymbol{\Sigma}_2$. We first review the univariate case and then consider the analogous multivariate situation.

3.9.1 Univariate Case

When $\sigma_1^2 \neq \sigma_2^2$, the usual t-statistic (3.54) does not have a t-distribution. Several approximate procedures for testing H_0: $\mu_1 = \mu_2$ have been proposed. The following solution is due to Welch (1937, 1947). If $\sigma_1^2 \neq \sigma_2^2$, then $\text{var}(\bar{y}_1 - \bar{y}_2) = \sigma_1^2/n_1 + \sigma_2^2/n_2$ (assuming two independent samples), which can be estimated by $s_1^2/n_1 + s_2^2/n_2$. When this is used,

$$t_\nu = \frac{\bar{y}_1 - \bar{y}_2}{\sqrt{s_1^2/n_1 + s_2^2/n_2}} \tag{3.79}$$

has an approximate t-distribution with degrees of freedom

$$\nu = \frac{s_1^2/n_1 + s_2^2/n_2}{(s_1^2/n_1)^2/(n_1+1) + (s_2^2/n_2)^2/(n_2+1)} - 2.$$

We reject H_0 if $|t_\nu| \geq t_{\alpha/2,\nu}$.

Nel, van der Merwe, and Moser (1990) derived the exact distribution of t_ν in (3.79) and compared it to Welch's approximation for two values of the ratio σ_1^2/σ_2^2. The differences were negligible, ranging from .000001 to .000054.

3.9.2 Multivariate Case

We assume two independent samples $\mathbf{y}_{11}, \mathbf{y}_{12}, \ldots, \mathbf{y}_{1n_1}$ and $\mathbf{y}_{21}, \mathbf{y}_{22}, \ldots, \mathbf{y}_{2n_2}$ from $N_p(\boldsymbol{\mu}_1, \boldsymbol{\Sigma}_1)$ and $N_p(\boldsymbol{\mu}_2, \boldsymbol{\Sigma}_2)$, respectively, where $\boldsymbol{\Sigma}_1 \neq \boldsymbol{\Sigma}_2$. When $\boldsymbol{\Sigma}_1 \neq \boldsymbol{\Sigma}_2$, the T^2-statistic in (3.57) does not have a T^2-distribution. We consider two cases: equal sample sizes and unequal sample sizes.

3.9.2a Equal Sample Sizes, $n_1 = n_2$

If $n_1 = n_2 = n$, say, we can use the paired observation test given in Section 3.8 since $\boldsymbol{\Sigma}_{yy} = \boldsymbol{\Sigma}_{xx}$ is not assumed in (3.74). We randomly pair each observation vector in the first sample with an observation vector in the second sample and then proceed as in Section 3.8. The procedure yields an exact T^2-test but has the disadvantage of having only $n - 1$ degrees of freedom instead of $2(n - 1)$. This loss of degrees of freedom makes the test somewhat less powerful, and a researcher may wish to use one of the approximations given in Section 3.9.2b. (These approximations can be used when $n_1 = n_2$.)

3.9.2b Unequal Sample Sizes, $n_1 \neq n_2$

The first solution to the multivariate Behrens–Fisher problem was given by Bennett (1951). Bennett's approach leads to an exact T^2-test but discards $n_2 - n_1$ (assuming $n_2 > n_1$) values of $\mathbf{y}_{2i}$ in making the computation. This procedure has two disadvantages: (1) there is a loss of power if n_1 is much less than n_2 and (2) different researchers will obtain different test results, depending on which values of $\mathbf{y}_{2i}$ are discarded. Because of these drawbacks, we do not recommend Bennett's procedure and will not provide computational details.

If $\boldsymbol{\Sigma}_1$ and $\boldsymbol{\Sigma}_2$ were known,

$$Z^2 = (\overline{\mathbf{y}}_1 - \overline{\mathbf{y}}_2)' \left(\frac{\boldsymbol{\Sigma}_1}{n_1} + \frac{\boldsymbol{\Sigma}_2}{n_2} \right)^{-1} (\overline{\mathbf{y}}_1 - \overline{\mathbf{y}}_2) \tag{3.80}$$

would have a χ_p^2-distribution under H_0. The sample version of (3.80),

$$T^{*2} = (\overline{\mathbf{y}}_1 - \overline{\mathbf{y}}_2)' \left(\frac{\mathbf{S}_1}{n_1} + \frac{\mathbf{S}_2}{n_2} \right)^{-1} (\overline{\mathbf{y}}_1 - \overline{\mathbf{y}}_2), \tag{3.81}$$

does not have Hotelling's T^2-distribution (or a χ^2-distribution). James (1954) obtained a correction to the approximate chi-square critical values for (3.81). Yao (1965) generalized Welch's (1937, 1947) univariate solution (see Section 3.9.1). Additional approximate solutions to the multivariate Behrens–Fisher problem have been given by Johansen (1980), Nel and van der Merwe (1986), and Kim (1992). The procedures of James, Yao, Johansen, and Nel and van der Merwe all use T^{*2} in (3.81). We use the following notation in the discussion of these four solutions:

$$\mathbf{V}_i = \frac{\mathbf{S}_i}{n_i}, \qquad i = 1, 2,$$

$$\mathbf{S}_e = \mathbf{V}_1 + \mathbf{V}_2,$$

$$\mathbf{V} = \mathbf{V}_1^{-1} + \mathbf{V}_2^{-1}.$$

James (1954) used the product $(A + B\chi^2_{\alpha,p})(\chi^2_{\alpha,p})$ as an approximate critical value for T^{*2} in (3.81), where

$$A = 1 + \frac{1}{2p} \sum_{i=1}^{2} \frac{1}{n_i - 1} \left[\mathrm{tr}(\mathbf{S}_e^{-1}\mathbf{V}_i)\right]^2,$$

$$B = \frac{1}{2p(p+2)} \sum_{i=1}^{2} \frac{1}{n_i - 1} \left\{\mathrm{tr}\left[2(\mathbf{S}_e^{-1}\mathbf{V}_i)^2\right] + \left[\mathrm{tr}(\mathbf{S}_e^{-1}\mathbf{V}_i)\right]^2\right\},$$

and $\chi^2_{\alpha,p}$ is the upper α percentile of the chi-square distribution with p degrees of freedom.

Yao (1965) approximated the distribution of T^{*2} in (3.81) by that of $T^2_{p,\nu}$, where

$$\frac{1}{\nu} = \frac{1}{(T^{*2})^2} \sum_{i=1}^{2} \frac{1}{n_i - 1} \left[(\overline{\mathbf{y}}_1 - \overline{\mathbf{y}}_2)'\mathbf{S}_e{}^{-1}\mathbf{V}_i\mathbf{S}_e{}^{-1}(\overline{\mathbf{y}}_1 - \overline{\mathbf{y}}_2)\right]^2. \tag{3.82}$$

The degrees of freedom ν obtained from (3.82) ranges between $\min(n_1 - 1, n_2 - 1)$ and $n_1 + n_2 - 2$ (Subrahmaniam and Subrahmaniam 1973). The test therefore tends to be conservative if $\boldsymbol{\Sigma}_1 = \boldsymbol{\Sigma}_2$. The discriminant function coefficient vector for Yao's approximate test is

$$\mathbf{a} = \mathbf{S}_e^{-1}(\overline{\mathbf{y}}_1 - \overline{\mathbf{y}}_2). \tag{3.83}$$

Bonferroni intervals for $\mu_{1j} - \mu_{2j}$, $j = 1, 2, \ldots, p$, are given by

$$\overline{y}_{1j} - \overline{y}_{2j} \pm t_{\alpha/2p,\nu_j} \sqrt{\frac{s^2_{1,jj}}{n_1} + \frac{s^2_{2,jj}}{n_2}}, \tag{3.84}$$

where $s^2_{1,jj}$ and $s^2_{2,jj}$ are the jth diagonal elements of $\mathbf{S}_1$ and $\mathbf{S}_2$, respectively, and

$$\frac{1}{\nu_j} = \frac{c_j^2}{n_1 - 1} + \frac{(1 - c_j)^2}{n_2 - 1}, \qquad c_j = \frac{s^2_{1,jj}/n_1}{s^2_{1,jj}/n_1 + s^2_{2,jj}/n_2}.$$

Johansen's (1980) test statistic is $\nu p T^{*2}/(\nu - p + 1)C$, which is approximately distributed as $T^2_{p,\nu}$, where T^{*2} is given in (3.81), and

$$\nu = \frac{p(p+2)}{3D}, \qquad C = p + 2D - \frac{6D}{p(p-1)+2},$$

$$D = \sum_{i=1}^{2} \frac{1}{2(n_i - 1)} \left\{ \mathrm{tr}(\mathbf{I} - \mathbf{V}^{-1}\mathbf{V}_i^{-1})^2 + \left[\mathrm{tr}(\mathbf{I} - \mathbf{V}^{-1}\mathbf{V}_i^{-1})\right]^2 \right\}$$

(see Tang and Algina 1993).

The approximate test by Nel and van der Merwe (1986) uses T^{*2} in (3.81), which is approximately distributed as $T^2_{p,\nu}$, with

$$\nu = \frac{\mathrm{tr}(\mathbf{S}_e)^2 + [\mathrm{tr}(\mathbf{S}_e)]^2}{(n_1 - 1)^{-1}\left\{\mathrm{tr}\left(\mathbf{V}_1^2\right) + [\mathrm{tr}(\mathbf{V}_1)]^2\right\} + (n_2 - 1)^{-1}\left\{\mathrm{tr}\left(\mathbf{V}_2^2\right) + [\mathrm{tr}\,(\mathbf{V}_2)]^2\right\}}.$$

For Kim's (1992) approximate test, we use the notation $\mathbf{A}^{-2} = (\mathbf{A}^{-1})^2$, $\mathbf{D} = \mathrm{diag}(d_1, d_2, \ldots, d_p)$, $\mathbf{Q} = (\mathbf{q}_1, \mathbf{q}_2, \ldots, \mathbf{q}_p)$, $\mathbf{w} = \mathbf{Q}'(\overline{\mathbf{y}}_1 - \overline{\mathbf{y}}_2)$, and $r = (\Pi_{j=1}^p d_j)^{1/2p}$, where the d_j's and $\mathbf{q}_j$'s are eigenvalues and eigenvectors of $\mathbf{V}_2^{-1}\mathbf{V}_1$. Then,

$$\frac{\nu - p + 1}{bc\nu}\mathbf{w}'(\mathbf{D}^{1/2} + r\mathbf{I})^{-2}\mathbf{w} \tag{3.85}$$

is approximately distributed as $F_{c,\nu-p+1}$, where $b = (\sum_{j=1}^p a_j^2)/(\sum_{j=1}^p a_j)$, $c = (\sum_j a_j)^2/(\sum_j a_j^2)$, $a_j = (d_j + 1)/(d_j^{1/2} + r)^2$, and

$$\frac{1}{\nu} = \frac{1}{n_1 - 1}\left[\frac{\mathbf{w}'\mathbf{D}(\mathbf{D}+\mathbf{I})^{-2}\mathbf{w}}{\mathbf{w}'(\mathbf{D}+\mathbf{I})^{-1}\mathbf{w}}\right]^2 + \frac{1}{n_2 - 1}\left[\frac{\mathbf{w}'(\mathbf{D}+\mathbf{I})^{-2}\mathbf{w}}{\mathbf{w}'(\mathbf{D}+\mathbf{I})^{-1}\mathbf{w}}\right]^2.$$

Kim (1992) noted that this expression for ν yields the same value as that given by (3.82).

Simulation studies by Algina and Tang (1988) and Yao (1965) have indicated that the Type I error rate τ for Yao's test is closer to the true α than is τ for James's test. Algina, Oshima, and Tang (1991) compared James's, Yao's, and Johansen's tests for robustness to heteroscedasticity and nonnormality by a simulation study. Subject to the limitations of their study, they concluded that the Type I error rate τ is close to α for symmetric distributions but that τ is too large for certain patterns of asymmetry. Under most conditions, James's test tended to perform more poorly than the other two procedures, neither of which had a clear advantage over the other.

Nel et al. (1990) obtained the exact distribution of T^{*2} in (3.81), but because of its computational intractability, they did not compare it to the various approximations. Hwang and Paulson (1986) suggested two additional solutions to the multivariate Behrens–Fisher problem, one based on the characteristic function and the other based on an "F-ratio-like-statistic." They provided limited tables of critical values but did not compare their procedures with other techniques.

Christensen and Rencher (1997) compared all the above procedures, as well as the T^2-statistic in (3.57), for Type I error rates and power when $\boldsymbol{\Sigma}_1 \neq \boldsymbol{\Sigma}_2$. Kim's (1992) method had the highest power among those with conservative Type I error rate, with Nel and van der Merwe's (1986) method only slightly less powerful. The Type I error rates for Hotelling's T^2 were inflated, even when the sample sizes were equal.

Example 3.9.2. Cox and Martin (1937) compared Iowa soils containing *Azotobacter* with soils not containing this organism. The variables measured were $y_1 =$ pH, $y_2 =$ amount of readily available phosphate, and $y_3 =$ total nitrogen content. An excerpt of the data is given in Table 3.6.

The mean vectors and covariance matrices are

$$\bar{\mathbf{y}}_1 = \begin{pmatrix} 7.81 \\ 108.77 \\ 44.92 \end{pmatrix}, \qquad \bar{\mathbf{y}}_2 = \begin{pmatrix} 5.89 \\ 41.90 \\ 20.80 \end{pmatrix},$$

$$\mathbf{S}_1 = \begin{pmatrix} .461 & 1.18 & 4.49 \\ 1.18 & 3776.4 & -17.35 \\ 4.49 & -17.35 & 147.24 \end{pmatrix}, \qquad \mathbf{S}_2 = \begin{pmatrix} .148 & -.679 & .209 \\ -.679 & 96.10 & 20.20 \\ .209 & 20.20 & 24.18 \end{pmatrix}.$$

An assumption of equal population covariance matrices seems unjustified in this case, and we apply the procedures of James, Yao, Johansen, Nel and van der Merwe, and Kim. From (3.81), we obtain

$$T^{*2} = (\bar{\mathbf{y}}_1 - \bar{\mathbf{y}}_2)' \left(\frac{\mathbf{S}_1}{13} + \frac{\mathbf{S}_2}{10} \right)^{-1} (\bar{\mathbf{y}}_1 - \bar{\mathbf{y}}_2) = 96.818.$$

Table 3.6. Comparison of Soils

With *Azotobacter*			Without *Azotobacter*		
y_1	y_2	y_3	y_1	y_2	y_3
8.0	60	58	6.2	49	30
8.0	156	68	5.6	31	23
8.0	90	37	5.8	42	22
6.1	44	27	5.7	42	14
7.4	207	31	6.2	40	23
7.4	120	32	6.4	49	18
8.4	65	43	5.8	31	17
8.1	237	45	6.4	31	19
8.3	57	60	5.4	62	26
7.0	94	43	5.4	42	16
8.5	86	40			
8.4	52	48			
7.9	146	52			

For James's approximation, we find $A = 1.08723$ and $B = .030014$. The approximate critical value is $\left(1.087 + .03\chi^2_{.05,3}\right)(\chi^2_{.05,3}) = 10.329$. Since $96.818 > 10.329$, we reject H_0: $\boldsymbol{\mu}_1 = \boldsymbol{\mu}_2$.

Yao's approximation uses (3.82) and rejects H_0 if $T^{*2} > T^2_{\alpha,p,\nu}$, where

$$\frac{1}{\nu} = \frac{1}{(96.818)^2}(515.8946 + 36.5483) = .0589355,$$

from which $\nu = 16.97$. We reject H_0: $\boldsymbol{\mu}_1 = \boldsymbol{\mu}_2$ because $T^{*2} = 96.818 > T^2_{.05,3,17} = 11.177$.

For Johansen's test, we first obtain $D = .35595$, $\nu = 14.0469$, and $C = 3.4449$. The test statistic is

$$\frac{\nu p T^{*2}}{(\nu - p + 1)C} = 98.311,$$

which exceeds $T^2_{.05,3,14} = 12.216$, and we reject H_0.

For the approximation by Nel and van der Merwe, we calculate $\nu = 12.874$ and reject H_0, since $T^{*2} = 96.818 > T^2_{.05,3,13} = 12.719$.

For Kim's approximate F-test, we have $\nu = 16.97$ (the same value of ν as obtained for Yao's procedure), and by (3.85), $F = 26.958$, with a p-value of 3.08×10^{-6}.

3.10 POWER AND SAMPLE SIZE

We define the *power* of a statistical test as the probability of rejecting H_0 when H_0 is false. For the tests we have considered, the power increases with an increase in any of the following:

1. the value of α,
2. the sample size or sizes,
3. the extent of departure of the true parameter values from the hypothesized parameter values.

The difference in number 3 is measured by the *noncentrality parameter*, an index of how far the actual distribution differs from the hypothesized one. For the T^2-tests, the noncentrality parameter is obtained from the test statistic by replacing sample estimates by the corresponding population parameters. Thus for the one-sample case, the noncentrality parameter is

$$\lambda = n\Delta^2 = n(\boldsymbol{\mu} - \boldsymbol{\mu}_0)'\boldsymbol{\Sigma}^{-1}(\boldsymbol{\mu} - \boldsymbol{\mu}_0), \tag{3.86}$$

and for the two-sample test it is given by

$$\lambda = \frac{n_1 n_2}{n_1 + n_2}\Delta^2 = \frac{n_1 n_2}{n_1 + n_2}(\boldsymbol{\mu}_1 - \boldsymbol{\mu}_2)'\boldsymbol{\Sigma}^{-1}(\boldsymbol{\mu}_1 - \boldsymbol{\mu}_2). \tag{3.87}$$

In (3.87), $\boldsymbol{\Sigma}$ is the common covariance matrix of the two populations.

Tiku (1967) has provided tables of $\beta = 1 - power$ for the F-test. Since T^2 is so readily converted to F by (3.17), we can use Tiku's tables to find the power for T^2-tests. (The noncentrality parameter for the F is the same as that of the T^2 since both are related to the noncentrality parameter of the chi-square.) Table B.3 (in Appendix B) presents the portion of Tiku's tables that corresponds to $\alpha = .05$. To use the tables, we calculate the noncentrality parameter according to (3.86) or (3.87) and use the degrees of freedom of the F-transformation (Theorem 3.3B),

$$\nu_1 = p$$

$$\nu_2 = \begin{cases} n - p & \text{for the one-sample test,} \\ n_1 + n_2 - p - 1 & \text{for the two-sample test.} \end{cases}$$

To avoid the complexity of a four-way table, λ and ν_1 are combined in a "standardized" form,

$$\phi = \sqrt{\frac{\lambda}{\nu_1 + 1}}. \tag{3.88}$$

Note that Table B.3 gives values of $\beta = 1 - power$; we therefore subtract the table entry from 1 to find the power.

We can use Table B.3 in at least two ways: (1) to find the power retrospectively in a particular experimental situation and (2) to find the sample size required to yield a desired power. To estimate the power for a test with a particular data set, we could use sample values in place of population parameters in the noncentrality parameter λ. This power estimate may be of special interest when the test does not reject the hypothesis. In this case, if the results indicate low power for the test, we are not very confident that $\boldsymbol{\mu}$ is close to $\boldsymbol{\mu}_0$ or that $\boldsymbol{\mu}_1$ is close to $\boldsymbol{\mu}_2$, as hypothesized.

The other use for Tiku's tables is to determine the sample size required to achieve a desired level of power, given a difference $\boldsymbol{\mu} - \boldsymbol{\mu}_0$ or $\boldsymbol{\mu}_1 - \boldsymbol{\mu}_2$ of interest to the researcher. An estimate of $\boldsymbol{\Sigma}$ might be obtained from previous data or from a pilot study. We use the same value of n for both samples in the two-group case. For a particular choice of n, we calculate ϕ from (3.88) and read $power = 1 - \beta$ from Table B.3. Other values of n are tried until the desired power is obtained.

For maximum power in the two-group case, one should always use $n_1 = n_2$. The noncentrality parameter given in (3.87) is a function of $n_1 n_2/(n_1 + n_2)$. For a given value of $N = n_1 + n_2$, the ratio $n_1 n_2/(n_1 + n_2)$ is maximized when $n_1 = n_2$.

Stevens (1980) converted some values from Tiku's tables to give power directly in terms of T^2 without the need to convert to F. Table 3.7 shows some of these values for the two-group case. This gives a good perspective of the power for various values of p, n, and $\Delta^2 = (\boldsymbol{\mu}_1 - \boldsymbol{\mu}_2)'\boldsymbol{\Sigma}^{-1}(\boldsymbol{\mu}_1 - \boldsymbol{\mu}_2)$. Note that for fixed values of Δ^2 and n, the power decreases as p increases. This corresponds to a feature of the T^2-table that can be readily noted in Table B.1, namely that the critical values T^2_α (for a given value of ν) increase with p.

Some computer software packages provide a noncentral F-distribution function that can be used in place of Tiku's tables. If this function is denoted by PROBF, as

Table 3.7. Power of the Two-Sample T^2-Test

Number of Variables	n	$\Delta^2 = .25$	$\Delta^2 = .64$	$\Delta^2 = 1$	$\Delta^2 = 2.25$
2	15	.26	.44	.65	.95
2	25	.33	.66	.86	.97
2	50	.60	.95	1.00	1.00
2	100	.90	1.00	1.00	1.00
3	15	.23	.37	.58	.91
3	25	.28	.58	.80	.95
3	50	.54	.93	1.00	1.00
3	100	.86	1.00	1.00	1.00
5	15	.21	.32	.42	.83
5	25	.26	.42	.72	.96
5	50	.44	.88	1.00	1.00
5	100	.78	1.00	1.00	1.00
7	15	.18	.27	.37	.77
7	25	.22	.38	.64	.94
7	50	.40	.82	.97	1.00
7	100	.72	1.00	1.00	1.00

Note: There are n observations in each sample.

for example, in the software package SAS, then

$$\text{Power} = 1 - \text{PROBF}(F_\alpha, \nu_1, \nu_2, \lambda) = P(F > F_\alpha),$$

where F_α is a critical value of the central F-distribution.

Example 3.10(a). We illustrate the estimation of power in a situation where the hypothesis is accepted. Frets (1921) measured head length and breadth (y_1 and y_2) on the first son in 25 families and head length and breadth (x_1 and x_2) on the second son in the same 25 families (see diskette for data). To compare mean vectors for the first and second sons using the paired-comparison T^2-test in Section 3.8, we calculate $\mathbf{d}_i = \mathbf{y}_i - \mathbf{x}_i$, $i = 1, 2, \ldots, 25$, where $\mathbf{y}_i$ = head measurements on the first son and $\mathbf{x}_i$ = head measurements on the second son. From $\mathbf{d}_1, \mathbf{d}_2, \ldots, \mathbf{d}_{25}$, we obtain

$$\bar{\mathbf{d}} = \begin{pmatrix} 1.880 \\ 1.880 \end{pmatrix}, \qquad \mathbf{S}_d = \begin{pmatrix} 56.78 & 11.98 \\ 11.98 & 29.28 \end{pmatrix},$$

$$T^2 = n\bar{\mathbf{d}}'\mathbf{S}_d^{-1}\bar{\mathbf{d}} = 3.6124.$$

Since $3.6124 < T^2_{.05,2,24} = 7.142$, we accept H_0: $\boldsymbol{\mu}_y = \boldsymbol{\mu}_x$. Note the unexpected result $\bar{d}_1 = \bar{d}_2 = 1.880$ for this data set.

To estimate the power, we use sample values in place of population parameters. For $\mathbf{d}_i$, (3.86) becomes $\lambda = n\boldsymbol{\mu}_d'\boldsymbol{\Sigma}_d^{-1}\boldsymbol{\mu}_d$. If we estimate this by $\hat{\lambda} = T^2$, then ϕ in

(3.88) becomes

$$\hat{\phi} = \sqrt{\frac{T^2}{p+1}} = \sqrt{\frac{3.6124}{3}} = 1.0973.$$

For ν_1 and ν_2 we have

$$\nu_1 = p = 2, \qquad \nu_2 = n - p = 25 - 2 = 23.$$

In Table B.3 we obtain (by interpolation)

$$\text{Power} = 1 - .6603 = .3397.$$

This level of power is not very reassuring. Having accepted H_0, we would be more confident that it is close to being true if the power were higher. A larger sample size would increase the power to a more convincing level.

Example 3.10(b). We illustrate the determination of sample size needed to achieve a desired value of power. Suppose we have a two-sample situation in which the common covariance matrix is known (or estimated) to be

$$\boldsymbol{\Sigma} = \begin{pmatrix} 6 & -3 & 3 \\ -3 & 5 & -6 \\ 3 & -6 & 9 \end{pmatrix},$$

and we wish to detect a difference of

$$\boldsymbol{\mu}_1 - \boldsymbol{\mu}_2 = \begin{pmatrix} 3 \\ -2 \\ 3 \end{pmatrix}.$$

Using $n_1 = n_2 = n$ in (3.87), we find that the noncentrality parameter in this case is (unexpectedly) equal to n:

$$\begin{aligned} \lambda &= \frac{n_1 n_2}{n_1 + n_2}(\boldsymbol{\mu}_1 - \boldsymbol{\mu}_2)'\boldsymbol{\Sigma}^{-1}(\boldsymbol{\mu}_1 - \boldsymbol{\mu}_2) \\ &= \frac{n^2}{2n}(3, -2, 3)\begin{pmatrix} 6 & -3 & 3 \\ -3 & 5 & -6 \\ 3 & -6 & 9 \end{pmatrix}^{-1}\begin{pmatrix} 3 \\ -2 \\ 3 \end{pmatrix} \\ &= \frac{n}{2}(3, -2, 3)\left(\frac{1}{12}\right)\begin{pmatrix} 3 & 3 & 1 \\ 3 & 15 & 9 \\ 1 & 9 & 7 \end{pmatrix}\begin{pmatrix} 3 \\ -2 \\ 3 \end{pmatrix} \\ &= \frac{n}{2}(2) = n. \end{aligned}$$

The other parameter values are

$$\nu_1 = p = 3,$$
$$\nu_2 = n_1 + n_2 - p - 1 = 2n - 4,$$
$$\phi = \sqrt{\frac{\lambda}{\nu_1 + 1}} = \sqrt{\frac{n}{4}}.$$

Using $\alpha = .05$, we find the power from Table B.3 for several choices of $\lambda = n$:

n	ϕ	ν_2	Power
10	1.5811	16	.6438
12	1.7321	20	.7520
14	1.8708	24	.8329
16	2.0000	28	.8936

We can then choose the sample size needed to achieve the desired probability of rejection.

3.11 TESTS ON A SUBVECTOR

If a large number of variables is available, the researcher may wish to delete those variables that are least useful for separating the two groups. In general, as more variables are measured, the unique information available in any single variable (information that is not available in the other variables) tends to decrease. We first discuss the two-sample case in Sections 3.11.1–3.11.3; then for completeness, we consider the one-sample case in Section 3.11.4.

3.11.1 Two-Sample Case

We begin with the two-sample case in which we compare two groups. We partition the basic observation vector into $\binom{\mathbf{y}}{\mathbf{x}}$, where $\mathbf{y}$ is $p \times 1$ and $\mathbf{x}$ is $q \times 1$. In some cases, $\mathbf{y}$ represents the original variables and $\mathbf{x}$ contains variables that are measured in addition to $\mathbf{y}$. In other cases, $\binom{\mathbf{y}}{\mathbf{x}}$ represents a partitioning of the original variables into two subsets of interest. We assume that the two populations we have sampled are multivariate normal with a common covariance matrix; that is, $\binom{\mathbf{y}_{1i}}{\mathbf{x}_{1i}}$, $i = 1, 2, \ldots, n_1$, are from $N_{p+q}(\boldsymbol{\mu}_1, \boldsymbol{\Sigma})$, and $\binom{\mathbf{y}_{2i}}{\mathbf{x}_{2i}}$, $i = 1, 2, \ldots, n_2$, are from $N_{p+q}(\boldsymbol{\mu}_2, \boldsymbol{\Sigma})$. We partition $\boldsymbol{\mu}_1$, $\boldsymbol{\mu}_2$, and $\boldsymbol{\Sigma}$ accordingly:

$$\boldsymbol{\mu}_1 = \begin{pmatrix} \boldsymbol{\mu}_{y1} \\ \boldsymbol{\mu}_{x1} \end{pmatrix}, \quad \boldsymbol{\mu}_2 = \begin{pmatrix} \boldsymbol{\mu}_{y2} \\ \boldsymbol{\mu}_{x2} \end{pmatrix}, \quad \boldsymbol{\Sigma} = \begin{pmatrix} \boldsymbol{\Sigma}_{yy} & \boldsymbol{\Sigma}_{yx} \\ \boldsymbol{\Sigma}_{xy} & \boldsymbol{\Sigma}_{xx} \end{pmatrix}.$$

These population parameters are estimated by corresponding sample values

$$\begin{pmatrix} \bar{\mathbf{y}}_1 \\ \bar{\mathbf{x}}_1 \end{pmatrix}, \quad \begin{pmatrix} \bar{\mathbf{y}}_2 \\ \bar{\mathbf{x}}_2 \end{pmatrix}, \quad \mathbf{S}_{\text{pl}} = \begin{pmatrix} \mathbf{S}_{yy} & \mathbf{S}_{yx} \\ \mathbf{S}_{xy} & \mathbf{S}_{xx} \end{pmatrix},$$

where $\mathbf{S}_{\text{pl}}$ is a pooled sample covariance matrix.

The hypothesis of interest is that $\mathbf{x}_1$ and $\mathbf{x}_2$ are superfluous for separating the two groups; in other words, T^2 based on $\mathbf{y}$ and $\mathbf{x}$ is not significantly larger than T^2 based on $\mathbf{y}$ alone. In order to render the hypothesis in more precise terminology, we make the following definitions. For the full vector $\binom{\mathbf{y}}{\mathbf{x}}$, the population Mahalanobis distance is

$$\Delta_{p+q}^2 = (\boldsymbol{\mu}_1 - \boldsymbol{\mu}_2)'\boldsymbol{\Sigma}^{-1}(\boldsymbol{\mu}_1 - \boldsymbol{\mu}_2) = \boldsymbol{\delta}'\boldsymbol{\Sigma}^{-1}\boldsymbol{\delta}, \tag{3.89}$$

and for the subvector $\mathbf{y}$,

$$\Delta_p^2 = (\boldsymbol{\mu}_{y1} - \boldsymbol{\mu}_{y2})'\boldsymbol{\Sigma}_{yy}^{-1}(\boldsymbol{\mu}_{y1} - \boldsymbol{\mu}_{y2}) = \boldsymbol{\delta}_y'\boldsymbol{\Sigma}_{yy}^{-1}\boldsymbol{\delta}_y, \tag{3.90}$$

where $\boldsymbol{\delta} = \boldsymbol{\mu}_1 - \boldsymbol{\mu}_2$, $\boldsymbol{\delta}_y = \boldsymbol{\mu}_{y1} - \boldsymbol{\mu}_{y2}$, and $\boldsymbol{\delta}_x = \boldsymbol{\mu}_{x1} - \boldsymbol{\mu}_{x2}$. Let the population discriminant function be $\boldsymbol{\alpha}'\binom{\mathbf{y}}{\mathbf{x}}$, where

$$\boldsymbol{\alpha} = \boldsymbol{\Sigma}^{-1}(\boldsymbol{\mu}_1 - \boldsymbol{\mu}_2) = \boldsymbol{\Sigma}^{-1}\boldsymbol{\delta}, \tag{3.91}$$

and let $\boldsymbol{\alpha}$ be partitioned as

$$\boldsymbol{\alpha} = \begin{pmatrix} \boldsymbol{\alpha}_y \\ \boldsymbol{\alpha}_x \end{pmatrix}.$$

(Note the use of $\boldsymbol{\alpha}$ as the population counterpart of the sample discriminant function coefficient vector $\mathbf{a}$. This vector $\boldsymbol{\alpha}$ is not related to α used as the probability of a Type I error.) With these definitions, the hypothesis of redundancy of $\mathbf{x}$ can be expressed in the following three equivalent forms:

$$H_0\colon \Delta_{p+q}^2 = \Delta_p^2, \tag{3.92}$$

$$H_0\colon \boldsymbol{\alpha}_x = \mathbf{0}, \tag{3.93}$$

$$H_0\colon \boldsymbol{\delta}_x - \boldsymbol{\Sigma}_{xy}\boldsymbol{\Sigma}_{yy}^{-1}\boldsymbol{\delta}_y = \mathbf{0} \text{ or } \boldsymbol{\delta}_x = \boldsymbol{\Sigma}_{xy}\boldsymbol{\Sigma}_{yy}^{-1}\boldsymbol{\delta}_y. \tag{3.94}$$

To compare these, note that (3.92) states that the additional variables represented by $\mathbf{x}$ do not increase the distance between the two populations; (3.93) asserts that the additional variables in $\mathbf{x}$ do not add anything to the linear combination of the variables in $\mathbf{y}$ that best separates the populations; and (3.94) relates $\boldsymbol{\delta}_x$ to $\boldsymbol{\delta}_y$; that is, the separation provided by the additional variables $x_1, x_2, \ldots, x_q$ is totally predictable from the separation already achieved by the variables $y_1, y_2, \ldots, y_p$.

To test for significance of the x's above and beyond the y's, we compare

$$T_{p+q}^2 = \frac{n_1 n_2}{n_1 + n_2} \left[\begin{pmatrix} \bar{\mathbf{y}}_1 \\ \bar{\mathbf{x}}_1 \end{pmatrix} - \begin{pmatrix} \bar{\mathbf{y}}_2 \\ \bar{\mathbf{x}}_2 \end{pmatrix} \right]' \mathbf{S}_{\text{pl}}^{-1} \left[\begin{pmatrix} \bar{\mathbf{y}}_1 \\ \bar{\mathbf{x}}_1 \end{pmatrix} - \begin{pmatrix} \bar{\mathbf{y}}_2 \\ \bar{\mathbf{x}}_2 \end{pmatrix} \right], \tag{3.95}$$

based on the y's and the x's (the "full model"), and

$$T_p^2 = \frac{n_1 n_2}{n_1 + n_2} (\bar{\mathbf{y}}_1 - \bar{\mathbf{y}}_2)' \mathbf{S}_{yy}^{-1} (\bar{\mathbf{y}}_1 - \bar{\mathbf{y}}_2), \tag{3.96}$$

based only on the p measurements in $\mathbf{y}$ (the "reduced model"). The statistic T_{p+q}^2 can be decomposed into (see Problem 3.33)

$$T_{p+q}^2 = T_p^2 + T_{x \cdot y}^2, \tag{3.97}$$

where $T_{x\cdot y}^2 = [n_1 n_2/(n_1 + n_2)] \hat{\boldsymbol{\delta}}_{x\cdot y}' \mathbf{S}_{x\cdot y}^{-1} \hat{\boldsymbol{\delta}}_{x\cdot y}$, $\hat{\boldsymbol{\delta}}_{x\cdot y} = \hat{\boldsymbol{\delta}}_x - \mathbf{S}_{xy}\mathbf{S}_{yy}^{-1}\hat{\boldsymbol{\delta}}_y$, $\mathbf{S}_{x\cdot y} = \mathbf{S}_{xx} - \mathbf{S}_{xy}\mathbf{S}_{yy}^{-1}\mathbf{S}_{yx}$, $\hat{\boldsymbol{\delta}}_x = \bar{\mathbf{x}}_1 - \bar{\mathbf{x}}_2$, and $\hat{\boldsymbol{\delta}}_y = \bar{\mathbf{y}}_1 - \bar{\mathbf{y}}_2$. However, $T_{x\cdot y}^2 = T_{p+q}^2 - T_p^2$ does not have Hotelling's T^2-distribution. We use the following modification: When H_0 is true, the statistic

$$T^2 = (\nu - p) \left(\frac{T_{p+q}^2 - T_p^2}{\nu + T_p^2} \right) \tag{3.98}$$

is distributed as $T_{q,\nu-p}^2$, and we reject H_0 if $T^2 \geq T_{\alpha,q,\nu-p}^2$, where $\nu = n_1 + n_2 - 2$. The degrees of freedom $\nu - p$ reflect the adjustment for $\mathbf{y}$. By Theorem 3.3B, T^2 can be transformed to

$$F = \left(\frac{\nu - p - q + 1}{q} \right) \left(\frac{T_{p+q}^2 - T_p^2}{\nu + T_p^2} \right), \tag{3.99}$$

which is distributed as $F_{q,\nu-p-q+1}$. We reject H_0 if $F \geq F_{\alpha,q,\nu-p-q+1}$. Either of (3.98) or (3.99) is the likelihood ratio test (Problem 3.34). The test statistic (3.99) was originally obtained by Rao (1952, pp. 252–253).

Another use for (3.98) or (3.99) is to remove the effect of one or more covariates (Section 4.10). If $\mathbf{y}$ is a set of covariates and $\mathbf{x}$ contains the variables of interest, then the ability of the x's to separate the groups after adjusting for the y's is tested by (3.98) or (3.99).

If there is only one x to be added, then $q = 1$, and using either (3.98) or (3.99), we reject H_0 if

$$(\nu - p) \left(\frac{T_{p+1}^2 - T_p^2}{\nu + T_p^2} \right) \geq t_{\alpha/2,\nu-p}^2 = F_{\alpha,1,\nu-p}. \tag{3.100}$$

3.11.2 Step-Down Test

In this section we do not use the partitioned-vector notation $\binom{\mathbf{y}}{\mathbf{x}}$ because we wish to test one variable at a time. We use $\mathbf{y}$ alone and define $\mathbf{d} = \mathbf{y}_1 - \mathbf{y}_2$ and $\boldsymbol{\delta} = \boldsymbol{\mu}_1 - \boldsymbol{\mu}_2$, where these vectors are $p \times 1$.

If there is an a priori ordering among the elements of $\boldsymbol{\delta}' = (\delta_1, \delta_2, \ldots, \delta_p)$, with a corresponding ordering on $\mathbf{d}' = (d_1, d_2 \ldots, d_p)$, we can test the variables sequentially using (3.100), which becomes

$$t_k^2 = (\nu - k + 1)\left(\frac{T_k^2 - T_{k-1}^2}{\nu + T_{k-1}^2}\right), \qquad k = 1, 2, \ldots, p. \tag{3.101}$$

The statistic t_k^2 is the square of a t-distributed random variable with $\nu - k + 1$ degrees of freedom, or alternatively, $t_k^2 = F_{1,\nu-k+1}$. For $k = 1$ we set $T_0^2 = 0$. The statistics $t_1^2, t_2^2, \ldots, t_p^2$ are calculated in that order.

The hypothesis tested by (3.101) is that each variable is redundant, given the preceding variables tested. If we reject this hypothesis in one of the p tests, we have in effect rejected H_0: $\boldsymbol{\delta} = \mathbf{0}$ or H_0: $\boldsymbol{\mu}_1 = \boldsymbol{\mu}_2$. If we accept this hypothesis in all p tests, we have accepted H_0. It can be shown that the test statistics $t_1^2, t_2^2, \ldots, t_p^2$ are independent. The overall α is therefore related to the α's for the individual tests by

$$1 - \alpha = \prod_{k=1}^{p}(1 - \alpha_k),$$

where α_k is the significance level of the kth test. The levels α_k, $k = 1, 2, \ldots, p$, can be adjusted to produce the desired overall α. If the first few variables in the ordering are more important than the later variables, the α_k's can be chosen accordingly.

3.11.3 Selection of Variables

If there is no natural a priori ordering of variables, we can let the data determine the order of entry of the variables in (3.101). A *forward selection* of variables would proceed as follows. In the first step, calculate t_1^2 for each of the p variables and choose the variable with maximum t_1^2. In the second step, calculate t_2^2 for each of the $p - 1$ variables not selected in the first step and choose the variable with maximum t_2^2, that is, the variable that adds the most separation to the variable selected in step 1. In the third step, calculate t_3^2 for each of the $p - 2$ remaining variables and choose the variable that maximizes t_3^2. Continue this process until for some step k the maximum t_k^2 fails to exceed a predetermined threshold value.

A *stepwise* procedure would proceed similarly, except that in each step the variables that entered previously would be reexamined to see if each still contributed a significant amount. If in some step an earlier variable is no longer significant in the presence of variables that have entered later, the variable is deleted.

Note that the statistic t_k^2 used in this way no longer has a t^2- (or F-) distribution because t_k^2 is maximized at each step. The use of critical values from the t- (or F-) table in a decision to stop entering variables would bias the results in favor of including too many variables. This problem is discussed further in Section 5.10 in connection with stepwise discriminant analysis.

3.11.4 One-Sample Case

We now consider the test for a subvector in the one-sample context of comparing a sample mean vector with a hypothesized mean vector, H_0: $\boldsymbol{\mu} = \boldsymbol{\mu}_0$.

Let $\mathbf{y}$ be $p \times 1$ and $\mathbf{x}$ be $q \times 1$ such that $\binom{\mathbf{y}}{\mathbf{x}}$ is $N_{p+q}(\boldsymbol{\mu}, \boldsymbol{\Sigma})$, where

$$\boldsymbol{\mu} = \begin{pmatrix} \boldsymbol{\mu}_y \\ \boldsymbol{\mu}_x \end{pmatrix} \quad \text{and} \quad \boldsymbol{\Sigma} = \begin{pmatrix} \boldsymbol{\Sigma}_{yy} & \boldsymbol{\Sigma}_{yx} \\ \boldsymbol{\Sigma}_{xy} & \boldsymbol{\Sigma}_{xx} \end{pmatrix},$$

with corresponding sample values $\overline{\mathbf{y}}, \overline{\mathbf{x}}, \mathbf{S}, \mathbf{S}_{yy}, \mathbf{S}_{yx}, \mathbf{S}_{xy}$, and $\mathbf{S}_{xx}$. The hypothesized mean vector $\boldsymbol{\mu}_0$ is similarly partitioned:

$$\boldsymbol{\mu}_0 = \begin{pmatrix} \boldsymbol{\mu}_{0y} \\ \boldsymbol{\mu}_{0x} \end{pmatrix}.$$

We wish to test the hypothesis H_0 that $\mathbf{x}$ is redundant in the presence of $\mathbf{y}$ for detecting a significant difference between the sample mean vector and the hypothesized mean vector. This hypothesis can be phrased in expressions analogous to (3.92), (3.93), and (3.94). First define T^2 based on $\binom{\mathbf{y}}{\mathbf{x}}$ and T^2 based on $\mathbf{y}$:

$$T_{p+q}^2 = n\left[\begin{pmatrix} \overline{\mathbf{y}} \\ \overline{\mathbf{x}} \end{pmatrix} - \begin{pmatrix} \boldsymbol{\mu}_{0y} \\ \boldsymbol{\mu}_{0x} \end{pmatrix}\right]' \mathbf{S}^{-1} \left[\begin{pmatrix} \overline{\mathbf{y}} \\ \overline{\mathbf{x}} \end{pmatrix} - \begin{pmatrix} \boldsymbol{\mu}_{0y} \\ \boldsymbol{\mu}_{0x} \end{pmatrix}\right],$$

$$T_p^2 = n(\overline{\mathbf{y}} - \boldsymbol{\mu}_{0y})' \mathbf{S}_{yy}^{-1} (\overline{\mathbf{y}} - \boldsymbol{\mu}_{0y}).$$

Then we reject H_0 if

$$T^2 = (n - p - 1)\left(\frac{T_{p+q}^2 - T_p^2}{n - 1 + T_p^2}\right) \geq T_{\alpha,q,n-p-1}^2 \tag{3.102}$$

or, alternatively, if

$$F = \frac{n - p - q}{q}\left(\frac{T_{p+q}^2 - T_p^2}{n - 1 + T_p^2}\right) \geq F_{\alpha,q,n-p-q}. \tag{3.103}$$

For $q = 1$ in the one-sample case, there are expressions that correspond to (3.100) and (3.101).

3.12 NONNORMAL APPROACHES TO HYPOTHESIS TESTING

3.12.1 Elliptically Contoured Distributions

It was noted in Section 2.4 that many estimates and tests derived for elliptically contoured distributions are similar to those based on the multivariate normal distribution. The random vector $\mathbf{y}$ has an elliptically contoured distribution if its density function is of the form

$$f(\mathbf{y}) = |\boldsymbol{\Sigma}|^{-1/2} g\left[(\mathbf{y} - \boldsymbol{\mu})'\boldsymbol{\Sigma}^{-1}(\mathbf{y} - \boldsymbol{\mu})\right],$$

where $g(\cdot)$ is a nonincreasing positive-valued function (Tong 1990, p. 63).

Fang and Zhang (1990, Section 5.2) and Anderson and Fang (1990) show that for the elliptically contoured distribution the likelihood ratio tests of $H_0\colon \boldsymbol{\mu} = \boldsymbol{\mu}_0$ and $H_0\colon \boldsymbol{\mu}_1 = \boldsymbol{\mu}_2$ are the same T^2-tests as in (3.14) and (3.57). Furthermore, the distribution of these T^2-statistics is the same as in the normal case.

3.12.2 Nonparametric Tests

In Section 3.7, we discussed the robustness of T^2 to nonnormality and to heterogeneity of covariance matrices. In Section 3.9.2, we discussed several approximate tests that could be used to test $H_0\colon \boldsymbol{\mu}_1 = \boldsymbol{\mu}_2$ when $\boldsymbol{\Sigma}_1 \neq \boldsymbol{\Sigma}_2$.

An alternative approach in these situations is offered by nonparametric tests that are based on ranks or normal scores (Puri and Sen 1971). To describe these tests, let

$$U = \frac{T^2}{n_1 + n_2 - 2}, \tag{3.104}$$

where T^2 is the usual two-sample version in (3.57). The nonparametric test is given by

$$v = (n_1 + n_2 - 1)\frac{U^*}{1 + U^*}, \tag{3.105}$$

where U^* is analogous to U in (3.104) but is based on ranks or normal scores. To obtain ranks, the $n_1 + n_2$ observations for each of the p variables are ranked from 1 to $n_1 + n_2$. In the ranking, the two groups (samples) are combined together. Alternatively, expected normal scores can be substituted for ranks. The test statistic v in (3.105) is asymptotically distributed as χ^2_p for both the rank and normal score procedures. For computational details of both tests and a description of available software, see Zwick (1985).

The properties of these two tests were investigated by Zwick (1986). Subject to the limitations of her simulation study, the results were as follows. The nonparametric tests tended to be conservative in Type I error rates. Under normality and homoscedasticity ($\boldsymbol{\Sigma}_1 = \boldsymbol{\Sigma}_2$), the T^2-test was more powerful, as we would expect, although the nonparametric tests were not far behind. Under nonnormal conditions and heteroscedasticity ($\boldsymbol{\Sigma}_1 \neq \boldsymbol{\Sigma}_2$), both nonparametric tests were more powerful than the T^2-test.

Jan and Randles (1994) reviewed several nonparametric procedures for the multivariate one-sample problem and proposed a new method based on signed sums. Under certain conditions, the signed-sum test outperformed Hotelling's T^2 when the observations arose from heavy-tailed distributions.

3.12.3 Robust Versions of T^2

Mudholkar and Srivastava (1996a) provided a robust version of the one-sample T^2 based on trimmed means. The proposed test is more powerful than T^2 if the true population is heavy tailed. Mudholkar and Srivastava (1996b) suggested two robust versions of the two-sample T^2 and showed them to be effective for heavy-tailed distributions.

3.13 APPLICATION OF T^2 IN MULTIVARIATE QUALITY CONTROL

In industrial processes, it is often necessary to monitor several variables simultaneously. In the spirit of the last paragraph of Section 3.6.4, one approach is to monitor the variables multivariately and, if an out-of-control condition is detected, to use univariate control charts for the individual variables to identify the variable(s) that caused the out-of-control signal.

One technique for monitoring a process for shifts in the means of several variables involves the use of T^2. To check one observation $\mathbf{y}$ at a time, we can use

$$T^2 = (\mathbf{y} - \mathbf{m})'\mathbf{S}^{-1}(\mathbf{y} - \mathbf{m}), \tag{3.106}$$

where $\mathbf{m}$ is a target value or a long-term average. In this case we have $n = 1$; that is, $\mathbf{y}$ is the "mean" of one observation. However, $\mathbf{S}$ cannot be obtained from a single observation vector. If k observation vectors are available from past experience (in control) to calculate $\mathbf{S}$, then T^2 in (3.106) is distributed as $T^2_{p,k-1}$. The value of T^2 for each $\mathbf{y}$ can be plotted with an upper control limit of T^2_α. There is no lower limit. For α, we can use .05 or possibly .0027, corresponding to the "3-sigma" limits of univariate $\bar{x}$-charts. Rigdon (1996) discussed the sample size k required to avoid a high false alarm rate, that is, a high probability of an out-of-control signal when the process has not shifted. Sullivan and Woodall (1996) discussed calculations of $\mathbf{S}$ in situations in which the mean vector is shifting. Griffith, Young, and Mason (1996) investigated the behavior of (3.106) for $p = 2$ when sampling is from a truncated bivariate normal distribution.

To interpret (3.106), Mason, Tracy, and Young (1995) recommended a decomposition of T^2 into the effect of each variable as in Section 3.3.5 (see also Rencher 1993). Hawkins (1993, 1996) discussed transformations of the variables to *regression adjustments* that improve the diagnostic capability of multivariate control charts. Timm (1996) proposed the use of finite intersection tests to detect shifts that occur for a subset of variables. Kourti and MacGregor (1996) suggested methods for detecting the variables that contribute to an out-of-control signal and discussed other approaches based on projection methods such as principal component analysis and partial least squares.

If $k - 1 < p$, then $\mathbf{S}^{-1}$ in (3.106) does not exist. In such a case, it is possible to reduce the dimensionality using principal components (Chapter 9) before calculating T^2. This may also be useful when p is large but does not exceed $k - 1$. An alternative approach based on constraints on the covariance matrix is advocated by Boyles (1996).

If we use the mean vector $\bar{\mathbf{y}}$ of a sample of size n, instead of an individual observation vector $\mathbf{y}$, then T^2 becomes

$$T^2 = n(\bar{\mathbf{y}} - \mathbf{m})'\mathbf{S}_{\text{pl}}^{-1}(\bar{\mathbf{y}} - \mathbf{m}), \tag{3.107}$$

which is distributed as $T^2_{p,k(n-1)}$, where $\mathbf{S}_{\text{pl}} = \sum_{i=1}^{k} \mathbf{S}_i/k$ (see Section 3.3.4).

For reviews of these and other multivariate quality control approaches such as CUSUM procedures and exponentially weighted average control charts, see Alt and Smith (1988), Healy (1987), Jackson (1985), Sparks (1992), Pignatiello and Runger (1990), Lowry et al. (1992), Roes and Does (1995), Nomikos and MacGregor (1995), Hamilton and Lesperance (1995), Liu (1995), Runger (1996), Runger and Prabhu (1996), and Prabhu and Runger (1997).

PROBLEMS

3.1 Show that the use of $Z^2 = n(\bar{\mathbf{y}} - \boldsymbol{\mu}_0)'\boldsymbol{\Sigma}^{-1}(\bar{\mathbf{y}} - \boldsymbol{\mu}_0)$ in (3.1) to test $H_0\colon \boldsymbol{\mu} = \boldsymbol{\mu}_0$ is equivalent to the likelihood ratio test.

3.2 If $z_i = ay_i + b, i = 1, 2, \ldots, n$, verify the following results, which were used to prove that $t_z^2 = t_y^2$ in (3.5):

(a) $\bar{z} = a\bar{y} + b$

(b) $s_z^2 = a^2 s_y^2$

3.3 Verify that the likelihood function (3.9) is maximized by $\hat{\mu} = \bar{y}$ and $\hat{\sigma}^2 = \sum_i (y_i - \bar{y})^2/n$.

3.4 Show that under $H_0\colon \mu = \mu_0$ the likelihood function (3.9) is maximized by $\hat{\sigma}_0^2 = \sum_i (y_i - \mu_0)^2/n$, as noted following (3.10).

3.5 Show that $\sum_i (y_i - \bar{y}) = 0$, as used in obtaining (3.13) from (3.12).

3.6 If a random sample $\mathbf{y}_1, \mathbf{y}_2, \ldots, \mathbf{y}_n$ is obtained from $N_p(\boldsymbol{\mu}, \boldsymbol{\Sigma})$, show that $\bar{\mathbf{v}} = \sqrt{n}(\bar{\mathbf{y}} - \boldsymbol{\mu}_0)$ is $N_p(\mathbf{0}, \boldsymbol{\Sigma})$ under $H_0\colon \boldsymbol{\mu} = \boldsymbol{\mu}_0$, that $\mathbf{W} = (n-1)\mathbf{S}$ is $W_p(n-1, \boldsymbol{\Sigma})$, and that $\bar{\mathbf{v}}$ and $\mathbf{W}$ are independent, thus verifying that $T^2 = n(\bar{\mathbf{y}} - \boldsymbol{\mu}_0)'\mathbf{S}^{-1}(\bar{\mathbf{y}} - \boldsymbol{\mu}_0)$ satisfies the formal definition (3.16).

3.7 Use (A.7.10) to verify (3.15), the alternative formula for T^2 based on determinants.

3.8 Prove Theorem 3.3C.

3.9 Show that $T^2 = n(\bar{\mathbf{y}} - \boldsymbol{\mu}_0)'\mathbf{S}^{-1}(\bar{\mathbf{y}} - \boldsymbol{\mu}_0)$ is unchanged by the transformation $\mathbf{z}_i = \mathbf{A}\mathbf{y}_i + \mathbf{b}$, where $\mathbf{A}$ is nonsingular, thus verifying property 1 in Section 3.3.6.

3.10 Show that when $\boldsymbol{\mu} = \boldsymbol{\mu}_0$, the value of $\boldsymbol{\Sigma}$ that maximizes the likelihood function (3.21) is $\hat{\boldsymbol{\Sigma}}_0 = \sum_i (\mathbf{y}_i - \boldsymbol{\mu}_0)(\mathbf{y}_i - \boldsymbol{\mu}_0)'/n$.

3.11 Show that (3.24) follows from $\max_{H_0} L$ in (3.23). Use $\hat{\boldsymbol{\Sigma}}_0 = \sum_i (\mathbf{y}_i - \boldsymbol{\mu}_0)(\mathbf{y}_i - \boldsymbol{\mu}_0)'/n$.

3.12 Show that application of (A.7.10) to (3.25) produces (3.26); that is, show that $|\mathbf{S} + n(\bar{\mathbf{y}} - \boldsymbol{\mu}_0)(\bar{\mathbf{y}} - \boldsymbol{\mu}_0)'/(n-1)| = |\mathbf{S}|[1 + n(\bar{\mathbf{y}} - \boldsymbol{\mu}_0)'\mathbf{S}^{-1}(\bar{\mathbf{y}} - \boldsymbol{\mu}_0)/(n-1)]$.

3.13 Show that for $\mathbf{a}' = (0, \ldots, 0, 1, 0, \ldots, 0)$ with a 1 in the jth position, $\mathbf{a}'\overline{\mathbf{y}} = \overline{y}_j$ and $\mathbf{a}'\mathbf{S}\mathbf{a} = s_{jj}$ as in (3.37).

3.14 Show that for $t(\mathbf{a})$ defined by (3.48), $\max_{\mathbf{a}} t^2(\mathbf{a})$ is distributed as T^2_{p,n_1+n_2-2}.

3.15 **(a)** Show that $T^2 = n(\mathbf{C}\overline{\mathbf{y}})'(\mathbf{CSC}')^{-1}(\mathbf{C}\overline{\mathbf{y}})$ in (3.51) has the $T^2_{p-1,n-1}$ distribution (under H_0).

(b) Show that $T^2 = n(\mathbf{C}\overline{\mathbf{y}} - \boldsymbol{\gamma})'(\mathbf{CSC}')^{-1}(\mathbf{C}\overline{\mathbf{y}} - \boldsymbol{\gamma})$ in (3.52) has the $T^2_{k,n-1}$ distribution (under H_0).

3.16 Let $(y_1, x_1), (y_2, x_2), \ldots, (y_n, x_n)$ be a random sample from the bivariate normal $N_2(\boldsymbol{\mu}, \boldsymbol{\Sigma})$, where $\boldsymbol{\mu}' = (\mu_y, \mu_x)$ and let $d_i = y_i - x_i$, $i = 1, 2, \ldots, n$. Show that the T^2-test of H_0: $\mu_y = \mu_x$ obtainable from (3.51) with appropriately defined $\mathbf{C}$ is equivalent to the paired comparison t-statistic $t = \sqrt{n}\overline{d}/s_d$, where $\overline{d}$ and s_d are the sample mean and standard deviation of the d_i's.

3.17 Show that $\overline{\mathbf{y}}'\mathbf{S}^{-1}\overline{\mathbf{y}} = (\mathbf{C}\overline{\mathbf{y}})'(\mathbf{CSC}')^{-1}\mathbf{C}\overline{\mathbf{y}} + (\overline{\mathbf{y}}'\mathbf{S}^{-1}\mathbf{j})^2/\mathbf{j}'\mathbf{S}^{-1}\mathbf{j}$, where $\mathbf{C}$ is $(p-1) \times p$ of rank $p-1$ such that $\mathbf{Cj} = \mathbf{0}$. From this it follows that $n\overline{\mathbf{y}}'\mathbf{S}^{-1}\overline{\mathbf{y}} > n(\mathbf{C}\overline{\mathbf{y}})'(\mathbf{CSC}')^{-1}\mathbf{C}\overline{\mathbf{y}}$ as in (3.52).

3.18 Show that the t-statistic in (3.54) satisfies the formal definition (3.3) and the corresponding assumptions.

3.19 Verify the likelihood ratio result in (3.56).

3.20 Show that $\mathbf{S}_{\text{pl}} = [(n_1 - 1)\mathbf{S}_1 + (n_2 - 1)\mathbf{S}_2]/(n_1 + n_2 - 2)$ is an unbiased estimator of the common population covariance matrix $\boldsymbol{\Sigma}$.

3.21 Show that the two-sample T^2-statistic given in (3.57) satisfies the formal definition of a T^2 random variable in (3.16).

3.22 Verify the alternative formula for T^2 in (3.59).

3.23 Prove Theorem 3.5B.

3.24 Prove Theorem 3.5C.

3.25 Verify that $\max_{\mathbf{a}} t^2(\mathbf{a}) = T^2$ as in Theorem 3.5D.

3.26 Show that the vector $\mathbf{a}$ that maximizes $t^2(\mathbf{a})$ in (3.63) is $\mathbf{a} = \mathbf{S}_{\text{pl}}^{-1}(\overline{\mathbf{y}}_1 - \overline{\mathbf{y}}_2)$ as in (3.65).

3.27 Show that if the discriminant function coefficient vector $\mathbf{a} = \mathbf{S}_{\text{pl}}^{-1}(\overline{\mathbf{y}}_1 - \overline{\mathbf{y}}_2)$ is substituted in $t^2(\mathbf{a})$ from (3.63), the result is $t^2(\mathbf{a}) = T^2$, where T^2 is given by (3.57).

3.28 Verify that the T^2 expression in (3.73) for testing H_0: $\mathbf{C}(\boldsymbol{\mu}_1 - \boldsymbol{\mu}_2) = \mathbf{0}$ is distributed as T^2_{p-1,n_1+n_2-2}.

3.29 Show that if $n_1 = n_2$, the approximate T^{*2} in (3.81) is the same as the usual T^2 in (3.57). (The degrees of freedom are different for the two cases.)

3.30 Show that for fixed $N = n_1 + n_2$ the ratio $n_1 n_2/(n_1 + n_2)$ is maximized when $n_1 = n_2$, as noted in Section 3.10.

3.31 Show that H_0: $\Delta^2_{p+q} = \Delta^2_p$ in (3.92) is equivalent to H_0: $\boldsymbol{\delta}_x - \boldsymbol{\Sigma}_{xy}\boldsymbol{\Sigma}_{yy}^{-1}\boldsymbol{\delta}_y = \mathbf{0}$ in (3.94).

3.32 Show that H_0: $\boldsymbol{\alpha}_x = \mathbf{0}$ in (3.93) is equivalent to H_0: $\boldsymbol{\delta}_x - \boldsymbol{\Sigma}_{xy}\boldsymbol{\Sigma}_{yy}^{-1}\boldsymbol{\delta}_y = \mathbf{0}$ in (3.94).

3.33 Show that T^2 can be decomposed as in (3.97), $T^2_{p+q} = T^2_p + T^2_{x\cdot y}$, where T^2_{p+q} and T^2_p are defined in (3.95) and (3.96), and $T^2_{x\cdot y} = [n_1 n_2/(n_1 + n_2)]\hat{\boldsymbol{\delta}}'_{x\cdot y}\mathbf{S}^{-1}_{x\cdot y}\hat{\boldsymbol{\delta}}_{x\cdot y}$, with $\hat{\boldsymbol{\delta}}_{x\cdot y} = \hat{\boldsymbol{\delta}}_x - \mathbf{S}_{xy}\mathbf{S}^{-1}_{yy}\hat{\boldsymbol{\delta}}_y$ and $\mathbf{S}_{x\cdot y} = \mathbf{S}_{xx} - \mathbf{S}_{xy}\mathbf{S}^{-1}_{yy}\mathbf{S}_{yx}$.

3.34 Show that the test statistic T^2 in (3.98) is a function of the likelihood ratio for H_0 given in (3.92)–(3.94).

3.35 Verify that the test statistic for a single x in (3.100) follows from (3.98) and (3.99).

3.36 Use (3.102) or (3.103) to find an expression analogous to (3.100) for $q = 1$ in the one-sample case.

3.37 Using the data in $\mathbf{Y}$ below, illustrate that T^2 for $H_0\colon \boldsymbol{\mu} = (8, 5)'$ remains unchanged, as in property 1 of Section 3.3.6, if each observation $\mathbf{y}_i, i = 1, 2, 3, 4, 5$, is transformed to $\mathbf{z}_i = \mathbf{A}\mathbf{y}_i + \mathbf{b}$, where

$$\mathbf{Y} = \begin{pmatrix} 7 & 8 \\ 12 & 7 \\ 9 & 4 \\ 4 & 5 \\ 6 & 4 \end{pmatrix}, \qquad \mathbf{A} = \begin{pmatrix} 1 & 1 \\ 1 & -1 \end{pmatrix}, \qquad \mathbf{b} = \begin{pmatrix} -3 \\ 7 \end{pmatrix}.$$

Note that the observations

$$\mathbf{z}_i = \mathbf{A}\mathbf{y}_i + \mathbf{b} = \begin{pmatrix} y_{1i} + y_{2i} - 3 \\ y_{1i} - y_{2i} + 7 \end{pmatrix}$$

yield the data matrix

$$\begin{pmatrix} 7+8-3 & 7-8+7 \\ 12+7-3 & 12-7+7 \\ 9+4-3 & 9-4+7 \\ 4+5-3 & 4-5+7 \\ 6+4-3 & 6-4+7 \end{pmatrix}$$

and that $\boldsymbol{\mu}_0$ should be transformed to $\boldsymbol{\mu}_{0z} = \mathbf{A}\boldsymbol{\mu}_0 + \mathbf{b}$.

3.38 Using the probe word data in Table 1.3, do the following:

(a) Test $H_0\colon \boldsymbol{\mu} = (30, 25, 40, 25, 30)'$.

(b) Obtain 95% simultaneous confidence intervals for $\mu_1, \mu_2, \ldots, \mu_5$.

(c) Obtain 95% Bonferroni confidence intervals for $\mu_1, \mu_2, \ldots, \mu_5$.

(d) Test the hypotheses $H_{0j}\colon \mu_j = \mu_{0j}$ for $j = 1, 2, \ldots, 5$ using t_j in (3.46) with a Bonferroni critical value.

(e) If the T^2-test in part (a) leads to rejection of $H_0\colon \boldsymbol{\mu} = \boldsymbol{\mu}_0$, test each $H_{0j}\colon \mu_j = \mu_{0j}$, $j = 1, 2, \ldots, 5$, using a critical value $t_{\alpha/2,n-1}$, in the spirit of the last paragraph in Section 3.6.4.

3.39 Using the probe word data in Table 1.3 and the hypothesis $H_0\colon \boldsymbol{\mu} = (30, 25, 40, 25, 30)'$, do the following:

(a) Check the effect of each variable on T^2 as in Section 3.3.5. Calculate $\hat{t}_x$, t_x, R^2, and the increase in T^2.

(b) Test the significance of y_4 and y_5 adjusted for y_1, y_2, and y_3.

(c) Test the significance of each variable adjusted for the other four.

3.40 Using the probe word data in Table 1.3, do the following:

(a) Test H_0: $\mu_1 = \mu_2 = \mu_3 = \mu_4 = \mu_5$.

(b) Test H_0: $\frac{1}{2}(\mu_1 + \mu_2) = \frac{1}{3}(\mu_3 + \mu_4 + \mu_5)$.

(c) Test H_0: $\mu_1 = \mu_4$.

(d) Test H_0: $\begin{pmatrix} \mu_1 - \mu_2 \\ \mu_3 - \mu_4 \end{pmatrix} = \begin{pmatrix} 0 \\ 0 \end{pmatrix}$.

3.41 Lubischew (1962) measured the following four variables on two species of flea beetles:

y_1 = the distance of the transverse groove from the posterior border of the prothorax (μm),

y_2 = the length of the elytra (in 0.01 mm),

y_3 = the length of the second antennal joint (μm),

y_4 = the length of the third antennal joint (μm).

The data are given in Table 3.8.

Table 3.8. Four Measurements on Two Species of Flea Beetles

	Haltica oleracea					*Haltica carduorum*			
Specimen Number	y_1	y_2	y_3	y_4	Specimen Number	y_1	y_2	y_3	y_4
1	189	245	137	163	1	181	305	184	209
2	192	260	132	217	2	158	237	133	188
3	217	276	141	192	3	184	300	166	231
4	221	299	142	213	4	171	273	162	213
5	171	239	128	158	5	181	297	163	224
6	192	262	147	173	6	181	308	160	223
7	213	278	136	201	7	177	301	166	221
8	192	255	128	185	8	198	308	141	197
9	170	244	128	192	9	180	286	146	214
10	201	276	146	186	10	177	299	171	192
11	195	242	128	192	11	176	317	166	213
12	205	263	147	192	12	192	312	166	209
13	180	252	121	167	13	176	285	141	200
14	192	283	138	183	14	169	287	162	214
15	200	294	138	188	15	164	265	147	192
16	192	277	150	177	16	181	308	157	204
17	200	287	136	173	17	192	276	154	209
18	181	255	146	183	18	181	278	149	235
19	192	287	141	198	19	175	271	140	192
					20	197	303	170	205

(a) Test $H_0\colon \boldsymbol{\mu}_1 = \boldsymbol{\mu}_2$ using T^2.

(b) Calculate the discriminant function coefficient vector $\mathbf{a} = \mathbf{S}_{\text{pl}}^{-1}(\overline{\mathbf{y}}_1 - \overline{\mathbf{y}}_2)$.

(c) Substitute the vector **a** found in part (b) into $t^2(\mathbf{a})$ from (3.63). Is the result the same as the value of T^2 found in part (a), as indicated in Theorem 3.5D?

(d) Test the significance of y_3 and y_4 adjusted for y_1 and y_2.

(e) Test the significance of each y_j adjusted for the other three.

3.42 Using the data in Table 3.8, check the effect of each variable on the T^2-statistic, as in Theorem 3.5B. Calculate $\hat{t}_x, t_x, R^2$, and the increase in T^2, as in Examples 3.5.3 (a) and (b).

3.43 Suppose there were evidence that $\boldsymbol{\Sigma}_1 \neq \boldsymbol{\Sigma}_2$ for the data in Table 3.8.

(a) Carry out James's approximate test (Section 3.9.2b).

(b) Carry out Yao's approximate test (Section 3.9.2b).

(c) Carry out Johansen's approximate test (Section 3.9.2b).

(d) Carry out Nel and van der Merwe's approximate test (Section 3.9.2b).

(e) Carry out Kim's approximate test (Section 3.9.2b).

3.44 Using the data in Table 3.8, do the following:

(a) Find 95% simultaneous confidence intervals for $\mu_{1j} - \mu_{2j}, j = 1, 2, 3, 4$, using (3.68).

(b) Find 95% Bonferroni confidence intervals for $\mu_{1j} - \mu_{2j}, j = 1, 2, 3, 4$, using (3.70).

3.45 Using the data in Table 3.8, do the following:

(a) Use (3.73) to test

$$H_0\colon \begin{pmatrix} \mu_{12} - \mu_{11} \\ \mu_{13} - \mu_{12} \\ \mu_{14} - \mu_{13} \end{pmatrix} = \begin{pmatrix} \mu_{22} - \mu_{21} \\ \mu_{23} - \mu_{22} \\ \mu_{24} - \mu_{23} \end{pmatrix},$$

which can be expressed as $H_0\colon \mathbf{C}\boldsymbol{\mu}_1 = \mathbf{C}\boldsymbol{\mu}_2$ with

$$\mathbf{C} = \begin{pmatrix} -1 & 1 & 0 & 0 \\ 0 & -1 & 1 & 0 \\ 0 & 0 & -1 & 1 \end{pmatrix}.$$

(b) Test $H_0\colon \mathbf{C}\boldsymbol{\mu}_1 = \mathbf{C}\boldsymbol{\mu}_2$ with

$$\mathbf{C} = \begin{pmatrix} 1 & -3 & 1 & 1 \\ 1 & 0 & 0 & -1 \end{pmatrix}.$$

3.46 Using the data in Table 1.4, do the following:

(a) Test $H_0\colon \boldsymbol{\mu}_y = \boldsymbol{\mu}_x$ using the paired comparison test (3.74).

(b) Test the significance of each variable adjusted for the other two. Use (3.100) adapted for the paired observation case.

(c) Test each variable by itself using (3.78).

(d) Calculate the discriminant function coefficients in (3.75).

3.47 Suppose we are going to collect data from two populations in order to compare them using two variables. We conjecture that the common covariance matrix is

$$\boldsymbol{\Sigma} = \begin{pmatrix} 13 & -15 \\ -15 & 25 \end{pmatrix}.$$

We wish to detect a difference of $\boldsymbol{\mu}_1 - \boldsymbol{\mu}_2 = \binom{-2}{6}$ with probability .8 using $\alpha = .05$ for the test. How large must the sample size be for each of the two samples?

3.48 Suppose we are going to take a sample from each of two populations and compare means on four variables. How large a sample should we take from each population in order to ensure a 90% chance of detecting a difference $\boldsymbol{\mu}_1 - \boldsymbol{\mu}_2 = (1, -2, 4, 3)'$. Assume the following common covariance matrix:

$$\boldsymbol{\Sigma} = \begin{pmatrix} 10 & -1 & -6 & 3 \\ -1 & 16 & 0 & -1 \\ -6 & 0 & 15 & 6 \\ 3 & -1 & 6 & 12 \end{pmatrix}.$$

3.49 Using the data in Table 3.8, calculate the test statistic v in (3.105) based on ranks and compare it to $\chi^2_{.05,4}$.

CHAPTER 4

Multivariate Analysis of Variance

In Chapter 3 we considered various tests that involve a single sample or two samples of observation vectors. We now extend to a comparison of several samples of observation vectors and to a two-way classification of observation vectors for both the balanced and unbalanced cases. In addition to multivariate analysis of variance, we also consider comparisons of covariance matrices (Section 4.3) and multivariate analysis of covariance (Section 4.10). For profile analysis, repeated measures, and growth curves, see Rencher (1995, Sections 6.8, 6.9, and 6.10).

4.1 ONE-WAY CLASSIFICATION

In this section we consider the balanced one-way classification in which the sample sizes are all equal. The unbalanced case with unequal sample sizes is discussed in Section 4.8.3.

4.1.1 Model for One-Way Multivariate Analysis of Variance

In a one-way multivariate analysis of variance (MANOVA), we assume that a random sample of n p-variate observations is available from each of k multivariate normal populations with equal covariance matrices $\boldsymbol{\Sigma}$. The model for each observation vector is

$$\begin{aligned} \mathbf{y}_{ij} &= \boldsymbol{\mu} + \boldsymbol{\alpha}_i + \boldsymbol{\varepsilon}_{ij} \\ &= \boldsymbol{\mu}_i + \boldsymbol{\varepsilon}_{ij}, \qquad i = 1, 2, \ldots, k, \quad j = 1, 2, \ldots, n, \end{aligned} \tag{4.1}$$

where $\boldsymbol{\mu} = (\mu_1, \mu_2, \ldots, \mu_p)'$ and $\boldsymbol{\alpha}_i = (\alpha_{i1}, \alpha_{i2}, \ldots, \alpha_{ip})'$ provide a mean and "treatment" effect, respectively, for all p variables. The hypothesis of interest in one-way MANOVA is H_0: $\boldsymbol{\mu}_1 = \boldsymbol{\mu}_2 = \cdots = \boldsymbol{\mu}_k$ versus H_1: at least two $\boldsymbol{\mu}_i$'s are unequal. Note that the notation for subscripts differs from that of previous chapters, where the subscript i represented the observation. In this chapter, we use the last subscript in a model such as (4.1) to represent the observation.

We define sample totals and means as follows:

$$\mathbf{y}_{i\cdot} = \textstyle\sum_{j=1}^{n} \mathbf{y}_{ij}, \qquad \mathbf{y}_{\cdot\cdot} = \sum_{i=1}^{k}\sum_{j=1}^{n} \mathbf{y}_{ij},$$

$$\overline{\mathbf{y}}_{i\cdot} = \frac{\mathbf{y}_{i\cdot}}{n}, \qquad \overline{\mathbf{y}}_{\cdot\cdot} = \frac{\mathbf{y}_{\cdot\cdot}}{kn}.$$

To summarize variation in the data, we use "between" and "within" matrices **H** and **E**, defined as

$$\mathbf{H} = n\sum_{i=1}^{k}(\overline{\mathbf{y}}_{i\cdot} - \overline{\mathbf{y}}_{\cdot\cdot})(\overline{\mathbf{y}}_{i\cdot} - \overline{\mathbf{y}}_{\cdot\cdot})' \tag{4.2}$$

$$= \sum_{i=1}^{k} \frac{1}{n}\mathbf{y}_{i\cdot}\mathbf{y}_{i\cdot}' - \frac{1}{kn}\mathbf{y}_{\cdot\cdot}\mathbf{y}_{\cdot\cdot}', \tag{4.3}$$

$$\mathbf{E} = \sum_{i=1}^{k}\sum_{j=1}^{n}(\mathbf{y}_{ij} - \overline{\mathbf{y}}_{i\cdot})(\mathbf{y}_{ij} - \overline{\mathbf{y}}_{i\cdot})' \tag{4.4}$$

$$= \sum_{i=1}^{k}\sum_{j=1}^{n}\mathbf{y}_{ij}\mathbf{y}_{ij}' - \sum_{i=1}^{k}\frac{1}{n}\mathbf{y}_{i\cdot}\mathbf{y}_{i\cdot}'. \tag{4.5}$$

The $p \times p$ "hypothesis" matrix **H** has on its diagonal the between sum of squares (SSH) for each of the p variables. The off-diagonal elements of **H** are analogous sums of products for all pairs of variables. The $p \times p$ "error" matrix **E** has on its diagonal the within sum of squares (SSE) for each variable with analogous sums of products off-diagonal. From (4.3) and (4.5), we can infer the degrees of freedom for hypothesis and error: $\nu_H = k - 1$ and $\nu_E = k(n-1)$. The rank of **H** is $\min(p, \nu_H)$, and the rank of **E** must be p in order to carry out a test; that is, we must have $\nu_E \geq p$.

Several test statistics have been proposed for H_0: $\boldsymbol{\mu}_1 = \boldsymbol{\mu}_2 = \cdots = \boldsymbol{\mu}_k$. We consider the likelihood ratio test in Section 4.1.2, the union–intersection test in Section 4.1.3, and two additional tests in Section 4.1.4.

4.1.2 Wilks' Likelihood Ratio Test

In Section 3.5.5 we showed that the two-sample T^2-test of H_0: $\boldsymbol{\mu}_1 = \boldsymbol{\mu}_2$ is both the likelihood ratio test and the union–intersection test. In MANOVA, these two approaches lead to two different test statistics. We first consider the likelihood ratio test.

Theorem 4.1A. If $\mathbf{y}_{ij}$, $i = 1, 2, \ldots, k$, $j = 1, 2, \ldots, n$, are independently observed from $N_p(\boldsymbol{\mu}_i, \boldsymbol{\Sigma})$, then the likelihood ratio test statistic for H_0: $\boldsymbol{\mu}_1 = \boldsymbol{\mu}_2 = \ldots = \boldsymbol{\mu}_k$ can be expressed as

$$\Lambda = \frac{|\mathbf{E}|}{|\mathbf{E} + \mathbf{H}|},$$

where **H** and **E** are defined in (4.2) and (4.4).

Proof. There is a likelihood function similar to (2.45) or (2.48) for the n observation vectors in each of the k samples. The overall likelihood function for the kn observation vectors $\mathbf{y}_{ij}$ is the product of these k likelihood functions and can be written in the form

$$L(\boldsymbol{\mu}_1, \ldots, \boldsymbol{\mu}_k, \boldsymbol{\Sigma}) = \frac{1}{(\sqrt{2\pi})^{knp} |\boldsymbol{\Sigma}|^{kn/2}} e^{-\frac{1}{2} \operatorname{tr}\left[\boldsymbol{\Sigma}^{-1} \sum_{ij} (\mathbf{y}_{ij} - \boldsymbol{\mu}_i)(\mathbf{y}_{ij} - \boldsymbol{\mu}_i)'\right]}. \tag{4.6}$$

When H_0: $\boldsymbol{\mu}_1 = \boldsymbol{\mu}_2 = \cdots = \boldsymbol{\mu}_k = \boldsymbol{\mu}$ is true, the k samples become a single sample from $N_p(\boldsymbol{\mu}, \boldsymbol{\Sigma})$, and by Theorem 2.3A, the maximum likelihood estimators of $\boldsymbol{\mu}$ and $\boldsymbol{\Sigma}$ are $\overline{\mathbf{y}}_{..}$ and $(\mathbf{E} + \mathbf{H})/kn$, since

$$\mathbf{E} + \mathbf{H} = \sum_{i=1}^{k} \sum_{j=1}^{n} (\mathbf{y}_{ij} - \overline{\mathbf{y}}_{..})(\mathbf{y}_{ij} - \overline{\mathbf{y}}_{..})'. \tag{4.7}$$

Substituting $\hat{\boldsymbol{\mu}} = \overline{\mathbf{y}}_{..}$ and $\hat{\boldsymbol{\Sigma}} = (\mathbf{E} + \mathbf{H})/kn$ into (4.6), we obtain

$$\max_{H_0} L = \frac{1}{(\sqrt{2\pi})^{knp} |(\mathbf{E} + \mathbf{H})/kn|^{kn/2}} e^{-knp/2}. \tag{4.8}$$

Under H_1, the maximum likelihood estimator of each $\boldsymbol{\mu}_i$ is $\overline{\mathbf{y}}_{i.}$ (ignoring the possibility of equality of some $\boldsymbol{\mu}_i$'s as long as at least two $\boldsymbol{\mu}_i$'s are unequal), and the maximum likelihood estimator of $\boldsymbol{\Sigma}$ is the pooled within-sample estimator $\mathbf{E}/kn$. With these estimates, we obtain

$$\max_{H_1} L = \frac{1}{(\sqrt{2\pi})^{knp} |\mathbf{E}/kn|^{kn/2}} e^{-knp/2}. \tag{4.9}$$

From (4.8) and (4.9), the likelihood ratio becomes

$$\text{LR} = \left[\frac{|\mathbf{E}|}{|\mathbf{E} + \mathbf{H}|}\right]^{kn/2}. \tag{4.10}$$

The test is usually conducted using $(\text{LR})^{2/kn}$,

$$\Lambda = \frac{|\mathbf{E}|}{|\mathbf{E} + \mathbf{H}|}, \tag{4.11}$$

which is known as Wilks' Λ. □

The statistic in (4.11), first introduced by Wilks (1932a), is also called Wilks' U. This statistic is distributed as the Wilks Λ-distribution if the following three conditions hold: (1) $\mathbf{E}$ has the Wishart distribution $W_p(\nu_E, \boldsymbol{\Sigma})$, (2) $\mathbf{H}$ is $W_p(\nu_H, \boldsymbol{\Sigma})$ under H_0, and (3) $\mathbf{E}$ and $\mathbf{H}$ are independent. We now show that (4.11) satisfies these three conditions.

Theorem 4.1B. If $\mathbf{y}_{ij}, i = 1, 2, \ldots, k, j = 1, 2 \ldots, n$, are independently distributed as $N_p(\boldsymbol{\mu}_i, \boldsymbol{\Sigma})$, and if **H** and **E** are defined as in (4.2) and (4.4), then **E** is $W_p(\nu_E, \boldsymbol{\Sigma})$, **H** is $W_p(\nu_H, \boldsymbol{\Sigma})$ under H_0: $\boldsymbol{\mu}_1 = \boldsymbol{\mu}_2 = \cdots = \boldsymbol{\mu}_k$, and **E** and **H** are independent.

Proof. Define the $kn \times p$ data matrix for the k samples as

$$\mathbf{Y} = \begin{pmatrix} \mathbf{Y}_1 \\ \mathbf{Y}_2 \\ \vdots \\ \mathbf{Y}_k \end{pmatrix}, \text{ where } \mathbf{Y}_i = \begin{pmatrix} \mathbf{y}'_{i1} \\ \mathbf{y}'_{i2} \\ \vdots \\ \mathbf{y}'_{in} \end{pmatrix}.$$

Define the $kn \times kn$ matrices **A** and **B** as follows:

$$\mathbf{A} = \frac{1}{n}\begin{pmatrix} \mathbf{J}_n & \mathbf{O} & \ldots & \mathbf{O} \\ \mathbf{O} & \mathbf{J}_n & \ldots & \mathbf{O} \\ \vdots & \vdots & & \vdots \\ \mathbf{O} & \mathbf{O} & \ldots & \mathbf{J}_n \end{pmatrix} - \frac{1}{kn}\mathbf{J}_{kn}$$

$$= \frac{1}{n}\mathbf{G} - \frac{1}{kn}\mathbf{J}_{kn},$$

$$\mathbf{B} = \begin{pmatrix} \mathbf{I}_n - \frac{1}{n}\mathbf{J}_n & \mathbf{O} & \ldots & \mathbf{O} \\ \mathbf{O} & \mathbf{I}_n - \frac{1}{n}\mathbf{J}_n & \ldots & \mathbf{O} \\ \vdots & \vdots & & \vdots \\ \mathbf{O} & \mathbf{O} & \ldots & \mathbf{I}_n - \frac{1}{n}\mathbf{J}_n \end{pmatrix}$$

$$= \mathbf{I}_{kn} - \frac{1}{n}\mathbf{G},$$

where $\mathbf{J}_n$ is an $n \times n$ matrix of 1s, $\mathbf{J}_{kn}$ is a $kn \times kn$ matrix of 1s, $\mathbf{I}_n$ is an $n \times n$ identity matrix, $\mathbf{I}_{kn}$ is a $kn \times kn$ identity matrix, and **O** is $n \times n$. The between and within matrices **H** and **E** in (4.3) and (4.5) can be expressed as $\mathbf{H} = \mathbf{Y}'\mathbf{A}\mathbf{Y}$ and $\mathbf{E} = \mathbf{Y}'\mathbf{B}\mathbf{Y}$ (see Problem 4.8). It can be shown that **A** and **B** are idempotent of rank $k - 1$ and $k(n - 1)$, respectively, and that $\mathbf{AB} = \mathbf{O}$ (Problem 4.9). The Wishart distributions and independence now follow from Theorem 2.3G and Theorem 2.3H. □

By Theorem 4.1B, Wilks' Λ is analogous to an F random variable, defined as the ratio of two independent chi-square random variables divided by their respective degrees of freedom. The Wilks Λ-distribution is indexed by three parameters: the dimension p and the degrees of freedom for hypothesis and error, ν_H and ν_E. For the one-way model in this section, $\nu_H = k - 1$ and $\nu_E = k(n - 1)$. We reject H_0: $\boldsymbol{\mu}_1 = \boldsymbol{\mu}_2 = \cdots = \boldsymbol{\mu}_k$ if $\Lambda \le \Lambda_{\alpha,p,\nu_H,\nu_E}$. Exact critical values $\Lambda_{\alpha,p,\nu_H,\nu_E}$ for Wilks' Λ have been tabulated by Wall (1967). An excerpt of Wall's tables appears in Table B.4 in Appendix B.

In the remainder of this section, we list some features of Wilks' Λ.

1. The matrices **E** and **E** + **H** are positive definite if $\nu_E \ge p$ and if the variables are not linearly related. Hence, Λ is defined only under the condition $\nu_E \ge p$.

2. The degrees of freedom ν_H and ν_E for Wilks' Λ are always the same as in the analogous univariate model. For example, in the balanced one-way model, it was noted above that $\nu_H = k - 1$ and $\nu_E = k(n - 1)$.
3. Wilks' Λ is invariant to full-rank linear (affine) transformations. Thus if $\mathbf{y}_{ij}$ is transformed to $\mathbf{z}_{ij} = \mathbf{F}\mathbf{y}_{ij} + \mathbf{g}$, where $\mathbf{F}$ is a $p \times p$ nonsingular matrix and $\mathbf{g}$ is a vector of constants, then

$$\Lambda_z = \Lambda_y. \tag{4.12}$$

4. The moments of Λ are given by

$$E(\Lambda^r) = \prod_{i=1}^{p} \frac{\Gamma\left[\frac{1}{2}(\nu_E - i + 1) + r\right] \Gamma\left[\frac{1}{2}(\nu_H + \nu_E - i + 1)\right]}{\Gamma\left[\frac{1}{2}(\nu_E - i + 1)\right] \Gamma\left[\frac{1}{2}(\nu_H + \nu_E - i + 1) + r\right]}, \tag{4.13}$$

where $\Gamma(\alpha)$ is the gamma function defined for $\alpha > 0$ by

$$\Gamma(\alpha) = \int_0^\infty e^{-x} x^{\alpha - 1}\, dx.$$

5. The distribution of Λ_{p,ν_H,ν_E} is the same as that of $\Lambda_{\nu_H,p,\nu_E+\nu_H-p}$.
6. Wilks' Λ can be written in terms of the eigenvalues $\lambda_1, \lambda_2, \ldots, \lambda_p$ of $\mathbf{E}^{-1}\mathbf{H}$:

$$\Lambda = \prod_{i=1}^{p} \frac{1}{1 + \lambda_i}. \tag{4.14}$$

We could also use the eigenvalues of $\mathbf{H}\mathbf{E}^{-1}$, which are the same as those of $\mathbf{E}^{-1}\mathbf{H}$ [see (A.10.9)]. However, the eigenvectors of the two matrices are different; we will use those of $\mathbf{E}^{-1}\mathbf{H}$ in Section 4.1.3.
7. The range of Λ is $0 \leq \Lambda \leq 1$, and we reject H_0 for small values of Λ. This follows because Λ is a simple function of the likelihood ratio in (4.10), small values of which lead to rejection of H_0.
8. For fixed values of ν_H and ν_E, the critical values of Λ decrease as p increases (see Table B.4). Thus the addition of variables that do not contribute to rejection of the hypothesis will lead to a reduction in power.
9. When $p = 1$, Wilks' Λ reduces to a simple function of an F-statistic, since in this case $|\mathbf{E}| = \text{SSE}$ and $|\mathbf{E} + \mathbf{H}| = \text{SSE} + \text{SSH}$. Thus for $p = 1$,

$$\Lambda = \frac{\text{SSE}}{\text{SSE} + \text{SSH}} = \frac{1}{1 + \text{SSH}/\text{SSE}} = \frac{1}{1 + \nu_H F/\nu_E}.$$

Solving for F gives

$$F = \frac{1 - \Lambda}{\Lambda} \frac{\nu_E}{\nu_H}. \tag{4.15}$$

Additional transformations from Λ to F are available when $p = 2$ and when $\nu_H = 1$ or 2. The four cases are summarized in Table 4.1.

Table 4.1. Transformations of Λ to Provide Exact Upper-Tail Tests Using F-Distributions

Parameters p, ν_H	Statistic Having F-distribution	Degrees of Freedom
$\nu_H = 1$, any p	$\dfrac{1-\Lambda}{\Lambda} \cdot \dfrac{\nu_E + \nu_H - p}{p}$	$p, \nu_E + \nu_H - p$
$\nu_H = 2$, any p	$\dfrac{1-\sqrt{\Lambda}}{\sqrt{\Lambda}} \cdot \dfrac{\nu_E + \nu_H - p - 1}{p}$	$2p, 2(\nu_E + \nu_H - p - 1)$
$p = 1$, any ν_H	$\dfrac{1-\Lambda}{\Lambda} \cdot \dfrac{\nu_E}{\nu_H}$	ν_H, ν_E
$p = 2$, any ν_H	$\dfrac{1-\sqrt{\Lambda}}{\sqrt{\Lambda}} \cdot \dfrac{\nu_E - 1}{\nu_H}$	$2\nu_H, 2(\nu_E - 1)$

10. If neither p nor ν_H is equal to 1 or 2, an approximate F-test can be performed. Rao (1951) showed that

$$F = \frac{1 - \Lambda^{1/t}}{\Lambda^{1/t}} \frac{\text{df}_2}{\text{df}_1} \tag{4.16}$$

is approximately distributed as an F with df_1 and df_2 degrees of freedom, where

$$\text{df}_1 = p\nu_H, \qquad \text{df}_2 = wt - \tfrac{1}{2}(p\nu_H - 2),$$

$$w = \nu_E + \nu_H - \tfrac{1}{2}(p + \nu_H + 1), \qquad t = \sqrt{\frac{p^2\nu_H^2 - 4}{p^2 + \nu_H^2 - 5}}.$$

When $p\nu_H = 2$, we set $t = 1$. When either ν_H or p is 1 or 2, the approximate F in (4.16) reduces in each case to the exact F-statistic given in Table 4.1.

A less accurate approximate test given by Bartlett (1938, 1947),

$$\chi^2 = -\left[\nu_E - \tfrac{1}{2}(p - \nu_H + 1)\right] \ln \Lambda = -f \ln \Lambda, \tag{4.17}$$

has an approximate χ^2-distribution with $p\nu_H$ degrees of freedom. We reject H_0 if $\chi^2 \geq \chi^2_\alpha$. If $p^2 + \nu_H^2 \leq f/3$, the approximation is accurate to three decimal places (Anderson 1984, p. 318).

11. Wilks' Λ is not associated directly with the univariate analysis of variance (ANOVA) F-tests on the individual variables in terms of acceptance and rejection. If the multivariate test based on Λ leads to rejection of $H_0\colon \boldsymbol{\mu}_1 = \boldsymbol{\mu}_2 = \cdots = \boldsymbol{\mu}_k$, it is possible that none of the F_i's on the individual variables will reject $H_{0i}\colon \mu_{1i} = \mu_{2i} = \cdots = \mu_{ki}, i = 1, 2, \ldots, p$. Conversely, when the Λ-test leads to acceptance of H_0, it is possible that one or more of the F's will lead to rejection of H_{0i}. A similar situation with Hotelling's T^2 was discussed in Section 3.4.5.

4.1.3 Roy's Union–Intersection Test

The union–intersection approach to test construction was introduced in Sections 3.3.8 and 3.5.5. To apply the method in the MANOVA setting, we define a linear combination $z_{ij} = \mathbf{a}'\mathbf{y}_{ij}$ for every $\mathbf{y}_{ij}$, $i = 1, 2, \ldots, k$, $j = 1, 2, \ldots, n$, thus reducing the observation vectors to scalars. We then find the vector $\mathbf{a}$ that maximizes the usual univariate F-statistic comparing the k groups,

$$F(\mathbf{a}) = \frac{ns_{\bar{z}}^2}{s_{e,z}^2} = \frac{n\sum_{i=1}^{k}(\bar{z}_{i\cdot} - \bar{z}_{\cdot\cdot})^2/(k-1)}{\sum_{i=1}^{k}\sum_{j=1}^{n}(z_{ij} - \bar{z}_{i\cdot})^2/(kn-k)}. \tag{4.18}$$

From (1.73), the sample variance of a transformed variable $z = \mathbf{a}'\mathbf{y}$ is $s_z^2 = \mathbf{a}'\mathbf{S}_y\mathbf{a}$. By analogy, (4.18) becomes

$$F(\mathbf{a}) = \frac{\mathbf{a}'\mathbf{H}\mathbf{a}/(k-1)}{\mathbf{a}'\mathbf{E}\mathbf{a}/(kn-k)}. \tag{4.19}$$

We establish the denominator directly (for the numerator, see Problem 4.13). The observation vectors $\mathbf{y}_{ij} = (y_{ij1}, y_{ij2}, \ldots, y_{ijp})'$ are transformed to scalars, $z_{ij} = \mathbf{a}'\mathbf{y}_{ij} = a_1y_{ij1} + a_2y_{ij2} + \cdots + a_py_{ijp}$, $i = 1, 2, \ldots, k$, $j = 1, 2, \ldots, n$. The mean of the ith sample can be expressed as

$$\begin{aligned}
\bar{z}_{i\cdot} &= \frac{1}{n}\sum_{j=1}^{n} z_{ij} \\
&= \frac{1}{n}\sum_{j=1}^{n} \mathbf{a}'\mathbf{y}_{ij} \\
&= \mathbf{a}'\sum_{j=1}^{n} \frac{\mathbf{y}_{ij}}{n} \qquad [\text{ by (A.2.17)}] \\
&= \mathbf{a}'\bar{\mathbf{y}}_{i\cdot},
\end{aligned}$$

and the within sum of squares for z is

$$\begin{aligned}
\text{SSE}(z) &= \sum_{i=1}^{k}\sum_{j=1}^{n}(z_{ij} - \bar{z}_{i\cdot})^2 \\
&= \sum_{ij}(\mathbf{a}'\mathbf{y}_{ij} - \mathbf{a}'\bar{\mathbf{y}}_{i\cdot})^2 \\
&= \mathbf{a}'\left[\sum_{ij}(\mathbf{y}_{ij} - \bar{\mathbf{y}}_{i\cdot})(\mathbf{y}_{ij} - \bar{\mathbf{y}}_{i\cdot})'\right]\mathbf{a} \qquad [\text{by (A.2.19)}] \\
&= \mathbf{a}'\mathbf{E}\mathbf{a} \qquad [\text{by (4.4)}].
\end{aligned}$$

As in Sections 3.3.8 and 3.5.5, we reject H_0: $\boldsymbol{\mu}_1 = \boldsymbol{\mu}_2 = \cdots = \boldsymbol{\mu}_k$ if $\max_{\mathbf{a}} F(\mathbf{a}) \geq c$, where $F(\mathbf{a})$ is given in (4.19). Maximizing $F(\mathbf{a})$ is equivalent to maximizing $\mathbf{a}'\mathbf{Ha}/\mathbf{a}'\mathbf{Ea} = \lambda$. The vector $\mathbf{a}$ that maximizes λ and the resulting maximum value of λ are given in the following theorem.

Theorem 4.1C. Under the assumptions of Theorem 4.1A, the maximum value of $\lambda = \mathbf{a}'\mathbf{Ha}/\mathbf{a}'\mathbf{Ea}$ and the vector $\mathbf{a}$ that produces the maximum are given by the largest eigenvalue λ_1 and the associated eigenvector of $\mathbf{E}^{-1}\mathbf{H}$, respectively.

Proof. To find $\mathbf{a}$, we differentiate λ with respect to $\mathbf{a}$ [see (A.13.3)] and set the result equal to $\mathbf{0}$:

$$\frac{\partial \lambda}{\partial \mathbf{a}} = \frac{\mathbf{a}'\mathbf{Ea}(2\mathbf{Ha}) - \mathbf{a}'\mathbf{Ha}(2\mathbf{Ea})}{(\mathbf{a}'\mathbf{Ea})^2} = \mathbf{0}.$$

Multiplying by $\mathbf{a}'\mathbf{Ea}$, we obtain

$$\left(\frac{\mathbf{a}'\mathbf{Ea}}{\mathbf{a}'\mathbf{Ea}}\right)\mathbf{Ha} - \left(\frac{\mathbf{a}'\mathbf{Ha}}{\mathbf{a}'\mathbf{Ea}}\right)\mathbf{Ea} = \mathbf{0}.$$

By the definition of λ, this can be written

$$\mathbf{Ha} - \lambda\mathbf{Ea} = \mathbf{0}$$

or

$$(\mathbf{H} - \lambda\mathbf{E})\mathbf{a} = \mathbf{0}. \tag{4.20}$$

Multiplying (on the left) by $\mathbf{E}^{-1}$ reduces (4.20) to the form

$$(\mathbf{E}^{-1}\mathbf{H} - \lambda\mathbf{I})\mathbf{a} = \mathbf{0}. \tag{4.21}$$

□

The function $z = \mathbf{a}'\mathbf{y}$, called the *discriminant function*, is discussed further in Sections 4.6 and 4.8.3 and in Chapter 5.

The product $\mathbf{E}^{-1}\mathbf{H}$ serves as a "standardized" version of $\mathbf{H}$, comparing between to within variation. Since the rank of $\mathbf{E}^{-1}\mathbf{H}$ is the same as the rank of $\mathbf{H}$ [see (A.5.2)], there are $s = \min(\nu_H, p)$ nonzero eigenvalues $\lambda_1 > \lambda_2 > \cdots > \lambda_s > 0$. (The probability is 0 that two eigenvalues would be equal.) By Theorem 4.1C, the union–intersection test uses only λ_1, the largest. The test based on λ_1 is usually referred to as *Roy's largest root test* after S. N. Roy, who first suggested union–intersection tests (Roy 1939, 1953, 1957). Roy's test, like Wilks', is invariant to full rank linear (affine) transformations.

To test H_0: $\boldsymbol{\mu}_1 = \boldsymbol{\mu}_2 = \cdots = \boldsymbol{\mu}_k$ using λ_1, we distinguish two cases. When $\nu_H = 1$ and $p \geq 2$, we use

$$F = \frac{\nu_E - p + 1}{p}\lambda_1, \tag{4.22}$$

which has an exact F-distribution with $|\nu_H - p| + 1$ and $\nu_E - p + 1$ degrees of freedom. When $\nu_H \geq 2$ and $p \geq 2$, we use Roy's test statistic

$$\theta = \frac{\lambda_1}{1 + \lambda_1}. \tag{4.23}$$

Critical values for θ are given in Table B.5 in Appendix B (Pearson and Hartley 1972, Pillai 1964, 1965). For tables that provide more extensive coverage of the parameters, see Pillai and Flury (1984). We reject H_0: $\boldsymbol{\mu}_1 = \boldsymbol{\mu}_2 = \cdots = \boldsymbol{\mu}_k$ if $\theta \geq \theta_{\alpha,s,m,N}$, where s, m, and N are defined by

$$s = \min(\nu_H, p), \qquad m = \tfrac{1}{2}(|\nu_H - p| - 1), \qquad N = \tfrac{1}{2}(\nu_E - p - 1).$$

Roy's test statistic θ does not have a transformation to an approximate F, but an "upper bound" on F that provides a "lower bound" on the p-value is given by

$$F^* = \frac{(\nu_E - d - 1)\lambda_1}{d}, \tag{4.24}$$

with degrees of freedom d and $\nu_E - d - 1$, where $d = \max(p, \nu_H)$. To partially justify (4.24), we note that if $\mathbf{a}$ is the eigenvector corresponding to the largest eigenvalue λ_1, then (4.19) becomes

$$\max_{\mathbf{a}} F(\mathbf{a}) = \frac{\mathbf{a}'\mathbf{H}\mathbf{a}/\nu_H}{\mathbf{a}'\mathbf{E}\mathbf{a}/\nu_E} = \lambda_1 \frac{\nu_E}{\nu_H}, \tag{4.25}$$

which resembles (4.24), with adjustment to degrees of freedom. Since (4.24) is related to the maximum value of F, it is referred to as an upper bound. Note that (4.19) would have an F-distribution for an arbitrary value of $\mathbf{a}$ (chosen before seeing the data) but that it does not have an F-distribution for the value of $\mathbf{a}$ that maximizes $F(\mathbf{a})$ based on the observed values of $\mathbf{E}$ and $\mathbf{H}$.

It is of interest to compare (4.25) with the F-value of each of the p variables, $F_i = ns^2_{\bar{y}_i}/s^2_{e,y_i}$, $i = 1, 2, \ldots, p$, as in (4.18). In (4.25) we have the largest F-value that could be obtained from any linear combination of the individual variables. If one of the F_i-values is close to the maximum possible F, then the other variables add little to separation of the mean vectors.

The matrix $\mathbf{E}^{-1}\mathbf{H}$ is not symmetric. However, it has the same eigenvalues as the symmetric matrix $(\mathbf{U}')^{-1}\mathbf{H}\mathbf{U}^{-1}$, where $\mathbf{U}'\mathbf{U} = \mathbf{E}$ is the Cholesky factorization of $\mathbf{E}$ (Section A.6). This follows from the result in (A.10.9) that the eigenvalues of $\mathbf{AB}$ are the same as those of $\mathbf{BA}$. Let $\mathbf{A} = \mathbf{U}^{-1}$ and $\mathbf{B} = (\mathbf{U}')^{-1}\mathbf{H}$. Then $\mathbf{AB} = \mathbf{U}^{-1}(\mathbf{U}')^{-1}\mathbf{H} = (\mathbf{U}'\mathbf{U})^{-1}\mathbf{H} = \mathbf{E}^{-1}\mathbf{H}$ has the same eigenvalues as $\mathbf{BA} = (\mathbf{U}')^{-1}\mathbf{H}\mathbf{U}^{-1}$. Note that $(\mathbf{U}')^{-1}\mathbf{H}\mathbf{U}^{-1}$ is positive semidefinite, so that $\lambda_i \geq 0$ for all i. The eigenvectors of $\mathbf{BA} = (\mathbf{U}')^{-1}\mathbf{H}\mathbf{U}^{-1}$ are of the form $\mathbf{Ba} = (\mathbf{U}')^{-1}\mathbf{Ha}$, where $\mathbf{a}$ is an eigenvector of $\mathbf{AB} = \mathbf{E}^{-1}\mathbf{H}$ (See Rencher 1995, Section 6.1.4).

4.1.4 The Pillai and Lawley–Hotelling Test Statistics

We consider two additional test statistics for H_0: $\boldsymbol{\mu}_1 = \boldsymbol{\mu}_2 = \cdots = \boldsymbol{\mu}_k$. *Pillai's trace statistic*

$$V^{(s)} = \mathrm{tr}\left[(\mathbf{E} + \mathbf{H})^{-1}\mathbf{H}\right] = \sum_{i=1}^{s} \frac{\lambda_i}{1 + \lambda_i} \tag{4.26}$$

was proposed by Pillai (1955). We reject H_0 for $V^{(s)} \geq V^{(s)}_{\alpha,s,m,N}$, where s, m, and N are defined as in Section 4.1.3 for Roy's test. Upper percentage points of $V^{(s)}$ for $s \geq 2$ were provided by Schuurmann et al. (1975); see Table B.6 in Appendix B. For $s = 1$, use (4.22), since Pillai's test statistic in (4.26) reduces to Roy's statistic $\theta = \lambda_1/(1 + \lambda_1)$ when $s = 1$. Good approximations for p-values of $V^{(s)}$ were given by Butler and Huzurbazar (1992).

An approximate F-statistic for $V^{(s)}$ (Pillai 1954, 1956a) is given by

$$F_1 = \frac{(2N + s + 1)V^{(s)}}{(2m + s + 1)(s - V^{(s)})} \tag{4.27}$$

with $s(2m + s + 1)$ and $s(2N + s + 1)$ degrees of freedom. Two alternative F-approximations were proposed by Pillai (1956a, 1960, p. 19):

$$F_2 = \frac{s(\nu_E - \nu_H + s)V^{(s)}}{p\nu_H(s - V^{(s)})} \tag{4.28}$$

with $p\nu_H$ and $s(\nu_E - \nu_H + s)$ degrees of freedom and

$$F_3 = \frac{(\nu_E - p + s)V^{(s)}}{d(s - V^{(s)})} \tag{4.29}$$

with sd and $s(\nu_E - p + s)$ degrees of freedom, where $d = \max(p, \nu_H)$. It can be shown that F_3 in (4.29) is the same as F_1 in (4.27).

The *Lawley–Hotelling trace statistic*, due to Lawley (1938) and Hotelling (1951), is given by

$$U^{(s)} = \mathrm{tr}(\mathbf{E}^{-1}\mathbf{H}) = \sum_{i=1}^{s} \lambda_i \tag{4.30}$$

and is also called *Hotelling's generalized* T^2*-statistic*. Upper percentage points of

$$\frac{\nu_E}{\nu_H}\,\mathrm{tr}(\mathbf{E}^{-1}\mathbf{H}) = \frac{\nu_E}{\nu_H}U^{(s)}, \tag{4.31}$$

by Davis (1970a, 1970b, 1980a), are given in Table B.7. An F-approximation (McKeon 1974) is given by

$$F_1 = \frac{U^{(s)}}{c} = \frac{\mathrm{tr}(\mathbf{E}^{-1}\mathbf{H})}{c}, \tag{4.32}$$

which is approximately distributed as $F_{a,b}$, where

$$a = p\nu_H, \qquad b = 4 + \frac{a+2}{B-1},$$

$$c = \frac{a(b-2)}{b(\nu_E - p - 1)}, \qquad B = \frac{(\nu_E + \nu_H - p - 1)(\nu_E - 1)}{(\nu_E - p - 3)(\nu_E - p)}.$$

Another F-approximation (Pillai 1954, 1956b) is

$$F_2 = \frac{2(sN+1)U^{(s)}}{s^2(2m+s+1)}, \tag{4.33}$$

with $s(2m + s + 1)$ and $2(sN + 1)$ degrees of freedom, and a third approximation (Pillai and Sampson 1959) is

$$F_3 = \frac{[s(\nu_E - s - 1) + 2]U^{(s)}}{s^2 \nu_H}, \tag{4.34}$$

with $s\nu_H$ and $s(\nu_E - s - 1) + 2$ degrees of freedom. When $p \leq \nu_H$, F_2 and F_3 are the same.

Swalberg (1995) compared the Type I error rates of the F-approximations for $V^{(s)}$ and $U^{(s)}$ using simulation methods. He found F_1 for $V^{(s)}$ to be superior to F_2. For $U^{(s)}$, F_1 and F_2 both do very well.

As with Wilks' Λ and Roy's θ, the Pillai and Lawley–Hotelling test statistics are invariant to full rank linear (affine) transformations of the form $\mathbf{z}_{ij} = \mathbf{F}\mathbf{y}_{ij} + \mathbf{g}$.

Other tests based on eigenvalues have been proposed (Gnanadesikan et al. 1965, Olson 1975, Pillai 1955, Roy 1945, 1953), but each is inferior to at least one of the above four that are in standard use.

4.1.5 Summary of the Four Test Statistics

We summarize the four test statistics in terms of $\mathbf{E}, \mathbf{H}$, and the eigenvalues $\lambda_1 > \lambda_2 > \cdots > \lambda_s$ of $\mathbf{E}^{-1}\mathbf{H}$, where $s = \min(\nu_H, p)$, p is the number of variables, ν_H is the hypothesis degrees of freedom, and ν_E is the error degrees of freedom:

Wilks' lambda $\quad \Lambda = \prod_{i=1}^{s} \frac{1}{1+\lambda_i} = \frac{|\mathbf{E}|}{|\mathbf{H}+\mathbf{E}|}$

Pillai $\quad V^{(s)} = \sum_{i=1}^{s} \frac{\lambda_i}{1+\lambda_i} = \operatorname{tr}\left[(\mathbf{E}+\mathbf{H})^{-1}\mathbf{H}\right]$

Lawley–Hotelling $\quad U^{(s)} = \sum_{i=1}^{s} \lambda_i = \operatorname{tr}(\mathbf{E}^{-1}\mathbf{H})$

Roy's largest root $\quad \theta = \frac{\lambda_1}{1+\lambda_1}$

Clearly, for all four tests we must have $\nu_E \geq p$.

When H_0 is true, all four tests have probability α of rejecting H_0. However, the power of each test depends on the configuration of $\boldsymbol{\mu}_1, \boldsymbol{\mu}_2, \ldots, \boldsymbol{\mu}_k$ in the s-dimensional space of mean vectors. In fact, for a given sample the four tests may lead to different conclusions even when H_0 is true; some may reject H_0 while others may accept it. For further discussion, see Section 4.2 and Rencher (1995, Section 6.1.7).

When $\nu_H = 1$, there is only one nonzero eigenvalue. In this case, all four test statistics are functions of each other and can be transformed to an F using (4.22). In terms of Λ, for example, the other three become

$$U^{(1)} = \lambda_1 = \frac{1-\Lambda}{\Lambda}, \tag{4.35}$$

$$V^{(1)} = \frac{\lambda_1}{1+\lambda_1} = 1 - \Lambda, \tag{4.36}$$

$$\theta = 1 - \Lambda. \tag{4.37}$$

4.1.6 Effect of an Additional Variable on Wilks' Λ

As with Hotelling's T^2 (see Section 3.3.5), the addition of variables to Wilks' Λ may either strengthen the evidence against the hypothesis or weaken it. A glance at the critical values of Λ in Table B.4 will show that when the number of variables increases, the power is reduced if the additional variables are not sufficiently informative. For example,

$$\Lambda_{.05,5,3,30} - \Lambda_{.05,6,3,30} = .4147 - .3547 = .0600.$$

In this case, the addition of a sixth variable must reduce Λ by .06 to avoid a loss of power. In Section 4.9.1, we give a test for the significance of the change in Λ due to an additional variable. In the present section, we examine the factors that determine the size of the change. The following discussion is based on Rencher (1993).

Using the notation of Sections 1.5, 3.3.5, and 3.5.3, we denote by x the variable that is to be added to $\mathbf{y}' = (y_1, y_2, \ldots, y_p)$. With this notation, the hypothesis and error matrices for $y_1, y_2, \ldots, y_p, x$ can be written as

$$\mathbf{H} = \begin{pmatrix} \mathbf{H}_{yy} & \mathbf{h}_{xy} \\ \mathbf{h}'_{xy} & h_{xx} \end{pmatrix}, \qquad \mathbf{E} = \begin{pmatrix} \mathbf{E}_{yy} & \mathbf{e}_{xy} \\ \mathbf{e}'_{xy} & e_{xx} \end{pmatrix}.$$

The precise nature of the reduction in Wilks' Λ due to the added variable x is given in the following theorem.

Theorem 4.1D. Wilks' Λ for $y_1, y_2, \ldots, y_p, x$ can be partitioned as

$$\Lambda_{y,x} = \Lambda_y \frac{1 - R_e^2}{(1 + cF_x)(1 - R_{e+h}^2)},$$

where Λ_y is the value of Λ based on the y's, R_e^2 is the within-groups squared multiple correlation of x regressed on the y's (based on $\mathbf{E}$), R_{e+h}^2 is the squared multiple correlation of x regressed on the y's in the combined sample ignoring groups (based on $\mathbf{E} + \mathbf{H}$), F_x is the univariate ANOVA F-statistic comparing the groups on x alone, and $c = \nu_H/\nu_E$.

Proof. Applying (A.7.11) to the partitioned form of $\mathbf{H}$ and $\mathbf{E}$ above, Wilks' Λ for the y's and x can be factored:

$$\begin{aligned}
\Lambda_{y,x} &= \frac{|\mathbf{E}|}{|\mathbf{E}+\mathbf{H}|} \\
&= \frac{|\mathbf{E}_{yy}|(e_{xx} - \mathbf{e}_{xy}'\mathbf{E}_{yy}^{-1}\mathbf{e}_{xy})}{|\mathbf{E}_{yy}+\mathbf{H}_{yy}|[e_{xx}+h_{xx} - (\mathbf{e}_{xy}+\mathbf{h}_{xy})'(\mathbf{E}_{yy}+\mathbf{H}_{yy})^{-1}(\mathbf{e}_{xy}+\mathbf{h}_{xy})]} \\
&= \frac{|\mathbf{E}_{yy}|}{|\mathbf{E}_{yy}+\mathbf{H}_{yy}|} \\
&\quad \times \frac{e_{xx}(1 - \mathbf{e}_{xy}'\mathbf{E}_{yy}^{-1}\mathbf{e}_{xy}/e_{xx})}{(e_{xx}+h_{xx})\left[1 - (\mathbf{e}_{xy}+\mathbf{h}_{xy})'(\mathbf{E}_{yy}+\mathbf{H}_{yy})^{-1}(\mathbf{e}_{xy}+\mathbf{h}_{xy})/(e_{xx}+h_{xx})\right]}.
\end{aligned} \tag{4.38}$$

In (4.38), the ratio $|\mathbf{E}_{yy}|/|\mathbf{E}_{yy}+\mathbf{H}_{yy}|$ is Wilks' Λ based on the y's and will be denoted by Λ_y. Since $h_{xx} = \text{SSH}_x$ and $e_{xx} = \text{SSE}_x$, the ratio $e_{xx}/(e_{xx}+h_{xx})$ can be written as

$$\frac{e_{xx}}{e_{xx}+h_{xx}} = \frac{1}{1+h_{xx}/e_{xx}} = \frac{1}{1+\nu_H F_x/\nu_E},$$

where F_x is the univariate ANOVA F-statistic comparing the groups on x alone. The second term in the parentheses in the numerator of (4.38) can be written as

$$\begin{aligned}
\frac{\mathbf{e}_{xy}'\mathbf{E}_{yy}^{-1}\mathbf{e}_{xy}}{e_{xx}} &= \frac{(\mathbf{e}_{xy}/\nu_E)'(\mathbf{E}_{yy}/\nu_E)^{-1}(\mathbf{e}_{xy}/\nu_E)}{e_{xx}/\nu_E} \\
&= \frac{\mathbf{s}_{xy}'\mathbf{S}_{yy}^{-1}\mathbf{s}_{xy}}{s_{xx}} = R_e^2,
\end{aligned}$$

which, by (7.66), is the within-groups squared multiple correlation of x regressed on the y's. Similarly, the second term in brackets in the denominator of (4.38) becomes R_{e+h}^2, the squared multiple correlation of x regressed on the y's in the combined sample ignoring groups. With these simplifications, (4.38) becomes

$$\Lambda_{y,x} = \Lambda_y \frac{1-R_e^2}{(1+cF_x)(1-R_{e+h}^2)}, \tag{4.39}$$

where $c = \nu_H/\nu_E$. □

By Theorem 4.1D, the reduction from Λ_y to $\Lambda_{y,x}$ will be important if F_x is relatively large or if $R_e^2 > R_{e+h}^2$, so that x is more correlated with the y's after fitting the MANOVA model than before. If x is less correlated with the y's after fitting the MANOVA model ($R_e^2 < R_{e+h}^2$), then the reduction from Λ_y to $\Lambda_{y,x}$ is less (but may still be substantial if F_x is large). Note that if x were orthogonal to the y's, we would have $R_e^2 = R_{e+h}^2 = 0$ because $\mathbf{e}_{xy} = \mathbf{h}_{xy} = \mathbf{0}$, and (4.39) would simplify to $\Lambda_{y,x} = \Lambda_y/(1 + cF_x)$.

4.1.7 Tests on Individual Variables

In the previous section, we presented a descriptive breakdown of the effect of each variable on Wilks' Λ in the presence of the other variables. In this section, we consider univariate tests on individual variables. These tests ignore correlations among the variables.

4.1.7a Tests on Individual Variables without a Multivariate Test

In this section we consider testing each variable separately without regard to a MANOVA test; that is, a MANOVA test is not performed or is ignored if performed.

An experimenter could test $H_0\colon \boldsymbol{\mu}_1 = \boldsymbol{\mu}_2 = \cdots = \boldsymbol{\mu}_k$ by testing each of $H_{0i}\colon \mu_{1i} = \mu_{2i} = \cdots = \mu_{ki}, i = 1, 2, \ldots, p$. This can be done by carrying out a univariate ANOVA F-test on each of the p variables. To protect against inflation of the experimentwise error rate (overall α-level; see Section 3.4.5), we could use a Bonferroni approach and test each variable at the α/p level of significance (see Sections 3.4.5 and 3.6.4). Values of $F_{\alpha/p,\nu_H,\nu_E}$ are available in Ludwig, Gottlieb, and Lienert (1986). Alternatively, we can compute the p-value of each test and reject H_{0i} if the p-value is less than α/p (see Section 3.4.5). (Note the distinction between the use of p as the number of variables and the use of p in the p-value defined in Section 3.4.5.)

4.1.7b Tests on Individual Variables following a Multivariate Test

If a MANOVA test such as Wilks' Λ is carried out prior to conducting F-tests on the individual variables, the result of the MANOVA test can be used as a guide in making decisions about use of the F-tests. Rencher and Scott (1990; see also Rencher 1995, Section 6.4) showed that if individual F-tests are conducted only when the Wilks Λ-test rejects $H_0\colon \boldsymbol{\mu}_1 = \boldsymbol{\mu}_2 = \cdots = \boldsymbol{\mu}_k$, then the experimentwise error rate remains close to the nominal value.

Tests conducted in this manner are reminiscent of Fisher's (1949) protected LSD (least significant difference) tests, conducted as follows. If an ANOVA F-test rejects $H_0\colon \mu_1 = \mu_2 = \cdots = \mu_k$, differences between means $\overline{y}_{i.}$ and $\overline{y}_{j.}$ are compared to

$$\text{LSD} = t_{\alpha/2}\sqrt{s_e^2\left(\frac{1}{n_i} + \frac{1}{n_j}\right)}.$$

Carmer and Swanson (1973) showed that the experimentwise error rate of this procedure is close to the nominal α-value.

4.2 POWER AND ROBUSTNESS COMPARISONS FOR THE FOUR MANOVA TEST STATISTICS

When $H_0: \boldsymbol{\mu}_1 = \boldsymbol{\mu}_2 = \cdots = \boldsymbol{\mu}_k$ is true, the mean vectors are coincident, and all four MANOVA test statistics have the same Type I error rate α, as noted in Section 4.1.5. When H_0 is false, however, the four tests have different levels of power.

In univariate ANOVA with $p = 1$, the F-test is uniformly most powerful. When $p = 1$, the means $\mu_1, \mu_2, \ldots, \mu_k$ can be uniquely ordered in one dimension. In MANOVA, the mean vectors are points in $s = \min(p, \nu_H)$ dimensions, and no single test is uniformly most powerful. The power of each of the four tests depends on the configuration of the mean vectors $\boldsymbol{\mu}_1, \boldsymbol{\mu}_2, \ldots, \boldsymbol{\mu}_k$ in the s-dimensional space, specifically on the dimension of the subspace occupied by the mean vectors.

In Section 4.1.5 (and Sections 4.1.2–4.1.4) we noted that the four test statistics are functions of the s eigenvalues of $\mathbf{E}^{-1}\mathbf{H}$. To consider the population eigenvalues that correspond to these sample eigenvalues, define the matrix

$$\boldsymbol{\Omega} = n \sum_{i=1}^{k} (\boldsymbol{\mu}_i - \boldsymbol{\mu})(\boldsymbol{\mu}_i - \boldsymbol{\mu})', \tag{4.40}$$

where $\boldsymbol{\mu} = \sum_{i=1}^{k} \boldsymbol{\mu}_i / k$ is the overall mean vector. Then, with an adjustment for degrees of freedom, $\boldsymbol{\Sigma}^{-1}\boldsymbol{\Omega}$ is a population counterpart of $\mathbf{E}^{-1}\mathbf{H}$. Among the p eigenvalues $\gamma_1, \gamma_2, \ldots, \gamma_p$ of $\boldsymbol{\Sigma}^{-1}\boldsymbol{\Omega}$, at most s can be nonzero, where $s = \min(\nu_H, p)$ is the rank of $\boldsymbol{\Omega}$. The number of nonzero eigenvalues of $\boldsymbol{\Sigma}^{-1}\boldsymbol{\Omega}$ reflects the number of dimensions occupied by $\boldsymbol{\mu}_1, \boldsymbol{\mu}_2, \ldots, \boldsymbol{\mu}_k$ (see illustrations in the next two paragraphs). If the mean vectors are *colinear*, there is only one nonzero eigenvalue. The case $\gamma_1 = \gamma_2 = \cdots = \gamma_s$, in which the mean vectors are evenly spread in s dimensions, was called *diffuse* by Schatzoff (1966). Between these two extremes lie many populations with intermediate structures for the arrangement of the mean vectors.

We now give two illustrations of the above statement that the rank of $\boldsymbol{\Omega}$ (the number of nonzero eigenvalues of $\boldsymbol{\Sigma}^{-1}\boldsymbol{\Omega}$) equals the number of dimensions occupied by $\boldsymbol{\mu}_1, \boldsymbol{\mu}_2, \ldots, \boldsymbol{\mu}_k$. For the one-dimensional case, let $k = 3$ and suppose that $\boldsymbol{\mu}_1, \boldsymbol{\mu}_2$, and $\boldsymbol{\mu}_3$ are colinear. We show that in this case $\text{rank}(\boldsymbol{\Omega}) = 1$. Without loss of generality, we assume that the origin is on the line. Then the overall mean $\boldsymbol{\mu} = \frac{1}{3}\sum_{i=1}^{3} \boldsymbol{\mu}_i$ is also on the same line, and we have

$$\boldsymbol{\mu}_1 = c_1\boldsymbol{\mu}, \qquad \boldsymbol{\mu}_2 = c_2\boldsymbol{\mu}, \qquad \boldsymbol{\mu}_3 = c_3\boldsymbol{\mu}.$$

By (4.40), $\frac{1}{n}\boldsymbol{\Omega}$ becomes

$$\begin{aligned}\frac{1}{n}\boldsymbol{\Omega} &= \sum_{i=1}^{3} (\boldsymbol{\mu}_i - \boldsymbol{\mu})(\boldsymbol{\mu}_i - \boldsymbol{\mu})' = \sum_{i=1}^{3} (c_i\boldsymbol{\mu} - \boldsymbol{\mu})(c_i\boldsymbol{\mu} - \boldsymbol{\mu})' \\ &= \left[\sum_{i=1}^{3} (c_i - 1)^2\right] \boldsymbol{\mu}\boldsymbol{\mu}',\end{aligned}$$

which clearly has rank 1.

For the two-dimensional case, suppose that four mean vectors $\boldsymbol{\mu}_1, \boldsymbol{\mu}_2, \boldsymbol{\mu}_3$, and $\boldsymbol{\mu}_4$ lie on a plane that also (without loss of generality) includes the origin. We show that in this case rank($\boldsymbol{\Omega}$) = 2. Since two vectors span a plane, $\boldsymbol{\mu}_3, \boldsymbol{\mu}_4$, and $\boldsymbol{\mu} = \frac{1}{4}\sum_{i=1}^{4}\boldsymbol{\mu}_i$ can be expressed in terms of $\boldsymbol{\mu}_1$ and $\boldsymbol{\mu}_2$ (assuming $\boldsymbol{\mu}_1$ and $\boldsymbol{\mu}_2$ are not colinear with the origin):

$$\boldsymbol{\mu}_3 = a_1\boldsymbol{\mu}_1 + a_2\boldsymbol{\mu}_2, \qquad \boldsymbol{\mu}_4 = b_1\boldsymbol{\mu}_1 + b_2\boldsymbol{\mu}_2,$$
$$\boldsymbol{\mu} = c_1\boldsymbol{\mu}_1 + c_2\boldsymbol{\mu}_2.$$

Then $\frac{1}{n}\boldsymbol{\Omega}$ becomes

$$\begin{aligned}\frac{1}{n}\boldsymbol{\Omega} &= \sum_{i=1}^{4}(\boldsymbol{\mu}_i - \boldsymbol{\mu})(\boldsymbol{\mu}_i - \boldsymbol{\mu})' \\ &= (\boldsymbol{\mu}_1 - c_1\boldsymbol{\mu}_1 - c_2\boldsymbol{\mu}_2)(\boldsymbol{\mu}_1 - c_1\boldsymbol{\mu}_1 - c_2\boldsymbol{\mu}_2)' \\ &\quad + (\boldsymbol{\mu}_2 - c_1\boldsymbol{\mu}_1 - c_2\boldsymbol{\mu}_2)(\boldsymbol{\mu}_2 - c_1\boldsymbol{\mu}_1 - c_2\boldsymbol{\mu}_2)' \\ &\quad + (a_1\boldsymbol{\mu}_1 + a_2\boldsymbol{\mu}_2 - c_1\boldsymbol{\mu}_1 - c_2\boldsymbol{\mu}_2)(a_1\boldsymbol{\mu}_1 + a_2\boldsymbol{\mu}_2 - c_1\boldsymbol{\mu}_1 - c_2\boldsymbol{\mu}_2)' \\ &\quad + (b_1\boldsymbol{\mu}_1 + b_2\boldsymbol{\mu}_2 - c_1\boldsymbol{\mu}_1 - c_2\boldsymbol{\mu}_2)(b_1\boldsymbol{\mu}_1 + b_2\boldsymbol{\mu}_2 - c_1\boldsymbol{\mu}_1 - c_2\boldsymbol{\mu}_2)' \\ &= \sum_{i=1}^{4}(g_{1i}\boldsymbol{\mu}_1 + g_{2i}\boldsymbol{\mu}_2)(g_{1i}\boldsymbol{\mu}_1 + g_{2i}\boldsymbol{\mu}_2)',\end{aligned}$$

where $g_{11} = 1 - c_1,\ g_{21} = -c_2, \ldots, g_{24} = b_2 - c_2$.

By (A.3.4), we can express $g_{1i}\boldsymbol{\mu}_1 + g_{2i}\boldsymbol{\mu}_2$ as

$$g_{1i}\boldsymbol{\mu}_1 + g_{2i}\boldsymbol{\mu}_2 = \mathbf{M}\mathbf{g}_i,$$

where $\mathbf{M} = (\boldsymbol{\mu}_1, \boldsymbol{\mu}_2)$ and $\mathbf{g}_i = (g_{1i}, g_{2i})'$. Thus, by (A.2.20) and (A.3.8), $\frac{1}{n}\boldsymbol{\Omega}$ becomes

$$\frac{1}{n}\boldsymbol{\Omega} = \sum_{i=1}^{4}(\mathbf{M}\mathbf{g}_i)(\mathbf{M}\mathbf{g}_i)' = \mathbf{M}\left(\sum_{i=1}^{4}\mathbf{g}_i\mathbf{g}_i'\right)\mathbf{M}' = \mathbf{M}\mathbf{G}\mathbf{G}'\mathbf{M}',$$

where $\mathbf{G} = (\mathbf{g}_1, \mathbf{g}_2, \mathbf{g}_3, \mathbf{g}_4)$. Now $\mathbf{G}$ is 2×4, and we demonstrate that its rank is 2. Suppose that rank($\mathbf{G}$) = 1; then $\mathbf{G}$ would have the form

$$\mathbf{G} = \begin{pmatrix} g_{11} & g_{12} & g_{13} & g_{14} \\ g_{21} & g_{22} & g_{23} & g_{24} \end{pmatrix} = \begin{pmatrix} g_{11} & g_{12} & g_{13} & g_{14} \\ dg_{11} & dg_{12} & dg_{13} & dg_{14} \end{pmatrix},$$

and $\boldsymbol{\mu}_i - \boldsymbol{\mu}$ would be expressible as

$$\boldsymbol{\mu}_i - \boldsymbol{\mu} = g_{1i}\boldsymbol{\mu}_1 + g_{2i}\boldsymbol{\mu}_2 = g_{1i}(\boldsymbol{\mu}_1 + d\boldsymbol{\mu}_2),$$

so that all four $\boldsymbol{\mu}_i - \boldsymbol{\mu}$, $i = 1, 2, 3, 4$, would lie on a line instead of in a plane.

Thus rank($\mathbf{G}$) = 2, and by (A.6.1), $\mathbf{G}\mathbf{G}'$ is positive definite. By (A.6.2), $\mathbf{G}\mathbf{G}'$ can be factored into $\mathbf{T}'\mathbf{T}$, where $\mathbf{T}$ is nonsingular. Then by (A.5.2), rank($\mathbf{T}\mathbf{M}'$) = 2, and by (A.4.6), rank($\mathbf{MT'TM'}$) = rank($\mathbf{MGG'M'}$) = 2, from which the rank of $\mathbf{\Omega}$ is 2, and there are two nonzero eigenvalues of $\mathbf{\Sigma}^{-1}\mathbf{\Omega}$.

Some indication of the pattern of the eigenvalues of $\mathbf{\Sigma}^{-1}\mathbf{\Omega}$ may be gained by examining the eigenvalues of $\mathbf{E}^{-1}\mathbf{H}$, which are available in the output of most MANOVA software programs. Roy's test is more powerful than the others when the mean vectors are colinear, because the test uses only the largest eigenvalue of $\mathbf{E}^{-1}\mathbf{H}$. The other three tests have greater power than Roy's when the mean vectors are more diffuse.

The tests based on Λ, $V^{(s)}$, and $U^{(s)}$ are asymptotically equivalent (tend to the same p-value as $\nu_E \to \infty$) but give different results for small or moderate ν_E. Olson (1974, 1975) suggested that these three tests may be considered equivalent when $\nu_E \geq 10p\nu_H$. For smaller samples, the power of the four tests has been compared by several authors (Lee 1971, Pillai and Jayachandran 1967, Roy, Gnanadesikan, and Srivastava 1971, Schatzoff 1966). For the colinear case, their results indicate that in terms of power the tests are ordered $\theta \geq U^{(s)} \geq \Lambda \geq V^{(s)}$. In the diffuse case and an intermediate structure between colinear and diffuse, the ordering of power is reversed, $V^{(s)} \geq \Lambda \geq U^{(s)} \geq \theta$. However, the differences in power among $V^{(s)}$, Λ, and $U^{(s)}$ are small, and it may be more appropriate to compare the tests for robustness to nonnormality and for heterogeneity of covariance matrices.

Davis (1980b) investigated the robustness of Λ to skewness and kurtosis and found that an increase in skewness increases α and an increase in kurtosis lowers α. Skewness has a greater effect for smaller ν_E, while the impact of kurtosis is greater for larger ν_E. The effect of skewness and kurtosis on θ was similar (Davis 1982). For both Λ and θ, Davis found the effects of moderate departures from multivariate normality to be small. Other robustness studies that compare the four tests were made by Ito (1969, 1980), Ito and Schull (1964), Korin (1972), Mardia (1971), and Olson (1974, 1975). In general, positive kurtosis (caused by extreme observations) has a slight conservative effect on α, and the tests are ordered $V^{(s)} \geq \Lambda \geq U^{(s)} \geq \theta$. Kurtosis reduces the power of all four tests; the ordering remains the same as that noted above for colinear mean vectors or for diffuse structures.

Departure from the basic assumption of equality of covariance matrices has a greater effect on α. In some cases with highly divergent covariance matrices, the Type I error rate for θ is unacceptably high, with $U^{(s)}$ and Λ not far behind. The increase in α for $V^{(s)}$ is generally less critical. The following examples from Olson (1976) illustrate the elevation of α with heterogeneity of covariance matrices. In the first example, the parameter values were $k = 3$, $n = 5$, and $p = 3$, where, as usual, k is the number of groups (populations), n is the number of observation vectors from each group, and p is the number of variables. In one of the three populations, all three standard deviations were three times as large as those of the other two populations. The target value for α was .05, but the resulting actual Type I error rates were .09 for $V^{(s)}$, .13 for Λ, .15 for $U^{(s)}$, and .17 for θ. In the second example with $k = 3$, $n = 5$, $p = 6$, all the standard deviations of one of the three populations were six times as large as those of the other two. The actual Type I error rates became .09 for $V^{(s)}$, .49 for Λ, .58 for $U^{(s)}$, and .62 for θ.

However, Stevens (1979) suggested that the extreme heterogeneity of covariance matrices (variances in one population 36 times as large as in the other populations) modeled by Olson (1976) would rarely be encountered in practice. He argued that for the amount of heterogeneity likely to occur in practice the three statistics $U^{(s)}$, Λ, and $V^{(s)}$ are essentially equally robust to heterogeneity. Based on additional simulations, Olson (1979) countered that $V^{(s)}$ is more robust even in less extreme cases.

O'Brien, Parente, and Schmitt (1982) simulated skewness and kurtosis by means of truncated sampling and concluded that all four test statistics are reasonably robust. Frost (1994) compared the four test statistics using a more general pattern for $\mathbf{\Sigma}$ than that of previous studies. He found all four statistics to be robust to heterogeneity of covariance matrices but not to skewness and kurtosis. Wilks' Λ was slightly more powerful than Pillai's $V^{(s)}$ in most cases, with or without violation of the assumptions.

4.3 TESTS FOR EQUALITY OF COVARIANCE MATRICES

For a one-way MANOVA with k groups ($k \geq 2$), the assumption of equality of covariance matrices can be stated as a hypothesis to be tested:

$$H_0: \mathbf{\Sigma}_1 = \mathbf{\Sigma}_2 = \cdots = \mathbf{\Sigma}_k \tag{4.41}$$

versus H_1: at least two $\mathbf{\Sigma}_i$'s are unequal. For independent samples of size $n_1, n_2, \ldots, n_k$ from multivariate normal distributions $N_p(\boldsymbol{\mu}_i, \mathbf{\Sigma}_i)$, the likelihood ratio for testing H_0 is

$$\mathrm{LR} = \frac{|\hat{\mathbf{\Sigma}}_1|^{n_1/2}|\hat{\mathbf{\Sigma}}_2|^{n_2/2} \cdots |\hat{\mathbf{\Sigma}}_k|^{n_k/2}}{|\hat{\mathbf{\Sigma}}_{\mathrm{pl}}|^{\sum_i n_i/2}}, \tag{4.42}$$

in which $\hat{\mathbf{\Sigma}}_{\mathrm{pl}}$ is the pooled estimator that combines the $\hat{\mathbf{\Sigma}}_i$'s, and $\hat{\mathbf{\Sigma}}_i$ is the (biased) maximum likelihood estimator

$$\hat{\mathbf{\Sigma}}_i = \frac{1}{n_i}\sum_{j=1}^{n_i}(\mathbf{y}_{ij} - \bar{\mathbf{y}}_{i\cdot})(\mathbf{y}_{ij} - \bar{\mathbf{y}}_{i\cdot})' = \frac{1}{n_i}\mathbf{W}_i,$$

where $\mathbf{W}_i = \sum_{j=1}^{n_i}(\mathbf{y}_{ij} - \bar{\mathbf{y}}_{i.})(\mathbf{y}_{ij} - \bar{\mathbf{y}}_{i.})'$. As usual, the likelihood ratio varies between 0 and 1, with values near 1 favoring H_0 and values near 0 indicating rejection of H_0.

However, the test based on (4.42) is biased (Perlman 1980). (A test is said to be unbiased if the power is at least α for all values of the parameters in H_1.) An unbiased version that has better χ^2- and F-approximations is obtained if LR in (4.42) is modified as follows:

$$M = \frac{|\mathbf{S}_1|^{\nu_1/2}|\mathbf{S}_2|^{\nu_2/2} \cdots |\mathbf{S}_k|^{\nu_k/2}}{|\mathbf{S}_{\mathrm{pl}}|^{\sum_i \nu_i/2}}, \tag{4.43}$$

where $\nu_i = n_i - 1$, $\mathbf{S}_i = \mathbf{W}_i/\nu_i$ is the unbiased sample covariance matrix, and $\mathbf{S}_{\mathrm{pl}}$ is the pooled sample covariance matrix,

$$\mathbf{S}_{\mathrm{pl}} = \frac{\sum_{i=1}^{k} \nu_i \mathbf{S}_i}{\sum_{i=1}^{k} \nu_i} = \frac{\mathbf{E}}{\nu_E}. \tag{4.44}$$

We must have all $\nu_i > p$ so that $M > 0$. We can easily extend (4.43) and (4.44) to compare covariance matrices for the cells of a higher order design using appropriate ν_i.

Box's M-test provides both χ^2- and F-approximations for the distribution of M (Box 1949, 1950). The statistic

$$u = -2(1 - c_1)\ln M \tag{4.45}$$

has an approximate χ^2-distribution with $\frac{1}{2}(k-1)p(p+1)$ degrees of freedom, where

$$c_1 = \left[\sum_{i=1}^{k} \frac{1}{\nu_i} - \frac{1}{\sum_{i=1}^{k} \nu_i}\right]\left[\frac{2p^2 + 3p - 1}{6(p+1)(k-1)}\right] \tag{4.46}$$

and

$$\ln M = \frac{1}{2}\sum_{i=1}^{k} \nu_i \ln|\mathbf{S}_i| - \frac{1}{2}\left(\sum_{i=1}^{k} \nu_i\right)\ln|\mathbf{S}_{\text{pl}}|. \tag{4.47}$$

We reject H_0 if $u > \chi^2_\alpha$. If $\nu_1 = \nu_2 = \cdots = \nu_k = \nu$, then c_1 can be written as

$$c_1 = \frac{(k+1)(2p^2 + 3p - 1)}{6k\nu(p+1)}. \tag{4.48}$$

The degrees-of-freedom parameter of the χ^2-approximation arises from the following consideration: under H_1 we estimate $\frac{1}{2}kp(p+1)$ parameters in $\mathbf{\Sigma}_1, \mathbf{\Sigma}_2, \ldots, \mathbf{\Sigma}_k$, and under H_0 we estimate $\frac{1}{2}p(p+1)$ parameters in $\mathbf{\Sigma}$. The difference is $\frac{1}{2}(k-1)p(p+1)$.

For the F-approximation, Box's M-test uses c_1 from (4.46) or (4.48) as well as the following:

$$c_2 = \frac{(p-1)(p+2)}{6(k-1)}\left[\sum_{i=1}^{k} \frac{1}{\nu_i^2} - \frac{1}{\left(\sum_{i=1}^{k} \nu_i\right)^2}\right], \tag{4.49}$$

$$a_1 = \frac{1}{2}(k-1)p(p+1), \qquad a_2 = \frac{a_1 + 2}{|c_2 - c_1^2|},$$

$$b_1 = \frac{1 - c_1 - a_1/a_2}{a_1}, \qquad b_2 = \frac{1 - c_1 - 2/a_2}{a_2}.$$

We then define

$$F = \begin{cases} -2b_1 \ln M & \text{if } c_2 > c_1^2 \quad (4.50) \\[2ex] \dfrac{-a_2 b_2 \ln M}{a_1(1 + 2b_2 \ln M)} & \text{if } c_2 < c_1^2 \quad (4.51) \end{cases}$$

In either case, F is approximately distributed as F_{a_1,a_2}. If $\nu_1 = \nu_2 = \cdots = \nu_k = \nu$, then c_2 in (4.49) becomes

$$c_2 = \frac{(p-1)(p+2)(k^2+k+1)}{6k^2\nu^2}. \tag{4.52}$$

Box recommended the χ^2-approximation if $p \leq 5, k \leq 5$, and each $n_i > 20$; otherwise, he preferred the F-approximation. Pearson (1969) compared the two approximations with the exact distribution for $p \leq 5$ and concluded that both were adequate, with the F-approximation generally more accurate. Box's M-test is available in many MANOVA software programs.

Pederzoli and Rathie (1983) obtained the exact distribution of M. Lee et al. (1977) provided exact upper percentage points of $-2\ln M = \nu\left[k \ln|\mathbf{S}_{\text{pl}}| - \sum_i \ln|\mathbf{S}_i|\right]$ for the case of equal ν_i. Critical values from Lee et al. are given in Table B.8 in Appendix B.

Olson (1974) showed that the M-test is sensitive to some forms of nonnormality, such as kurtosis, that have only minor effects on the MANOVA tests. Manly and Rayner (1987) partitioned the likelihood ratio statistic (4.42) into three components that test equality of correlations, equality of variances, and equality of covariance matrices. Hawkins (1981) provided an approach for simultaneously testing normality and homogeneity of covariance matrices. This test does not require that all $\nu_i > p$. Sen (1968) provided a nonparametric test of $H_0\colon \boldsymbol{\Sigma}_1 = \boldsymbol{\Sigma}_2 = \cdots = \boldsymbol{\Sigma}_k$ based on ranks. Tiku and Balakrishnan (1985) and O'Brien (1992) proposed robust tests for $H_0\colon \boldsymbol{\Sigma}_1 = \boldsymbol{\Sigma}_2 = \cdots = \boldsymbol{\Sigma}_k$. These tests have good power against certain alternative hypotheses.

Graphical methods for comparing covariance matrices have been given by Gnanadesikan and Lee (1970), Gnanadesikan (1977), and Campbell (1981). These procedures provide some information as to how the covariance matrices differ. For example, Campbell discussed the interaction of groups by variances and groups by correlations. Flury (1983) used the eigenvalues and eigenvectors of $\mathbf{S}_1^{-1}\mathbf{S}_2$ to characterize the differences between $\mathbf{S}_1$ and $\mathbf{S}_2$. Rao (1982) gave likelihood ratio tests for the hypotheses $\boldsymbol{\Sigma}_2 = \sigma^2\boldsymbol{\Sigma}_1, \boldsymbol{\Sigma}_2 = \boldsymbol{\Gamma} + \sigma^2\boldsymbol{\Sigma}_1$, and $\boldsymbol{\Sigma}_2 = \boldsymbol{\Gamma} + \boldsymbol{\Sigma}_1$ along with a discussion of applications of these tests.

4.4 POWER AND SAMPLE SIZE FOR THE FOUR MANOVA TESTS

In Section 4.2, we made some comparisons of the power of the four tests for various configurations of the mean vectors $\boldsymbol{\mu}_1, \boldsymbol{\mu}_2, \ldots, \boldsymbol{\mu}_k$. In the present section, we discuss methods for computation of power in a given experimental framework. As suggested in Section 3.10 in connection with T^2-tests, we may wish to (1) find the power in a particular experimental situation or (2) find the sample size required to yield a desired level of power.

Lauter (1978) presented extensive tables (14 pages) of the power of the Lawley–Hotelling trace statistic $U^{(s)}$. The tables indicate the sample size that is needed in a one-way MANOVA to achieve a given level of power. However, the tables have limited usefulness because they are based on an overly simplified alternative hypothesis. Lauter considers three alternative hypotheses, for each of which the k mean vectors differ in only one of the p variables. In this restricted case, the mean vectors are colinear, and Roy's test statistic θ would be more powerful than $U^{(s)}$.

The distributions of the four test statistics when H_0 is false are not known in complete generality. Partial results are given by Nagarsenker (1979) and Carter (1989). Sarkar (1984) showed that if $\boldsymbol{\Sigma} = \text{cov}(\mathbf{y}_{ij})$ and $\boldsymbol{\Omega} = n\Sigma_{i=1}^{k}(\boldsymbol{\mu}_i - \boldsymbol{\mu})(\boldsymbol{\mu}_i - \boldsymbol{\mu})'$ remain constant, then the power of Wilks' Λ decreases as either p or ν_H increases.

Muller and Peterson (1984) suggested a method for calculating approximate power that can be readily implemented using straightforward matrix calculations. Their technique is cast in the framework of the general linear hypothesis (see Sections 7.2.4c and 7.5.3). We present an adaptation to the one-way MANOVA model using the F-approximations of Sections 4.1.2 and 4.1.4. For most purposes, the resulting power approximations are sufficiently accurate.

In a manner analogous to the procedure in Section 3.10, the noncentrality parameter λ is defined as $\lambda = p\nu_H F_{\text{pn}}$, where F_{pn} indicates that the sample estimators in the approximate F are replaced by population parameters. To justify this form for λ, note that a noncentral F random variable is defined as the ratio of a noncentral χ^2 random variable and a central χ^2 random variable, each divided by its degrees of freedom. Then λ is the noncentrality parameter associated with the noncentral χ^2. Thus $\lambda = \text{SSH}_{\text{pn}}/\sigma^2$, where SSH_{pn} is the population sum of squares (quadratic form) for the hypothesis (treatment), and σ^2 is the error variance, MSE_{pn}. This can be converted to a form involving F,

$$\lambda = \frac{\text{df}_1(\text{SSH}_{\text{pn}}/\text{df}_1)}{\text{MSE}_{\text{pn}}} = \text{df}_1 F_{\text{pn}} = p\nu_H F_{\text{pn}}, \tag{4.53}$$

since $\text{df}_1 = p\nu_H$ for many of the F-approximations in Sections 4.1.2 and 4.1.4.

For Wilks' Λ, the noncentrality parameter is obtained from (4.16) and (4.53) as

$$\lambda = p\nu_H F_{\text{pn}} = \frac{p\nu_H[1 - (\Lambda_{\text{pn}})^{1/t}]\text{df}_2}{(\Lambda_{\text{pn}})^{1/t}\text{df}_1}, \tag{4.54}$$

where t, df_1, and df_2 are defined following (4.16); Λ_{pn} is obtained from the Λ-statistic (4.11) by using population values in place of $\mathbf{E}$ and $\mathbf{H}$,

$$\Lambda_{\text{pn}} = \frac{|\nu_E\boldsymbol{\Sigma}|}{|\nu_E\boldsymbol{\Sigma} + \nu_H\boldsymbol{\Omega}|}; \tag{4.55}$$

and the parameter matrix $\boldsymbol{\Omega}$ corresponding to $\mathbf{H}$ is given in (4.40) as

$$\boldsymbol{\Omega} = n\sum_{i=1}^{k}(\boldsymbol{\mu}_i - \boldsymbol{\mu})(\boldsymbol{\mu}_i - \boldsymbol{\mu})'.$$

For $\boldsymbol{\Sigma}$, we might use prior experience, an estimated value from the present sample, or a value obtained from a pilot study. For $\boldsymbol{\Omega}$, we could use a value of interest or a value from the present data set. We then obtain the power from Table B.3, with $\nu_1 = \text{df}_1$,

$\nu_2 = \text{df}_2$, and

$$\phi = \sqrt{\frac{\lambda}{\nu_1 + 1}},$$

as in (3.88). Note that Table B.3 is limited to $\alpha = .05$ and gives $\beta = 1 - \text{power}$; we therefore use power $= 1 - \beta$.

As an alternative to using Table B.3, some software packages provide a noncentral F-distribution function, denoted, for example, by PROBF in SAS, which gives β directly:

$$\beta = \text{PROBF}(F_\alpha, \text{df}_1, \text{df}_2, \lambda) = P(F < F_\alpha),$$

where F_α is a critical value of the central F-distribution (under H_0).

For Pillai's trace statistic, the power is found from $\lambda = p\nu_H F_{\text{pn}}$ in (4.53), using any of the approximate F-statistics (4.27), (4.28), or (4.29), with degrees of freedom as specified in each case and with the substitution of population parameters in $V^{(s)} = \text{tr}\left[(\mathbf{E} + \mathbf{H})^{-1}\mathbf{H}\right]$:

$$V_{\text{pn}}^{(s)} = \text{tr}\left[(\nu_E \boldsymbol{\Sigma} + \nu_H \boldsymbol{\Omega})^{-1}(\nu_H \boldsymbol{\Omega})\right].$$

Similarly, for the Lawley–Hotelling trace statistic, we find λ in (4.53) from the approximate F's given by (4.32), (4.33), or (4.34), with corresponding degrees of freedom, and

$$U_{\text{pn}}^{(s)} = \text{tr}\left[(\nu_E \boldsymbol{\Sigma})^{-1}(\nu_H \boldsymbol{\Omega})\right].$$

Betz (1987) has given an alternative approximation for the power of the Hotelling–Lawley trace statistic $U^{(s)}$. Muller et al. (1992) extended the approximate power methods of Muller and Peterson (1984) to additional models.

4.5 CONTRASTS AMONG MEAN VECTORS

In this section we discuss contrasts among several mean vectors $\boldsymbol{\mu}_1, \boldsymbol{\mu}_2, \ldots, \boldsymbol{\mu}_k$. Contrasts among the elements of a vector $\boldsymbol{\mu}$ are discussed in Sections 3.4.6, 3.6.5, and 4.8.4. We consider only the balanced model with $n_1 = n_2 = \cdots = n_k = n$. For contrasts in the unbalanced case, see Sections 4.8.2 and 4.8.3.

4.5.1 Univariate Contrasts

To set the stage for a discussion of multivariate contrasts, we first review contrasts in a univariate framework ($p = 1$). In univariate ANOVA, the model is $y_{ij} = \mu + \alpha_i + \varepsilon_{ij} = \mu_i + \varepsilon_{ij}$, with the assumptions that y_{ij} is distributed as $N(\mu_i, \sigma^2)$ for $i = 1, 2, \ldots, k$ and $j = 1, 2, \ldots, n$ and that all the y_{ij}'s are independent. We estimate μ_i by $\overline{y}_{i.} = \sum_j y_{ij}/n$ and σ^2 by MSE $= \sum_{ij}(y_{ij} - \overline{y}_{i.})^2/k(n-1)$.

A *contrast* in the population means $\mu_1, \mu_2, \ldots, \mu_k$ is a linear combination

$$\delta = c_1\mu_1 + c_2\mu_2 + \cdots + c_k\mu_k, \tag{4.56}$$

where

$$\sum_{i=1}^{k} c_i = 0. \tag{4.57}$$

An unbiased estimator of δ is obtained as the same linear combination of the sample means:

$$\hat{\delta} = c_1\overline{y}_{1.} + c_2\overline{y}_{2.} + \cdots + c_k\overline{y}_{k.}. \tag{4.58}$$

Since the $\overline{y}_{i.}$'s are independent with $\text{var}(\overline{y}_{i.}) = \sigma^2/n$, we have

$$\text{var}(\hat{\delta}) = \frac{\sigma^2}{n}\sum_{i=1}^{k} c_i^2$$

and an unbiased estimator

$$s_{\hat{\delta}}^2 = \frac{\text{MSE}}{n}\sum_{i=1}^{k} c_i^2. \tag{4.59}$$

A typical hypothesis is

$$H_0\colon \delta = c_1\mu_1 + c_2\mu_2 + \cdots + c_k\mu_k = 0,$$

and therefore a contrast is also called a *comparison* among the treatment means. This can be tested using

$$t = \frac{\hat{\delta} - 0}{s_{\hat{\delta}}}, \tag{4.60}$$

which is distributed as t_{ν_E}, since MSE in (4.59) is independent of each $\overline{y}_{i.}$ in $\hat{\delta}$. As usual, $\nu_E = k(n-1)$ in the one-way balanced case. We can also use

$$\begin{aligned} F = t^2 = \frac{\hat{\delta}^2}{s_{\hat{\delta}}^2} &= \frac{\left(\sum_{i=1}^{k} c_i\overline{y}_{i.}\right)^2}{\text{MSE}\sum_{i=1}^{k} c_i^2/n} \\ &= \frac{n\left(\sum_i c_i\overline{y}_{i.}\right)^2/\sum_i c_i^2}{\text{MSE}}, \end{aligned} \tag{4.61}$$

with 1 and ν_E degrees of freedom. We reject $H_0\colon \delta = 0$ if $F > F_{\alpha,1,\nu_E}$. The numerator of (4.61) is the sum of squares for the contrast.

Two contrasts $\delta = \sum_i a_i \mu_i$ and $\gamma = \sum_i b_i \mu_i$ are said to be *orthogonal* if $\sum_i a_i b_i = 0$. In the balanced case, the two corresponding sums of squares are independent.

Theorem 4.5A. Let $\hat{\delta} = \sum_{i=1}^k a_i \overline{y}_{i\cdot} = \mathbf{a}'\overline{\mathbf{y}}$ and $\hat{\gamma} = \sum_{i=1}^k b_i \overline{y}_{i\cdot} = \mathbf{b}'\overline{\mathbf{y}}$, where $\overline{\mathbf{y}} = (\overline{y}_{1.}, \overline{y}_{2.}, \ldots, \overline{y}_{k.})'$. Assuming normality and independence of all y_{ij}, the corresponding sums of squares $n(\sum_{i=1}^k a_i \overline{y}_{i\cdot})^2 / \sum_{i=1}^k a_i^2$ and $n(\sum_{i=1}^k b_i \overline{y}_{i\cdot})^2 / \sum_{i=1}^k b_i^2$ are independent if $\sum_{i=1}^k a_i b_i = 0$, that is, if the two contrasts are orthogonal.

Proof. It is sufficient to show that the two contrasts $\mathbf{a}'\overline{\mathbf{y}}$ and $\mathbf{b}'\overline{\mathbf{y}}$ are independent. We note that $\text{cov}(\overline{\mathbf{y}}) = \sigma^2 \mathbf{I}/n$, because the $\overline{y}_{i\cdot}$'s are independent with $\text{var}(\overline{y}_{i\cdot}) = \sigma^2/n$. Then, by (1.91),

$$\text{cov}(\hat{\delta}, \hat{\gamma}) = \text{cov}(\mathbf{a}'\overline{\mathbf{y}}, \mathbf{b}'\overline{\mathbf{y}}) = \mathbf{a}'\,\text{cov}(\overline{\mathbf{y}})\mathbf{b}$$

$$= \mathbf{a}'\left[\frac{\sigma^2 \mathbf{I}}{n}\right]\mathbf{b} = \frac{\sigma^2}{n}\mathbf{a}'\mathbf{b} = \frac{\sigma^2}{n}\sum_{i=1}^k a_i b_i.$$

The two estimated contrasts $\hat{\delta}$ and $\hat{\gamma}$ will therefore be independent if $\text{cov}(\hat{\delta}, \hat{\gamma}) = 0$ or $\sum_i a_i b_i = 0$ (assuming normality). □

Any $k - 1$ orthogonal contrasts will partition the treatment sum of squares SSH into $k - 1$ independent sums of squares, each with one degree of freedom.

If a single contrast of interest is specified before seeing the data, we can test $H_0\colon \delta = 0$ by the F-test in (4.61). If several, say $m < k - 1$, contrasts are to be tested (all specified a priori), we can protect against inflation of α by using a Bonferroni approach and testing each at the α/m level of significance. Values of $F_{\alpha/m,1,\nu_E}$ could be obtained by squaring the $t_{\alpha/2m,\nu_E}$ values in Table B.2, or we could compare the p-value of each test against α/m. This procedure guarantees an overall (experimentwise) α-level. The global hypothesis is that all m hypotheses are true, and we reject this overall hypothesis if one (or more) of the m hypotheses is rejected at the α/m level.

In the above Bonferroni procedure, we assume that the contrasts are chosen before seeing the data. If we wish to select a contrast after seeing the means $\overline{y}_{1.}, \overline{y}_{2.}, \ldots, \overline{y}_{k.}$, we would naturally emphasize the largest differences, and the critical value for the F-test would no longer hold. For example, the F-value based on $\overline{y}_{1.} - \overline{y}_{2.}$ will usually be much less than the F based on $\overline{y}_{\text{max}} - \overline{y}_{\text{min}}$. The latter F does not have an F-distribution and should not be compared with F_α or even with $F_{\alpha/m}$. If the F based on $\overline{y}_{\text{max}} - \overline{y}_{\text{min}}$ were compared with F_α, its rejection rate (under H_0) would be much greater than α.

Scheffé (1953) provided a method of testing any and all contrasts, including those selected after viewing the sample means (post hoc). For any contrast, the F-statistic (4.61) is calculated and H_0 is rejected if $F > \nu_H F_{\alpha,\nu_H,\nu_E}$ [see Theorem 7.2B and (7.44)]. This allows for any number of contrasts to be tested with an experimentwise Type I error rate of at most α. Scheffe's procedure has the property that if the overall F is significant, there exists at least one contrast that is also significant. However,

Scheffe's approach is very conservative, since it is a simultaneous technique that allows for all possible contrasts to be tested. It is therefore not as powerful for detecting differences of prior interest.

4.5.2 Multivariate Contrasts

In this section we continue to use the MANOVA model (4.1) with accompanying assumptions of normality and independence of the $\mathbf{y}_{ij}$'s.

A contrast among the population mean vectors $\boldsymbol{\mu}_1, \boldsymbol{\mu}_2, \ldots, \boldsymbol{\mu}_k$ is defined as

$$\boldsymbol{\delta} = c_1\boldsymbol{\mu}_1 + c_2\boldsymbol{\mu}_2 + \cdots + c_k\boldsymbol{\mu}_k, \tag{4.62}$$

where $\sum_{i=1}^{k} c_i = 0$. Note that (4.62) is a contrast among vectors, and as such it differs from contrasts among the elements $\mu_1, \mu_2, \ldots, \mu_p$ of a single mean vector $\boldsymbol{\mu}$, expressed as $\mathbf{C}\boldsymbol{\mu}$ in Sections 3.4.6 and 3.6.5.

The contrast $\boldsymbol{\delta}$ is estimated by the corresponding contrast in the sample mean vectors:

$$\hat{\boldsymbol{\delta}} = c_1\overline{\mathbf{y}}_{1.} + c_2\overline{\mathbf{y}}_{2.} + \cdots + c_k\overline{\mathbf{y}}_{k.}. \tag{4.63}$$

Because the $\overline{\mathbf{y}}_{i.}$'s are independent with covariance matrix $\boldsymbol{\Sigma}/n$, by (2.56), we obtain

$$\text{cov}(\hat{\boldsymbol{\delta}}) = \sum_{i=1}^{k} c_i^2 \frac{\boldsymbol{\Sigma}}{n}.$$

An unbiased estimator of $\text{cov}(\hat{\boldsymbol{\delta}})$ is given by

$$\widehat{\text{cov}}(\hat{\boldsymbol{\delta}}) = \sum_{i=1}^{k} c_i^2 \frac{\mathbf{S}_{\text{pl}}}{n},$$

where $\mathbf{S}_{\text{pl}} = \mathbf{E}/\nu_E$.

Since the $\mathbf{y}_{ij}$'s are distributed as $N_p(\boldsymbol{\mu}_i, \boldsymbol{\Sigma})$ for all i, j, $\mathbf{S}_{\text{pl}}$ is independent of each $\overline{\mathbf{y}}_{i.}$ in $\hat{\boldsymbol{\delta}}$ (Theorem 2.3D), and H_0: $c_1\boldsymbol{\mu}_1 + c_2\boldsymbol{\mu}_2 + \cdots + c_k\boldsymbol{\mu}_k = \mathbf{0}$ can therefore be tested with

$$\begin{aligned} T^2 &= \hat{\boldsymbol{\delta}}'\left(\sum_{i=1}^{k} c_i^2 \frac{\mathbf{S}_{\text{pl}}}{n}\right)^{-1} \hat{\boldsymbol{\delta}} \\ &= \frac{n}{\sum_{i=1}^{k} c_i^2}\left(\sum_{i=1}^{k} c_i\overline{\mathbf{y}}_{i.}\right)'\left(\frac{\mathbf{E}}{\nu_E}\right)^{-1}\left(\sum_{i=1}^{k} c_i\overline{\mathbf{y}}_{i.}\right), \end{aligned} \tag{4.64}$$

which is distributed as T^2_{p,ν_E} if a single a priori contrast is of interest. If several prespecified contrasts are to be tested, appropriate Bonferroni adjustments to the α_i for each test should be made.

Alternatively, H_0 can be tested with Wilks' Λ. By analogy with the numerator of (4.61), the hypothesis matrix for the contrast is given by

$$\mathbf{H}_1 = \frac{n}{\sum_{i=1}^{k} c_i^2} \left(\sum_{i=1}^{k} c_i \overline{\mathbf{y}}_{i\cdot} \right) \left(\sum_{i=1}^{k} c_i \overline{\mathbf{y}}_{i\cdot} \right)', \tag{4.65}$$

and the test statistic is

$$\Lambda = \frac{|\mathbf{E}|}{|\mathbf{E} + \mathbf{H}_1|}, \tag{4.66}$$

which is distributed as $\Lambda_{p,1,\nu_E}$. The other three MANOVA test statistics, $V^{(s)}$, $U^{(s)}$, and θ, can also be obtained from the single nonzero eigenvalue of $\mathbf{E}^{-1}\mathbf{H}_1$. In this case $\nu_H = 1$, and all four MANOVA statistics and T^2 give the same results [see (4.35)–(4.37) in Section 4.1.5].

A condition for independence of two contrasts (in the balanced case) is given in the following theorem.

Theorem 4.5B. Assuming normality and independence of the $\mathbf{y}_{ij}$'s, the contrasts $\hat{\boldsymbol{\delta}} = \sum_{i=1}^{k} a_i \overline{\mathbf{y}}_{i\cdot}$ and $\hat{\boldsymbol{\gamma}} = \sum_{i=1}^{k} b_i \overline{\mathbf{y}}_{i\cdot}$ are independent if $\sum_{i=1}^{k} a_i b_i = 0$.

Proof. By Problem 2.29, $\text{cov}(\hat{\boldsymbol{\delta}}, \hat{\boldsymbol{\gamma}}) = \sum_{i=1}^{k} a_i b_i \boldsymbol{\Sigma}/n$, which equals $\mathbf{O}$ if $\sum_i a_i b_i = 0$. □

Two contrasts $\hat{\boldsymbol{\delta}} = \sum_{i=1}^{k} a_i \overline{\mathbf{y}}_{i\cdot}$ and $\hat{\boldsymbol{\gamma}} = \sum_{i=1}^{k} b_i \overline{\mathbf{y}}_{i\cdot}$, where $\sum_{i=1}^{k} a_i b_i = 0$, are said to be orthogonal. If $k - 1$ orthogonal contrasts are used, they partition the $\mathbf{H}$ matrix into $k - 1$ independent matrices, $\mathbf{H}_1, \mathbf{H}_2, \ldots, \mathbf{H}_{k-1}$, each with one degree of freedom.

For simultaneous tests that can be used after seeing the mean vectors, we turn to Roy's union–intersection principle (Sections 3.3.8, 3.5.5, and 4.1.3), in which the multivariate hypothesis is restated in terms of all possible linear combinations. Thus $H_0\colon \boldsymbol{\mu}_1 = \boldsymbol{\mu}_2 = \cdots = \boldsymbol{\mu}_k$ is true if and only if all contrasts of linear combinations $\mathbf{a}'\boldsymbol{\mu}_i$ are equal to 0, that is, if

$$H_0\colon \sum_{i=1}^{k} c_i \mathbf{a}'\boldsymbol{\mu}_i = 0 \tag{4.67}$$

is true for all possible sets $c_1, c_2, \ldots, c_k$ with $\sum_{i=1}^{k} c_i = 0$ and all possible $\mathbf{a}$. Maximization of $\sum_i c_i \mathbf{a}'\overline{\mathbf{y}}$ with respect to $\mathbf{a}$ leads directly to λ_1, the largest eigenvalue of $\mathbf{E}^{-1}\mathbf{H}$, as a test criterion. The test of H_0 in (4.67) can be carried out in terms of confidence intervals given in the following theorem.

Theorem 4.5C. Simultaneous confidence intervals for all contrasts $\sum_{i=1}^{k} c_i \mathbf{a}'\boldsymbol{\mu}_i$ that allow for all possible $\mathbf{a}$ are given by

$$\sum_{i=1}^{k} c_i \mathbf{a}'\overline{\mathbf{y}}_{i\cdot} \pm \sqrt{\lambda_\alpha \sum_{i=1}^{k} \frac{c_i^2}{n} \mathbf{a}'\mathbf{E}\mathbf{a}}, \tag{4.68}$$

where $\lambda_\alpha = \theta_\alpha/(1-\theta_\alpha)$ is the upper α critical value of λ_1, and θ_α is a critical value of Roy's statistic (see Table B.5).

Proof. For a particular $\mathbf{a}$, the F-statistic for a contrast $\sum_{i=1}^{k} c_i \mathbf{a}'\overline{\mathbf{y}}_{i\cdot}$ is given by (4.61) as

$$F_c(\mathbf{a}) = \frac{n\left(\sum_{i=1}^{k} c_i \mathbf{a}'\overline{\mathbf{y}}_{i\cdot}\right)^2 / \sum_{i=1}^{k} c_i^2}{\text{MSE}_{\mathbf{a}'\mathbf{y}}} = \frac{n\left(\sum_{i=1}^{k} c_i \mathbf{a}'\overline{\mathbf{y}}_{i\cdot}\right)^2 / \sum_{i=1}^{k} c_i^2}{\mathbf{a}'\mathbf{E}\mathbf{a}/\nu_E}.$$

By Scheffe's approach in the last paragraph of Section 4.5.1, $F_c(\mathbf{a}) \leq \nu_H F_\alpha(\mathbf{a})$ with probability $1-\alpha$ for any and all contrasts, where $F_\alpha(\mathbf{a})$ is the upper critical value of $F(\mathbf{a})$ in (4.19). From Roy's union–intersection approach in Section 4.1.3, $F(\mathbf{a})$ is maximized by $\nu_E\lambda_1/\nu_H$. Thus $F_\alpha(\mathbf{a}) = \nu_E\lambda_\alpha/\nu_H$, where λ_α is the upper critical value of λ_1. Hence, with probability $1-\alpha$,

$$F_c(\mathbf{a}) = \frac{n\left(\sum_{i=1}^{k} c_i \mathbf{a}'\overline{\mathbf{y}}_{i\cdot}\right)^2 / \sum_{i=1}^{k} c_i^2}{\mathbf{a}'\mathbf{E}\mathbf{a}/\nu_E} \leq \frac{\nu_H \lambda_\alpha \nu_E}{\nu_H} = \nu_E \lambda_\alpha,$$

or

$$\left(\sum_{i=1}^{k} c_i \mathbf{a}'\overline{\mathbf{y}}_{i\cdot}\right)^2 \leq \frac{1}{n} \sum_{i=1}^{k} c_i^2 \mathbf{a}'\mathbf{E}\mathbf{a}\lambda_\alpha,$$

which reduces to (4.68). □

For any given $\mathbf{a}$, we can test the hypothesis (4.67) by observing whether the interval (4.68) contains 0. Any contrast suggested by the data can be examined. However, the intervals tend to be extremely wide and may not be very useful.

If we use $\mathbf{a}' = (0, \ldots, 0, 1, 0, \ldots, 0)$ with a 1 in the rth position in (4.68), we obtain a confidence interval for $\sum_{i=1}^{k} c_i \mu_{ir}$, a contrast in the means of the rth variable. In particular, confidence intervals for $\mu_{ir} - \mu_{i'r}$, the difference between the ith and i'th mean on the rth variable, are given by

$$\overline{y}_{i.r} - \overline{y}_{i'.r} \pm \sqrt{\frac{\theta_\alpha}{1-\theta_\alpha}\left(\frac{2}{n}\right) e_{rr}}, \tag{4.69}$$

where $\overline{y}_{i.r}$ is the rth element of $\overline{\mathbf{y}}_{i\cdot}$ and e_{rr} is the rth diagonal element of $\mathbf{E}$.

If we were interested in only a few, say m, preselected contrasts on linear combinations $\mathbf{a}'\boldsymbol{\mu}_i$, we could use Bonferroni coefficients to obtain narrower intervals:

$$\sum_i c_i \mathbf{a}'\bar{\mathbf{y}}_{i.} \pm t_{\alpha/2m,\nu_E} \sqrt{\mathbf{a}'\mathbf{E}\mathbf{a} \sum_i c_i^2 / n\nu_E}. \tag{4.70}$$

4.6 TWO-WAY MULTIVARIATE ANALYSIS OF VARIANCE

A balanced two-way fixed-effects MANOVA model for p dependent variables can be expressed as

$$\begin{gathered} \mathbf{y}_{ijk} = \boldsymbol{\mu} + \boldsymbol{\alpha}_i + \boldsymbol{\beta}_j + \boldsymbol{\gamma}_{ij} + \boldsymbol{\varepsilon}_{ijk} = \boldsymbol{\mu}_{ij} + \boldsymbol{\varepsilon}_{ijk}, \\ i = 1, 2, \ldots, a, \qquad j = 1, 2, \ldots, b, \qquad k = 1, 2, \ldots, n, \end{gathered} \tag{4.71}$$

with side conditions $\sum_{i=1}^a \boldsymbol{\alpha}_i = \sum_{j=1}^b \boldsymbol{\beta}_j = \mathbf{0}; \sum_{i=1}^a \boldsymbol{\gamma}_{ij} = \mathbf{0}, j = 1, 2, \ldots, b$; and $\sum_{j=1}^b \boldsymbol{\gamma}_{ij} = \mathbf{0}, i = 1, 2, \ldots, a$. We assume that the $\boldsymbol{\varepsilon}_{ijk}$'s are independently distributed as $N_p(\mathbf{0}, \boldsymbol{\Sigma})$.

In the presence of interaction, we define the main effect of A as an average over the levels of B: $\boldsymbol{\alpha}_i = \bar{\boldsymbol{\mu}}_{i.} - \bar{\boldsymbol{\mu}}_{..}, i = 1, 2, \ldots, a$, where $\bar{\boldsymbol{\mu}}_{i.} = \sum_{j=1}^b \boldsymbol{\mu}_{ij}/b$ and $\bar{\boldsymbol{\mu}}_{..} = \sum_{ij} \boldsymbol{\mu}_{ij}/ab$. This definition of $\boldsymbol{\alpha}_i$ follows from the side condition $\sum_{i=1}^a \boldsymbol{\alpha}_i = \mathbf{0}$. The effects of B and AB are defined similarly: $\boldsymbol{\beta}_j = \bar{\boldsymbol{\mu}}_{.j} - \bar{\boldsymbol{\mu}}_{..}$ and $\boldsymbol{\gamma}_{ij} = \boldsymbol{\mu}_{ij} - \bar{\boldsymbol{\mu}}_{i.} - \bar{\boldsymbol{\mu}}_{.j} + \bar{\boldsymbol{\mu}}_{..}$.

The interaction $\boldsymbol{\gamma}_{ij}$ can be expressed as $\boldsymbol{\gamma}_{ij} = \boldsymbol{\mu}_{ij} - \bar{\boldsymbol{\mu}}_{i.} - \boldsymbol{\beta}_j$ or as $\boldsymbol{\gamma}_{ij} = \boldsymbol{\mu}_{ij} - \bar{\boldsymbol{\mu}}_{.j} - \boldsymbol{\alpha}_i$. If the B effect at the ith level of A, $\boldsymbol{\mu}_{ij} - \bar{\boldsymbol{\mu}}_{i.}$, is the same for all i, then $\boldsymbol{\mu}_{ij} - \bar{\boldsymbol{\mu}}_{i.} = \boldsymbol{\beta}_j$, and we have $\boldsymbol{\gamma}_{ij} = \mathbf{0}$. Similarly, if the A effect at the jth level of B, $\boldsymbol{\mu}_{ij} - \bar{\boldsymbol{\mu}}_{.j}$, is the same for all j, then $\boldsymbol{\mu}_{ij} - \bar{\boldsymbol{\mu}}_{.j} = \boldsymbol{\alpha}_i$, and $\boldsymbol{\gamma}_{ij} = \mathbf{0}$.

In the presence of interaction, the hypothesis for factor A can be expressed as

$$H_{0A}: \bar{\boldsymbol{\mu}}_{1.} = \bar{\boldsymbol{\mu}}_{2.} = \cdots = \bar{\boldsymbol{\mu}}_{a.},$$

or equivalently as

$$H_{0A}: \boldsymbol{\alpha}_1 = \boldsymbol{\alpha}_2 = \cdots = \boldsymbol{\alpha}_{a-1} = \mathbf{0}.$$

For factor B, we have

$$H_{0B}: \bar{\boldsymbol{\mu}}_{.1} = \bar{\boldsymbol{\mu}}_{.2} = \cdots = \bar{\boldsymbol{\mu}}_{.b},$$

or equivalently

$$H_{0B}: \boldsymbol{\beta}_1 = \boldsymbol{\beta}_2 = \cdots = \boldsymbol{\beta}_{b-1} = \mathbf{0}.$$

For the interaction AB, the hypothesis can be expressed as

$$H_{0AB}: \boldsymbol{\gamma}_{ij} = \boldsymbol{\mu}_{ij} - \bar{\boldsymbol{\mu}}_{i.} - \bar{\boldsymbol{\mu}}_{.j} + \bar{\boldsymbol{\mu}}_{..} = \mathbf{0}$$

for $i = 1, 2, \ldots, a - 1$ and $j = 1, 2, \ldots, b - 1$.

If the interaction is absent, the hypothesis for factor A becomes

$$H_{0A}: \boldsymbol{\mu}_{1j} = \boldsymbol{\mu}_{2j} = \cdots = \boldsymbol{\mu}_{aj}, \qquad j = 1, 2, \ldots, b;$$

that is, the effect of A is the same for all levels of B. There is a similar expression for H_{0B}.

The sums of squares and products matrices for making the tests are given in Table 4.2. As in Section 4.1.1, $\overline{\mathbf{y}}_{i..} = \sum_{jk} \mathbf{y}_{ijk}/nb$, with analogous definitions for $\overline{\mathbf{y}}_{.j.}$, $\overline{\mathbf{y}}_{ij.}$, and $\overline{\mathbf{y}}_{...}$. In the balanced case considered in this section, the total sum of squares and products matrix is partitioned as

$$\mathbf{T} = \mathbf{H}_A + \mathbf{H}_B + \mathbf{H}_{AB} + \mathbf{E}. \tag{4.72}$$

The main effects A and B and the interaction AB can be tested using all four MANOVA test statistics. For Wilks' Λ, we have

$$\Lambda_A = \frac{|\mathbf{E}|}{|\mathbf{E} + \mathbf{H}_A|} \quad \text{is} \quad \Lambda_{p,a-1,ab(n-1)};$$

$$\Lambda_B = \frac{|\mathbf{E}|}{|\mathbf{E} + \mathbf{H}_B|} \quad \text{is} \quad \Lambda_{p,b-1,ab(n-1)};$$

$$\Lambda_{AB} = \frac{|\mathbf{E}|}{|\mathbf{E} + \mathbf{H}_{AB}|} \quad \text{is} \quad \Lambda_{p,(a-1)(b-1),ab(n-1)}.$$

In each of Λ_A, Λ_B, and Λ_{AB}, the indicated distribution holds when H_0 is true. The other three MANOVA test statistics can be obtained using the eigenvalues of $\mathbf{E}^{-1}\mathbf{H}_A, \mathbf{E}^{-1}\mathbf{H}_B$, and $\mathbf{E}^{-1}\mathbf{H}_{AB}$.

If the interaction is significant, we can still test the main effects A and B by defining each main effect as the average effect over the levels of the other factor. (The side conditions $\sum_i \boldsymbol{\alpha}_i = \mathbf{0}$ and $\sum_j \boldsymbol{\beta}_j = \mathbf{0}$ effectively impose this definition.)

Table 4.2. Two-Way MANOVA

Source	Sum of Squares and Products Matrix	df
A	$\mathbf{H}_A = nb\, \Sigma_i(\overline{\mathbf{y}}_{i..} - \overline{\mathbf{y}}_{...})(\overline{\mathbf{y}}_{i..} - \overline{\mathbf{y}}_{...})'$	$a - 1$
B	$\mathbf{H}_B = na\, \Sigma_j(\overline{\mathbf{y}}_{.j.} - \overline{\mathbf{y}}_{...})(\overline{\mathbf{y}}_{.j.} - \overline{\mathbf{y}}_{...})'$	$b - 1$
AB	$\mathbf{H}_{AB} = n\, \Sigma_{ij}(\overline{\mathbf{y}}_{ij.} - \overline{\mathbf{y}}_{i..} - \overline{\mathbf{y}}_{.j.} + \overline{\mathbf{y}}_{...})$ $\times (\overline{\mathbf{y}}_{ij.} - \overline{\mathbf{y}}_{i..} - \overline{\mathbf{y}}_{.j.} + \overline{\mathbf{y}}_{...})'$	$(a - 1)(b - 1)$
Error	$\mathbf{E} = \Sigma_{ijk}(\mathbf{y}_{ijk} - \overline{\mathbf{y}}_{ij.})(\overline{\mathbf{y}}_{ijk} - \overline{\mathbf{y}}_{ij.})'$	$ab(n - 1)$
Total	$\mathbf{T} = \Sigma_{ijk}(\mathbf{y}_{ijk} - \overline{\mathbf{y}}_{...})(\mathbf{y}_{ijk} - \overline{\mathbf{y}}_{...})'$	$abn - 1$

Interpretation of the main effects is not as simple in the presence of interaction, but there is information to be gained. Novak and Cramer (1987) advocate plotting the group means of the first two discriminant functions to give additional insights into multivariate main effects and interactions.

A contrast among the levels of factor A is defined as $\sum_{i=1}^{a} c_i \bar{\boldsymbol{\mu}}_{i\cdot}$, where $\sum_{i=1}^{a} c_i = 0$ and $\bar{\boldsymbol{\mu}}_{i\cdot} = \sum_{j=1}^{b} \boldsymbol{\mu}_{ij}/b$. Similarly, a contrast among the levels of B is given by $\sum_{j=1}^{b} c_j \bar{\boldsymbol{\mu}}_{\cdot j}$. Hypotheses associated with these contrasts can be tested by T^2 or any of the four MANOVA test statistics, as in (4.64)–(4.66). Thus to test H_0: $\sum_{i=1}^{a} c_i \bar{\boldsymbol{\mu}}_{i\cdot} = \mathbf{0}$, we can use

$$T^2 = \frac{nb}{\sum_{i=1}^{a} c_i^2} \left(\sum_{i=1}^{a} c_i \bar{\mathbf{y}}_{i..} \right)' \left(\frac{\mathbf{E}}{\nu_E} \right)^{-1} \left(\sum_{i=1}^{a} c_i \bar{\mathbf{y}}_{i..} \right), \tag{4.73}$$

which is distributed as T^2_{p,ν_E} (under H_0), or alternatively, we can test H_0 with

$$\Lambda = \frac{|\mathbf{E}|}{|\mathbf{E} + \mathbf{H}_1|},$$

which is distributed as $\Lambda_{p,1,\nu_E}$ (under H_0), where

$$\mathbf{H}_1 = \frac{nb}{\sum_{i=1}^{a} c_i^2} \left(\sum_{i=1}^{a} c_i \bar{\mathbf{y}}_{i..} \right) \left(\sum_{i=1}^{a} c_i \bar{\mathbf{y}}_{i..} \right)' \tag{4.74}$$

[see (4.65)]. The other three MANOVA test statistics can be obtained from the single nonzero eigenvalue of $\mathbf{E}^{-1}\mathbf{H}_1$. Since $\nu_H = 1$, all five test statistics will give equivalent results (see Section 4.1.5).

For the levels of factor A, simultaneous confidence intervals for all possible contrasts on all possible linear combinations of mean vectors (including comparisons made after seeing the data) are given by Theorem 4.5C as

$$\sum_{i=1}^{a} c_i \mathbf{a}' \bar{\mathbf{y}}_{i..} \pm \sqrt{\frac{\theta_\alpha}{1-\theta_\alpha} \sum_{i=1}^{a} \frac{c_i^2}{nb} \mathbf{a}'\mathbf{E}\mathbf{a}} \tag{4.75}$$

with a similar expression for the levels of B,

$$\sum_{j=1}^{b} c_j \mathbf{a}' \bar{\mathbf{y}}_{.j.} \pm \sqrt{\frac{\theta_\alpha}{1-\theta_\alpha} \sum_{j=1}^{b} \frac{c_j^2}{na} \mathbf{a}'\mathbf{E}\mathbf{a}}. \tag{4.76}$$

If we are interested in only a few, say m, preselected contrasts on linear combinations of the levels of A, we can obtain shorter intervals by using the Bonferroni method in (4.70),

$$\sum_i c_i \mathbf{a}'\overline{\mathbf{y}}_{i..} \pm t_{\alpha/2m,\nu_E}\sqrt{\frac{\mathbf{a}'\mathbf{E}\mathbf{a}}{\nu_E}\sum_i \frac{c_i^2}{nb}}. \tag{4.77}$$

There is an analogous expression for contrasts on the levels of B.

The discriminant function $z = \mathbf{a}'\mathbf{y}$ obtained from the first eigenvector of $\mathbf{E}^{-1}\mathbf{H}_A$ and of $\mathbf{E}^{-1}\mathbf{H}_B$ can be examined for the contribution of each variable in the presence of the others toward separation of the levels of A and B, respectively (see Section 4.1.3 and Chapter 5). The discriminant function obtained from $\mathbf{E}^{-1}\mathbf{H}_A$ will not have the same pattern as the one from $\mathbf{E}^{-1}\mathbf{H}_B$. This is not surprising, since we expect that the relative contribution of the variables toward separating the levels of factor A will be different from the relative contribution toward separating the levels of B. For tests on individual variables by themselves, see Section 4.1.7b.

For a randomized block design or a two-way MANOVA without replication, the analysis is similar to that given above for the two-way model with replication; therefore we will provide no specific details.

4.7 HIGHER ORDER MODELS

The two-way fixed-effects models in Section 4.6 generalize to higher order fixed-effects models with no difficulty. Test construction parallels that for the two-way model using the matrix for error to test all factors.

Every univariate ANOVA design has a multivariate counterpart, including fixed, random, and mixed models, and experimental structures that are crossed, nested, or a combination of the two. Roebruck (1982) proved that univariate mixed models generalize to multivariate mixed models. For the one-way multivariate random-effects model, Schott and Saw (1984) showed that the likelihood ratio approach leads to the same test statistics as in the fixed-effects model.

For random-effects models and mixed models, a table of expected mean squares for the terms in the analogous univariate model can be used to find the appropriate error matrix to test each term in the model, since the expected mean square matrices have the same pattern as that of the corresponding univariate model. However, the matrix indicated for "error" in each case must have degrees of freedom greater than p or a test cannot be made. For terms with no appropriate error matrix, a multivariate Satterthwaite approximation has been considered by Boik (1988), Tan and Gupta (1983), Naik and Rao (1994), and Khuri, Mathew, and Nel (1994).

Zhou and Mathew (1993) compared various approaches to testing in multivariate mixed models. They obtained a locally best invariant test that turned out to be equivalent to Pillai's test for the balanced case but not for the unbalanced case.

4.8 UNBALANCED DATA

4.8.1 Introduction

All experimental structures considered up to this point have involved an equal number of observations in each cell (that is, each treatment combination; for example, first level of A and second level of B). We now turn to the unbalanced case, in which the number of replications may vary from one cell to another.

The overparameterized or non-full-rank model that we have used for balanced data allows for main effects, interactions, nested terms, split plots, and so forth. The models in (4.1) and (4.71) are examples of overparameterized models. Such models have served well in the balanced case, but unfortunately, various attempts to generalize the overparameterized model to the unbalanced case have led to contradictory results.

To set the stage for unbalanced MANOVA models, we first treat unbalanced univariate ANOVA models. Univariate one-way and two-way models are covered in Sections 4.8.2 and 4.8.4. The corresponding one-way and two-way MANOVA models are treated in Sections 4.8.3 and 4.8.5. In the remainder of this section, we review various approaches to (univariate) unbalanced data analysis.

Speed, Hocking, and Hackney (1978) reviewed the most common methods of analysis of unbalanced data for univariate two-way models $y_{ijk} = \mu + \alpha_i + \beta_j + \gamma_{ij} + \varepsilon_{ijk}$ and delineated the hypothesis that was being tested in each case. Some of these methods use the *reduction in sum of squares*, or $R(\cdot)$, notation (Speed and Hocking 1976). This notation is used in expressions such as

$$R(\alpha|\mu, \beta) = R(\mu, \alpha, \beta) - R(\mu, \beta)$$

to indicate the sum of squares due to α "adjusted" for μ and β. This notation serves well for balanced data analysis, but in the unbalanced case, it may fail to indicate the hypothesis being tested. The usual hypothesis of interest for the main effects is that the levels of a factor do not differ when they are averaged over the levels of the other factors. The only approaches that test this hypothesis are the method of *weighted squares of means* (originally proposed by Yates 1934; see also Morrison 1983, pp. 407–412) and a modernization called the *cell means model*. The hypotheses tested by all other methods are functions of the cell means weighted by the cell frequencies n_{ij}. This seems very unnatural unless the n_{ij}'s from the sample are somehow characteristic of the population. The following popular methods test hypotheses that are misleading or at best unintended:

1. The method of unweighted means was a popular computational scheme before the days of modern computing power (Searle 1971, Winer 1971).
2. The method of fitting constants (Rao 1965, pp. 211–214; Searle 1971, p. 139; Snedecor and Cochran 1967) produces "adjusted main effect" sums of squares that do not test main effects in the usual sense.
3. Overall and Spiegel (1969) presented three methods, one of which is equivalent to the weighted squares of means approach. The other two methods test

inappropriate hypotheses, but unfortunately, these are the two that Overall and Spiegel recommended as the most correct.

4. The regression, or dummy-variable, method obtains a full-rank model by deleting redundant terms in the overparameterized model. The method is deceptively simple because it obtains sums of squares denoted $R(\alpha|\mu, \beta, \gamma)$ and $R(\beta|\mu, \alpha, \gamma)$, where γ is the interaction effect, but these sums of squares test hypotheses that are concerned only with the first level of A and the first level of B.

4.8.2 Univariate One-Way Model

The univariate one-way unbalanced model is

$$y_{ij} = \mu + \alpha_i + \varepsilon_{ij} \tag{4.78}$$

$$= \mu_i + \varepsilon_{ij}, \tag{4.79}$$

$$i = 1, 2, \ldots, k, \qquad j = 1, 2, \ldots, n_i.$$

We also assume that the y_{ij}'s are independently distributed as $N(\mu_i, \sigma^2)$. The appropriate sums of squares are given in Table 4.3, where $y_{i\cdot} = \sum_{j=1}^{n_i} y_{ij}$; $y_{\cdot\cdot} = \sum_{i=1}^{k}\sum_{j=1}^{n_i} y_{ij} = \sum_i y_{i\cdot}$; $\overline{y}_{i\cdot} = \sum_j y_{ij}/n_i = y_{i\cdot}/n_i$; $\overline{y}_{\cdot\cdot} = \sum_{ij} y_{ij}/N = y_{\cdot\cdot}/N$; and $N = \sum_i n_i$. An F-statistic for testing H_0: $\mu_1 = \mu_2 = \cdots = \mu_k$ is given by

$$F = \frac{\text{SSB}/(k-1)}{\text{SSE}/(N-k)} = \frac{\text{MSB}}{\text{MSE}}, \tag{4.80}$$

with $\nu_H = k - 1$ and $\nu_E = N - k$ degrees of freedom.

A contrast in the population means is defined as in (4.56), $\delta = c_1\mu_1 + c_2\mu_2 + \cdots + c_k\mu_k$, where $\sum_i c_i = 0$. An estimate is given by $\hat{\delta} = c_1\overline{y}_{1.} + c_2\overline{y}_{2.} + \cdots + c_k\overline{y}_{k.}$, with variance $\sigma^2 \sum_{i=1}^{k} c_i^2/n_i$. The F-statistic for testing H_0: $\delta = 0$ that corresponds to (4.61) is

$$F = \frac{(\sum_{i=1}^{k} c_i\overline{y}_{i\cdot})^2/(\sum_{i=1}^{k} c_i^2/n_i)}{\text{MSE}}, \tag{4.81}$$

with 1 and $\nu_E = N - k$ degrees of freedom. We reject H_0: $\delta = 0$ if $F > F_{\alpha,1,\nu_E}$. The numerator of (4.81) is the sum of squares for the contrast.

Table 4.3. One-Way Unbalanced Univariate ANOVA

Source	Sum of Squares	df
Between	$\text{SSB} = \sum_i n_i(\overline{y}_{i.} - \overline{y}_{..})^2 = \sum_i y_{i.}^2/n_i - y_{..}^2/N$	$k - 1$
Error	$\text{SSE} = \sum_{ij}(y_{ij} - \overline{y}_{i.})^2 = \sum_{ij} y_{ij}^2 - \sum_i y_{i.}^2/n_i$	$N - k$
Total	$\text{SST} = \sum_{ij}(y_{ij} - \overline{y}_{..})^2 = \sum_{ij} y_{ij}^2 - y_{..}^2/N$	$N - 1$

A researcher is typically interested in orthogonal contrasts $\hat{\delta} = \sum_{i=1}^{k} a_i \overline{y}_{i\cdot}$ and $\hat{\gamma} = \sum_{i=1}^{k} b_i \overline{y}_{i\cdot}$, where $\sum_{i=1}^{k} a_i b_i = 0$. However, in the case of unbalanced data, two orthogonal contrasts of this type are not independent.

Theorem 4.8A. In the unbalanced model (4.78), two contrasts $\hat{\delta} = \sum_{i=1}^{k} a_i \overline{y}_{i\cdot}$ and $\hat{\gamma} = \sum_{i=1}^{k} b_i \overline{y}_{i\cdot}$ are independent if and only if $\sum_{i=1}^{k} a_i b_i / n_i = 0$.

Proof. We express the two contrasts in vector notation as $\hat{\delta} = \mathbf{a}'\overline{\mathbf{y}}$ and $\hat{\gamma} = \mathbf{b}'\overline{\mathbf{y}}$, where $\overline{\mathbf{y}} = (\overline{y}_{1\cdot}, \overline{y}_{2\cdot}, \ldots, \overline{y}_{k\cdot})'$. Since $\text{var}(\overline{y}_{i\cdot}) = \sigma^2/n_i$ and the $\overline{y}_{i\cdot}$'s are independent, the covariance matrix for $\overline{\mathbf{y}}$ is

$$\text{cov}(\overline{\mathbf{y}}) = \sigma^2 \begin{pmatrix} 1/n_1 & 0 & \cdots & 0 \\ 0 & 1/n_2 & \cdots & 0 \\ \vdots & \vdots & & \vdots \\ 0 & 0 & \cdots & 1/n_k \end{pmatrix} = \sigma^2 \mathbf{D}.$$

Then by (1.91),

$$\text{cov}(\hat{\delta}, \hat{\gamma}) = \text{cov}(\mathbf{a}'\overline{\mathbf{y}}, \mathbf{b}'\overline{\mathbf{y}}) = \mathbf{a}'\text{cov}(\overline{\mathbf{y}})\mathbf{b} = \sigma^2 \mathbf{a}'\mathbf{D}\mathbf{b}$$

$$= \sigma^2 \sum_{i=1}^{k} \frac{a_i b_i}{n_i}. \tag{4.82}$$

Hence (assuming normality) the two contrast coefficient vectors must satisfy $\sum_i a_i b_i / n_i = 0$ to be independent. □

We refer to contrasts whose coefficients satisfy $\sum_i a_i b_i / n_i = 0$ as *weighted contrasts*. If we have $k - 1$ contrasts of this type, they partition the treatment sum of squares into $k - 1$ independent sums of squares, each with one degree of freedom. Orthogonal contrasts that satisfy only $\sum_i a_i b_i = 0$ do not do this (for unbalanced data).

This requirement of dividing by n_i in order for the contrasts to be independent imposes an unnatural restriction. In practice, weighted contrasts are often of less interest than orthogonal contrasts. The n_i's seldom reflect population characteristics that we wish to take into account. However, it is not necessary that the tests be independent. If we use ordinary orthogonal contrasts with $\sum_i a_i b_i = 0$, then (4.81) tests each contrast adjusted for the other contrasts because (4.81) is essentially a general linear hypothesis test [see a comment following (7.34)].

Example 4.8.2. Suppose that we wish to compare the means of three treatments and that the orthogonal contrasts of interest are the rows in the matrix

$$\begin{pmatrix} 2 & -1 & -1 \\ 0 & 1 & -1 \end{pmatrix}$$

with corresponding hypotheses

$$H_{01}\colon \mu_1 = \tfrac{1}{2}(\mu_2 + \mu_3) \qquad H_{02}\colon \mu_2 = \mu_3.$$

If the sample sizes for the three treatments are 10, 20, and 5, the two estimated contrasts,

$$\hat{\delta} = 2\overline{y}_{1.} - \overline{y}_{2.} - \overline{y}_{3.} \qquad \text{and} \qquad \hat{\gamma} = \overline{y}_{2.} - \overline{y}_{3.},$$

are not independent, and the corresponding sums of squares do not partition the treatment sum of squares.

To obtain two independent (weighted) contrasts whose coefficients satisfy $\sum_i a_i b_i / n_i = 0$, we could use, for example, the rows of

$$\begin{pmatrix} 25 & -20 & -5 \\ 0 & 1 & -1 \end{pmatrix},$$

but the comparison

$$H_{01}\colon 25\mu_1 = 20\mu_2 + 5\mu_3 \quad \text{or} \quad H_{01}\colon \mu_1 = \tfrac{4}{5}\mu_2 + \tfrac{1}{5}\mu_3$$

is not what we were initially interested in.

4.8.3 Multivariate One-Way Model

Analysis of the multivariate one-way unbalanced model is an immediate extension of the univariate case. The model for each p-variate observation vector is

$$\mathbf{y}_{ij} = \boldsymbol{\mu} + \boldsymbol{\alpha}_i + \boldsymbol{\varepsilon}_{ij} \tag{4.83}$$

$$= \boldsymbol{\mu}_i + \boldsymbol{\varepsilon}_{ij}, \tag{4.84}$$

$$i = 1, 2, \ldots, k, \qquad j = 1, 2, \ldots, n_i.$$

The $\boldsymbol{\varepsilon}_{ij}$'s are assumed to be independently distributed as $N_p(\mathbf{0}, \boldsymbol{\Sigma})$. The model (4.84) involves the cell mean vectors. The appropriate matrices for making the test of $H_0\colon \boldsymbol{\mu}_1 = \boldsymbol{\mu}_2 = \cdots = \boldsymbol{\mu}_k$ are given in Table 4.4, where $\overline{\mathbf{y}}_{i.} = \sum_{j=1}^{n_i} \mathbf{y}_{ij}/n_i$, $\overline{\mathbf{y}}_{..} = \sum_{i=1}^{k} \sum_{j=1}^{n_i} \mathbf{y}_{ij}/N$, and $N = \sum_{i=1}^{k} n_i$. The test of H_0 is conducted with $\mathbf{H}$ and $\mathbf{E}$ using any of the four MANOVA test statistics; for example, $\Lambda = |\mathbf{E}|/|\mathbf{E} + \mathbf{H}|$ is distributed as Λ_{p,ν_H,ν_E}, where $\nu_H = k - 1$ and $\nu_E = N - k$.

Table 4.4. One-Way Unbalanced MANOVA

Source	Sum of Squares and Products Matrix	df
Treatments	$\mathbf{H} = \sum_i n_i(\overline{\mathbf{y}}_{i.} - \overline{\mathbf{y}}_{..})(\overline{\mathbf{y}}_{i.} - \overline{\mathbf{y}}_{..})'$	$\nu_H = k - 1$
Error	$\mathbf{E} = \sum_{ij}(\mathbf{y}_{ij} - \overline{\mathbf{y}}_{i.})(\mathbf{y}_{ij} - \overline{\mathbf{y}}_{i.})'$	$\nu_E = N - k$
Total	$\mathbf{T} = \sum_{ij}(\mathbf{y}_{ij} - \overline{\mathbf{y}}_{..})(\mathbf{y}_{ij} - \overline{\mathbf{y}}_{..})'$	$N - 1$

As in (4.62), a contrast in the population mean vectors is defined as $\boldsymbol{\delta} = c_1\boldsymbol{\mu}_1 + c_2\boldsymbol{\mu}_2 + \cdots + c_k\boldsymbol{\mu}_k$, where $\sum_{i=1}^k c_i = 0$. The hypothesis $H_{01}\colon \boldsymbol{\delta} = \mathbf{0}$ can be tested by

$$T^2 = \hat{\boldsymbol{\delta}}'\left(\sum_{i=1}^k \frac{c_i^2}{n_i}\mathbf{S}_{\text{pl}}\right)^{-1}\hat{\boldsymbol{\delta}}$$

$$= \frac{1}{\sum_{i=1}^k (c_i^2/n_i)}\left(\sum_{i=1}^k c_i\overline{\mathbf{y}}_{i\cdot}\right)'\left(\frac{\mathbf{E}}{\nu_E}\right)^{-1}\left(\sum_{i=1}^k c_i\overline{\mathbf{y}}_{i\cdot}\right), \tag{4.85}$$

which is distributed as T^2_{p,ν_E} if a single a priori contrast is tested, where $\nu_E = N - k$. If several prespecified contrasts are to be tested, Bonferroni adjustments to the α_i for each test can be made.

The hypothesis $H_{01}\colon \sum_{i=1}^k c_i\boldsymbol{\mu}_i = \mathbf{0}$ can also be tested using any of the four MANOVA statistics. By (4.65) and (4.81), a hypothesis matrix due to the contrast is

$$\mathbf{H}_1 = \frac{1}{\sum_{i=1}^k (c_i^2/n_i)}\left(\sum_{i=1}^k c_i\overline{\mathbf{y}}_{i\cdot}\right)\left(\sum_{i=1}^k c_i\overline{\mathbf{y}}_{i\cdot}\right)', \tag{4.86}$$

and Wilks' statistic is given by

$$\Lambda_1 = \frac{|\mathbf{E}|}{|\mathbf{E} + \mathbf{H}_1|}, \tag{4.87}$$

which is distributed as $\Lambda_{p,1,\nu_E}$ under H_{01}, where $\nu_E = N - k$. The other three MANOVA tests can be obtained using the eigenvalues of $\mathbf{E}^{-1}\mathbf{H}_1$. All five tests give equivalent results in this case, since $\nu_H = 1$.

In the following theorem, a condition is given for two contrasts to be independent.

Theorem 4.8B. Assuming normality and independence of the $\overline{\mathbf{y}}_{i\cdot}$'s, two contrasts $\hat{\boldsymbol{\delta}} = \sum_{i=1}^k a_i\overline{\mathbf{y}}_{i\cdot}$ and $\hat{\boldsymbol{\gamma}} = \sum_{i=1}^k b_i\overline{\mathbf{y}}_{i\cdot}$ are independent if $\sum_{i=1}^k a_ib_i/n_i = 0$.

Proof. As in the hints for Problems 2.28 and 2.29 (Appendix C), define $\overline{\mathbf{y}}' = (\overline{\mathbf{y}}_{1\cdot}', \overline{\mathbf{y}}_{2\cdot}', \ldots, \overline{\mathbf{y}}_{k\cdot}')$, $\mathbf{A} = (a_1\mathbf{I}, a_2\mathbf{I}, \ldots, a_k\mathbf{I})$, and $\mathbf{B} = (b_1\mathbf{I}, b_2\mathbf{I}, \ldots, b_k\mathbf{I})$. Then

$$\boldsymbol{\Sigma}_{\overline{y}} = \begin{pmatrix} \boldsymbol{\Sigma}/n_1 & \mathbf{O} & \ldots & \mathbf{O} \\ \mathbf{O} & \boldsymbol{\Sigma}/n_2 & \ldots & \mathbf{O} \\ \vdots & \vdots & & \vdots \\ \mathbf{O} & \mathbf{O} & \ldots & \boldsymbol{\Sigma}/n_k \end{pmatrix},$$

and $\text{cov}(\hat{\boldsymbol{\delta}}, \hat{\boldsymbol{\gamma}}) = \mathbf{A}\boldsymbol{\Sigma}_{\overline{y}}\mathbf{B}' = \boldsymbol{\Sigma}\sum_{i=1}^k a_ib_i/n_i$, which is equal to $\mathbf{O}$ if $\sum_i a_ib_i/n_i = 0$. □

By an extension of Theorem 4.5C, simultaneous confidence intervals on $\sum_{i=1}^k c_i\mathbf{a}'\boldsymbol{\mu}_i$ for all contrasts and all possible $\mathbf{a}$ are given by

$$\sum_{i=1}^k c_i\mathbf{a}'\overline{\mathbf{y}}_{i\cdot} \pm \sqrt{\frac{\theta_\alpha}{1-\theta_\alpha}\sum_{i=1}^k \frac{c_i^2}{n_i}\mathbf{a}'\mathbf{E}\mathbf{a}}. \tag{4.88}$$

See Table B.5 for the critical values θ_α. From (4.88) we can obtain confidence intervals for $\mu_{ir} - \mu_{i'r}$, the difference between the ith and i'th mean on the rth variable,

$$\overline{y}_{i.r} - \overline{y}_{i'.r} \pm \sqrt{\frac{\theta_\alpha}{1-\theta_\alpha}\left(\frac{1}{n_i} + \frac{1}{n_{i'}}\right) e_{rr}}, \tag{4.89}$$

where $\overline{y}_{i.r}$ is the rth element of $\overline{\mathbf{y}}_{i\cdot}$ and e_{rr} is the rth diagonal element of $\mathbf{E}$.

If only a few, say m, preselected contrasts on linear combinations $\mathbf{a}'\boldsymbol{\mu}_i$ are desired, Bonferroni coefficients can be used to obtain shorter intervals,

$$\sum_{i=1}^{k} c_i \mathbf{a}'\overline{\mathbf{y}}_{i\cdot} \pm t_{\alpha/2m,\nu_E}\sqrt{\frac{\mathbf{a}'\mathbf{E}\mathbf{a}}{\nu_E}\sum_{i=1}^{k}\frac{c_i^2}{n_i}}. \tag{4.90}$$

If we are interested in the individual variables following a significant MANOVA test, we can do an F-test on each variable with significance level α on each F-test. Rencher and Scott (1990) showed that this procedure will preserve an experimentwise error rate of α as long as the univariate tests are carried out only if the MANOVA test rejects H_0: $\boldsymbol{\mu}_1 = \boldsymbol{\mu}_2 = \cdots = \boldsymbol{\mu}_k$.

Information on the effect of the individual variables in the presence of one another can be obtained from the coefficients of the first discriminant function, which are obtained from the first eigenvector of $\mathbf{E}^{-1}\mathbf{H}$. (For a method of standardization, see Section 5.4.) The discriminant coefficients take into account the correlations among the variables and show the joint contribution of the variables to group separation.

Example 4.8.3. Table 4.5 contains data reported by Allison, Zappasodi, and Lurie (1962) and analyzed by Kramer and Jensen (1969b). The study compared four groups of rabbits infected with human tuberculosis:

Table 4.5. Metabolic Comparisons on Four Groups of Rabbits

	G_1		G_2		G_3		G_4	
	y_1	y_2	y_1	y_2	y_1	y_2	y_1	y_2
	24.0	3.5	7.4	3.5	16.4	3.2	25.1	2.7
	13.3	3.5	13.2	3.0	24.0	2.5	5.9	2.3
	12.2	4.0	8.5	3.0	53.0	1.5		
	14.0	4.0	10.1	3.0	32.7	2.6		
	22.2	3.6	9.3	2.0	42.8	2.0		
	16.1	4.3	8.5	2.5				
	27.9	5.2	4.3	1.5				
Total	129.7	28.1	61.3	18.5	168.9	11.8	31.0	5.0
Mean	18.53	4.01	8.76	2.64	33.78	2.36	15.50	2.50

G_1 = unvaccinated control
G_2 = infected during metabolic depression
G_3 = infected during heightened metabolic activity
G_4 = infected during normal activity

The variables were y_1 = number of bacilli inhaled per tubercle formed and y_2 = tubercle size (in millimeters). Thus we have $p = 2,\ k = 4,\ N = 21,\ \nu_H = k-1 = 3$, and $\nu_E = N - k = 17$.

To illustrate the computation, we convert the matrices in Table 4.4 into computational format using totals analogous to those in Table 4.3; for example, $\mathbf{H} = \sum_{i=1}^{k} \mathbf{y}_{i\cdot}\mathbf{y}_{i\cdot}'/n_i - \mathbf{y}_{\cdot\cdot}\mathbf{y}_{\cdot\cdot}'/N$. The totals $\mathbf{y}_{i\cdot}$ for each group are given in Table 4.5, and the grand total vector is $\mathbf{y}_{\cdot\cdot} = (390.9,\ 63.4)'$. The elements of $\mathbf{H}$ are therefore given by

$$h_{11} = \frac{(129.7)^2}{7} + \frac{(61.3)^2}{7} + \frac{(168.9)^2}{5} + \frac{(31.0)^2}{2} - \frac{(390.9)^2}{21} = 1849.59,$$

$$h_{12} = \frac{(129.7)(28.1)}{7} + \frac{(61.3)(18.5)}{7} + \frac{(168.9)(11.8)}{5} + \frac{(31.0)(5.0)}{2} - \frac{(390.9)(63.4)}{21}$$
$$= -21.38,$$

$$h_{22} = \frac{(28.1)^2}{7} + \frac{(18.5)^2}{7} + \frac{(11.8)^2}{5} + \frac{(5.0)^2}{2} - \frac{(63.4)^2}{21} = 10.63.$$

The elements of $\mathbf{E} = \sum_{i=1}^{k}\sum_{j=1}^{n_i} \mathbf{y}_{ij}\mathbf{y}_{ij}' - \sum_{i=1}^{k} \mathbf{y}_{i\cdot}\mathbf{y}_{i\cdot}'/n_i$ are given by

$$e_{11} = (24.0)^2 + \cdots + (5.9)^2 - \frac{(390.0)^2}{21} - 1849.58 = 1302.68,$$

$$e_{12} = (24.0)(3.5) + \cdots + (5.9)(2.3) - \frac{(390.9)(63.4)}{21} - 21.38 = -17.64,$$

$$e_{22} = 6.78.$$

Thus we have

$$\mathbf{H} = \begin{pmatrix} 1849.59 & -21.38 \\ -21.38 & 10.63 \end{pmatrix}, \qquad \mathbf{E} = \begin{pmatrix} 1302.6795 & -17.6447 \\ -17.6447 & 6.7778 \end{pmatrix},$$

$$\mathbf{E} + \mathbf{H} = \begin{pmatrix} 3152.2657 & -39.0257 \\ -39.0257 & 17.4124 \end{pmatrix}.$$

For Wilks' Λ, we obtain

$$\Lambda = \frac{|\mathbf{E}|}{|\mathbf{E} + \mathbf{H}|} = \frac{8517.9675}{53{,}365.5060}$$
$$= 0.1596 \leq \Lambda_{.05,2,3,17} = .317.$$

We obtain the other three test statistics from the eigenvalues of $\mathbf{E}^{-1}\mathbf{H}$, 1.599 and 1.411:

$$V^{(s)} = \sum_{i=1}^{s} \frac{\lambda_i}{1+\lambda_i} = 1.200,$$

$$U^{(s)} = \sum_{i=1}^{s} \lambda_i = 3.010,$$

$$\theta = \frac{\lambda_1}{1+\lambda_1} = .615.$$

With $s = 2, m = 0$, and $N = 7$, the .05 critical values for $V^{(s)}$ and θ are .573 and .489, respectively. [Note the use of $N = \frac{1}{2}(\nu_E - p - 1) = 7$ as a parameter for $V^{(s)}$ and θ to be distinguished from the total sample size $N = \sum_{i=1}^{k} n_i = 21$.] We compare $\nu_E U^{(s)}/\nu_H = 17.059$ with its .05 critical value, 5.6802. Thus all four tests lead to rejection of H_0.

From a description of the four groups, we surmise that three (a priori) contrasts of interest might be the rows of

$$\begin{pmatrix} 3 & -1 & -1 & -1 \\ 0 & -1 & -1 & 2 \\ 0 & 1 & -1 & 0 \end{pmatrix}.$$

We can calculate a hypothesis matrix $\mathbf{H}_i$ for each of these contrasts using (4.86). For the first, we obtain

$$\begin{aligned} \mathbf{H}_1 &= \frac{1}{\sum_{i=1}^{k} c_i^2/n_i} \left(\sum_{i=1}^{k} c_i \bar{\mathbf{y}}_{i\cdot} \right) \left(\sum_{i=1}^{k} c_i \bar{\mathbf{y}}_{i\cdot} \right)' \\ &= \frac{1}{2.129} \begin{pmatrix} -2.451 \\ 4.540 \end{pmatrix} (-2.451, 4.540) \\ &= \begin{pmatrix} 2.823 & -5.229 \\ -5.229 & 9.683 \end{pmatrix}. \end{aligned}$$

By (4.87) the Wilks Λ-test statistic is

$$\Lambda_1 = \frac{|\mathbf{E}|}{|\mathbf{E} + \mathbf{H}_1|} = \frac{8517.9}{20966.7} = .406 < \Lambda_{.05,2,1,17} = .688.$$

For the second contrast, we have

$$\mathbf{H}_2 = \frac{1}{2.343} \begin{pmatrix} -11.537 \\ -.00286 \end{pmatrix} (-11.537, -.00286) = \begin{pmatrix} 56.813 & .0141 \\ .0141 & 3.48 \times 10^{-6} \end{pmatrix},$$

$$\Lambda_2 = \frac{|\mathbf{E}|}{|\mathbf{E} + \mathbf{H}_2|} = \frac{8517.9}{8903.4} = .957 > \Lambda_{.05,2,1,17} = .688.$$

For the third contrast, we obtain

$$\mathbf{H}_3 = \frac{1}{.343}\begin{pmatrix} -25.023 \\ .283 \end{pmatrix}(-25.023, .283) = \begin{pmatrix} 1826.3 & -20.64 \\ -20.64 & .233 \end{pmatrix},$$

$$\Lambda_3 = \frac{8517.9}{20471.2} = .416 < \Lambda_{.05,2,1,17} = .688.$$

Thus the first and third contrasts are significant, but the second is not.

These three contrasts are orthogonal in the sense that $\sum_{i=1}^{k} a_i b_i = 0$ for each pair of contrasts, but they are not weighted contrasts that satisfy $\sum_{i=1}^{k} a_i b_i / n_i = 0$. Hence the three matrices $\mathbf{H}_1, \mathbf{H}_2$, and $\mathbf{H}_3$ are not independent and do not sum to $\mathbf{H}$.

To illustrate the use of weighted contrasts that satisfy $\sum_{i=1}^{k} a_i b_i / n_i = 0$, consider the rows of

$$\begin{pmatrix} 14 & -7 & -5 & -2 \\ 0 & -7 & -5 & 12 \\ 0 & 1 & -1 & 0 \end{pmatrix}.$$

These three contrasts partition $\mathbf{H}$ into independent single-degree-of-freedom components $\mathbf{H}_1$, $\mathbf{H}_2$, and $\mathbf{H}_3$, such that $\mathbf{H} = \mathbf{H}_1 + \mathbf{H}_2 + \mathbf{H}_3$. However, except for the third, the comparisons may not be as interesting as the more direct ones considered above. For these three weighted contrasts, we obtain

$$\Lambda_1 = \frac{8517.9}{22035.0} = .387, \qquad \Lambda_2 = \frac{8517.9}{8682.5} = .981, \qquad \Lambda_3 = \frac{8517.9}{20471.2} = .416.$$

4.8.4 Univariate Two-Way Model

The unbalanced two-way model in the univariate case is

$$y_{ijk} = \mu + \alpha_i + \beta_j + \gamma_{ij} + \varepsilon_{ijk} \tag{4.91}$$

$$= \mu_{ij} + \varepsilon_{ijk}, \tag{4.92}$$

$$i = 1, 2, \ldots, a, \qquad j = 1, 2, \ldots, b, \qquad k = 1, 2, \ldots, n_{ij}.$$

The ε_{ijk}'s are assumed to be independently distributed as $N(0, \sigma^2)$. We consider only the case in which all $n_{ij} > 0$. As discussed in Section 4.8.1, analysis of the overparameterized model (4.91) results in inconsistencies and nonuniqueness. On the other hand, the *cell means model* (4.92) provides a simple and unambiguous approach that leads to tests of clearly identifiable hypotheses. This approach to analyzing unbalanced data was first proposed by Yates (1934), but by the early 1950s it had been supplanted by the overparameterized model. More recently, a return to the cell means model has been advocated by Speed (1969), Urquhart, Weeks, and Henderson (1973), Nelder (1974), Hocking and Speed (1975), Bryce (1975), Bryce, Carter, and Reader (1976), Searle (1977), Speed et al. (1978), Bryce, Scott, and Carter (1980), Searle, Speed, and Henderson (1981), and Hocking (1985, 1996). Turner (1990) discusses the relationship between (4.91) and (4.92). In our development we follow Bryce, Scott, and Carter (1980) and Hocking (1985, 1996).

		B 1	2	3
A	1	$n_{11}=2$	$n_{12}=1$	$n_{13}=2$
	2	$n_{21}=1$	$n_{22}=3$	$n_{23}=2$

Figure 4.1. Cell counts for unbalanced data illustration.

4.8.4a Unconstrained Model

We first consider the *unconstrained model* in which the μ_{ij}'s are unrestricted. The constrained model, which can accommodate a no-interaction model, for example, is discussed in Section 4.8.4b.

To illustrate the cell means model (4.92), we use a 2×3 design with the cell counts n_{ij} in Figure 4.1. This example will be referred to throughout the remainder of this section.

For the 11 observations, the model $y_{ijk} = \mu_{ij} + \varepsilon_{ijk}$ is

$$\begin{aligned} y_{111} &= \mu_{11} + \varepsilon_{111} \\ y_{112} &= \mu_{11} + \varepsilon_{112} \\ y_{121} &= \mu_{12} + \varepsilon_{121} \\ &\vdots \\ y_{231} &= \mu_{23} + \varepsilon_{231} \\ y_{232} &= \mu_{23} + \varepsilon_{232}, \end{aligned}$$

or in matrix form

$$\mathbf{y} = \mathbf{W}\boldsymbol{\mu} + \boldsymbol{\varepsilon}, \tag{4.93}$$

where

$$\mathbf{y} = \begin{pmatrix} y_{111} \\ y_{112} \\ \vdots \\ y_{232} \end{pmatrix}, \quad \mathbf{W} = \begin{pmatrix} 1 & 0 & 0 & \cdots & 0 \\ 1 & 0 & 0 & \cdots & 0 \\ 0 & 1 & 0 & \cdots & 0 \\ 0 & 0 & 1 & \cdots & 0 \\ 0 & 0 & 1 & \cdots & 0 \\ \vdots & \vdots & \vdots & & \vdots \\ 0 & 0 & 0 & \cdots & 1 \end{pmatrix},$$

$$\boldsymbol{\mu} = \begin{pmatrix} \mu_{11} \\ \mu_{12} \\ \mu_{13} \\ \mu_{21} \\ \mu_{22} \\ \mu_{23} \end{pmatrix}, \quad \boldsymbol{\varepsilon} = \begin{pmatrix} \varepsilon_{111} \\ \varepsilon_{112} \\ \vdots \\ \varepsilon_{232} \end{pmatrix}.$$

Note that $\mathbf{y}$ and $\boldsymbol{\varepsilon}$ are 11×1 and $\mathbf{W}$ is 11×6. We assume that $\mathbf{y}$ is $N_{11}(\mathbf{W}\boldsymbol{\mu}, \sigma^2\mathbf{I})$.

Since $\mathbf{W}$ is full rank, we can use a regression approach. The analysis is further simplified because $\mathbf{W}'\mathbf{W} = \text{diag}(n_{11}, n_{12}, n_{13}, n_{21}, n_{22}, n_{23})$. A review of estimation and hypothesis testing in the full-rank linear model can be found in Section 7.2. A least squares estimator of $\boldsymbol{\mu}$ [see (7.6)] is given by

$$\hat{\boldsymbol{\mu}} = (\mathbf{W}'\mathbf{W})^{-1}\mathbf{W}'\mathbf{y} = \bar{\mathbf{y}}, \tag{4.94}$$

where $\bar{\mathbf{y}} = (\bar{y}_{11.}, \bar{y}_{12.}, \bar{y}_{13.}, \bar{y}_{21.}, \bar{y}_{22.}, \bar{y}_{23.})'$ contains the sample means of the cells. The covariance matrix for $\hat{\boldsymbol{\mu}}$ [see (7.8)] is

$$\text{cov}(\hat{\boldsymbol{\mu}}) = \sigma^2(\mathbf{W}'\mathbf{W})^{-1}, \tag{4.95}$$

and an unbiased estimator of σ^2 [see (7.10)] is given by

$$s^2 = \frac{\text{SSE}}{\nu_E} = \frac{(\mathbf{y} - \mathbf{W}\hat{\boldsymbol{\mu}})'(\mathbf{y} - \mathbf{W}\hat{\boldsymbol{\mu}})}{\nu_E}, \tag{4.96}$$

where $\nu_E = \sum_{i=1}^a \sum_{j=1}^b (n_{ij} - 1) = N - ab$, with $N = \sum_{ij} n_{ij}$. Two alternative forms of SSE are

$$\text{SSE} = \mathbf{y}'[\mathbf{I} - \mathbf{W}(\mathbf{W}'\mathbf{W})^{-1}\mathbf{W}']\mathbf{y}, \tag{4.97}$$

$$\text{SSE} = \sum_{i=1}^a \sum_{j=1}^b \left[\sum_{k=1}^{n_{ij}} (y_{ijk} - \bar{y}_{ij.})^2 \right]. \tag{4.98}$$

From (4.98), we see that s^2 is a pooled estimator and can be expressed as

$$s^2 = \frac{\sum_{i=1}^a \sum_{j=1}^b (n_{ij} - 1)s_{ij}^2}{N - ab}, \tag{4.99}$$

where s_{ij}^2 is the variance estimator in the ijth cell.

The overparameterized model (4.91) shows main effects and interactions, but the means model (4.92) does not. To carry out tests in the cell means model, we use contrasts to express the main effects and the interaction as functions of the μ_{ij}'s in $\boldsymbol{\mu}$.

In the vector $\boldsymbol{\mu} = (\mu_{11}, \mu_{12}, \mu_{13}, \mu_{21}, \mu_{22}, \mu_{23})'$, the first three elements correspond to the first level of A and the last three to the second level, as seen in Figure 4.2. Therefore, the main effect of A is obtained by comparing μ_{11}, μ_{12}, and μ_{13} with

		B		
		1	2	3
A	1	μ_{11}	μ_{12}	μ_{13}
	2	μ_{21}	μ_{22}	μ_{23}

Figure 4.2. Cell means.

μ_{21}, μ_{22}, and μ_{23}. We can make this comparison with the contrast

$$\mathbf{a}'\boldsymbol{\mu} = \mu_{11} + \mu_{12} + \mu_{13} - \mu_{21} - \mu_{22} - \mu_{23},$$

where $\mathbf{a}' = (1 \quad 1 \quad 1 \quad -1 \quad -1 \quad -1)$. The hypothesis that we would typically wish to test is $H_0\colon \mathbf{a}'\boldsymbol{\mu} = 0$, which can be written as $H_0\colon (\mu_{11} - \mu_{21}) + (\mu_{12} - \mu_{22}) + (\mu_{13} - \mu_{23}) = 0$. This states that the effect of A averaged over the levels of B is 0.

Factor B has three levels that correspond to the three columns of Figure 4.2. For three levels there are two degrees of freedom, which will require two contrasts. Suppose we wish to compare the first column with the other two and then compare the second column with the third. This can be done with two orthogonal contrasts as follows:

$$\begin{aligned}
\mathbf{b}_1'\boldsymbol{\mu} &= 2(\mu_{11} + \mu_{21}) - (\mu_{12} + \mu_{22}) - (\mu_{13} + \mu_{23}) \\
&= 2\mu_{11} - \mu_{12} - \mu_{13} + 2\mu_{21} - \mu_{22} - \mu_{23} \\
&= (2 \quad -1 \quad -1 \quad 2 \quad -1 \quad -1)\boldsymbol{\mu}, \\
\mathbf{b}_2'\boldsymbol{\mu} &= (\mu_{12} + \mu_{22}) - (\mu_{13} + \mu_{23}) \\
&= \mu_{12} - \mu_{13} + \mu_{22} - \mu_{23} \\
&= (0 \quad 1 \quad -1 \quad 0 \quad 1 \quad -1)\boldsymbol{\mu}.
\end{aligned}$$

Note that the contrasts $\mathbf{a}'\boldsymbol{\mu}$, $\mathbf{b}_1'\boldsymbol{\mu}$, and $\mathbf{b}_2'\boldsymbol{\mu}$ compare the elements μ_{ij} within the mean vector $\boldsymbol{\mu}$, whereas the contrasts in Section 4.8.3 involved comparisons of entire mean vectors.

We can combine $\mathbf{b}_1'$ and $\mathbf{b}_2'$ into the matrix

$$\mathbf{B} = \begin{pmatrix} 2 & -1 & -1 & 2 & -1 & -1 \\ 0 & 1 & -1 & 0 & 1 & -1 \end{pmatrix},$$

and the hypothesis becomes $H_0\colon \mathbf{B}\boldsymbol{\mu} = \mathbf{0}$, which is equivalent to $H_0\colon \mu_{11} + \mu_{21} = \mu_{12} + \mu_{22} = \mu_{13} + \mu_{23}$. This states that the three levels of B do not differ when averaged over the levels of A.

The interaction AB has two degrees of freedom, and the two associated contrasts can be found by taking products of corresponding elements of $\mathbf{a}$ and $\mathbf{b}_1$ and of $\mathbf{a}$ and $\mathbf{b}_2$. This procedure gives

$$\begin{aligned}
\mathbf{c}_1' &= [\,(1)(2) \quad (1)(-1) \quad (1)(-1) \quad (-1)(2) \quad (-1)(-1) \quad (-1)(-1)\,] \\
&= (2 \quad -1 \quad -1 \quad -2 \quad 1 \quad 1), \\
\mathbf{c}_2' &= [\,(1)(0) \quad (1)(1) \quad (1)(-1) \quad (-1)(0) \quad (-1)(1) \quad (-1)(-1)\,] \\
&= (0 \quad 1 \quad -1 \quad 0 \quad -1 \quad 1),
\end{aligned}$$

which can be combined into

$$\mathbf{C} = \begin{pmatrix} 2 & -1 & -1 & -2 & 1 & 1 \\ 0 & 1 & -1 & 0 & -1 & 1 \end{pmatrix}.$$

This elementwise multiplication of two vectors is called the *Hadamard product*. In this case, it produces interaction contrasts that are orthogonal to each other and to the main-effect contrasts. To see that these contrasts define meaningful interaction effects in this example, note that $\mathbf{c}_1'\boldsymbol{\mu}$ and $\mathbf{c}_2'\boldsymbol{\mu}$ can be expressed as

$$\mathbf{c}_1'\boldsymbol{\mu} = 2(\mu_{11} - \mu_{21}) - (\mu_{12} - \mu_{22}) - (\mu_{13} - \mu_{23}),$$
$$\mathbf{c}_2'\boldsymbol{\mu} = (\mu_{12} - \mu_{22}) - (\mu_{13} - \mu_{23}).$$

The hypothesis $H_0\colon \mathbf{C}\boldsymbol{\mu} = \mathbf{0}$ is equivalent to $H_0\colon \mu_{11} - \mu_{21} = \mu_{12} - \mu_{22} = \mu_{13} - \mu_{23}$, which is a comparison of the "A effect" across the levels of B. If these A effects differ, we have an interaction.

We now construct tests for the hypotheses $H_0\colon \mathbf{a}'\boldsymbol{\mu} = 0, H_0\colon \mathbf{B}\boldsymbol{\mu} = \mathbf{0}$, and $H_0\colon \mathbf{C}\boldsymbol{\mu} = \mathbf{0}$, corresponding to the two main effects and the interaction. The hypothesis for the main effect of A, $H_0\colon \mathbf{a}'\boldsymbol{\mu} = 0$, is easily tested using a t-statistic similar to (4.60). The contrast $\mathbf{a}'\boldsymbol{\mu}$ is estimated by $\mathbf{a}'\hat{\boldsymbol{\mu}}$, which, by (1.87) and (4.95), has variance equal to

$$\text{var}(\mathbf{a}'\hat{\boldsymbol{\mu}}) = \mathbf{a}'\text{cov}(\hat{\boldsymbol{\mu}})\mathbf{a} = \sigma^2\mathbf{a}'(\mathbf{W}'\mathbf{W})^{-1}\mathbf{a} = \sigma^2\sum_{i=1}^{a}\sum_{j=1}^{b}\frac{a_{ij}^2}{n_{ij}}.$$

This variance can estimated by

$$\widehat{\text{var}}(\mathbf{a}'\hat{\boldsymbol{\mu}}) = s^2\mathbf{a}'(\mathbf{W}'\mathbf{W})^{-1}\mathbf{a},$$

where s^2 is given by (4.96). Hence, the t-statistic for testing $H_0\colon \mathbf{a}'\boldsymbol{\mu} = 0$ is

$$t = \frac{\mathbf{a}'\hat{\boldsymbol{\mu}} - 0}{\sqrt{\widehat{\text{var}}(\mathbf{a}'\hat{\boldsymbol{\mu}})}} = \frac{\mathbf{a}'\hat{\boldsymbol{\mu}}}{s\sqrt{\mathbf{a}'(\mathbf{W}'\mathbf{W})^{-1}\mathbf{a}}}, \tag{4.100}$$

which is distributed as t_{ν_E} (when H_0 is true). This tests for factor A above and beyond the B effect and the interaction.

If the t-statistic in (4.100) is squared, we obtain an F-statistic with 1 and ν_E degrees of freedom:

$$F = \frac{(\mathbf{a}'\hat{\boldsymbol{\mu}})^2/[\mathbf{a}'(\mathbf{W}'\mathbf{W})^{-1}\mathbf{a}]}{s^2}$$

$$= \frac{(\mathbf{a}'\hat{\boldsymbol{\mu}})'[\mathbf{a}'(\mathbf{W}'\mathbf{W})^{-1}\mathbf{a}]^{-1}(\mathbf{a}'\hat{\boldsymbol{\mu}})}{\text{SSE}/\nu_E} \qquad \text{[by (4.96)]} \tag{4.101}$$

$$= \frac{\text{SSA}}{\text{SSE}/\nu_E}. \tag{4.102}$$

The F-statistic in (4.101) for factor A is a squared standardized distance from $\mathbf{a}'\hat{\boldsymbol{\mu}}$ to 0. Similarly, to test the hypothesis for the B main effect, $H_0\colon \mathbf{B}\boldsymbol{\mu} = \mathbf{0}$, we work with the distance from $\mathbf{B}\hat{\boldsymbol{\mu}}$ to $\mathbf{0}$ (see Section 1.8). The covariance matrix for $\mathbf{B}\hat{\boldsymbol{\mu}}$ is

$$\text{cov}(\mathbf{B}\hat{\boldsymbol{\mu}}) = \mathbf{B}\,\text{cov}(\hat{\boldsymbol{\mu}})\mathbf{B}' = \sigma^2\mathbf{B}(\mathbf{W}'\mathbf{W})^{-1}\mathbf{B}',$$

and the squared standardized distance from $\mathbf{B}\hat{\boldsymbol{\mu}}$ to $\mathbf{0}$ is

$$(\mathbf{B}\hat{\boldsymbol{\mu}}-\mathbf{0})'[\sigma^2\mathbf{B}(\mathbf{W}'\mathbf{W})^{-1}\mathbf{B}']^{-1}(\mathbf{B}\hat{\boldsymbol{\mu}}-\mathbf{0}) = \frac{1}{\sigma^2}(\mathbf{B}\hat{\boldsymbol{\mu}})'[\mathbf{B}(\mathbf{W}'\mathbf{W})^{-1}\mathbf{B}']^{-1}\mathbf{B}\hat{\boldsymbol{\mu}}. \quad (4.103)$$

By Theorems 2.2B(ii) and 2.2F, the expression in (4.103) has a χ^2-distribution with $\nu_B = 2$ degrees of freedom. In general, ν_B is the number of rows of $\mathbf{B}$.

To extend (4.101) to an F-statistic that involves $\mathbf{B}\hat{\boldsymbol{\mu}}$ as in (4.103), we first note that SSE/σ^2 is distributed as χ^2 with degrees of freedom $\nu_E = \sum_{i=1}^{a}\sum_{j=1}^{b}(n_{ij}-1) = N - ab$, SSE and $\mathbf{B}\hat{\boldsymbol{\mu}}$ are independent, and therefore SSE and the sum of squares due to B in (4.103) are independent. (For a justification of these assertions in a multivariate setting, see Theorem 7.4G.) By definition, an F random variable is the ratio of two independent χ^2 random variables, each divided by its degrees of freedom. Hence a test statistic for the factor B main effect hypothesis $H_0\colon \mathbf{B}\boldsymbol{\mu} = \mathbf{0}$ is given by

$$F = \frac{(\mathbf{B}\hat{\boldsymbol{\mu}})'[\mathbf{B}(\mathbf{W}'\mathbf{W})^{-1}\mathbf{B}']^{-1}\mathbf{B}\hat{\boldsymbol{\mu}}/\sigma^2\nu_B}{\text{SSE}/\sigma^2\nu_E}$$

$$= \frac{(\mathbf{B}\hat{\boldsymbol{\mu}})'[\mathbf{B}(\mathbf{W}'\mathbf{W})^{-1}\mathbf{B}']^{-1}\mathbf{B}\hat{\boldsymbol{\mu}}/\nu_B}{\text{SSE}/\nu_E} \quad (4.104)$$

$$= \frac{\text{SSB}/\nu_B}{\text{SSE}/\nu_E}. \quad (4.105)$$

When H_0 is true, F is distributed as F_{ν_B,ν_E}. Alternatively, (4.104) can be obtained directly as the test statistic for the general linear hypothesis $H_0\colon \mathbf{B}\boldsymbol{\mu} = \mathbf{0}$ (see Section 7.2.4c).

A test statistic for the interaction hypothesis $H_0\colon \mathbf{C}\boldsymbol{\mu} = \mathbf{0}$ is obtained similarly:

$$F = \frac{(\mathbf{C}\hat{\boldsymbol{\mu}})'[\mathbf{C}(\mathbf{W}'\mathbf{W})^{-1}\mathbf{C}']^{-1}\mathbf{C}\hat{\boldsymbol{\mu}}/\nu_{AB}}{\text{SSE}/\nu_E} \quad (4.106)$$

$$= \frac{\text{SSAB}/\nu_{AB}}{\text{SSE}/\nu_E}, \quad (4.107)$$

which is distributed as F_{ν_{AB},ν_E}, where ν_{AB} is the degrees of freedom for interaction.

Because of the unequal n_{ij}'s, the three sums of squares SSA, SSB, and SSAB do not add up to the total sum of squares for treatments, and therefore they are not statistically independent. In each case, the effect being tested is adjusted for the other effects; that is, the given effect is tested "above and beyond" the others [see a comment following (7.34)].

4.8.4b Constrained Model

Sometimes constraints must be added to the cell means model (4.92) or (4.93) to allow for missing interactions or other restrictions. For example, the model

$$y_{ijk} = \mu_{ij} + \varepsilon_{ijk}$$

cannot represent the no-interaction model

$$y_{ijk} = \mu + \alpha_i + \beta_j + \varepsilon_{ijk} \tag{4.108}$$

unless we specify some relationships among the μ_{ij}'s.

In our 2×3 illustration, the two interaction contrasts were expressible as

$$\mathbf{C}\boldsymbol{\mu} = \begin{pmatrix} 2 & -1 & -1 & -2 & 1 & 1 \\ 0 & 1 & -1 & 0 & -1 & 1 \end{pmatrix} \boldsymbol{\mu}.$$

If we wish to operate with a model in which interaction is absent, then $\mathbf{C}\boldsymbol{\mu} = \mathbf{0}$ is not a hypothesis to be tested but an assumption to be included in the statement of the model. In general, for constraints $\mathbf{G}\boldsymbol{\mu} = \mathbf{0}$, the model can be stated as

$$\mathbf{y} = \mathbf{W}\boldsymbol{\mu} + \boldsymbol{\varepsilon} \text{ subject to } \mathbf{G}\boldsymbol{\mu} = \mathbf{0}. \tag{4.109}$$

The estimate of $\boldsymbol{\mu}$ will be somewhat different in order to allow for the constraints. We incorporate the constraints $\mathbf{G}\boldsymbol{\mu} = \mathbf{0}$ into the model by defining a matrix

$$\mathbf{A} = \begin{pmatrix} \mathbf{K} \\ \mathbf{G} \end{pmatrix}. \tag{4.110}$$

The first row of $\mathbf{K}$ typically corresponds to a test of the overall mean, H_0: $\mu = 0$, with μ defined in (4.108). The remaining rows of $\mathbf{K}$ include the contrasts for the hypotheses of interest.

In our continuing illustration, suppose we wish to test the A and B main effects in the no-interaction model (4.108). Then we would define $\mathbf{G} = \mathbf{C}$ and

$$\mathbf{K} = \begin{pmatrix} 1 & 1 & 1 & 1 & 1 & 1 \\ 1 & 1 & 1 & -1 & -1 & -1 \\ 2 & -1 & -1 & 2 & -1 & -1 \\ 0 & 1 & -1 & 0 & 1 & -1 \end{pmatrix}.$$

The second row of $\mathbf{K}$ is $\mathbf{a}'$ and corresponds to the effect of A. The third and fourth rows are from $\mathbf{B}$ and represent the B-effect.

If the rows of $\mathbf{G}$ are orthogonal to $\mathbf{K}$, then $\mathbf{A}$ will be full rank and have an inverse. This holds true in our example, in which $\mathbf{G} = \mathbf{C}$. We incorporate the constraints into the model by inserting $\mathbf{A}^{-1}\mathbf{A}$ into (4.109):

$$\begin{aligned} \mathbf{y} &= \mathbf{W}\mathbf{A}^{-1}\mathbf{A}\boldsymbol{\mu} + \boldsymbol{\varepsilon} \qquad (4.111) \\ &= \mathbf{Z}\boldsymbol{\delta} + \boldsymbol{\varepsilon}, \end{aligned}$$

where $\mathbf{Z} = \mathbf{W}\mathbf{A}^{-1}$ and $\boldsymbol{\delta} = \mathbf{A}\boldsymbol{\mu}$.

By (A.3.5), we partition $\boldsymbol{\delta}$ into

$$\boldsymbol{\delta} = \mathbf{A}\boldsymbol{\mu} = \begin{pmatrix} \mathbf{K} \\ \mathbf{G} \end{pmatrix} \boldsymbol{\mu} = \begin{pmatrix} \mathbf{K}\boldsymbol{\mu} \\ \mathbf{G}\boldsymbol{\mu} \end{pmatrix} = \begin{pmatrix} \boldsymbol{\delta}_1 \\ \boldsymbol{\delta}_2 \end{pmatrix},$$

and with a corresponding partitioning on the columns of $\mathbf{Z}$, we obtain

$$\begin{aligned} \mathbf{y} = \mathbf{Z}\boldsymbol{\delta} + \boldsymbol{\varepsilon} &= (\mathbf{Z}_1, \mathbf{Z}_2)\begin{pmatrix} \boldsymbol{\delta}_1 \\ \boldsymbol{\delta}_2 \end{pmatrix} + \boldsymbol{\varepsilon} \\ &= \mathbf{Z}_1\boldsymbol{\delta}_1 + \mathbf{Z}_2\boldsymbol{\delta}_2 + \boldsymbol{\varepsilon} \qquad \text{[by (A.3.2)]}. \end{aligned} \tag{4.112}$$

Since $\boldsymbol{\delta}_2 = \mathbf{G}\boldsymbol{\mu} = \mathbf{0}$, the constrained model in (4.112) simplifies to

$$\mathbf{y} = \mathbf{Z}_1\boldsymbol{\delta}_1 + \boldsymbol{\varepsilon}, \tag{4.113}$$

and an estimate of $\boldsymbol{\delta}_1$ [see (7.6)] is given by

$$\hat{\boldsymbol{\delta}}_1 = (\mathbf{Z}_1'\mathbf{Z}_1)^{-1}\mathbf{Z}_1'\mathbf{y}.$$

To solve for $\boldsymbol{\mu}$ subject to the constraints, we multiply

$$\mathbf{A}\boldsymbol{\mu} = \begin{pmatrix} \boldsymbol{\delta}_1 \\ \boldsymbol{\delta}_2 \end{pmatrix}$$

by

$$\mathbf{A}^{-1} = [\mathbf{K}'(\mathbf{K}\mathbf{K}')^{-1},\ \mathbf{G}'(\mathbf{G}\mathbf{G}')^{-1}] \tag{4.114}$$

to obtain

$$\boldsymbol{\mu}_c = \mathbf{K}'(\mathbf{K}\mathbf{K}')^{-1}\boldsymbol{\delta}_1,$$

since $\boldsymbol{\delta}_2 = \mathbf{0}$. We then estimate $\boldsymbol{\mu}_c$ by

$$\hat{\boldsymbol{\mu}}_c = \mathbf{K}'(\mathbf{K}\mathbf{K}')^{-1}\hat{\boldsymbol{\delta}}_1 = \mathbf{K}'(\mathbf{K}\mathbf{K}')^{-1}(\mathbf{Z}_1'\mathbf{Z}_1)^{-1}\mathbf{Z}_1'\mathbf{y}. \tag{4.115}$$

The covariance matrix of $\hat{\boldsymbol{\mu}}_c$ is given by

$$\text{cov}(\hat{\boldsymbol{\mu}}_c) = \sigma^2\mathbf{K}'(\mathbf{K}\mathbf{K}')^{-1}(\mathbf{Z}_1'\mathbf{Z}_1)^{-1}(\mathbf{K}\mathbf{K}')^{-1}\mathbf{K}. \tag{4.116}$$

To test the hypothesis $H_0\colon \mathbf{B}\boldsymbol{\mu}_c = \mathbf{0}$ in the constrained model, we need the covariance matrix of $\mathbf{B}\hat{\boldsymbol{\mu}}_c$, which is obtained from (4.116) as

$$\text{cov}(\mathbf{B}\hat{\boldsymbol{\mu}}_c) = \sigma^2\mathbf{B}\mathbf{K}'(\mathbf{K}\mathbf{K}')^{-1}(\mathbf{Z}_1'\mathbf{Z}_1)^{-1}(\mathbf{K}\mathbf{K}')^{-1}\mathbf{K}\mathbf{B}'.$$

Thus for the constrained model, the test statistic in (4.104) becomes

$$F = \frac{(\mathbf{B}\hat{\boldsymbol{\mu}}_c)'[\mathbf{B}\mathbf{K}'(\mathbf{K}\mathbf{K}')^{-1}(\mathbf{Z}_1'\mathbf{Z}_1)^{-1}(\mathbf{K}\mathbf{K}')^{-1}\mathbf{K}\mathbf{B}']^{-1}\mathbf{B}\hat{\boldsymbol{\mu}}_c/\nu_B}{\text{SSE}_c/\nu_{Ec}}, \tag{4.117}$$

where SSE_c (in this case, with $\mathbf{G}\boldsymbol{\mu} = \mathbf{0}$) is obtained using $\hat{\boldsymbol{\mu}}_c$ from (4.115) in (4.96), (4.97), or (4.98). (In our example, where $\mathbf{G} = \mathbf{C}$ for interaction, SSE_c effectively pools SSE and SSAB from the unconstrained model.) The degrees of freedom ν_{Ec} is obtained as $\nu_{Ec} = \nu_E + \text{rank}(\mathbf{G})$, where $\nu_E = N - ab$ is for the unconstrained model, as defined following (4.96). [In our example, adding rank(**G**) adds two degrees of freedom for SSAB.] We reject H_0 if $F > F_{\alpha,\nu_B,\nu_{Ec}}$. Expressions similar to (4.101) or (4.106) would result for hypotheses H_0: $\mathbf{a}'\boldsymbol{\mu}_c = 0$ or H_0: $\mathbf{C}\boldsymbol{\mu}_c = \mathbf{0}$ in the constrained model.

4.8.5 Multivariate Two-Way Model

The results in the previous section extend immediately to the multivariate case. Corresponding to (4.92), the two-way model for p-variate observation vectors becomes

$$\mathbf{y}_{ijk} = \boldsymbol{\mu}_{ij} + \boldsymbol{\varepsilon}_{ijk}, \tag{4.118}$$
$$i = 1, 2, \ldots, a, \quad j = 1, 2, \ldots, b, \quad k = 1, 2, \ldots, n_{ij}.$$

The error vectors $\boldsymbol{\varepsilon}_{ijk}$ are assumed to be independently distributed as $N_p(\mathbf{0}, \boldsymbol{\Sigma})$. Corresponding to (4.93), the model for all observation vectors becomes

$$\mathbf{Y} = \mathbf{W}\mathbf{M} + \boldsymbol{\Xi}, \tag{4.119}$$

where

$$\mathbf{Y} = \begin{pmatrix} \mathbf{y}'_{111} \\ \mathbf{y}'_{112} \\ \vdots \\ \mathbf{y}'_{abn_{ab}} \end{pmatrix}, \quad \mathbf{M} = \begin{pmatrix} \boldsymbol{\mu}'_{11} \\ \boldsymbol{\mu}'_{12} \\ \vdots \\ \boldsymbol{\mu}'_{ab} \end{pmatrix}, \quad \boldsymbol{\Xi} = \begin{pmatrix} \boldsymbol{\varepsilon}'_{111} \\ \boldsymbol{\varepsilon}'_{112} \\ \vdots \\ \boldsymbol{\varepsilon}'_{abn_{ab}} \end{pmatrix},$$

and **W** has the same form as in the univariate cell means model that was illustrated following (4.93).

If we denote the columns of **Y** by $\mathbf{y}_{(i)}$, as in (A.2.22) and (1.27), we have

$$\mathbf{Y} = (\mathbf{y}_{(1)}, \mathbf{y}_{(2)}, \ldots, \mathbf{y}_{(p)}).$$

Thus $\mathbf{y}_{(1)}$ contains all the $N = \sum_{ij} n_{ij}$ observations on the first variable, $\mathbf{y}_{(2)}$ consists of all the N observations on the second variable, and so on. We can likewise express **M** in the form

$$\mathbf{M} = (\boldsymbol{\mu}_{(1)}, \boldsymbol{\mu}_{(2)}, \ldots, \boldsymbol{\mu}_{(p)}),$$

where $\boldsymbol{\mu}_{(1)}$ contains the cell means for the first variable, $\boldsymbol{\mu}_{(2)}$ has the cell means for the second variable, and so on. With a similar notation for $\boldsymbol{\varepsilon}_{(i)}$, we can write (4.119) as

$$\begin{aligned}(\mathbf{y}_{(1)}, \mathbf{y}_{(2)}, \ldots, \mathbf{y}_{(p)}) &= \mathbf{W}(\boldsymbol{\mu}_{(1)}, \boldsymbol{\mu}_{(2)}, \ldots, \boldsymbol{\mu}_{(p)}) + (\boldsymbol{\varepsilon}_{(1)}, \boldsymbol{\varepsilon}_{(2)}, \ldots, \boldsymbol{\varepsilon}_{(p)}) \\ &= (\mathbf{W}\boldsymbol{\mu}_{(1)}, \mathbf{W}\boldsymbol{\mu}_{(2)}, \ldots, \mathbf{W}\boldsymbol{\mu}_{(p)}) + (\boldsymbol{\varepsilon}_{(1)}, \boldsymbol{\varepsilon}_{(2)}, \ldots, \boldsymbol{\varepsilon}_{(p)}),\end{aligned}$$

by (A.2.26) . The multivariate model thus consists of the juxtaposition of the p univariate models $\mathbf{y}_{(i)} = \mathbf{W}\boldsymbol{\mu}_{(i)} + \boldsymbol{\varepsilon}_{(i)}$, $i = 1, 2, \ldots, p$.

4.8.5a Unconstrained Model

Using (4.94), we can estimate each column of $\mathbf{M}$ in the unconstrained model as

$$\hat{\boldsymbol{\mu}}_{(i)} = (\mathbf{W}'\mathbf{W})^{-1}\mathbf{W}'\mathbf{y}_{(i)}.$$

Hence $\hat{\mathbf{M}}$ is given by

$$\begin{aligned}\hat{\mathbf{M}} &= (\hat{\boldsymbol{\mu}}_{(1)}, \hat{\boldsymbol{\mu}}_{(2)}, \ldots, \hat{\boldsymbol{\mu}}_{(p)}) \\ &= (\mathbf{W}'\mathbf{W})^{-1}\mathbf{W}'(\mathbf{y}_{(1)}, \mathbf{y}_{(2)}, \ldots, \mathbf{y}_{(p)}) \\ &= (\mathbf{W}'\mathbf{W})^{-1}\mathbf{W}'\mathbf{Y}. \end{aligned} \tag{4.120}$$

For an alternative derivation of $\hat{\mathbf{M}}$ using multivariate regression, see Theorem 7.4A. The matrix that corresponds to SSE in (4.96) is

$$\mathbf{E} = (\mathbf{Y} - \mathbf{W}\hat{\mathbf{M}})'(\mathbf{Y} - \mathbf{W}\hat{\mathbf{M}}) \tag{4.121}$$

[see also (7.82)]. The diagonal of $\mathbf{E}$ contains the SSE for each of the p variables, and the off-diagonal elements of $\mathbf{E}$ are analogous cross-product terms.

To test hypotheses, we use contrasts as in Section 4.8.4, in which the procedure was illustrated with two levels of A and three levels of B. If $\mathbf{B}$ is the matrix of contrasts used to test the main effect of factor B, a hypothesis matrix analogous to the sum of squares in the numerator of (4.104) is obtained as

$$\mathbf{H}_B = (\mathbf{B}\hat{\mathbf{M}})'[\mathbf{B}(\mathbf{W}'\mathbf{W})^{-1}\mathbf{B}']^{-1}\mathbf{B}\hat{\mathbf{M}}, \tag{4.122}$$

and Wilks' test statistic for $H_0\colon \mathbf{BM} = \mathbf{O}$ is

$$\Lambda = \frac{|\mathbf{E}|}{|\mathbf{E} + \mathbf{H}_B|},$$

which is distributed as Λ_{p,ν_B,ν_E} when H_0 is true. The other three MANOVA test statistics for factor B can be calculated using the eigenvalues of $\mathbf{E}^{-1}\mathbf{H}_B$. Tests for the A effect ($H_0\colon \mathbf{a}'\mathbf{M} = \mathbf{0}'$, assuming two levels of A as in Section 4.8.4) and the AB interaction ($H_0\colon \mathbf{CM} = \mathbf{O}$) can be obtained similarly using

$$\mathbf{H}_A = (\mathbf{a}'\hat{\mathbf{M}})'[\mathbf{a}'(\mathbf{W}'\mathbf{W})^{-1}\mathbf{a}]^{-1}(\mathbf{a}'\hat{\mathbf{M}}), \tag{4.123}$$

$$\mathbf{H}_{AB} = (\mathbf{C}\hat{\mathbf{M}})'[\mathbf{C}(\mathbf{W}'\mathbf{W})^{-1}\mathbf{C}']^{-1}(\mathbf{C}\hat{\mathbf{M}}). \tag{4.124}$$

4.8.5b Constrained Model

Estimation and testing in the multivariate constrained model are likewise immediate extensions of the univariate case in Section 4.8.4. The model becomes

$$\mathbf{Y} = \mathbf{WM} + \boldsymbol{\Xi} \text{ subject to } \mathbf{GM} = \mathbf{O}, \tag{4.125}$$

where $\mathbf{GM} = \mathbf{O}$ may include interactions assumed to be zero or other constraints of interest. We define $\mathbf{A}$ as in (4.110),

$$\mathbf{A} = \begin{pmatrix} \mathbf{K} \\ \mathbf{G} \end{pmatrix},$$

where $\mathbf{K}$ contains contrasts associated with hypotheses of interest. The constrained model becomes

$$\begin{aligned} \mathbf{Y} &= \mathbf{W}\mathbf{A}^{-1}\mathbf{A}\mathbf{M} + \boldsymbol{\Xi} \\ &= \mathbf{Z}\boldsymbol{\Delta} + \boldsymbol{\Xi}, \end{aligned} \tag{4.126}$$

where $\mathbf{Z} = \mathbf{W}\mathbf{A}^{-1}$ and $\boldsymbol{\Delta} = \mathbf{A}\mathbf{M}$. By (A.3.6), we partition $\boldsymbol{\Delta}$ into

$$\boldsymbol{\Delta} = \mathbf{A}\mathbf{M} = \begin{pmatrix} \mathbf{K} \\ \mathbf{G} \end{pmatrix}\mathbf{M} = \begin{pmatrix} \mathbf{KM} \\ \mathbf{GM} \end{pmatrix} = \begin{pmatrix} \boldsymbol{\Delta}_1 \\ \boldsymbol{\Delta}_2 \end{pmatrix},$$

and with an analogous partitioning on the columns of $\mathbf{Z}$, we obtain

$$\mathbf{Y} = \mathbf{Z}_1\boldsymbol{\Delta}_1 + \mathbf{Z}_2\boldsymbol{\Delta}_2 + \boldsymbol{\Xi}, \tag{4.127}$$

which simplifies to

$$\mathbf{Y} = \mathbf{Z}_1\boldsymbol{\Delta}_1 + \boldsymbol{\Xi} \tag{4.128}$$

because $\boldsymbol{\Delta}_2 = \mathbf{GM} = \mathbf{O}$. From the constrained model (4.128) we obtain

$$\hat{\boldsymbol{\Delta}}_1 = (\mathbf{Z}_1'\mathbf{Z}_1)^{-1}\mathbf{Z}_1'\mathbf{Y}.$$

By analogy with (4.115), we estimate $\mathbf{M}_c = \mathbf{K}'(\mathbf{KK}')^{-1}\boldsymbol{\Delta}_1$ by

$$\hat{\mathbf{M}}_c = \mathbf{K}'(\mathbf{KK}')^{-1}(\mathbf{Z}_1'\mathbf{Z}_1)^{-1}\mathbf{Z}_1'\mathbf{Y}. \tag{4.129}$$

To test the hypothesis $H_0\colon \mathbf{BM}_c = \mathbf{O}$ in the constrained model, we construct a hypothesis matrix analogous to the sum of squares in the numerator of (4.117):

$$\mathbf{H}_c = (\mathbf{B}\hat{\mathbf{M}}_c)'[\mathbf{BK}'(\mathbf{KK}')^{-1}(\mathbf{Z}_1'\mathbf{Z}_1)^{-1}(\mathbf{KK}')^{-1}\mathbf{KB}']^{-1}\mathbf{B}\hat{\mathbf{M}}_c. \tag{4.130}$$

With the constraints $\mathbf{GM} = \mathbf{O}$, the within matrix $\mathbf{E}_c$ is obtained by using $\hat{\mathbf{M}}_c$ in (4.121). (If $\mathbf{GM} = \mathbf{O}$ represents interaction, then $\mathbf{E}_c$ effectively pools $\mathbf{E}$ and $\mathbf{H}_{AB}$ from the unconstrained model.) For Wilks' Λ-test of $H_0\colon \mathbf{BM}_c = \mathbf{O}$, we compute

$$\Lambda = \frac{|\mathbf{E}_c|}{|\mathbf{E}_c + \mathbf{H}_c|}$$

and compare it to $\Lambda_{\alpha,p,\nu_B,\nu_{Ec}}$, where $\nu_{Ec} = \nu_E + \text{rank}(\mathbf{G}) = N - ab + \text{rank}(\mathbf{G})$, with $N = \Sigma_{ij}n_{ij}$. [If $\mathbf{GM} = \mathbf{O}$ represents interaction, then rank($\mathbf{G}$) adds degrees of freedom corresponding to $\mathbf{H}_{AB}$.] We can also use any of the other three MANOVA test statistics based on the eigenvalues of $\mathbf{E}_c^{-1}\mathbf{H}_c$. Expressions for $\mathbf{H}_c$ similar to (4.123) and (4.124) would result for hypotheses $H_0: \mathbf{a}'\mathbf{M}_c = \mathbf{0}'$ and $H_0: \mathbf{CM}_c = \mathbf{O}$ in the constrained model.

Example 4.8.5. Table 4.6 contains partial data on healing time for patients with fractures of the jaw (Woodard 1931; also analyzed by Jensen 1972). Factor A is gender (1 = male, 2 = female) and factor B is type of fracture (1 = one compound fracture, 2 = two compound fractures, 3 = one simple fracture). The variables are y_1 = age of patient, y_2 = blood lymphocytes, and y_3 = blood polymorphonuclears.

The cell counts are given in Figure 4.3, and the $\mathbf{W}$ and $\mathbf{W}'\mathbf{W}$ matrices are also provided:

Table 4.6. Two-Way Unbalanced Data for Fractures of the Jaw

Gender	Type of Fracture	y_1	y_2	y_3
1	1	42	35	61
1	1	42	43	55
1	1	48	35	64
1	1	35	33	65
1	1	25	31	64
1	1	45	36	58
1	2	23	27	64
1	2	22	32	64
1	2	25	30	64
1	2	28	39	56
1	2	24	31	69
1	2	52	28	60
1	2	17	30	64
1	2	24	42	57
1	3	32	37	54
1	3	52	34	62
1	3	53	45	51
1	3	49	35	60
1	3	55	32	60
1	3	30	34	62
2	1	22	56	43
2	2	22	29	68
2	2	38	25	73
2	2	21	37	59
2	2	42	42	56
2	3	43	30	67
2	3	30	36	60

		B: 1	2	3
A	1	$n_{11} = 6$	$n_{12} = 8$	$n_{13} = 6$
	2	$n_{21} = 1$	$n_{22} = 4$	$n_{23} = 2$

Figure 4.3. Cell counts for fracture data.

$$\mathbf{W} = \begin{bmatrix} 1 & 0 & 0 & 0 & 0 & 0 \\ 1 & 0 & 0 & 0 & 0 & 0 \\ 1 & 0 & 0 & 0 & 0 & 0 \\ 1 & 0 & 0 & 0 & 0 & 0 \\ 1 & 0 & 0 & 0 & 0 & 0 \\ 1 & 0 & 0 & 0 & 0 & 0 \\ 0 & 1 & 0 & 0 & 0 & 0 \\ 0 & 1 & 0 & 0 & 0 & 0 \\ 0 & 1 & 0 & 0 & 0 & 0 \\ 0 & 1 & 0 & 0 & 0 & 0 \\ 0 & 1 & 0 & 0 & 0 & 0 \\ 0 & 1 & 0 & 0 & 0 & 0 \\ 0 & 1 & 0 & 0 & 0 & 0 \\ 0 & 1 & 0 & 0 & 0 & 0 \\ 0 & 0 & 1 & 0 & 0 & 0 \\ 0 & 0 & 1 & 0 & 0 & 0 \\ 0 & 0 & 1 & 0 & 0 & 0 \\ 0 & 0 & 1 & 0 & 0 & 0 \\ 0 & 0 & 1 & 0 & 0 & 0 \\ 0 & 0 & 1 & 0 & 0 & 0 \\ 0 & 0 & 0 & 1 & 0 & 0 \\ 0 & 0 & 0 & 0 & 1 & 0 \\ 0 & 0 & 0 & 0 & 1 & 0 \\ 0 & 0 & 0 & 0 & 1 & 0 \\ 0 & 0 & 0 & 0 & 1 & 0 \\ 0 & 0 & 0 & 0 & 0 & 1 \\ 0 & 0 & 0 & 0 & 0 & 1 \end{bmatrix}, \qquad \mathbf{W}'\mathbf{W} = \begin{pmatrix} 6 & 0 & 0 & 0 & 0 & 0 \\ 0 & 8 & 0 & 0 & 0 & 0 \\ 0 & 0 & 6 & 0 & 0 & 0 \\ 0 & 0 & 0 & 1 & 0 & 0 \\ 0 & 0 & 0 & 0 & 4 & 0 \\ 0 & 0 & 0 & 0 & 0 & 2 \end{pmatrix}.$$

We then obtain

$$\hat{\mathbf{M}} = (\mathbf{W}'\mathbf{W})^{-1}\mathbf{W}'\mathbf{Y} = \begin{pmatrix} 39.50 & 35.50 & 61.17 \\ 26.88 & 32.38 & 62.25 \\ 45.17 & 36.17 & 58.17 \\ 22.00 & 56.00 & 43.00 \\ 30.75 & 33.25 & 64.00 \\ 36.50 & 33.00 & 63.50 \end{pmatrix},$$

$$\mathbf{E} = (\mathbf{Y} - \mathbf{W}\hat{\mathbf{M}})'(\mathbf{Y} - \mathbf{W}\hat{\mathbf{M}}) = \begin{pmatrix} 2192.46 & 47.96 & -144.92 \\ 47.96 & 582.96 & -475.42 \\ -144.92 & -475.42 & 523.67 \end{pmatrix}.$$

For factor A, we use the contrast vector $\mathbf{a}' = (1\ 1\ 1\ -1\ -1\ -1)$ to obtain

$$\mathbf{H}_A = (\mathbf{a}'\hat{\mathbf{M}})'[\mathbf{a}'(\mathbf{W}'\mathbf{W})^{-1}\mathbf{a}]^{-1}(\mathbf{a}'\hat{\mathbf{M}})$$
$$= \begin{pmatrix} 225.02 & -183.80 & 111.88 \\ -183.80 & 150.13 & -91.39 \\ 111.88 & -91.39 & 55.63 \end{pmatrix}.$$

To test the A effect using Wilks' Λ, we have

$$\Lambda_A = \frac{|\mathbf{E}|}{|\mathbf{E} + \mathbf{H}_A|} = \frac{1.6692 \times 10^8}{2.3341 \times 10^8} = .715 > \Lambda_{.05,3,1,21} = .669,$$

and we cannot reject $H_0\colon \mathbf{a}'\mathbf{M} = \mathbf{0}'$. The other three MANOVA tests are equivalent to Λ because $\nu_H = 1$.

To test the B effect, we use the contrast matrix

$$\mathbf{B} = \begin{pmatrix} -1 & -1 & 2 & -1 & -1 & 2 \\ 1 & -1 & 0 & 1 & -1 & 0 \end{pmatrix},$$

in which the first row compares the two compound fracture groups with the simple fracture group. Using $\mathbf{B}$, we obtain

$$\mathbf{H}_B = (\mathbf{B}\hat{\mathbf{M}})'[\mathbf{B}(\mathbf{W}'\mathbf{W})^{-1}\mathbf{B}']^{-1}\mathbf{B}\hat{\mathbf{M}}$$
$$= \begin{pmatrix} 571.12 & -1.85 & -36.35 \\ -1.85 & 442.25 & -372.92 \\ -36.35 & -372.92 & 316.98 \end{pmatrix}.$$

To test the B effect with Wilks' Λ, we have

$$\Lambda_B = \frac{|\mathbf{E}|}{|\mathbf{E} + \mathbf{H}_B|} = \frac{1.6692 \times 10^8}{3.7157 \times 10^8} = .449 < \Lambda_{.05,3,2,21} = .532,$$

and we therefore reject $H_0\colon \mathbf{BM} = \mathbf{O}$. To obtain the other three MANOVA tests, we use the eigenvalues of $\mathbf{E}^{-1}\mathbf{H}_B$, .764 and .262:

$$V^{(s)} = \sum_{i=1}^{2} \frac{\lambda_i}{1 + \lambda_i} = .641, \qquad U^{(s)} = \sum_{i=1}^{2} \lambda_i = 1.026,$$

$$\frac{\nu_E}{\nu_H} U_{(s)} = 10.811,$$

$$\theta = \frac{\lambda_1}{1 + \lambda_1} = .433.$$

With $s = 2, m = 0$, and $N = 8.5$, we obtain $V_{.05}^{(s)} = .506$, $\theta_{.05} = .431$, and a critical value of 5.405 for $\nu_E U^{(s)}/\nu_H$. Thus all four tests lead to rejection of H_0: $\mathbf{BM} = \mathbf{O}$, and the B effect is significant.

To test the interaction, we use the contrast matrix

$$\mathbf{C} = \begin{pmatrix} -1 & -1 & 2 & 1 & 1 & -2 \\ 1 & -1 & 0 & -1 & 1 & 0 \end{pmatrix},$$

which gives

$$\mathbf{H}_{AB} = \begin{pmatrix} 353.08 & -204.00 & 211.04 \\ -204.00 & 331.58 & -331.70 \\ 211.04 & -331.70 & 332.03 \end{pmatrix}$$

and

$$\Lambda_{AB} = \frac{|\mathbf{E}|}{|\mathbf{E} + \mathbf{H}_{AB}|} = \frac{1.6692 \times 10^8}{3.2563 \times 10^8} = .513 < \Lambda_{.05,3,2,21} = .532.$$

The eigenvalues of $\mathbf{E}^{-1}\mathbf{H}_{AB}$ are .786 and .092, from which we obtain

$$V^{(s)} = \sum_{i=1}^{2} \frac{\lambda_i}{1 + \lambda_i} = .524, \qquad U^{(s)} = \sum_{i=1}^{2} \lambda_i = .878,$$

$$\frac{\nu_E}{\nu_H} U^{(s)} = 9.223,$$

$$\theta = \frac{\lambda_1}{1 + \lambda_1} = .440.$$

In this case, ν_H for AB is the same as ν_H for B, and the critical values for these test statistics are the same as those given above for B. Thus all four tests reject H_0: $\mathbf{CM} = \mathbf{O}$, and the interaction is significant.

4.9 TESTS ON A SUBVECTOR

4.9.1 Testing a Single Subvector

In Section 3.11, we considered tests of significance of a subvector in the T^2-test of H_0: $\boldsymbol{\mu}_1 = \boldsymbol{\mu}_2$. We now extend our discussion to several groups in a one-way MANOVA setting, in which we wish to test H_0: $\boldsymbol{\mu}_1 = \boldsymbol{\mu}_2 = \cdots = \boldsymbol{\mu}_k$. The results could be extended to higher order designs. We use Wilks' Λ because it can be partitioned in some convenient ways.

Let the observation vector be partitioned into two subsets of interest, $\mathbf{y}$ and $\mathbf{x}$, where $\mathbf{y}$ is $p \times 1$ and $\mathbf{x}$ is $q \times 1$. We wish to determine whether $\mathbf{x}$ makes a significant contribution to the test of H_0 above and beyond $\mathbf{y}$; that is, we wish to see if Λ for $\binom{\mathbf{y}}{\mathbf{x}}$ is significantly less than Λ for $\mathbf{y}$ alone.

For the one-way (unbalanced) model, the parameter matrix corresponding to **H** [see (4.40)] is

$$\boldsymbol{\Omega} = \sum_{i=1}^{k} n_i(\boldsymbol{\mu}_i - \boldsymbol{\mu})(\boldsymbol{\mu}_i - \boldsymbol{\mu})',$$

where $\boldsymbol{\mu} = \sum_{i=1}^{k} n_i\boldsymbol{\mu}_i/N$ and $N = \sum_{i=1}^{k} n_i$. Let $\boldsymbol{\alpha}_i$, $i = 1, 2, \ldots, s$, be the population discriminant function coefficient vectors, found as eigenvectors of $\boldsymbol{\Sigma}^{-1}\boldsymbol{\Omega}$, where $s = \min(p, k-1)$. Let $\boldsymbol{\alpha}_i$ $(i = 1, 2, \ldots, s)$, $\boldsymbol{\mu}_i$ $(i = 1, 2, \ldots, k)$, $\boldsymbol{\Omega}$, and $\boldsymbol{\Sigma}$ be partitioned to conform to $\binom{\mathbf{y}}{\mathbf{x}}$:

$$\boldsymbol{\mu}_i = \begin{pmatrix} \boldsymbol{\mu}_{yi} \\ \boldsymbol{\mu}_{xi} \end{pmatrix}, \qquad \boldsymbol{\alpha}_i = \begin{pmatrix} \boldsymbol{\alpha}_{yi} \\ \boldsymbol{\alpha}_{xi} \end{pmatrix},$$

$$\boldsymbol{\Omega} = \begin{pmatrix} \boldsymbol{\Omega}_{yy} & \boldsymbol{\Omega}_{yx} \\ \boldsymbol{\Omega}_{xy} & \boldsymbol{\Omega}_{xx} \end{pmatrix}, \quad \boldsymbol{\Sigma} = \begin{pmatrix} \boldsymbol{\Sigma}_{yy} & \boldsymbol{\Sigma}_{yx} \\ \boldsymbol{\Sigma}_{xy} & \boldsymbol{\Sigma}_{xx} \end{pmatrix},$$

where $\boldsymbol{\mu}_{yi}$ and $\boldsymbol{\alpha}_{yi}$ are $p \times 1$, $\boldsymbol{\Omega}_{yy}$ and $\boldsymbol{\Sigma}_{yy}$ are $p \times p$, $\boldsymbol{\mu}_{xi}$ and $\boldsymbol{\alpha}_{xi}$ are $q \times 1$, and $\boldsymbol{\Omega}_{xx}$ and $\boldsymbol{\Sigma}_{xx}$ are $q \times q$. The hypothesis of interest is that **x** does not contribute to separation of groups in the presence of **y**. This hypothesis can be stated in three equivalent ways, analogous to (3.92)–(3.94) (see also the attendant comments):

$$H_{01}\colon \boldsymbol{\alpha}_{x1} = \boldsymbol{\alpha}_{x2} = \cdots = \boldsymbol{\alpha}_{xs} = \mathbf{0},$$

$$H_{02}\colon \boldsymbol{\mu}_{x1} - \boldsymbol{\Sigma}_{xy}\boldsymbol{\Sigma}_{yy}^{-1}\boldsymbol{\mu}_{y1} = \cdots = \boldsymbol{\mu}_{xk} - \boldsymbol{\Sigma}_{xy}\boldsymbol{\Sigma}_{yy}^{-1}\boldsymbol{\mu}_{yk},$$

$$H_{03}\colon \mathrm{tr}(\boldsymbol{\Sigma}^{-1}\boldsymbol{\Omega}) = \mathrm{tr}(\boldsymbol{\Sigma}_{yy}^{-1}\boldsymbol{\Omega}_{yy}).$$

The implication of H_{01} is clear. In H_{02} we see that any difference $\boldsymbol{\mu}_{xi} - \boldsymbol{\mu}_{xj}$ can be predicted from the corresponding $\boldsymbol{\mu}_{yi} - \boldsymbol{\mu}_{yj}$; that is, $\boldsymbol{\mu}_{xi} - \boldsymbol{\mu}_{xj} = \boldsymbol{\Sigma}_{xy}\boldsymbol{\Sigma}_{yy}^{-1}(\boldsymbol{\mu}_{yi} - \boldsymbol{\mu}_{yj})$. Since $\mathrm{tr}(\boldsymbol{\Sigma}^{-1}\boldsymbol{\Omega})$ is a measure of the distances among the k populations, H_{03} states that a subset of variables preserves the intergroup distances of the full set.

To test H_{01}, H_{02}, and H_{03}, we compute **E** and **H** based on the p y's and the q x's and then partition them:

$$\mathbf{E} = \begin{pmatrix} \mathbf{E}_{yy} & \mathbf{E}_{yx} \\ \mathbf{E}_{xy} & \mathbf{E}_{xx} \end{pmatrix}, \qquad \mathbf{H} = \begin{pmatrix} \mathbf{H}_{yy} & \mathbf{H}_{yx} \\ \mathbf{H}_{xy} & \mathbf{H}_{xx} \end{pmatrix},$$

where **E** and **H** are $(p+q) \times (p+q)$ and $\mathbf{E}_{yy}$ and $\mathbf{H}_{yy}$ are $p \times p$. Then Wilks' Λ based on the y's and x's,

$$\Lambda(\mathbf{y}, \mathbf{x}) = \frac{|\mathbf{E}|}{|\mathbf{E} + \mathbf{H}|}, \tag{4.131}$$

is distributed as $\Lambda_{q+p,\nu_H,\nu_E}$, where $\nu_H = k - 1$ and $\nu_E = N - k$. For the reduced vector **y**, Wilks' Λ is given by

$$\Lambda(\mathbf{y}) = \frac{|\mathbf{E}_{yy}|}{|\mathbf{E}_{yy} + \mathbf{H}_{yy}|}, \tag{4.132}$$

which is distributed as Λ_{p,ν_H,ν_E}. A test for the significance of the change from $\Lambda(\mathbf{y})$ to $\Lambda(\mathbf{y}, \mathbf{x})$ is given in the following theorem.

Theorem 4.9A. With $\Lambda(\mathbf{y}, \mathbf{x})$ and $\Lambda(\mathbf{y})$ defined in (4.131) and (4.132), the three equivalent hypotheses H_{01}, H_{02}, and H_{03} can be tested by

$$\Lambda(\mathbf{x}|\mathbf{y}) = \frac{\Lambda(\mathbf{y}, \mathbf{x})}{\Lambda(\mathbf{y})},$$

which is distributed as $\Lambda_{q,\nu_H,\nu_E-p}$.

Proof. For notational convenience, we define $\mathbf{T} = \mathbf{E} + \mathbf{H}$. Then by (A.7.8), $|\mathbf{E}|$ and $|\mathbf{T}|$ can be factored as

$$|\mathbf{E}| = |\mathbf{E}_{yy}||\mathbf{E}_{xx} - \mathbf{E}_{xy}\mathbf{E}_{yy}^{-1}\mathbf{E}_{yx}| = |\mathbf{E}_{yy}||\mathbf{E}_{x\cdot y}|,$$

$$|\mathbf{T}| = |\mathbf{T}_{yy}||\mathbf{T}_{xx} - \mathbf{T}_{xy}\mathbf{T}_{yy}^{-1}\mathbf{T}_{yx}| = |\mathbf{T}_{yy}||\mathbf{T}_{x\cdot y}|,$$

where $\mathbf{E}_{x\cdot y} = \mathbf{E}_{xx} - \mathbf{E}_{xy}\mathbf{E}_{yy}^{-1}\mathbf{E}_{yx}$ and $\mathbf{T}_{x\cdot y} = \mathbf{T}_{xx} - \mathbf{T}_{xy}\mathbf{T}_{yy}^{-1}\mathbf{T}_{yx}$. We can now express $\Lambda(\mathbf{y}, \mathbf{x})$ as

$$\begin{aligned} \Lambda(\mathbf{y}, \mathbf{x}) &= \frac{|\mathbf{E}|}{|\mathbf{T}|} = \frac{|\mathbf{E}_{yy}||\mathbf{E}_{xx} - \mathbf{E}_{xy}\mathbf{E}_{yy}^{-1}\mathbf{E}_{yx}|}{|\mathbf{T}_{yy}||\mathbf{T}_{xx} - \mathbf{T}_{xy}\mathbf{T}_{yy}^{-1}\mathbf{T}_{yx}|} = \frac{|\mathbf{E}_{yy}||\mathbf{E}_{x\cdot y}|}{|\mathbf{T}_{yy}||\mathbf{T}_{x\cdot y}|} \\ &= \Lambda(\mathbf{y})\Lambda(\mathbf{x}|\mathbf{y}), \end{aligned} \tag{4.133}$$

where

$$\Lambda(\mathbf{x}|\mathbf{y}) = \frac{|\mathbf{E}_{x\cdot y}|}{|\mathbf{T}_{x\cdot y}|}.$$

It can be shown that $\mathbf{E}_{x\cdot y}$ and $\mathbf{H}_{x\cdot y}$ have Wishart distributions and are independent of $\mathbf{E}_{yy}$ and $\mathbf{H}_{yy}$, respectively (Seber 1984, p. 50). Hence $\Lambda(\mathbf{x}|\mathbf{y})$ is distributed as $\Lambda_{q,\nu_H,\nu_E-p}$ and is independent of $\Lambda(\mathbf{y})$. From (4.133), $\Lambda(\mathbf{x}|\mathbf{y})$ can be expressed as

$$\Lambda(\mathbf{x}|\mathbf{y}) = \frac{\Lambda(\mathbf{y}, \mathbf{x})}{\Lambda(\mathbf{y})}. \tag{4.134}$$

□

Note that the dimension q of $\Lambda(\mathbf{x}|\mathbf{y})$ is the number of x's, not the combined number of y's and x's. The error degrees of freedom $\nu_E - p$ reflects the adjustment for the p y's. The right side of (4.134) provides an easy computational device.

If there is only one x, then (4.134) becomes

$$\Lambda(x|y_1, \ldots, y_p) = \frac{\Lambda(y_1, \ldots, y_p, x)}{\Lambda(y_1, \ldots, y_p)}, \tag{4.135}$$

which is distributed as $\Lambda_{1,\nu_H,\nu_E-p}$. Since $q = 1$, this Λ-statistic has an exact F-transformation (see Table 4.1):

$$F = \frac{1-\Lambda}{\Lambda}\frac{\nu_E - p}{\nu_H}. \tag{4.136}$$

The statistic (4.135) can be called a *partial* Λ*-statistic*, and correspondingly, (4.136) becomes a *partial F-statistic*.

From (4.135), we obtain

$$\Lambda(y_1, \ldots, y_p, x) = \Lambda(x|y_1, \ldots, y_p)\Lambda(y_1, \ldots, y_p) \le \Lambda(y_1, \ldots, y_p), \tag{4.137}$$

which demonstrates that Wilks' Λ decreases with an additional variable.

4.9.2 Step-Down Test

If the variables $y_1, y_2, \ldots, y_p$ are arranged in order of importance to the researcher, they can be tested sequentially using (4.135) . The hypothesis at the kth step, $k = 1, 2, \ldots, p$, is that y_k is redundant given $y_1, y_2, \ldots, y_{k-1}$. This is tested with

$$\Lambda(y_k|y_1, y_2, \ldots, y_{k-1}) = \frac{\Lambda(y_1, \ldots, y_k)}{\Lambda(y_1, \ldots, y_{k-1})}, \tag{4.138}$$

which is distributed as $\Lambda_{1,\nu_H,\nu_E-k+1}$. The degrees of freedom for error will decrease by 1 at each step. For $k = 1$, we have $k - 1 = 0$, and we set Λ_0 equal to 1 in the denominator.

If we reject one of the p tests, we have in effect rejected the overall MANOVA hypothesis, H_0: $\boldsymbol{\mu}_1 = \boldsymbol{\mu}_2 = \cdots = \boldsymbol{\mu}_k$. If we accept all p tests, we have accepted H_0. Since the p test statistics in (4.138) are independent, the overall α-level is related to the α-levels $\alpha_1, \alpha_2, \ldots, \alpha_p$ of the individual tests by

$$1 - \alpha = \prod_{i=1}^{p}(1 - \alpha_i),$$

and the levels α_i can be adjusted to obtain the overall α-level desired. If the first variables in the ordering are more important than the later variables, the α_i's can be chosen accordingly.

4.9.3 Stepwise Selection of Variables

If there is no natural a priori ordering of variables, one could still use (4.138) by letting the data determine the order of entry (or exit) of the variables. In the kth step of a *forward selection procedure*, we choose the variable y_k that minimizes (4.138). In the kth step of a *stepwise discriminant analysis*, a variable is selected as in a forward selection procedure, and then the variables previously selected are reexamined to see if any one of them has now become redundant and can be deleted. The stepwise approach, similar to that outlined in Section 3.11.3 for two groups, is available in many software packages. Even though this procedure is commonly called stepwise discriminant analysis, no discriminant functions are calculated as part of the selection process. A more apt designation would be "stepwise MANOVA."

When the statistic $\Lambda(y_k | y_1, y_2, \ldots, y_{k-1})$ is minimized at each step, it is not distributed as $\Lambda_{1,\nu_H,\nu_E-k+1}$, and the corresponding F-statistic does not have an F-distribution. Thus using a critical value from the F-table as a stopping rule will bias the results in favor of including too many variables. This problem is discussed further in Sections 5.10 and 6.7 in connection with stepwise discriminant analysis and classification analysis.

4.10 MULTIVARIATE ANALYSIS OF COVARIANCE

4.10.1 Introduction

Analysis of covariance is sometimes described as a blend of ANOVA and regression. In addition to the dependent variable(s), there are one or more quantitative variables known as *covariates* or *concomitant variables*. If the experimental units (subjects) have characteristics that may affect the outcome, these characteristics can possibly be treated as covariates. The primary motivation for inclusion of covariates in an experiment is to gain precision by reducing the error variance. A secondary goal is to decrease the effect of factors the experimenter cannot control. In certain cases, an attempt to include various levels of a variable as a full factor will cause the design to become unwieldy. If such a factor is quantitative, it can be included as a covariate, with a resulting adjustment to the dependent variable(s) before comparing groups. Variables of this type may occur in experimental situations in which the subjects cannot be randomly assigned to treatments. In such cases, we forfeit the causality implication of a designed experiment, and analysis of covariance is closer in spirit to descriptive model building.

Heuristically speaking, in analysis of covariance we regress the dependent variable(s) on the covariate(s) and carry out ANOVA on the residuals. In terms of one dependent variable and one covariate, analysis of covariance will be successful if the following three assumptions hold.

1. *Dependent Variable is Linearly Related to the Covariate.* If this assumption holds, part of the error in the model is predictable and can be removed to reduce the error variance. This assumption can be checked by testing $H_0: \beta = 0$, where β is the slope from the regression of the dependent variable on the covariate (see Section 4.10.2). Since the estimated slope $\hat{\beta}$ will never be exactly zero, analysis of covariance will always give a smaller sum of squares for error than ANOVA. If $\hat{\beta}$ is close to zero, however, the small reduction in error sum of squares may not offset the loss of a degree of freedom. This problem is more likely to arise with multiple covariates, especially if they are highly correlated.

2. *Groups (Treatments) Have the Same Slope.* In assumption 1 above, a common slope β is assumed. We can check this assumption by testing $H_0: \beta_1 = \beta_2 = \cdots = \beta_k$

(assuming a one-way model with k groups), where β_i is the slope in the ith group (see Section 4.10.2).

3. *Covariate Is Unaffected by Differences Among the Means of the Groups (Treatments).* If the group differences are reduced when the dependent variable is adjusted for the covariate, the test for differences will be less powerful. This assumption can be checked by running an ANOVA on the covariate.

Covariates can be either fixed constants (values chosen by the researcher) or random variables. The model we consider below involves a fixed covariate, but initially we treat it as random. However, the estimation and testing procedures are the same in both cases, although the properties of estimators and tests are somewhat different for fixed and random covariates. For example, in the fixed-covariate case, the power of the test depends on the actual values chosen for the covariates, whereas in the random case, the power of the test depends on the population covariance matrix of the covariates. In general, we need not be concerned about such distinctions.

As an illustration of the use of analysis of covariance, suppose we wish to compare three methods of teaching language. Three classes are available, and we assign a class to each of the teaching methods. The students are free to sign up for any one of the three classes and are therefore not randomly assigned. One of the classes may end up with a disproportionate share of the best students, in which case we cannot claim that teaching methods have produced a significant difference in final grades. However, we can use previous grades or other measures of performance as covariates and then compare the students' adjusted scores for the three methods.

We review univariate analysis of covariance in Sections 4.10.2–4.10.4 before proceeding to multivariate analysis of covariance in Section 4.10.5.

4.10.2 Univariate Analysis of Covariance: One-Way Model with One Covariate

We begin with the case of one dependent variable and one covariate. We use the analogy to regression suggested in Section 4.10.1. The model can be written as

$$y_{ij} = \mu + \tau_i + \beta x_{ij} + \varepsilon_{ij}, \quad i = 1, 2, \ldots, k, \qquad j = 1, 2, \ldots, n_i, \tag{4.139}$$

where the τ_i are treatment effects, x_{ij} is a covariate observed on the same sampling unit as y_{ij}, and β is a slope relating x_{ij} to y_{ij}. The parameters $\mu + \tau_i$, $i = 1, 2, \ldots, k$, also serve as regression intercepts for the k groups. In analysis of covariance, the treatment means are adjusted so that they are compared for the same value of x, say the grand mean $\overline{x}_{..}$.

If we rewrite (4.139) as

$$y_{ij} - \beta x_{ij} = \mu + \tau_i + \varepsilon_{ij},$$

then (informally) we can approach analysis of covariance as ANOVA of $y_{ij} - \beta x_{ij}$. The variance of $y_{ij} - \beta x_{ij}$ is equal to $\sigma_y^2 - 2\beta\sigma_{xy} + \beta^2\sigma_x^2$, and (for known β) a within-sample estimator of this is given by

$$\begin{aligned} s_{y-\beta x}^2 &= s_y^2 - 2\beta s_{xy} + \beta^2 s_x^2 \\ &= \frac{\text{SSE}_y}{N-k} - 2\beta \frac{\text{SPE}}{N-k} + \beta^2 \frac{\text{SSE}_x}{N-k}, \end{aligned} \tag{4.140}$$

where $\text{SSE}_y = \sum_{i=1}^k \sum_{j=1}^{n_i} (y_{ij} - \overline{y}_{i\cdot})^2$, $\text{SSE}_x = \sum_{ij} (x_{ij} - \overline{x}_{i\cdot})^2$, $N = \sum_{i=1}^k n_i$, and the sum of products for error is $\text{SPE} = \sum_{ij} (x_{ij} - \overline{x}_{i\cdot})(y_{ij} - \overline{y}_{i\cdot})$. To estimate β, we write (4.140) in the form

$$\text{SSE}_{y-\beta x} = \text{SSE}_y - 2\beta \text{SPE} + \beta^2 \text{SSE}_x$$

and complete the square on β to obtain

$$\text{SSE}_{y-\beta x} = \text{SSE}_x \left(\beta - \frac{\text{SPE}}{\text{SSE}_x} \right)^2 + \text{SSE}_y - \frac{(\text{SPE})^2}{\text{SSE}_x}. \tag{4.141}$$

It is clear that

$$\hat{\beta} = \frac{\text{SPE}}{\text{SSE}_x} \tag{4.142}$$

minimizes $\text{SSE}_{y-\beta x}$ in (4.141). Substituting $\hat{\beta}$ in (4.142) for β in (4.141) gives the minimum value

$$\text{SSE}_{y\cdot x} = \text{SSE}_{y-\hat{\beta}x} = \text{SSE}_y - \frac{(\text{SPE})^2}{\text{SSE}_x}, \tag{4.143}$$

with $N - k - 1$ degrees of freedom. Note that SSE_y has $N - k$ degrees of freedom and $(\text{SPE})^2/\text{SSE}_x$ has one degree of freedom. Thus the degrees of freedom of $\text{SSE}_{y\cdot x}$ are reduced by 1 for estimation of β. In using analysis of covariance, the researcher expects the reduction from SSE_y to $\text{SSE}_{y\cdot x}$ to at least offset the loss of a degree of freedom.

By analogy with (4.143), the total sum of squares for $y_{ij} - \hat{\beta} x_{ij}$ is given by

$$\text{SST}_{y\cdot x} = \text{SST}_y - \frac{(\text{SPT})^2}{\text{SST}_x}, \tag{4.144}$$

where SST_y and SST_x are the total sums of squares for y and x, respectively, and SPT is the total sum of products:

$$\text{SST}_y = \sum_{i=1}^k \sum_{j=1}^{n_i} (y_{ij} - \overline{y}_{\cdot\cdot})^2, \qquad \text{SST}_x = \sum_{ij} (x_{ij} - \overline{x}_{\cdot\cdot})^2,$$

$$\text{SPT} = \sum_{ij} (x_{ij} - \overline{x}_{\cdot\cdot})(y_{ij} - \overline{y}_{\cdot\cdot}).$$

In (4.144), $\text{SST}_{y\cdot x}$ has $N - 2$ degrees of freedom. Again, one degree of freedom has been lost due to estimation of β.

Alternatively, we can obtain $\text{SST}_{y\cdot x}$ in (4.144) as the error sum of squares in the simple linear regression model

$$y_{ij} = \beta_0 + \beta_1 x_{ij} + \varepsilon_{ij}, \tag{4.145}$$

which is obtained from (4.139) by removing the treatment effect τ_i:

$$\begin{aligned} \text{SST}_{y\cdot x} = \text{SSE}_{\text{reg}} &= \sum_{i=1}^{k}\sum_{j=1}^{n_i}(y_{ij} - \hat{y}_{ij})^2 = \sum_{ij}(y_{ij} - \hat{\beta}_0 - \hat{\beta}_1 x_{ij})^2 \\ &= \sum_{ij}(y_{ij} - \overline{y}_{..})^2 - \hat{\beta}_1 \sum_{ij}(x_{ij} - \overline{x}_{..})(y_{ij} - \overline{y}_{..}) \\ &= \sum_{ij}(y_{ij} - \overline{y}_{..})^2 - \frac{\left[\sum_{ij}(x_{ij} - \overline{x}_{..})(y_{ij} - \overline{y}_{..})\right]^2}{\sum_{ij}(x_{ij} - \overline{x}_{..})^2} \qquad \text{[by (4.142)]} \\ &= \text{SST}_y - \frac{(\text{SPT})^2}{\text{SST}_x}. \end{aligned}$$

If we consider (4.139) as the full model and (4.145) as the reduced model obtained by deleting the treatment effect, then the treatment sum of squares is calculated as

$$\begin{aligned} \text{SSTR}_{y\cdot x} &= \text{SSE(reduced)} - \text{SSE(full)} \\ &= \text{SSE[for(4.145)]} - \text{SSE[for(4.139)]} \\ &= \text{SST}_{y\cdot x} - \text{SSE}_{y\cdot x}, \end{aligned} \tag{4.146}$$

which has $k - 1$ degrees of freedom. Note that the sum of squares $\text{SSTR}_{y\cdot x}$ in (4.146) is obtained by subtraction, not by an adjustment to $\text{SSTR}_y = \sum_i n_i(\overline{y}_{i\cdot} - \overline{y}_{..})^2$, analogous to the adjusted $\text{SSE}_{y\cdot x}$ and $\text{SST}_{y\cdot x}$ in (4.143) and (4.144). We use the full and reduced model approach, which computes $\text{SSTR}_{y\cdot x}$ by subtraction, because we do not have the same covariate values for each treatment, and the design is therefore unbalanced (even if the n_i's are equal). If $\text{SSTR}_{y\cdot x}$ were computed in an adjusted manner as in (4.143) or (4.144), then $\text{SSE}_{y\cdot x} + \text{SSTR}_{y\cdot x}$ would not equal $\text{SST}_{y\cdot x}$ because of lack of balance caused by the covariate. We will follow a similar computational scheme to that of (4.146) for each term in the two-way (balanced) design in Section 4.10.3.

The test statistic for the treatment effect hypothesis

$$H_{01}: \tau_1 = \tau_2 = \cdots = \tau_k,$$

adjusted for the covariate, is thus given by

$$F = \frac{\text{SSTR}_{y\cdot x}/(k - 1)}{\text{SSE}_{y\cdot x}/(N - k - 1)}, \tag{4.147}$$

Table 4.7. Analysis of Covariance for One-Way Model

Source	SS and SP Corrected for Mean			SS Adjusted for Covariate	Adjusted df
	y	x	xy		
Treatments	SSTR_y	SSTR_x	SPTR	$\text{SSTR}_{y\cdot x} = \text{SST}_{y\cdot x} - \text{SSE}_{y\cdot x}$	$k-1$
Error	SSE_y	SSE_x	SPE	$\text{SSE}_{y\cdot x} = \text{SSE}_y - \dfrac{(\text{SPE})^2}{\text{SSE}_x}$	$N-k-1$
Total	SST_y	SST_x	SPT	$\text{SST}_{y\cdot x} = \text{SST}_y - \dfrac{(\text{SPT})^2}{\text{SST}_x}$	$N-2$

which is distributed as $F_{k-1,N-k-1}$ under H_{01}. The calculations leading to the test statistic in (4.147) are summarized in Table 4.7.

To test

$$H_{02}\colon \beta = 0,$$

we note that in (4.143) the sum of squares due to β is $(\text{SPE})^2/\text{SSE}_x$. This sum of squares can be divided by $\text{SSE}_{y\cdot x}/(N-k-1)$ to obtain the test statistic

$$F = \frac{(\text{SPE})^2/\text{SSE}_x}{\text{SSE}_{y\cdot x}/(N-k-1)}, \tag{4.148}$$

which under H_{02} (and H_{03} below) is distributed as an F with 1 and $N-k-1$ degrees of freedom.

The test of $H_{02}\colon \beta = 0$ using (4.148) assumes a common slope for all k groups. To check this assumption, we can test the hypothesis of *homogeneity of slopes* in the groups,

$$H_{03}\colon \beta_1 = \beta_2 = \cdots = \beta_k,$$

that is, that the k regression lines are parallel. In many cases, it would be advisable to test H_{03} before testing H_{02}.

We obtain a test of $H_{03}\colon \beta_1 = \beta_2 = \cdots = \beta_k$ by comparing the full model, with a different slope estimated for each group, to the reduced model, which has a common slope, β. Let

$$\text{SSE}_{xi} = \sum_{j=1}^{n_i}(x_{ij} - \overline{x}_{i\cdot})^2$$

be the error sum of squares for x within the ith group, and define SPE_i in an analogous fashion. Then the slope estimate for the ith group is

$$\hat{\beta}_i = \frac{\text{SPE}_i}{\text{SSE}_{xi}},$$

and the sum of squares due to β_i is $(\text{SPE}_i)^2/\text{SSE}_{xi}$. We sum this expression over the k groups to obtain the sum of squares due to the β_i's in the full model,

$$\text{SS}_\text{F} = \sum_{i=1}^{k} \frac{(\text{SPE}_i)^2}{\text{SSE}_{xi}}.$$

The sum of squares due to the common slope β in the reduced model is

$$\text{SS}_\text{R} = \frac{(\text{SPE})^2}{\text{SSE}_x}.$$

Accordingly, our sum of squares for testing $H_{03}\colon \beta_1 = \beta_2 = \cdots = \beta_k$ is

$$\text{SS}_\text{F} - \text{SS}_\text{R} = \sum_{i=1}^{k} \frac{(\text{SPE}_i)^2}{\text{SSE}_{xi}} - \frac{(\text{SPE})^2}{\text{SSE}_x},$$

which has $k - 1$ degrees of freedom. The error sum of squares for the test statistic is based on the full model:

$$\text{SSE}_{(F)y\cdot x} = \text{SSE}_y - \sum_{i=1}^{k} \frac{(\text{SPE}_i)^2}{\text{SSE}_{xi}}, \tag{4.149}$$

with $N - 2k$ degrees of freedom. The test statistic for H_{03} is therefore

$$F = \frac{(\text{SS}_\text{F} - \text{SS}_\text{R})/(k-1)}{\text{SSE}_{(F)y\cdot x}/(N-2k)} \tag{4.150}$$

$$= \frac{\left[\sum_{i=1}^{k}(\text{SPE}_i)^2/\text{SSE}_{xi} - (\text{SPE})^2/\text{SSE}_x\right]/(k-1)}{\left[\text{SSE}_y - \sum_{i=1}^{k}(\text{SPE}_i)^2/\text{SSE}_{xi}\right]/(N-2k)},$$

which is distributed as $F_{k-1,N-2k}$ when H_{03} is true.

4.10.3 Univariate Analysis of Covariance: Two-Way Model with One Covariate

Analysis of covariance can be applied to any ANOVA design with nested, crossed, fixed, or random effects or any combination of these. In a repeated measures design, there could be covariates on the subjects and possibly on each level of within-subject treatment. In this section, we discuss the two-way fixed-effects model with one covariate.

Consider a balanced fixed-effects model,

$$y_{ijk} = \mu + \alpha_i + \gamma_j + (\alpha\gamma)_{ij} + \beta x_{ijk} + \varepsilon_{ijk}, \tag{4.151}$$

$$i = 1, 2, \ldots, a, \qquad j = 1, 2, \ldots, c, \qquad k = 1, 2, \ldots, n,$$

where α_i is the effect of factor A, γ_j is the effect of factor C, $(\alpha\gamma)_{ij}$ is the AC interaction effect, and x_{ijk} is a covariate measured on the same experimental unit as y_{ijk}. The sums of squares and products used in testing main effects and interaction are given in Table 4.8.

Table 4.8. Sums of Squares and Products for x and y in a Two-Way Model

	SS and SP Corrected for the Mean		
Source	y	x	xy
A	SSA_y	SSA_x	SPA
C	SSC_y	SSC_x	SPC
AC	SSAC_y	SSAC_x	SPAC
Error	SSE_y	SSE_x	SPE
$A+E$	$\text{SSA}_y + \text{SSE}_y$	$\text{SSA}_x + \text{SSE}_x$	SPA + SPE
$C+E$	$\text{SSC}_y + \text{SSE}_y$	$\text{SSC}_x + \text{SSE}_x$	SPC + SPE
$AC+E$	$\text{SSAC}_y + \text{SSE}_y$	$\text{SSAC}_x + \text{SSE}_x$	SPAC + SPE

The sums of squares for x and y in Table 4.8 are analogous to those for ordinary two-way ANOVA. For example,

$$\text{SSA}_y = nc\sum_{i=1}^{a}(\overline{y}_{i..} - \overline{y}_{...})^2,$$

$$\text{SSAC}_y = n\sum_{i=1}^{a}\sum_{j=1}^{c}(\overline{y}_{ij.} - \overline{y}_{i..} - \overline{y}_{.j.} + \overline{y}_{...})^2$$

$$= n\sum_{ij}(y_{ij.} - \overline{y}_{...})^2 - \text{SSA}_y - \text{SSC}_y,$$

with analogous expressions for SSA_x and SSAC_x. The corresponding sums of products are

$$\text{SPA} = nc\sum_{i}(\overline{x}_{i..} - \overline{x}_{...})(\overline{y}_{i..} - \overline{y}_{...}),$$

$$\text{SPAC} = n\sum_{ij}(\overline{x}_{ij.} - \overline{x}_{i..} - \overline{x}_{.j.} - \overline{x}_{...})(\overline{y}_{ij.} - \overline{y}_{i..} - \overline{y}_{.j.} - \overline{y}_{...})$$

$$= n\sum_{ij}(\overline{x}_{ij.} - \overline{x}_{..})(\overline{y}_{ij.} - \overline{y}_{..}) - \text{SPA} - \text{SPC}.$$

The orthogonality of the balanced design is lost when adjustments are made for the covariate (Bingham and Feinberg 1982). We therefore obtain a "total" for each term (A, C, or AC) by adding the error SS or SP to the term SS or SP. These totals are then used to obtain sums of squares adjusted for the covariate in a manner analogous to that employed in the one-way model. For example, the adjusted sum of squares for factor A is obtained as follows:

$$\text{SS}(A + E)_{y\cdot x} = \text{SSA}_y + \text{SSE}_y - \frac{(\text{SPA} + \text{SPE})^2}{\text{SSA}_x + \text{SSE}_x}, \tag{4.152}$$

$$\text{SSE}_{y\cdot x} = \text{SSE}_y - \frac{(\text{SPE})^2}{\text{SSE}_x}, \tag{4.153}$$

$$\text{SSA}_{y\cdot x} = \text{SS}(A + E)_{y\cdot x} - \text{SSE}_{y\cdot x}. \tag{4.154}$$

The statistic for testing $H_{01}\colon \alpha_1 = \alpha_2 = \cdots = \alpha_a$, corresponding to the main effect of A, is then given by

$$F = \frac{\text{SSA}_{y\cdot x}/(a-1)}{\text{SSE}_{y\cdot x}/[ac(n-1)-1]},$$

which (under H_{01}) is distributed as an F with $a - 1$ and $ac(n - 1) - 1$ degrees of freedom. The number of degrees of freedom for $\text{SSE}_{y\cdot x}$ has been reduced by 1 for the covariate adjustment. Tests for factor C and the interaction AC are developed in an entirely analogous fashion.

To test the hypothesis $H_{02}\colon \beta = 0$, the sum of squares due to β is $(\text{SPE})^2/\text{SSE}_x$, and the F-statistic is given by

$$F = \frac{(\text{SPE})^2/\text{SSE}_x}{\text{SSE}_{y\cdot x}/[ac(n-1)-1]}, \tag{4.155}$$

which (under H_{02} and H_{03} below) is F with 1 and $ac(n - 1) - 1$ degrees of freedom.

The test for homogeneity of slopes can be carried out separately for factor A, factor C, and the interaction AC. We describe the test for A. The hypothesis is

$$H_{03}\colon \beta_1 = \beta_2 = \cdots = \beta_a;$$

that is, the regression lines for the a levels of A are parallel. The intercepts, of course, may be different. To obtain a slope estimate $\hat{\beta}_i$ for each level of A, consider the within-cells error sum of squares for x pooled across the levels of C,

$$\text{SSE}_{xi} = \sum_{j=1}^{c} \sum_{k=1}^{n} (x_{ijk} - \overline{x}_{ij.})^2, \tag{4.156}$$

with an analogous expression for SPE_i. Then the slope estimate for the ith level of A is

$$\hat{\beta}_i = \frac{\text{SPE}_i}{\text{SSE}_{xi}},$$

and the sum of squares due to β_i is $(\text{SPE}_i)^2/\text{SSE}_{xi}$. The sum of squares for the full model in which the β_i's are different is given by

$$\text{SS}_\text{F} = \sum_{i=1}^{a} \frac{(\text{SPE}_i)^2}{\text{SSE}_{xi}},$$

and the sum of squares in the reduced model with a common slope is

$$\text{SS}_\text{R} = \frac{(\text{SPE})^2}{\text{SSE}_x}.$$

Our test statistic for H_{03} is then similar to (4.150):

$$F = \frac{\left[\sum_{i=1}^{a}(\text{SPE}_i)^2/\text{SSE}_{xi} - (\text{SPE})^2/\text{SSE}_x\right]/(a-1)}{\left[\text{SSE}_y - \sum_{i=1}^{a}(\text{SPE}_i)^2/\text{SSE}_{xi}\right]/\left[ac(n-1)-a\right]}, \tag{4.157}$$

with $a-1$ and $ac(n-1)-a$ degrees of freedom, where SSE_y is obtained from the acn cells of the model, as seen in Table 4.8. The tests for homogeneity of slopes for C and AC are constructed in a similar fashion.

4.10.4 Additional Topics in Univariate Analysis of Covariance

Extensions to higher order models can easily be obtained by following the pattern in Section 4.10.3. Another straightforward extension is to include more than one covariate. A good treatment of this case has been provided by Morrison (1983, pp. 466–474). Since each covariate decreases the error degrees of freedom by 1, the inclusion of too many covariates may lead to loss of power.

The results in Section 4.10.3 are for balanced ANOVA models. The case in which the ANOVA model is itself unbalanced before the addition of a covariate was treated by Hendrix, Carter, and Scott (1982), whose main emphasis was on dealing with heterogeneity of slopes. Searle (1979) also discussed the problem of heterogeneity. Reader (1973) showed that much information can be recovered in the presence of various types of heterogeneity of slopes. Urquhart (1982) suggested some adjustments that can be made when the covariate is affected by one of the factors in the experiment.

For an alternative development of analysis of covariance, we can consider the cell means model of Section 4.8.4. This model allows for imbalance in the n_{ij}'s as well as the inherent imbalance in analysis of covariance models [see a comment following (4.146)]. For an analysis of covariance model with a single covariate for which there is a common slope β, we extend the cell means model (4.93) as

$$\mathbf{y} = (\mathbf{W}, \mathbf{x})\begin{pmatrix}\boldsymbol{\mu} \\ \beta\end{pmatrix} + \boldsymbol{\varepsilon} = \mathbf{W}\boldsymbol{\mu} + \beta\mathbf{x} + \boldsymbol{\varepsilon}. \tag{4.158}$$

The vector $\boldsymbol{\mu}$ contains the means for a one-way model as in (4.139), a two-way model as in (4.151), or some other model. We can test hypotheses about main effects, interactions, the covariate, or other effects, by using contrasts on $\binom{\boldsymbol{\mu}}{\beta}$ as in Section 4.8.4.

The hypothesis $H_{02}\colon \beta = 0$ can be expressed in the form $H_{02}\colon (0, \ldots, 0, 1)\binom{\boldsymbol{\mu}}{\beta} = 0$. To test H_{02}, we use a statistic analogous to (4.101).

To test $H_{03}\colon \beta_1 = \beta_2 = \cdots = \beta_k$ for a one-way model (or $H_{03}\colon \beta_1 = \beta_2 = \cdots = \beta_a$ for the slopes of the a levels of factor A, and so on), we expand the model (4.158) to include the β_i's:

$$\mathbf{y} = (\mathbf{W}, \mathbf{W}_x)\begin{pmatrix}\boldsymbol{\mu} \\ \boldsymbol{\beta}\end{pmatrix} + \boldsymbol{\varepsilon} = \mathbf{W}\boldsymbol{\mu} + \mathbf{W}_x\boldsymbol{\beta} + \boldsymbol{\varepsilon}, \tag{4.159}$$

where $\boldsymbol{\beta} = (\beta_1, \beta_2, \ldots, \beta_k)'$, and $\mathbf{W}_x$ has a value of x_{ij} in each row and the rest 0s. Then $H_{03}: \beta_1 = \beta_2 = \cdots = \beta_k$ can be expressed as $H_{03}: \mathbf{C}\boldsymbol{\beta} = \mathbf{0}$, where $\mathbf{C}$ is a $(k-1) \times k$ matrix of rank $k-1$ such that $\mathbf{Cj} = \mathbf{0}$. We can test $H_{03}: \mathbf{C}\boldsymbol{\beta} = \mathbf{0}$ using a statistic analogous to (4.104).

Constraints on the μ's or the β's can be introduced by inserting nonsingular matrices $\mathbf{A}$ and $\mathbf{A}_x$ into (4.159):

$$\mathbf{y} = \mathbf{WA}^{-1}\mathbf{A}\boldsymbol{\mu} + \mathbf{W}_x\mathbf{A}_x^{-1}\mathbf{A}_x\boldsymbol{\beta} + \boldsymbol{\varepsilon}. \tag{4.160}$$

The matrix $\mathbf{A}$ has the form illustrated in (4.110). The matrix $\mathbf{A}_x$ provides constraints on the β's. For example, if

$$\mathbf{A}_x = \begin{pmatrix} \mathbf{j}' \\ \mathbf{C} \end{pmatrix},$$

where $\mathbf{C}$ is a $(k-1) \times k$ matrix of rank $k-1$ such that $\mathbf{Cj} = \mathbf{0}$ as above, then the model (4.160) has a common slope. The matrices $\mathbf{A}$ and $\mathbf{A}_x$ would often be the same.

4.10.5 Multivariate Analysis of Covariance

In *multivariate analysis of covariance*, we have several dependent variables $y_1, y_2, \ldots, y_p$ and one or more accompanying *covariates* or *concomitant variables*, $x_1, x_2, \ldots, x_q$. As in the univariate analog, the primary motivation is the reduction of error variance by removing the variation due to the relationship with the covariate(s). We present the results in the context of a one-way model, but the test statistics have a general form applicable to any balanced fixed-effects model. Analogous to (4.139), the model for a multivariate one-way analysis of covariance with q covariates is

$$\mathbf{y}_{ij} = \boldsymbol{\mu} + \boldsymbol{\tau}_i + \mathbf{Bx}_{ij} + \boldsymbol{\varepsilon}_{ij}, \quad i = 1, 2, \ldots, k, \qquad j = 1, 2, \ldots, n_i, \tag{4.161}$$

where

$$\mathbf{B} = \begin{pmatrix} \boldsymbol{\beta}_1' \\ \boldsymbol{\beta}_2' \\ \vdots \\ \boldsymbol{\beta}_p' \end{pmatrix}$$

is a $p \times q$ matrix of regression coefficients of $\mathbf{y}$ on $\mathbf{x}$. Thus for the p variables in $\mathbf{y}_{ij}$, we have

$$\begin{aligned} E(y_{ij1}) &= \mu_1 + \tau_{i1} + \boldsymbol{\beta}_1'\mathbf{x}_{ij}, \\ E(y_{ij2}) &= \mu_2 + \tau_{i2} + \boldsymbol{\beta}_2'\mathbf{x}_{ij}, \\ &\vdots \\ E(y_{ijp}) &= \mu_p + \tau_{ip} + \boldsymbol{\beta}_p'\mathbf{x}_{ij}. \end{aligned}$$

The parameter vectors $\boldsymbol{\mu} + \boldsymbol{\tau}_i$, $i = 1, 2, \ldots, k$, also serve as regression intercepts for the k groups. For multivariate two-way analysis of covariance, the model that corresponds to (4.151) is

$$\mathbf{y}_{ijk} = \boldsymbol{\mu} + \boldsymbol{\alpha}_i + \boldsymbol{\gamma}_j + (\boldsymbol{\alpha\gamma})_{ij} + \mathbf{Bx}_{ijk} + \boldsymbol{\varepsilon}_{ijk}, \tag{4.162}$$
$$i = 1, 2, \ldots, a, \qquad j = 1, 2, \ldots, c, \qquad k = 1, 2, \ldots, n.$$

The assumptions for multivariate analysis of covariance (MANCOVA) are similar to those for univariate analysis of covariance (Section 4.10.1) and for MANOVA in general (Section 4.1.1).

We describe hypotheses and tests that correspond to those in Section 4.10.2 for the univariate one-way model and include a few comments about extensions to the two-way model (4.162) corresponding to Section 4.10.3. For the multivariate one-way model (4.161), the hypotheses take the form

$$H_{01}: \boldsymbol{\tau}_1 = \boldsymbol{\tau}_2 = \cdots = \boldsymbol{\tau}_k \text{ (adjusted for } \mathbf{x}),$$
$$H_{02}: \mathbf{B} = \mathbf{O},$$
$$H_{03}: \mathbf{B}_1 = \mathbf{B}_2 = \cdots = \mathbf{B}_k,$$

where $\mathbf{B}_i$ is the matrix of regression coefficients in the ith group. For the two-way model (4.162), we would have analogous hypotheses involving either of the main effects or the interaction effect.

To make the tests, we start with the $\mathbf{E}$ and $\mathbf{H}$ matrices for the combined vector $\binom{\mathbf{x}}{\mathbf{y}}$, which can easily be computed by running a one-way MANOVA on all variables $x_1, x_2, \ldots, x_q, y_1, y_2, \ldots, y_p$. The $\mathbf{E}$ and $\mathbf{H}$ matrices are partitioned into blocks corresponding to $\mathbf{x}$ and $\mathbf{y}$:

$$\mathbf{E} = \begin{pmatrix} \mathbf{E}_{xx} & \mathbf{E}_{xy} \\ \mathbf{E}_{yx} & \mathbf{E}_{yy} \end{pmatrix}, \qquad \mathbf{H} = \begin{pmatrix} \mathbf{H}_{xx} & \mathbf{H}_{xy} \\ \mathbf{H}_{yx} & \mathbf{H}_{yy} \end{pmatrix}, \tag{4.163}$$

where $\mathbf{E}_{xx}$ is $q \times q$, $\mathbf{E}_{xy}$ is $q \times p$, $\mathbf{E}_{yx}$ is $p \times q$, and $\mathbf{E}_{yy}$ is $p \times p$. In the one-way model, the actual structure of the submatrices $\mathbf{E}_{xx}, \mathbf{E}_{yy}$, and $\mathbf{E}_{xy}$ would be

$$\mathbf{E}_{xx} = \sum_{i=1}^{k}\sum_{j=1}^{n_i}(\mathbf{x}_{ij} - \overline{\mathbf{x}}_{i\cdot})(\mathbf{x}_{ij} - \overline{\mathbf{x}}_{i\cdot})',$$

$$\mathbf{E}_{yy} = \sum_{ij}(\mathbf{y}_{ij} - \overline{\mathbf{y}}_{i\cdot})(\mathbf{y}_{ij} - \overline{\mathbf{y}}_{i\cdot})',$$

$$\mathbf{E}_{xy} = \sum_{ij}(\mathbf{x}_{ij} - \overline{\mathbf{x}}_{i\cdot})(\mathbf{y}_{ij} - \overline{\mathbf{y}}_{i\cdot})'.$$

Note that in multivariate analysis of covariance, we are reversing the subvectors $\mathbf{x}$ and $\mathbf{y}$ from the order used in Section 4.9.1. This is because in H_{01} above, we wish to check the effect of $\mathbf{y}$ adjusted for the covariates in $\mathbf{x}$. This is the reverse of the order in Section 4.9.1, where we tested for the effect of $\mathbf{x}$ adjusted for $\mathbf{y}$.

To test H_{01}, we use, by analogy with (4.152)–(4.154),

$$\Lambda_1 = \frac{|\mathbf{E}_{yy} - \mathbf{E}_{yx}\mathbf{E}_{xx}^{-1}\mathbf{E}_{xy}|}{|\mathbf{E}_{yy} + \mathbf{H}_{yy} - (\mathbf{E}_{yx} + \mathbf{H}_{yx})(\mathbf{E}_{xx} + \mathbf{H}_{xx})^{-1}(\mathbf{E}_{xy} + \mathbf{H}_{xy})|}, \quad (4.164)$$

which is distributed as $\Lambda_{p,\nu_H,\nu_E-q}$, where $\nu_H = k - 1$ and $\nu_E = N - k$, with $N = \sum_{i=1}^{k} n_i$. The degrees of freedom for error, $\nu_E - q$, is reduced by q because of adjustment for q covariates. The statistic Λ_1 in (4.164) can be extended for use in a two-way model by using appropriately partitioned forms of $\mathbf{H}_A, \mathbf{H}_C, \mathbf{H}_{AC}$, and $\mathbf{E}$, along with suitable ν_H and ν_E.

To calculate the Pillai, Lawley–Hotelling, and Roy test statistics corresponding to Λ_1 in (4.164), we use eigenvalues of $\mathbf{E}_{y\cdot x}^{-1}\mathbf{H}_{y\cdot x}$, where $\mathbf{E}_{y\cdot x}$ is the numerator matrix in (4.164) and $\mathbf{H}_{y\cdot x}$ is obtained by subtracting the numerator matrix from the denominator matrix:

$$\mathbf{E}_{y\cdot x} = \mathbf{E}_{yy} - \mathbf{E}_{yx}\mathbf{E}_{xx}^{-1}\mathbf{E}_{xy},$$

$$\mathbf{H}_{y\cdot x} = \mathbf{H}_{yy} - (\mathbf{E}_{yx} + \mathbf{H}_{yx})(\mathbf{E}_{xx} + \mathbf{H}_{xx})^{-1}(\mathbf{E}_{xy} + \mathbf{H}_{xy}) + \mathbf{E}_{yx}\mathbf{E}_{xx}^{-1}\mathbf{E}_{xy}.$$

We can also examine the discriminant functions obtained from eigenvectors of $\mathbf{E}_{y\cdot x}^{-1}\mathbf{H}_{y\cdot x}$.

The test statistic Λ_1 in (4.164) can be calculated directly from the basic matrices involved. However, a simpler approach is to use the test statistic for a subvector in (4.134) to test for the significance of $\mathbf{y}$ adjusted for $\mathbf{x}$:

$$\Lambda_1 = \Lambda(\mathbf{y}|\mathbf{x}) = \frac{\Lambda(\mathbf{x}, \mathbf{y})}{\Lambda(\mathbf{x})}, \quad (4.165)$$

where

$$\Lambda(\mathbf{x}, \mathbf{y}) = \frac{|\mathbf{E}|}{|\mathbf{E} + \mathbf{H}|}, \qquad \Lambda(\mathbf{x}) = \frac{|\mathbf{E}_{xx}|}{|\mathbf{E}_{xx} + \mathbf{H}_{xx}|},$$

with $\mathbf{E}$, $\mathbf{H}$, $\mathbf{E}_{xx}$, and $\mathbf{H}_{xx}$ defined in (4.163) .

In practice we would not ordinarily compare $\Lambda(\mathbf{x})$ to its critical value $\Lambda_{\alpha,q,\nu_H,\nu_E}$, but if we did so and found significance, then use of the covariate model may be questionable. [A test using $\Lambda(\mathbf{x})$ is analogous to the ANOVA test on a single covariate discussed in assumption 1 in Section 4.10.1.] The covariates are intended to take up some of the within-cells variance, not the between-cells variation. If $\Lambda(\mathbf{x})$ is significant, we should compare (4.165) to

$$\Lambda(\mathbf{y}) = \frac{|\mathbf{E}_{yy}|}{|\mathbf{E}_{yy} + \mathbf{H}_{yy}|}. \quad (4.166)$$

If $\Lambda(\mathbf{y}|\mathbf{x})$ based on the adjusted y's in (4.165) is not appreciably less than $\Lambda(\mathbf{y})$ based on the unadjusted y's in (4.166), then the use of covariates has not fulfilled our primary purpose. A formal test for the efficacy of the covariates is given in the next paragraph.

To test H_{02}: $\mathbf{B} = \mathbf{O}$, we extend the univariate sum of squares due to β, $(\text{SPE})^2/\text{SSE}_x$, to obtain the analogous hypothesis matrix in the multivariate case, $\mathbf{H}_2 = \mathbf{E}_{yx}\mathbf{E}_{xx}^{-1}\mathbf{E}_{xy}$, which is a $p \times p$ matrix with q degrees of freedom due to fitting q x's. The error matrix is $\mathbf{E}_2 = \mathbf{E}_{y \cdot x} = \mathbf{E}_{yy} - \mathbf{E}_{yx}\mathbf{E}_{xx}^{-1}\mathbf{E}_{xy}$. The sum of these two for use in the denominator of Λ is

$$\mathbf{E}_2 + \mathbf{H}_2 = \mathbf{E}_{yy} - \mathbf{E}_{yx}\mathbf{E}_{xx}^{-1}\mathbf{E}_{xy} + \mathbf{E}_{yx}\mathbf{E}_{xx}^{-1}\mathbf{E}_{xy} = \mathbf{E}_{yy}.$$

The test statistic for H_{02}: $\mathbf{B} = \mathbf{O}$ is therefore

$$\Lambda_2 = \frac{|\mathbf{E}_2|}{|\mathbf{E}_2 + \mathbf{H}_2|} = \frac{|\mathbf{E}_{yy} - \mathbf{E}_{yx}\mathbf{E}_{xx}^{-1}\mathbf{E}_{xy}|}{|\mathbf{E}_{yy}|}, \tag{4.167}$$

which is distributed as $\Lambda_{p,q,\nu_E - q}$ under H_{02} (and H_{03} below).

Note that Λ_2 can be written in the form

$$\begin{aligned}
\Lambda_2 &= |\mathbf{E}_{yy}^{-1}||\mathbf{E}_{yy} - \mathbf{E}_{yx}\mathbf{E}_{xx}^{-1}\mathbf{E}_{xy}| && \text{[by (A.7.6)]} \\
&= |\mathbf{E}_{yy}^{-1}(\mathbf{E}_{yy} - \mathbf{E}_{yx}\mathbf{E}_{xx}^{-1}\mathbf{E}_{xy})| && \text{[by (A.7.5)]} \\
&= |\mathbf{I} - \mathbf{E}_{yy}^{-1}\mathbf{E}_{yx}\mathbf{E}_{xx}^{-1}\mathbf{E}_{xy}| \\
&= \prod_{i=1}^{s}(1 - r_i^2), && \text{[by (A.10.8)]},
\end{aligned}$$

where $s = \min(p, q)$ and the r_i^2's are the nonzero eigenvalues of $\mathbf{E}_{yy}^{-1}\mathbf{E}_{yx}\mathbf{E}_{xx}^{-1}\mathbf{E}_{xy}$. The eigenvalues $r_1^2, r_2^2, \ldots, r_s^2$ are the (within-group) *squared canonical correlations* between $\mathbf{y}$ and $\mathbf{x}$ (see Chapter 8).

To construct a test for H_{03}: $\mathbf{B}_1 = \mathbf{B}_2 = \cdots = \mathbf{B}_k$ (for a one-way model), we first find

$$\mathbf{E}_i = \begin{pmatrix} \mathbf{E}_{xxi} & \mathbf{E}_{xyi} \\ \mathbf{E}_{yxi} & \mathbf{E}_{yyi} \end{pmatrix}, \qquad i = 1, 2, \ldots, k,$$

that correspond to $\mathbf{B}_1, \mathbf{B}_2, \ldots, \mathbf{B}_k$ for groups $1, 2, \ldots, k$. For a one-way model, there would be a $\mathbf{B}_i$ and an $\mathbf{E}_i$ for each group. To test H_{03}: $\mathbf{B}_1 = \mathbf{B}_2 = \cdots = \mathbf{B}_k$, we compare the full model in which H_{03} is not true to the reduced model under H_{03}. The hypothesis matrix for regression is computed separately in each group and the results are summed. Thus $\mathbf{H}_{3F}$ for the full model is an extension of $\text{SS}_F = \sum_{i=1}^{k}(\text{SPE}_i)^2/\text{SSE}_{xi}$:

$$\mathbf{H}_{3F} = \sum_{i=1}^{k} \mathbf{E}_{yxi}\mathbf{E}_{xxi}^{-1}\mathbf{E}_{xyi}. \tag{4.168}$$

For the reduced model under H_{03}, there is a single $\mathbf{B}$, and by extension of $\text{SS}_R = (\text{SPE})^2/\text{SSE}_x$, the hypothesis matrix is

$$\mathbf{H}_{3R} = \mathbf{E}_{yx}\mathbf{E}_{xx}^{-1}\mathbf{E}_{xy}.$$

The difference between full and reduced models gives

$$\mathbf{H}_3 = \mathbf{H}_{3F} - \mathbf{H}_{3R} = \sum_{i=1}^{k} \mathbf{E}_{yxi}\mathbf{E}_{xxi}^{-1}\mathbf{E}_{xyi} - \mathbf{E}_{yx}\mathbf{E}_{xx}^{-1}\mathbf{E}_{xy}, \tag{4.169}$$

which has $qk - q = q(k-1)$ degrees of freedom, since each of the k terms in the sum in (4.169) has q degrees of freedom (equal to its rank), as does $\mathbf{E}_{yx}\mathbf{E}_{xx}^{-1}\mathbf{E}_{xy}$. Note that there must be more than q observation vectors in each group so that $\mathbf{E}_{xxi}$ will have an inverse.

For the error matrix $\mathbf{E}_3$ (based on the full model), we pool the error matrices for the groups [by extension of $\text{SSE}_{(F)y\cdot x} = \text{SSE}_y - \sum_{i=1}^{k}(\text{SPE}_i)^2/\text{SSE}_{xi}$ in (4.149)] to obtain

$$\begin{aligned}
\mathbf{E}_3 &= \sum_{i=1}^{k} \mathbf{E}_{(y\cdot x)i} = \sum_{i=1}^{k} (\mathbf{E}_{yyi} - \mathbf{E}_{yxi}\mathbf{E}_{xxi}^{-1}\mathbf{E}_{xyi}) \\
&= \sum_i \mathbf{E}_{yyi} - \sum_i \mathbf{E}_{yxi}\mathbf{E}_{xxi}^{-1}\mathbf{E}_{xyi} \\
&= \mathbf{E}_{yy} - \sum_i \mathbf{E}_{yxi}\mathbf{E}_{xxi}^{-1}\mathbf{E}_{xyi},
\end{aligned}$$

since $\sum_{i=1}^{k} \mathbf{E}_{yyi} = \mathbf{E}_{yy}$ by definition. The degrees of freedom of $\mathbf{E}_3$ are $\nu_E - qk$. Our test statistic for H_{03} is thus

$$\Lambda_3 = \frac{|\mathbf{E}_3|}{|\mathbf{E}_3 + \mathbf{H}_3|} = \frac{|\mathbf{E}_{yy} - \sum_{i=1}^{k} \mathbf{E}_{yxi}\mathbf{E}_{xxi}^{-1}\mathbf{E}_{xyi}|}{|\mathbf{E}_{yy} - \mathbf{E}_{yx}\mathbf{E}_{xx}^{-1}\mathbf{E}_{xy}|}, \tag{4.170}$$

which is distributed as $\Lambda_{p,q(k-1),\nu_E - qk}$ when H_{03} is true. In many cases, it would be advisable to test H_{03} before testing H_{02}.

For the two-way model, k would be a, c, and ac for factor A, C, and AC, respectively. If we are comparing the $\mathbf{B}_i$ matrices for one of the main effects, say A, then $\mathbf{E}_i$ would be calculated for all observation vectors belonging to the ith level of A. If we wish to compare the $\mathbf{B}_{ij}$ matrices for the interaction, we would compute an $\mathbf{E}_{ij}$ for each "cell," that is, for all observation vectors that correspond to the ith level of A and the jth level of C.

For a two-way or higher order unbalanced model, we can use the unbalanced MANOVA approach in Section 4.8 to extend the analysis of covariance formulation of Hendrix et al. (1982) to the multivariate case. Some of the other references in Section 4.10.4 are also directly applicable to the multivariate case.

Example 4.10.5. To illustrate multivariate analysis of covariance, consider the partial data in Table 4.9 from Brown and Beerstecher (1951) and Smith, Gnanadesikan, and Hughes (1962). The variables are y_1 = modified creatinine coefficient, y_2 = pigment creatinine, y_3 = phosphate (mg/ml), x_1 = volume (ml), and x_2 = specific gravity.

Table 4.9. Biochemical Measurements on Four Weight Groups

y_1	y_2	y_3	x_1	x_2
		Group 1		
4.67	17.6	1.50	205	24
4.67	13.4	1.65	160	32
2.70	20.3	.90	480	17
3.49	22.3	1.75	230	30
3.49	20.5	1.40	235	30
3.49	18.5	1.20	215	27
4.84	12.1	1.90	215	25
4.84	12.0	1.65	190	30
		Group 2		
3.48	18.1	1.50	220	31
3.48	19.7	1.65	300	23
3.48	16.9	1.40	305	32
2.63	23.7	1.65	275	20
2.63	19.2	.90	405	18
2.63	18.0	1.60	210	23
4.46	14.8	2.45	170	31
4.46	15.6	1.65	235	28
		Group 3		
1.26	17.0	.70	350	18
1.26	12.5	.80	475	10
2.52	21.5	1.80	195	33
2.52	22.2	1.05	375	25
2.52	13.0	2.20	160	35
3.24	13.0	3.55	240	33
3.24	10.9	3.30	205	31
3.24	12.0	3.65	270	34
		Group 4		
4.12	12.5	5.90	105	32
4.12	8.7	4.25	115	25
4.12	9.4	3.85	97	28
2.14	15.0	2.45	325	27
2.14	12.9	1.70	310	23
2.03	12.1	1.80	245	25
2.03	13.2	3.65	170	26
2.03	11.5	2.25	220	34

Groups 1 and 2 are underweight groups, while groups 3 and 4 represent overweight groupings. The **H** and **E** matrices, partitioned according to the y's and the x's, are as follows:

$$\mathbf{H} = \begin{pmatrix} \mathbf{H}_{yy} & \mathbf{H}_{yx} \\ \mathbf{H}_{xy} & \mathbf{H}_{xx} \end{pmatrix}$$

$$= \left(\begin{array}{rrr|rr} 11.00 & 26.73 & -8.24 & -81.75 & -6.550 \\ 26.73 & 183.5 & -51.28 & 1648.3 & -43.64 \\ -8.24 & -51.28 & 15.18 & -475.0 & 10.28 \\ \hline -81.75 & 1648.3 & -475.0 & 32575.8 & -258.1 \\ -6.550 & -43.64 & 10.28 & -258.1 & 15.25 \end{array}\right),$$

$$\mathbf{E} = \begin{pmatrix} \mathbf{E}_{yy} & \mathbf{E}_{yx} \\ \mathbf{E}_{xy} & \mathbf{E}_{xx} \end{pmatrix}$$

$$= \left(\begin{array}{rrr|rr} 21.40 & -47.21 & 18.04 & -1612.8 & 83.12 \\ -47.21 & 336.2 & -36.10 & 3300.4 & -125.6 \\ 18.04 & -36.10 & 27.47 & -1718.2 & 88.32 \\ \hline -1612.8 & 3300.4 & -1718.2 & 245262.9 & -11466.5 \\ 83.12 & -125.6 & 88.32 & -11466.5 & 1044.2 \end{array}\right).$$

To test equality of group effects, $H_{01}: \boldsymbol{\tau}_1 = \boldsymbol{\tau}_2 = \boldsymbol{\tau}_3 = \boldsymbol{\tau}_4$, adjusted for the covariates, we use (4.165):

$$\Lambda_1 = \frac{\Lambda(\mathbf{x}, \mathbf{y})}{\Lambda(\mathbf{x})} = \frac{|\mathbf{E}|/|\mathbf{E} + \mathbf{H}|}{|\mathbf{E}_{xx}|/|\mathbf{E}_{xx} + \mathbf{H}_{xx}|}$$

$$= \frac{3.3775 \times 10^{12}/2.8515 \times 10^{13}}{124,635,135/156,903,291} = .1491 < \Lambda_{.05,3,3,26} = .514.$$

To test $H_{02}: \mathbf{B} = \mathbf{O}$, we have

$$\mathbf{E}_{yx}\mathbf{E}_{xx}^{-1}\mathbf{E}_{xy} = \begin{pmatrix} 10.722 & -21.266 & 11.420 \\ -21.266 & 46.039 & -22.668 \\ 11.420 & -22.668 & 12.162 \end{pmatrix},$$

$$\Lambda_2 = \frac{|\mathbf{E}_{yy} - \mathbf{E}_{yx}\mathbf{E}_{xx}^{-1}\mathbf{E}_{xy}|}{|\mathbf{E}_{yy}|} = \frac{27099.4}{60621.3}$$

$$= .4470 < \Lambda_{.05,3,2,26} = .604.$$

To test $H_{03}: \mathbf{B}_1 = \mathbf{B}_2 = \mathbf{B}_3 = \mathbf{B}_4$, we have

$$\sum_{i=1}^{4} \mathbf{E}_{yxi}\mathbf{E}_{xxi}^{-1}\mathbf{E}_{yxi} = \begin{pmatrix} 14.853 & -27.964 & 14.246 \\ -27.964 & 84.922 & -18.933 \\ 14.246 & -18.933 & 18.049 \end{pmatrix},$$

$$\Lambda_3 = \frac{|\mathbf{E}_{yy} - \sum_{i=1}^{4} \mathbf{E}_{yxi}\mathbf{E}_{xxi}^{-1}\mathbf{E}_{xyi}|}{|\mathbf{E}_{yy} - \mathbf{E}_{yx}\mathbf{E}_{xx}^{-1}\mathbf{E}_{xy}|}$$

$$= \frac{8970.9372}{27,099.403} = .3310 > \Lambda_{.05,3,6,20} = .250.$$

4.11 ALTERNATIVE APPROACHES TO TESTING $H_0\colon \boldsymbol{\mu}_1 = \boldsymbol{\mu}_2 = \cdots = \boldsymbol{\mu}_k$

In Section 4.2, we discussed the robustness (or lack thereof) of the four MANOVA test statistics to nonnormality and heterogeneity of the covariance matrices. In this section, we consider alternative approaches to testing $H_0\colon \boldsymbol{\mu}_1 = \boldsymbol{\mu}_2 = \cdots = \boldsymbol{\mu}_k$ that do not require the multivariate normal assumption.

Nath and Pavur (1985) proposed a simple rank transformation test statistic for use with data from a one-way MANOVA model with k groups. The observed values of each variable are ranked from 1 to N, where $N = n_1 + n_2 + \cdots + n_k$. This is done separately for each of the p variables. The ranking is carried out across groups. In the case of tied observations, each is assigned the average rank. A rank-transformed Wilks Λ, denoted by Λ_R, is obtained by computing Λ on the ranks. Let $\mathbf{E}_R$ and $\mathbf{H}_R$ be the $\mathbf{E}$ and $\mathbf{H}$ matrices obtained as in (4.2) and (4.4) by substituting ranks in place of the observations. Then Λ_R is defined as

$$\Lambda_R = \frac{|\mathbf{E}_R|}{|\mathbf{E}_R + \mathbf{H}_R|}.$$

The distribution of Λ_R is approximated by that of Λ_{p,ν_H,ν_E}.

Nath and Pavur (1985) investigated the properties of Λ_R in a simulation study. For multivariate normal populations, Λ_R yielded Type I error rates close to the nominal α-values. To illustrate nonnormal populations, they used uniform and Cauchy distributions, the latter of which is heavy tailed. For these distributions, Λ_R was closer to the nominal α-values than was the ordinary Wilks Λ. In terms of power, Λ_R was a little below Λ for the normal case, but Λ_R was much better than Λ for the Cauchy distribution.

Robust estimates of $\boldsymbol{\mu}$ and $\boldsymbol{\Sigma}$ were considered in Section 1.10. Shiraishi (1990) proposed a test procedure for $H_0\colon \boldsymbol{\mu}_1 = \boldsymbol{\mu}_2 = \cdots = \boldsymbol{\mu}_k$ based on Huber's (1964) M-estimators and showed that the test statistic has an asymptotic χ^2-distribution. The test is robust to various departures from normality. It is not necessary, for example, that the underlying distribution be symmetric.

PROBLEMS

4.1 Show that the two forms given in (4.2) and (4.3) for **H** are equivalent and the forms for **E** in (4.4) and (4.5) are equivalent; that is, show that

(a) $n\sum_{i=1}^{k}(\bar{\mathbf{y}}_{i.} - \bar{\mathbf{y}}_{..})(\bar{\mathbf{y}}_{i.} - \bar{\mathbf{y}}_{..})' = \sum_{i=1}^{k} \frac{1}{n}\mathbf{y}_{i.}\mathbf{y}_{i.}' - \frac{1}{kn}\mathbf{y}_{..}\mathbf{y}_{..}'$,

(b) $\sum_{i=1}^{k}\sum_{j=1}^{n}(\mathbf{y}_{ij} - \bar{\mathbf{y}}_{i.})(\mathbf{y}_{ij} - \bar{\mathbf{y}}_{i.})' = \sum_{ij}\mathbf{y}_{ij}\mathbf{y}_{ij}' - \frac{1}{n}\sum_{i}\mathbf{y}_{i.}\mathbf{y}_{i.}'$

4.2 Show that the likelihood function can be written in the form given in (4.6).

4.3 Show that $\sum_{ij}(\mathbf{y}_{ij} - \bar{\mathbf{y}}_{..})(\mathbf{y}_{ij} - \bar{\mathbf{y}}_{..})' = \mathbf{E} + \mathbf{H}$ as in (4.7).

4.4 Show that $\max_{H_0} L$ and $\max_{H_1} L$ assume the forms given in (4.8) and (4.9).

4.5 Show that (4.8) and (4.9) lead to the likelihood ratio given in (4.10).

4.6 Show that $\mathbf{H}$ and $\mathbf{E}$ can be expressed as $\mathbf{H} = \mathbf{Y}'\mathbf{AY}$ and $\mathbf{E} = \mathbf{Y}'\mathbf{BY}$, where $\mathbf{Y}, \mathbf{A}$, and $\mathbf{B}$ are defined in the proof of Theorem 4.1B.

4.7 Show that $\mathbf{A}$ and $\mathbf{B}$ are idempotent of rank $k - 1$ and $k(n - 1)$, respectively, and that $\mathbf{AB} = \mathbf{O}$, where $\mathbf{A}$ and $\mathbf{B}$ are defined in the proof of Theorem 4.1B.

4.8 Verify (4.12), that is, that Wilks' Λ is invariant under full-rank linear transformations.

4.9 Prove property 6 in Section 4.1.2; that is, show that Λ can be expressed in terms of the eigenvalues of $\mathbf{E}^{-1}\mathbf{H}$ as in (4.14).

4.10 Show that the eigenvalues of $\mathbf{HE}^{-1}$ are the same as those for $\mathbf{E}^{-1}\mathbf{H}$; thus the four MANOVA test statistics summarized in Section 4.1.5 can be expressed in terms of eigenvalues of $\mathbf{HE}^{-1}$.

4.11 **(a)** Use (A.10.9) to show that if $\mathbf{a}$ is an eigenvector of $\mathbf{E}^{-1}\mathbf{H}$, then $\mathbf{Ha}$ is an eigenvector of $\mathbf{HE}^{-1}$.

(b) Insert $\mathbf{E}^{-1}\mathbf{E}$ in (4.20) to obtain $(\mathbf{H} - \lambda\mathbf{E})\mathbf{E}^{-1}\mathbf{Ea} = \mathbf{0}$, which gives $\mathbf{Ea}$ as an eigenvector of $\mathbf{HE}^{-1}$.

(c) Since $\mathbf{Ha} \neq \mathbf{Ea}$, how can both serve as eigenvectors of $\mathbf{HE}^{-1}$?

4.12 Show that the approximate F-statistic given by (4.16) reduces to the exact F-transformations in Table 4.1 when either ν_H or p is 1 or 2, as noted in property 10 of Section 4.1.2.

4.13 Show that, for $z_{ij} = \mathbf{a}'\mathbf{y}_{ij}$, the between sum of squares $n\sum_{i=1}^{k}(\bar{z}_{i.} - \bar{z}_{..})^2$ in (4.18) is equal to $\mathbf{a}'\mathbf{Ha}$, thus establishing the numerator of (4.19).

4.14 At the end of Section 4.1.3, we noted that the eigenvalues of $\mathbf{E}^{-1}\mathbf{H}$ are the same as those of $(\mathbf{U}')^{-1}\mathbf{HU}^{-1}$, where $\mathbf{E} = \mathbf{U}'\mathbf{U}$ is the Cholesky factorization of $\mathbf{E}$. We also noted that the eigenvectors of $(\mathbf{U}')^{-1}\mathbf{HU}^{-1}$ are of the form $(\mathbf{U}')^{-1}\mathbf{Ha}$, where $\mathbf{a}$ is an eigenvector of $\mathbf{E}^{-1}\mathbf{H}$.

(a) Show that the eigenvectors of $(\mathbf{U}')^{-1}\mathbf{HU}^{-1}$ can also be expressed as $\mathbf{Ua}$.

(b) The two vectors $(\mathbf{U}')^{-1}\mathbf{Ha}$ and $\mathbf{Ua}$ are not equal. How can they both serve as eigenvectors?

4.15 Show that F_3 in (4.29) is the same as F_1 in (4.27).

4.16 Show that the Lawley–Hotelling statistic $U^{(s)}$ in (4.30) can be expressed as a linear combination of Hotelling T^2-statistics.

4.17 Show that the eigenvalues of $\mathbf{E}^{-1}\mathbf{H}$ are unchanged under full rank linear transformations, $\mathbf{z}_{ij} = \mathbf{Fy}_{ij} + \mathbf{g}$, where $\mathbf{F}$ is nonsingular. Thus all four MANOVA test statistics are invariant to such transformations, as noted at the end of Section 4.1.4.

4.18 Show that the likelihood ratio for testing $H_0\colon \boldsymbol{\Sigma}_1 = \boldsymbol{\Sigma}_2 = \cdots = \boldsymbol{\Sigma}_k$ is given by LR in (4.42).

4.19 Show that if some $\nu_i < p$, then $M = 0$ in (4.43).

4.20 Show that if all $\nu_i = \nu$, then c_1 in (4.46) reduces to the simpler form in (4.48).

4.21 Show that if all $\nu_i = \nu$, then c_2 in (4.49) reduces to the form in (4.52).

4.22 Show that T^2 in (4.64) has the T^2_{p,ν_E}-distribution.

4.23 Show that the confidence interval for $\mu_{ir} - \mu_{i'r}$ given in (4.69) can be obtained from (4.68) in Theorem 4.5C.

4.24 Justify the Bonferroni intervals in (4.70).

4.25 Show that the simultaneous confidence intervals in (4.75) follow from Theorem 4.5C.

4.26 Show that T^2 in (4.85) has the T^2_{p,ν_E}-distribution.

4.27 Show that $\mathbf{A}\boldsymbol{\Sigma}_{\bar{y}}\mathbf{B}' = \boldsymbol{\Sigma} \sum_{i=1}^{k} a_i b_i / n_i$, where $\mathbf{A}$, $\boldsymbol{\Sigma}_{\bar{y}}$, and $\mathbf{B}$ are defined in the proof of Theorem 4.8B.

4.28 Obtain the confidence interval in (4.89) from (4.88).

4.29 Verify (4.95), $\text{cov}(\hat{\boldsymbol{\mu}}) = \sigma^2(\mathbf{W}'\mathbf{W})^{-1}$.

4.30 Obtain SSE in (4.97) from SSE in (4.96).

4.31 Obtain SSE in (4.98) from (4.96).

4.32 Obtain s^2 in (4.99) from (4.98).

4.33 Evaluate $\mathbf{a}'(\mathbf{W}'\mathbf{W})^{-1}\mathbf{a}$ in (4.100) for $\mathbf{a}' = (1 \quad 1 \quad 1 \quad -1 \quad -1 \quad -1)$. Use the $\mathbf{W}$ matrix for the 11 observations in the illustration in Section 4.8.4.

4.34 Evaluate $\mathbf{B}(\mathbf{W}'\mathbf{W})^{-1}\mathbf{B}'$ in (4.103) for the matrices $\mathbf{B}$ and $\mathbf{W}$ used in the illustration in Section 4.8.4.

4.35 Show that $\mathbf{A}\mathbf{A}^{-1} = \mathbf{I}$, where $\mathbf{A} = \binom{\mathbf{K}}{\mathbf{G}}$ as in (4.110) and

$$\mathbf{A}^{-1} = [\mathbf{K}'(\mathbf{K}\mathbf{K}')^{-1}, \ \mathbf{G}'(\mathbf{G}\mathbf{G}')^{-1}]$$

as in (4.114).

4.36 Show that $\text{cov}(\hat{\boldsymbol{\mu}}_c) = \sigma^2\mathbf{K}'(\mathbf{K}\mathbf{K}')^{-1}(\mathbf{Z}_1'\mathbf{Z}_1)^{-1}(\mathbf{K}\mathbf{K}')^{-1}\mathbf{K}$, thus verifying (4.116).

4.37 Show that the hypotheses H_{01}, H_{02}, and H_{03} in Section 4.9.1 are equivalent.

4.38 Verify that completing the square on β leads to the result in (4.141).

4.39 Show that the Wilks Λ-test for equality of adjusted means in analysis of covariance in (4.164) is equal to the test for a subvector in (4.165).

4.40 In an experiment to assess the effect of four nitrogen treatments and four blocks, several variables were measured for Jonathan apples (Ratkowsky and Martin 1974):

y_1 = total nitrogen	y_5 = calcium
y_2 = protein nitrogen	y_6 = magnesium
y_3 = phosphorus	x_1 = mean fruit weight (g)
y_4 = potassium	x_2 = incidence of bitter pit (%)

All measurements are in parts per million except for x_1 and x_2. The data are given in Table 4.10. Ignore the blocks and treat the data as a one-way (unbalanced) MANOVA. Use $y_1, y_2, \ldots, y_6$.

Table 4.10. Mineral Content, Mean Fruit Weight, and Incidence of Bitter Pit for Jonathan Apples

Block	y_1	y_2	y_3	y_4	y_5	y_6	x_1	x_2
				Treatment A: Control				
1	3580	1790	932	8220	244	410	85 · 3	0.0
	2880	1670	836	9840	142	367	113 · 8	3.2
	3260	1530	740	8180	269	387	92 · 9	0.0
2	2870	1700	926	7550	272	332	48 · 9	0.0
	3430	1800	899	9520	202	370	99 · 4	3.6
	2930	1490	847	8310	272	413	79 · 1	0.0
3	3110	1700	770	8180	297	389	70 · 0	2.7
	3300	1840	891	8970	225	362	86 · 9	1.8
4	3370	1780	899	9420	212	403	87 · 7	6.5
	3290	1730	879	7240	206	330	67 · 3	4.3
				Treatment B: Urea				
1	3040	1810	798	10760	138	414	117 · 5	47.0
	4470	2020	886	9990	151	401	98 · 9	39.6
2	5810	2400	1037	11340	165	479	108 · 5	44.2
	4610	2070	840	9070	151	351	104 · 4	19.0
	4690	2070	914	9730	199	429	96 · 8	10.0
3	3010	1780	813	9830	159	403	94 · 5	18.5
	6740	2310	1111	11150	158	458	90 · 6	7.3
	4510	2320	912	10360	163	401	100 · 8	23.6
4	4890	2040	925	9550	239	397	96 · 0	6.5
	4340	1990	915	10440	180	428	99 · 9	20.4
	4130	1870	710	9040	199	363	84 · 6	0.0
				Treatment C: Calcium and Potassium Nitrates				
1	4250	2040	932	11830	169	408	127 · 1	9.5
	3710	1810	792	10530	210	392	108 · 5	3.9
	4640	2340	883	11210	172	393	99 · 9	1.6
2	6950	2300	1202	12910	148	510	124 · 8	27.2
	4880	1800	829	11210	219	411	94 · 5	2.0
3	4680	1940	850	11010	224	411	99 · 4	2.7
	5170	2130	862	11750	152	419	117 · 5	13.9
	5730	2560	1161	12440	160	454	135 · 0	50.0
4	5360	2000	898	10960	211	428	85 · 6	3.6
	6310	2420	984	12210	178	428	102 · 5	14.3
	4370	2080	874	12560	183	404	110 · 8	10.0
				Treatment D: Ammonium Sulfate				
1	4700	1990	938	8830	148	349	77 · 4	50.0
	5930	2720	1211	11430	128	449	91 · 3	54.0
	4840	2360	1038	10370	132	406	91 · 3	89.5
2	7230	3280	1233	10840	120	437	81 · 7	70.5
	7650	2670	1289	10800	124	455	89 · 2	37.5
3	5760	2610	1137	9200	147	378	69 · 6	64.0
	7140	2240	1074	9300	255	420	69 · 0	16.0
	7950	2730	1200	10630	172	425	73 · 7	39.5
4	5040	2270	869	9120	140	334	75 · 1	36.1
	3850	1880	823	8520	181	334	87 · 0	58.6

(a) Compare the four treatments using all four MANOVA tests (Section 4.8.3).

(b) Examine the eigenvalues of $\mathbf{E}^{-1}\mathbf{H}$ for an indication of the dimensionality of the mean vectors (Section 4.2).

(c) Using contrasts, test the following comparisons of treatments (Section 4.8.3): 1 versus 2, 3, 4; 2 versus 3, 4; 3 versus 4.

(d) If any of the four tests in part (a) are significant, run an individual ANOVA F-test on each of $y_1, y_2, \ldots, y_6$ (Sections 4.1.7b and 4.8.2).

4.41 Use $y_1, y_2, \ldots, y_6$ from the apple data in Table 4.10; ignore blocks.

(a) Using Wilks' Λ, test the significance of y_1 and y_2 adjusted for y_3, y_4, y_5, and y_6 (Section 4.9.1).

(b) Using Wilks' Λ, test the significance of each y_i adjusted for the other five variables (Section 4.9.1).

(c) Check the effect of each variable on Wilks' Λ as in Theorem 4.1D. Calculate F_x, R_e^2, R_{e+h}^2, and the ratio $\Lambda_{y,x}/\Lambda_y$.

(d) Test $H_0\colon \mathbf{\Sigma}_1 = \mathbf{\Sigma}_2 = \mathbf{\Sigma}_3 = \mathbf{\Sigma}_4$ (Section 4.3). Use both χ^2- and F-approximations.

4.42 Use $y_1, y_2, \ldots, y_6$ in the apple data in Table 4.10 and consider blocks as a second factor. Analyze as a two-way unbalanced design, testing for treatments, blocks, and interaction (Section 4.8.5).

4.43 Use $y_1, y_2, \ldots, y_6, x_1, x_2$ from the apple data in Table 4.10; ignore blocks. Analyze as a one-way (unbalanced) multivariate analysis of covariance with x_1 and x_2 as covariates (Section 4.10.5). Test the following three hypotheses using (4.165), (4.167), and (4.170):

$$H_{01}\colon \boldsymbol{\tau}_1 = \boldsymbol{\tau}_2 = \cdots = \boldsymbol{\tau}_k \quad \text{(adjusted for } \mathbf{x}\text{)},$$

$$H_{02}\colon \mathbf{B} = \mathbf{O},$$

$$H_{03}\colon \mathbf{B}_1 = \mathbf{B}_2 = \cdots = \mathbf{B}_k.$$

4.44 Use $y_1, y_2, \ldots, y_6, x_1, x_2$ from the apple data in Table 4.10 and consider blocks to be a second factor. Analyze as a two-way unbalanced multivariate analysis of covariance with x_1 and x_2 as covariates (Sections 4.8.5 and 4.10.5). Test H_{01}, H_{02}, and H_{03} for treatments, for blocks, and for interaction.

4.45 Table 4.11 from Smith et al. (1962) gives data from a one-way (unbalanced) multivariate analysis of covariance. A subset of these data was used in Example 4.10.5.

The variables are

y_1 = pH	y_2 = modified creatinine coefficient
y_3 = pigment creatinine	y_4 = phosphate (mg/ml)
y_5 = calcium (mg/ml)	y_6 = phosphorus (mg/ml)

Table 4.11. Biochemical Data on Four Weight Groups

y_1	y_2	y_3	y_4	y_5	y_6	y_7	y_8	y_9	y_{10}	y_{11}	x_1	x_2
Group 1												
5.7	4.67	17.6	1.50	.104	1.50	1.88	5.15	8.40	7.5	.11	205	24
5.5	4.67	13.4	1.65	.245	1.32	2.24	5.75	4.50	7.1	.11	160	32
4.6	2.70	20.3	.90	.097	.89	1.28	4.35	1.20	2.3	.11	480	17
5.7	3.49	22.3	1.75	.174	1.50	2.24	7.55	2.75	4.0	.12	230	30
5.6	3.49	20.5	1.40	.210	1.19	2.00	8.50	3.30	2.0	.12	235	30
4.0	3.49	18.5	1.20	.275	1.03	1.84	10.25	2.00	2.0	.12	215	27
5.3	4.84	12.1	1.90	.170	1.87	2.40	5.95	2.60	14.8	.14	215	25
5.4	4.84	12.0	1.65	.164	1.68	3.00	4.30	2.72	14.5	.14	190	30
5.4	4.84	10.1	2.30	.275	2.08	2.68	5.45	2.40	.9	.20	190	28
5.6	4.48	14.7	2.35	.210	2.55	3.00	3.75	7.00	2.0	.21	175	24
5.6	4.48	14.8	2.35	.050	1.32	2.84	5.10	4.00	.4	.12	145	24
5.6	4.48	14.4	2.50	.143	2.38	2.84	4.05	8.00	3.8	.18	155	27
Group 2												
5.2	3.48	18.1	1.50	.153	1.20	2.60	9.00	2.35	14.5	.13	220	31
5.2	3.48	19.7	1.65	.203	1.73	1.88	5.30	2.52	12.5	.20	300	23
5.6	3.48	16.9	1.40	.074	1.15	1.72	9.85	2.45	8.0	.07	305	32
5.8	2.63	23.7	1.65	.155	1.58	1.60	3.60	3.75	4.9	.10	275	20
4.0	2.63	19.2	.90	.155	.96	1.20	4.05	3.30	.2	.10	405	18
5.3	2.63	18.0	1.60	.129	1.68	2.00	4.40	3.00	3.6	.18	210	23
5.4	4.46	14.8	2.45	.245	2.15	3.12	7.15	1.81	12.0	.13	170	31
5.6	4.46	15.6	1.65	.422	1.42	2.56	7.25	1.92	5.2	.15	235	28
5.3	2.80	14.2	1.65	.063	1.62	2.04	5.30	3.90	10.2	.12	185	21
5.4	2.80	14.1	1.25	.042	1.62	1.84	3.10	4.10	8.5	.30	255	20
5.5	2.80	17.5	1.05	.030	1.56	1.48	2.40	2.10	9.6	.20	265	15
5.4	2.57	14.1	2.70	.194	2.77	2.56	4.25	2.60	6.9	.17	305	24
5.4	2.57	19.1	1.60	.139	1.59	1.88	5.80	2.30	4.7	.16	440	24
5.2	2.57	22.5	.85	.046	1.65	1.20	1.55	1.50	3.5	.21	430	14
Group 3												
5.5	1.26	17.0	.70	.094	.97	1.24	4.55	2.90	1.9	.12	350	18
5.9	1.26	12.5	.80	.039	.80	.64	2.65	.72	.7	.13	475	10
5.6	2.52	21.5	1.80	.142	1.77	2.60	4.50	2.48	8.3	.17	195	33
5.6	2.52	22.2	1.05	.080	1.17	1.48	4.85	2.20	9.3	.14	375	25
5.3	2.52	13.0	2.20	.215	1.85	3.84	8.75	2.40	13.0	.11	160	35
5.6	3.24	13.0	3.55	.166	3.18	3.48	5.20	3.50	18.3	.22	240	33
5.5	3.24	10.9	3.30	.111	2.79	3.04	4.75	2.52	10.5	.21	205	31
5.6	3.24	12.0	3.65	.180	2.40	3.00	5.85	3.00	14.5	.21	270	34
5.4	1.56	22.8	.55	.069	1.00	1.14	2.85	2.90	3.3	.15	475	14
5.3	1.56	14.5	2.05	.222	1.49	2.40	4.55	3.90	6.3	.11	430	31
5.2	1.56	18.4	1.05	.267	1.17	1.36	4.60	2.00	4.9	.11	490	28
Group 4												
5.8	4.12	12.5	5.90	.093	3.80	3.84	2.90	3.00	22.5	.24	105	32
5.7	4.12	8.7	4.25	.147	3.62	5.32	3.00	3.55	19.5	.20	115	25
5.7	4.12	9.4	3.85	.217	3.36	5.52	3.40	5.20	1.3	.31	97	28
5.4	2.14	15.0	2.45	.418	2.38	2.40	5.40	1.81	20.0	.17	325	27
5.4	2.14	12.9	1.70	.323	1.74	2.48	4.45	1.88	1.0	.15	310	23
4.9	2.03	12.1	1.80	.205	2.00	2.24	4.30	3.70	5.0	.19	245	25
5.0	2.03	13.2	3.65	.348	1.95	2.12	5.00	1.80	3.0	.15	170	26
4.9	2.03	11.5	2.25	.320	2.25	3.12	3.40	2.50	5.1	.18	220	34

y_7 = creatinine (mg/ml) y_8 = chloride (mg/ml)
y_9 = boron (μg/ml) y_{10} = choline (μg/ml)
y_{11} = copper (μg/ml) x_1 = volume (ml)
x_2 = (specific gravity $-$ 1) $\times 10^3$

Using x_1 and x_2 as covariates in the model (4.161), test the following three hypotheses using (4.165), (4.167), and (4.170):

$$H_{01}\colon \boldsymbol{\tau}_1 = \boldsymbol{\tau}_2 = \cdots = \boldsymbol{\tau}_k \quad \text{(adjusted for } \mathbf{x}\text{)},$$

$$H_{02}\colon \mathbf{B} = \mathbf{O},$$

$$H_{03}\colon \mathbf{B}_1 = \mathbf{B}_2 = \cdots = \mathbf{B}_k.$$

4.46 Use $y_1, y_2, \ldots, y_{11}$ from the biochemical data in Table 4.11. Analyze as a one-way (unbalanced) MANOVA design.

(a) Compare the four groups using all four MANOVA tests (Section 4.8.3).

(b) Examine the eigenvalues of $\mathbf{E}^{-1}\mathbf{H}$ for an indication of the dimensionality of the mean vectors (Section 4.2).

(c) Using contrasts, test the following comparisons of treatments (Section 4.8.3): 1, 2 versus 3, 4; 1 versus 2; 3 versus 4.

(d) If any of the four tests in part (a) are significant, run an individual ANOVA F-test on each of $y_1, y_2, \ldots, y_{11}$ (Sections 4.1.7b and 4.8.2).

(e) Test the significance of $y_1, y_2, \ldots, y_5$ adjusted for $y_6, y_7, \ldots, y_{11}$.

4.47 Test $H_0\colon \boldsymbol{\Sigma}_1 = \boldsymbol{\Sigma}_2$ (Section 4.4) for the soils data of Table 3.6. Use both χ^2- and F-approximations.

CHAPTER 5

Discriminant Functions for Descriptive Group Separation

5.1 INTRODUCTION

Discriminant functions are used to describe or elucidate the differences between two or more groups. These linear functions of the variables were introduced in Section 3.3.8 for one group, in Section 3.5.5 for two groups, in Section 4.1.3 for several groups, and in Section 4.6 for two-way models. In those sections, interest was centered on follow-up to Hotelling's T^2-tests and MANOVA tests. In this chapter, we expand that focus and cover other aspects of these useful functions.

Chapter 6 treats *classification analysis*, in which linear or quadratic functions of the variables are employed to assign a subject or individual sampling unit to one of the groups.

Reviews of the issues in discriminant analysis and classification analysis have been given by Huberty (1984, 1994, 1996) and Williams (1983).

5.2 TWO GROUPS

The discriminant function $z = \mathbf{a}'\mathbf{y}$ with coefficient vector $\mathbf{a} = \mathbf{S}_{\text{pl}}^{-1}(\overline{\mathbf{y}}_1 - \overline{\mathbf{y}}_2)$ was obtained in Section 3.5.5 [see (3.65)] as a byproduct of the union–intersection derivation of Hotelling's T^2-test for two groups (see Theorem 3.5D). We now obtain the same result using a nonparametric approach in which the normality assumption is not needed.

We assume the availability of two random samples of p-variate observation vectors, $\mathbf{y}_{11}, \mathbf{y}_{12}, \ldots, \mathbf{y}_{1n_1}$ and $\mathbf{y}_{21}, \mathbf{y}_{22}, \ldots, \mathbf{y}_{2n_2}$, from populations with mean vectors $\boldsymbol{\mu}_1$ and $\boldsymbol{\mu}_2$ and common covariance matrix $\boldsymbol{\Sigma}$. We seek the linear combination $z = \mathbf{a}'\mathbf{y}$ of these p variables that maximizes the distance between the two means $\overline{z}_1$ and $\overline{z}_2$. By (1.72), $\overline{z}_1 - \overline{z}_2 = \mathbf{a}'\overline{\mathbf{y}}_1 - \mathbf{a}'\overline{\mathbf{y}}_2 = \mathbf{a}'(\overline{\mathbf{y}}_1 - \overline{\mathbf{y}}_2)$.

Without restrictions on $\mathbf{a}$, the difference $\overline{z}_1 - \overline{z}_2$ has no maximum. We therefore standardize by dividing by the standard deviation, $s_z = (\mathbf{a}'\mathbf{S}_{\text{pl}}\mathbf{a})^{1/2}$, where $\mathbf{S}_{\text{pl}}$ is the pooled covariance matrix from the two samples [see (3.41)]. We then seek the vector

a that maximizes $(\bar{z}_1 - \bar{z}_2)/s_z$, or for mathematical convenience

$$\frac{(\bar{z}_1 - \bar{z}_2)^2}{s_z^2} = \frac{[\mathbf{a}'(\bar{\mathbf{y}}_1 - \bar{\mathbf{y}}_2)]^2}{\mathbf{a}'\mathbf{S}_{\text{pl}}\mathbf{a}}. \tag{5.1}$$

Theorem 5.2A. For two groups, the discriminant function coefficient vector **a** that maximizes the discriminant criterion (5.1) is given by

$$\mathbf{a} = \mathbf{S}_{\text{pl}}^{-1}(\bar{\mathbf{y}}_1 - \bar{\mathbf{y}}_2) \tag{5.2}$$

or by any multiple of $\mathbf{S}_{\text{pl}}^{-1}(\bar{\mathbf{y}}_1 - \bar{\mathbf{y}}_2)$. □

The maximizing vector **a** in (5.2) is not unique, but its "direction" is unique; that is, the relative values or ratios of $a_1, a_2, \ldots, a_p$ are unique. Note that in order for $\mathbf{S}_{\text{pl}}^{-1}$ to exist, we must have $n_1 + n_2 - 2 > p$. This discriminant function coefficient vector **a** for separating two groups was originally proposed by Fisher (1936).

The discriminant function $z = \mathbf{a}'\mathbf{y}$ is invariant to nonsingular linear transformations. Let **y** be transformed to $\mathbf{u} = \mathbf{A}\mathbf{y}$, which has sample mean vectors $\bar{\mathbf{u}}_1$ and $\bar{\mathbf{u}}_2$, sample pooled covariance matrix $\mathbf{S}_u$, and discriminant function $z_u = \mathbf{b}'\mathbf{u}$, where $\mathbf{b} = \mathbf{S}_u^{-1}(\bar{\mathbf{u}}_1 - \bar{\mathbf{u}}_2)$. Then if **A** is nonsingular, it can easily be shown that

$$z_u = \mathbf{b}'\mathbf{u} = \mathbf{a}'\mathbf{y}. \tag{5.3}$$

The vector $\mathbf{a} = \mathbf{S}_{\text{pl}}^{-1}(\bar{\mathbf{y}}_1 - \bar{\mathbf{y}}_2)$ can be expressed in terms of regression coefficients from a multiple regression of dummy grouping variables on the y's (Rencher 1995, Sections 5.6.2 and 8.3).

5.3 SEVERAL GROUPS

5.3.1 Discriminant Functions

In discriminant analysis for several groups, we seek linear combinations that best separate the transformed mean vectors of the groups. The resulting discriminant functions for several groups were first introduced in Section 4.1.3 in connection with Roy's union–intersection test for one-way MANOVA. We now obtain the same functions without the assumption of a multivariate normal distribution. Geometrically, we have k groups of observations in p-dimensional space, and our goal is to represent group separation in fewer than p dimensions. Note that if $k < p$ the k mean vectors $\bar{\mathbf{y}}_{1\cdot}, \bar{\mathbf{y}}_{2\cdot}, \ldots, \bar{\mathbf{y}}_{k\cdot}$ occupy a subspace of dimension $k - 1$ or less.

Extending Fisher's (1936) criterion for two groups in (5.1), we seek the vector **a** that maximizes the ratio of between to within sums of squares for $z_{ij} = \mathbf{a}'\mathbf{y}_{ij}$, $i = 1, 2, \ldots, k,\ j = 1, 2, \ldots, n_i$:

$$\lambda = \frac{\text{SSB}_z}{\text{SSE}_z} = \frac{\sum_{i=1}^k n_i(\bar{z}_{i\cdot} - \bar{z}_{\cdot\cdot})^2}{\sum_{i=1}^k \sum_{j=1}^{n_i}(z_{ij} - \bar{z}_{i\cdot})^2}, \tag{5.4}$$

where $\bar{z}_{i\cdot} = \mathbf{a}'\bar{\mathbf{y}}_{i\cdot}$ is the mean of the ith group and $\bar{z}_{\cdot\cdot}$ is the overall mean. It was shown (for the balanced case) in Section 4.1.3 that this ratio is equal to

$$\lambda = \frac{\mathbf{a}'\mathbf{H}\mathbf{a}}{\mathbf{a}'\mathbf{E}\mathbf{a}}. \tag{5.5}$$

By Theorem 4.1C, the maximum value of λ is λ_1, the largest eigenvalue of $\mathbf{E}^{-1}\mathbf{H}$, and the vector $\mathbf{a}$ that produces the maximum is the associated eigenvector $\mathbf{a}_1$. The rank of $\mathbf{E}^{-1}\mathbf{H}$ is $s = \min(k-1, p)$. [Note the distinction in notation between $s = \text{rank}\left(\mathbf{E}^{-1}\mathbf{H}\right)$ and the standard deviation s_z or covariance $s_{z_1 z_2}$.] Thus there are s (nonzero) eigenvalues $\lambda_1, \lambda_2, \ldots, \lambda_s$, typically arranged in descending order of magnitude, and s associated eigenvectors $\mathbf{a}_1, \mathbf{a}_2, \ldots, \mathbf{a}_s$, from which we obtain the s discriminant functions $z_1 = \mathbf{a}_1'\mathbf{y}$, $z_2 = \mathbf{a}_2'\mathbf{y}, \ldots, z_s = \mathbf{a}_s'\mathbf{y}$. The eigenvectors are of course determined only up to an arbitrary multiplier; that is, if $\mathbf{a}$ is an eigenvector, $c\mathbf{a}$ is also an eigenvector. We typically choose c so that $\mathbf{a}'\mathbf{a} = 1$.

The relationship among the s discriminant functions obtained from the eigenvectors of $\mathbf{E}^{-1}\mathbf{H}$ is given in the following theorem.

Theorem 5.3A. The first discriminant function $z_1 = \mathbf{a}_1'\mathbf{y} = a_{11}y_1 + a_{12}y_2 + \cdots + a_{1p}y_p$ has the largest discriminant criterion $\lambda_1 = \mathbf{a}_1'\mathbf{H}\mathbf{a}_1/\mathbf{a}_1'\mathbf{E}\mathbf{a}_1$ achievable by any linear combination of the y's. The second discriminant function $z_2 = \mathbf{a}_2'\mathbf{y} = a_{21}y_1 + a_{22}y_2 + \cdots + a_{2p}y_p$ has the largest discriminant criterion $\lambda_2 = \mathbf{a}_2'\mathbf{H}\mathbf{a}_2/\mathbf{a}_2'\mathbf{E}\mathbf{a}_2$ achievable by any linear combination of the y's that is uncorrelated with z_1. The third discriminant function $z_3 = \mathbf{a}_3'\mathbf{y}$ has the largest discriminant criterion value λ_3 among all linear combinations of the y's that are uncorrelated with both z_1 and z_2. This property extends to $z_4, z_5, \ldots, z_s$.

Proof. By Theorem 4.1C, $z_1 = \mathbf{a}_1'\mathbf{y}$ maximizes $\mathbf{a}'\mathbf{H}\mathbf{a}/\mathbf{a}'\mathbf{E}\mathbf{a}$. We first show that z_1 and z_2 are uncorrelated. By (1.75) we can express the within-sample correlation in terms of $\mathbf{a}_1'\mathbf{E}\mathbf{a}_2$:

$$\begin{aligned} r_{z_1 z_2} &= \frac{s_{z_1 z_2}}{s_{z_1} s_{z_2}} = \frac{\mathbf{a}_1'\mathbf{S}\mathbf{a}_2}{\sqrt{(\mathbf{a}_1'\mathbf{S}\mathbf{a}_1)(\mathbf{a}_2'\mathbf{S}\mathbf{a}_2)}} \\ &= \frac{\mathbf{a}_1'\mathbf{E}\mathbf{a}_2}{\sqrt{(\mathbf{a}_1'\mathbf{E}\mathbf{a}_1)(\mathbf{a}_2'\mathbf{E}\mathbf{a}_2)}}, \end{aligned} \tag{5.6}$$

where $\mathbf{S} = \mathbf{E}/(N-k)$ is the pooled estimator of $\boldsymbol{\Sigma}$, with $N = \sum_{i=1}^{k} n_i$. To show that $\mathbf{a}_1'\mathbf{E}\mathbf{a}_2 = 0$, we begin with the defining expressions from (4.20),

$$\mathbf{H}\mathbf{a}_1 = \lambda_1\mathbf{E}\mathbf{a}_1, \qquad \mathbf{H}\mathbf{a}_2 = \lambda_2\mathbf{E}\mathbf{a}_2,$$

and multiply them by $\mathbf{a}_2'$ and $\mathbf{a}_1'$ to obtain

$$\mathbf{a}_2'\mathbf{H}\mathbf{a}_1 = \lambda_1\mathbf{a}_2'\mathbf{E}\mathbf{a}_1, \qquad \mathbf{a}_1'\mathbf{H}\mathbf{a}_2 = \lambda_2\mathbf{a}_1'\mathbf{E}\mathbf{a}_2.$$

Since $\mathbf{a}_2'\mathbf{H}\mathbf{a}_1$ is a scalar and $\mathbf{H}$ is symmetric, the left-hand sides are equal and we can equate the right-hand sides:

$$\lambda_1\mathbf{a}_2'\mathbf{E}\mathbf{a}_1 = \lambda_2\mathbf{a}_1'\mathbf{E}\mathbf{a}_2 \quad \text{or} \quad (\lambda_1 - \lambda_2)\mathbf{a}_2'\mathbf{E}\mathbf{a}_1 = 0.$$

Since the eigenvalues of $\mathbf{E}^{-1}\mathbf{H}$ are all distinct (with probability 1), $\lambda_1 - \lambda_2 \neq 0$. Therefore, $\mathbf{a}_2'\mathbf{E}\mathbf{a}_1 = 0$, and $r_{z_1 z_2}$ in (5.6) is 0.

We next show that $z_2 = \mathbf{a}_2'\mathbf{y}$ has the largest discriminant criterion

$$\lambda_2 = \mathbf{a}_2'\mathbf{H}\mathbf{a}_2/\mathbf{a}_2'\mathbf{E}\mathbf{a}_2$$

subject to the constraint $r_{z_1 z_2} = 0$. We incorporate $r_{z_1 z_2} = 0$ or $\mathbf{a}_1'\mathbf{E}\mathbf{a}_2 = 0$ by using a Lagrange multiplier γ, and we then find the maximum by differentiating with respect to $\mathbf{a}_2$ and setting the result equal to $\mathbf{0}$:

$$\frac{\partial}{\partial \mathbf{a}_2}\left(\frac{\mathbf{a}_2'\mathbf{H}\mathbf{a}_2}{\mathbf{a}_2'\mathbf{E}\mathbf{a}_2} + \gamma \mathbf{a}_1'\mathbf{E}\mathbf{a}_2\right) = \mathbf{0},$$

$$\frac{\mathbf{a}_2'\mathbf{E}\mathbf{a}_2 2\mathbf{H}\mathbf{a}_2 - \mathbf{a}_2'\mathbf{H}\mathbf{a}_2 2\mathbf{E}\mathbf{a}_2}{(\mathbf{a}_2'\mathbf{E}\mathbf{a}_2)^2} + \gamma \mathbf{E}\mathbf{a}_1 = \mathbf{0} \qquad \text{[by (A.12.3)]}.$$

Substituting λ_2 for $\mathbf{a}_2'\mathbf{H}\mathbf{a}_2/\mathbf{a}_2'\mathbf{E}\mathbf{a}_2$ gives

$$\frac{2\mathbf{H}\mathbf{a}_2 - 2\lambda_2\mathbf{E}\mathbf{a}_2}{\mathbf{a}_2'\mathbf{E}\mathbf{a}_2} + \gamma\mathbf{E}\mathbf{a}_1 = \mathbf{0}. \tag{5.7}$$

We now multiply by $\mathbf{a}_1'$ to obtain

$$\frac{2\mathbf{a}_1'\mathbf{H}\mathbf{a}_2 - 2\lambda_2\mathbf{a}_1'\mathbf{E}\mathbf{a}_2}{\mathbf{a}_2'\mathbf{E}\mathbf{a}_2} + \gamma\mathbf{a}_1'\mathbf{E}\mathbf{a}_1 = \mathbf{a}_1'\mathbf{0} = 0.$$

By differentiating with respect to γ, we obtain $\mathbf{a}_1'\mathbf{E}\mathbf{a}_2 = 0$, and this implies that $\mathbf{a}_1'\mathbf{H}\mathbf{a}_2 = 0$ as well (see Problem 5.4). Thus

$$\gamma\mathbf{a}_1'\mathbf{E}\mathbf{a}_1 = 0 \quad \text{or} \quad \gamma = 0.$$

Hence (5.7) becomes

$$\mathbf{H}\mathbf{a}_2 - \lambda_2\mathbf{E}\mathbf{a}_2 = \mathbf{0},$$

and $\mathbf{a}_2$, the "second" eigenvector of $\mathbf{E}^{-1}\mathbf{H}$, maximizes $\lambda_2 = \mathbf{a}_2'\mathbf{H}\mathbf{a}_2/\mathbf{a}_2'\mathbf{E}\mathbf{a}_2$ subject to $r_{z_1 z_2} = 0$. The analogous property of z_3 can be shown similarly. □

Thus discriminant analysis reveals the uncorrelated "dimensions" of group differences in decreasing order of importance. We emphasize that z_1 and z_2 are uncorrelated, not orthogonal ($\mathbf{a}_1'\mathbf{a}_2 \neq 0$ because $\mathbf{E}^{-1}\mathbf{H}$ is not symmetric). The correlation in (5.6) is a within-sample correlation. It can be shown that the total-sample correlation is also zero; that is, if all $N = \sum_{i=1}^{k} n_i$ of the $\mathbf{y}_{ij}$'s are combined into a single sample and z_1 and z_2 are calculated in the same manner, their correlation is zero.

In Section 5.5, tests of significance for discriminant functions are provided that can be used as a guide to how many of the $s = \min(k-1, p)$ discriminant functions should be retained to adequately separate groups. Sometimes, however, not all functions that are statistically significant are of practical importance in describing group differences. A useful measure of the relative importance of the ith discriminant function $z_i = \mathbf{a}_i'\mathbf{y}$ is given by

$$\frac{\lambda_i}{\sum_{j=1}^{s} \lambda_j}. \tag{5.8}$$

A few discriminant functions will often suffice to describe the group differences. The proportion in (5.8) is often referred to as the "percent of discriminating variance," but this is a misnomer, since λ_i is not the variance of z_i.

The discriminant functions are invariant to full-rank linear transformations $\mathbf{u} = \mathbf{A}\mathbf{y}$, where $\mathbf{A}$ is nonsingular. Let $\mathbf{H}_u$ and $\mathbf{E}_u$ be the hypothesis and error matrices of $\mathbf{u}$ and let $\mathbf{b}_1, \mathbf{b}_2, \ldots, \mathbf{b}_s$ be the eigenvectors of $\mathbf{E}_u^{-1}\mathbf{H}_u$. Then the discriminant functions based on $\mathbf{u}$ are the same as those based on $\mathbf{y}$; that is,

$$z_i = \mathbf{b}_i'\mathbf{u} = \mathbf{a}_i'\mathbf{y}, \qquad i = 1, 2, \ldots, s, \tag{5.9}$$

where $\mathbf{b}_i$ and $\mathbf{a}_i$ correspond to the ith largest eigenvalue in each case.

5.3.2 Assumptions

The assumptions for discriminant analysis are similar to those for MANOVA (see Section 4.1.1), including equal population covariance matrices, independence of the observation vectors, and multivariate normality. The most important of these assumptions is independence of the observation vectors. Reliable results cannot be expected in the absence of independence.

Since discriminant function coefficient vectors are obtained as eigenvectors of $\mathbf{E}^{-1}\mathbf{H}$, it is clear that we have assumed equal population covariance matrices; we use $\mathbf{S} = \mathbf{E}/\nu_E$ as the estimator of the common covariance matrix $\boldsymbol{\Sigma}$. Tests in discriminant analysis (Section 5.5) are fairly robust to the assumption of homogeneity of covariance matrices when the sample sizes n_i are large or equal. If the sample sizes are small or unequal, the significance tests can be seriously affected by heterogeneity of covariance matrices. Similarly, the descriptive representation of the data vectors by discriminant functions will be distorted in an unpredictable way in the presence of heterogeneity. The homogeneity assumption can be checked by Box's M-test (Section 4.3). Another check on the homogeneity assumption is to calculate the first two discriminant functions z_1 and z_2 for each observation vector and plot these in a bivariate scatterplot. If the k groups are roughly equal in size and shape, we have some assurance of homogeneity of covariance matrices.

We would expect the discriminant functions to be well behaved for multivariate normal data, since linear combinations of multivariate normal variables are normally distributed. Very little is known about the effect of nonnormality on the behavior of discriminant function coefficients.

The sampling variability, var(a_i), of the coefficients a_i in $z = \mathbf{a}'\mathbf{y} = a_1y_1 + a_2y_2 + \cdots + a_py_p$ increases with p and k and with distances between group means and decreases with the sample sizes n_i. When sample sizes are small relative to p, the pattern in the coefficients may be an artifact of the particular sample and therefore not representative of the population. Whenever possible, the procedure should be repeated on new data. If $N = \sum_{i=1}^{k} n_i$ is large, the sample could be partitioned into two parts, with one part reserved for verification.

Discriminant analysis is sensitive to outliers. Rencher (1995, Section 4.5.2) reviewed methods for detecting multivariate outliers. Campbell (1978) presented an influence function that is useful for detecting outliers for two groups. For the several-group case, Radhakrishnan (1983, 1985) and Romanazzi (1991) proposed influence functions that can be used in techniques for outlier detection.

5.4 STANDARDIZED COEFFICIENTS

It was noted at the end of Section 3.5 that if the variables are not commensurate, the coefficients $a_1, a_2, \ldots, a_p$ in a discriminant function $z = \sum_{i=1}^{p} a_i y_i$ should be standardized. This standardization of each a_i is easily carried out by multiplying by s_i:

$$a_i^* = s_i a_i, \qquad i = 1, 2, \ldots, p, \tag{5.10}$$

where s_i is the within-sample standard deviation of the ith variable. The standardized coefficients a_i^* apply to standardized variables that have been rendered scale free by dividing by the standard deviation. Mueller and Cozad (1993) compared the use of within-sample standard deviation and total-sample standard deviation. They concluded that the within-sample estimate is preferred.

For the two-group case, Rencher and Scott (1990) partitioned the standardized discriminant function coefficient a_i^* in (5.10) into a part due to the multiple correlation of y_i with the other variables and a part due to the influence of the other variables on y_i as it contributes to the discriminant function. This result is given in the following theorem.

Theorem 5.4A. Let u be the variable of interest, where u is any one of $y_1, y_2, \ldots, y_p$, and denote the other $p - 1$ y's collectively by y or $\mathbf{y}$. Then the standardized discriminant function coefficient a_u^* as defined in (5.10) can be expressed as

$$a_u^* = \frac{D_u - \hat{D}_u}{1 - R^2}, \tag{5.11}$$

where

$$D_u = \frac{\overline{u}_1 - \overline{u}_2}{s_u}, \qquad \hat{D}_u = \frac{\hat{\boldsymbol{\beta}}'(\overline{\mathbf{y}}_1 - \overline{\mathbf{y}}_2)}{s_u},$$

$$R^2 = \frac{\mathbf{s}_{uy}'\mathbf{S}_{yy}^{-1}\mathbf{s}_{uy}}{s_u^2}, \qquad \hat{\boldsymbol{\beta}} = \mathbf{S}_{yy}^{-1}\mathbf{s}_{uy}. \qquad \square$$

Note that in Theorem 5.4A, $\hat{\boldsymbol{\beta}}$ is the vector of regression coefficients of u regressed on the other $p-1$ y's and R^2 is the corresponding squared multiple correlation [see (7.54) and (7.66)].

In (5.11) we see that the contribution of each u (each y_i) to separation of the two groups is due to

1. the relationship of u to the other y's, as exhibited by R^2, and
2. how well the standardized distance $D_u = (\overline{u}_1 - \overline{u}_2)/s_u$ can be predicted (linearly) from the other y's by $\hat{D}_u = \hat{\boldsymbol{\beta}}'(\overline{\mathbf{y}}_1 - \overline{\mathbf{y}}_2)/s_u$. If either D_u or $\hat{D}_u$ is substantially larger in absolute value than the other, the addition of u will be important.

We note from the linearity inherent in both $\hat{D}_u$ and R^2 that, in general, discriminant functions are more meaningful for multivariate normal data (or at least elliptically contoured data; see Section 2.4) than for other types of data. We also see in (5.11) that if the variables were uncorrelated (orthogonal) in the sample, a_u^* would reduce to $D_u = (\overline{u}_1 - \overline{u}_2)/s_u$.

5.5 TESTS OF HYPOTHESES

In this section we assume that the $\mathbf{y}_{ij}$'s are independently distributed as $N_p(\boldsymbol{\mu}_i, \boldsymbol{\Sigma})$.

5.5.1 Two Groups

If we denote the population discriminant function by $\boldsymbol{\alpha}'\mathbf{y} = (\boldsymbol{\mu}_1 - \boldsymbol{\mu}_2)'\boldsymbol{\Sigma}^{-1}\mathbf{y}$ as in (3.91), then the most basic hypothesis is $H_0: \boldsymbol{\alpha} = \mathbf{0}$. Since $\boldsymbol{\alpha} = \mathbf{0}$ is equivalent to $\boldsymbol{\mu}_1 - \boldsymbol{\mu}_2 = \mathbf{0}$, the test of H_0 can be carried out using the two-sample Hotelling T^2 in (3.57),

$$T^2 = \frac{n_1 n_2}{n_1 + n_2}(\overline{\mathbf{y}}_1 - \overline{\mathbf{y}}_2)'\mathbf{S}_{\text{pl}}^{-1}(\overline{\mathbf{y}}_1 - \overline{\mathbf{y}}_2). \tag{5.12}$$

To test the hypothesis that some of the population discriminant coefficients are zero, say $H_0: \boldsymbol{\alpha}_2 = \mathbf{0}$, where $\boldsymbol{\alpha}' = (\boldsymbol{\alpha}_1', \boldsymbol{\alpha}_2')$, we can use (3.98) or (3.99), because $H_0: \boldsymbol{\alpha}_2 = \mathbf{0}$ is equivalent to (3.93).

A test for a hypothetical discriminant function $\boldsymbol{\alpha}_0'\mathbf{y}$ was given by Siotani et al. (1985, pp. 402–403). The coefficient vector $\boldsymbol{\alpha}_0$ might represent previous experience or a simplified version we wish to test for adequate fit to the data. The hypothesis can be stated as

$$H_0: \boldsymbol{\alpha} \text{ is proportional to } \boldsymbol{\alpha}_0. \tag{5.13}$$

We do not express (5.13) as $\boldsymbol{\alpha} = \boldsymbol{\alpha}_0$ because discriminant function coefficients are unique only up to a constant multiplier [see a remark following (A.10.3)]. We compare the observed discriminant function coefficient vector $\mathbf{a} = \mathbf{S}_{\text{pl}}^{-1}(\overline{\mathbf{y}}_1 - \overline{\mathbf{y}}_2)$ to

$\boldsymbol{\alpha}_0$ by comparing $t^2(\mathbf{a})$ to $t^2(\boldsymbol{\alpha}_0)$, where, by (3.63),

$$t^2(\mathbf{a}) = \frac{[\mathbf{a}'(\overline{\mathbf{y}}_1 - \overline{\mathbf{y}}_2)]^2}{[(n_1 + n_2)/n_1 n_2]\mathbf{a}'\mathbf{S}_{\text{pl}}\mathbf{a}}.$$

For $\mathbf{a} = \mathbf{S}_{\text{pl}}^{-1}(\overline{\mathbf{y}}_1 - \overline{\mathbf{y}}_2)$, $t^2(\mathbf{a})$ is equal to T^2, the usual two-sample test statistic in (5.12) [see Theorem 3.5D and (3.65)]. We denote this T^2 by T_a^2 to distinguish it from the test statistic T^2 below in (5.14). Since $t^2(\boldsymbol{\alpha}_0)$ is distributed as t_ν^2 and thereby as $T_{1,\nu}^2$, we can compare T_a^2 with $t^2(\boldsymbol{\alpha}_0)$ by means of (3.98) or (3.99). We reject H_0 in (5.13) if

$$T^2 = (\nu - 1)\left(\frac{T_a^2 - t^2(\boldsymbol{\alpha}_0)}{\nu + t^2(\boldsymbol{\alpha}_0)}\right) \geq T^2_{\alpha,p-1,\nu-1} \tag{5.14}$$

or if

$$F = \left(\frac{\nu - p + 1}{p - 1}\right)\left(\frac{T_a^2 - t^2(\boldsymbol{\alpha}_0)}{\nu + t^2(\boldsymbol{\alpha}_0)}\right) \geq F_{\alpha,p-1,\nu-p+1}, \tag{5.15}$$

where $\nu = n_1 + n_2 - 2$.

Example 5.5.1. For the psychological data summarized in Example 3.5.3(a), the discriminant function coefficient vector is given by

$$\mathbf{a} = \begin{pmatrix} .5104 \\ -.2033 \\ .4660 \\ -.3097 \end{pmatrix}.$$

For illustrative purposes we consider the hypothesis $H_0 : \boldsymbol{\alpha}$ is proportional to $\boldsymbol{\alpha}_0 = (2, -1, 2, -1)'$. If the four variables were more commensurate (the ratio of the largest variance in $\mathbf{S}_{\text{pl}}$ to the smallest is $29.36/7.164 = 4.1$), this might represent a pattern of simplified coefficients for interpretation purposes. We first calculate

$$t^2(\boldsymbol{\alpha}_0) = \frac{[\boldsymbol{\alpha}_0'(\overline{\mathbf{y}}_1 - \overline{\mathbf{y}}_2)]^2}{[(n_1 + n_2)/n_1 n_2]\boldsymbol{\alpha}_0'\mathbf{S}_{\text{pl}}\boldsymbol{\alpha}_0}$$

$$= \frac{(25.500)^2}{[(32 + 32)/(32)^2](108.668)} = 95.741$$

and compare this to $T_a^2 = 97.601$ from Example 3.5.3(a). Using (5.14) to test the significance of the difference, where $\nu = n_1 + n_2 - 2 = 62$, we have

$$T^2 = (62 - 1)\frac{97.601 - 95.741}{62 + 95.741} = .720,$$

which, of course, is not significant. Thus there is no evidence to reject the hypothesis that the population discriminant function coefficients have the pattern $(2, -1, 2, -1)'$.

5.5.2 Several Groups

If the s population discriminant functions are denoted by $\boldsymbol{\alpha}_1'\mathbf{y}, \boldsymbol{\alpha}_2'\mathbf{y}, \ldots, \boldsymbol{\alpha}_s'\mathbf{y}$, then the hypothesis for the test of significance of the discriminant functions can be written as $H_0: \boldsymbol{\alpha}_1 = \boldsymbol{\alpha}_2 = \cdots = \boldsymbol{\alpha}_s = \mathbf{0}$. By analogy to (4.40), the $\boldsymbol{\alpha}_i$'s are eigenvectors of $\boldsymbol{\Sigma}^{-1}\boldsymbol{\Omega}$, where $\boldsymbol{\Omega} = \sum_{i=1}^k n_i(\boldsymbol{\mu}_i - \boldsymbol{\mu})(\boldsymbol{\mu}_i - \boldsymbol{\mu})'$. Since all the eigenvectors of a nonzero matrix are nonzero (even when some of the eigenvalues are zero), $\boldsymbol{\alpha}_i = \mathbf{0}$ for all i implies that $\boldsymbol{\Sigma}^{-1}\boldsymbol{\Omega} = \mathbf{O}$, in which case $\boldsymbol{\Omega} = \mathbf{O}$ and $\boldsymbol{\mu}_1 = \boldsymbol{\mu}_2 = \ldots = \boldsymbol{\mu}_k$. Thus $H_0: \boldsymbol{\alpha}_1 = \boldsymbol{\alpha}_2 = \cdots = \boldsymbol{\alpha}_s = \mathbf{0}$ is equivalent to $H_0: \boldsymbol{\mu}_1 = \boldsymbol{\mu}_2 = \cdots = \boldsymbol{\mu}_k$, and the s discriminant functions can be tested for significance by the overall MANOVA likelihood ratio test using (4.14),

$$\Lambda_1 = \prod_{i=1}^{s} \frac{1}{1 + \lambda_i}, \tag{5.16}$$

which is distributed as $\Lambda_{p,k-1,N-k}$, where p is the number of variables, k is the number of groups, $N = \sum_i n_i$, and the λ_i's are eigenvalues of $\mathbf{E}^{-1}\mathbf{H}$.

The statistic Λ_1 in (5.16) tests the significance of all the discriminant functions $z_i = \mathbf{a}_i'\mathbf{y}$, $i = 1, 2, \ldots, s$. If the test rejects H_0, we are only sure of the significance of $z_1 = \mathbf{a}_1'\mathbf{y}$. To test the significance of the discriminant functions corresponding to some of the smaller λ's, say $\lambda_m, \lambda_{m+1}, \ldots, \lambda_s$, we use the following function of the likelihood ratio,

$$\Lambda_m = \prod_{i=m}^{s} \frac{1}{1 + \lambda_i}, \tag{5.17}$$

which is distributed as $\Lambda_{p-m+1,k-m,N-k-m+1}$. Bartlett (1947) showed that

$$\begin{aligned} V_m &= -\left(N - 1 - \frac{p+k}{2}\right)\ln \Lambda_m \\ &= \left(N - 1 - \frac{p+k}{2}\right)\sum_{i=m}^{s} \ln(1 + \lambda_i) \end{aligned} \tag{5.18}$$

has an approximate χ^2-distribution with $(p - m + 1)(k - m)$ degrees of freedom.

We can also use the F-approximation in (4.16) for Λ_m in (5.17),

$$F = \frac{1 - \Lambda_m^{1/t}}{\Lambda_m^{1/t}} \frac{\text{df}_2}{\text{df}_1},$$

where

$$t = \sqrt{\frac{(p - m + 1)^2(k - m)^2 - 4}{(p - m + 1)^2 + (k - m)^2 - 5}}, \qquad w = N - 1 - \frac{1}{2}(p + k),$$

$$\text{df}_1 = (p - m + 1)(k - m), \qquad \text{df}_2 = wt - \frac{1}{2}[(p - m + 1)(k - m) - 2].$$

To test the hypothesis that the coefficients in a subset of each population discriminant function are zero, say $H_0 : \boldsymbol{\alpha}_{21} = \boldsymbol{\alpha}_{22} = \cdots = \boldsymbol{\alpha}_{2s} = \mathbf{0}$, where each $\boldsymbol{\alpha}_i'$ is partitioned as $\boldsymbol{\alpha}_i' = (\boldsymbol{\alpha}_{1i}', \boldsymbol{\alpha}_{2i}')$, we use (4.134) because H_0 is equivalent to H_{01} in Section 4.9.1. For a test of the hypothesis that a discriminant function has a specified value, $H_0 : \boldsymbol{\alpha}$ is proportional to $\boldsymbol{\alpha}_0$; see Siotani et al. (1985, pp. 439–440).

5.6 DISCRIMINANT ANALYSIS FOR HIGHER ORDER DESIGNS

Discriminant analysis for higher order fixed-effects designs is an immediate extension of discriminant analysis for the one-way model presented in Section 5.3. For example, in a balanced two-way fixed-effects design (Section 4.6), the total sum of squares and products matrix can be partitioned as in (4.72),

$$\mathbf{T} = \mathbf{H}_A + \mathbf{H}_B + \mathbf{H}_{AB} + \mathbf{E}.$$

We therefore have three sets of discriminant functions, a set for each of the two main effects and a set for the interaction. For the A effect, the coefficient vectors are eigenvectors of $\mathbf{E}^{-1}\mathbf{H}_A$, and there are $s = \min(p, a - 1)$ discriminant functions, where a is the number of levels of factor A. Similarly, for the B effect, the coefficient vectors are the $s = \min(p, b - 1)$ eigenvectors of $\mathbf{E}^{-1}\mathbf{H}_B$. For the AB interaction, the coefficients for the discriminant functions are the $s = \min[p, (a - 1)(b - 1)]$ eigenvectors of $\mathbf{E}^{-1}\mathbf{H}_{AB}$.

We would not expect the discriminant functions for the three effects to be the same. The pattern of coefficients in each case would indicate the contribution of the variables to the corresponding main effect or interaction. The coefficients could be standardized as in (5.10).

5.7 INTERPRETATION OF DISCRIMINANT FUNCTIONS

A distinction can be made between interpreting discriminant functions and determining the contribution of each variable, but we will not always distinguish between the two cases. In interpretation, we wish to describe or name the discriminant function, and therefore the signs of the coefficients are important. In a determination of the contribution of each variable to the discriminant function, the coefficients are simply ranked in absolute value. In either case, the coefficients show the contribution of the variables in the presence of each other; that is, the coefficients provide a multivariate interpretation that allows for the correlations among the variables.

In Sections 5.7.1 and 5.7.2, we discuss three common approaches to assessing the contribution of each variable to separating the groups: (1) standardized discriminant function coefficients, (2) partial F-tests, and (3) correlations between the variables and the discriminant function. The correlations are widely recommended, but we show that they are not useful. For a more comprehensive treatment of interpretation of discriminant functions, see Rencher (1992a).

5.7.1 Standardized Coefficients and Partial F-Values

As noted in Section 5.4, the standardized coefficients $a_1^*, a_2^*, \ldots, a_p^*$ reveal the joint contribution of the p variables to the discriminant function. The precise sense in which the coefficients provide this information was given in Theorem 5.4A for the two-group case.

A partial F-test for each variable shows its contribution to Hotelling's T^2 or Wilks' Λ after adjusting for the other variables. For two groups, the partial F-statistic is given by (3.100); for several groups, see (4.136).

In the case of several groups, the partial F-values are not associated with a single discriminant function. They indicate the overall unique contribution of each variable to group separation. Thus, for interpretation of a particular discriminant function, the standardized discriminant function coefficients are more useful.

5.7.2 Correlations between Variables and Discriminant Functions

The correlation between each variable and a discriminant function is widely recommended as a useful measure of variable importance in the discriminant function (see, for example, Huberty 1996; 1994, pp. 209, 218, 231, 246). These correlations, sometimes called structure coefficients, are provided in many software packages. However, Rencher (1988, 1992b) has shown that these correlations do not show the multivariate contribution of each variable, but rather provide only univariate information, showing how each variable by itself separates the groups, ignoring the presence of the other variables. These results are given in Sections 5.7.2a and 5.7.2b.

5.7.2a Two-Group Case

Theorem 5.7A. In the two-group case, the within-group sample correlation $r_{y_i z}$ of each variable y_i with the discriminant function $z = \mathbf{a}'\mathbf{y}$ is directly proportional to the two-sample t-statistic for that variable:

$$r_{y_i z} = ct_i = c\frac{\overline{y}_{1i} - \overline{y}_{2i}}{\sqrt{(1/n_1 + 1/n_2)s_{ii}}}, \tag{5.19}$$

$i = 1, 2, \ldots, p$, where $\mathbf{a} = \mathbf{S}_{\text{pl}}^{-1}(\overline{\mathbf{y}}_1 - \overline{\mathbf{y}}_2)$, $c = \sqrt{(n_1 + n_2)/(n_1 n_2 D^2)}$, s_{ii} is the ith diagonal element of the pooled covariance matrix $\mathbf{S}_{\text{pl}}$, and $D^2 = (\overline{\mathbf{y}}_1 - \overline{\mathbf{y}}_2)'\mathbf{S}_{\text{pl}}^{-1}(\overline{\mathbf{y}}_1 - \overline{\mathbf{y}}_2)$.

Proof. Let $\mathbf{g}$ be a vector with a 1 in the ith position and 0s in all other positions. Then $y_i = \mathbf{g}'\mathbf{y}$, and by (1.75), the correlation between y_i and z is given by

$$r_{y_i z} = \frac{s_{y_i z}}{\sqrt{s_{y_i}^2 s_z^2}} = \frac{\mathbf{g}'\mathbf{S}_{\text{pl}}\mathbf{a}}{\sqrt{(\mathbf{g}'\mathbf{S}_{\text{pl}}\mathbf{g})(\mathbf{a}'\mathbf{S}_{\text{pl}}\mathbf{a})}}$$

$$= \frac{\mathbf{g}'\mathbf{S}_{\text{pl}}\mathbf{S}_{\text{pl}}^{-1}(\overline{\mathbf{y}}_1 - \overline{\mathbf{y}}_2)}{\sqrt{s_{ii}(\overline{\mathbf{y}}_1 - \overline{\mathbf{y}}_2)'\mathbf{S}_{\text{pl}}^{-1}\mathbf{S}_{\text{pl}}\mathbf{S}_{\text{pl}}^{-1}(\overline{\mathbf{y}}_1 - \overline{\mathbf{y}}_2)}} = \frac{\overline{y}_{1i} - \overline{y}_{2i}}{\sqrt{s_{ii}D^2}}. \qquad \square$$

By Theorem 5.7A, the correlation $r_{y_i z}$ is simply a univariate measure of how well the ith variable by itself separates the two groups. Contrary to what is claimed for it, $r_{y_i z}$ conveys no information about the contribution of y_i to the discriminant function z.

5.7.2b *Multiple-Group Case*

In the case of k groups, with n_i observation vectors in the ith group, the correlations again provide only univariate information. They do not depict the contribution of each variable in a multivariate context. Since each variable y_i is correlated with all s of the discriminant functions, the relationship between the correlations and the univariate F for y_i is not as simple in the multiple-group case as it is in the two-group case. A weighted sum of squares of the correlations of y_i with the s discriminant functions is proportional to the F for y_i alone, as noted in the following theorem.

Theorem 5.7B. Let $r_{y_i z_j}$, $i = 1, 2, \ldots, p$, $j = 1, 2, \ldots, s$, be the within-sample correlations between $y_1, y_2, \ldots, y_p$ and $z_1 = \mathbf{a}_1'\mathbf{y}, \ldots, z_s = \mathbf{a}_s'\mathbf{y}$:

$$\begin{pmatrix} r_{y_1z_1} & r_{y_1z_2} & \cdots & r_{y_1z_s} \\ r_{y_2z_1} & r_{y_2z_2} & \cdots & r_{y_2z_s} \\ \vdots & \vdots & & \vdots \\ r_{y_pz_1} & r_{y_pz_2} & \cdots & r_{y_pz_s} \end{pmatrix}.$$

Then the correlations in the ith row of this array relate to F_i, the test statistic that compares the k groups on y_i, as follows:

$$r_{y_iz_1}^2\lambda_1 + r_{y_iz_2}^2\lambda_2 + \cdots + r_{y_iz_s}^2\lambda_s = \frac{h_{ii}}{e_{ii}} = \frac{k-1}{N-k}F_i, \tag{5.20}$$

where the λ_j's are the eigenvalues of $\mathbf{E}^{-1}\mathbf{H}$, h_{ii} and e_{ii} are diagonal elements of $\mathbf{E}$ and $\mathbf{H}$, respectively, $N = \sum_{i=1}^k n_i$, and $\mathbf{E}$ and $\mathbf{H}$ are defined in Table 4.4.

Proof. The jth column of the above array of $r_{y_iz_j}$ is the vector of within-sample correlations between the discriminant function $z_j = \mathbf{a}_j'\mathbf{y}$ and the variables in $\mathbf{y}$,

$$\mathbf{r}_{yz_j} = \begin{pmatrix} r_{y_1z_j} \\ r_{y_2z_j} \\ \vdots \\ r_{y_pz_j} \end{pmatrix}.$$

Using (1.53) and (1.79), this vector can be obtained as

$$\mathbf{r}_{yz_j} = \mathbf{R}\mathbf{a}_j^*, \tag{5.21}$$

where $\mathbf{R}$ is the within-sample correlation matrix of the y's and $\mathbf{a}_j^* = (a_{j1}^*, a_{j2}^*, \ldots, a_{jp}^*)'$ is the vector of standardized discriminant function coefficients defined in (5.10). By analogy to (1.51), $\mathbf{R}$ can be obtained from $\mathbf{E}$ as

$$\mathbf{R} = \mathbf{D}^{-1}\mathbf{E}\mathbf{D}^{-1}, \tag{5.22}$$

where $\mathbf{D}^2 = \text{diag}(\mathbf{E}) = \text{diag}(\sqrt{e_{11}}, \sqrt{e_{22}}, \ldots, \sqrt{e_{pp}})$. Similarly, $\mathbf{a}_j^*$ can be expressed as

$$\mathbf{a}_j^* = \mathbf{D}\mathbf{a}_j. \tag{5.23}$$

Using (5.22) and (5.23) in (5.21), we have

$$\mathbf{r}_{yz_j} = \mathbf{R}\mathbf{a}_j^* = \mathbf{D}^{-1}\mathbf{E}\mathbf{D}^{-1}\mathbf{D}\mathbf{a}_j = \mathbf{D}^{-1}\mathbf{E}\mathbf{a}_j,$$

from which the ith element of $\mathbf{r}_{yz_j}$ is

$$r_{y_i z_j} = \frac{\mathbf{e}_i'\mathbf{a}_j}{\sqrt{e_{ii}}}, \tag{5.24}$$

where $\mathbf{e}_i'$ is the ith row of $\mathbf{E}$. Using (5.24), the left side of (5.20) can be written as

$$\sum_{j=1}^{s} r_{y_i z_j}^2 \lambda_j = \sum_{j=1}^{s} \left(\frac{\mathbf{e}_i'\mathbf{a}_j}{\sqrt{e_{ii}}} \right)^2 \lambda_j = \left(\frac{1}{e_{ii}} \right) \mathbf{e}_i' \left[\sum_{j=1}^{s} \mathbf{a}_j\mathbf{a}_j'\lambda_j \right] \mathbf{e}_i. \tag{5.25}$$

By the last paragraph in Section 4.1.3, the eigenvalues of $\mathbf{E}^{-1}\mathbf{H}$ are the same as those of the symmetric matrix $(\mathbf{U}')^{-1}\mathbf{H}\mathbf{U}^{-1}$, where $\mathbf{U}'\mathbf{U} = \mathbf{E}$ is the Cholesky factorization of $\mathbf{E}$ (Section A.6). The eigenvectors $\mathbf{a}_j$ of $\mathbf{E}^{-1}\mathbf{H}$ are related to the eigenvectors $\mathbf{c}_j$ of $(\mathbf{U}')^{-1}\mathbf{H}\mathbf{U}^{-1}$ by $\mathbf{a}_j = \mathbf{U}^{-1}\mathbf{c}_j$. Using $\mathbf{a}_j = \mathbf{U}^{-1}\mathbf{c}_j$ in $\sum_{j=1}^{s} \mathbf{a}_j\mathbf{a}_j'\lambda_j$, we obtain

$$\sum_{j=1}^{s} \mathbf{a}_j\mathbf{a}_j'\lambda_j = \mathbf{U}^{-1} \left[\sum_{j=1}^{s} \mathbf{c}_j\mathbf{c}_j'\lambda_j \right] (\mathbf{U}^{-1})'.$$

The sum $\sum_{j=1}^{s} \mathbf{c}_j\mathbf{c}_j'\lambda_j$ is the spectral decomposition of $(\mathbf{U}^{-1})'\mathbf{H}\mathbf{U}^{-1}$ [see (A.11.4)], and we have

$$\begin{aligned} \sum_{j=1}^{s} \mathbf{a}_j\mathbf{a}_j'\lambda_j &= \mathbf{U}^{-1}\left[(\mathbf{U}^{-1})'\mathbf{H}\mathbf{U}^{-1}\right](\mathbf{U}^{-1})' = (\mathbf{U}'\mathbf{U})^{-1}\mathbf{H}(\mathbf{U}'\mathbf{U})^{-1} \\ &= \mathbf{E}^{-1}\mathbf{H}\mathbf{E}^{-1}. \end{aligned} \tag{5.26}$$

Substituting (5.26) into (5.25) gives

$$\begin{aligned} \sum_{j=1}^{s} r_{y_i z_j}^2 \lambda_j &= \frac{1}{e_{ii}} \mathbf{e}_i' \left[\mathbf{E}^{-1}\mathbf{H}\mathbf{E}^{-1} \right] \mathbf{e}_i \\ &= \frac{1}{e_{ii}} (0, \ldots, 0, 1, 0, \ldots, 0)\mathbf{H}(0, \ldots, 0, 1, 0, \ldots, 0)' \\ &= \frac{h_{ii}}{e_{ii}} \\ &= \frac{k-1}{N-k} F_i \qquad \text{[by (4.19)]}. \end{aligned}$$

□

Thus in the multiple-group case, the squared correlations for a variable are related directly to the F for that variable. The correlations therefore provide only univariate information and are not useful for interpreting discriminant functions.

If λ_1 is a large fraction of $\sum_{j=1}^{s} \lambda_j$, then by (5.20), $r_{y_i z_1}^2 \lambda_1$ will be almost equal to $(k-1)F_i/(N-k)$; that is,

$$F_i \cong \frac{\lambda_1(N-k)}{k-1} r_{y_i z_1}^2. \tag{5.27}$$

If λ_1 does not constitute a large portion of $\sum_{j=1}^{s} \lambda_j$, then $r_{y_i z_1}^2, r_{y_i z_2}^2, \ldots, r_{y_i z_s}^2$ are still not informative in a multivariate context because they are merely additive components of $(k-1)F_i/(N-k)$, which pertains only to y_i. For interpretation, we therefore recommend the use of standardized coefficients or partial F's.

Example 5.7.2. In a preliminary study of a possible link between football helmet design and neck injuries, six head measurements were made on 30 individuals in each of three player groups (Bryce and Barker, Brigham Young University, personal communication; see diskette for data set). The pooled within-groups correlation matrix is

$$\mathbf{R} = \begin{bmatrix} 1.0 & .497 & .327 & .115 & .242 & .569 \\ .497 & 1.0 & .776 & .331 & .243 & .464 \\ .327 & .776 & 1.0 & .094 & .221 & .351 \\ .115 & .331 & .094 & 1.0 & .361 & .062 \\ .242 & .243 & .221 & .361 & 1.0 & .019 \\ .569 & .464 & .351 & .062 & .019 & 1.0 \end{bmatrix}.$$

The eigenvalues of $\mathbf{E}^{-1}\mathbf{H}$ are $\lambda_1 = 1.9178$ and $\lambda_2 = .1159$. Since $\lambda_1/(\lambda_1+\lambda_2) = .94$ [see (5.8)], the three mean vectors lie largely in one dimension. We therefore consider only the first discriminant function $z_1 = \mathbf{a}_1'\mathbf{y}$.

In Table 5.1, we give the standardized discriminant function coefficients, the correlations between variables and discriminant functions, and the univariate F_i's. As expected, the correlations and univariate F_i's rank the variables in the same

Table 5.1. Standardized Coefficients a_{1i}^* and Correlations $r_{y_i z_1}$ for $z_1 = \mathbf{a}_1'\mathbf{y}$ Compared to Univariate F_i's and $cr_{y_i z_1}^2$

Variable	a_{1i}^*	Rank	$r_{y_i z_1}$	Rank	F_i	Rank	$cr_{y_i z_1}^2$
WDIM	−.621	2	−.148	5	2.55	5	1.83
CIRCUM	.006	5	.271	3	6.23	3	6.15
FBEYE	.005	6	.141	6	1.67	6	1.65
EYEHD	.719	1	.824	1	58.16	1	56.71
EARHD	.397	4	.518	2	22.43	2	22.36
JAW	.508	3	.212	4	4.51	4	3.75

$c = (N-k)\lambda_1/(k-1)$

order of importance. In fact, the squared correlations with the first discriminant function largely reproduce the F's, as seen in the last column of Table 5.1, where $[(N-k)\lambda_1/(k-1)]r^2_{y_i z_1}$ can be compared with F_i [see (5.27)].

5.7.3 Other Approaches

Several other approaches have been suggested for interpretation of discriminant functions or for assessing the contribution of each variable to separation of groups. We mention these for completeness; it appears that most of them are not as informative as are standardized coefficients or partial F-values.

1. A univariate t- or F-test can be carried out on each variable, but such a test tells only how well the variable by itself separates the groups, not what it contributes to the discriminant function. A univariate test does not take into account how a variable behaves in the presence of other variables and is therefore of little interest in a multivariate context. These tests are equivalent to the correlations discussed in Section 5.7.2.
2. When the variables are entered one by one in a forward selection or stepwise MANOVA procedure (Sections 3.11.3, 4.9.3, 5.9.2), the order of entry is sometimes used as a ranking of the variables. However, the variable entered at each step is compared only with the variables previously entered rather than with all the other $p-1$ variables. Thus this ranking is less informative than the partial F.
3. Gabriel (1968) proposed comparing all groups and subsets of groups using all subsets of variables. However, even when the computational and expositional load of this approach is feasible, the results provide only partial information about variable importance.
4. Some authors recommend rotation of the coefficients (see Section 10.5) to obtain a pattern with (absolute values of) coefficients closer to zero or one. Discriminant functions with such coefficients may be more interpretable, but they are correlated and they no longer maximize group separation (Rencher 1992a).
5. Harris (1985, pp. 101–109) suggested replacing the discriminant function coefficients with integers to obtain simplified forms that are easier to interpret. The simplified versions can be tested for significance by (5.14) or (5.15). However, this approach cannot readily be used with standardized coefficients and is therefore largely limited to variables that are already commensurate.
6. Darroch and Mosimann (1985) proposed a method of partitioning the sum of the variances of the discriminant functions into *size* and *shape* components.
7. Thomas (1992) proposed new indices called *discriminant ratio coefficients* for the quantification of the relative importance of the individual variables and compared these indices with other techniques.

5.8 CONFIDENCE INTERVALS

To better visualize the configuration of the mean vectors $\overline{\mathbf{y}}_{1.}, \overline{\mathbf{y}}_{2.}, \ldots, \overline{\mathbf{y}}_{k.}$, we can plot these vectors in a two-dimensional projection formed by the first two discriminant functions. The transformed mean vectors are

$$\overline{\mathbf{z}}_{i\cdot} = \begin{pmatrix} \overline{z}_{1i.} \\ \overline{z}_{2i.} \end{pmatrix} = \begin{pmatrix} \mathbf{a}_1'\overline{\mathbf{y}}_{i\cdot} \\ \mathbf{a}_2'\overline{\mathbf{y}}_{i\cdot} \end{pmatrix} = \mathbf{A}\overline{\mathbf{y}}_{i\cdot}, \qquad i = 1, 2, \ldots, k.$$

If the eigenvectors are normalized so that $\mathbf{a}_i'\mathbf{S}\mathbf{a}_i = 1$, then

$$\widehat{\text{cov}}(\overline{\mathbf{z}}_{i\cdot}) = \frac{\mathbf{I}}{n_i}, \tag{5.28}$$

since, by Theorem 5.3A, $\mathbf{a}_1'\mathbf{S}\mathbf{a}_2 = 0$, where $\mathbf{S} = \mathbf{E}/\nu_E$. Then, assuming that $\mathbf{y}_{ij}$ is $N_p(\boldsymbol{\mu}_i, \boldsymbol{\Sigma})$, a $100(1-\alpha)\%$ confidence region for $\boldsymbol{\mu}_{z_i} = \mathbf{A}\boldsymbol{\mu}_i$ is given by

$$(\overline{\mathbf{z}}_{i\cdot} - \boldsymbol{\mu}_{z_i})' \left(\frac{\mathbf{I}}{n_i}\right)^{-1} (\overline{\mathbf{z}}_{i\cdot} - \boldsymbol{\mu}_{z_i}) \leq T^2_{\alpha,2,\nu_E}$$

or

$$n_i(\overline{\mathbf{z}}_{i\cdot} - \boldsymbol{\mu}_{z_i})'(\overline{\mathbf{z}}_{i\cdot} - \boldsymbol{\mu}_{z_i}) \leq T^2_{\alpha,2,\nu_E}. \tag{5.29}$$

For large ν_E (≥ 50), a good approximation of (5.29) is given by

$$n_i(\overline{\mathbf{z}}_{i\cdot} - \boldsymbol{\mu}_{z_i})'(\overline{\mathbf{z}}_{i\cdot} - \boldsymbol{\mu}_{z_i}) \leq \chi^2_{\alpha,2}, \tag{5.30}$$

since there are only two dimensions. (For larger values of p, $\chi^2_{\alpha,p}$ is not a good approximation for $T^2_{\alpha,p,50}$; see Table B.1.) Both (5.29) and (5.30) have the form of a circle, which is relatively simple to add to a plot of the mean vectors $\overline{\mathbf{z}}_{i\cdot}$.

An assumption for either (5.29) or (5.30) is that $\mathbf{a}_1$ and $\mathbf{a}_2$ are constant vectors from sample to sample. However, $\mathbf{a}_1$ and $\mathbf{a}_2$ are sample dependent, and thus (5.29) and (5.30) are approximations. Based on an empirical study of several data sets, Krzanowski (1989) concluded that the circular regions in (5.30) are too small. He proposed elliptical regions that come closer to the nominal confidence level. Schott (1990) presented improved confidence regions for the case in which the eigenvalues are close together.

5.9 SUBSET SELECTION

In many applications, the experimenter measures a large number of dependent variables to be assured of including any that might prove effective in separating the groups. For a given sample size, however, an increase in the number of noninformative variables (variables whose population means do not differ across groups) leads to reduced power of tests and increased variability of estimates. Hence a search for redundant variables that can be discarded is very desirable in many cases.

We now discuss three major procedures for selecting dependent variables that best separate the groups: (1) delete variables with small coefficients in the standardized discriminant functions, (2) delete or add variables one at a time using stepwise methods, and (3) examine all possible subsets. (Note that methods 2 and 3 have a different purpose here than in Section 5.7.3.) McKay and Campbell (1982a) give a good survey of subset selection techniques. Other useful reviews are given by Huberty (1975; 1984; 1994, Section VIII-2), Krishnaiah (1982), and Berk (1980).

5.9.1 Discriminant Function Approach to Selection

In Sections 3.11.1, 4.9.1, 5.5.1, and 5.5.2, we noted that testing the significance of a subset of population discriminant function coefficients is equivalent to testing for significance of the corresponding subset of variables. We can therefore delete variables whose coefficients are small (in absolute value) in the discriminant functions. The coefficients should first be standardized as in Section 5.4 to better reflect the relative contribution of the variables. Since the coefficients change as the number of dependent variables changes, the discriminant functions should be recomputed each time a variable is deleted.

To be a candidate for deletion in the several-group case, a variable should have small coefficients in all discriminant functions of practical importance. A weighted average of (absolute values of) the coefficients for a given variable across the discriminant functions could be used in an attempt to devise an objective screening procedure.

5.9.2 Stepwise Selection

In stepwise selection, the variable entered (selected) at each step is the one that maximizes the partial F-statistic based on the partial T^2 or partial Wilks' Λ (Sections 3.11.3 and 4.9.3). After a new variable has entered, the other variables are reexamined to see if one of them has become redundant and can be deleted. The procedure continues until the largest partial F among the variables available for entry is less than a preset threshold value. The problem of choosing a threshold value is discussed in Section 5.10. Computationally efficient algorithms for stepwise selection (Jennrich 1977) are widely available.

Berk (1980) showed that in stepwise selection a return to a subset of size k will give a smaller value of Wilks' Λ. For example, if y_1, y_2, and y_3 have entered and then y_2 is deleted and replaced by y_4, the subset $\{y_1, y_3, y_4\}$ has a smaller overall Λ-value than does $\{y_1, y_2, y_3\}$.

Other criteria have been proposed to replace a partial Wilks' Λ at each step. For example, we could compute the Mahalanobis distance D^2 (Section 1.8) between the means of all pairs of groups and select the variable that gives maximum increase in D^2 between the two closest groups. Miller (1962) suggested entering the variable that maximizes the increase in the Lawley–Hotelling trace statistic $U^{(s)} = \text{tr}(\mathbf{E}^{-1}\mathbf{H})$. Farmer and Freund (1975) discussed a backward elimination technique in which at each step the variable is deleted that has the largest R^2 with the remaining variables, where R^2 is computed from the $\mathbf{E}$ matrix.

Todorov, Neykov, and Neytchev (1990) compared several methods of robust estimation applied to Wilks' Λ for variable selection, including the M-estimators and

MVE estimators discussed in Section 1.10. In the presence of multiple outliers, Wilks' Λ based on MVE-type estimators performed better than the ordinary Wilks' Λ or Wilks' Λ based on M-estimators. Krusinska and Liebhart (1989) discussed a robust version of Wilks' Λ for variable selection based on M-estimators. They found it to work well for three sets of medical data, in each of which there were outliers.

5.9.3 All Possible Subsets

Because variables are included or excluded one at a time, the stepwise procedure will often fail to find the subset with minimum Λ. However, if p is large, computation of Λ for all $2^p - 1$ subsets could be prohibitive. McCabe (1975) provided a program that finds the subset with minimum Λ-value for each subset size without processing all possible subsets. The procedure is based on the "leaps-and-bounds" algorithm of multiple regression (Furnival 1971, Furnival and Wilson 1974). However, even with this gain in efficiency, McCabe's program cannot readily handle as many variables as a stepwise discriminant program can process.

Since the overall Wilks Λ decreases as the subset size increases [see (4.138)], a plot may be useful in determining what subset size to use. An illustrative plot of the minimum value of Λ for each subset size is given in Figure 5.1. An appropriate subset size can be chosen where the plot levels off.

5.9.4 Selection in Higher Order Designs

McCabe's (1975) program (Section 5.9.3) and most stepwise programs (Section 5.9.2) assume a one-way MANOVA structure and use raw data as input. For a higher way MANOVA situation, McHenry (1978) provided a program that accepts **E** and

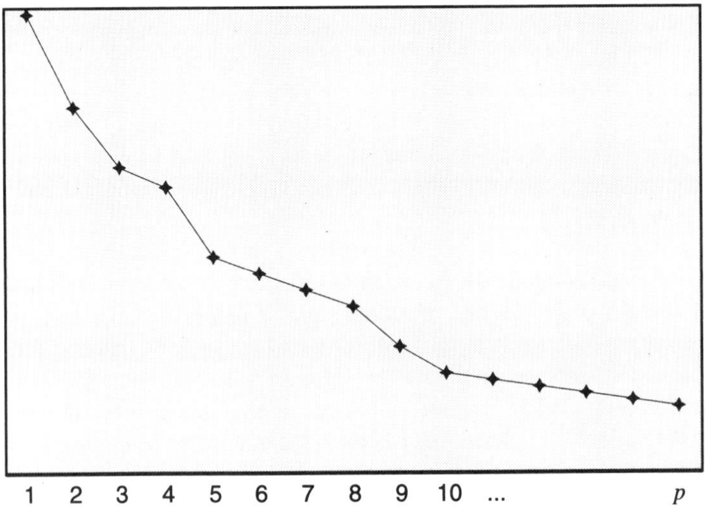

Figure 5.1. Minimum value of Wilks' Λ for each subset size.

H matrices as input. Thus, for example, the **H** matrix for a main effect or interaction in a factorial experiment along with the **E** matrix can be obtained from a MANOVA program and used as input to McHenry's program, which then finds the best subset of each size. The algorithm involves a combination of an all-possible-subsets approach and a stepwise approach. The program is efficient and can handle at least as many variables as McCabe's (1975) program can process.

5.10 BIAS IN SUBSET SELECTION

When the best or near best subset of a given size is found using procedures discussed in Section 5.9, the subset is optimal only in the sample, as noted in Section 3.11.3. We cannot be sure that the same subset of variables will be optimal in the population.

The optimality dilemma is related to the problem of determining significance levels in subset selection. When the partial Λ-statistic (4.135) or (4.138) is minimized at each step of a stepwise routine, it obviously does not have Wilks' Λ-distribution. A similar statement can be made about the corresponding maximum partial F in (4.136). This phenomenon is illustrated in Figure 5.2, where we have denoted the maximum partial F by F_{max}. The density of F_{max} is unknown.

As typified in Figure 5.2, the proportion of partial F-values exceeding F_α is α, while the proportion of F_{max}-values exceeding F_α is much greater than α. This bias

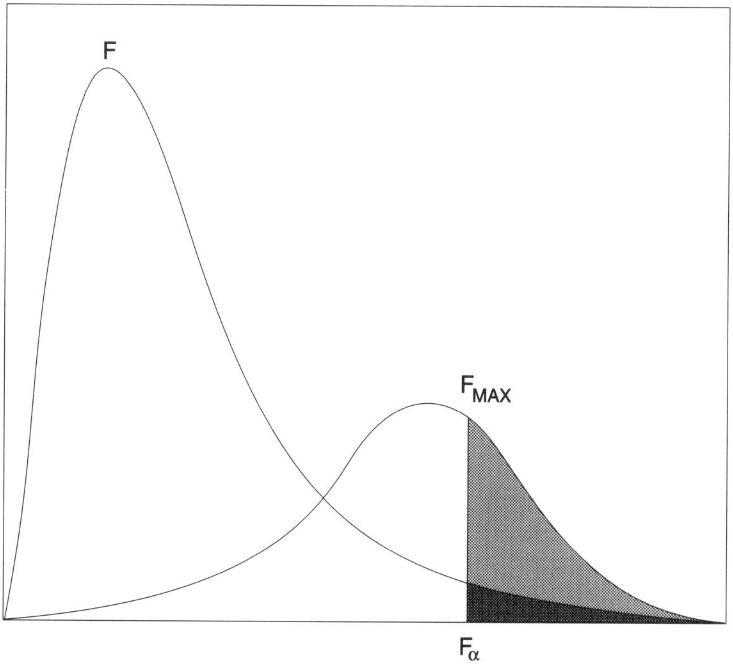

Figure 5.2. Density of F and of F_{max}.

is analogous to the bias in R^2 in best subset regression discussed by Flack and Chang (1987), Rencher and Pun (1980), Berk (1978), Pope and Webster (1972), and Diehr and Hoflin (1974).

The bias in F and Wilks' Λ may lead to inclusion of too many variables in the subset or even selection of an entirely spurious subset. These two problems are especially likely to occur if the significance levels for F are used in a stopping rule. If the actual significance levels of $F_{\max}$ were available to be used, the final subset would be smaller in many cases.

The above two problems become especially acute when the number of variables p is large and the total sample size $N = \sum_{i=1}^{n} n_i$ is relatively small. In some studies p exceeds N. In such cases, a value of Wilks' Λ cannot be computed for the entire set of variables, and hence it is not known whether the full set of variables can significantly separate the groups. However, stepwise discriminant analysis can still be used because it starts with a single variable and stops before the number of variables reaches the sample size N. On such shaky ground, a stepwise procedure should be used with caution, if at all. However, the researcher may feel that there is little choice but to eliminate some variables and will proceed to do so.

The results of a study in which $p > N$ should be presented with appropriate caveats. Clearly, a subset of variables chosen under such circumstances is likely to be unstable; that is, a different subset would typically emerge from a new sample. If possible, therefore, the researcher should redo the study with new data or reserve part of the sample to be used for verification of the results.

Rencher and Larson (1980) investigated the bias in stepwise discriminant analysis. The question of interest was the following: When there is a relatively large number of variables available, can the separation shown by the overall Wilks' Λ for the best subset be spurious? The null case of no differences between population means ($\boldsymbol{\mu}_1 = \boldsymbol{\mu}_2 = \cdots = \boldsymbol{\mu}_g$) was used so as to ascertain the levels Wilks' Λ may reach when there is no real separation from group to group. An excerpt of the results of Rencher and Larson for $p = 4$ is shown in Table 5.2.

Table 5.2. Simulation Results Showing Bias in Overall Wilks' Λ in Subset Selection

			$p = 4$									
			$k = 5$		$k = 10$		$k = 20$		$k = 30$		$k = 40$	
g	n	$E(\Lambda_0)$	$\overline{\Lambda}_s$	$\Lambda_{.05}$	$\overline{\Lambda}_s$	$\Lambda_{.05}$	$\overline{\Lambda}_s$	$\Lambda_{.05}$	$\overline{\Lambda}_s$	$\Lambda_{.05}$	$\overline{\Lambda}_s$	$\Lambda_{.05}$
2	5	.556	.454	.131	.215	.039	.081	.013	.046	.008	.029	.005
	10	.789	.747	.513	.577	.348	.422	.231	.349	.198	.297	.165
4	5	.470	.400	.198	.242	.111	.153	.070	.119	.055	.096	.045
	10	.716	.673	.515	.554	.397	.461	.332	.411	.291	.375	.274
6	5	.447	.392	.236	.272	.151	.192	.111	.159	.091	.137	.079
	10	.695	.658	.528	.565	.368	.489	.386	.447	.360	.418	.330
8	5	.437	.396	.245	.293	.177	.219	.137	.187	.117	.166	.105
	10	.685	.658	.520	.573	.449	.508	.412	.469	.358	.450	.364

The notation is as follows:

p = number of variables selected,
k = total number of variables available,
g = number of groups,
n = number of observation vectors in each group,
$\overline{\Lambda}_s$ = average value of Wilks' Λ for the p variables selected out of k,
$\Lambda_{.05}$ = lower .05 percentage point of Λ for the p variables selected out of k,
$E(\Lambda_0)$ = average value of Wilks' Λ if the p variables were not selected out of k [see (4.13)].

As Table 5.2 shows, the average value of Λ when the variables were obtained by selection, $\overline{\Lambda}_s$, is less than the average for the same number of variables not obtained by selection, $E(\Lambda_0)$; that is, $E(\Lambda_0)$ is the average that we would obtain if only p variables were available. If k is much larger than the sample size ng, the values of $\overline{\Lambda}_s$ are much smaller than $E(\Lambda_0)$. For example, compare $\overline{\Lambda}_s$ and $E(\Lambda_0)$ when $k = 40$ and $n = 5$ for all four values of g.

The values of $\Lambda_{.05}$ provide a critical value that one must attain for "significance." These values are surprisingly low in many cases.

In an attempt to control the significance level in stepwise selection, Hawkins (1976) suggested that a variable be entered only if its partial Λ or F is significant at the $\alpha/(k - p)$ level, where α is the desired level of significance, p is the number of variables already entered, and k is the total number of variables. A similar guideline is used for possible deletion of a variable at each step. Hawkins conjectured that this rule for inclusion and deletion is conservative.

McKay (1977) presented a simultaneous test procedure for subset selection in discriminant analysis. He suggested that Wilks' Λ for each subset be compared to Λ_α for the full set of variables and showed that the significance level for the test on each subset will not exceed the α-value for the full set. Since the critical values Λ_α increase as p decreases (Section 4.1.2, property 9), the test becomes more conservative for smaller subsets. Because we do not know how conservative it is for particular subset sizes, a small subset may fail to show significant separation when in fact the variables are important discriminators.

To apply McKay's (1977) technique, we would first test the full set of variables, and if significance is achieved, we could proceed to test all subsets of interest. If the initial test fails to reject H_0, there is no point in continuing, since all Λ's for subsets are greater than the Λ obtained for the full set.

5.11 OTHER ESTIMATORS OF DISCRIMINANT FUNCTIONS

In this section, we discuss two approaches to estimation of discriminant function coefficients as alternatives to the traditional methods in Sections 5.2 and 5.3. These methods give improved properties under certain distributional conditions.

5.11.1 Ridge Discriminant Analysis and Related Techniques

In the two-group case, the sample discriminant function coefficient vector $\mathbf{a} = \mathbf{S}_{\text{pl}}^{-1}(\overline{\mathbf{y}}_1 - \overline{\mathbf{y}}_2)$ is an estimate of the population discriminant function coefficient vector $\boldsymbol{\alpha} = \boldsymbol{\Sigma}^{-1}(\boldsymbol{\mu}_1 - \boldsymbol{\mu}_2)$. When $\boldsymbol{\Sigma}$ is near singular, $\mathbf{S}_{\text{pl}}$ will often be near singular, and small changes in $\mathbf{S}_{\text{pl}}$ will produce large changes in $\mathbf{S}_{\text{pl}}^{-1}$. The a_i's in $\mathbf{a} = \mathbf{S}_{\text{pl}}^{-1}(\overline{\mathbf{y}}_1 - \overline{\mathbf{y}}_2)$ will then have large variances.

This increase in variance when $\boldsymbol{\Sigma}$ is near singular can be demonstrated more formally as follows. Das Gupta (1968) showed that

$$\text{cov}(\mathbf{a}) = c_1\left[\Delta^2\boldsymbol{\Sigma}^{-1} + c_2\boldsymbol{\Sigma}^{-1} + c_3\boldsymbol{\Sigma}^{-1}\boldsymbol{\delta}\boldsymbol{\delta}'\boldsymbol{\Sigma}^{-1}\right], \tag{5.31}$$

where $\Delta^2 = (\boldsymbol{\mu}_1 - \boldsymbol{\mu}_2)'\boldsymbol{\Sigma}^{-1}(\boldsymbol{\mu}_1 - \boldsymbol{\mu}_2)$; $\boldsymbol{\delta} = \boldsymbol{\mu}_1 - \boldsymbol{\mu}_2$; and c_1, c_2, and c_3 are functions of n_1, n_2, and p. Using (A.8.1) and (A.8.2), the sum of the variances of the a_i's becomes

$$\begin{aligned}\sum_{i=1}^{p} \text{var}(a_i) &= \text{tr}[\text{cov}(\mathbf{a})] \\ &= c_1\left[\Delta^2 \,\text{tr}\,\boldsymbol{\Sigma}^{-1} + c_2 \,\text{tr}\,\boldsymbol{\Sigma}^{-1} + c_3\boldsymbol{\delta}'(\boldsymbol{\Sigma}^{-1})^2\boldsymbol{\delta}\right].\end{aligned} \tag{5.32}$$

Now, by (A.10.5) and (A.10.6),

$$\text{tr}\,\boldsymbol{\Sigma}^{-1} = \sum_{i=1}^{p}\frac{1}{\theta_i}, \tag{5.33}$$

where $\theta_1, \theta_2, \ldots, \theta_p$ are the eigenvalues of $\boldsymbol{\Sigma}$. If $\boldsymbol{\Sigma}$ is near singular, one or more of the θ_i's will be small and $\text{tr}\,\boldsymbol{\Sigma}^{-1}$ will be large, whereupon $\sum_i \text{var}(a_i)$ will be large.

In an attempt to reduce the inflated variances of the a_i's, Smidt and McDonald (1976) proposed a *ridge-adjusted estimator* of $\mathbf{a}$. This approach is patterned after *ridge regression*, a biased estimation technique that produces improved estimates of the regression coefficients when $\mathbf{X}'\mathbf{X}$ is near singular (Hoerl and Kennard 1970a, 1970b).

To obtain ridge discriminant function coefficients, consider the spectral decomposition of $\mathbf{S}_{\text{pl}}^{-1}$ obtained from (A.10.5) and (A.11.4),

$$\mathbf{S}_{\text{pl}}^{-1} = \sum_{i=1}^{p}\frac{1}{\lambda_i}\mathbf{c}_i\mathbf{c}_i', \tag{5.34}$$

where $\lambda_1, \lambda_2, \ldots, \lambda_p$ are the eigenvalues of $\mathbf{S}_{\text{pl}}$ and $\mathbf{c}_1, \mathbf{c}_2, \ldots, \mathbf{c}_p$ are the normalized eigenvectors. The *ridge discriminant function* of Smidt and McDonald (1976) is defined as

$$\mathbf{a}_k = \mathbf{S}_k^{-1}(\overline{\mathbf{y}}_1 - \overline{\mathbf{y}}_2) = (\mathbf{S}_{\text{pl}} + k\mathbf{I})^{-1}(\overline{\mathbf{y}}_1 - \overline{\mathbf{y}}_2), \tag{5.35}$$

where (see Problem 5.16)

$$(\mathbf{S}_{\text{pl}} + k\mathbf{I})^{-1} = \sum_{i=1}^{p} \frac{1}{\lambda_i + k} \mathbf{c}_i \mathbf{c}_i'. \tag{5.36}$$

The scalar constant k is a positive quantity that is chosen to stabilize $\mathbf{S}_k^{-1}$ and thus reduce the variances of the elements in $\mathbf{a}_k$. Smidt and McDonald suggested using a value of k proportional to λ_p, where λ_p is the smallest eigenvalue of $\mathbf{S}_{\text{pl}}$. In a Monte Carlo study Smidt and McDonald found that for various choices of near-singular $\boldsymbol{\Sigma}$, 80% of the ridge coefficients in $\mathbf{a}_k = \mathbf{S}_k^{-1}(\overline{\mathbf{y}}_1 - \overline{\mathbf{y}}_2)$ from (5.35) were closer to the population coefficients in $\boldsymbol{\alpha} = \boldsymbol{\Sigma}^{-1}(\boldsymbol{\mu}_1 - \boldsymbol{\mu}_2)$ than were the classical coefficients from $\mathbf{a} = \mathbf{S}_{\text{pl}}^{-1}(\overline{\mathbf{y}}_1 - \overline{\mathbf{y}}_2)$.

One cause of near singularity of $\boldsymbol{\Sigma}$ with attendant instability of $\mathbf{S}_{\text{pl}}^{-1}$ is the presence of highly intercorrelated variables. An alternative to the ridge approach is to delete some of the most redundant variables by means of a subset selection as in Section 5.9. Another option would be to delete m of the smallest eigenvalues in the spectral decomposition of $\mathbf{S}_{\text{pl}}$ so that (5.34) becomes

$$\mathbf{S}_m^{-1} = \sum_{i=1}^{p-m} \frac{1}{\lambda_i} \mathbf{c}_i \mathbf{c}_i'. \tag{5.37}$$

Of course $\mathbf{S}_m^{-1}$ is no longer an inverse of $\mathbf{S}_{\text{pl}}$, but by deleting one or more of the smallest λ_i, the diagonal elements of $\mathbf{S}_m^{-1}$ are smaller, and the resulting discriminant function coefficients in

$$\mathbf{a}_m = \mathbf{S}_m^{-1}(\overline{\mathbf{y}}_1 - \overline{\mathbf{y}}_2) \tag{5.38}$$

have smaller variances.

In the several-group case, highly intercorrelated variables lead to near singularity of $\mathbf{E}$ with a resulting instability of the eigenvectors of $\mathbf{E}^{-1}\mathbf{H}$. The techniques involved in (5.36) or (5.37) can be readily adapted to obtain a version of $\mathbf{E}^{-1}$ that will result in smaller variances of the discriminant function coefficients.

5.11.2 Robust Discriminant Analysis

As with many other multivariate procedures, discriminant functions are adversely affected by outliers and other manifestations of nonnormal data. Outliers claim too much attention from the discriminant functions, and the result is a poorer description of the rest of the data. Lachenbruch, Sneeringer, and Revo (1973) noted that the distortion was particularly marked for data from a long-tailed distribution.

Ahmed and Lachenbruch (1977) and Randles et al. (1978) suggested the use of robust procedures for two-group discriminant analysis. Larson (1980) and Campbell (1982) have discussed the several-group case. Both Larson and Campbell used M-estimators with weights similar to (1.119)–(1.121) to adjust the $\mathbf{E}$ matrix. The total sum-of-squares-and-products matrix $\mathbf{T} = \mathbf{E} + \mathbf{H}$ can be adjusted using the same weights. If the adjusted **E** and **T** matrices are denoted by $\mathbf{E}_c$ and $\mathbf{T}_c$, an adjusted hypothesis matrix can be found by $\mathbf{H}_c = \mathbf{T}_c - \mathbf{E}_c$. The robust discriminant functions are then obtained from the eigenvectors of $\mathbf{E}_c^{-1}\mathbf{H}_c$.

Example 5.11.2. Larson (1980) calculated robust discriminant functions for a data set from Pollock, Jackson, and Pate (1980), who used discriminant analysis in an attempt to determine if physiological attributes could separate good distance runners from elite distance runners. The 20 elite runners, all of whom had national or

Table 5.3. Physiological Data

	Runner	Fat	Lean	Maximum VO_2	Lactic Acid	Submaximum VO_2
			Middle-Long Distance Runners			
1.	D. Brown	8.3	65.8	5.77	30	4.71
2.	Anonymous	3.3	57.6	4.51	33	4.15
3.	T. Canstaneda	2.9	59.9	4.82	46	4.20
4.	J. Crawford	1.4	56.6	4.38	37	3.99
5.	P. Geis	7.1	59.2	5.30	26	4.29
6.	J. Johnson	3.2	58.9	4.96	24	4.07
7.	D. Kardong	4.1	65.8	5.41	19	4.57
8.	M. Manley	1.8	67.6	5.29	32	4.44
9.	P. Ndoo	4.0	49.6	4.33	39	3.53
10.	S. Prefontaine	3.8	62.5	5.60	16	4.79
11.	N. Rose	4.7	54.5	4.67	24	3.96
12.	G. Tuttle	0.9	60.1	5.04	31	4.12
			Marathon Runners			
13.	N. Cusack	5.5	58.7	5.07	17	4.32
14.	J. Galloway	4.8	60.2	4.74	34	4.00
15.	D. Kennedy	2.1	53.7	4.04	38	3.40
16.	K. Moore	2.8	61.5	4.77	40	4.20
17.	R. Pate	1.4	55.6	4.39	44	3.99
18.	F. Shorter	2.0	59.3	4.37	19	3.71
19.	R. Wayne	5.8	55.5	4.47	27	3.85
20.	D. Williams	1.9	64.2	4.84	29	4.10
			Good Runners			
21.		3.5	64.0	4.43	77	4.34
22.		9.2	56.3	4.45	83	4.35
23.		5.6	60.2	4.49	94	4.35
24.		4.2	67.5	4.75	83	4.64
25.		3.3	59.7	4.72	36	4.34
26.		4.7	63.9	4.96	32	4.31
27.		0.8	61.4	4.54	67	4.22
28.		5.9	64.7	4.62	80	4.36

international reputations, were divided into two groups: middle-long distance runners and marathon runners. The eight good runners were all in training but did not belong to the elite class.

The variables were (1) fat weight, (2) lean weight, (3) VO_2 during maximal running, (4) blood lactic acid content after submaximal running, and (5) VO_2 during submaximal running. The term VO_2 represents aerobic efficiency, a measure of the ability of the body to take in and process oxygen. The data appear in Table 5.3.

The eigenvalues of $\mathbf{E}^{-1}\mathbf{H}$ and the standardized discriminant function coefficients from (5.10) are in Table 5.4. A plot of (z_1, z_2) for each observation vector is given in Figure 5.3.

The first discriminant function separates the good runners from the elite, while the second separates the middle-long distance runners from the marathon runners. The runners with the two extreme scores on the second discriminant function are Frank Shorter and Steve Prefontaine, well known as outstanding runners even among those considered to be elite. Could these two exceptional runners be considered as outliers, with undue influence on the results?

Robust discriminant functions were obtained using the MLT procedure described in Section 1.10. Robust estimates $\tilde{\boldsymbol{\mu}}_i$, $\tilde{\mathbf{E}}$, and $\tilde{\mathbf{H}}$, similar to those in (1.119) and (1.120), were used. The weights w_i in (1.121) for Prefontaine and Shorter were .240 and .778, respectively. Eigenvalues and eigenvectors of $\tilde{\mathbf{E}}^{-1}\tilde{\mathbf{H}}$ were obtained, and the resulting discriminant functions are given in Table 5.5, with a plot of (z_1, z_2) for each observation vector in Figure 5.4.

Table 5.4. Means, Eigenvalues, and Discriminant Function Coefficients from a Classical Discriminant Analysis

	Group Means				
Group	Fat	Lean	Maximum VO_2	Lactic Acid	Submaximum VO_2
1	3.79	59.83	4.235	29.8	5.005
2	3.28	58.58	3.947	31.0	4.585
3	4.65	62.20	4.363	69.0	5.620

Function	Eigenvalue	Percent of Variance
1	2.635	80.53
2	.637	19.47

	Standardized Coefficients	
	1	2
Fat	−.273	−.361
Lean	−.445	−.761
Maximum VO_2	−.620	.217
Lactic Acid	−.239	.448
Submaximum VO_2	1.000	1.000

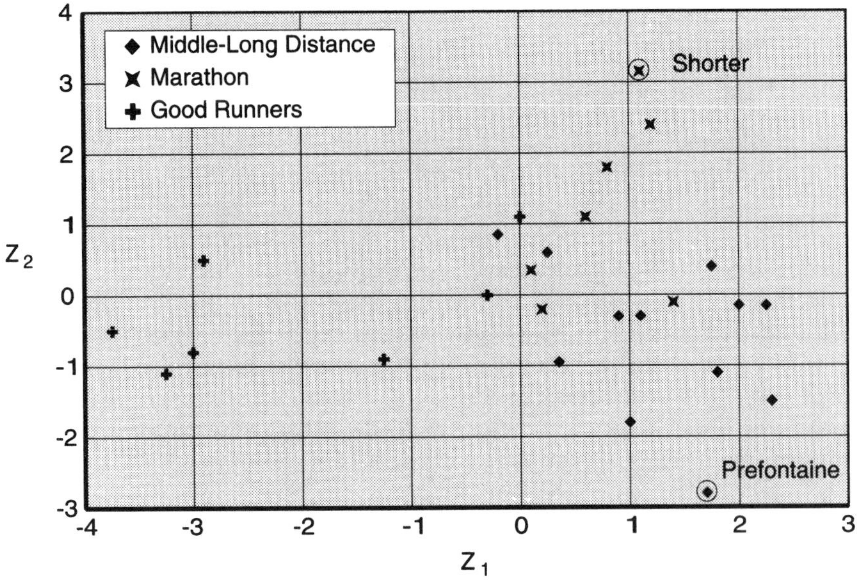

Figure 5.3. Discriminant scores based on classical estimates.

Table 5.5. Means, Eigenvalues, and Discriminant Function Coefficients Using Robust Estimates

	Group Means				
Group	Fat	Lean	Maximum VO_2	Lactic Acid	Submaximum VO_2
1	3.84	59.50	4.200	30.2	4.953
2	3.19	58.93	3.955	32.3	4.582
3	4.66	61.81	4.344	68.4	4.608

Function	Eigenvalue	Percent of Variance
1	47.68	76.45
2	14.67	23.55

Standardized Coefficients	1	2
Fat	−.300	.354
Lean	−.496	1.000
Maximum VO_2	−.571	−.656
Lactic Acid	−.292	−.254
Submaximum VO_2	1.000	−.701

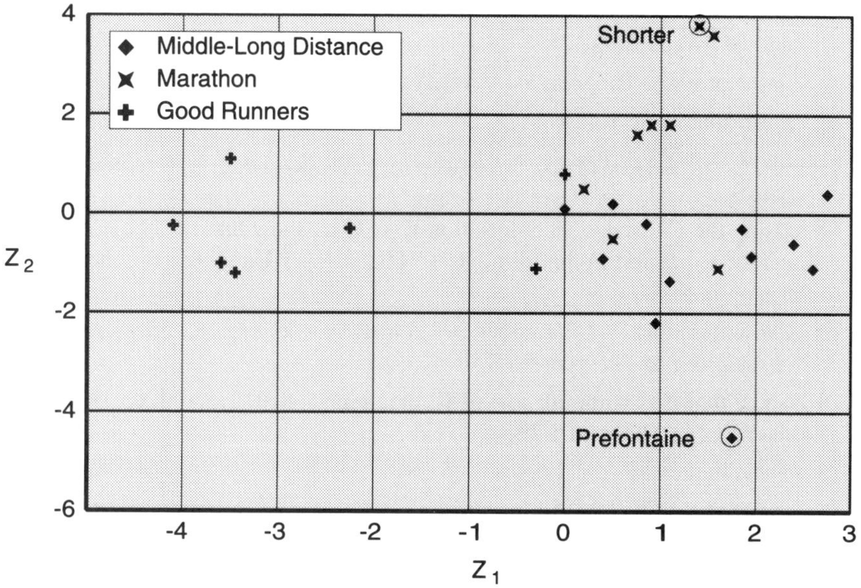

Figure 5.4. Discriminant scores based on robust estimates.

The robust results are similar to the classical analysis in Table 5.4, but with some notable differences. In Figure 5.4, Shorter and Prefontaine are now more isolated from the remaining observations, indicating that these two are perhaps not representative of the rest of the elite group. The first discriminant function is virtually unchanged in the robust version, but the coefficients in the second function show substantial differences between the robust and classical approaches.

PROBLEMS

5.1 Prove Theorem 5.2A.

5.2 Show that the discriminant function for two groups is invariant to nonsingular linear transformations $\mathbf{u} = \mathbf{Ay}$ as in (5.3).

5.3 Obtain $\lambda = \mathbf{a}'\mathbf{Ha}/\mathbf{a}'\mathbf{Ea}$ in (5.5) from $\lambda = \text{SSB}_z/\text{SSE}_z$ in (5.4).

5.4 Show that $\mathbf{a}_1'\mathbf{Ea}_2 = 0$ implies $\mathbf{a}_1'\mathbf{Ha}_2 = 0$, as used in the proof of Theorem 5.3A.

5.5 By Theorem 5.3A the within-sample correlation between z_1 and z_2 is 0, which implies $\mathbf{a}_1'\mathbf{Ea}_2 = 0$. Show that the total-sample correlation is also 0, that is, $\mathbf{a}_1'(\mathbf{E} + \mathbf{H})\mathbf{a}_2 = 0$.

5.6 Show that for the several-group case, the discriminant functions are invariant to nonsingular linear transformations as in (5.9).

5.7 Prove Theorem 5.4A.

5.8 Show that the hypothesis $H_0 : \boldsymbol{\alpha}$ is proportional to $\boldsymbol{\alpha}_0$ in (5.13) can be tested with (5.14) or (5.15).

5.9 In the proof of Theorem 5.7A, verify that $\mathbf{g}'\mathbf{S}_{\text{pl}}\mathbf{g} = s_{ii}$.

5.10 Show that $\mathbf{r}_{yz} = \mathbf{Ra}^*$ as in (5.21).

5.11 Show that $\sum_{j=1}^{s}(\mathbf{e}_i'\mathbf{a}_j/\sqrt{e_{ii}})^2\lambda_j = (1/e_{ii})\mathbf{e}_i'\left(\sum_{j=1}^{s}\mathbf{a}_j\mathbf{a}_j'\lambda_j\right)\mathbf{e}_i$, thus verifying (5.25).

5.12 Using the Cholesky decomposition $\mathbf{E} = \mathbf{U}'\mathbf{U}$, show that $(\mathbf{E}^{-1}\mathbf{H} - \lambda\mathbf{I})\mathbf{a} = \mathbf{0}$ can be expressed in the form $[(\mathbf{U}')^{-1}\mathbf{H}\mathbf{U}^{-1} - \lambda\mathbf{I}]\mathbf{U}\mathbf{a} = \mathbf{0}$, as in the proof of Theorem 5.7B.

5.13 Show that if $\mathbf{a}_j = \mathbf{U}^{-1}\mathbf{c}_j$, then $\sum_{j=1}^{s}\mathbf{a}_j\mathbf{a}_j'\lambda_j = \mathbf{U}^{-1}(\sum_{j=1}^{s}\mathbf{c}_j\mathbf{c}_j'\lambda_j)(\mathbf{U}^{-1})'$, as in the proof of Theorem 5.7B.

5.14 Show that if $\mathbf{e}_i'$ is the ith row of $\mathbf{E}$, then $\mathbf{e}_i'\mathbf{E}^{-1} = (0, \ldots, 0, 1, 0, \ldots, 0)$, as in the proof of Theorem 5.7B.

5.15 Show that $\widehat{\text{cov}}(\bar{\mathbf{z}}_{i\cdot}) = \mathbf{I}/n_i$ as in (5.28).

5.16 Show that $\text{tr}(\boldsymbol{\Sigma}^{-1}\boldsymbol{\delta}\boldsymbol{\delta}'\boldsymbol{\Sigma}^{-1}) = \boldsymbol{\delta}'(\boldsymbol{\Sigma}^{-1})^2\boldsymbol{\delta}$, as used in obtaining (5.32) from (5.31).

5.17 If $\mathbf{A}$ has eigenvalues $\lambda_1, \lambda_2, \ldots, \lambda_p$, show that $\mathbf{A} + k\mathbf{I}$ has eigenvalues $\lambda_i + k$, as used in (5.36).

5.18 For the beetles data in Table 3.8, do the following:

(a) Find the discriminant function coefficient vector.

(b) Find the standardized coefficients.

(c) Find the partial F for each variable (Section 5.7.1). Do the partial F's rank the variables in the same order of importance as the standardized coefficients?

5.19 Use the beetles data in Table 3.8.

(a) Show the breakdown of each standardized discriminant function coefficient as in Theorem 5.4A. Calculate D_u, $\hat{D}_u$, and R^2 for each coefficient.

(b) Test the hypothesis $H_0 : \boldsymbol{\alpha}' = (6, -3, -2, -3)$ as in Section 5.5.1.

5.20 Use $y_1, y_2, \ldots, y_6$ in the first two groups (treatments A and B) from the apple data in Table 4.10. Ignore blocks.

(a) Find the discriminant function coefficient vector.

(b) Find the standardized coefficients.

(c) Find the partial F for each variable (Section 5.7.1). Do the partial F's rank the variables in the same order of importance as the standardized coefficients?

5.21 Use $y_1, y_2, \ldots, y_6$ in all four groups (treatments) in the apple data in Table 4.10. Ignore blocks.

(a) Find the eigenvectors of $\mathbf{E}^{-1}\mathbf{H}$.

(b) Find the relative importance of each discriminant function as in (5.8), $\lambda_i/\sum_{j=1}^{s}\lambda_j$.

(c) Carry out tests of significance for the discriminant functions using Λ_1, Λ_2, Λ_3 as in Section 5.5.2. Do the test results agree with part (b) as to the number of important discriminant functions?

(d) Find the standardized coefficients and comment on the contribution of the variables to separation of groups.

(e) Find the partial F for each variable (Section 5.7.1). Do the partial F's rank the variables in the same order as the standardized coefficients for the first discriminant function?

5.22 Use $y_1, y_2, \ldots, y_{11}$ from the biochemical data in Table 4.11.

(a) Find the eigenvectors of $\mathbf{E}^{-1}\mathbf{H}$.

(b) Find the relative importance of each discriminant function as in (5.8), $\lambda_i / \sum_{j=1}^{s} \lambda_j$.

(c) Carry out tests of significance for the discriminant functions using Λ_1, Λ_2, Λ_3 as in Section 5.5.2. Do the test results agree with part (b) as to the number of important discriminant functions?

(d) Find the standardized coefficients and comment on the contribution of the variables to separation of groups.

(e) Find the partial F for each variable (Section 5.7.1). Do the partial F's rank the variables in the same order as the standardized coefficients for the first discriminant function?

5.23 Carry out a stepwise selection of variables for $y_1, y_2, \ldots, y_6$ in all four groups in the apple data of Table 4.10.

5.24 Carry out a stepwise selection of variables for the biochemical data in Table 4.11 using $y_1, y_2, \ldots, y_{11}$.

5.25 Use $y_1, y_2, \ldots, y_6$ in the apple data of Table 4.10 and consider blocks as a second factor.

(a) Find discriminant functions for the two main effects (treatments and blocks) and for the interaction, as in Section 5.6.

(b) Standardize the coefficients in part (a) and comment on how the variables contribute to the first discriminant function in each of the three cases.

CHAPTER 6

Classification of Observations into Groups

6.1 INTRODUCTION

In Chapter 5 we considered descriptive group separation, in which we use discriminant functions to reveal the dimensionality of separation and the contribution of the variables to separation. In this chapter, we treat *classification analysis*, the *predictive* aspect of discriminant analysis. In classification analysis, we use the variables measured on a sampling unit (subject or object) to assign the unit to a group.

Classification analysis literature is abundant; numerous papers discuss applications and theoretical developments. Useful monographs and reviews have been given by Lachenbruch (1975), Lachenbruch and Goldstein (1979), Hand (1981), James (1985), Huberty (1975, 1984, 1994), and McLachlan (1992). Hand (1992) reviewed many of the methods discussed in this chapter and discussed their application to medical diagnosis. In the literature, classification is often referred to as discriminant analysis.

6.2 TWO GROUPS

The majority of the papers in the literature are concerned with the case of two populations. Based on a vector $\mathbf{y}$ of variables measured on a sampling unit, we wish to classify the unit into one of the two populations. For convenience we will speak of "classifying $\mathbf{y}$."

6.2.1 Equal Population Covariance Matrices

We first consider the case in which $\boldsymbol{\Sigma}_1 = \boldsymbol{\Sigma}_2$. Let $f(\mathbf{y} \mid G_1)$ represent the density function for $\mathbf{y}$ from G_1 and $f(\mathbf{y} \mid G_2)$ be the density for $\mathbf{y}$ from G_2, where G_1 and G_2 are populations. [The notation $f(\mathbf{y} \mid G_1)$ does not represent a conditional distribution in the usual sense.] Let p_1 and p_2 be the *prior probabilities*, that is, the probability (before observing $\mathbf{y}$) that $\mathbf{y}$ will come from G_1 and G_2, respectively, where $p_2 = 1 - p_1$. The following result is due to Welch (1939).

Theorem 6.2A. The optimal classification rule, namely, the rule that minimizes the total probability of misclassification [see (6.27)], is: Assign y to G_1 if

$$p_1 f(\mathbf{y} \mid G_1) > p_2 f(\mathbf{y} \mid G_2) \tag{6.1}$$

and assign **y** to G_2 otherwise. □

If $p_1 = p_2$, or if the p_i's are unknown, the optimal rule in (6.1) becomes: Assign **y** to G_1 if

$$f(\mathbf{y} \mid G_1) > f(\mathbf{y} \mid G_2) \tag{6.2}$$

and to G_2 otherwise. The optimal rule in (6.2) is also called the *maximum likelihood* rule, since $f(\mathbf{y} \mid G_i)$ is the likelihood function of **y** for the ith group.

Let us apply Welch's optimal rule in Theorem 6.2A to the multivariate normal distribution with $\boldsymbol{\Sigma}_1 = \boldsymbol{\Sigma}_2 = \boldsymbol{\Sigma}$, so that $f(\mathbf{y} \mid G_1) = N_p(\boldsymbol{\mu}_1, \boldsymbol{\Sigma})$ and $f(\mathbf{y} \mid G_2) = N_p(\boldsymbol{\mu}_2, \boldsymbol{\Sigma})$, or

$$f(\mathbf{y} \mid G_i) = (2\pi)^{-p/2} |\boldsymbol{\Sigma}|^{-1/2} e^{-(\mathbf{y}-\boldsymbol{\mu}_i)'\boldsymbol{\Sigma}^{-1}(\mathbf{y}-\boldsymbol{\mu}_i)/2}, \qquad i = 1, 2. \tag{6.3}$$

If we write the inequality in (6.1) as

$$\frac{f(\mathbf{y} \mid G_1)}{f(\mathbf{y} \mid G_2)} > \frac{p_2}{p_1}, \tag{6.4}$$

then using $f(\mathbf{y} \mid G_i)$ in (6.3) for the multivariate normal, we obtain

$$\frac{f(\mathbf{y} \mid G_1)}{f(\mathbf{y} \mid G_2)} = e^{-(\mathbf{y}-\boldsymbol{\mu}_1)'\boldsymbol{\Sigma}^{-1}(\mathbf{y}-\boldsymbol{\mu}_1)/2 + (\mathbf{y}-\boldsymbol{\mu}_2)'\boldsymbol{\Sigma}^{-1}(\mathbf{y}-\boldsymbol{\mu}_2)/2}. \tag{6.5}$$

This can be shown to be equal to

$$\frac{f(\mathbf{y} \mid G_1)}{f(\mathbf{y} \mid G_2)} = e^{(\boldsymbol{\mu}_1-\boldsymbol{\mu}_2)'\boldsymbol{\Sigma}^{-1}\mathbf{y} - (\boldsymbol{\mu}_1-\boldsymbol{\mu}_2)'\boldsymbol{\Sigma}^{-1}(\boldsymbol{\mu}_1+\boldsymbol{\mu}_2)/2} \tag{6.6}$$

(see Problem 6.2). Taking logarithms in (6.4) and (6.6), the optimal rule in (6.1) becomes: Assign **y** to G_1 if

$$(\boldsymbol{\mu}_1 - \boldsymbol{\mu}_2)'\boldsymbol{\Sigma}^{-1}\mathbf{y} > \frac{1}{2}(\boldsymbol{\mu}_1 - \boldsymbol{\mu}_2)'\boldsymbol{\Sigma}^{-1}(\boldsymbol{\mu}_1 + \boldsymbol{\mu}_2) + \ln\left(\frac{p_2}{p_1}\right) \tag{6.7}$$

and to G_2 otherwise.

Since $\boldsymbol{\mu}_1$, $\boldsymbol{\mu}_2$, and $\boldsymbol{\Sigma}$ are ordinarily unknown, we use estimators $\overline{\mathbf{y}}_1$, $\overline{\mathbf{y}}_2$, and $\mathbf{S}_{\text{pl}}$ based on samples of size n_1 and n_2 that are known to have come from G_1 and G_2. Then an observation **y** of unknown origin can be classified as follows: Assign **y** to G_1 if

$$(\overline{\mathbf{y}}_1 - \overline{\mathbf{y}}_2)'\mathbf{S}_{\text{pl}}^{-1}\mathbf{y} > \frac{1}{2}(\overline{\mathbf{y}}_1 - \overline{\mathbf{y}}_2)'\mathbf{S}_{\text{pl}}^{-1}(\overline{\mathbf{y}}_1 + \overline{\mathbf{y}}_2) + \ln\left(\frac{p_2}{p_1}\right) \tag{6.8}$$

and assign $\mathbf{y}$ to G_2 otherwise. Since the left side of (6.8) is a linear function of $\mathbf{y}$, (6.8) is called a *linear classification rule*. (It is sometimes referred to as a linear discriminant rule.) Note that the linear function $(\overline{\mathbf{y}}_1 - \overline{\mathbf{y}}_2)'\mathbf{S}_{\text{pl}}^{-1}\mathbf{y}$ is the same as the discriminant function for two groups in Theorem 5.2A. However, for the case of unequal covariance matrices, $\boldsymbol{\Sigma}_1 \neq \boldsymbol{\Sigma}_2$, in Section 6.2.2, we do not use the discriminant functions of Section 5.3.1 for classification.

The sample-based rule in (6.8) that uses $\overline{\mathbf{y}}_1, \overline{\mathbf{y}}_2$, and $\mathbf{S}_{\text{pl}}$ is not optimal as is (6.7), which is based on $\boldsymbol{\mu}_1, \boldsymbol{\mu}_2$, and $\boldsymbol{\Sigma}$. The sample-based rule is only *asymptotically optimal*; that is, it approaches optimality (minimizes the total probability of misclassification) as n_1 and n_2 increase (Siotani et al. 1985, p. 391). For classification rules that use estimated parameters, at present there is no general guideline that is analogous to Welch's (1939) optimal rule in (6.1) and (6.2) for completely known densities. We therefore use the asymptotically optimal rule. In the remainder of this chapter, we will not always distinguish between optimality and asymptotic optimality.

6.2.2 Unequal Population Covariance Matrices

If $\boldsymbol{\Sigma}_1 \neq \boldsymbol{\Sigma}_2$ for the two populations, the logarithm of the density ratio for the multivariate normal is

$$Q(\mathbf{y}) = \ln \frac{f(\mathbf{y} \mid G_1)}{f(\mathbf{y} \mid G_2)}$$

$$= \frac{1}{2}\ln\left(\frac{|\boldsymbol{\Sigma}_2|}{|\boldsymbol{\Sigma}_1|}\right) - \frac{1}{2}(\mathbf{y} - \boldsymbol{\mu}_1)'\boldsymbol{\Sigma}_1^{-1}(\mathbf{y} - \boldsymbol{\mu}_1) + \frac{1}{2}(\mathbf{y} - \boldsymbol{\mu}_2)'\boldsymbol{\Sigma}_2^{-1}(\mathbf{y} - \boldsymbol{\mu}_2) \quad (6.9)$$

$$= \frac{1}{2}\ln\left(\frac{|\boldsymbol{\Sigma}_2|}{|\boldsymbol{\Sigma}_1|}\right) - \frac{1}{2}(\boldsymbol{\mu}_1'\boldsymbol{\Sigma}_1^{-1}\boldsymbol{\mu}_1 - \boldsymbol{\mu}_2'\boldsymbol{\Sigma}_2^{-1}\boldsymbol{\mu}_2)$$

$$+(\boldsymbol{\mu}_1'\boldsymbol{\Sigma}_1^{-1} - \boldsymbol{\mu}_2'\boldsymbol{\Sigma}_2^{-1})\mathbf{y} - \frac{1}{2}\mathbf{y}'(\boldsymbol{\Sigma}_1^{-1} - \boldsymbol{\Sigma}_2^{-1})\mathbf{y}. \quad (6.10)$$

Since the last term, $\mathbf{y}'(\boldsymbol{\Sigma}_1^{-1} - \boldsymbol{\Sigma}_2^{-1})\mathbf{y}$, involves squares and products of the y's, $Q(\mathbf{y})$ is a *quadratic classification function*. Many writers refer to $Q(\mathbf{y})$ as a quadratic discriminant function.

By Theorem 6.2A the optimal classification rule for $Q(\mathbf{y})$ is: Assign $\mathbf{y}$ to G_1 if

$$Q(\mathbf{y}) > \ln\left(\frac{p_2}{p_1}\right) \quad (6.11)$$

and to G_2 otherwise. If samples are available from G_1 and G_2, the sample analogue of $Q(\mathbf{y})$ is obtained by replacing $\boldsymbol{\mu}_i$ by $\overline{\mathbf{y}}_i$ and $\boldsymbol{\Sigma}_i$ by $\mathbf{S}_i$ (the sample covariance matrix for the ith group):

$$Q_s(\mathbf{y}) = \frac{1}{2}\ln\left(\frac{|\mathbf{S}_2|}{|\mathbf{S}_1|}\right) - \frac{1}{2}(\overline{\mathbf{y}}_1'\mathbf{S}_1^{-1}\overline{\mathbf{y}}_1 - \overline{\mathbf{y}}_2'\mathbf{S}_2^{-1}\overline{\mathbf{y}}_2) + (\overline{\mathbf{y}}_1'\mathbf{S}_1^{-1} - \overline{\mathbf{y}}_2'\mathbf{S}_2^{-1})\mathbf{y}$$

$$- \frac{1}{2}\mathbf{y}'(\mathbf{S}_1^{-1} - \mathbf{S}_2^{-1})\mathbf{y}, \quad (6.12)$$

which is of the form

$$Q_s(\mathbf{y}) = b + \mathbf{c}'\mathbf{y} - \mathbf{y}'\mathbf{A}\mathbf{y}.$$

The classification rule for $Q_s(\mathbf{y})$ that corresponds to (6.11) is: Assign $\mathbf{y}$ to G_1 if

$$Q_s(\mathbf{y}) > \ln\left(\frac{p_2}{p_1}\right) \tag{6.13}$$

and to G_2 otherwise. Note that $\ln(p_2/p_1) = 0$ if $p_1 = p_2$.

When $\boldsymbol{\Sigma}_1 \neq \boldsymbol{\Sigma}_2$, the quadratic classification function $Q_s(\mathbf{y})$ in (6.12) is asymptotically optimal. However, for small samples, $\mathbf{S}_i$ is not a stable estimator of $\boldsymbol{\Sigma}_i$; that is, $\mathbf{S}_i$ would vary widely in repeated sampling from the same population (group). In such cases, the linear classification function will often do better than $Q_s(\mathbf{y})$. For large sample sizes and large differences between $\boldsymbol{\Sigma}_1$ and $\boldsymbol{\Sigma}_2$, the quadratic classification function is preferred. These conclusions are supported in studies by Marks and Dunn (1974), Wahl and Kronmal (1977), Van Ness (1979), Aitchison, Habbema, and Kay (1977), and Remme, Habbema, and Hermans (1980). Han and Huang (1987) studied the following procedure: test $H_0 : \boldsymbol{\Sigma}_1 = \boldsymbol{\Sigma}_2$; if H_0 is accepted, use the linear classification function; otherwise, use the quadratic classification function. For $p = 1$ or 2 variables, the preliminary test had little effect on error rates. For $p = 5$ variables, Han and Huang suggested using α equal to .10 and .50, where α is the probability of a Type I error in testing H_0. It would be of interest to investigate this question further with additional values of p and α.

Wakaki (1990) compared the error rates (see Section 6.4) of linear and quadratic classification functions for two groups and gave some guidelines in terms of sample size for choosing between the two methods.

6.2.3 Unequal Costs of Misclassification

If misclassification into one of the groups is more serious than into the other, we can assign different costs to these two misclassifications. Suppose, for example, that G_1 consists of those persons who have a certain type of cancer and that G_2 indicates those who do not have the disease. Then misclassifying a member of G_1 into G_2 is a more serious error than misclassifying a member of G_2 into G_1.

Let c_1 and c_2 be the costs of misclassifying a member of G_1 and G_2, respectively. Then the rule that minimizes the total expected costs of misclassification is: Assign $\mathbf{y}$ to G_1 if

$$c_1 p_1 f(\mathbf{y} \mid G_1) > c_2 p_2 f(\mathbf{y} \mid G_2) \tag{6.14}$$

and to G_2 otherwise. Since c_1 and c_2 are usually unknown and experimenters may be hesitant to assign them subjectively, this method is probably not used as often as it could be. Note that (6.14) reduces to (6.1) if $c_1 = c_2$.

Rudolph and Karson (1988) investigated the consequences of assuming equal prior probabilities and equal misclassification costs when these assumptions are not true.

6.2.4 Posterior Probability Approach

The prior probabilities p_1 and p_2 are defined in Section 6.2.1 as the probability, before observing $\mathbf{y}$, that $\mathbf{y}$ will arise from G_1 and G_2, respectively. The *posterior probability* is defined as the conditional probability, given an observed value of $\mathbf{y}$, that $\mathbf{y}$ came from G_1 or G_2. The posterior probability of G_2 can be obtained from an application of Bayes' Theorem:

$$\begin{aligned} P(G_2 \mid \mathbf{y}) &= \frac{P(G_2) f(\mathbf{y} \mid G_2)}{P(G_1) f(\mathbf{y} \mid G_1) + P(G_2) f(\mathbf{y} \mid G_2)} \\ &= \frac{p_2 f(\mathbf{y} \mid G_2)}{p_1 f(\mathbf{y} \mid G_1) + p_2 f(\mathbf{y} \mid G_2)}, \end{aligned} \tag{6.15}$$

where $p_1 = P(G_1)$ and $p_2 = P(G_2)$; also we have used $f(\mathbf{y} \mid G_i)$ in place of $P(\mathbf{y} \mid G_i)$. Similarly,

$$P(G_1 \mid \mathbf{y}) = \frac{p_1 f(\mathbf{y} \mid G_1)}{p_1 f(\mathbf{y} \mid G_1) + p_2 f(\mathbf{y} \mid G_2)}. \tag{6.16}$$

In terms of posterior probabilities, the classification rule is: Assign $\mathbf{y}$ to G_1 if

$$P(G_1 \mid \mathbf{y}) > P(G_2 \mid \mathbf{y})$$

and to G_2 otherwise. Since the denominators of (6.15) and (6.16) are the same, this rule becomes: Assign $\mathbf{y}$ to G_1 if

$$p_1 f(\mathbf{y} \mid G_1) > p_2 f(\mathbf{y} \mid G_2),$$

which is the same as (6.1).

6.2.5 Robustness to Departures from the Assumptions

Under the assumptions of multivariate normality and $\mathbf{\Sigma}_1 = \mathbf{\Sigma}_2$, the linear classification rule (6.8) is optimal. Lachenbruch (1975, Chapter 3), Krzanowski (1977), Seber (1984, pp. 279–300), and Huberty (1994, pp. 63–65) have reviewed the robustness of the linear rule to violation of various assumptions.

In the case of nonnormal continuous data, linear classification functions are moderately robust to a low level of skewness but not to highly skewed distributions (Lachenbruch et al. 1973, Crawley 1979, Chinganda and Subrahmaniam 1979, Subrahmaniam and Chinganda 1978). The linear functions are fairly robust to long-tailed symmetric distributions and to mixtures of normals (Ashikaga and Chang 1981). Beauchamp, Folkert, and Robson (1980) and Beauchamp and Robson (1986) discussed the use of transformations of nonnormal data to achieve approximate normality; in particular, they considered the power transformation of Box and Cox (1964). In their studies, which were limited to univariate and bivariate distributions for two groups, they found virtually no improvement in error rates after transforming the data.

When the data are discrete, the normality assumption does not hold. Lachenbruch (1975) reviewed several studies and concluded that "the general indications seem to be that the linear discriminant function performs fairly well on discrete data of various types"(p. 45). Further support for this conclusion is given by Titterington et al. (1981) and Gilbert (1968). However, Moore (1973) noted a problem with binary data in certain configurations.

In this chapter we have assumed that there are no misclassifications in the initial samples used to compute the classification functions. If there are initial misclassifications, the performance of the classification function in future samples may be seriously impaired. Lachenbruch (1966, 1974, 1979) and McLachlan (1972, 1992, pp. 35–37) have studied the problem for two groups, and Aitchison and Begg (1976) have studied it for several groups. Lachenbruch obtained the optimum error rate in terms of α_1 and α_2, the proportions of initial misclassifications in the two groups, and showed that if $\alpha_1 = \alpha_2$, the optimum error rate is unaffected. However, the apparent error rate is greatly influenced by initial misclassification. (The optimum error and apparent error rate are defined in Section 6.4.) The error rates for quadratic classification functions are even more adversely affected by initial misclassifications than are those for the linear rule.

6.2.6 Robust Procedures

Robust classification functions can easily be constructed based on the robust estimators of $\boldsymbol{\mu}_i$ and $\boldsymbol{\Sigma}$ from Section 1.10. If $\tilde{\boldsymbol{\mu}}_1, \tilde{\boldsymbol{\mu}}_2$, and $\tilde{\mathbf{S}}_{\text{pl}}$ are the robust estimators for the two-sample case, analogous to those defined in (1.119) and (1.120) for a single sample, a robust linear classification rule based on (6.8) is: Assign $\mathbf{y}$ to G_1 if

$$(\tilde{\boldsymbol{\mu}}_1 - \tilde{\boldsymbol{\mu}}_2)'\tilde{\mathbf{S}}_{\text{pl}}^{-1}\mathbf{y} > \frac{1}{2}(\tilde{\boldsymbol{\mu}}_1 - \tilde{\boldsymbol{\mu}}_2)'\tilde{\mathbf{S}}_{\text{pl}}^{-1}(\tilde{\boldsymbol{\mu}}_1 + \tilde{\boldsymbol{\mu}}_2) + \ln\left(\frac{p_1}{p_2}\right) \tag{6.17}$$

and to G_2 otherwise. A robust quadratic classification rule analogous to (6.12) can be defined similarly. Delaney (1987) found that a robust approach performed somewhat better than a classical method on a sample of failed and healthy banks. Another approach (Broffitt 1982, pp. 157–159) is to find a robust estimator of $\mathbf{a}$ in Fisher's ratio $[\mathbf{a}'(\bar{\mathbf{y}}_1 - \bar{\mathbf{y}}_2)]^2/\mathbf{a}'\mathbf{S}_{\text{pl}}\mathbf{a}$ in (5.1) and to use this version of $\mathbf{a}$ in place of $(\tilde{\boldsymbol{\mu}}_1 - \tilde{\boldsymbol{\mu}}_2)'\tilde{\mathbf{S}}_{\text{pl}}^{-1}$ in (6.17). Randles et al. (1978) compared some robust rules in a Monte Carlo study.

Krusinska (1988) reviewed robust methods that had been applied to linear and quadratic classification functions and to selection of variables. Most of her discussion centered on the two-group case, but she also considered the several-group setting. Hu et al. (1988) applied robust methods to classification of some marketing data and found improved error rates. Krusinska and Liebhart (1990b) applied robust methods to medical data that consisted of three groups (disease categories). The data contained outliers, which were downweighted by the robust approach. The robust methods improved diagnosis.

6.3 SEVERAL GROUPS

We now turn to the case of sampling from k groups $G_1, G_2, \ldots, G_k$. In Section 6.3.1 we assume that $\mathbf{\Sigma}_1 = \mathbf{\Sigma}_2 = \cdots = \mathbf{\Sigma}_k$, and in Section 6.3.2 we consider unequal covariance matrices.

6.3.1 Equal Population Covariance Matrices

When sampling from several normal populations with equal covariance matrices, the optimal classification rule leads to linear classification functions. We now obtain these linear classification functions.

If $p_1, p_2, \ldots, p_k$ are the prior probabilities that an observation $\mathbf{y}$ arises from $G_1, G_2, \ldots, G_k$, respectively, the optimum classification rule for known densities is an immediate extension of (6.1): Assign $\mathbf{y}$ to G_i if

$$p_i f(\mathbf{y} \mid G_i) \geq p_j f(\mathbf{y} \mid G_j) \text{ for all } j = 1, 2, \ldots, k, \tag{6.18}$$

that is, if $p_i f(\mathbf{y} \mid G_i) = \max_j p_j f(\mathbf{y} \mid G_j)$. Maximizing $p_i f(\mathbf{y} \mid G_i)$ is equivalent to maximizing $\ln[p_i f(\mathbf{y} \mid G_i)]$. Using $N_p(\boldsymbol{\mu}_i, \mathbf{\Sigma})$ for $f(\mathbf{y} \mid G_i)$ and taking logarithms, we obtain

$$\ln[p_i f(\mathbf{y} \mid G_i)] = \ln p_i - \tfrac{1}{2} p \ln(2\pi) - \tfrac{1}{2} \ln |\mathbf{\Sigma}| - \tfrac{1}{2}(\mathbf{y} - \boldsymbol{\mu}_i)' \mathbf{\Sigma}^{-1} (\mathbf{y} - \boldsymbol{\mu}_i), \tag{6.19}$$

where $\mathbf{\Sigma}$ is the common covariance matrix. Note that p_i is a prior probability, while p represents the number of variables. If we expand the last term of (6.19) and delete terms common to all groups (terms that do not involve i), we obtain $\ln p_i + \boldsymbol{\mu}_i' \mathbf{\Sigma}^{-1} \mathbf{y} - \frac{1}{2} \boldsymbol{\mu}_i' \mathbf{\Sigma}^{-1} \boldsymbol{\mu}_i$. Assigning $\mathbf{y}$ to the group that maximizes this expression provides the optimal rule. Substituting estimators of $\boldsymbol{\mu}_i$ and $\mathbf{\Sigma}$, we have the linear function

$$L_i(\mathbf{y}) = \ln p_i + \overline{\mathbf{y}}_i' \mathbf{S}_{\text{pl}}^{-1} \mathbf{y} - \tfrac{1}{2} \overline{\mathbf{y}}_i' \mathbf{S}_{\text{pl}}^{-1} \overline{\mathbf{y}}_i, \tag{6.20}$$

where $\mathbf{S}_{\text{pl}} = \mathbf{E}/(N - k)$ is defined in (4.85). We assign $\mathbf{y}$ to the group with maximum value of $L_i(\mathbf{y})$. This rule is asymptotically optimal. Note that the linear functions $L_i(\mathbf{y})$ are not the same as the linear discriminant functions $z_i = \mathbf{a}_i' \mathbf{y}$ in Section 5.3 based on eigenvectors of $\mathbf{E}^{-1}\mathbf{H}$.

If $p_1 = p_2 = \cdots = p_k$, the rule in (6.18) can be called a *maximum likelihood rule*, since $f(\mathbf{y} \mid G_i)$ is the likelihood function for $\mathbf{y}$ for the ith group.

The posterior probability rule for two groups discussed in Section 6.2.4 can be readily extended to several groups. The posterior probability of the ith group, given the observed value of $\mathbf{y}$, is

$$P(G_i \mid \mathbf{y}) = \frac{p_i f(\mathbf{y} \mid G_i)}{\sum_{j=1}^{k} p_j f(\mathbf{y} \mid G_j)}. \tag{6.21}$$

Note that assigning $\mathbf{y}$ to the group with maximum posterior probability is clearly equivalent to maximizing $p_i f(\mathbf{y} \mid G_i)$, as in the optimum rule in (6.18).

Some software programs provide a posterior probability for each observation $\mathbf{y}_{ij}$. These are typically computed using (6.21) with estimators of $\boldsymbol{\mu}_i$ and $\boldsymbol{\Sigma}$ substituted in the multivariate normal density $f(\mathbf{y} \mid G_i) = N_p(\boldsymbol{\mu}_i, \boldsymbol{\Sigma})$:

$$P(G_i \mid \mathbf{y}) = \frac{p_i e^{-D_i^2/2}}{\sum_{j=1}^{k} p_j e^{-D_j^2/2}}, \tag{6.22}$$

where $D_i^2 = (\mathbf{y} - \overline{\mathbf{y}}_i)'\mathbf{S}_{\text{pl}}^{-1}(\mathbf{y} - \overline{\mathbf{y}}_i)$.

Example 6.3.1. For the football data discussed in Example 5.7.2, we illustrate the posterior probability (6.22) with $p_1 = p_2 = p_3$ for the first and third observations in group 1. For the first observation, $\mathbf{y}_{11}$, we obtain

$$P(G_1 \mid \mathbf{y}_{11}) = \frac{e^{-D_1^2/2}}{\sum_{j=1}^{3} e^{-D_j^2/2}} = .9503,$$

$$P(G_2 \mid \mathbf{y}_{11}) = \frac{e^{-D_2^2/2}}{\sum_{j=1}^{3} e^{-D_j^2/2}} = .0169,$$

$$P(G_3 \mid \mathbf{y}_{11}) = \frac{e^{-D_3^2/2}}{\sum_{j=1}^{3} e^{-D_j^2/2}} = .0328.$$

We thus classify $\mathbf{y}_{11}$ as a member of group 1. For the third observation, $\mathbf{y}_{13}$, we obtain

$$P(G_1 \mid \mathbf{y}_{13}) = .0291, \quad P(G_2 \mid \mathbf{y}_{13}) = .7380, \quad P(G_3 \mid \mathbf{y}_{13}) = .2329,$$

and misclassify $\mathbf{y}_{13}$ as a member of group 2.

6.3.2 Unequal Population Covariance Matrices

If linear classification procedures are used for populations with unequal covariance matrices, observations tend to be classified too frequently into groups with larger variances. However, in the case of unequal population covariance matrices, the classification rules can easily be altered to preserve optimality of classification rates.

Assuming p-variate normal populations with unequal covariance matrices, $\boldsymbol{\Sigma}_1, \boldsymbol{\Sigma}_2, \ldots, \boldsymbol{\Sigma}_k$, we use $N_p(\boldsymbol{\mu}_i, \boldsymbol{\Sigma}_i)$ as $f(\mathbf{y} \mid G_i)$ in $\ln[p_i f(\mathbf{y} \mid G_i)]$:

$$\ln[p_i f(\mathbf{y} \mid G_i)] = \ln p_i - \tfrac{1}{2} p \ln(2\pi) - \tfrac{1}{2} \ln |\boldsymbol{\Sigma}_i| - \tfrac{1}{2}(\mathbf{y} - \boldsymbol{\mu}_i)'\boldsymbol{\Sigma}_i^{-1}(\mathbf{y} - \boldsymbol{\mu}_i).$$

Using the sample mean vector $\overline{\mathbf{y}}_i$ and the sample covariance matrix $\mathbf{S}_i$ from the ith group in place of $\boldsymbol{\mu}_i$ and $\boldsymbol{\Sigma}_i$ and deleting $-(p/2)\ln(2\pi)$, we obtain the *quadratic classification function*,

$$\begin{aligned} Q_i(\mathbf{y}) &= \ln p_i - \tfrac{1}{2} \ln |\mathbf{S}_i| - \tfrac{1}{2}(\mathbf{y} - \overline{\mathbf{y}}_i)'\mathbf{S}_i^{-1}(\mathbf{y} - \overline{\mathbf{y}}_i) \\ &= \ln p_i - \tfrac{1}{2} \ln |\mathbf{S}_i| - \tfrac{1}{2}\overline{\mathbf{y}}_i'\mathbf{S}_i^{-1}\overline{\mathbf{y}}_i + \overline{\mathbf{y}}_i'\mathbf{S}_i^{-1}\mathbf{y} - \tfrac{1}{2}\mathbf{y}'\mathbf{S}_i^{-1}\mathbf{y}. \end{aligned} \tag{6.23}$$

The classification rule is: Assign $\mathbf{y}$ to the group for which $Q_i(\mathbf{y})$ is largest. If $p_1 = p_2 = \cdots = p_k$ or if the p_i's are unknown, the term $\ln p_i$ is deleted. In order for $\mathbf{S}_i^{-1}$ to exist, we must have $n_i > p$, $i = 1, 2, \ldots, k$.

For multivariate normal populations with unequal $\boldsymbol{\Sigma}_i$, the posterior probability (6.21) (using estimates of $\boldsymbol{\mu}_i$ and $\boldsymbol{\Sigma}_i$) becomes

$$P(G_i \mid \mathbf{y}) = \frac{p_i |\mathbf{S}_i|^{-1/2} e^{-D_i^2/2}}{\sum_{j=1}^k p_j |\mathbf{S}_j|^{-1/2} e^{-D_j^2/2}}, \tag{6.24}$$

where $D_i^2 = (\mathbf{y} - \overline{\mathbf{y}}_i)' \mathbf{S}_i^{-1} (\mathbf{y} - \overline{\mathbf{y}}_i)$.

The normal-based classification rules in this section and Section 6.3.1 allow for the use of prior probabilities. If the p_i's are known, the error rates will generally be improved, but in the majority of applications the p_i's are not available. Many programs provide an option to estimate each p_i as proportional to the sample size n_i. However, this option should not be used unless the sample proportions represent the population proportions.

For large samples, quadratic classification functions will often perform better than linear functions; that is, they will have smaller error rates (see Section 6.4 for a discussion of error rates). For small samples, however, the results from quadratic rules will be less stable across repeated sampling than the results from linear rules, because far more parameters are estimated in $\mathbf{S}_1, \mathbf{S}_2, \ldots, \mathbf{S}_k$ than in $\mathbf{S}_{\text{pl}}$ and because each $\mathbf{S}_i$ is based on fewer degrees of freedom than $\mathbf{S}_{\text{pl}}$. Quadratic rules are also more sensitive to nonnormality (Michaelis 1973, Huberty and Curry 1978).

Velilla and Barrio (1994) suggested a transformation of the data to approximate normality for either the linear classification rules $L_i(\mathbf{y})$ in (6.20) or the quadratic rules $Q_i(\mathbf{y})$ in (6.23). They noted that this is especially appropriate for long-tailed marginal distributions.

In an attempt to improve on the performance of quadratic rules, Greene and Rayens (1989) proposed replacing $\mathbf{S}_i$ in (6.23) by

$$\mathbf{S}_i(w_i) = w_i \mathbf{S}_i + (1 - w_i) \mathbf{S}_{\text{pl}}, \tag{6.25}$$

where $0 \leq w_i \leq 1$. The w_i's for (6.25) are chosen by an empirical Bayes method in which a large variability among $\mathbf{S}_1, \mathbf{S}_2, \ldots, \mathbf{S}_k$ or a large n_i leads to large w_i, whereas a small variability of the $\mathbf{S}_i$'s or a small n_i favors small w_i. In a simulation study, Greene and Rayens found that their method based on (6.25) often performed better than either the linear or quadratic classification functions.

Friedman (1989) proposed a further adjustment for $\mathbf{S}_i$ to be used in (6.23):

$$\mathbf{S}_i(\lambda, \gamma) = (1 - \gamma) \mathbf{S}_i(\lambda) + \frac{\gamma}{p} [\operatorname{tr} \mathbf{S}_i(\lambda)] \mathbf{I}, \tag{6.26}$$

where $\mathbf{S}_i(\lambda) = (1 - \lambda) \mathbf{S}_i + \lambda \mathbf{S}_{\text{pl}}$ as in (6.25). This approach is similar to ridge regression; for a given value of λ, the additional parameter γ shrinks the estimate toward a multiple of the identity matrix $\mathbf{I}$. The coefficient of $\mathbf{I}$ is $\text{tr}\,\mathbf{S}_i(\lambda)/p$, the average eigenvalue of $\mathbf{S}_i(\lambda)$. This shrinkage toward $\mathbf{I}$ has the effect of reducing the variation in $\mathbf{S}_i^{-1}$ in (6.23) due to a small sample size n_i relative to the number of variables p. Friedman called the method based on (6.26) *regularized discriminant analysis* (note the use of *discriminant* in place of *classification*). The parameters λ and γ are chosen jointly to minimize future rates of misclassification estimated by the holdout method (see Section 6.5.2). Friedman recommended this procedure when the n_i's are small relative to p. In a simulation study, he found his method to be better than linear or quadratic classification in many situations.

Rayens and Greene (1991) compared their 1989 procedure (reviewed above) with Friedman's (1989) method. They found that in some cases the additional complexity of Friedman's method does not lead to improvement. Higbee (1994) further generalized Friedman's method by allowing the weighting and regularization parameters to vary for each group. Higbee et al. (1996) compared several classification methods using various criteria such as simplicity, robustness, applicability, and performance.

Feiveson (1983) suggested a method for reducing the computational time for large data sets such as satellite or aircraft remote-sensing multispectral images. If $Q_i(\mathbf{y})$ in (6.23) exceeds a *threshold*, certain other $Q_i(\mathbf{y})$ values need not be calculated. Yau and Manry (1992) proposed the use of a different threshold T_i, $i = 1, 2, \ldots, k$, for each group. The classification rule associated with (6.23) is modified as follows: Assign $\mathbf{y}$ to the group for which $Q_i(\mathbf{y})$ is largest, provided that $Q_i(\mathbf{y}) \geq T_i$. In an example with a large data set, Yau and Manry's procedure had fewer misclassifications than the use of $Q_i(\mathbf{y})$ without thresholds.

6.3.3 Use of Linear Discriminant Functions for Classification

Green (1979) and Williams (1982) have demonstrated a direct mathematical relationship between the linear discriminant functions $z_i = \mathbf{a}_i'\mathbf{y}$ based on eigenvectors of $\mathbf{E}^{-1}\mathbf{H}$ (Section 5.3.1) and the linear classification functions in (6.20) based on the assumption of multivariate normality, homogeneous covariance matrices, and equal prior probabilities. Some writers (for example, Johnson and Wichern 1992, pp. 547–552) suggest that the linear discriminant functions $z_i = \mathbf{a}_i'\mathbf{y}$ be used in classification. This is done indirectly by calculating the $s = \min(p, k - 1)$ discriminant function scores $z_1, z_2, \ldots, z_s$ for each observation vector and then using these as raw data input in a classification procedure. A possible advantage of the use of discriminant functions is that s will often be much less than p, with a resulting computational efficiency. Some software programs have adopted this approach.

Kshirsagar and Arseven (1975) showed that if all s linear discriminant functions are used in the above procedure, the classification results will be identical to the results using the original variables. If fewer than s discriminant functions are used in the classification analysis, the results may not be the same. We can also expect the results to differ if linear discriminant scores are used in place of original variables as input to the quadratic classification functions (6.23).

6.4 ESTIMATION OF ERROR RATES

After obtaining a classification rule, it is natural to inquire how well the rule will classify observations into the correct groups. That is, we wish to know the *correct classification rate*, the probability that the rule will classify an observation into the correct group. Alternatively, we could work with its complement, the misclassification rate, commonly called the *error rate*. To be more precise, our interest is in the probability that our classification functions based on a set of data will misclassify a future observation $\mathbf{y}$. This probability is usually referred to as the *actual error rate* but is also known as the *conditional error rate*. Our discussion in this section is largely limited to the case of two groups.

For two groups, misclassification can occur in two ways. Hence

$$\begin{aligned}\text{Actual error rate} &= P(\text{misclassification into either group})\\ &= p_1 P(\text{classify as } G_2 \,|\, G_1) + p_2 P(\text{classify as } G_1 \,|\, G_2),\end{aligned} \tag{6.27}$$

where p_1 and p_2 are the prior probabilities and $P(\text{classify as } G_2 \,|\, G_1)$ means the probability of classifying $\mathbf{y}$ as a member of G_2 when $\mathbf{y}$ really came from G_1, with an analogous definition for $P(\text{classify as } G_1 \,|\, G_2)$. The classification procedure could be (6.8), (6.13), or a similar sample-based method that involves estimates of parameters.

The definition in (6.27) is for a classification procedure based on one particular sample. We might also be interested in the average error rate based on all possible samples:

$$\begin{aligned}\text{Expected actual error rate} &= p_1 E\left[P(\text{classify as } G_2 \,|\, G_1)\right]\\ &\quad + p_2 E\left[P(\text{classify as } G_1 \,|\, G_2)\right].\end{aligned} \tag{6.28}$$

The *expected actual error rate* is called the *true error rate* by some writers (Hand 1981, p. 187).

To evaluate the probabilities in (6.27) and (6.28), we would need to know the population parameters and assume a density, such as the normal. In most cases population parameters are unknown; we therefore consider some estimators of error rates.

The *plug-in* estimator of error rate is an estimator of the *optimum error rate* based on the optimum rule in (6.7),

$$\begin{aligned}\text{Optimum error rate (known parameters)} &= p_1 P(\text{classify as } G_2 \,|\, G_1)\\ &\quad + p_2 P(\text{classify as } G_1 \,|\, G_2).\end{aligned} \tag{6.29}$$

We assume the multivariate normal distribution with known $\boldsymbol{\mu}_1$, $\boldsymbol{\mu}_2$, and $\boldsymbol{\Sigma}$ in (6.29) and then "plug in" estimators of these parameters. Note that the optimum error rate in (6.29) has the same appearance as (6.27). However, in (6.27), $P(\text{classify as } G_2 \,|\, G_1)$ and $P(\text{classify as } G_1 \,|\, G_2)$ are based on (6.8) using estimates from a sample, whereas in (6.29) they are based on the optimum rule (6.7) using known parameters.

To evaluate (6.29) for the multivariate normal using (6.7), we have

$$P(\text{classify as } G_1 \mid G_2) = P\left[\boldsymbol{\alpha}'\mathbf{y} > \frac{1}{2}(\boldsymbol{\mu}_1 - \boldsymbol{\mu}_2)'\boldsymbol{\Sigma}^{-1}(\boldsymbol{\mu}_1 + \boldsymbol{\mu}_2) + \ln\left(\frac{p_2}{p_1}\right)\right], \tag{6.30}$$

where $\boldsymbol{\alpha}'\mathbf{y} = (\boldsymbol{\mu}_1 - \boldsymbol{\mu}_2)'\boldsymbol{\Sigma}^{-1}\mathbf{y}$. By Theorem 2.2B, if $\mathbf{y}$ is from G_2, then $\boldsymbol{\alpha}'\mathbf{y}$ is distributed as $N(\boldsymbol{\alpha}'\boldsymbol{\mu}_2, \boldsymbol{\alpha}'\boldsymbol{\Sigma}\boldsymbol{\alpha})$, which can be expressed as $N(\boldsymbol{\alpha}'\boldsymbol{\mu}_2, \Delta^2)$, since

$$\begin{aligned} \boldsymbol{\alpha}'\boldsymbol{\Sigma}\boldsymbol{\alpha} &= (\boldsymbol{\mu}_1 - \boldsymbol{\mu}_2)'\boldsymbol{\Sigma}^{-1}\boldsymbol{\Sigma}\boldsymbol{\Sigma}^{-1}(\boldsymbol{\mu}_1 - \boldsymbol{\mu}_2) \\ &= (\boldsymbol{\mu}_1 - \boldsymbol{\mu}_2)'\boldsymbol{\Sigma}^{-1}(\boldsymbol{\mu}_1 - \boldsymbol{\mu}_2) = \Delta^2, \end{aligned}$$

where Δ^2 is the Mahalanobis distance defined in Section 1.8. We can therefore evaluate (6.30) using the univariate standard normal distribution. We standardize by subtracting the mean $\boldsymbol{\alpha}'\boldsymbol{\mu}_2$ and dividing by the standard deviation Δ:

$$\begin{aligned} P&\left[\boldsymbol{\alpha}'\mathbf{y} > \frac{1}{2}(\boldsymbol{\mu}_1 - \boldsymbol{\mu}_2)'\boldsymbol{\Sigma}^{-1}(\boldsymbol{\mu}_1 + \boldsymbol{\mu}_2) + \ln\left(\frac{p_2}{p_1}\right)\right] \\ &= P\left[\frac{\boldsymbol{\alpha}'\mathbf{y} - \boldsymbol{\alpha}'\boldsymbol{\mu}_2}{\Delta} > \frac{\frac{1}{2}(\boldsymbol{\mu}_1 - \boldsymbol{\mu}_2)'\boldsymbol{\Sigma}^{-1}(\boldsymbol{\mu}_1 + \boldsymbol{\mu}_2) + \ln(p_2/p_1) - \boldsymbol{\alpha}'\boldsymbol{\mu}_2}{\Delta}\right] \\ &= P\left[w > \frac{\frac{1}{2}(\boldsymbol{\mu}_1 - \boldsymbol{\mu}_2)'\boldsymbol{\Sigma}^{-1}(\boldsymbol{\mu}_1 + \boldsymbol{\mu}_2 - 2\boldsymbol{\mu}_2) + \ln(p_2/p_1)}{\Delta}\right] \\ &= P\left[w > \frac{\frac{1}{2}\Delta^2 + \ln(p_2/p_1)}{\Delta}\right], \end{aligned}$$

where $w = (\boldsymbol{\alpha}'\mathbf{y} - \boldsymbol{\alpha}'\boldsymbol{\mu}_2)/\Delta$. By the symmetry of the standard normal distribution, this probability becomes

$$\begin{aligned} P(\text{classify as } G_1 \mid G_2) &= P\left[w < \frac{-\frac{1}{2}\Delta^2 - \ln(p_2/p_1)}{\Delta}\right] \\ &= \Phi\left[\frac{-\frac{1}{2}\Delta^2 - \ln(p_2/p_1)}{\Delta}\right], \end{aligned} \tag{6.31}$$

where $\Phi[\cdot]$ is the standard normal distribution function, which can be evaluated by tables or by numerical integration. Similarly (see Problem 6.6),

$$P(\text{classify as } G_2 \mid G_1) = \Phi\left[\frac{-\frac{1}{2}\Delta^2 + \ln(p_2/p_1)}{\Delta}\right]. \tag{6.32}$$

We can now use (6.31) and (6.32) to evaluate (6.29):

$$\text{Optimum error rate} = p_1\Phi\left[\frac{-\frac{1}{2}\Delta^2 + \ln(p_2/p_1)}{\Delta}\right] + p_2\Phi\left[\frac{-\frac{1}{2}\Delta^2 - \ln(p_2/p_1)}{\Delta}\right]. \tag{6.33}$$

If $p_1 = p_2 = \frac{1}{2}$, this reduces to

$$\begin{aligned}\text{Optimum error rate} &= \tfrac{1}{2}\Phi\left(-\tfrac{1}{2}\Delta\right) + \tfrac{1}{2}\Phi\left(-\tfrac{1}{2}\Delta\right) \\ &= \Phi\left(-\tfrac{1}{2}\Delta\right).\end{aligned} \tag{6.34}$$

The plug-in estimator is readily obtained by substituting $D^2 = (\bar{\mathbf{y}}_1 - \bar{\mathbf{y}}_2)'\mathbf{S}_{\text{pl}}^{-1}(\bar{\mathbf{y}}_1 - \bar{\mathbf{y}}_2)$ in place of Δ^2 in (6.33):

$$\text{Plug-in error rate} = p_1\Phi\left[\frac{-\frac{1}{2}D^2 + \ln(p_2/p_1)}{D}\right] + p_2\Phi\left[\frac{-\frac{1}{2}D^2 - \ln(p_2/p_1)}{D}\right]. \tag{6.35}$$

When $p_1 = p_2 = \frac{1}{2}$, this reduces to

$$\text{Plug-in error rate} = \Phi\left(-\tfrac{1}{2}D\right). \tag{6.36}$$

The plug-in estimators given by (6.35) and (6.36) can also be shown to be the *maximum likelihood estimators* of the error rate and are sometimes referred to in this way.

A modified plug-in method (also known as the *shrunken D* method) proposed by Lachenburch and Mickey (1968) uses

$$D_s^2 = \frac{n_1 + n_2 - p - 3}{n_1 + n_2 - 2}D^2 \tag{6.37}$$

in place of D^2 in (6.35) or (6.36):

$$\begin{aligned}\text{Modified plug-in error rate} = {} & p_1\Phi\left[\frac{-\frac{1}{2}D_s^2 + \ln(p_2/p_1)}{D_s}\right] \\ & + p_2\Phi\left[\frac{-\frac{1}{2}D_s^2 - \ln(p_2/p_1)}{D_s}\right].\end{aligned} \tag{6.38}$$

When $p_1 = p_2 = \frac{1}{2}$, this reduces to

$$\text{Modified plug-in error rate} = \Phi\left(-\tfrac{1}{2}D_s\right). \tag{6.39}$$

The basis for the definition of D_s^2 given in (6.37) is that (Problem 6.7)

$$E\left[\frac{(n_1 + n_2 - p - 3)\mathbf{S}^{-1}}{n_1 + n_2 - 2}\right] = \boldsymbol{\Sigma}^{-1}. \tag{6.40}$$

A comparison of error rates for the above two methods is given by Dorveo (1993).

A simple nonparametric estimator of error rate (Smith 1947) can be obtained by applying the classification procedure (6.8) to all observations in the two samples used to compute (6.8). This method is called *resubstitution*. After the observation vectors $\mathbf{y}_{1j}$, $j = 1, 2, \ldots, n_1$, and $\mathbf{y}_{2j}$, $j = 1, 2, \ldots, n_2$, are classified into the two groups, the proportion of misclassifications is called the *apparent error rate*. The *apparent correct classification rate* is defined similarly. The method of resubstitution can be readily extended to the case of several groups using any of the classification procedures in Section 6.3.

The apparent error rate is biased because we are classifying the same observations that are used to compute the classification functions (Hills 1966). The apparent error rate also has a large variance for small sample sizes (Glick 1978). In Section 6.5, we consider some approaches to reducing the bias.

Houshmand (1993) suggested an approximate method of calculating the error rates for quadratic classification functions, assuming diagonal covariance matrices. Hand (1994) discussed other approaches to evaluating the performance of classification techniques.

Example 6.4. We illustrate the plug-in estimate of error rate for the psychological data first considered in Example 3.5.3(a). Using values of $\overline{\mathbf{y}}_1, \overline{\mathbf{y}}_2$, and $\mathbf{S}_{\text{pl}}$ from Example 3.5.3(a), we obtain

$$D^2 = (\overline{\mathbf{y}}_1 - \overline{\mathbf{y}}_2)'\mathbf{S}_{\text{pl}}^{-1}(\overline{\mathbf{y}}_1 - \overline{\mathbf{y}}_2) = 6.1001.$$

With $p_1 = p_2$, (6.36) yields

$$\text{Plug-in error rate} = \Phi\left(-\tfrac{1}{2}D\right) = \Phi\left(-\tfrac{1}{2}\sqrt{6.1001}\right) = .1084.$$

For the apparent error rate, we classify each observation in the two samples and show the results in Table 6.1. The "predicted group" shows the number of

Table 6.1. Classification Table for the Psychological Data

Actual Group	Number of Observations	Predicted Group 1	Predicted Group 2
Male	32	28	4
Female	32	4	28

observations classified into each group. For example, of the 32 observations in G_1, 28 were classified into G_1 and four were classified into G_2. The apparent error rate is given by

$$\text{Apparent error rate} = \frac{4 + 4}{32 + 32} = .125.$$

6.5 CORRECTING FOR BIAS IN THE APPARENT ERROR RATE

The asymptotic (large-sample) bias of the apparent error rate was derived by McLachlan (1976) for the case of two multivariate normal populations (groups). For large samples, the bias is small, but for small samples, the bias may be substantial (Rencher 1992a). Several techniques have been proposed for reducing the bias in the apparent error rate. We review and compare three of these.

6.5.1 Partitioning the Sample

In the several-group case, an unbiased estimator of error rate could be obtained by splitting the sample of $N = \sum_{i=1}^{k} n_i$ observations into two parts, then constructing the classification rule on one part and evaluating it on the other part. Partitioning the sample has two drawbacks: Partitioning requires very large samples to be effective, and the classification functions it evaluates are not based on the entire sample. Estimators of error rate based on part of the sample will have a larger variance than those based on the entire sample.

6.5.2 Holdout Method

The above sample-splitting procedure can be improved by using $N - 1$ observations to compute the classification rule and then classifying the omitted observation. This procedure is repeated for each observation. Referred to as the *holdout method*, it produces a nearly unbiased estimator of the expected actual error rate.

The holdout procedure, proposed by Lachenbruch (1965, 1967, 1975), is also called the *leaving-one-out method* and *cross-validation*. Another label that has been applied to the procedure is *jackknife*, but this characterization is incorrect (Seber 1984, p. 289). Efron and Gong (1983) described the true jackknife estimator and compared it to the holdout method and to the *bootstrap* estimator (Section 6.5.3).

Using Monte Carlo methods, Lachenbruch and Mickey (1968; see also Lachenbruch 1975) compared several methods of estimating error rates in the two-group case. Among normal-based methods, the best was a technique that used Okamoto's (1963) asymptotic expansion of the linear classification function in (6.8). The holdout method was better than the apparent error rate and also outperformed the plug-in method.

As noted previously, the apparent error rate obtained from resubstitution is satisfactory only for large samples. Lachenbruch (see Seber 1984, p. 292) recommends that each n_i exceed twice the number of parameters estimated for the corresponding G_i. For the multinormal case, the total number of parameters in $\boldsymbol{\mu}_i$ and $\boldsymbol{\Sigma}$ is $p(p+3)/2$. Therefore, according to Lachenbruch's rule of thumb, we should prefer the holdout method over resubstitution unless each $n_i > p(p+3)$. Glick (1978), however, showed that the holdout estimator has a relatively large variance from sample to sample. The bootstrap method (Section 6.5.3) is apparently more stable.

6.5.3 Bootstrap Estimator

The *bootstrap estimator* of error rate (Efron 1979, 1981, Efron and Tibshirani 1993, McLachlan 1980a) is essentially a bias correction for the apparent error rate, based on *resampling* the original samples. We describe the procedure for the two-group case with samples of size n_1 and n_2. From the first sample we take a random sample of size n_1 *with replacement*. Typically, some of the observations from the original sample will not appear in the new sample, while others will show up more than once. The probability is approximately .368 that an observation vector will not be included in the bootstrap sample (Problem 6.9). We similarly resample from the second group.

The bootstrap procedure is thus a method of attenuating the samples so that the classification functions will not perform so optimistically. We recompute the classification rule using the two new samples and then classify both the original samples and the new samples. Let

$$d_i = \frac{m_{i,\text{old}} - m_{i,\text{new}}}{n_i}, \qquad i = 1, 2, \tag{6.41}$$

where $m_{i,\text{old}}$ is the number of observations out of n_i that are misclassified in the original sample and $m_{i,\text{new}}$ is the number misclassified in the new sample. We carry out this resampling procedure repeatedly (100 or 200 replications have been suggested) and use $\overline{d}_i$ as a bias correction term:

$$\text{Bootstrap error rate} = \text{apparent error rate} + \overline{d}_1 + \overline{d}_2. \tag{6.42}$$

6.5.4 Comparison of Error Estimators

Good reviews of various estimators of error rates and the issues involved have been given by Hand (1981, pp. 186–190) and Ganeshanandam and Krzanowski (1990). Efron (1979) showed that the bootstrap estimator of error is almost unbiased and that its variance is less than the variance of the holdout estimator but greater than that of the apparent error rate. Efron (1983) made further comparisons of the bootstrap and holdout methods, including several improved variations of the bootstrap. Davison and Hall (1992) showed that for two populations the bootstrap and holdout estimators of error rate do not differ appreciably in bias or variability except when the two populations are close, in which case the bootstrap has less variability but greater bias than the holdout method.

Snapinn and Knoke (1984) used numerical integration and simulation to compare four methods of estimating the error rate in the two-group case: the plug-in, modified plug-in, resubstitution, and holdout methods. Their criterion for comparison was the mean squared error of the actual error rate. For normal data with small or moderate sample sizes, they recommend the modified plug-in estimate when $p \leq 3$ and the holdout method when $p > 3$. Based on a very limited study of robustness involving univariate nonnormal distributions, they recommend the resubstitution method when $p \leq 3$ and the holdout method when $p > 3$.

For the two-group case, Glick (1978) introduced a smoothed version of the apparent error rate designed to reduce the variance. Snapinn and Knoke (1985) proposed the following modification of Glick's estimator of error rate for the ith group, $i = 1, 2$:

$$P_i = \Phi\left[(-1)^i \frac{1}{2} D \sqrt{\frac{n_i}{n_i B + n_i - 1}}\right],$$

where

$$D = \sqrt{(\overline{\mathbf{y}}_1 - \overline{\mathbf{y}}_2)' \mathbf{S}_{\text{pl}}^{-1} (\overline{\mathbf{y}}_1 - \overline{\mathbf{y}}_2)}, \qquad B = \frac{(p+2)(n_1 - 1) + n_2 - 1}{n_1(n_1 + n_2 - p - 3)},$$

and n_1 and n_2 are the sample sizes of the two groups. They compared this estimator to those based on the resubstitution method, the holdout method, and an idealized bootstrap. The holdout error rate was better than the smoothed estimator for small $\Delta = [(\boldsymbol{\mu}_1 - \boldsymbol{\mu}_2)' \boldsymbol{\Sigma}^{-1} (\boldsymbol{\mu}_1 - \boldsymbol{\mu}_2)]^{1/2}$, but the smoothed estimator was the best among all four when Δ was of intermediate or large size. In a very limited simulation involving univariate nonnormality, all the methods appeared to be robust to departures from normality. Snapinn and Knoke (1988) suggested two modifications of the smoothed estimator, both of which showed improvements in bias and variance. Looney (1988) proposed a method based on repeated measures for testing the statistical significance of the differences in error rates of several classification methods applied to the same set of data. Hirst (1996) generalized Snapinn and Knoke's (1985) technique to the several-group case.

Page (1985) compared the following eight parametric methods of estimating error rates for the two-group case: the plug-in and modified plug-in methods, Okamoto's method (Okamoto 1963, Lachenburch and Mickey 1968) using D^2 and D_s^2 [see (6.37)], a method suggested by Lachenbruch (1968) that was based on the mean and variance of the discriminant function, two methods by Sorum (1971), and a method due to McLachlan (1975). For estimation of the actual error rate, the Lachenbruch, McLachlan, and modified Okamoto methods (using D_s^2) worked best. For large samples with a moderate number of variables, the modified plug-in method also worked very well. Because of its computational simplicity compared to the modified Okamoto method, the modified plug-in method becomes a viable alternative.

Chernick, Murthy, and Nealy (1985) compared seven nonparametric estimators of the expected actual error rate for two and three groups: resubstitution, holdout, and five variations of the bootstrap. Three of these variations, the standard bootstrap, the ".632 estimator," and the e_0 estimator, are by Efron (1983). The other two are the "MC estimator" by Chernick and Murthy (1985) and the convex bootstrap introduced by Chernick et al. (1985). The .632 estimator ranked first most often, followed by the MC and e_0 methods.

For the two-group case, Ganeshanandam and Krzanowski (1990) compared 11 estimators of error rates, many of which are described above: resubstitution, plug-in, modified plug-in, a modified D^2 method by Lachenbruch (1967), McLachlan's (1974) method, the modified smoothed estimator by Snapinn and Knoke (1985), the holdout method, the method given by Lachenburch and Mickey (1968), a modified jackknife technique, the .632 variation of the bootstrap, and the holdout method applied to quadratic classification functions. Ganeshanandam and Krzanowski found that for both normal and nonnormal data the best estimators were the holdout method, the modified jackknife, Lachenbruch's method, McLachlan's estimator, and Lachenbruch and Mickey's method. The smoothed, resubstitution, and plug-in estimators performed poorly.

For two and three groups, Huberty, Wisenbaker, and Smith (1987) compared the holdout estimator to one based on posterior probabilities (Sections 6.2.4 and 6.3.1) proposed by Fukunaga and Kessell (1973) and Glick (1978). Based on their results, Huberty, Wisenbaker, and Smith favored use of the holdout method.

Hand (1994) discussed error rates and two additional criteria for evaluating and comparing classification rules.

6.6 SUBSET SELECTION

We consider three types of variable selection in classification analysis: (1) techniques designed for separation of groups as described in Sections 3.11.3, 4.9.3, and 5.9.2; (2) procedures based on allocation, such as including at each step the variable with maximum improvement in error rates; and (3) methods for the case of heteroscedastic (unequal) covariance matrices. Good reviews have been given by McKay and Campbell (1982b), Krishnaiah (1982), Schaafsman (1982), McLachlan (1976), Farver and Dunn (1979), Huberty (1984), and McCulloch (1986). Biscay, Valdes, and Pascual (1990) proposed a method of subset selection based on principal components (see Chapter 9).

6.6.1 Selection Based on Separation of Groups

Selection schemes for classification analysis have often been based on a discriminant analysis approach (Section 5.9.2). Stepwise discriminant analysis or a similar procedure is used to find the subset that best separates groups, and then these variables are used to construct classification functions. This method is available in most major statistical software packages.

Clearly, there is some association between separation and classification. If a subset of variables can be found using Wilks' Λ, for example, that separates the groups as well as (or almost as well as) the full set of variables, we would expect that subset to also classify the observations well. However, the subset that best separates the groups according to Wilks' Λ may not be the subset that best allocates observations to groups (see comments in the first paragraph of Section 6.6.2). Krusinska and Liebhart (1990a) considered the use of Wilks' Λ for selection when both continuous and discrete variables are present.

Eisenbis, Gilbert, and Avery (1973) compared six separation-based methods of subset selection in terms of apparent error rate. Their study was limited to the two-group case. Two of the methods, one based on univariate (not partial) F-statistics and another based on standardized discriminant function coefficients, were noticeably inferior to the others. Among the other four, which included three stepwise procedures and an all-subsets approach, none stood out as clearly superior.

In a Monte Carlo study, Farver and Dunn (1979) compared two forward selection procedures for the two-group case. Both of the selection methods produced subsets with lower error rates than the original full set of variables.

Huberty and Smith (1976) compared six methods for ranking variables by their contribution to classification in the several-group case: a univariate F-statistic approach, a stepwise technique, two methods based on correlations between the variables and the discriminant functions, and two methods that use standardized discriminant function coefficients. The results did not indicate that any one of the six methods was uniformly superior, and Huberty and Smith recommended that further studies be carried out.

Example 6.6.1. Espahbodi (1991) used stepwise discriminant analysis to select four variables out of 13 financial ratios available to discriminate between failed and nonfailed banks. The 13 variables were observed two years before failure and again one year before failure. The same four variables emerged in both cases. For the data observed one year before failure, Table 6.2 shows the partial F for each variable [see (3.100)] and the value of T^2 that compares the two groups using all four variables.

Using the method of resubstitution to estimate the error rate, the observations were classified by linear classification functions based on the four variables selected in Table 6.2. The classification results are in Table 6.3, where the apparent correct classification rate is $(32 + 31)/(38 + 35) = .863$.

Table 6.2. Stepwise Discriminant Analysis for One Year before Failure

Step	Variable Number	Variable Name	Partial F
1	6	$\dfrac{\text{Total loan revenue}}{\text{Total operating income}}$	11.04
2	9	$\dfrac{\text{Interest paid on deposits}}{\text{Total operating income}}$	24.78
3	8	$\dfrac{\text{Interest on government obligations}}{\text{Total operating income}}$	14.88
4	11	$\dfrac{\text{Total time and saving deposits}}{\text{Total demand deposits}}$	10.78

Note: Overall T^2 for the four variables is 73.90.

Table 6.3. Classification Table for One Year before Failure.[a]

Actual Group	Number of Observations	Predicted Group	
		Failed	Solvent
Failed	38	32	6
Solvent	35	4	31

Note: Apparent correct classification rate .863; apparent error rate 1 − .863 = .137.
[a] Linear classification functions based on the four variables in Table 6.2.

Table 6.4. Stepwise Discriminant Analysis for Two Years before Failure

Step	Variable	Partial F
1	8	18.58
2	9	10.70
3	11	7.23
4	6	4.51

Note: Overall T^2 for the four variables is 43.33.

Table 6.5. Classification Table for Two Years before Failure[a]

Actual Group	Number of Observations	Predicted Group	
		Failed	Solvent
Failed	37	32	5
Solvent	33	6	27

Note: Apparent correct classification rate .843; apparent error rate 1 − .843 = .157.
[a] Linear classification functions based on the four variables in Table 6.4.

Tables 6.4 and 6.5 contain the selection and classification results for the data obtained two years before failure.

6.6.2 Selection Based on Allocation

When our goal is to find the optimal subset for classification, we would expect that a subset selection procedure based on misclassification rates would be superior to one based on separation of groups. We therefore consider choosing the subset of variables (of a given size) that has the lowest error rate. For several groups, the partial Wilks' Λ-criterion (4.135) or corresponding partial F in (4.136) for entry of a variable in stepwise discriminant analysis (Sections 5.9 and 5.10) may not select the variable that minimizes the error rate. Habbema and Hermans (1977) gave an example in which the variable with the smallest partial F yielded better classification results than the one with the largest F. This apparent anomaly is possible because the partial-F approach to subset selection tends to select variables that best separate the groups that are furthest apart rather than variables that best separate the closest groups. For some data sets, at least, a procedure designed to find variables that maximally separate the two closest groups would lead to better classification rates.

For the two-group case, McLachlan (1976, 1980b) developed an approximate probability that deletion of a given subset will not increase the actual error rate; he then compared this approximate probability method with the partial F [see (3.99)] for significance of a subset to be deleted. Constanza and Afifi (1980) discussed stopping rules for subset selection in two-group classification analysis using forward selection. They compared seven methods for determining how many variables to include and recommended two: (1) the partial F-test with nominal significance levels between .10 and .25 and (2) an estimator of the probability of correct classification based on the Mahalanobis distance between groups (Lachenburch and Mickey 1968).

For the several-group case, Habbema and Hermans (1977) proposed a forward selection procedure based on error rates. They compared their procedure with the usual partial-F method on two data sets and found their approach to be superior for the two examples. The procedure is available in the programs ALLOC-1 (Hermans and Habbema 1976) and ALLOC 80 (Hermans, Habbema, and Schafer 1982).

Ganeshanandam and Krzanowski (1989) considered several methods of variable selection based on error rates computed by the holdout method. Based on a simulation study, they recommended a (forward) stepwise procedure in which each observation is left out (one at a time), the subset of variables is selected, and then the omitted unit is classified using linear classification functions computed from the selected variables. The resulting error rates are used to choose the variable to be entered at each step. This is very computer intensive, since a different subset of variables may be selected for each omitted unit.

Seaman and Young (1990) proposed an allocation-based backward-elimination method that utilizes a holdout approach. They used an efficient method of upgrading the inverse of a matrix in order to reduce the computational load.

6.6.3 Selection in the Heteroscedastic Case

If the population covariance matrices are not assumed to be equal, we can consider using the quadratic classification functions (6.12) or (6.23). In (6.23), we must estimate $\frac{1}{2}kp(p+3)$ parameters, which requires large sample sizes if p is large. In this case, it would be advantageous to reduce the value of p by judicious subset selection, if possible.

Young and Odell (1986) proposed three methods of subset selection for the heteroscedastic case (unequal covariance matrices). In a limited simulation for a particular pattern of heteroscedasticity, the three methods were shown to have smaller error rates than standard methods of classification. Fatti and Hawkins (1986) proposed a method of variable selection in the heteroscedastic case based on a partitioning of the test statistic into three terms.

Pynnonen (1987) adapted the test for additional information (Section 4.9.1) to obtain a selection procedure that is applicable to the heteroscedastic case. In a limited simulation, this method yielded subsets that classified better than the full set of variables. Schott (1993) noted that, in general, reducing the number of variables reduces the error rates, and he proposed a test to determine the minimum number of variables that will yield good results.

6.7 BIAS IN STEPWISE CLASSIFICATION ANALYSIS

The bias in Wilks' Λ in stepwise discriminant analysis for separation of groups was discussed in Section 5.10. A similar bias in error rates is evident in subset selection for classification. This bias due to subset selection is in addition to the bias in apparent error rates due to the data reuse effect (Section 6.5).

The likelihood of bias is of course greater in situations in which the number of variables is large relative to the sample size. Habbema and Hermans (1977, p. 452) asserted that "sample sizes of say 10–40 are not unusual, with a number of variables ranging from 10–200." Mucciardi and Gose (1971) presented a classification analysis of 242 observations on 157 variables. Farver and Dunn (1979) note a study by Carpenter, Strauss, and Bartko (1973) in which 12 variables were selected out of 415 with a sample size of 560. Kwan and Kowalski (1978) used stepwise selection for classification on a data set with 47 observations on 256 variables. We discuss this data set and the results of Kwan and Kowalski in Example 6.7.

An analysis of two industrial data sets (Center for Collaborative Research and Statistical Consultation, Brigham Young University) illustrates the bias due to selection. The first data set consisted of two groups with $n_1 + n_2 = 26$ observations and $p = 28$ variables. With stepwise discriminant analysis, 2 variables were selected out of 28, and using these 2 variables, only 3 of the 26 observations were misclassified. To check these results, the 26 observation vectors were reassigned at random to the two groups (with n_1 and n_2 the same as before). Then a new subset of variables was selected with which to classify the observations. This was repeated 10 times. In 4 of the replications, 3 or fewer observations were misclassified, and in the other 6, there were 4 misclassifications. These results cast serious doubt on the ability of the original subset of variables to classify future observations.

The second data set had $n_1 + n_2 = 42$ observations and $p = 300$ variables. The observations were similarly shuffled 10 times, and in all 10 replications the new subsets of variables classified as least as well as the original subset.

In cases in which the number of variables exceeds the total sample size, variables should be selected with caution, if at all. It is in just such a situation, however, that the researcher has the greatest need to reduce the number of variables. The resulting subset should be considered suspect (in terms of classifying future observations) unless confirmed by an independent procedure such as a repetition of the study.

The bias in error rates due to selection in the two-group case was discussed by Murray (1977), who noted a tendency for the error rate to decline as the subset size increases from 1 to $\frac{1}{2}k$, where k is the total number of variables. Murray found a corresponding tendency for the error rate to increase as the subset size increases from $\frac{1}{2}k$ to k. Murray attributed this to the fact that the number of subsets $\binom{k}{r}$ increases as r approaches $\frac{1}{2}k$, then decreases again as r goes from $\frac{1}{2}k$ to k. Hecker and Wegener

(1978) considered the subset selection bias for two groups and found this bias to be substantial in many cases. However, Hecker and Wegener's results were based on 10 or fewer Monte Carlo replications, which led to large standard errors in estimation of classification rates.

Snapinn and Knoke (1989) compared the bootstrap and smoothed estimators of error rates (Sections 6.5.3 and 6.5.4) in stepwise discriminant analysis for two groups. Their results favored the smoothed estimator. Rutter, Flack, and Lachenbruch (1991) compared six estimators of error rates for stepwise discriminant analysis in the two-group case: resubstitution (Section 6.4), two levels of partitioning the sample (Section 6.5.1), two versions of the plug-in method (Section 6.4), and a smoothed estimator (Section 6.5.4). The least bias in error rates was found for the sample partitioning methods followed by the smoothed estimator. The others exhibited substantial bias.

For the several-group case, Rencher (1992a) investigated the distribution of apparent correct classification rates under subset selection. The results are described in the remainder of this section and in Example 6.7. The study considered only the null case of no differences between groups (populations) in order to provide guidelines to the optimistic levels of correct classification attainable when the variables have no real ability to separate groups. The selection procedure was based on stepwise discriminant analysis (Section 5.9) using partial Wilks Λ values converted to partial F's. Both the bias due to selection and the bias due to resubstitution were obtained.

The observation vectors were simulated from a multivariate normal distribution using correlated variables. Threshold F-values of 1.0, 2.0, 3.0, and 4.0 were used for entry (or deletion) of variables in the stepwise selection procedure. The number of variables selected from the k available was denoted by p. There were n observations in each of the g groups. The proportions of observations correctly classified under selection and without selection were denoted by C_s and C_0, respectively. For each combination of F, g, n, and k, 1000 replications were used to compute the average values $\overline{p}$, $\overline{C}_s$, and $\overline{C}_0$, and the upper 95th percentage point $C_{.95}$.

Because there was no difference between populations, the expected correct classification rate is $1/g$. The bias due solely to resubstitution is $\overline{B}_0 = \overline{C}_0 - 1/g$, and the bias due to stepwise selection is $\overline{B}_s = \overline{C}_s - \overline{C}_0$. An excerpt from Rencher's (1992a) results is given in Table 6.6, in which both types of bias are readily discernible: $\overline{B}_0$ varies from .06 to .73, and $\overline{B}_s$ varies from .02 to .23.

Example 6.7. This example illustrates the misleading results that may be obtained in selection of variables when the sample size is small relative to the number of variables. Dyer and Ansher (1977) measured concentration levels of 16 major chemical and spectrographic components in 1970 and 1971 vintage Rhine and Moselle German white wines to determine if vintage year and wine regions could be distinguished by these variables. From an examination of range, mean, and standard deviation for each of the 16 variables, Dyer and Ansher concluded that the variables did not separate vintage years and wine regions.

Kwan and Kowalski (1978) added all possible ratios of the original 16 variables of Dyer and Ansher to obtain a total of 256 variables. There were four groups (region–year combinations): Moselle 1970, Moselle 1971, Rhine 1970, and Rhine

Table 6.6. Average and 95th Percent Apparent Correct Classification Rates and Bias with Stepwise Selection and without Selection

k	F	$\overline{p}$	$\overline{C}_0$	$\overline{C}_s$	$C_{.95}$	$\overline{B}_0$	$\overline{B}_s$	$\overline{p}$	$\overline{C}_0$	$\overline{C}_s$	$C_{.95}$	$\overline{B}_0$	$\overline{B}_s$
		$n = 5, g = 4, 1/g = .25$						$n = 5, g = 8, 1/g = .125$					
10	4	1.2	.38	.52	.70	.13	.14						
	3	1.4	.40	.52	.75	.15	.12	1.1	.21	.27	.40	.08	.06
	2	2.0	.43	.56	.80	.18	.13	1.4	.22	.28	.45	.10	.06
	1	4.3	.57	.70	.95	.32	.13	4.2	.36	.44	.60	.24	.08
20	4	1.4	.40	.55	.80	.15	.15						
	3	1.9	.42	.59	.90	.17	.17	1.1	.21	.28	.42	.08	.07
	2	3.8	.53	.72	1.00	.28	.19	1.9	.25	.34	.60	.12	.09
	1	10.2	.84	.95	1.00	.59	.11	7.9	.53	.67	.90	.40	.14
30	4	1.6	.40	.59	.85	.15	.19						
	3	2.5	.46	.65	.95	.21	.19	1.2	.21	.29	.42	.08	.08
	2	6.4	.66	.85	1.00	.41	.19	2.7	.29	.41	.65	.16	.12
	1	13.6	.96	1.00	1.00	.71	.04	12.4	.70	.86	1.00	.58	.16
40	4	1.8	.41	.61	.90	.16	.20						
	3	3.4	.50	.73	1.00	.25	.23	1.3	.22	.30	.48	.10	.08
	2	9.6	.80	.94	1.00	.55	.14	3.3	.33	.47	.72	.20	.14
	1	14.0	.98	1.00	1.00	.73	.02	17.1	.84	.95	1.00	.72	.11
		$n = 10, g = 4, 1/g = .25$						$n = 10, g = 8, 1/g = .125$					
10	4	1.1	.33	.41	.52	.08	.08						
	3	1.2	.34	.41	.52	.09	.07	1.1	.18	.21	.29	.06	.03
	2	1.7	.36	.43	.60	.11	.07	1.3	.19	.22	.31	.06	.03
	1	4.0	.44	.51	.68	.19	.07	4.1	.27	.31	.42	.14	.04
20	4	1.1	.33	.42	.55	.08	.09						
	3	1.5	.35	.44	.60	.10	.09	1.1	.18	.22	.30	.06	.04
	2	3.0	.40	.51	.70	.15	.11	1.8	.20	.25	.35	.08	.05
	1	8.7	.57	.71	.88	.32	.14	8.7	.38	.46	.59	.26	.08
30	4	1.2	.33	.43	.60	.08	.10						
	3	1.8	.36	.47	.65	.11	.11	1.1	.18	.23	.31	.06	.05
	2	4.5	.45	.60	.85	.20	.15	2.2	.21	.28	.39	.08	.07
	1	14.4	.71	.87	1.00	.46	.16	13.2	.47	.59	.74	.34	.12
40	4	1.3	.34	.44	.60	.09	.10						
	3	2.2	.37	.50	.72	.12	.13	1.2	.18	.23	.31	.06	.05
	2	6.5	.51	.70	.95	.26	.19	2.7	.23	.30	.44	.10	.07
	1	23.4	.89	.98	1.00	.64	.09	18.0	.57	.71	.85	.44	.14

1971. Kwan and Kowalski also made eight two-group comparisons: the six possible pairings among the above four groups and comparisons of the two regions and of the two years.

The four groups had 9, 16, 13, and 9 observation vectors, respectively, with 256 variables measured in each vector. Using a stepwise discriminant analysis procedure, three variables were selected out of 256, yielding an apparent correct classification rate for the four groups of "about 50%." For the eight two-group comparisons, the average apparent classification rate with three variables selected in each case was 84.6%. Kwan and Kowalski concluded that "separations by vintage years and wine regions are made possible by these methods."

To check these results, Rencher (1992a) used the simulation program described above to evaluate the following two sets of (balanced) parameter values: $g = 4$, $n = 12$, $k = 256$, $F = 4.3$; and $g = 2$, $n = 12$, $k = 256$, $F = 9.7$. These threshold F-values yielded values of $\overline{p}$ close to 3, the number of variables selected by Kwan and Kowalski. The results for $g = 4$ were $\overline{p} = 3.12$, $\overline{C}_s = .59$, and $C_{.95} = .84$. For the two-group comparisons ($g = 2$), the results were $\overline{p} = 3.08$, $\overline{C}_s = .88$, and $C_{.95} = 1.00$. When these values are compared to those of 50% and 84.6% obtained by Kwan and Kowalski for $g = 4$ and $g = 2$, respectively, it appears that the best three variables out of 256 in the data set cannot classify any better than variables generated randomly from similar populations with no differences in means.

6.8 LOGISTIC AND PROBIT CLASSIFICATION

We now consider two approaches that can be used with variables that are discrete or a mixture of discrete and continuous. We consider logistic classification in Sections 6.8.1–6.8.5 and probit classification in Section 6.8.6.

6.8.1 The Logistic Model for Two Groups with $\boldsymbol{\Sigma}_1 = \boldsymbol{\Sigma}_2$

For the multivariate normal distribution with $\boldsymbol{\Sigma}_1 = \boldsymbol{\Sigma}_2$, the logarithm of the density ratio (6.6) is a linear function of the observation vector $\mathbf{y}$:

$$\begin{aligned}\ln \frac{f(\mathbf{y} \mid G_1)}{f(\mathbf{y} \mid G_2)} &= -\tfrac{1}{2}(\boldsymbol{\mu}_1 - \boldsymbol{\mu}_2)'\boldsymbol{\Sigma}^{-1}(\boldsymbol{\mu}_1 + \boldsymbol{\mu}_2) + (\boldsymbol{\mu}_1 - \boldsymbol{\mu}_2)'\boldsymbol{\Sigma}^{-1}\mathbf{y} \\ &= \alpha + \boldsymbol{\beta}'\mathbf{y}. \end{aligned} \tag{6.43}$$

In addition to the multivariate normal, other multivariate distributions also satisfy (6.43), some of which involve discrete random variables or mixtures of continuous and discrete variables. The model in (6.43) is known as the *logistic model*, and from (6.4) the classification rule becomes: Assign $\mathbf{y}$ to G_1 if

$$\alpha + \boldsymbol{\beta}'\mathbf{y} > \ln\left(\frac{p_2}{p_1}\right) \tag{6.44}$$

and to G_2 otherwise. *Logistic classification* is also referred to as *logistic discrimination* and as *logistic multiple regression*. For a description of logistic regression (also called logit regression) and accompanying weighted least squares estimation techniques, see Kleinbaum (1994), Hosmer, Jovanovic, and Lemeshow (1989), Myers (1990, pp. 317–320), or Neter et al. (1996, pp. 567–609). Van Houwelingen and LeCessie (1988) reviewed logistic regression as applied to classification analysis.

We can also express posterior probabilities in terms of the logistic model. From (6.43) and (6.16), we obtain

$$\begin{aligned} P(G_1 \mid \mathbf{y}) &= \frac{p_1 f(\mathbf{y} \mid G_1)}{p_1 f(\mathbf{y} \mid G_1) + p_2 f(\mathbf{y} \mid G_2)} \\ &= \frac{e^{\ln(p_1/p_2)+\alpha+\boldsymbol{\beta}'\mathbf{y}}}{1 + e^{\ln(p_1/p_2)+\alpha+\boldsymbol{\beta}'\mathbf{y}}} \end{aligned} \tag{6.45}$$

$$= \frac{e^{\alpha_0+\boldsymbol{\beta}'\mathbf{y}}}{1 + e^{\alpha_0+\boldsymbol{\beta}'\mathbf{y}}}, \tag{6.46}$$

where $\alpha_0 = \ln(p_1/p_2) + \alpha$ (see Problem 6.11). It follows immediately that

$$P(G_2 \,|\, \mathbf{y}) = 1 - P(G_1 \,|\, \mathbf{y}) = \frac{1}{1 + e^{\alpha_0+\boldsymbol{\beta}'\mathbf{y}}}. \tag{6.47}$$

Models (6.43), (6.46), and (6.47) were suggested by Cornfield (1962), Cox (1966), Day and Kerridge (1967), and Anderson (1972). Anderson (1982) reviewed logistic classification and provided a stepwise variable selection procedure. Veldhuisen and Temkin (1994) compared logistic classification with CART (classification and regression trees; see Breiman et al. 1984), a computer-intensive partitioning model that handles both categorical and continuous variables.

To estimate α and $\boldsymbol{\beta}$, we can use logistic regression with weighted least squares or a maximum likelihood approach, expressing the appropriate likelihood function in terms of (6.46) and (6.47). To maximize the likelihood function, an iterative procedure is usually required, such as the standard Newton–Raphson technique or a quasi-Newton method (Gill and Murray 1972).

Example 6.8.1. Logistic classification has been successfully applied to medical data that consist of both continuous and discrete variables. The following example is from Clayton, Anderson, and McNicol (1976) as reported by Anderson (1982). The goal was to predict postoperative deep vein thrombosis, a condition that is best treated preoperatively with anticoagulant therapy. However, this treatment produces bleeding problems in some patients, and it is important to identify those patients most at risk, if possible.

Of the 124 patients in the study, none showed preoperative evidence of deep vein thrombosis. After the operation, 20 patients developed the condition. These 20 constituted G_1, with the remaining 104 in G_2. From 26 variables available on each patient, preliminary screening and a stepwise procedure selected five variables; four of these were continuous (y_1, y_2, y_4, y_5), and the other (y_3) was discrete.

The logistic model (6.43) was estimated as

$$\begin{aligned} w &= \hat{\alpha} + \hat{\boldsymbol{\beta}}'\mathbf{y} \\ &= -11.3 + .009y_1 + .22y_2 + .085y_3 + .043y_4 + 2.19y_5. \end{aligned} \tag{6.48}$$

A value of w was calculated for each of the 124 patients; these values are plotted in Figure 6.1. Generally, the two groups are well separated. If the classification rule in (6.44) is applied with $p_1 = p_2$, those patients with $w > 0$ would be assigned to G_1, the deep vein thrombosis group. Using this procedure, 11 of the 124 would be misclassified, for an apparent error rate of 9%. However, using $w > 0$ would miss some of the high-risk patients. Therefore, a recommendation was made to give preoperative anticoagulant therapy to patients with $w > -2.5$. Anderson (1982) reported that this decision system had very satisfactory results, reducing the incidence of thrombosis to 3%, as compared with 16% (20 out of 124) in this study.

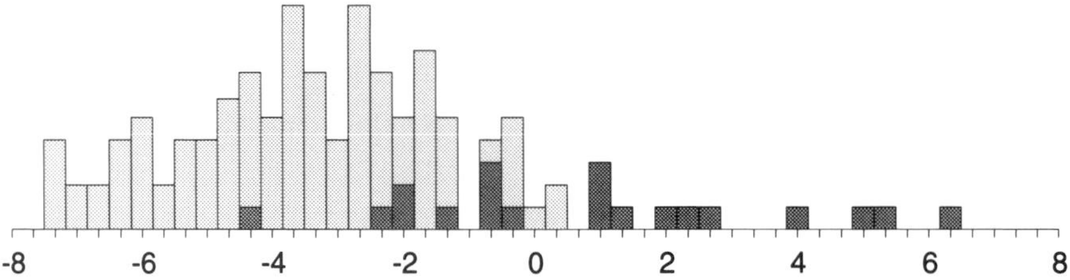

Figure 6.1. Prognostic index (w) for patients at risk for deep vein thrombosis (DVT). Dark shading, patients with DVT; light shading, patients with no DVT.

6.8.2 Comparison of Logistic Classification with Linear Classification Functions

For multivariate normal data with $\mathbf{\Sigma}_1 = \mathbf{\Sigma}_2$, linear classification functions will be superior to logistic classification, since the linear functions are optimal for such data. However, for binary data, these assumptions usually do not hold, and the logistic approach may be better. Efron (1975) showed that for normal data with $\mathbf{\Sigma}_1 = \mathbf{\Sigma}_2$ the asymptotic efficiency of the linear classification approach relative to the logistic approach declines rapidly for $\Delta \geq 2.5$, where $\Delta = [(\boldsymbol{\mu}_1 - \boldsymbol{\mu}_2)'\mathbf{\Sigma}^{-1}(\boldsymbol{\mu}_1 - \boldsymbol{\mu}_2)]^{1/2}$ is the square root of the Mahalanobis distance defined in Section 1.8. For further discussion of the asymptotic efficiency of logistic regression relative to linear classification, see Ruiz-Velasco (1991).

Crawley (1979) compared the logistic approach to the linear classification method for small samples under three conditions. For normal data with $\mathbf{\Sigma}_1 = \mathbf{\Sigma}_2$, the logistic functions classified almost as well as the linear classification functions. For normal distributions with $\mathbf{\Sigma}_1 \neq \mathbf{\Sigma}_2$, the logistic approach was slightly superior. For nonnormal data, the logistic method performed much better than the linear classification functions.

Harrell and Lee (1985) compared the logistic approach to the linear classification method for two multivariate normal distributions with equal covariance matrices. They used sample sizes of 50 and 130 and six values of $\Delta = [(\boldsymbol{\mu}_1 - \boldsymbol{\mu}_2)'\mathbf{\Sigma}^{-1}(\boldsymbol{\mu}_1 - \boldsymbol{\mu}_2)]^{1/2}$ ranging from .94 to 4.68. Using several measures of predictive accuracy, they found that while the linear classification functions were superior to the logistic approach as expected, the difference was never great. Harrell and Lee concluded that "even when the conditions under which the linear classification method was optimized are satisfied, the performance of the logistic model is nearly as good as that of the linear classification method for reasonable sample sizes and values of Δ (p. 341)." They further argued that "the logistic model is the tool of first choice among these two competitors. With the availability of efficient computers and computer programs, the issue of the computational requirements of the logistic model becomes unimportant" (p. 341).

Cherkaoui and Cleroux (1991) compared six classification methods on two groups of data involving a mixture of continuous, binary, ordinal, and nominal variables. The six methods were linear classification as in (6.8), quadratic classification as in (6.12) and (6.13), logistic classification as in (6.44), and three nonparametric methods. Cherkaoui and Cleroux discuss the conditions under which the various methods give superior performance.

O'Gorman and Woolson (1991) compared stepwise logistic regression (Hosmer et al. 1989) and stepwise discriminant analysis (Sections 3.11.3, 4.9.3, and 5.9.2). The simulation involved eight variables; four of the variables were related to group separation, and the other four were not related. Stepwise discriminant analysis selected the correct variables more often than did stepwise logistic regression.

Example 6.8.2. Espahbodi (1991) compared the logistic method with linear classification functions for data on bank failures. The results for four variables selected by stepwise discriminant analysis were reported in Example 6.6.1. We now give Espahbodi's results for the logistic approach.

The logistic model used by Espahbodi that corresponds to (6.43) was

$$\ln\left(\frac{P_i}{1 - P_i}\right) = \alpha + \beta_1 y_{i1} + \beta_2 y_{i2} + \cdots + \beta_q y_{iq},$$

where P_i is the probability that the ith bank fails. For both the one-year-before-failure data and the two-years-before-failure data, a stepwise procedure selected the same four variables as did the stepwise discriminant approach reported in Example 6.6.1.

Using posterior probabilities as in (6.46) and (6.47), Espahbodi used a cutoff point of .5. Thus if $P(\text{Failure} \mid \mathbf{y}) > .5$, a bank is classified as failed, where $\mathbf{y}$ is the bank's measurements on the four variables. The results are given in Tables 6.7 and 6.8.

Comparing the results in Example 6.6.1 with those in Tables 6.7 and 6.8, we see that the logistic method gives a slightly better error rate for one year before failure, but the linear classification functions do better for two years before failure.

Table 6.7. Classification Table for One Year before Failure Based on Logistic Regression

Actual Group	Number of Observations	Predicted Group: Failed	Predicted Group: Solvent
Failed	38	33	5
Solvent	35	4	31

Note: Apparent correct classification rate .877; apparent error rate $1 - .877 = .123$.

Table 6.8. Classification Table for Two Years before Failure Based on Logistic Regression

Actual Group	Number of Observations	Predicted Group: Failed	Predicted Group: Solvent
Failed	37	29	8
Solvent	33	9	24

Note: Apparent correct classification rate .757; apparent error rate $1 - .757 = .243$.

6.8.3 Quadratic Logistic Functions When $\boldsymbol{\Sigma}_1 \neq \boldsymbol{\Sigma}_2$

If $\boldsymbol{\Sigma}_1 \neq \boldsymbol{\Sigma}_2$, the logistic function for the multivariate normal in (6.43) is no longer a linear function of $\mathbf{y}$. By (6.10), we have

$$\begin{aligned}\ln \frac{f(\mathbf{y}\,|\,G_1)}{f(\mathbf{y}\,|\,G_2)} &= c_0 + (\boldsymbol{\mu}_1'\boldsymbol{\Sigma}_1^{-1} - \boldsymbol{\mu}_2'\boldsymbol{\Sigma}_2^{-1})\mathbf{y} + \tfrac{1}{2}\mathbf{y}'(\boldsymbol{\Sigma}_2^{-1} - \boldsymbol{\Sigma}_1^{-1})\mathbf{y} \\ &= c_0 + \boldsymbol{\gamma}'\mathbf{y} + \mathbf{y}'\boldsymbol{\Omega}\mathbf{y}, \qquad (6.49)\end{aligned}$$

where $c_0 = \frac{1}{2}\ln(|\boldsymbol{\Sigma}_2|/|\boldsymbol{\Sigma}_1|) - \frac{1}{2}(\boldsymbol{\mu}_1'\boldsymbol{\Sigma}_1^{-1}\boldsymbol{\mu}_1 - \boldsymbol{\mu}_2'\boldsymbol{\Sigma}_2^{-1}\boldsymbol{\mu}_2)$, $\boldsymbol{\gamma}' = \boldsymbol{\mu}_1'\boldsymbol{\Sigma}_1^{-1} - \boldsymbol{\mu}_2'\boldsymbol{\Sigma}_2^{-1}$, and $\boldsymbol{\Omega} = \boldsymbol{\Sigma}_2^{-1} - \boldsymbol{\Sigma}_1^{-1}$.

Even though the function in (6.49) is not linear in the y's, it is linear in the parameters. For example, if $p = 3$, we have

$$\begin{aligned}\ln \frac{f(\mathbf{y}\,|\,G_1)}{f(\mathbf{y}\,|\,G_2)} &= c_0 + \gamma_1 y_1 + \gamma_2 y_2 + \gamma_3 y_3 + \omega_{11} y_1^2 + \omega_{22} y_2^2 + \omega_{33} y_3^2 \\ &\quad + 2\omega_{12} y_1 y_2 + 2\omega_{13} y_1 y_3 + 2\omega_{23} y_2 y_3 \\ &= c_0 + \boldsymbol{\beta}'\mathbf{z},\end{aligned}$$

where

$$\begin{aligned}\mathbf{z} &= (y_1, y_2, y_3, y_1^2, y_2^2, y_3^2, 2y_1y_2, 2y_1y_3, 2y_2y_3)', \\ \boldsymbol{\beta} &= (\gamma_1, \gamma_2, \gamma_3, \omega_{11}, \omega_{22}, \omega_{33}, \omega_{12}, \omega_{13}, \omega_{23})'.\end{aligned}$$

Thus we have a linear logistic model as before, and in principle, the parameters can be estimated using the same iterative technique. However, the number of parameters is $1 + p(p + 3)/2$, which becomes large as p increases. For example, if $p = 5$ there are 21 parameters, and if $p = 10$ there are 66. Unless the sample size is very large, the parameters cannot be estimated at all or cannot be estimated with reasonable precision. Anderson (1975) discussed quadratic logistic classification in detail and suggested some approximations for $\mathbf{y}'\boldsymbol{\Omega}\mathbf{y}$ that reduce the number of parameters to be estimated.

6.8.4 Logistic Classification for Several Groups

The logistic approach has been extended to the several-group case by Cox (1966), Anderson (1972), Marshall and Chisholm (1985), and Albert and Lesaffre (1986). Bull and Donner (1987a, 1987b) compared the logistic approach with the usual linear classification functions in the several-group setting. Gong (1986) employed a forward logistic procedure to select a subset of variables and compared three methods of estimating error rates: leaving-one-out, the bootstrap, and the jackknife (see Efron 1982). Gong concluded that the bootstrap was superior to the other two approaches.

Lesaffre, Willems, and Albert (1989) showed by simulation that for logistic classification the resubstitution estimator yields even more optimistic estimators of the error rate than does resubstitution with ordinary linear or quadratic classification. They proposed an approximate holdout technique, noting that the ordinary holdout method applied to logistic classification would require a prohibitive amount of computer time.

6.8.5 Additional Topics in Logistic Classification

Cox and Ferry (1991) proposed a model for logistic classification for two groups that is robust against outlying observations. The logistic model (6.43) can be written as

$$f(\mathbf{y} \mid G_1) = e^{\alpha+\boldsymbol{\beta}'\mathbf{y}} f(\mathbf{y} \mid G_2). \tag{6.50}$$

From (6.50) we see that either $f(\mathbf{y} \mid G_1)$ or $f(\mathbf{y} \mid G_2)$ can be much greater than the other, depending on the values of α and $\boldsymbol{\beta}$. Outliers will affect the estimation of α and $\boldsymbol{\beta}$. To guard against this influence of outliers, Cox and Ferry suggested other forms of (6.50), one of which is

$$f(\mathbf{y} \mid G_1) = \frac{e^{\gamma} + e^{\alpha+\boldsymbol{\beta}'\mathbf{y}}}{1 + e^{\gamma} e^{\alpha+\boldsymbol{\beta}'\mathbf{y}}} f(\mathbf{y} \mid G_2), \tag{6.51}$$

so that $f(\mathbf{y} \mid G_1)/f(\mathbf{y} \mid G_2)$ is bounded above and below by e^{γ} and $e^{-\gamma}$. In a simulation as well as an example, the robust model (6.51) outperformed ordinary logistic classification in terms of error rate.

Pregibon (1981), Cook and Weisberg (1982, Chapter 2), and Johnson (1985) provided methods for identifying observations in the data that are influential in any of three ways: (1) the estimation of α and $\boldsymbol{\beta}$ in (6.43); (2) the estimation of $P(G_1 \mid \mathbf{y})$ and $P(G_2 \mid \mathbf{y})$ in (6.46) and (6.47); and (3) the classification of future observations. Bhattacharajee and Dunsmore (1991) considered the influence of the y's on $P(G_1 \mid \mathbf{y})$ and $P(G_2 \mid \mathbf{y})$.

Le Cessie and Van Houwelingen (1992) proposed the use of ridge estimators (see Sections 5.11.1 and 6.9) in logistic classification to improve the parameter estimates and to reduce the error rates in classification. Hosmer et al. (1989) discussed subset selection for logistic regression.

Lesaffre et al. (1991) discussed the following techniques as applied to logistic classification for several groups: (1) significance tests for the logistic assumption (6.43); (2) procedures to detect outliers, influential observations, or high leverage points; and (3) partial residual plots (Larsen and McCleary 1972) to check whether a variable y_i should be included and how the functional form should be altered accordingly.

6.8.6 Probit Classification

In some cases the groups are defined *quantitatively* rather than *qualitatively*. For example, we could partition a single group of students into two groups on the basis of a continuous variable such as grade point average (GPA) so that we have a "high-GPA" group and a "low-GPA" group. Using a vector $\mathbf{y}$ of test scores and other measurements, we wish to predict membership in the two GPA groups. Albert and Anderson (1981) discussed *probit* classification functions for this model, based on a similar approach in regression (Finney 1971, Neter et al. 1996, pp. 572–573). Klemm and Gust (1982) and Grablowsky and Talley (1981) compared the performance of the probit method of classification with that of the linear classification functions. Albert and Anderson (1981) showed that the probit method can often be satisfactorily modeled using the logistic approach.

In the following discussion, we follow Albert and Anderson (1981). Let z be the continuous random variable such as GPA in the above illustration. If t is a threshold value, then an individual belongs to G_2 (for example, low GPA) if $z \leq t$ and to G_1 (high GPA) if $z > t$. (We have designated G_1 and G_2 in this way so as to correspond to the use of G_1 and G_2 in logistic classification in Section 6.8.1.)

To begin, we assume that $(z, \mathbf{y}')$ is $N_{p+1}(\boldsymbol{\mu}, \boldsymbol{\Sigma})$, where

$$\boldsymbol{\mu} = \begin{pmatrix} \mu_z \\ \boldsymbol{\mu}_y \end{pmatrix} \quad \text{and} \quad \boldsymbol{\Sigma} = \begin{pmatrix} \sigma_z^2 & \boldsymbol{\sigma}'_{zy} \\ \boldsymbol{\sigma}_{zy} & \boldsymbol{\Sigma}_{yy} \end{pmatrix}.$$

Then by Theorem 2.2F, the conditional distribution of z given $\mathbf{y}$ is normal with

$$E(z \mid \mathbf{y}) = \mu_z + \boldsymbol{\sigma}'_{zy}\boldsymbol{\Sigma}_{yy}^{-1}(\mathbf{y} - \boldsymbol{\mu}_y), \tag{6.52}$$

$$\text{var}(z \mid \mathbf{y}) = \sigma_z^2 - \boldsymbol{\sigma}'_{zy}\boldsymbol{\Sigma}_{yy}^{-1}\boldsymbol{\sigma}_{zy}. \tag{6.53}$$

Hence

$$\begin{aligned} P(G_1 \mid \mathbf{y}) &= P(z > t \mid \mathbf{y}) \\ &= P\left(\frac{z - \mu_{z|y}}{\sigma_{z|y}} > \frac{t - \mu_{z|y}}{\sigma_{z|y}}\right) \\ &= 1 - \Phi\left(\frac{t - \mu_{z|y}}{\sigma_{z|y}}\right) \\ &= \Phi\left(\frac{-t + \mu_{z|y}}{\sigma_{z|y}}\right), \end{aligned} \tag{6.54}$$

where $\Phi(\cdot)$ is the standard normal distribution function. Using (6.52) and (6.53) for $\mu_{z|y}$ and $\sigma^2_{z|y}$, (6.54) becomes

$$\begin{aligned} P(G_1 \mid \mathbf{y}) &= \Phi\left[\frac{-t + \mu_z + \boldsymbol{\sigma}'_{zy}\boldsymbol{\Sigma}_{yy}^{-1}(\mathbf{y} - \boldsymbol{\mu}_y)}{\sqrt{\sigma_z^2 - \boldsymbol{\sigma}'_{zy}\boldsymbol{\Sigma}_{yy}^{-1}\boldsymbol{\sigma}_{zy}}}\right] \\ &= \Phi(\gamma_o + \boldsymbol{\gamma}'_1\mathbf{y}), \end{aligned} \tag{6.55}$$

where

$$\gamma_0 = -(t - \mu_z + \boldsymbol{\sigma}'_{zy}\boldsymbol{\Sigma}_{yy}^{-1}\boldsymbol{\mu}_y)/\sqrt{\sigma_z^2 - \boldsymbol{\sigma}'_{zy}\boldsymbol{\Sigma}_{yy}^{-1}\boldsymbol{\sigma}_{zy}}$$

and

$$\boldsymbol{\gamma}_1' = \boldsymbol{\sigma}_{zy}'\boldsymbol{\Sigma}_{yy}^{-1}\Big/\sqrt{\sigma_z^2 - \boldsymbol{\sigma}_{zy}'\boldsymbol{\Sigma}_{yy}^{-1}\boldsymbol{\sigma}_{zy}}.$$

The classification rule is: Assign $\mathbf{y}$ to G_1 if $P(z > t \mid \mathbf{y}) \geq P(z < t \mid \mathbf{y})$, that is, if $P(G_1 \mid \mathbf{y}) \geq P(G_2 \mid \mathbf{y})$. By (6.55), this rule becomes: Assign $\mathbf{y}$ to G_1 if $\Phi(\gamma_0 + \boldsymbol{\gamma}_1'\mathbf{y}) \geq 1 - \Phi(\gamma_0 + \boldsymbol{\gamma}_1'\mathbf{y})$, which is equivalent to $\Phi(\gamma_0 + \boldsymbol{\gamma}_1'\mathbf{y}) \geq \frac{1}{2}$. In terms of $\gamma_0 + \boldsymbol{\gamma}_1'\mathbf{y}$, the rule can be expressed as: Assign $\mathbf{y}$ to G_1 if

$$\gamma_0 + \boldsymbol{\gamma}_1'\mathbf{y} \geq 0 \tag{6.56}$$

and to G_2 otherwise.

Maximum likelihood estimates of γ_0 and $\boldsymbol{\gamma}_1$ (requiring iterative solution) can be obtained using a dichotomized form of $z : w = 0$ if $z \leq t$, $w = 1$ if $z > t$. It is not necessary that $\mathbf{y}$ be multivariate normal, only that the conditional distribution of z given $\mathbf{y}$ be normal. This permits the use of discrete y's.

Example 6.8.6. The probit classification method was applied to the data of Example 6.8.1 by regarding the formation of deep vein thrombosis as varying continuously. The linear function $\gamma_0 + \boldsymbol{\gamma}_1'\mathbf{y}$ of (6.56) was estimated as

$$\begin{aligned} u &= \hat{\gamma}_0 + \hat{\boldsymbol{\gamma}}_1'\mathbf{y} \\ &= -6.22 + .0049y_1 + .112y_2 + .0485y_3 + .020y_4 + 1.19y_5. \end{aligned} \tag{6.57}$$

Using the classification rule in (6.56), the apparent error rate based on resubstitution was the same as that of the logistic method in Example 6.8.1.

Comparing (6.57) to (6.48) in Example 6.8.1, we see that the coefficients in (6.57) are slightly over $\frac{1}{2}$ those in (6.48). This relationship, along with the identical error rates obtained, illustrates the correspondence between logistic classification and probit classification.

6.9 ADDITIONAL TOPICS IN CLASSIFICATION

A ridge-type adjustment for discriminant function coefficients was discussed in Section 5.11.1 for the two-group case. A ridge adjustment can also be used in classification analysis when $\boldsymbol{\Sigma}$ is near singular. DiPillo (1976, 1979) and Smidt and McDonald (1976) have shown that the use of $(\mathbf{S} + k\mathbf{I})^{-1}$ in place of $\mathbf{S}^{-1}$ in classification rules gives improved error rates when $\boldsymbol{\Sigma}$ is near singular and appears not to do worse otherwise. This improvement is, of course, dependent on finding an appropriate value of the biasing parameter k. In practice, this may be difficult, and several procedures are suggested in the three papers. Loh (1993) gave an expression for the biasing parameter in a ridge classification scheme and showed by simulation that the resulting classification rule has reasonably good properties.

Fatti (1983) proposed a random-effects model for classification analysis in which the group means $\boldsymbol{\mu}_1, \boldsymbol{\mu}_2, \ldots, \boldsymbol{\mu}_k$ have been randomly selected from some population.

He investigated the behavior of the usual classification rule given in (6.20) under the random-effects model.

Krzanowski (1980, 1983) and Wernecke et al. (1988) described classification methods for categorical data. Wernecke (1992) proposed a coupling procedure for use with data that include both continuous and categorical variables. In some examples the coupling procedure showed improved error rates over procedures that use the continuous variables alone or the categorical variables alone. Krzanowski (1993) discussed the use of the "conditional Gaussian distribution" for mixtures of categorical and continuous variables. This distribution is defined as the conditional distribution (multivariate normal) of the continuous variables (given the values of the categorical variables) multiplied by the marginal distribution (multinomial) of the latter.

Back, Gray, and Woodward (1994) presented a classification rule (for two groups) in which one of the misclassification probabilities can be controlled. The method, which is obtained by applying the bootstrap to the generalized likelihood ratio, can be used with any mixture of continuous and discrete variables. Feldmann (1993) proposed a classification technique that allows for ordered or partially ordered categories of the response variable in the logistic and other approaches. He illustrated with a data set for which the three possible outcomes were (1) the postoperative hospital stay is two weeks or less, (2) the postoperative hospital stay is more than two weeks, and (3) the patient dies during the hospital stay. Villarroya, Rios, and Oller (1995) proposed a classification algorithm (for several groups) based on a distance function. In many cases, their method resulted in improved error rates when compared with linear or quadratic classification.

As noted in Section 6.2.5, the linear classification functions are fairly robust to nonnormality. However, they can be sensitive to heavy tails or outliers in the data. One approach to remedy this deficiency is the method of classification based on linear programming proposed by Freed and Glover (1981a, 1981b, 1982). Lee and Ord (1990) proposed a modification based on least absolute deviations. Koehler and Erenguc (1990) considered another programming approach, and Celeux and Mkhadri (1992) proposed a method of regularized "discriminant" analysis (Section 6.3.2) for binary data. Hastie and Tibshirani (1996) used Gaussian mixtures to improve classfication of nonnormal data.

Campbell (1978) and Fung (1992, 1995) discussed the influence that individual observations have on the estimation of parameters, and Johnson (1987) considered the influence of individual observations on the classification of future observations. Critchley and Vitiello (1991), Fung (1992), and Whitcomb and Lahiff (1993) examined the influence of observations on error rates. In all six of the above papers, only the two-group case was considered.

Kshirsagar, Kocherlakota, and Kocherlakota (1990) proposed a method of classification based on principal components (Chapter 9) for the case in which the number of variables exceeds the sample size. They considered both the two-group and several-group cases.

Twedt and Gill (1992) considered the effect of imputation for missing data in classification analysis for two groups. They compared three methods of replacing missing data and found that the three performed similarly and that all three were better than not replacing missing observations. Dubuisson and Masson (1993) presented a classification technique for use when the number of groups is not known or when no observations are available from some of the known groups.

Patuwo, Hu, and Hung (1993) discussed the use of *neural networks* for two-group classification for "dirty" data sets with unknown distributions and unknown covariance matrices. The method is based on models for biological cognition and learning. It involves a layered network of nodes and arcs whose values and weights are determined using a training sample. For a detailed development of the technique, see Wythoff (1993).

The parametric classification rules in (6.1) and (6.18) require an assumption about the density $f(\mathbf{y} \mid G_i)$. This density can be estimated from the data in a *nonparametric* approach for use in classification and estimation of error rates. Another nonparametric approach is the *k-nearest-neighbor* method, in which an observation $\mathbf{y}_i$ is classified according to the group membership of the majority of its nearest neighbors. A review of these procedures is given by Rencher (1995, Section 9.7). Buturovic (1993) suggested modifications of these methods for improved error rates. Granville and Rasson (1995) proposed classification methods (for several groups) based on penalized likelihood and adaptive bandwith. The method uses the density estimation and nearest neighbor approaches and is suitable for high-dimensional data.

PROBLEMS

6.1 Prove Theorem 6.2A.

6.2 Show that the expression in (6.6) for $f(\mathbf{y} \mid G_1)/f(\mathbf{y} \mid G_2)$ can be obtained from (6.5).

6.3 Show that if $p_1 = p_2$, the classification rule in (6.8) is equivalent to the following: Assign $\mathbf{y}$ to G_1 if $D_1^2 < D_2^2$, where $D_i^2 = (\mathbf{y} - \bar{\mathbf{y}}_i)'\mathbf{S}_{\text{pl}}^{-1}(\mathbf{y} - \bar{\mathbf{y}}_i)$, $i = 1, 2$.

6.4 Show that $Q(\mathbf{y})$ in (6.10) follows from (6.9).

6.5 Verify that $P(G_i \mid \mathbf{y})$ in (6.22) can be obtained from (6.21).

6.6 Verify the expression in (6.32) for $P(\text{classify as } G_2 \mid G_1)$.

6.7 Show that $E[(n_1 + n_2 - p - 3)\mathbf{S}^{-1}/(n_1 + n_2 - 2)] = \mathbf{\Sigma}^{-1}$ as in (6.40).

6.8 Show that $2p + \binom{p}{2} = \frac{1}{2}p(p + 3)$. This was given in Section 6.5.2 for the number of parameters in $\boldsymbol{\mu}_i$ and $\mathbf{\Sigma}$.

6.9 Verify that the probability is approximately .368 that a given observation will not appear when sampling with replacement, as in Section 6.5.3.

6.10 Verify the logistic model in (6.43).

6.11 Verify the expression for $P(G_1 \mid \mathbf{y})$ in (6.45).

6.12 Show that for the logistic model in (6.43) we have

$$P(G_2 \mid \mathbf{y}) = \frac{1}{1 + e^{\alpha_0 + \boldsymbol{\beta}'\mathbf{y}}}$$

as in (6.47).

6.13 Do a classification analysis on the beetle data in Table 3.8 as follows:

(a) Find the classification rule (6.8). Assume $p_1 = p_2$. Use the method of resubstitution to estimate the error rate.

(b) Find the quadratic classification function (6.12) (assuming $\mathbf{\Sigma}_1 \neq \mathbf{\Sigma}_2$). Use the method of resubstitution to estimate the error rate.

6.14 Do a classification analysis on the soils data in Table 3.6 as follows:

(a) Find the classification rule (6.8). Assume $p_1 = p_2$. Use the method of resubstitution to estimate the error rate.

(b) Find the quadratic classification function (6.12) (assuming $\mathbf{\Sigma}_1 \neq \mathbf{\Sigma}_2$). Use the method of resubstitution to estimate the error rate.

(c) Repeat part (a) using p_1 and p_2 proportional to sample sizes.

6.15 Use $y_1, y_2, \ldots, y_6$ and the first two groups (treatments *A* and *B*) from the apple data in Table 4.10 (ignore blocks).

(a) Find the classification rule (6.8). Assume $p_1 = p_2$. Use the method of resubstitution to estimate the error rate.

(b) Use the plug-in method to estimate the error rate.

(c) Use the modified plug-in method to estimate the error rate.

(d) Use the method of logistic regression to estimate the error rate.

6.16 Use $y_1, y_2, \ldots, y_6$ and the first two groups (treatments *A* and *B*) from the apple data in Table 4.10 (ignore blocks).

(a) Carry out a stepwise discriminant selection of variables (see Section 3.11.3).

(b) For the variables selected in part (a), find the classification rule (6.8), assuming $p_1 = p_2$. Use the method of resubstitution to estimate the error rate. Compare with the results of part (a) of Problem 6.15.

6.17 Do a classification analysis on all four groups from the apple data in Table 4.10 using $y_1, y_2, \ldots, y_6$. Assume $p_1 = p_2 = p_3 = p_4$.

(a) Find the linear classification functions (assuming $\mathbf{\Sigma}_1 = \mathbf{\Sigma}_2 = \mathbf{\Sigma}_3 = \mathbf{\Sigma}_4$).

(b) Use the method of resubstitution to estimate the error rate.

(c) Find the quadratic classification functions in (6.23) (assuming population covariance matrices are not equal).

(d) Use the method of resubstitution to estimate the error rate.

6.18 Do a classification analysis on $y_1, y_2, \ldots, y_{11}$ of the biochemical data of Table 4.11 as follows:

(a) Find the linear classification functions (assuming $\mathbf{\Sigma}_1 = \mathbf{\Sigma}_2 = \mathbf{\Sigma}_3 = \mathbf{\Sigma}_4$).

(b) Use the method of resubstitution to estimate the error rate.

(c) Use the holdout method (Section 6.5.2) to estimate the error rate.

6.19 Use $y_1, y_2, \ldots, y_{11}$ of the biochemical data of Table 4.11.

(a) For the variables selected by stepwise discriminant analysis in Problem 5.24, find the linear classification functions.

(b) Use the method of resubstitution to estimate the error rate.

(c) Compare the results of part (b) with the results in part (b) of Problem 6.18.

6.20 Use $y_1, y_2, \ldots, y_6$ in all four groups of the apple data of Table 4.10.

(a) For the variables selected by stepwise discriminant analysis in Problem 5.23, find the classification table using linear classification functions.

(b) Compare the results of part (a) with the results in part (b) of Problem 6.17.

CHAPTER 7

Multivariate Regression

7.1 INTRODUCTION

In ordinary *multiple regression*, we attempt to predict a *dependent* or *response* variable y by assuming that it has a linear relationship with several *independent* variables x_1, $x_2, \ldots, x_q$. In *multivariate multiple regression* we have several dependent variables $y_1, y_2, \ldots, y_p$ to be predicted by linear relationships with $x_1, x_2, \ldots, x_q$.

The y's are random variables. In some cases the x's are fixed, that is, controlled by the experimenter, and in other cases the x's are random variables. Multiple regression with fixed x's is reviewed in Section 7.2, and the random-x case is covered in Section 7.3. Multivariate multiple regression with both fixed and random x's is treated in Sections 7.4–7.8.

In addition to constructing a model for prediction, we may wish to measure the strength of the relationship between one or more y's and one or more x's. For one y and one x, we use the ordinary Pearson correlation coefficient r. For one y and several x's, we use the multiple correlation coefficient R (Sections 7.2.5 and 7.3.3). For measures of correlation between several y's and several x's, we use *canonical correlations* (Chapter 8).

In the random-x case in Sections 7.3 and 7.7, the x's are assumed to be continuous. Fixed x's (Sections 7.2 and 7.4–7.6) are either discrete or continuous. In ANOVA (Chapter 4), the x's are fixed and discrete. In this chapter, all y's are treated as continuous. The logistic classification model in Section 6.8 is related to a logistic regression model with binary y-values. *Generalized linear models* involve y's that are categorical, discrete, or continuous. Various aspects of generalized linear models have been treated as follows: Dobson (1990) and McCullagh and Nelder (1989) provided basic development, Percy (1993) discussed prediction, Cordeiro and McCullagh (1991) gave corrections for bias, and Thomas and Cook (1989) proposed diagnostics to check for various types of influential observations.

7.2 MULTIPLE REGRESSION: FIXED x's

Useful applied expositions of multiple regression for the fixed-x case can be found in Draper and Smith (1981), Morrison (1983), Myers (1990), and Neter et al. (1996). Theoretical treatments are given by Graybill (1976), Seber (1977), Guttman (1982), Wang and Chow (1994), and Hocking (1976, 1985, 1996). We give a review of the main results.

7.2.1 Least Squares Estimators and Properties

In the fixed-x regression model, we assume that the mean of y can be expressed as a linear function (linear in the β's) of q x's:

$$E(y) = \beta_0 + \beta_1 x_1 + \beta_2 x_2 + \cdots + \beta_q x_q. \tag{7.1}$$

To estimate the β's, we use data from n independent observations on y and the associated x's. The model for the ith observation is

$$y_i = \beta_0 + \beta_1 x_{i1} + \beta_2 x_{i2} + \cdots + \beta_q x_{iq} + \varepsilon_i, \quad i = 1, 2, \ldots, n, \tag{7.2}$$

where ε_i is a random error term.

The models for the n observations in (7.2) can be written in matrix form:

$$\begin{pmatrix} y_1 \\ y_2 \\ \vdots \\ y_n \end{pmatrix} = \begin{pmatrix} 1 & x_{11} & x_{12} & \cdots & x_{1q} \\ 1 & x_{21} & x_{22} & \cdots & x_{2q} \\ \vdots & \vdots & \vdots & & \vdots \\ 1 & x_{n1} & x_{n2} & \cdots & x_{nq} \end{pmatrix} \begin{pmatrix} \beta_0 \\ \beta_1 \\ \vdots \\ \beta_q \end{pmatrix} + \begin{pmatrix} \varepsilon_1 \\ \varepsilon_2 \\ \vdots \\ \varepsilon_n \end{pmatrix}$$

or

$$\mathbf{y} = \mathbf{X}\boldsymbol{\beta} + \boldsymbol{\varepsilon}. \tag{7.3}$$

Desirable assumptions on $\boldsymbol{\varepsilon}$ or $\mathbf{y}$ that lead to good estimators of the β_j's are

1. $E(\boldsymbol{\varepsilon}) = \mathbf{0}$ or $E(\mathbf{y}) = \mathbf{X}\boldsymbol{\beta}$ and
2. $\text{cov}(\boldsymbol{\varepsilon}) = \sigma^2\mathbf{I}$ or $\text{cov}(\mathbf{y}) = \sigma^2\mathbf{I}$.

We now review the *least squares approach* to estimation of the β's in the fixed-x model. No distributional assumptions are required.

For the parameters $\beta_0, \beta_1, \ldots, \beta_q$, we seek estimators that minimize the sum of squares of deviations of the n observed y's from their predicted values. Thus we seek $\hat{\beta}_0, \hat{\beta}_1, \ldots, \hat{\beta}_q$ that minimize

$$\begin{aligned} \sum_{i=1}^{n} \hat{\varepsilon}_i^2 &= \sum_{i=1}^{n} (y_i - \hat{y}_i)^2 \\ &= \sum_{i=1}^{n} (y_i - \hat{\beta}_0 - \hat{\beta}_1 x_{i1} - \hat{\beta}_2 x_{i2} - \cdots - \hat{\beta}_q x_{iq})^2. \end{aligned} \tag{7.4}$$

We can express (7.4) as

$$\hat{\boldsymbol{\varepsilon}}'\hat{\boldsymbol{\varepsilon}} = \sum_{i=1}^{n}(y_i - \mathbf{x}_i'\hat{\boldsymbol{\beta}})^2 = (\mathbf{y} - \mathbf{X}\hat{\boldsymbol{\beta}})'(\mathbf{y} - \mathbf{X}\hat{\boldsymbol{\beta}}), \tag{7.5}$$

where $\mathbf{x}_i' = (1, x_{i1}, \ldots, x_{iq})$ is the ith row of $\mathbf{X}$.

The value of $\hat{\boldsymbol{\beta}} = (\hat{\beta}_0, \hat{\beta}_1, \ldots, \hat{\beta}_q)'$ that minimizes (7.4) or (7.5) is given by

$$\hat{\boldsymbol{\beta}} = (\mathbf{X}'\mathbf{X})^{-1}\mathbf{X}'\mathbf{y}. \tag{7.6}$$

The least squares estimator $\hat{\boldsymbol{\beta}} = (\mathbf{X}'\mathbf{X})^{-1}\mathbf{X}'\mathbf{y}$ in (7.6) is obtained without the use of the assumptions $E(\mathbf{y}) = \mathbf{X}\boldsymbol{\beta}$ and $\text{cov}(\mathbf{y}) = \sigma^2\mathbf{I}$. However, if $E(\mathbf{y}) = \mathbf{X}\boldsymbol{\beta}$ and $\text{cov}(\mathbf{y}) = \sigma^2\mathbf{I}$ hold, $\hat{\boldsymbol{\beta}}$ has the following three properties [(7.7), (7.8), and Theorem 7.2A]:

1. If $E(\mathbf{y}) = \mathbf{X}\boldsymbol{\beta}$, then $\hat{\boldsymbol{\beta}}$ is an unbiased estimator for $\boldsymbol{\beta}$; that is,

$$E(\hat{\boldsymbol{\beta}}) = \boldsymbol{\beta}. \tag{7.7}$$

2. If $\text{cov}(\mathbf{y}) = \sigma^2\mathbf{I}$, then

$$\text{cov}(\hat{\boldsymbol{\beta}}) = \sigma^2(\mathbf{X}'\mathbf{X})^{-1}. \tag{7.8}$$

Theorem 7.2A (Gauss–Markov Theorem). If $E(\mathbf{y}) = \mathbf{X}\boldsymbol{\beta}$ and $\text{cov}(\mathbf{y}) = \sigma^2\mathbf{I}$, the least squares estimators $\hat{\beta}_j$, $j = 0, 1, \ldots, q$, have minimum variance among all linear unbiased estimators. □

The Gauss–Markov theorem is sometimes stated as follows: If $E(\mathbf{y}) = \mathbf{X}\boldsymbol{\beta}$ and $\text{cov}(\mathbf{y}) = \sigma^2\mathbf{I}$, the least squares estimators are "best linear unbiased estimators" (BLUE). In this statement, "best" means minimum variance, and "linear" indicates that the estimators are linear functions of $\mathbf{y}$.

7.2.2 An Estimator for σ^2

By assumption 2 following (7.3), $\sigma^2 = \text{var}(y_i) = E[y_i - E(y_i)]^2$ is the same for all i. We estimate σ^2 by an average from the sample,

$$s^2 = \frac{1}{n-q-1}\sum_{i=1}^{n}(y_i - \mathbf{x}_i'\hat{\boldsymbol{\beta}})^2, \tag{7.9}$$

$$= \frac{1}{n-q-1}(\mathbf{y} - \mathbf{X}\hat{\boldsymbol{\beta}})'(\mathbf{y} - \mathbf{X}\hat{\boldsymbol{\beta}}) \qquad \text{[by (7.5)]} \tag{7.10}$$

$$= \frac{\mathbf{y}'\mathbf{y} - \hat{\boldsymbol{\beta}}'\mathbf{X}'\mathbf{y}}{n-q-1} \qquad \text{[by (A.2.7)]} \tag{7.11}$$

$$= \frac{\text{SSE}}{n-q-1},$$

where $\text{SSE} = (\mathbf{y} - \mathbf{X}\hat{\boldsymbol{\beta}})'(\mathbf{y} - \mathbf{X}\hat{\boldsymbol{\beta}}) = \mathbf{y}'\mathbf{y} - \hat{\boldsymbol{\beta}}'\mathbf{X}'\mathbf{y}$. It can be shown that

$$E(\text{SSE}) = \sigma^2[n - (q + 1)] = \sigma^2(n - q - 1) \tag{7.12}$$

(see Problem 7.3), and therefore $E(s^2) = \sigma^2$.

7.2.3 The Model in Centered Form

The model in (7.2) for each y_i can be written in terms of centered x's as

$$y_i = \alpha + \beta_1(x_{i1} - \bar{x}_1) + \beta_2(x_{i2} - \bar{x}_2) + \cdots + \beta_q(x_{iq} - \bar{x}_q) + \varepsilon_i, \tag{7.13}$$

$i = 1, 2, \ldots, n$, where

$$\alpha = \beta_0 + \beta_1\bar{x}_1 + \beta_2\bar{x}_2 + \cdots + \beta_q\bar{x}_q, \tag{7.14}$$

and $\bar{x}_j = \sum_{i=1}^n x_{ij}/n$, $j = 1, 2, \ldots, q$. In matrix form corresponding to (7.3), the centered model (7.13) for $y_1, y_2, \ldots, y_n$ becomes

$$\mathbf{y} = (\mathbf{j}, \mathbf{X}_c)\begin{pmatrix} \alpha \\ \boldsymbol{\beta}_1 \end{pmatrix} + \boldsymbol{\varepsilon}, \tag{7.15}$$

where

$$\mathbf{X}_c = \begin{pmatrix} x_{11} - \bar{x}_1 & x_{12} - \bar{x}_2 & \cdots & x_{1q} - \bar{x}_q \\ x_{21} - \bar{x}_1 & x_{22} - \bar{x}_2 & \cdots & x_{2q} - \bar{x}_q \\ \vdots & \vdots & & \vdots \\ x_{n1} - \bar{x}_1 & x_{n2} - \bar{x}_2 & \cdots & x_{nq} - \bar{x}_q \end{pmatrix}, \qquad \boldsymbol{\beta}_1 = \begin{pmatrix} \beta_1 \\ \beta_2 \\ \vdots \\ \beta_q \end{pmatrix}, \tag{7.16}$$

and $\mathbf{j}$ is a vector of 1s. As in (7.6), the least squares estimator for $\binom{\alpha}{\boldsymbol{\beta}_1}$ is

$$\begin{pmatrix} \hat{\alpha} \\ \hat{\boldsymbol{\beta}}_1 \end{pmatrix} = [(\mathbf{j}, \mathbf{X}_c)'(\mathbf{j}, \mathbf{X}_c)]^{-1}(\mathbf{j}, \mathbf{X}_c)'\mathbf{y}. \tag{7.17}$$

The product $(\mathbf{j}, \mathbf{X}_c)'(\mathbf{j}, \mathbf{X}_c)$ in (7.17) becomes

$$\begin{aligned}(\mathbf{j}, \mathbf{X}_c)'(\mathbf{j}, \mathbf{X}_c) &= \begin{pmatrix} \mathbf{j}' \\ \mathbf{X}_c' \end{pmatrix}(\mathbf{j}, \mathbf{X}_c) = \begin{pmatrix} \mathbf{j}'\mathbf{j} & \mathbf{j}'\mathbf{X}_c \\ \mathbf{X}_c'\mathbf{j} & \mathbf{X}_c'\mathbf{X}_c \end{pmatrix} \\ &= \begin{pmatrix} n & \mathbf{0}' \\ \mathbf{0} & \mathbf{X}_c'\mathbf{X}_c \end{pmatrix}, \end{aligned} \tag{7.18}$$

where $\mathbf{j}'\mathbf{X}_c = \mathbf{0}'$ because the columns of $\mathbf{X}_c$ sum to zero (Problem 7.4). Similarly,

$$(\mathbf{j}, \mathbf{X}_c)'\mathbf{y} = \begin{pmatrix} \mathbf{j}' \\ \mathbf{X}_c' \end{pmatrix}\mathbf{y} = \begin{pmatrix} n\bar{y} \\ \mathbf{X}_c'\mathbf{y} \end{pmatrix}.$$

The least squares estimators (7.17) are now given by

$$\begin{pmatrix} \hat{\alpha} \\ \hat{\boldsymbol{\beta}}_1 \end{pmatrix} = \left[(\mathbf{j}, \mathbf{X}_c)'(\mathbf{j}, \mathbf{X}_c)\right]^{-1} (\mathbf{j}, \mathbf{X}_c)'\mathbf{y} = \begin{pmatrix} n & \mathbf{0}' \\ \mathbf{0} & \mathbf{X}_c'\mathbf{X}_c \end{pmatrix}^{-1} \begin{pmatrix} n\bar{y} \\ \mathbf{X}_c'\mathbf{y} \end{pmatrix}$$

$$= \begin{pmatrix} 1/n & \mathbf{0}' \\ \mathbf{0} & (\mathbf{X}_c'\mathbf{X}_c)^{-1} \end{pmatrix} \begin{pmatrix} n\bar{y} \\ \mathbf{X}_c'\mathbf{y} \end{pmatrix} = \begin{pmatrix} \bar{y} \\ (\mathbf{X}_c'\mathbf{X}_c)^{-1}\mathbf{X}_c'\mathbf{y} \end{pmatrix},$$

or

$$\hat{\alpha} = \bar{y}, \tag{7.19}$$

$$\hat{\boldsymbol{\beta}}_1 = (\mathbf{X}_c'\mathbf{X}_c)^{-1}\mathbf{X}_c'\mathbf{y}. \tag{7.20}$$

These estimators are the same as the usual least squares estimators in (7.6), with the adjustment

$$\hat{\beta}_0 = \hat{\alpha} - \hat{\beta}_1\bar{x}_1 - \hat{\beta}_2\bar{x}_2 - \cdots - \hat{\beta}_q\bar{x}_q \tag{7.21}$$

obtained from an estimator of α as defined in (7.14).

We can use (7.19)–(7.21) to express $\hat{\boldsymbol{\beta}}_1$ and $\hat{\beta}_0$ in terms of sample variances and covariances. By (1.43), the centered matrix $\mathbf{X}_c$ can be written as

$$\mathbf{X}_c = \left(\mathbf{I} - \frac{1}{n}\mathbf{J}\right)\mathbf{X}_1, \tag{7.22}$$

where $\mathbf{J}$ is a square matrix of 1s and

$$\mathbf{X}_1 = \begin{pmatrix} x_{11} & x_{12} & \cdots & x_{1q} \\ x_{21} & x_{22} & \cdots & x_{2q} \\ \vdots & \vdots & & \vdots \\ x_{n1} & x_{n2} & \cdots & x_{nq} \end{pmatrix}. \tag{7.23}$$

By (1.42),

$$\mathbf{X}_c'\mathbf{X}_c = \mathbf{X}_1'\left(\mathbf{I} - \frac{1}{n}\mathbf{J}\right)'\left(\mathbf{I} - \frac{1}{n}\mathbf{J}\right)\mathbf{X}_1 = \mathbf{X}_1'\left(\mathbf{I} - \frac{1}{n}\mathbf{J}\right)\mathbf{X}_1.$$

By (1.40) and (1.42), $\mathbf{X}_c'\mathbf{X}_c$ is the numerator of the sample covariance matrix of the x's:

$$\mathbf{S}_{xx} = \frac{1}{n-1}\mathbf{X}_c'\mathbf{X}_c = \frac{1}{n-1}\left[\mathbf{X}_1'\left(\mathbf{I} - \frac{1}{n}\mathbf{J}\right)\mathbf{X}_1\right]. \tag{7.24}$$

Note that because $\mathbf{X}_1$ is fixed, $\mathbf{S}_{xx}$ is not a random matrix that estimates a population covariance matrix. The product $\mathbf{X}_c'\mathbf{y}$ is also related to covariances:

$$\mathbf{s}_{yx} = \frac{1}{n-1}\mathbf{X}_c'\mathbf{y}, \tag{7.25}$$

where $\mathbf{s}_{yx} = (s_{y1}, s_{y2}, \ldots, s_{yq})'$ contains the sample covariances s_{yj} of y with each x_j.

Using (7.24) and (7.25), $\hat{\boldsymbol{\beta}}_1$ in (7.20) becomes

$$\hat{\boldsymbol{\beta}}_1 = (n-1)(\mathbf{X}_c'\mathbf{X}_c)^{-1}\frac{\mathbf{X}_c'\mathbf{y}}{n-1} = \left(\frac{\mathbf{X}_c'\mathbf{X}_c}{n-1}\right)^{-1}\frac{\mathbf{X}_c'\mathbf{y}}{n-1}$$

$$= \mathbf{S}_{xx}^{-1}\mathbf{s}_{yx}, \tag{7.26}$$

and from (7.19) and (7.21), we have

$$\hat{\beta}_0 = \hat{\alpha} - \hat{\boldsymbol{\beta}}_1'\overline{\mathbf{x}} = \overline{y} - \mathbf{s}_{yx}'\mathbf{S}_{xx}^{-1}\overline{\mathbf{x}}. \tag{7.27}$$

7.2.4 Hypothesis Tests and Confidence Intervals

To obtain tests and confidence intervals, we assume that $\mathbf{y}$ in (7.3) is distributed as $N_n(\mathbf{X}\boldsymbol{\beta}, \sigma^2\mathbf{I})$.

7.2.4a Test for Significance of Regression

We first describe a test for the overall regression hypothesis that none of the x's predict y in (7.2). This hypothesis can be expressed as $H_0: \boldsymbol{\beta}_1 = \mathbf{0}$, where $\boldsymbol{\beta}_1 = (\beta_1, \beta_2, \ldots, \beta_q)'$ as in (7.16). To obtain a test of H_0, we begin by partitioning the (corrected) total sum of squares $\mathbf{y}'\mathbf{y} - n\overline{y}^2$ as

$$\mathbf{y}'\mathbf{y} - n\overline{y}^2 = (\mathbf{y}'\mathbf{y} - \hat{\boldsymbol{\beta}}'\mathbf{X}'\mathbf{y}) + (\hat{\boldsymbol{\beta}}'\mathbf{X}'\mathbf{y} - n\overline{y}^2)$$

$$= \text{SSE} + \text{SSR}, \tag{7.28}$$

where $\text{SSR} = \hat{\boldsymbol{\beta}}'\mathbf{X}'\mathbf{y} - n\overline{y}^2$ is the overall regression sum of squares corrected for the mean.

It can be shown that SSE and SSR are independent and have χ^2-distributional properties. We can therefore construct an F-statistic for testing $H_0: \boldsymbol{\beta}_1 = \mathbf{0}$,

$$F = \frac{\text{SSR}/q}{\text{SSE}/(n-q-1)}, \tag{7.29}$$

which is distributed as $F_{q,n-q-1}$ when H_0 is true.

7.2.4b Test on a Subset of the β's

In some cases, a subset of the β's is of interest (for a single β_j, see Section 7.2.4d). Without loss of generality, we assume that the variables in the subset of interest have been arranged last in $\boldsymbol{\beta}$ and in the columns of $\mathbf{X}$. Then $\boldsymbol{\beta}$ and $\mathbf{X}$ can be partitioned accordingly, and by (A.3.2), the model (7.3) for all n observations becomes

$$\mathbf{y} = \mathbf{X}\boldsymbol{\beta} + \boldsymbol{\varepsilon} = (\mathbf{X}_r, \mathbf{X}_d)\begin{pmatrix}\boldsymbol{\beta}_r \\ \boldsymbol{\beta}_d\end{pmatrix} + \boldsymbol{\varepsilon}$$

$$= \mathbf{X}_r\boldsymbol{\beta}_r + \mathbf{X}_d\boldsymbol{\beta}_d + \boldsymbol{\varepsilon}, \tag{7.30}$$

where $\mathbf{X}_d$ and $\boldsymbol{\beta}_d$ pertain to the subset to be *deleted* if it is not significant, and $\mathbf{X}_r$ and $\boldsymbol{\beta}_r$ correspond to the variables *remaining* in the *reduced* model. (The intercept β_0

would ordinarily be included in $\boldsymbol{\beta}_r$.) Thus the hypothesis of interest is $H_0: \boldsymbol{\beta}_d = \mathbf{0}$, and the reduced model becomes

$$\mathbf{y} = \mathbf{X}_r\boldsymbol{\beta}_r + \boldsymbol{\varepsilon}. \tag{7.31}$$

If we designate the number of parameters in $\boldsymbol{\beta}_d$ by h, then $\boldsymbol{\beta}_r$ is $(q + 1 - h) \times 1$, $\mathbf{X}_d$ is $n \times h$, and $\mathbf{X}_r$ is $n \times (q - h + 1)$. The estimator of $\boldsymbol{\beta}_r$ in the reduced model (7.31) is $\hat{\boldsymbol{\beta}}_r = (\mathbf{X}_r'\mathbf{X}_r)^{-1}\mathbf{X}_r'\mathbf{y}$.

We now partition the total sum of squares as

$$\mathbf{y}'\mathbf{y} = (\mathbf{y}'\mathbf{y} - \hat{\boldsymbol{\beta}}'\mathbf{X}'\mathbf{y}) + (\hat{\boldsymbol{\beta}}'\mathbf{X}'\mathbf{y} - \hat{\boldsymbol{\beta}}_r'\mathbf{X}_r'\mathbf{y}) + \hat{\boldsymbol{\beta}}_r'\mathbf{X}_r'\mathbf{y}, \tag{7.32}$$

where $\hat{\boldsymbol{\beta}}_r'\mathbf{X}_r'\mathbf{y}$ is from the reduced model (7.31), and $\hat{\boldsymbol{\beta}}'\mathbf{X}'\mathbf{y} - \hat{\boldsymbol{\beta}}_r'\mathbf{X}_r'\mathbf{y}$ is the "extra" regression sum of squares due to $\boldsymbol{\beta}_d$ adjusted for $\boldsymbol{\beta}_r$. It can be shown that SSH $= \hat{\boldsymbol{\beta}}'\mathbf{X}'\mathbf{y} - \hat{\boldsymbol{\beta}}_r'\mathbf{X}_r'\mathbf{y}$ has a χ^2-distribution (if divided by σ^2) and that SSH is independent of SSE $= \mathbf{y}'\mathbf{y} - \hat{\boldsymbol{\beta}}'\mathbf{X}'\mathbf{y}$. We can therefore construct an F-statistic for testing $H_0: \boldsymbol{\beta}_d = \mathbf{0}$:

$$F = \frac{(\hat{\boldsymbol{\beta}}'\mathbf{X}'\mathbf{y} - \hat{\boldsymbol{\beta}}_r'\mathbf{X}_r'\mathbf{y})/h}{(\mathbf{y}'\mathbf{y} - \hat{\boldsymbol{\beta}}'\mathbf{X}'\mathbf{y})/(n - q - 1)}, \tag{7.33}$$

which is distributed as $F_{h,n-q-1}$ if H_0 is true.

7.2.4c *The General Linear Hypothesis* $H_0: \mathbf{C}\boldsymbol{\beta} = \mathbf{0}$

Using appropriate choices for $\mathbf{C}$, the hypotheses $H_0: \boldsymbol{\beta}_1 = \mathbf{0}$ and $H_0: \boldsymbol{\beta}_d = \mathbf{0}$ in Sections 7.2.4a and 7.2.4b can be expressed in the form $H_0: \mathbf{C}\boldsymbol{\beta} = \mathbf{0}$. The formulation $H_0: \mathbf{C}\boldsymbol{\beta} = \mathbf{0}$ also allows for more general hypotheses such as

$$H_0: \beta_1 - \beta_2 = \beta_2 - 2\beta_3 + \beta_4 = \beta_1 - \beta_4 = 0,$$

which can be expressed in the form $\mathbf{C}\boldsymbol{\beta} = \mathbf{0}$ as follows:

$$\begin{pmatrix} 0 & 1 & -1 & 0 & 0 \\ 0 & 0 & 1 & -2 & 1 \\ 0 & 1 & 0 & 0 & -1 \end{pmatrix} \begin{pmatrix} \beta_0 \\ \beta_1 \\ \beta_2 \\ \beta_3 \\ \beta_4 \end{pmatrix} = \begin{pmatrix} 0 \\ 0 \\ 0 \end{pmatrix}.$$

To construct an F-test for $H_0: \mathbf{C}\boldsymbol{\beta} = \mathbf{0}$, where $\mathbf{C}$ is $m \times (q + 1)$ of rank m, we estimate the linear functions $\mathbf{C}\boldsymbol{\beta}$ by $\mathbf{C}\hat{\boldsymbol{\beta}}$. By (1.94) and (7.8), $\text{cov}(\mathbf{C}\hat{\boldsymbol{\beta}}) = \sigma^2\mathbf{C}(\mathbf{X}'\mathbf{X})^{-1}\mathbf{C}'$. Then by Theorems 2.2B and 2.2F, SSH$/\sigma^2 = (\mathbf{C}\hat{\boldsymbol{\beta}})'[\mathbf{C}(\mathbf{X}'\mathbf{X})^{-1}\mathbf{C}']^{-1}\mathbf{C}\hat{\boldsymbol{\beta}}/\sigma^2$ is χ^2_m when H_0 is true. It can be shown that SSH and SSE $= \mathbf{y}'\mathbf{y} - \hat{\boldsymbol{\beta}}'\mathbf{X}'\mathbf{y}$ are independent.

We can therefore test $H_0: \mathbf{C}\boldsymbol{\beta} = \mathbf{0}$ using the F-statistic

$$F = \frac{\text{SSH}/m}{\text{SSE}/(n-q-1)}$$
$$= \frac{(\mathbf{C}\hat{\boldsymbol{\beta}})'[\mathbf{C}(\mathbf{X}'\mathbf{X})^{-1}\mathbf{C}']^{-1}(\mathbf{C}\hat{\boldsymbol{\beta}})/m}{(\mathbf{y}'\mathbf{y} - \hat{\boldsymbol{\beta}}'\mathbf{X}'\mathbf{y})/(n-q-1)}, \tag{7.34}$$

which is distributed as $F_{m,n-q-1}$ when H_0 is true. It can be shown that the F-statistic in (7.34) tests the hypothesis $H_0: \mathbf{C}\boldsymbol{\beta} = \mathbf{0}$ above and beyond other effects; that is, the F-test based on (7.34) is a full-and-reduced-model test.

7.2.4d Tests and Confidence Intervals for an Individual β_j

To test $H_0: \beta_j = 0$ using the general linear hypothesis test of Section 7.2.4c, the matrix $\mathbf{C}$ in (7.34) becomes the single row vector $\mathbf{a}' = (0, \ldots, 0, 1, 0, \ldots, 0)$ with a 1 in the jth position, and we have $\mathbf{a}'\hat{\boldsymbol{\beta}} = \hat{\beta}_j$. Let $s^2 = (\mathbf{y}'\mathbf{y} - \hat{\boldsymbol{\beta}}'\mathbf{X}'\mathbf{y})/(n-q-1)$ and g_{jj} be the jth diagonal element of $(\mathbf{X}'\mathbf{X})^{-1}$. Then (7.34) becomes

$$F = \frac{(\mathbf{a}'\hat{\boldsymbol{\beta}})'\left[\mathbf{a}'(\mathbf{X}'\mathbf{X})^{-1}\mathbf{a}\right]^{-1}(\mathbf{a}'\hat{\boldsymbol{\beta}})}{s^2} \tag{7.35}$$

$$= \frac{\hat{\beta}_j^2}{s^2 g_{jj}}, \tag{7.36}$$

which is distributed as $F_{1,n-q-1}$ if H_0 is true. The F-statistic in (7.36) tests the significance of $\hat{\beta}_j$ adjusted for the other $\hat{\beta}$'s.

Since the F-statistic in (7.36) has 1 and $n - q - 1$ degrees of freedom, we can equivalently use the t-statistic

$$t = \frac{\hat{\beta}_j}{s\sqrt{g_{jj}}} \tag{7.37}$$

to test the effect of β_j above and beyond the other β's. We reject $H_0: \beta_j = 0$ if $|t| > t_{\alpha/2,n-q-1}$. We find $100(1-\alpha)\%$ confidence limits for β_j by

$$\hat{\beta}_j \pm t_{\alpha/2,n-q-1} s\sqrt{g_{jj}}. \tag{7.38}$$

7.2.4e Simultaneous Tests and Confidence Intervals for Regression Coefficients

If the test in (7.37) or the confidence interval in (7.38) is carried out for each of $\hat{\beta}_1, \ldots, \hat{\beta}_q$, the overall α-level may be inflated. One way to preserve the α-level is to use the Bonferroni or simultaneous procedures introduced in Section 3.4. For Bonferroni tests of $H_{0j}: \beta_j = 0$, $j = 1, 2, \ldots, q$, we use (7.37),

$$t_j = \frac{\hat{\beta}_j}{s\sqrt{g_{jj}}},$$

and reject H_{0j} if $|t_j| \geq t_{\alpha/2q,n-q-1}$, where g_{jj} is the jth diagonal element of $(\mathbf{X}'\mathbf{X})^{-1}$. The corresponding confidence limits are

$$\hat{\beta}_j \pm t_{\alpha/2q,n-q-1} s \sqrt{g_{jj}} \tag{7.39}$$

for $j = 1, 2, \ldots, q$. Bonferroni critical values $t_{\alpha/2q}$ are given in Table B.2.

Simultaneous confidence limits are given in the following theorem.

Theorem 7.2B. Confidence limits that hold simultaneously with probability $1-\alpha$ for all possible linear combinations $\mathbf{a}'\boldsymbol{\beta}$ are given by

$$\mathbf{a}'\hat{\boldsymbol{\beta}} \pm \sqrt{(q+1)\mathbf{a}'(\mathbf{X}'\mathbf{X})^{-1}\mathbf{a}s^2 F_{\alpha,q+1,n-q-1}}. \tag{7.40}$$

Proof. By an adaptation of (3.33), we obtain

$$\max_{\mathbf{a}} \frac{(\mathbf{a}'\hat{\boldsymbol{\beta}})^2}{\mathbf{a}'[s^2(\mathbf{X}'\mathbf{X})^{-1}]\mathbf{a}} = \hat{\boldsymbol{\beta}}'[s^2(\mathbf{X}'\mathbf{X})^{-1}]^{-1}\hat{\boldsymbol{\beta}} \tag{7.41}$$

$$= \frac{\hat{\boldsymbol{\beta}}'\mathbf{X}'\mathbf{X}\hat{\boldsymbol{\beta}}}{s^2}. \tag{7.42}$$

By (7.34), the expression $\hat{\boldsymbol{\beta}}'\mathbf{X}'\mathbf{X}\hat{\boldsymbol{\beta}}/s^2$ in (7.42) has an F-distribution if it is divided by $q+1$. Thus

$$P\left[\frac{(\mathbf{a}'\hat{\boldsymbol{\beta}})^2}{(q+1)s^2\mathbf{a}'(\mathbf{X}'\mathbf{X})^{-1}\mathbf{a}} \leq F_{\alpha,q+1,n-q-1}\right] = 1-\alpha \tag{7.43}$$

simultaneously for all possible $\mathbf{a}$, and the result in (7.40) follows. □

For individual β_j's, we use $\mathbf{a}' = (0, \ldots, 0, 1, 0, \ldots, 0)$ in (7.40) to obtain

$$\hat{\beta}_j \pm \sqrt{(q+1)F_{\alpha,q+1,n-q-1}}\, s\sqrt{g_{jj}}. \tag{7.44}$$

The simultaneous intervals in (7.40) or (7.44) are known as Scheffé intervals or S-intervals (Scheffé 1953; 1959, p. 68). Equivalently, we can test $H_{0j}: \beta_j = 0$ for all $j = 0, 1, 2, \ldots, q$ with

$$t_j = \frac{\hat{\beta}_j}{s\sqrt{g_{jj}}}$$

and reject H_{0j} for $|t_j| \geq \sqrt{(q+1)F_{\alpha,q+1,n-q-1}}$. However, since

$$t_{\alpha/2q,n-q-1} < \sqrt{(q+1)F_{\alpha,q+1,n-q-1}}, \tag{7.45}$$

the Bonferroni tests and confidence intervals for individual β_j's are better than the simultaneous tests and intervals for individual β_j's.

7.2.5 R^2 in Fixed-x Regression

In (7.28) the corrected total sum of squares of the y's is partitioned into sums of squares due to regression and error,

$$\begin{aligned}\mathbf{y}'\mathbf{y} - n\bar{y}^2 &= \text{SSE} + \text{SSR} \\ &= (\mathbf{y}'\mathbf{y} - \hat{\boldsymbol{\beta}}'\mathbf{X}'\mathbf{y}) + (\hat{\boldsymbol{\beta}}'\mathbf{X}'\mathbf{y} - n\bar{y}^2).\end{aligned}$$

The *coefficient of multiple determination*, also called the *squared multiple correlation*, is defined as

$$R^2 = \frac{\text{SSR}}{\text{SSR} + \text{SSE}} = \frac{\hat{\boldsymbol{\beta}}'\mathbf{X}'\mathbf{y} - n\bar{y}^2}{\mathbf{y}'\mathbf{y} - n\bar{y}^2}. \tag{7.46}$$

The positive square root, R, is called the *multiple correlation.* In (7.46), R^2 is the proportion of total variation in the y's that is accounted for by regression on the x's, and thus R^2 is a measure of model fit. In (7.66) we have another form of R^2 for the case of x's that are random variables. The two forms can be shown to be equal.

7.2.6 Model Validation

There has been much interest in model validation (verification that the assumptions hold in a given data set); see, for example, Myers (1990, Chapters 5 and 6); Neter et al. (1996, Chapters 9 and 10); Beckman and Cook (1983); Cook and Weisberg (1982); Draper and Smith (1981, Chapter 6); Belsley, Kuh, and Welsch (1980); Weisberg (1985, Chapters 5 and 6); Cook (1977, 1979); Snee (1977); and Hocking (1996, Chapter 9).

By Theorem 7.2A, the assumptions $E(\mathbf{y}) = \mathbf{X}\boldsymbol{\beta}$ and $\text{cov}(\mathbf{y}) = \sigma^2\mathbf{I}$ lead to best linear unbiased estimators. The normality assumption was added in Section 7.2.4. If some of the rows of the $\mathbf{X}$ matrix are repeated, a lack-of-fit test can be made to check the assumption $E(\mathbf{y}) = \mathbf{X}\boldsymbol{\beta}$ (Myers 1990, pp. 116–120). If replicate rows are not available in $\mathbf{X}$, various procedures have been proposed for grouping the observations into near replicates for a lack-of-fit test; see, for example, Green (1971), Shillington (1979), Christensen (1989), Joglekar et al. (1989), and Miller et al. (1996).

A plot of the residuals $\hat{\varepsilon}_i = y_i - \hat{y}_i$, $i = 1, 2, \ldots, n$, against the fitted values $\hat{y}_i$ may reveal deficiencies in the assumption $E(\mathbf{y}) = \mathbf{X}\boldsymbol{\beta}$ (for example, the need for quadratic terms) or failure of the assumption $\text{cov}(\mathbf{y}) = \sigma^2\mathbf{I}$. The residuals can also be plotted against predictor variables not included in the model or transformations of variables in the model (for example, x_j^2). A systematic, nonrandom pattern in such a plot would reveal a need for the indicated variable to be added to the model in some form (see Cook 1996). Berk and Booth (1993) discuss these and other residual plots designed to detect a curve, and they illustrate with several real data sets. The ranked residuals can be plotted, as in Rencher (1995, Section 4.4.1), to detect a departure from normality. The residual plots also aid in detection of outliers and other influential observations.

If the data have a natural chronological ordering, a plot of residuals against time may show a lack of independence. A formal test of significance can be carried out using the *Durbin-Watson test*. See Neter et al. (1996, pp. 504–507, 1349–1350) for the test statistic and tables of critical values.

7.3 MULTIPLE REGRESSION: RANDOM x's

Throughout Section 7.2, we assumed that the x's were fixed, that is, that they would remain constant in repeated sampling. However, in many regression applications, the x's are random variables. Our estimation and testing results follow from the assumption that $y, x_1, x_2, \ldots, x_q$ are jointly distributed as a multivariate normal.

7.3.1 Model for Random x's

We assume that $(y, x_1, \ldots, x_q) = (y, \mathbf{x}')$ is distributed as $N_{q+1}(\boldsymbol{\mu}, \boldsymbol{\Sigma})$ with

$$\boldsymbol{\mu} = \begin{pmatrix} \mu_y \\ \hline \mu_1 \\ \vdots \\ \mu_q \end{pmatrix} = \begin{pmatrix} \mu_y \\ \boldsymbol{\mu}_x \end{pmatrix}, \tag{7.47}$$

$$\boldsymbol{\Sigma} = \left(\begin{array}{c|ccc} \sigma_{yy} & \sigma_{y1} & \cdots & \sigma_{yq} \\ \hline \sigma_{1y} & \sigma_{11} & \cdots & \sigma_{1q} \\ \vdots & \vdots & & \vdots \\ \sigma_{qy} & \sigma_{q1} & \cdots & \sigma_{qq} \end{array}\right) = \begin{pmatrix} \sigma_{yy} & \boldsymbol{\sigma}'_{yx} \\ \boldsymbol{\sigma}_{yx} & \boldsymbol{\Sigma}_{xx} \end{pmatrix}. \tag{7.48}$$

From Theorem 2.2E we have

$$E(y \,|\, \mathbf{x}) = \mu_y + \boldsymbol{\sigma}'_{yx}\boldsymbol{\Sigma}_{xx}^{-1}(\mathbf{x} - \boldsymbol{\mu}_x) \tag{7.49}$$

$$= \beta_0 + \boldsymbol{\beta}_1'\mathbf{x}, \tag{7.50}$$

where

$$\beta_0 = \mu_y - \boldsymbol{\sigma}'_{yx}\boldsymbol{\Sigma}_{xx}^{-1}\boldsymbol{\mu}_x, \tag{7.51}$$

$$\boldsymbol{\beta}_1 = \boldsymbol{\Sigma}_{xx}^{-1}\boldsymbol{\sigma}_{yx}. \tag{7.52}$$

From Theorem 2.2E, we also obtain

$$\mathrm{var}(y \,|\, \mathbf{x}) = \sigma_{yy} - \boldsymbol{\sigma}'_{yx}\boldsymbol{\Sigma}_{xx}^{-1}\boldsymbol{\sigma}_{yx} = \sigma^2. \tag{7.53}$$

We denote $\sigma_{yy} - \boldsymbol{\sigma}'_{yx}\boldsymbol{\Sigma}_{xx}^{-1}\boldsymbol{\sigma}_{yx}$ by σ^2 to indicate that $\mathrm{var}(y \,|\, \mathbf{x})$ is not a function of $\mathbf{x}$. Thus under the multivariate normal assumption, we have a linear model with constant variance, analogous to the fixed-x case.

7.3.2 Estimation of β_0, $\boldsymbol{\beta}_1$, and σ^2

Maximum likelihood estimators of β_0, $\boldsymbol{\beta}_1$, and σ^2 are given in the following theorem.

Theorem 7.3A. If $(y_1, \mathbf{x}_1')$, $(y_2, \mathbf{x}_2'), \ldots, (y_n, \mathbf{x}_n')$ is a random sample from $N_{q+1}(\boldsymbol{\mu}, \boldsymbol{\Sigma})$, where $\boldsymbol{\mu}$ and $\boldsymbol{\Sigma}$ are given by (7.47) and (7.48), the maximum likelihood estimators for β_0, $\boldsymbol{\beta}_1$, and σ^2 are as follows:

$$\hat{\boldsymbol{\beta}}_1 = \mathbf{S}_{xx}^{-1}\mathbf{s}_{yx}, \tag{7.54}$$

$$\hat{\beta}_0 = \bar{y} - \mathbf{s}_{yx}'\mathbf{S}_{xx}^{-1}\bar{\mathbf{x}}, \tag{7.55}$$

$$s^2 = s_{yy} - \mathbf{s}_{yx}'\mathbf{S}_{xx}^{-1}\mathbf{s}_{yx}. \tag{7.56}$$

The variance estimator s^2 has been corrected for bias.

Proof. By Theorem 2.3A, the maximum likelihood estimators of $\boldsymbol{\mu}$ and $\boldsymbol{\Sigma}$ are

$$\hat{\boldsymbol{\mu}} = \begin{pmatrix} \hat{\mu}_y \\ \hat{\boldsymbol{\mu}}_x \end{pmatrix} = \begin{pmatrix} \bar{y} \\ \bar{\mathbf{x}} \end{pmatrix}, \tag{7.57}$$

$$\hat{\boldsymbol{\Sigma}} = \frac{n-1}{n}\mathbf{S} = \frac{n-1}{n}\begin{pmatrix} s_{yy} & \mathbf{s}_{yx}' \\ \mathbf{s}_{yx} & \mathbf{S}_{xx} \end{pmatrix}, \tag{7.58}$$

where the partitioning on $\hat{\boldsymbol{\mu}}$ and $\hat{\boldsymbol{\Sigma}}$ is analogous to the partitioning of $\boldsymbol{\mu}$ and $\boldsymbol{\Sigma}$ in (7.47) and (7.48). By the invariance property of maximum likelihood estimators (Theorem 2.3B), we insert (7.57) and (7.58) into (7.51), (7.52), and (7.53) to obtain the desired results (using the unbiased estimator $\mathbf{S}$ in place of $\hat{\boldsymbol{\Sigma}}$). □

By comparison with the centered form of the least squares estimators given in (7.26) and (7.27), we see that the maximum likelihood estimators of $\boldsymbol{\beta}_1$ and β_0 in (7.54) and (7.55) are the same as the least squares estimators in the fixed-x case (Section 7.2.3).

The regression coefficient vector $\hat{\boldsymbol{\beta}}_1$ can be expressed in terms of correlations. By analogy to (7.48) and (7.58), the correlation matrix can be written in partitioned form as

$$\mathbf{R} = \begin{pmatrix} 1 & \mathbf{r}_{yx}' \\ \mathbf{r}_{yx} & \mathbf{R}_{xx} \end{pmatrix}, \tag{7.59}$$

where $\mathbf{r}_{yx}$ is the vector of correlations between y and the x's, and $\mathbf{R}_{xx}$ is the correlation matrix for the x's. By (1.52), the submatrices of $\mathbf{R}$ can be converted to corresponding submatrices of $\mathbf{S}$:

$$\mathbf{S}_{xx} = \mathbf{D}_x\mathbf{R}_{xx}\mathbf{D}_x, \tag{7.60}$$

$$\mathbf{s}_{yx} = s_y\mathbf{D}_x\mathbf{r}_{yx}, \tag{7.61}$$

where $s_y = \sqrt{s_{yy}}$ is the standard deviation of y and $\mathbf{D}_x$ is the diagonal matrix of standard deviations of the x's, $\mathbf{D}_x = \text{diag}(\sqrt{s_{11}}, \sqrt{s_{22}}, \ldots, \sqrt{s_{pp}})$. When (7.60) and (7.61) are substituted into (7.54), we obtain an expression for $\hat{\boldsymbol{\beta}}_1$ in terms of correlations,

$$\hat{\boldsymbol{\beta}}_1 = s_y \mathbf{D}_x^{-1} \mathbf{R}_{xx}^{-1} \mathbf{r}_{yx}. \tag{7.62}$$

The regression coefficients can be standardized in order to show the effect of standardized x-values (z-scores). For example, for $q = 2$ the model in centered form [see (7.13)] is

$$\hat{y}_i = \bar{y} + \hat{\beta}_1(x_{i1} - \bar{x}_1) + \hat{\beta}_2(x_{i2} - \bar{x}_2).$$

This can be expressed in the form

$$\frac{\hat{y}_i - \bar{y}}{s_y} = \frac{s_1}{s_y}\hat{\beta}_1 \frac{(x_{i1} - \bar{x}_1)}{s_1} + \frac{s_2}{s_y}\hat{\beta}_2 \frac{(x_{i2} - \bar{x}_2)}{s_2},$$

where s_j is the standard deviation of x_j. We therefore define the standardized coefficients as

$$\hat{\beta}_j^* = \frac{s_j}{s_y}\hat{\beta}_j;$$

these are often referred to as *beta weights* or *beta coefficients*. The beta weights can be expressed in vector form as

$$\hat{\boldsymbol{\beta}}_1^* = \frac{1}{s_y}\mathbf{D}_x \hat{\boldsymbol{\beta}}_1.$$

Using (7.62), $\hat{\boldsymbol{\beta}}_1^*$ becomes

$$\hat{\boldsymbol{\beta}}_1^* = \mathbf{R}_{xx}^{-1} \mathbf{r}_{yx}. \tag{7.63}$$

7.3.3 R^2 in Random-x Regression

The *population multiple correlation* $\rho_{y|\mathbf{x}}$ is defined as the correlation between y and $z = \mu_y + \boldsymbol{\sigma}_{yx}' \boldsymbol{\Sigma}_{xx}^{-1}(\mathbf{x} - \boldsymbol{\mu}_x)$. By (7.49), z is the population predicted value of y, that is, the population analogue of $\hat{y}$. It is easily established that cov(y, z) and var(z) have the same value:

$$\text{cov}(y, z) = \text{var}(z) = \boldsymbol{\sigma}_{yx}' \boldsymbol{\Sigma}_{xx}^{-1} \boldsymbol{\sigma}_{yx}. \tag{7.64}$$

Therefore

$$\rho_{y|\mathbf{x}} = \frac{\text{cov}(y, z)}{\sqrt{\text{var}(y)\,\text{var}(z)}} = \sqrt{\frac{\boldsymbol{\sigma}_{yx}' \boldsymbol{\Sigma}_{xx}^{-1} \boldsymbol{\sigma}_{yx}}{\sigma_{yy}}},$$

from which the *population coefficient of determination* or *population squared multiple correlation* $\rho_{y|\mathbf{x}}^2$ is

$$\rho_{y|\mathbf{x}}^2 = \frac{\boldsymbol{\sigma}_{yx}' \boldsymbol{\Sigma}_{xx}^{-1} \boldsymbol{\sigma}_{yx}}{\sigma_{yy}}. \tag{7.65}$$

We can estimate $\rho^2_{y|\mathbf{x}}$ by substituting estimators for the parameters in (7.65):

$$R^2 = \frac{\mathbf{s}'_{yx}\mathbf{S}^{-1}_{xx}\mathbf{s}_{yx}}{s_{yy}}. \tag{7.66}$$

We have used the notation R^2 rather than $\hat{\rho}^2_{y|\mathbf{x}}$ because (7.66) can be shown to be the same as R^2 for the fixed-x case in (7.46).

When $\rho^2_{y|\mathbf{x}} = 0, R^2$ is biased:

$$E(R^2|\rho^2_{y|\mathbf{x}} = 0) = \frac{q}{n-1}. \tag{7.67}$$

If the same sample size n is small relative to the number of variables q, the bias can be substantial.

7.3.4 Tests and Confidence Intervals

Likelihood ratio tests can be obtained for the random-x case corresponding to all hypotheses for fixed x's in Section 7.2.4. In every case, the resulting test statistic is the same as that found in Section 7.2.4 for the analogous fixed-x case. When H_0 is true, the distribution of the test statistic, say (7.34), is the same for fixed x's as for random x's, namely, central F in both cases. However, when H_0 is false, the distributions differ. For fixed x's, the distribution is noncentral F, but for random x's, the test statistic does not have a noncentral F-distribution.

The confidence intervals for the β's in Sections 7.2.4d and 7.2.4e also remain valid for the random-x case because the conditional distribution of y given $\mathbf{x}$ is normal. Thus the confidence coefficient, say 95%, for the fixed-x case holds as well for random x's. However, the expected width of the interval differs in the two cases.

A confidence interval for $\rho^2_{y|\mathbf{x}}$ was given by Helland (1987).

7.4 ESTIMATION IN THE MULTIVARIATE MULTIPLE REGRESSION MODEL: FIXED x's

7.4.1 The Multivariate Model

In the *multivariate multiple regression model*, we have *multivariate* dependent variables and *multiple* independent variables; that is, $y_1, y_2, \ldots, y_p$ will be predicted by $x_1, x_2, \ldots, x_q$. We assume that the x's are fixed.

In the sample, there are n observed values of the vector of y's. These observation vectors can be listed as rows in the data matrix

$$\mathbf{Y} = \begin{pmatrix} y_{11} & y_{12} & \cdots & y_{1p} \\ y_{21} & y_{22} & \cdots & y_{2p} \\ \vdots & \vdots & & \vdots \\ y_{n1} & y_{n2} & \cdots & y_{np} \end{pmatrix} = \begin{pmatrix} \mathbf{y}'_1 \\ \mathbf{y}'_2 \\ \vdots \\ \mathbf{y}'_n \end{pmatrix}.$$

The columns of $\mathbf{Y}$ are also of interest. Each column consists of the n observations on one of the p variables and therefore corresponds to the $\mathbf{y}$-vector for that variable in the univariate multiple regression model (7.3). We will indicate the column vectors of $\mathbf{Y}$ by $\mathbf{y}_{(j)}$ to distinguish them from the row vectors $\mathbf{y}_i'$. Hence $\mathbf{Y}$ can be expressed in two forms:

$$\mathbf{Y} = \begin{pmatrix} \mathbf{y}_1' \\ \mathbf{y}_2' \\ \vdots \\ \mathbf{y}_n' \end{pmatrix} = (\mathbf{y}_{(1)}, \mathbf{y}_{(2)}, \ldots, \mathbf{y}_{(p)}).$$

Thus, for example, $\mathbf{y}_2' = (y_{21}, y_{22}, \ldots, y_{2p})$, while

$$\mathbf{y}_{(2)} = \begin{pmatrix} y_{12} \\ y_{22} \\ \vdots \\ y_{n2} \end{pmatrix}.$$

Each column of y's will depend on the x's in its own way, different from that of the other columns of y's. We can indicate this by writing the model (7.3) for each $\mathbf{y}_{(j)}$:

$$\mathbf{y}_{(1)} = \mathbf{X}\boldsymbol{\beta}_{(1)} + \boldsymbol{\varepsilon}_{(1)}, \qquad \mathbf{y}_{(2)} = \mathbf{X}\boldsymbol{\beta}_{(2)} + \boldsymbol{\varepsilon}_{(2)}, \qquad \ldots, \qquad \mathbf{y}_{(p)} = \mathbf{X}\boldsymbol{\beta}_{(p)} + \boldsymbol{\varepsilon}_{(p)}.$$

These columns can be juxtaposed to form a combined model for $\mathbf{Y}$,

$$\begin{aligned}(\mathbf{y}_{(1)}, \ldots, \mathbf{y}_{(p)}) &= (\mathbf{X}\boldsymbol{\beta}_{(1)}, \ldots, \mathbf{X}\boldsymbol{\beta}_{(p)}) + (\boldsymbol{\varepsilon}_{(1)}, \ldots, \boldsymbol{\varepsilon}_{(p)}) \\ &= \mathbf{X}(\boldsymbol{\beta}_{(1)}, \ldots, \boldsymbol{\beta}_{(p)}) + (\boldsymbol{\varepsilon}_{(1)}, \ldots, \boldsymbol{\varepsilon}_{(p)}) \qquad \text{[by (A.2.26)]}.\end{aligned}$$

This model for $\mathbf{Y} = (\mathbf{y}_{(1)}, \ldots, \mathbf{y}_{(p)})$ can be written as

$$\mathbf{Y} = \mathbf{XB} + \boldsymbol{\Xi}, \tag{7.68}$$

where $\mathbf{B} = (\boldsymbol{\beta}_{(1)}, \boldsymbol{\beta}_{(2)}, \ldots, \boldsymbol{\beta}_{(p)})$; $\boldsymbol{\Xi} = (\boldsymbol{\varepsilon}_{(1)}, \boldsymbol{\varepsilon}_{(2)}, \ldots, \boldsymbol{\varepsilon}_{(p)})$; and $\mathbf{X}$ is the usual matrix of the x's from multiple regression,

$$\mathbf{X} = \begin{pmatrix} 1 & x_{11} & x_{12} & \cdots & x_{1q} \\ 1 & x_{21} & x_{22} & \cdots & x_{2q} \\ \vdots & \vdots & \vdots & & \vdots \\ 1 & x_{n1} & x_{n2} & \cdots & x_{nq} \end{pmatrix}.$$

Thus $\mathbf{Y}$ is $n \times p$, $\mathbf{X}$ is $n \times (q + 1)$, and $\mathbf{B}$ is $(q + 1) \times p$. In row form, the model $\mathbf{Y} = \mathbf{XB} + \boldsymbol{\Xi}$ in (7.68) becomes

$$\begin{pmatrix} \mathbf{y}_1' \\ \mathbf{y}_2' \\ \vdots \\ \mathbf{y}_n' \end{pmatrix} = \begin{pmatrix} \mathbf{x}_1' \\ \mathbf{x}_2' \\ \vdots \\ \mathbf{x}_n' \end{pmatrix} \mathbf{B} + \begin{pmatrix} \boldsymbol{\varepsilon}_1' \\ \boldsymbol{\varepsilon}_2' \\ \vdots \\ \boldsymbol{\varepsilon}_n' \end{pmatrix} = \begin{pmatrix} \mathbf{x}_1'\mathbf{B} \\ \mathbf{x}_2'\mathbf{B} \\ \vdots \\ \mathbf{x}_n'\mathbf{B} \end{pmatrix} + \begin{pmatrix} \boldsymbol{\varepsilon}_1' \\ \boldsymbol{\varepsilon}_2' \\ \vdots \\ \boldsymbol{\varepsilon}_n' \end{pmatrix}. \tag{7.69}$$

We assume that $\mathbf{X}$ is fixed from sample to sample. Additional assumptions on $\mathbf{Y}$ or $\boldsymbol{\Xi}$ and their rows that lead to the desirable properties of $\hat{\mathbf{B}}$ in Sections 7.4.2 and 7.4.3 are the following:

1. $E(\mathbf{Y}) = \mathbf{XB}$ or $E(\boldsymbol{\Xi}) = \mathbf{O}$.
2. $\text{cov}(\mathbf{y}_i) = \boldsymbol{\Sigma}$ or $\text{cov}(\boldsymbol{\varepsilon}_i) = \boldsymbol{\Sigma}$ for all $i = 1, 2, \ldots, n$.
3. $\text{cov}(\mathbf{y}_i, \mathbf{y}_j) = \mathbf{O}$ or $\text{cov}(\boldsymbol{\varepsilon}_i, \boldsymbol{\varepsilon}_j) = \mathbf{O}$ for all $i \neq j$.

7.4.2 Least Squares Estimator for B

In the univariate case, the least squares estimator $\hat{\boldsymbol{\beta}} = (\mathbf{X}'\mathbf{X})^{-1}\mathbf{X}'\mathbf{y}$ in (7.6) minimizes $\hat{\boldsymbol{\varepsilon}}'\hat{\boldsymbol{\varepsilon}} = \sum_{i=1}^n \hat{\varepsilon}_i^2$ in (7.4) or (7.5). In the multivariate model, $\hat{\boldsymbol{\Xi}}$ has a column for each y_j, $j = 1, 2, \ldots, p$: $\hat{\boldsymbol{\Xi}} = (\hat{\boldsymbol{\varepsilon}}_{(1)}, \hat{\boldsymbol{\varepsilon}}_{(2)}, \ldots, \hat{\boldsymbol{\varepsilon}}_{(p)})$. We seek a least squares estimator $\hat{\mathbf{B}}$ that minimizes $\hat{\boldsymbol{\varepsilon}}_{(j)}'\hat{\boldsymbol{\varepsilon}}_{(j)}$ for each $j = 1, 2, \ldots, p$, that is, minimizes $\sum_{j=1}^p \hat{\boldsymbol{\varepsilon}}_{(j)}'\hat{\boldsymbol{\varepsilon}}_{(j)} = \sum_{i=1}^n \sum_{j=1}^p \hat{\varepsilon}_{ij}^2$. By (A.8.4), $\sum_{j=1}^p \hat{\boldsymbol{\varepsilon}}_{(j)}'\hat{\boldsymbol{\varepsilon}}_{(j)} = \text{tr}(\hat{\boldsymbol{\Xi}}'\hat{\boldsymbol{\Xi}})$. The same estimator $\hat{\mathbf{B}}$ also "minimizes" the (positive definite) matrix $\hat{\boldsymbol{\Xi}}'\hat{\boldsymbol{\Xi}}$ itself in the sense that any change in $\hat{\mathbf{B}}$ will add a positive semidefinite matrix to $\hat{\boldsymbol{\Xi}}'\hat{\boldsymbol{\Xi}}$. The following theorem gives this "least squares" estimator.

Theorem 7.4A. The estimator

$$\hat{\mathbf{B}} = (\mathbf{X}'\mathbf{X})^{-1}\mathbf{X}'\mathbf{Y} \tag{7.70}$$

is a least squares estimator for $\mathbf{B}$ in the following three ways:

(i) $\hat{\mathbf{B}}$ minimizes $\hat{\boldsymbol{\Xi}}'\hat{\boldsymbol{\Xi}} = (\mathbf{Y} - \mathbf{X}\hat{\mathbf{B}})'(\mathbf{Y} - \mathbf{X}\hat{\mathbf{B}})$ in the sense defined above.
(ii) $\hat{\mathbf{B}}$ minimizes $\text{tr}(\hat{\boldsymbol{\Xi}}'\hat{\boldsymbol{\Xi}}) = \text{tr}(\mathbf{Y} - \mathbf{X}\hat{\mathbf{B}})'(\mathbf{Y} - \mathbf{X}\hat{\mathbf{B}})$.
(iii) $\hat{\mathbf{B}}$ minimizes $|\hat{\boldsymbol{\Xi}}'\hat{\boldsymbol{\Xi}}| = |(\mathbf{Y} - \mathbf{X}\hat{\mathbf{B}})'(\mathbf{Y} - \mathbf{X}\hat{\mathbf{B}})|$.

Proof. We prove parts (i) and (ii).

(i) Let $\mathbf{B}_0$ be an estimator that may possibly be better than $\hat{\mathbf{B}}$. We write $\hat{\boldsymbol{\Xi}}'\hat{\boldsymbol{\Xi}}$ in terms of $\mathbf{B}_0$ and $\hat{\mathbf{B}}$:

$$\begin{aligned} \hat{\boldsymbol{\Xi}}'\hat{\boldsymbol{\Xi}} &= (\mathbf{Y} - \mathbf{XB}_0)'(\mathbf{Y} - \mathbf{XB}_0) \\ &= (\mathbf{Y} - \mathbf{X}\hat{\mathbf{B}} + \mathbf{X}\hat{\mathbf{B}} - \mathbf{XB}_0)'(\mathbf{Y} - \mathbf{X}\hat{\mathbf{B}} + \mathbf{X}\hat{\mathbf{B}} - \mathbf{XB}_0) \\ &= \left[\mathbf{Y} - \mathbf{X}\hat{\mathbf{B}} + \mathbf{X}(\hat{\mathbf{B}} - \mathbf{B}_0)\right]'\left[\mathbf{Y} - \mathbf{X}\hat{\mathbf{B}} + \mathbf{X}(\hat{\mathbf{B}} - \mathbf{B}_0)\right] \end{aligned}$$

$$= (\mathbf{Y} - \mathbf{X}\hat{\mathbf{B}})'(\mathbf{Y} - \mathbf{X}\hat{\mathbf{B}}) + (\hat{\mathbf{B}} - \mathbf{B}_0)'\mathbf{X}'\mathbf{X}(\hat{\mathbf{B}} - \mathbf{B}_0)$$
$$+ (\mathbf{Y} - \mathbf{X}\hat{\mathbf{B}})'\mathbf{X}(\hat{\mathbf{B}} - \mathbf{B}_0) + (\hat{\mathbf{B}} - \mathbf{B}_0)'\mathbf{X}'(\mathbf{Y} - \mathbf{X}\hat{\mathbf{B}}). \tag{7.71}$$

The last two terms on the right side of (7.71) vanish with the substitution $\hat{\mathbf{B}} = (\mathbf{X}'\mathbf{X})^{-1}\mathbf{X}'\mathbf{Y}$. The second term on the right side of (7.71) is a positive semidefinite matrix that reduces to $\mathbf{O}$ if $\mathbf{B}_0 = \hat{\mathbf{B}}$. Therefore $\hat{\boldsymbol{\Xi}}'\hat{\boldsymbol{\Xi}}$ is minimized when $\mathbf{B}_0 = \hat{\mathbf{B}} = (\mathbf{X}'\mathbf{X})^{-1}\mathbf{X}'\mathbf{Y}$.

(ii) To show that $\text{tr}(\hat{\boldsymbol{\Xi}}'\hat{\boldsymbol{\Xi}}) = \text{tr}(\mathbf{Y} - \mathbf{X}\mathbf{B}_0)'(\mathbf{Y} - \mathbf{X}\mathbf{B}_0)$ is minimized by $\mathbf{B}_0 = \hat{\mathbf{B}}$, we first write $\mathbf{B}_0$ in column form, $\mathbf{B}_0 = (\mathbf{b}_{0(1)}, \mathbf{b}_{0(2)}, \ldots, \mathbf{b}_{0(p)})$, and then use (A.8.4) to write

$$\text{tr}(\mathbf{Y} - \mathbf{X}\mathbf{B}_0)'(\mathbf{Y} - \mathbf{X}\mathbf{B}_0) = \sum_{j=1}^{p}(\mathbf{y}_{(j)} - \mathbf{X}\mathbf{b}_{0(j)})'(\mathbf{y}_{(j)} - \mathbf{X}\mathbf{b}_{0(j)}). \tag{7.72}$$

We now show that each of the p terms on the right side of (7.72) is minimized by $\mathbf{b}_{0(j)} = \hat{\boldsymbol{\beta}}_{(j)}$:

$$\hat{\boldsymbol{\varepsilon}}_{(j)}'\hat{\boldsymbol{\varepsilon}}_{(j)} = (\mathbf{y}_{(j)} - \mathbf{X}\mathbf{b}_{0(j)})'(\mathbf{y}_{(j)} - \mathbf{X}\mathbf{b}_{0(j)})$$
$$= (\mathbf{y}_{(j)} - \mathbf{X}\hat{\boldsymbol{\beta}}_{(j)} + \mathbf{X}\hat{\boldsymbol{\beta}}_{(j)} - \mathbf{X}\mathbf{b}_{0(j)})'(\mathbf{y}_{(j)} - \mathbf{X}\hat{\boldsymbol{\beta}}_{(j)} + \mathbf{X}\hat{\boldsymbol{\beta}}_{(j)} - \mathbf{X}\mathbf{b}_{0(j)}) \tag{7.73}$$
$$= (\mathbf{y} - \mathbf{X}\hat{\boldsymbol{\beta}}_{(j)})'(\mathbf{y} - \mathbf{X}\hat{\boldsymbol{\beta}}_{(j)}) + (\hat{\boldsymbol{\beta}}_{(j)} - \mathbf{b}_{0(j)})'\mathbf{X}'\mathbf{X}(\hat{\boldsymbol{\beta}}_{(j)} - \mathbf{b}_{0(j)})$$
$$+2(\hat{\boldsymbol{\beta}}_{(j)} - \mathbf{b}_{0(j)})'(\mathbf{X}'\mathbf{y}_{(j)} - \mathbf{X}'\mathbf{X}\hat{\boldsymbol{\beta}}_{(j)}), \tag{7.74}$$

where $\hat{\boldsymbol{\beta}}_{(j)}$ is the jth column of $\hat{\mathbf{B}}$ [see (7.75) below]. The third term on the right side of (7.74) vanishes for $\hat{\boldsymbol{\beta}}_{(j)} = (\mathbf{X}'\mathbf{X})^{-1}\mathbf{X}'\mathbf{y}_{(j)}$. The second term is a positive definite quadratic form, and thus $\hat{\boldsymbol{\varepsilon}}_{(j)}'\hat{\boldsymbol{\varepsilon}}_{(j)}$ is minimized when $\mathbf{b}_{0(j)} = \hat{\boldsymbol{\beta}}_{(j)}$.

The left side of (7.72) is therefore minimized by $\mathbf{B}_0 = \hat{\mathbf{B}}$. For a proof that involves direct differentiation of $\text{tr}(\mathbf{Y} - \mathbf{X}\mathbf{B}_0)'(\mathbf{Y} - \mathbf{X}\mathbf{B}_0)$, see Timm (1975, p. 100). □

If $\mathbf{Y}$ is written in column form, $\hat{\mathbf{B}}$ can be expressed as

$$\hat{\mathbf{B}} = (\mathbf{X}'\mathbf{X})^{-1}\mathbf{X}'\mathbf{Y} = (\mathbf{X}'\mathbf{X})^{-1}\mathbf{X}'(\mathbf{y}_{(1)}, \mathbf{y}_{(2)}, \ldots, \mathbf{y}_{(p)})$$
$$= \left[(\mathbf{X}'\mathbf{X})^{-1}\mathbf{X}'\mathbf{y}_{(1)}, \ldots, (\mathbf{X}'\mathbf{X})^{-1}\mathbf{X}'\mathbf{y}_{(p)}\right]$$
$$= (\hat{\boldsymbol{\beta}}_{(1)}, \hat{\boldsymbol{\beta}}_{(2)}, \ldots, \hat{\boldsymbol{\beta}}_{(p)}), \tag{7.75}$$

which is simply a juxtaposition of the p least squares estimators that correspond to $y_1, y_2, \ldots, y_p$. Thus, each $\hat{\boldsymbol{\beta}}_{(j)}$ is obtained exclusively from $\mathbf{y}_{(j)}$, and the other columns of $\mathbf{Y}$ do not enter into the computation. (If they did, $\hat{\boldsymbol{\beta}}_{(j)}$ would be biased; see the proof of Theorem 7.4D.) The elements of $\hat{\boldsymbol{\beta}}_{(j)}$ and $\hat{\boldsymbol{\beta}}_{(k)}$ are correlated, however (Theorem 7.4C), and we need multivariate tests for hypotheses about $\mathbf{B}$ (Section 7.5).

7.4.3 Properties of $\hat{\mathbf{B}}$

The least squares estimator $\hat{\mathbf{B}}$ was obtained without imposing the assumptions $E(\mathbf{Y}) = \mathbf{XB}$, $\text{cov}(\mathbf{y}_i) = \mathbf{\Sigma}$, and $\text{cov}(\mathbf{y}_i, \mathbf{y}_j) = \mathbf{O}$ given at the end of Section 7.4.1. However, when these assumptions hold, $\hat{\mathbf{B}}$ has some good properties, which are given in the following three theorems. In the proofs, $\mathbf{X}$ is considered to be fixed (constant) from sample to sample.

Theorem 7.4B. If $E(\mathbf{Y}) = \mathbf{XB}$, then $\hat{\mathbf{B}}$ is an unbiased estimator of $\mathbf{B}$.

Proof.

$$\begin{aligned} E(\hat{\mathbf{B}}) &= E\left[(\mathbf{X}'\mathbf{X})^{-1}\mathbf{X}'\mathbf{Y}\right] \\ &= (\mathbf{X}'\mathbf{X})^{-1}\mathbf{X}'E(\mathbf{Y}) \qquad \text{[by (1.90)]} \\ &= (\mathbf{X}'\mathbf{X})^{-1}\mathbf{X}'\mathbf{XB} = \mathbf{B}. \end{aligned} \tag{7.76}$$

□

In the next theorem, we consider the covariance structure of the $\hat{\beta}_{jk}$'s in $\hat{\mathbf{B}}$.

Theorem 7.4C. Under the assumptions $\text{cov}(\mathbf{y}_i) = \mathbf{\Sigma}$ and $\text{cov}(\mathbf{y}_i, \mathbf{y}_j) = \mathbf{O}$, all elements $\hat{\beta}_{jk}$ of $\hat{\mathbf{B}}$ are intercorrelated. Specifically,

$$\text{cov}(\hat{\boldsymbol{\beta}}_{(j)}) = \sigma_{jj}(\mathbf{X}'\mathbf{X})^{-1}, \tag{7.77}$$

$$\text{cov}(\hat{\boldsymbol{\beta}}_{(j)}, \hat{\boldsymbol{\beta}}_{(k)}) = \sigma_{jk}(\mathbf{X}'\mathbf{X})^{-1}, \tag{7.78}$$

where σ_{jj} and σ_{jk} are elements of $\mathbf{\Sigma}$.

Proof. Equation (7.77) follows from (7.75) and (7.8). To prove (7.78), we first note that $\text{cov}(\mathbf{y}_{(j)}, \mathbf{y}_{(k)}) = \sigma_{jk}\mathbf{I}$. This follows directly from the two assumptions $\text{cov}(\mathbf{y}_i) = \mathbf{\Sigma}$ and $\text{cov}(\mathbf{y}_i, \mathbf{y}_j) = \mathbf{O}$, which state that the y's in a row of $\mathbf{Y}$ are correlated but y's in different rows are uncorrelated. We illustrate $\text{cov}(\mathbf{y}_{(j)}, \mathbf{y}_{(k)}) = \sigma_{jk}\mathbf{I}$ for $\mathbf{y}_{(1)}$ and $\mathbf{y}_{(2)}$:

$$\text{cov}(\mathbf{y}_{(1)}, \mathbf{y}_{(2)}) = \text{cov}\left[\begin{pmatrix} y_{11} \\ y_{21} \\ \vdots \\ y_{n1} \end{pmatrix}, \begin{pmatrix} y_{12} \\ y_{22} \\ \vdots \\ y_{n2} \end{pmatrix}\right]$$

$$= \begin{pmatrix} \text{cov}(y_{11}, y_{12}) & \text{cov}(y_{11}, y_{22}) & \cdots & \text{cov}(y_{11}, y_{n2}) \\ \text{cov}(y_{21}, y_{12}) & \text{cov}(y_{21}, y_{22}) & \cdots & \text{cov}(y_{21}, y_{n2}) \\ \vdots & \vdots & & \vdots \\ \text{cov}(y_{n1}, y_{12}) & \text{cov}(y_{n1}, y_{22}) & \cdots & \text{cov}(y_{n1}, y_{n2}) \end{pmatrix}$$

$$= \begin{pmatrix} \sigma_{12} & 0 & \cdots & 0 \\ 0 & \sigma_{12} & \cdots & 0 \\ \vdots & \vdots & & \vdots \\ 0 & 0 & \cdots & \sigma_{12} \end{pmatrix}.$$

Now, by (1.99),

$$\begin{aligned} \text{cov}(\hat{\boldsymbol{\beta}}_{(j)}, \hat{\boldsymbol{\beta}}_{(k)}) &= \text{cov}\left[(\mathbf{X}'\mathbf{X})^{-1}\mathbf{X}'\mathbf{y}_{(j)}, (\mathbf{X}'\mathbf{X})^{-1}\mathbf{X}'\mathbf{y}_{(k)}\right] \\ &= (\mathbf{X}'\mathbf{X})^{-1}\mathbf{X}' \text{cov}(\mathbf{y}_{(j)}, \mathbf{y}_{(k)})\mathbf{X}(\mathbf{X}'\mathbf{X})^{-1} \\ &= (\mathbf{X}'\mathbf{X})^{-1}\mathbf{X}'(\sigma_{jk}\mathbf{I})\mathbf{X}(\mathbf{X}'\mathbf{X})^{-1} \\ &= \sigma_{jk}(\mathbf{X}'\mathbf{X})^{-1}. \end{aligned} \tag{7.79}$$

□

Additional covariance expressions for the elements of $\hat{\mathbf{B}}$ are

$$\text{cov}(\hat{\beta}_{ij}, \hat{\beta}_{kl}) = \sigma_{jl} g_{ik}, \tag{7.80}$$

$$\text{cov}(\hat{\boldsymbol{\beta}}_i, \hat{\boldsymbol{\beta}}_k) = g_{ik}\boldsymbol{\Sigma}, \tag{7.81}$$

where $\hat{\beta}_{ij}$ and $\hat{\beta}_{kl}$ are individual elements of $\hat{\mathbf{B}}$; $\hat{\boldsymbol{\beta}}_i'$ and $\hat{\boldsymbol{\beta}}_k'$ are rows of $\hat{\mathbf{B}}$; and g_{ik} is from $\mathbf{G} = (\mathbf{X}'\mathbf{X})^{-1}$ (see Problem 7.11).

The Gauss–Markov theorem for the univariate y case (Theorem 7.2A) has a multivariate generalization.

Theorem 7.4D (Gauss–Markov Theorem). If $E(\mathbf{Y}) = \mathbf{XB}$, $\text{cov}(\mathbf{y}_i) = \boldsymbol{\Sigma}$, $i = 1, 2, \ldots, n$, and $\text{cov}(\mathbf{y}_i, \mathbf{y}_j) = \mathbf{O}$, $i \neq j$, then the least squares estimators $\hat{\beta}_{jk}$ in $\hat{\mathbf{B}}$ are BLUE for β_{jk}.

Proof. Without loss of generality, we prove this for the elements $\hat{\beta}_{02}, \hat{\beta}_{12}, \ldots, \hat{\beta}_{q2}$ of $\hat{\boldsymbol{\beta}}_{(2)}$, the second column of $\hat{\mathbf{B}}$. Let $\mathbf{b}_2 = \sum_{j=1}^{p} \mathbf{A}_j\mathbf{y}_{(j)}$ be a linear estimator of $\boldsymbol{\beta}_{(2)}$, based on all the columns in $\mathbf{Y}$. To be unbiased, $\mathbf{b}_2$ must satisfy

$$\boldsymbol{\beta}_{(2)} = E(\mathbf{b}_2) = E\left(\sum_{j=1}^{p} \mathbf{A}_j\mathbf{y}_{(j)}\right) = \sum_{j=1}^{p} \mathbf{A}_j\mathbf{X}\boldsymbol{\beta}_{(j)}.$$

For the sum on the right to equal $\boldsymbol{\beta}_{(2)}$, we must have

$$\mathbf{A}_j = \mathbf{O} \text{ for } j \neq 2.$$

Hence $\mathbf{b}_2$ reduces to $\mathbf{A}_2\mathbf{y}_{(2)}$. Since $\boldsymbol{\beta}_{(2)} = \mathbf{A}_2\mathbf{X}\boldsymbol{\beta}_{(2)}$ must hold for all possible values of $\boldsymbol{\beta}_{(2)}$, we have, by (A.4.5), the condition

$$\mathbf{A}_2\mathbf{X} = \mathbf{I}.$$

The variances of the b_{j2}'s are found on the diagonal of

$$\text{cov}(\mathbf{b}_2) = \sigma_{22}\mathbf{A}_2\mathbf{A}_2',$$

and the condition $\mathbf{A}_2\mathbf{X} = \mathbf{I}$ leads to $\mathbf{A}_2 = (\mathbf{X}'\mathbf{X})^{-1}\mathbf{X}'$ as the value that minimizes the diagonal elements of $\mathbf{A}_2\mathbf{A}_2'$ (see Problem 7.12). □

The remarkable feature of the Gauss–Markov theorem is its distributional generality. The result holds for any distribution of $\mathbf{y}_i$; normality is not required. The only assumptions used in the proof are $E(\mathbf{Y}) = \mathbf{XB}$, $\text{cov}(\mathbf{y}_i) = \mathbf{\Sigma}$, and $\text{cov}(\mathbf{y}_i, \mathbf{y}_j) = \mathbf{O}$ for $i \neq j$. If these assumptions do not hold, $\hat{\mathbf{B}}$ may be biased or each $\hat{\beta}_{jk}$ may have a larger variance than that of some other estimator.

7.4.4 An Estimator for $\mathbf{\Sigma}$

To estimate $\text{cov}(\mathbf{y}_i) = \mathbf{\Sigma}$, we use the multivariate analogue of (7.10) and (7.11):

$$\mathbf{S}_e = \frac{\mathbf{E}}{n-q-1} = \frac{(\mathbf{Y} - \mathbf{X}\hat{\mathbf{B}})'(\mathbf{Y} - \mathbf{X}\hat{\mathbf{B}})}{n-q-1} \tag{7.82}$$

$$= \frac{\mathbf{Y}'\mathbf{Y} - \hat{\mathbf{B}}'\mathbf{X}'\mathbf{Y}}{n-q-1}. \tag{7.83}$$

With the denominator $n - q - 1$, $\mathbf{S}_e$ is unbiased.

Theorem 7.4E. If $\mathbf{S}_e$, is defined as in (7.82), then

$$E(\mathbf{S}_e) = \mathbf{\Sigma}. \tag{7.84}$$

Proof. Consider the individual elements of the $p \times p$ matrix $(\mathbf{Y} - \mathbf{X}\hat{\mathbf{B}})'(\mathbf{Y} - \mathbf{X}\hat{\mathbf{B}})$:

$$\begin{pmatrix} (\mathbf{y}_{(1)} - \mathbf{X}\hat{\boldsymbol{\beta}}_{(1)})'(\mathbf{y}_{(1)} - \mathbf{X}\hat{\boldsymbol{\beta}}_{(1)}) & \cdots & (\mathbf{y}_{(1)} - \mathbf{X}\hat{\boldsymbol{\beta}}_{(1)})'(\mathbf{y}_{(p)} - \mathbf{X}\hat{\boldsymbol{\beta}}_{(p)}) \\ \vdots & & \vdots \\ (\mathbf{y}_{(p)} - \mathbf{X}\hat{\boldsymbol{\beta}}_{(p)})'(\mathbf{y}_{(1)} - \mathbf{X}\hat{\boldsymbol{\beta}}_{(1)}) & \cdots & (\mathbf{y}_{(p)} - \mathbf{X}\hat{\boldsymbol{\beta}}_{(p)})'(\mathbf{y}_{(p)} - \mathbf{X}\hat{\boldsymbol{\beta}}_{(p)}) \end{pmatrix}.$$

By (7.12), $E(\mathbf{y}_{(j)} - \mathbf{X}\hat{\boldsymbol{\beta}}_{(j)})'(\mathbf{y}_{(j)} - \mathbf{X}\hat{\boldsymbol{\beta}}_{(j)}) = (n - q - 1)\sigma_{jj}$, and by analogy,

$$E(\mathbf{y}_{(j)} - \mathbf{X}\hat{\boldsymbol{\beta}}_{(j)})'(\mathbf{y}_{(k)} - \mathbf{X}\hat{\boldsymbol{\beta}}_{(k)}) = (n - q - 1)\sigma_{jk} \tag{7.85}$$

(see Problem 7.14). Hence the expected value of $\mathbf{E} = (\mathbf{Y} - \mathbf{X}\hat{\mathbf{B}})'(\mathbf{Y} - \mathbf{X}\hat{\mathbf{B}})$ is

$$E(\mathbf{Y} - \mathbf{X}\hat{\mathbf{B}})'(\mathbf{Y} - \mathbf{X}\hat{\mathbf{B}}) = (n - q - 1)\begin{pmatrix} \sigma_{11} & \sigma_{12} & \cdots & \sigma_{1p} \\ \sigma_{21} & \sigma_{22} & \cdots & \sigma_{2p} \\ \vdots & \vdots & & \vdots \\ \sigma_{p1} & \sigma_{p2} & \cdots & \sigma_{pp} \end{pmatrix}$$

$$= (n - q - 1)\mathbf{\Sigma}, \tag{7.86}$$

and (7.84) follows. □

7.4.5 Normal Model for the $\mathbf{y}_i$'s

By (7.69), each row in the model $\mathbf{Y} = \mathbf{XB} + \boldsymbol{\Xi}$ can be expressed as

$$\mathbf{y}_i' = \mathbf{x}_i'\mathbf{B} + \boldsymbol{\varepsilon}_i', \quad i = 1, 2, \ldots, n,$$

or in transposed form,

$$\mathbf{y}_i = \mathbf{B}'\mathbf{x}_i + \boldsymbol{\varepsilon}_i.$$

To the three assumptions in Section 7.4.1, we now add multivariate normality of $\mathbf{y}_i$, $i = 1, 2, \ldots, n$:

$$\mathbf{y}_i \text{ is } N_p(\mathbf{B}'\mathbf{x}_i, \boldsymbol{\Sigma}),$$

or equivalently,

$$\boldsymbol{\varepsilon}_i \text{ is } N_p(\mathbf{0}, \boldsymbol{\Sigma}).$$

With the normality assumption, we can obtain maximum likelihood estimators for $\mathbf{B}$ and $\boldsymbol{\Sigma}$.

Theorem 7.4F. If $\mathbf{y}_i$ is $N_p(\mathbf{B}'\mathbf{x}_i, \boldsymbol{\Sigma})$, $i = 1, 2, \ldots, n$, where $\mathbf{y}_i'$ is the ith row of $\mathbf{Y}$ in the model $\mathbf{Y} = \mathbf{XB} + \boldsymbol{\Xi}$, and if $\mathbf{y}_1, \mathbf{y}_2, \ldots, \mathbf{y}_n$ are independent, then the maximum likelihood estimators for $\mathbf{B}$ and $\boldsymbol{\Sigma}$ are given by

$$\hat{\mathbf{B}} = (\mathbf{X}'\mathbf{X})^{-1}\mathbf{X}'\mathbf{Y}, \tag{7.87}$$

$$\hat{\boldsymbol{\Sigma}} = \frac{1}{n}(\mathbf{Y} - \mathbf{X}\hat{\mathbf{B}})'(\mathbf{Y} - \mathbf{X}\hat{\mathbf{B}}) = \frac{\mathbf{E}}{n}. \tag{7.88}$$

Proof. Since $\mathbf{y}_1, \mathbf{y}_2, \ldots, \mathbf{y}_n$ are independent, the likelihood function is the product of their densities:

$$\begin{aligned} L(\mathbf{B}, \boldsymbol{\Sigma}) &= \prod_{i=1}^{n} \frac{1}{(\sqrt{2\pi})^p |\boldsymbol{\Sigma}|^{1/2}} e^{-\frac{1}{2}(\mathbf{y}_i - \mathbf{B}'\mathbf{x}_i)'\boldsymbol{\Sigma}^{-1}(\mathbf{y}_i - \mathbf{B}'\mathbf{x}_i)} \\ &= \frac{1}{(\sqrt{2\pi})^{np} |\boldsymbol{\Sigma}|^{n/2}} e^{-\frac{1}{2}\sum_{i=1}^{n}(\mathbf{y}_i - \mathbf{B}'\mathbf{x}_i)'\boldsymbol{\Sigma}^{-1}(\mathbf{y}_i - \mathbf{B}'\mathbf{x}_i)}. \end{aligned}$$

We follow the approach used in the proof of Theorem 2.3A, noting first that, because $(\mathbf{y}_i - \mathbf{B}'\mathbf{x}_i)'\boldsymbol{\Sigma}^{-1}(\mathbf{y}_i - \mathbf{B}'\mathbf{x}_i)$ is a scalar, it is equal to its trace. Using (A.8.1) and (A.8.2), we have

$$\begin{aligned} \sum_{i=1}^{n}(\mathbf{y}_i - \mathbf{B}'\mathbf{x}_i)'\boldsymbol{\Sigma}^{-1}(\mathbf{y}_i - \mathbf{B}'\mathbf{x}_i) &= \sum_{i=1}^{n} \mathrm{tr}\left[(\mathbf{y}_i - \mathbf{B}'\mathbf{x}_i)'\boldsymbol{\Sigma}^{-1}(\mathbf{y}_i - \mathbf{B}'\mathbf{x}_i)\right] \\ &= \sum_{i=1}^{n} \mathrm{tr}\left[\boldsymbol{\Sigma}^{-1}(\mathbf{y}_i - \mathbf{B}'\mathbf{x}_i)(\mathbf{y}_i - \mathbf{B}'\mathbf{x}_i)'\right] \end{aligned}$$

$$= \text{tr}\left[\boldsymbol{\Sigma}^{-1}\sum_{i=1}^{n}(\mathbf{y}_i - \mathbf{B}'\mathbf{x}_i)(\mathbf{y}_i - \mathbf{B}'\mathbf{x}_i)'\right]$$

$$= \text{tr}\left[\boldsymbol{\Sigma}^{-1}(\mathbf{Y} - \mathbf{XB})'(\mathbf{Y} - \mathbf{XB})\right] \qquad \text{[by (A.3.8)].}$$

Now, by adding and subtracting $\mathbf{X}\hat{\mathbf{B}}$, it can be shown that

$$(\mathbf{Y} - \mathbf{XB})'(\mathbf{Y} - \mathbf{XB}) = (\mathbf{Y} - \mathbf{X}\hat{\mathbf{B}})'(\mathbf{Y} - \mathbf{X}\hat{\mathbf{B}}) + (\hat{\mathbf{B}} - \mathbf{B})'\mathbf{X}'\mathbf{X}(\hat{\mathbf{B}} - \mathbf{B}), \tag{7.89}$$

which leads to

$$L(\mathbf{B}, \boldsymbol{\Sigma}) = \frac{1}{(\sqrt{2\pi})^{np}|\boldsymbol{\Sigma}|^{n/2}} e^{-\frac{1}{2}\text{tr}(\boldsymbol{\Sigma}^{-1}\mathbf{E}) - \frac{1}{2}\text{tr}[\mathbf{X}(\hat{\mathbf{B}}-\mathbf{B})\boldsymbol{\Sigma}^{-1}(\hat{\mathbf{B}}-\mathbf{B})'\mathbf{X}']}, \tag{7.90}$$

where $\mathbf{E} = (\mathbf{Y} - \mathbf{X}\hat{\mathbf{B}})'(\mathbf{Y} - \mathbf{X}\hat{\mathbf{B}})$.

We find it convenient to work with $\ln L(\mathbf{B}, \boldsymbol{\Sigma})$:

$$\ln L(\mathbf{B}, \boldsymbol{\Sigma}) = -np \ln\sqrt{2\pi} - \frac{n}{2}\ln|\boldsymbol{\Sigma}| - \frac{1}{2}\text{tr}(\boldsymbol{\Sigma}^{-1}\mathbf{E}) - \frac{1}{2}\text{tr}\left[\mathbf{X}(\hat{\mathbf{B}} - \mathbf{B})\boldsymbol{\Sigma}^{-1}(\hat{\mathbf{B}} - \mathbf{B})'\mathbf{X}'\right].$$

Only the last term involves $\mathbf{B}$, and since the matrix $\mathbf{X}(\hat{\mathbf{B}} - \mathbf{B})\boldsymbol{\Sigma}^{-1}(\hat{\mathbf{B}} - \mathbf{B})'\mathbf{X}'$ is positive semidefinite, $\ln L(\mathbf{B}, \boldsymbol{\Sigma})$ is maximized (with respect to $\mathbf{B}$) by $\hat{\mathbf{B}}$, thus verifying (7.87).

To find the maximum likelihood estimator for $\boldsymbol{\Sigma}$, we differentiate

$$\ln L(\hat{\mathbf{B}}, \boldsymbol{\Sigma}) = -np \ln\sqrt{2\pi} - \frac{n}{2}\ln|\boldsymbol{\Sigma}| - \frac{1}{2}\text{tr}(\boldsymbol{\Sigma}^{-1}\mathbf{E})$$

with respect to $\boldsymbol{\Sigma}^{-1}$ [using (A.13.5) and (A.13.6)] and set the result equal to $\mathbf{O}$ [see also (2.49) and (2.50) and the discussion following] to obtain

$$\hat{\boldsymbol{\Sigma}} = \frac{1}{n}\mathbf{E} = \frac{1}{n}(\mathbf{Y}'\mathbf{Y} - \hat{\mathbf{B}}'\mathbf{X}'\mathbf{Y}). \qquad \square$$

For use in Section 7.5, we note that the maximum value of $L(\mathbf{B}, \boldsymbol{\Sigma})$ is given by

$$L(\hat{\mathbf{B}}, \hat{\boldsymbol{\Sigma}}) = \frac{e^{-np/2}}{(\sqrt{2\pi})^{np}|\hat{\boldsymbol{\Sigma}}|^{n/2}}. \tag{7.91}$$

The maximum likelihood estimator $\hat{\mathbf{B}}$ in (7.87) is the same as the least squares estimator (7.70). The estimator $\hat{\boldsymbol{\Sigma}}$ in (7.88) is biased, and we often use the unbiased version $\mathbf{S}_e = \mathbf{E}/(n - q - 1)$ as given by (7.82) or (7.83). Some of the properties of $\hat{\mathbf{B}}$ and $\hat{\boldsymbol{\Sigma}}$ are given in the following theorem (see Problem 7.19).

Theorem 7.4G. If $\mathbf{y}_i$ is $N_p(\mathbf{B}'\mathbf{x}_i, \boldsymbol{\Sigma})$, $i = 1, 2, \ldots, n$, the maximum likelihood estimators $\hat{\mathbf{B}}$ and $\hat{\boldsymbol{\Sigma}} = \mathbf{E}/n$ have the following properties:

(i) Each column of $\hat{\mathbf{B}}$ is normal: $\hat{\boldsymbol{\beta}}_{(j)}$ is $N_{q+1}\left[\boldsymbol{\beta}_{(j)}, \sigma_{jj}(\mathbf{X}'\mathbf{X})^{-1}\right]$, $j = 1, 2, \ldots, p$.

(ii) $n\hat{\boldsymbol{\Sigma}} = \mathbf{E}$ is distributed as $W_p(n - q - 1, \boldsymbol{\Sigma})$.

(iii) $\hat{\mathbf{B}}$ and $\hat{\boldsymbol{\Sigma}}$ are independent.

(iv) $\hat{\mathbf{B}}$ and $\hat{\boldsymbol{\Sigma}}$ are jointly sufficient for $\mathbf{B}$ and $\boldsymbol{\Sigma}$.

(v) Each $\hat{\beta}_{jk}$ and $[n/(n-q-1)]\hat{\sigma}_{jk}$ is the minimum variance unbiased estimator of β_{jk} and σ_{jk}, respectively. □

7.4.6 The Multivariate Model in Centered Form

Corresponding to (7.15) in Section 7.2.3, there is a centered multivariate model,

$$\mathbf{Y} = (\mathbf{j}, \mathbf{X}_c)\begin{pmatrix} \boldsymbol{\alpha}' \\ \mathbf{B}_1 \end{pmatrix} + \boldsymbol{\Xi}, \tag{7.92}$$

where $\mathbf{X}_c$ is given in (7.16), $\boldsymbol{\alpha}' = (\alpha_1, \alpha_2, \ldots, \alpha_p)$, and $\mathbf{B}_1$ consists of all rows of $\mathbf{B}$ except the first:

$$\mathbf{B} = \begin{pmatrix} \boldsymbol{\beta}_0' \\ \mathbf{B}_1 \end{pmatrix} = \left(\begin{array}{cccc} \beta_{01} & \beta_{02} & \cdots & \beta_{0p} \\ \hline \beta_{11} & \beta_{12} & \cdots & \beta_{1p} \\ \vdots & \vdots & & \vdots \\ \beta_{q1} & \beta_{q2} & \cdots & \beta_{qp} \end{array}\right). \tag{7.93}$$

By an extension of (7.19) and (7.20), we obtain the least squares estimators

$$\hat{\boldsymbol{\alpha}}' = \overline{\mathbf{y}}' = (\overline{y}_1, \overline{y}_2, \ldots, \overline{y}_p), \tag{7.94}$$

$$\hat{\mathbf{B}}_1 = (\mathbf{X}_c'\mathbf{X}_c)^{-1}\mathbf{X}_c'\mathbf{Y}, \tag{7.95}$$

where $\overline{y}_j$ is the mean of the jth column of $\mathbf{Y}$. These estimators give the same results as $\hat{\mathbf{B}} = (\mathbf{X}'\mathbf{X})^{-1}\mathbf{X}'\mathbf{Y}$ in (7.70), with an adjustment corresponding to (7.21) for each $\hat{\beta}_{0j}$,

$$\hat{\beta}_{0j} = \hat{\alpha}_j - \hat{\beta}_{1j}\overline{x}_1 - \hat{\beta}_{2j}\overline{x}_2 - \cdots - \hat{\beta}_{qj}\overline{x}_q. \tag{7.96}$$

We can express $\hat{\mathbf{B}}_1$ in (7.95) in terms of sample covariance matrices,

$$\begin{aligned} \hat{\mathbf{B}}_1 &= (n-1)(\mathbf{X}_c'\mathbf{X}_c)^{-1}\frac{\mathbf{X}_c'\mathbf{Y}}{n-1} = \left(\frac{\mathbf{X}_c'\mathbf{X}_c}{n-1}\right)^{-1}\frac{\mathbf{X}_c'\mathbf{Y}}{n-1} \\ &= \mathbf{S}_{xx}^{-1}\mathbf{S}_{xy}, \end{aligned} \tag{7.97}$$

where $\mathbf{S}_{xx}$ and $\mathbf{S}_{xy}$ are from the partitioned sample covariance matrix

$$\mathbf{S} = \begin{pmatrix} \mathbf{S}_{yy} & \mathbf{S}_{yx} \\ \mathbf{S}_{xy} & \mathbf{S}_{xx} \end{pmatrix}. \tag{7.98}$$

Note that since the y's are random variables and the x's are fixed, $\mathbf{S}_{yy}$ is the sample covariance matrix of the y's in the usual sense, while $\mathbf{S}_{xx}$ consists of an analogous mathematical expression involving the constant x's. Similarly, $\mathbf{S}_{xy}$ contains expressions such as $\sum_{i=1}^{n}(x_{ij} - \bar{x}_j)(y_{ik} - \bar{y}_k)/(n-1)$ for $j = 1, 2, \ldots, q$; $k = 1, 2, \ldots, p$.

By (7.96) and (7.97), we have

$$\hat{\boldsymbol{\beta}}_0 = \begin{pmatrix} \hat{\alpha}_1 - \hat{\boldsymbol{\beta}}'_{1(1)}\bar{\mathbf{x}} \\ \vdots \\ \hat{\alpha}_p - \hat{\boldsymbol{\beta}}'_{1(p)}\bar{\mathbf{x}} \end{pmatrix} \tag{7.99}$$

$$= \bar{\mathbf{y}} - \hat{\mathbf{B}}'_1\bar{\mathbf{x}} = \bar{\mathbf{y}} - \mathbf{S}_{yx}\mathbf{S}_{xx}^{-1}\bar{\mathbf{x}}, \tag{7.100}$$

where $\hat{\mathbf{B}}_1 = (\hat{\boldsymbol{\beta}}_{1(1)}, \hat{\boldsymbol{\beta}}_{1(2)}, \ldots, \hat{\boldsymbol{\beta}}_{1(p)})$ and $\hat{\boldsymbol{\beta}}_{1(j)} = (\hat{\beta}_{j1}, \hat{\beta}_{j2}, \ldots, \hat{\beta}_{jp})'$, $j = 1, 2, \ldots, p$.

7.4.7 Measures of Multivariate Association

To define measures of association between the y's and the x's, we first define a matrix analogous to R^2 in (7.46):

$$\mathbf{R}^2 = (\mathbf{Y}'\mathbf{Y} - n\bar{\mathbf{y}}\bar{\mathbf{y}}')^{-1}(\hat{\mathbf{B}}'\mathbf{X}'\mathbf{Y} - n\bar{\mathbf{y}}\bar{\mathbf{y}}'). \tag{7.101}$$

Measures of association that range between 0 and 1 are given by

$$r_t^2 = \frac{\text{tr}(\mathbf{R}^2)}{p} \quad \text{and} \quad r_d^2 = |\mathbf{R}^2|. \tag{7.102}$$

However, both r_t^2 and r_d^2 are poor measures of association (see Problem 7.22); both underestimate the "true" amount of relationship between the y's and the x's. This is especially true for r_d^2.

A better measure of association is provided by the largest canonical correlation r_1 (Chapter 8); for other measures of association, see Rencher (1995, Section 10.6).

7.5 HYPOTHESIS TESTS IN THE MULTIVARIATE MULTIPLE REGRESSION MODEL: FIXED x's

In this section, we consider tests that are analogous to those for univariate y's in Section 7.2.4. We assume that $\mathbf{X}$ is fixed and that $\mathbf{y}_i$ is $N_p(\mathbf{B}'\mathbf{x}_i, \boldsymbol{\Sigma})$, $i = 1, 2, \ldots, n$.

7.5.1 Test for Significance of Regression

The hypothesis that none of the q x's predict any of the p y's can be expressed as $H_0: \mathbf{B}_1 = \mathbf{O}$, where $\mathbf{B}_1$ is defined as in (7.93):

$$\mathbf{B} = \begin{pmatrix} \boldsymbol{\beta}_0' \\ \mathbf{B}_1 \end{pmatrix} = \begin{pmatrix} \beta_{01} & \beta_{02} & \cdots & \beta_{0p} \\ \hline \beta_{11} & \beta_{12} & \cdots & \beta_{1p} \\ \vdots & \vdots & & \vdots \\ \beta_{q1} & \beta_{q2} & \cdots & \beta_{qp} \end{pmatrix}.$$

To obtain a test of H_0, we begin by partitioning the (corrected) total sum of squares and products matrix $\mathbf{Y}'\mathbf{Y} - n\bar{\mathbf{y}}\bar{\mathbf{y}}'$ as

$$\begin{aligned} \mathbf{Y}'\mathbf{Y} - n\bar{\mathbf{y}}\bar{\mathbf{y}}' &= (\mathbf{Y}'\mathbf{Y} - \hat{\mathbf{B}}'\mathbf{X}'\mathbf{Y}) + (\hat{\mathbf{B}}'\mathbf{X}'\mathbf{Y} - n\bar{\mathbf{y}}\bar{\mathbf{y}}') \\ &= \mathbf{E} + \mathbf{H}, \end{aligned} \tag{7.103}$$

where $\bar{\mathbf{y}}$ is defined in (7.94) and $\hat{\mathbf{B}} = (\mathbf{X}'\mathbf{X})^{-1}\mathbf{X}'\mathbf{Y}$. By subtracting $n\bar{\mathbf{y}}\bar{\mathbf{y}}'$, the overall regression sum of squares and products matrix $\mathbf{H} = \hat{\mathbf{B}}'\mathbf{X}'\mathbf{Y} - n\bar{\mathbf{y}}\bar{\mathbf{y}}'$ is adjusted for the intercepts $\beta_{01}, \beta_{02}, \ldots, \beta_{0p}$ (Problem 7.23) and can therefore be used to test $H_0\colon \mathbf{B}_1 = \mathbf{O}$. By Theorem 7.4G(ii), $\mathbf{E} = \mathbf{Y}'\mathbf{Y} - \hat{\mathbf{B}}'\mathbf{X}'\mathbf{Y}$ has the Wishart distribution $W_p(n-q-1, \boldsymbol{\Sigma})$. It can be shown that (under H_0) $\mathbf{H} = \hat{\mathbf{B}}'\mathbf{X}'\mathbf{Y} - n\bar{\mathbf{y}}\bar{\mathbf{y}}'$ is $W_p(q, \boldsymbol{\Sigma})$ and that $\mathbf{H}$ is independent of $\mathbf{E}$ (Problem 7.24). Therefore, by property 1 of Section 4.1.2,

$$\Lambda = \frac{|\mathbf{E}|}{|\mathbf{E} + \mathbf{H}|} = \frac{|\mathbf{Y}'\mathbf{Y} - \hat{\mathbf{B}}'\mathbf{X}'\mathbf{Y}|}{|\mathbf{Y}'\mathbf{Y} - n\bar{\mathbf{y}}\bar{\mathbf{y}}'|} \tag{7.104}$$

is distributed as $\Lambda_{p,q,n-q-1}$ when $H_0 : \mathbf{B}_1 = \mathbf{O}$ is true. We reject H_0 if $\Lambda \leq \Lambda_{\alpha,p,q,n-q-1}$. Critical values for Wilks' Λ are given in Table B.4. We can also use the F- and χ^2-approximations for Λ in (4.16) and (4.17), with $\nu_H = q$ and $\nu_E = n - q - 1$. By (4.14), Wilks' Λ in (7.104) can be expressed in terms of the eigenvalues $\lambda_1, \lambda_2, \ldots, \lambda_s$ of $\mathbf{E}^{-1}\mathbf{H}$:

$$\Lambda = \prod_{i=1}^{s} \frac{1}{1 + \lambda_i}, \tag{7.105}$$

where $s = \min(p, q)$.

Wilks' Λ in (7.104) is a simple function of the likelihood ratio.

Theorem 7.5A. The likelihood ratio test of $H_0\colon \mathbf{B}_1 = \mathbf{O}$ is based on the Wilks' Λ test statistic in (7.104).

Proof. Under $H_0\colon \mathbf{B}_1 = \mathbf{O}$, the maximum likelihood estimator of $\boldsymbol{\Sigma}$ (see Problem 7.25) is

$$\hat{\boldsymbol{\Sigma}}_0 = \frac{1}{n}(\mathbf{Y}'\mathbf{Y} - n\bar{\mathbf{y}}\bar{\mathbf{y}}'). \tag{7.106}$$

By (7.91), the likelihood ratio reduces to

$$\mathrm{LR} = \frac{\max_{H_0} L}{\max_{H_1} L} = \left(\frac{|\hat{\boldsymbol{\Sigma}}|}{|\hat{\boldsymbol{\Sigma}}_0|} \right)^{n/2},$$

and the likelihood ratio test can therefore be based on

$$\Lambda = (\mathrm{LR})^{2/n} = \frac{|\hat{\boldsymbol{\Sigma}}|}{|\hat{\boldsymbol{\Sigma}}_0|} = \frac{|\mathbf{Y}'\mathbf{Y} - \hat{\mathbf{B}}'\mathbf{X}'\mathbf{Y}|}{|\mathbf{Y}'\mathbf{Y} - n\bar{\mathbf{y}}\bar{\mathbf{y}}'|}. \qquad \square$$

Using the centered form of the model (Section 7.4.6), the Wilks Λ-statistic in (7.104) can be written in terms of the partitioned covariance matrix in (7.98) by noting that

$$\frac{\mathbf{Y}'\mathbf{Y} - n\bar{\mathbf{y}}\bar{\mathbf{y}}'}{n-1} - \frac{\hat{\mathbf{B}}'\mathbf{X}'\mathbf{Y} - n\bar{\mathbf{y}}\bar{\mathbf{y}}'}{n-1} = \mathbf{S}_{yy} - \mathbf{S}_{yx}\mathbf{S}_{xx}^{-1}\mathbf{S}_{xy}, \tag{7.107}$$

which gives

$$\Lambda = \frac{|\mathbf{Y}'\mathbf{Y} - \hat{\mathbf{B}}'\mathbf{X}'\mathbf{Y}|}{|\mathbf{Y}'\mathbf{Y} - n\bar{\mathbf{y}}\bar{\mathbf{y}}'|} = \frac{|\mathbf{S}_{yy} - \mathbf{S}_{yx}\mathbf{S}_{xx}^{-1}\mathbf{S}_{xy}|}{|\mathbf{S}_{yy}|}. \tag{7.108}$$

By (A.7.9), the determinant of $\mathbf{S}$ can be expressed in the form $|\mathbf{S}| = |\mathbf{S}_{xx}||\mathbf{S}_{yy} - \mathbf{S}_{yx}\mathbf{S}_{xx}^{-1}\mathbf{S}_{xy}|$, which can be used to simplify (7.108):

$$\Lambda = \frac{|\mathbf{S}_{xx}||\mathbf{S}_{yy} - \mathbf{S}_{yx}\mathbf{S}_{xx}^{-1}\mathbf{S}_{xy}|}{|\mathbf{S}_{xx}||\mathbf{S}_{yy}|} = \frac{|\mathbf{S}|}{|\mathbf{S}_{xx}||\mathbf{S}_{yy}|}. \tag{7.109}$$

This same form of Λ is obtained in the test for independence of $\mathbf{y}$ and $\mathbf{x}$ in Theorem 8.4A, where $\mathbf{x}$ is a random vector.

We now consider the union–intersection test of $H_0: \mathbf{B}_1 = \mathbf{O}$.

Theorem 7.5B. The union–intersection test of $H_0: \mathbf{B}_1 = \mathbf{O}$ uses Roy's statistic

$$\theta = \frac{\lambda_1}{1 + \lambda_1},$$

where λ_1 is the largest eigenvalue of $\mathbf{E}^{-1}\mathbf{H}$. The hypothesis is rejected if $\theta \geq \theta_\alpha$.

Proof. We reduce each row $\mathbf{y}_i'$ in $\mathbf{Y}$ to univariate $z_i = \mathbf{y}_i'\mathbf{a}$ by the transformation

$$\mathbf{Ya} = \mathbf{XBa} + \boldsymbol{\Xi}\mathbf{a},$$

which we write as

$$\mathbf{z} = \mathbf{X}\boldsymbol{\beta} + \boldsymbol{\varepsilon}.$$

The hypothesis $H_0 : \mathbf{B}_1 = \mathbf{O}$ then reduces to $H_0 : \boldsymbol{\beta}_1 = \mathbf{0}$, where $\boldsymbol{\beta}_1$ includes all of $\boldsymbol{\beta} = \mathbf{Ba}$ except the intercept β_0. By (7.28), the sums of squares for testing $H_0 : \boldsymbol{\beta}_1 = \mathbf{0}$ can be obtained from the partitioning

$$\begin{aligned}\mathbf{z}'\mathbf{z} - n\bar{z}^2 &= (\mathbf{z}'\mathbf{z} - \hat{\boldsymbol{\beta}}'\mathbf{X}'\mathbf{z}) + (\hat{\boldsymbol{\beta}}'\mathbf{X}'\mathbf{z} - n\bar{z}^2) \\ &= \text{SSE}_z + \text{SSR}_z.\end{aligned}$$

From (7.29), the F-ratio is

$$F(\mathbf{a}) = \frac{\text{SSR}_z/q}{\text{SSE}_z/(n-q-1)} \tag{7.110}$$

$$= \frac{\mathbf{a}'\mathbf{Ha}/q}{\mathbf{a}'\mathbf{Ea}/(n-q-1)} \qquad \text{[by (4.19)]}, \tag{7.111}$$

where $\mathbf{H} = \hat{\mathbf{B}}'\mathbf{X}'\mathbf{Y} - n\bar{\mathbf{y}}\bar{\mathbf{y}}'$ and $\mathbf{E} = \mathbf{Y}'\mathbf{Y} - \hat{\mathbf{B}}'\mathbf{X}'\mathbf{Y}$. By Theorem 4.1C, the value of $\mathbf{a}$ that maximizes $F(\mathbf{a})$ is the eigenvector of $\mathbf{E}^{-1}\mathbf{H}$ that corresponds to the largest eigenvalue, λ_1. Roy's test statistic using λ_1 was given in (4.23) as

$$\theta = \frac{\lambda_1}{1+\lambda_1}. \tag{7.112}$$

□

Critical values θ_α for Roy's test are given in Table B.5, with

$$s = \min(p, q), \qquad m = \tfrac{1}{2}(|q-p|-1), \qquad N = \tfrac{1}{2}(n-q-p-2). \tag{7.113}$$

By (4.26) and (4.30), Pillai's $V^{(s)}$ and the Lawley–Hotelling $U^{(s)}$ are defined as

$$V^{(s)} = \text{tr}\left[(\mathbf{E}+\mathbf{H})^{-1}\mathbf{H}\right] = \sum_{i=1}^{s} \frac{\lambda_i}{1+\lambda_i}, \tag{7.114}$$

$$U^{(s)} = \text{tr}(\mathbf{E}^{-1}\mathbf{H}) = \sum_{i=1}^{s} \lambda_i, \tag{7.115}$$

where $\lambda_1, \lambda_2, \ldots, \lambda_s$ are the eigenvalues of $\mathbf{E}^{-1}\mathbf{H}$. For $V^{(s)}$, tables and F-approximations (Section 4.1.4) are indexed by s, m, and N as defined in (7.113). For $U^{(s)}$, the parameters are $\nu_H = q$ and $\nu_E = n - q - 1$. Critical values for $V^{(s)}$ and $\nu_E U^{(s)}/\nu_H$ are given in Tables B.6 and B.7, respectively.

When H_0 is true, all four test statistics have the same Type I error rate, α. When H_0 is false, the relative power of the tests depends on the configuration of the columns or rows of $\mathbf{B}_1$, which is reflected by the population eigenvalues (Section 4.2). To examine these, we write $|\mathbf{E}^{-1}\mathbf{H} - \lambda\mathbf{I}|$ in the form $|\mathbf{H} - \lambda\mathbf{E}| = 0$, for which the population counterpart is $|E(\mathbf{H}) - \phi E(\mathbf{E})| = 0$, where ϕ is an eigenvalue of $[E(\mathbf{E})]^{-1}E(\mathbf{H})$. From (7.86),

$$E(\mathbf{E}) = (n-q-1)\boldsymbol{\Sigma},$$

and it can be shown that

$$E(\mathbf{H}) = q\mathbf{\Sigma} + \mathbf{B}_1'\mathbf{X}_c'\mathbf{X}_c\mathbf{B}_1, \tag{7.116}$$

where $\mathbf{X}_c$ is the centered $\mathbf{X}$ matrix in (7.16). Thus

$$\begin{aligned} 0 &= |E(\mathbf{H}) - \phi E(\mathbf{E})| \\ &= |q\mathbf{\Sigma} + \mathbf{B}_1'\mathbf{X}_c'\mathbf{X}_c\mathbf{B}_1 - (n-q-1)\phi\mathbf{\Sigma}| \\ &= |\mathbf{B}_1'\mathbf{X}_c'\mathbf{X}_c\mathbf{B}_1 - \gamma\mathbf{\Sigma}|, \end{aligned} \tag{7.117}$$

where $\gamma = (n-q-1)\phi - q$ is an eigenvalue of $\mathbf{\Sigma}^{-1}\mathbf{B}_1'\mathbf{X}_c'\mathbf{X}_c\mathbf{B}_1$. If the eigenvalues $\gamma_1, \gamma_2, \ldots, \gamma_s$ are equal or not too disparate, then $V^{(s)}$, Λ, and $U^{(s)}$ are more powerful than θ. If there is only one nonzero γ, then θ is more powerful than $U^{(s)}$, Λ, and $V^{(s)}$. Since $\mathbf{\Sigma}$ and $\mathbf{X}_c'\mathbf{X}_c$ are assumed to be nonsingular, a single nonzero γ would imply rank $(\mathbf{B}_1) = 1$. There are various possible patterns in the β_{jk}'s that could produce a rank of 1.

Some indication of the pattern of the eigenvalues of $\mathbf{\Sigma}^{-1}\mathbf{B}_1'\mathbf{X}_c'\mathbf{X}_c\mathbf{B}_1$ is furnished by the eigenvalues of $\mathbf{E}^{-1}\mathbf{H}$. See additional comments in Section 4.2.

7.5.2 Test on a Subset of the Rows of B

We now obtain a test of significance for a subset of the rows of $\mathbf{B}$, say the last h rows of $\mathbf{B}$, corresponding to the last h columns of $\mathbf{X}$. The model for all n observations can be partitioned as

$$\mathbf{Y} = (\mathbf{X}_r, \mathbf{X}_d)\begin{pmatrix} \mathbf{B}_r \\ \mathbf{B}_d \end{pmatrix} + \mathbf{\Xi} = \mathbf{X}_r\mathbf{B}_r + \mathbf{X}_d\mathbf{B}_d + \mathbf{\Xi}, \tag{7.118}$$

where $\mathbf{X}_d$ has h columns and $\mathbf{B}_d$ has h rows. The hypothesis of interest is

$$H_0\colon \mathbf{B}_d = \mathbf{O},$$

which reduces (7.118) to

$$\mathbf{Y} = \mathbf{X}_r\mathbf{B}_r + \mathbf{\Xi}. \tag{7.119}$$

The total sum of squares and products matrix can be partitioned to reflect interest in $\mathbf{B}_d$:

$$\mathbf{Y}'\mathbf{Y} = (\mathbf{Y}'\mathbf{Y} - \hat{\mathbf{B}}'\mathbf{X}'\mathbf{Y}) + (\hat{\mathbf{B}}'\mathbf{X}'\mathbf{Y} - \hat{\mathbf{B}}_r'\mathbf{X}_r'\mathbf{Y}) + \hat{\mathbf{B}}_r'\mathbf{X}_r'\mathbf{Y}, \tag{7.120}$$

where $\hat{\mathbf{B}} = (\mathbf{X}'\mathbf{X})^{-1}\mathbf{X}'\mathbf{Y}$ is from the full model (7.118), $\hat{\mathbf{B}}_r = (\mathbf{X}_r'\mathbf{X}_r)^{-1}\mathbf{X}_r'\mathbf{Y}$ is from the reduced model (7.119), and $\hat{\mathbf{B}}'\mathbf{X}'\mathbf{Y} - \hat{\mathbf{B}}_r'\mathbf{X}_r'\mathbf{Y}$ is the "extra" regression sum of squares and products matrix due to $\mathbf{B}_d$ after adjusting for $\mathbf{B}_r$. We define

$$\mathbf{E} = \mathbf{Y}'\mathbf{Y} - \hat{\mathbf{B}}'\mathbf{X}'\mathbf{Y} \quad \text{and} \quad \mathbf{H} = \hat{\mathbf{B}}'\mathbf{X}'\mathbf{Y} - \hat{\mathbf{B}}_r'\mathbf{X}_r'\mathbf{Y}. \tag{7.121}$$

By Theorem 7.4G(ii), $\mathbf{E}$ is $W_p(n-q-1, \mathbf{\Sigma})$. It can be shown that (under H_0) $\mathbf{H} = \hat{\mathbf{B}}'\mathbf{X}'\mathbf{Y} - \hat{\mathbf{B}}_r'\mathbf{X}_r'\mathbf{Y}$ is $W_p(h, \mathbf{\Sigma})$ and that $\mathbf{H}$ is independent of $\mathbf{E}$ (see Problem

7.30). We can therefore test H_0 by Wilks' Λ:

$$\Lambda = \frac{|\mathbf{E}|}{|\mathbf{E}+\mathbf{H}|} = \frac{|\mathbf{Y}'\mathbf{Y} - \hat{\mathbf{B}}'\mathbf{X}'\mathbf{Y}|}{|\mathbf{Y}'\mathbf{Y} - \hat{\mathbf{B}}_r'\mathbf{X}_r'\mathbf{Y}|}, \tag{7.122}$$

which is distributed as $\Lambda_{p,h,n-q-1}$ when $H_0: \mathbf{B}_d = \mathbf{O}$ is true. For critical values of Λ in Table B.4, we use $\nu_H = h$ and $\nu_E = n - q - 1$. Alternatively, the F- and χ^2-approximations in (4.16) or (4.17) can be used.

In (7.122), the error matrix for the full model appears in the numerator, and the error matrix for the reduced model is in the denominator. Thus (7.122) can be rewritten as the ratio of Wilks' Λ for the full model to Wilks' Λ for the reduced model,

$$\Lambda = \frac{\Lambda_f}{\Lambda_r}, \tag{7.123}$$

where Λ_f is the overall regression test statistic in (7.104), and

$$\Lambda_r = \frac{|\mathbf{Y}'\mathbf{Y} - \hat{\mathbf{B}}_r'\mathbf{X}_r'\mathbf{Y}|}{|\mathbf{Y}'\mathbf{Y} - n\bar{\mathbf{y}}\bar{\mathbf{y}}'|} \tag{7.124}$$

is the overall regression test statistic for the reduced model.

The ratio Λ_f/Λ_r in (7.123) is similar in appearance to (4.134). However, (7.123) tests a subset of x's, whereas (4.134) tests a subset of y's. [Note that in (4.134) the *dependent* variables are partitioned into $\mathbf{y}$ and $\mathbf{x}$, where, for expositional convenience, $\mathbf{x}$ represents a subset of the y's.] Thus Λ in (4.134) could be written as $\Lambda(y_{p+1}, \ldots, y_{p+q}|y_1, \ldots, y_p)$, whereas Λ in (7.123) could be expressed as $\Lambda(x_{q-h+1}, \ldots, x_q|x_1, \ldots, x_{q-h})$.

The likelihood ratio approach leads to the Wilks Λ-test.

Theorem 7.5C. The likelihood ratio for testing $H_0: \mathbf{B}_d = \mathbf{O}$ is a function of the Wilks Λ test statistic in (7.122).

Proof. As in the proof of Theorem 7.5A, the likelihood ratio has the form

$$\mathrm{LR} = \left(\frac{|\hat{\mathbf{\Sigma}}|}{|\hat{\mathbf{\Sigma}}_0|}\right)^{n/2},$$

where $\hat{\mathbf{\Sigma}} = (\mathbf{Y}'\mathbf{Y} - \hat{\mathbf{B}}'\mathbf{X}'\mathbf{Y})/n$ is the estimator of $\mathbf{\Sigma}$ under H_1, and $\hat{\mathbf{\Sigma}}_0 = (\mathbf{Y}'\mathbf{Y} - \hat{\mathbf{B}}_r'\mathbf{X}_r'\mathbf{Y})/n$ is the estimator of $\mathbf{\Sigma}$ subject to $H_0: \mathbf{B}_d = \mathbf{O}$, that is, in the reduced model $\mathbf{Y} = \mathbf{X}_r\mathbf{B}_r + \mathbf{\Xi}$. With these substitutions for $\hat{\mathbf{\Sigma}}$ and $\hat{\mathbf{\Sigma}}_0$, $\Lambda = (\mathrm{LR})^{2/n}$ is the same as (7.122). □

The union–intersection approach leads to Roy's test.

Theorem 7.5D. The union–intersection test of $H_0: \mathbf{B}_d = \mathbf{O}$ uses Roy's statistic

$$\theta = \frac{\lambda_1}{1 + \lambda_1},$$

where λ_1 is the maximum eigenvalue of $\mathbf{E}^{-1}\mathbf{H} = (\mathbf{Y}'\mathbf{Y} - \hat{\mathbf{B}}'\mathbf{X}'\mathbf{Y})^{-1}(\hat{\mathbf{B}}'\mathbf{X}'\mathbf{Y} - \hat{\mathbf{B}}_r'\mathbf{X}_r'\mathbf{Y})$. □

Critical values θ_α for Roy's test are given in Table B.5, with

$$s = \min(p, h), \qquad m = \tfrac{1}{2}(|h - p| - 1), \qquad N = \tfrac{1}{2}(n - p - h - 2). \tag{7.125}$$

Pillai's $V^{(s)}$ and the Lawley–Hotelling $U^{(s)}$ can also be obtained in the usual way from the eigenvalues of $\mathbf{E}^{-1}\mathbf{H}$. Critical values in Tables B.6 and B.7 are based on s, m, and N in (7.125) and on $\nu_H = h$ and $\nu_E = n - q - 1$.

An approximation for the power of tests for multivariate regression in Sections 7.5.1 and 7.5.2 can be obtained using approximate F-tests; for details, see Muller et al. (1992), O'Brien and Muller (1992), and Muller and Peterson (1984).

7.5.3 General Linear Hypotheses CB = O and CBM = O

The hypothesis $H_0: \mathbf{CB} = \mathbf{O}$ provides for contrasts in the elements of each column of $\mathbf{B}$ and includes the hypotheses in Sections 7.5.1 and 7.5.2 as special cases. The hypothesis $H_0: \mathbf{B}_1 = \mathbf{O}$ in Section 7.5.1, for example, can be expressed as

$$H_0: \mathbf{CB} = (\mathbf{0}, \mathbf{I}_q)\begin{pmatrix} \boldsymbol{\beta}_0' \\ \mathbf{B}_1 \end{pmatrix} = \mathbf{O},$$

where $\mathbf{0}$ is $q \times 1$ and $\mathbf{O}$ is $q \times p$. Similarly, $H_0: \mathbf{B}_d = \mathbf{O}$ in Section 7.5.2 can be written in the form

$$H_0: \mathbf{CB} = (\mathbf{O}, \mathbf{I}_h)\begin{pmatrix} \mathbf{B}_r \\ \mathbf{B}_d \end{pmatrix} = \mathbf{O},$$

where the $\mathbf{O}$ matrix in $(\mathbf{O}, \mathbf{I}_h)$ is $h \times (q - h + 1)$ and the $\mathbf{O}$ matrix on the right side is $h \times p$.

We use the full-and-reduced-model approach to develop a test statistic for $H_0: \mathbf{CB} = \mathbf{O}$. The reduced model is

$$\mathbf{Y} = \mathbf{XB} + \boldsymbol{\Xi} \text{ subject to } \mathbf{CB} = \mathbf{O}, \tag{7.126}$$

where $\mathbf{C}$ is $w \times (q + 1)$ of rank w. Using Lagrange multipliers, it can be shown (Lunneborg and Abbott 1983, p. 354) that the estimator of $\mathbf{B}$ in the reduced model (7.126) is

$$\hat{\mathbf{B}}_c = \hat{\mathbf{B}} - (\mathbf{X}'\mathbf{X})^{-1}\mathbf{C}'\left[\mathbf{C}(\mathbf{X}'\mathbf{X})^{-1}\mathbf{C}'\right]^{-1}\mathbf{C}\hat{\mathbf{B}}, \tag{7.127}$$

where $\hat{\mathbf{B}} = (\mathbf{X}'\mathbf{X})^{-1}\mathbf{X}'\mathbf{Y}$ is from the full model unrestricted by the hypothesis. The $\mathbf{X}$ matrix is unchanged in the reduced model (7.126), and substitution of (7.127) into

$\mathbf{H} = \hat{\mathbf{B}}'\mathbf{X}'\mathbf{Y} - \hat{\mathbf{B}}_c'\mathbf{X}'\mathbf{Y}$ [see (7.121)] leads to

$$\mathbf{H} = (\mathbf{C}\hat{\mathbf{B}})' \left[\mathbf{C}(\mathbf{X}'\mathbf{X})^{-1}\mathbf{C}'\right]^{-1} \mathbf{C}\hat{\mathbf{B}}, \tag{7.128}$$

which is analogous to SSH $= (\mathbf{C}\hat{\boldsymbol{\beta}})'[\mathbf{C}(\mathbf{X}'\mathbf{X})^{-1}\mathbf{C}']^{-1}(\mathbf{C}\hat{\boldsymbol{\beta}})$ in (7.34) for the univariate-y case. (For a more formal justification of $\hat{\mathbf{B}}_c'\mathbf{X}'\mathbf{Y}$ as the sum of squares and products matrix for the reduced model, see Problem 7.35.) The matrices $\mathbf{H}$ and $\mathbf{E} = \mathbf{Y}'\mathbf{Y} - \hat{\mathbf{B}}'\mathbf{X}'\mathbf{Y}$ can be shown to be independent and to have Wishart distributions (Problem 7.36); therefore H_0: $\mathbf{CB} = \mathbf{O}$ can be tested by

$$\Lambda = \frac{|\mathbf{E}|}{|\mathbf{E} + \mathbf{H}|},$$

which is distributed as $\Lambda_{p,w,n-q-1}$. The likelihood ratio principle leads to this same value of Wilks' Λ.

Roy's θ, Pillai's $V^{(s)}$, and the Lawley–Hotelling statistic $U^{(s)}$ can readily be obtained from the eigenvalues of $\mathbf{E}^{-1}\mathbf{H}$, where $\mathbf{E} = \mathbf{Y}'\mathbf{Y} - \hat{\mathbf{B}}'\mathbf{X}'\mathbf{Y}$ and $\mathbf{H}$ is given by (7.128). Critical values and approximate tests are based on $\nu_H = w$ and $\nu_E = n - q - 1$ and on

$$s = \min(p, w), \qquad m = \tfrac{1}{2}(|w - p| - 1), \qquad N = n - q - p - 2.$$

The hypothesis H_0 : $\mathbf{CB} = \mathbf{O}$ can be extended to H_0 : $\mathbf{CBM} = \mathbf{O}$, where $\mathbf{C}$ is $w \times (q + 1)$ of rank $w \leq q + 1$, $\mathbf{B}$ is $(q + 1) \times p$, and $\mathbf{M}$ is $p \times k$ of rank $k \leq p$. The columns of $\mathbf{M}$ provide coefficients for linear combinations of the elements in each row of $\mathbf{B}$. Essentially, $\mathbf{C}$ transforms the x's and $\mathbf{M}$ transforms the y's. Some hypotheses in profile analysis and repeated measures analysis can be cast in the form $\mathbf{CBM} = \mathbf{O}$ (Williams 1970, Morrison 1972).

To obtain a test statistic for H_0 : $\mathbf{CBM} = \mathbf{O}$, we first multiply the model $\mathbf{Y} = \mathbf{XB} + \boldsymbol{\Xi}$ by $\mathbf{M}$, to obtain

$$\mathbf{YM} = \mathbf{XBM} + \boldsymbol{\Xi}\mathbf{M},$$

which we write as

$$\mathbf{Z} = \mathbf{X}\boldsymbol{\Gamma} + \boldsymbol{\Xi}_0.$$

By Theorem 2.2B, each row of $\boldsymbol{\Xi}_0 = \boldsymbol{\Xi}\mathbf{M}$ is distributed as $N_k(\mathbf{0}', \mathbf{M}'\boldsymbol{\Sigma}\mathbf{M})$, where k is the number of columns of $\mathbf{M}$. The hypothesis H_0: $\mathbf{CBM} = \mathbf{O}$ becomes H_0: $\mathbf{C}\boldsymbol{\Gamma} = \mathbf{O}$, where $\boldsymbol{\Gamma} = \mathbf{BM}$. By analogy with (7.128) we obtain

$$\begin{aligned}
\mathbf{H}_M &= (\mathbf{C}\hat{\boldsymbol{\Gamma}})' \left[\mathbf{C}(\mathbf{X}'\mathbf{X})^{-1}\mathbf{C}'\right]^{-1} (\mathbf{C}\hat{\boldsymbol{\Gamma}}) \\
&= (\mathbf{C}\hat{\mathbf{B}}\mathbf{M})' \left[\mathbf{C}(\mathbf{X}'\mathbf{X})^{-1}\mathbf{C}'\right]^{-1} (\mathbf{C}\hat{\mathbf{B}}\mathbf{M}) \\
&= \mathbf{M}'\mathbf{H}\mathbf{M},
\end{aligned}$$

where $\mathbf{H}$ is given by (7.128). Similarly,

$$\begin{aligned}\mathbf{E}_M &= \mathbf{Z}'\mathbf{Z} - \hat{\mathbf{\Gamma}}'\mathbf{X}'\mathbf{Z} \\ &= (\mathbf{YM})'\mathbf{YM} - (\hat{\mathbf{B}}\mathbf{M})'\mathbf{X}'\mathbf{YM} \\ &= \mathbf{M}'(\mathbf{Y}'\mathbf{Y} - \hat{\mathbf{B}}'\mathbf{X}'\mathbf{Y})\mathbf{M} \\ &= \mathbf{M}'\mathbf{E}\mathbf{M}.\end{aligned}$$

Thus Wilks' Λ for $H_0: \mathbf{CBM} = \mathbf{O}$ becomes

$$\Lambda = \frac{|\mathbf{M}'\mathbf{E}\mathbf{M}|}{|\mathbf{M}'(\mathbf{E}+\mathbf{H})\mathbf{M}|}, \tag{7.129}$$

which is distributed as $\Lambda_{k,m,n-q-1}$. The other three test statistics can be obtained from the eigenvalues of $(\mathbf{M}'\mathbf{E}\mathbf{M})^{-1}(\mathbf{M}'\mathbf{H}\mathbf{M})$. Alternatively, the four tests can be carried out by transforming to $\mathbf{Z} = \mathbf{YM}$ and constructing the tests directly using $\mathbf{Z}$ as input.

7.5.4 Tests and Confidence Intervals for a Single β_{jk} and a Bilinear Function a′Bb

A test or confidence interval for a single β_{jk} in $\mathbf{B}$ is easily obtained. By Theorem 7.4G(i), $\hat{\boldsymbol{\beta}}_{(k)}$ is $N_{q+1}[\boldsymbol{\beta}_{(k)}, \sigma_{kk}(\mathbf{X}'\mathbf{X})^{-1}]$, where $\boldsymbol{\beta}_{(k)}$ is the kth column of $\mathbf{B}$. By Theorem 2.2C, the estimator $\hat{\beta}_{jk}$ is therefore distributed as the univariate normal $N(\beta_{jk}, \sigma_{kk}g_{jj})$, where g_{jj} is the jth diagonal element of $(\mathbf{X}'\mathbf{X})^{-1}$. We estimate σ_{kk} by s_k^2, the kth diagonal element of $\mathbf{S}_e = (\mathbf{Y}'\mathbf{Y} - \hat{\mathbf{B}}'\mathbf{X}'\mathbf{Y})/(n-q-1)$. By Theorem 7.4G(iii), $\hat{\beta}_{jk}$ and s_k^2 are independent. Hence we can test $H_0: \beta_{jk} = 0$ by

$$t = \frac{\hat{\beta}_{jk}}{s_k\sqrt{g_{jj}}}. \tag{7.130}$$

We reject H_0 if $|t| \geq t_{\alpha/2,n-q-1}$. The corresponding confidence interval for β_{jk} is

$$\hat{\beta}_{jk} \pm t_{\alpha/2,n-q-1}s_k\sqrt{g_{jj}}. \tag{7.131}$$

A bilinear function of $\mathbf{B}$, say $\mathbf{a}'\mathbf{B}\mathbf{b}$, can be used to express various linear functions of the β_{jk}'s. For example, let

$$\mathbf{B} = \begin{pmatrix} \beta_{01} & \beta_{02} \\ \beta_{11} & \beta_{12} \\ \beta_{21} & \beta_{22} \\ \beta_{31} & \beta_{32} \end{pmatrix}.$$

Then $(0,0,0,1)\mathbf{B}\binom{1}{-1} = \beta_{31} - \beta_{32}$; $(0,0,1,-1)\mathbf{B}\binom{1}{0} = \beta_{21} - \beta_{31}$; and $(0,0,1,-1) \times \mathbf{B}\binom{1}{-1} = \beta_{21} - \beta_{31} - \beta_{22} + \beta_{32}$. To construct a confidence interval for a single bilinear function $\mathbf{a}'\mathbf{B}\mathbf{b}$, we first find the mean and variance of the estimator $\mathbf{a}'\hat{\mathbf{B}}\mathbf{b}$. From (1.89), we obtain for the mean

$$E(\mathbf{a}'\hat{\mathbf{B}}\mathbf{b}) = \mathbf{a}'E(\hat{\mathbf{B}})\mathbf{b} = \mathbf{a}'\mathbf{B}\mathbf{b}. \tag{7.132}$$

It can be shown (Problem 7.37) that

$$\text{var}(\mathbf{a}'\hat{\mathbf{B}}\mathbf{b}) = (\mathbf{b}'\mathbf{\Sigma}\mathbf{b})\mathbf{a}'(\mathbf{X}'\mathbf{X})^{-1}\mathbf{a}. \tag{7.133}$$

A confidence interval for $\mathbf{a}'\mathbf{B}\mathbf{b}$ is thus given by

$$\mathbf{a}'\hat{\mathbf{B}}\mathbf{b} \pm t_{\alpha/2,n-q-1}\sqrt{(\mathbf{b}'\mathbf{S}_e\mathbf{b})\mathbf{a}'(\mathbf{X}'\mathbf{X})^{-1}\mathbf{a}}, \tag{7.134}$$

where $\mathbf{S}_e = (\mathbf{Y}'\mathbf{Y} - \hat{\mathbf{B}}'\mathbf{X}'\mathbf{Y})/(n-q-1)$.

7.5.5 Simultaneous Tests and Confidence Intervals for the β_{jk}'s and Bilinear Functions $\mathbf{a}'\mathbf{B}\mathbf{b}$

If several tests or confidence intervals are to be made, some protection is needed against inflation of the experimentwise error rate, α. If a few, say r, tests or intervals on the β_{jk}'s are desired, we can use Bonferroni critical values (Section 3.4.4). To test $H_0\colon \beta_{jk} = 0$ for r values of β_{jk}, we use (7.130),

$$t = \frac{\hat{\beta}_{jk}}{s_k\sqrt{g_{jj}}},$$

and reject H_0 if $|t| \geq t_{\alpha/2r,n-q-1}$. The corresponding confidence intervals are

$$\hat{\beta}_{jk} \pm t_{\alpha/2r,n-q-1}s_k\sqrt{g_{jj}}. \tag{7.135}$$

Similarly, for r choices of $\mathbf{a}$ and $\mathbf{b}$, the Bonferroni intervals for $\mathbf{a}'\mathbf{B}\mathbf{b}$ are given by

$$\mathbf{a}'\hat{\mathbf{B}}\mathbf{b} \pm t_{\alpha/2r,n-q-1}\sqrt{(\mathbf{b}'\mathbf{S}_e\mathbf{b})\mathbf{a}'(\mathbf{X}'\mathbf{X})^{-1}\mathbf{a}}. \tag{7.136}$$

In the following theorem, we obtain simultaneous confidence intervals for $\mathbf{a}'\mathbf{B}\mathbf{b}$ for all possible $\mathbf{a}$ and all possible $\mathbf{b}$.

Theorem 7.5E. Simultaneous confidence intervals for all possible bilinear functions $\mathbf{a}'\mathbf{B}\mathbf{b}$ are given by

$$\mathbf{a}'\hat{\mathbf{B}}\mathbf{b} \pm \sqrt{\frac{\theta_\alpha}{1-\theta_\alpha}(\mathbf{b}'\mathbf{E}\mathbf{b})\mathbf{a}'(\mathbf{X}'\mathbf{X})^{-1}\mathbf{a}}, \tag{7.137}$$

where θ_α is the upper α-level critical value of Roy's statistic.

Proof. The proof is an extension of the proof of Theorem 7.2B. We first transform to the model $\mathbf{Y}\mathbf{b} = \mathbf{X}\mathbf{B}\mathbf{b} + \mathbf{\Xi}\mathbf{b}$, or $\mathbf{z} = \mathbf{X}\boldsymbol{\beta} + \boldsymbol{\varepsilon}$, where $\boldsymbol{\beta} = \mathbf{B}\mathbf{b}$, as in the proof of

Theorem 7.5B. Then $\mathbf{a}'\mathbf{B}\mathbf{b}$ becomes $\mathbf{a}'\boldsymbol{\beta}$, and by (7.34) and (7.35), $H_0: \mathbf{a}'\boldsymbol{\beta} = 0$ can be tested with

$$F(\mathbf{a}, \mathbf{b}) = \frac{(\mathbf{a}'\hat{\boldsymbol{\beta}})' \left[\mathbf{a}'(\mathbf{X}'\mathbf{X})^{-1}\mathbf{a}\right]^{-1} (\mathbf{a}'\hat{\boldsymbol{\beta}})}{s_z^2} = \frac{(\mathbf{a}'\hat{\boldsymbol{\beta}})^2}{s_z^2 \mathbf{a}'(\mathbf{X}'\mathbf{X})^{-1}\mathbf{a}}, \qquad (7.138)$$

where $\hat{\boldsymbol{\beta}} = \hat{\mathbf{B}}\mathbf{b}$ and

$$s_z^2 = \frac{\mathbf{z}'\mathbf{z} - \hat{\boldsymbol{\beta}}'\mathbf{X}'\mathbf{z}}{n - q - 1} = \frac{\mathbf{b}'\mathbf{E}\mathbf{b}}{n - q - 1}. \qquad (7.139)$$

By (7.42) and (7.138)

$$\max_{\mathbf{a}} F(\mathbf{a}, \mathbf{b}) = \frac{\hat{\boldsymbol{\beta}}'\mathbf{X}'\mathbf{X}\hat{\boldsymbol{\beta}}}{s_z^2} = \frac{\mathbf{b}'\hat{\mathbf{B}}'\mathbf{X}'\mathbf{X}\hat{\mathbf{B}}\mathbf{b}}{\mathbf{b}'\mathbf{E}\mathbf{b}/(n - q - 1)}. \qquad (7.140)$$

By Theorem 4.1C, $\max_{\mathbf{b}}[\mathbf{b}'(\hat{\mathbf{B}}'\mathbf{X}'\mathbf{X}\hat{\mathbf{B}})\mathbf{b}/\mathbf{b}'\mathbf{E}\mathbf{b}] = \lambda_1$, the largest eigenvalue of $\mathbf{E}^{-1}(\hat{\mathbf{B}}'\mathbf{X}'\mathbf{X}\hat{\mathbf{B}})$. Thus

$$P\left[F(\mathbf{a}, \mathbf{b}) \le (n - q - 1)\lambda_\alpha\right] = 1 - \alpha \qquad (7.141)$$

for all possible $\mathbf{a}$ and $\mathbf{b}$, where λ_α is the upper critical value of λ_1. Using (7.138) with the substitution $\hat{\boldsymbol{\beta}} = \hat{\mathbf{B}}\mathbf{b}$ and $s_z^2 = \mathbf{b}'\mathbf{E}\mathbf{b}/(n - q - 1)$, (7.141) becomes

$$P\left[\frac{(n - q - 1)(\mathbf{a}'\hat{\mathbf{B}}\mathbf{b})^2}{(\mathbf{b}'\mathbf{E}\mathbf{b})\mathbf{a}'(\mathbf{X}'\mathbf{X})^{-1}\mathbf{a}} \le (n - q - 1)\lambda_\alpha\right] = 1 - \alpha.$$

With $\lambda_\alpha = \theta_\alpha/(1 - \theta_\alpha)$, we obtain the simultaneous intervals for $\mathbf{a}'\mathbf{B}\mathbf{b}$ for all possible $\mathbf{a}$ and $\mathbf{b}$ given in (7.137). □

Critical values θ_α are found in Table B.5 with

$$s = \min(q + 1, p), \qquad m = \tfrac{1}{2}(|q + 1 - p| - 1), \qquad N = \tfrac{1}{2}(n - q - p - 2).$$

We can use (7.137) to find intervals for each β_{ij} in $\mathbf{B}$ by choosing $\mathbf{a}$ to have a 1 in the ith position, with 0s elsewhere, and similarly defining $\mathbf{b}$ to have a 1 in the jth position. Generally, for all $p(q + 1)$ β_{jk}'s, the Bonferroni intervals in (7.135) or in (7.136) with $r = p(q + 1)$ will be shorter than the simultaneous intervals. Other simultaneous intervals are given by Schmidhammer (1982).

7.5.6 Tests in the Presence of Missing Data

In many studies, there will be missing values in the response vectors; that is, the observed values for one or more of $y_{i1}, y_{i2}, \ldots, y_{ip}$ will be missing in some of the observation vectors $\mathbf{y}_i'$ (rows of $\mathbf{Y}$). In most software packages, the default procedure is to delete the entire row of $\mathbf{Y}$ (and the corresponding row of $\mathbf{X}$) if there is a missing value anywhere in the row. This procedure, sometimes called *listwise deletion*, is satisfactory if only a few values are missing and if these values are missing at random (Section 1.9).

Barton and Cramer (1989) compared tests based on (1) the EM algorithm for imputation of the missing values (Section 1.9); (2) listwise deletion; and (3) the full set of data (with no missing values). The model was $\mathbf{Y} = \mathbf{XB} + \mathbf{\Xi}$, as in (7.68); the hypothesis of interest was $H_0: \mathbf{B}_1 = \mathbf{O}$, as in Section 7.5.1; and the test statistic was Wilks' Λ, as in (7.104) or (7.105). The parameters $\mathbf{B}$ and $\mathbf{\Sigma}$ were estimated by the EM algorithm, and Wilks' Λ was based on these estimates. This was compared to Wilks' Λ based on the listwise deleted data and to Wilks' Λ based on the full data without missing values. The results showed that the α-levels of the tests based on EM estimates were slightly conservative, ranging from .040 to .056 (nominal $\alpha = .05$). In most cases, the power for the EM approach was closer to the power for the full data than it was to the power for the listwise deleted data.

7.6 MULTIVARIATE MODEL VALIDATION: FIXED x's

A brief review of a few of the available methods for checking the assumptions in the univariate-y case was given in Section 7.2.6. For the multivariate model $\mathbf{Y} = \mathbf{XB} + \mathbf{\Xi}$, with $\mathbf{X}$ fixed, the usual assumptions are (1) $E(\mathbf{Y}) = \mathbf{XB}$; (2) the rows $\mathbf{y}_i'$ of $\mathbf{Y}$ are independent; (3) $\text{cov}(\mathbf{y}_i) = \mathbf{\Sigma}$ for $i = 1, 2, \ldots, n$; and (4) $\mathbf{y}_i$ is $N_p(\mathbf{B}'\mathbf{x}_i, \mathbf{\Sigma})$. Certain univariate methods for checking the assumptions can be adapted for the multivariate-y case.

7.6.1 Lack-of-Fit Tests

For the case in which some of the rows of $\mathbf{X}$ are repeated, Khuri (1985) provided a multivariate lack-of-fit test (see Section 7.2.6) and suggested a method for identifying which of the p y's contribute to lack of fit. This multivariate approach is preferable to testing each y univariately for lack of fit. Levy and Neill (1990) discussed the test of Khuri (1985) and proposed two additional tests for lack of fit in a multivariate regression setting. In a simulation study, no one of these three tests was found to dominate the other two in terms of power. Levy and Neill (1990) generalized the three lack-of-fit tests to the case in which the $\mathbf{X}$ matrix has no duplicate rows but some of the rows are "near replicates" of each other. Their generalization is based on Christensen's (1989) analogous generalization of univariate lack-of-fit tests using near replicates.

7.6.2 Residuals

If a lack-of-fit test is not feasible, the residuals (Section 7.2.6) can be used to check for departures from the assumptions. Multivariate residuals are defined as

$$\hat{\mathbf{\Xi}} = \mathbf{Y} - \hat{\mathbf{Y}} = \mathbf{Y} - \mathbf{X}\hat{\mathbf{B}}, \tag{7.142}$$

or

$$
\begin{aligned}
(\hat{\boldsymbol{\varepsilon}}_{(1)}, \ldots, \hat{\boldsymbol{\varepsilon}}_{(p)}) &= (\mathbf{y}_{(1)}, \ldots, \mathbf{y}_{(p)}) - \mathbf{X}(\hat{\boldsymbol{\beta}}_{(1)}, \ldots, \hat{\boldsymbol{\beta}}_{(p)}) \\
&= (\mathbf{y}_{(1)} - \mathbf{X}\hat{\boldsymbol{\beta}}_{(1)}, \ldots, \mathbf{y}_{(p)} - \mathbf{X}\hat{\boldsymbol{\beta}}_{(p)}) \qquad \text{[by (A.2.26)]}.
\end{aligned}
$$

Each column of $\hat{\boldsymbol{\Xi}} = (\hat{\boldsymbol{\varepsilon}}_{(1)}, \hat{\boldsymbol{\varepsilon}}_{(2)}, \ldots, \hat{\boldsymbol{\varepsilon}}_{(p)})$ can be examined separately using the plots and tests in Section 7.2.6. For example, the n residuals $\hat{\varepsilon}_{i2} = y_{i2} - \hat{y}_{i2}, i = 1, 2, \ldots, n$, in $\hat{\boldsymbol{\varepsilon}}_{(2)} = \mathbf{y}_{(2)} - \mathbf{X}\hat{\boldsymbol{\beta}}_{(2)}$ can be plotted against $\hat{y}_{i2}$, $i = 1, 2, \ldots, n$. The rows of $\hat{\boldsymbol{\Xi}}$ are discussed in Section 7.6.3.

7.6.3 Influence and Outliers

An influential observation vector $\mathbf{y}_i$ is one that has an unusually large impact on predicted values or parameter estimates. An outlier is an observation vector $\mathbf{y}_i$ whose residual $\mathbf{y}_i - \hat{\mathbf{y}}_i$ is unusually distant from the other residuals, where $\hat{\mathbf{y}}_i'$ is the ith row of $\hat{\mathbf{Y}} = \mathbf{X}\hat{\mathbf{B}}$. In the following discussion of influential observations, we follow Hossain and Naik (1989). For additional discussion of influence measures in multivariate regression, see Caroni (1987) and Barrett and Ling (1992).

The projection matrix $\mathbf{P} = \mathbf{X}(\mathbf{X}'\mathbf{X})^{-1}\mathbf{X}'$ projects $\mathbf{Y}$ onto $\hat{\mathbf{Y}}$, since $\hat{\mathbf{Y}} = \mathbf{X}\hat{\mathbf{B}} = \mathbf{P}\mathbf{Y}$. The residual matrix $\hat{\boldsymbol{\Xi}}$ in (7.142) can be written in terms of $\mathbf{P}$ as

$$
\hat{\boldsymbol{\Xi}} = \mathbf{Y} - \mathbf{X}\hat{\mathbf{B}} = \mathbf{Y} - \mathbf{P}\mathbf{Y} = (\mathbf{I} - \mathbf{P})\mathbf{Y}. \tag{7.143}
$$

It can also be shown that $\hat{\boldsymbol{\Xi}} = (\mathbf{I} - \mathbf{P})\boldsymbol{\Xi}$ (Problem 7.39). If we denote the rows of $\hat{\mathbf{Y}}$ as $\hat{\mathbf{y}}_i'$, $i = 1, 2, \ldots, n$, then $\hat{\mathbf{Y}} = \mathbf{P}\mathbf{Y}$ can be written as

$$
\begin{pmatrix} \hat{\mathbf{y}}_1' \\ \hat{\mathbf{y}}_2' \\ \vdots \\ \hat{\mathbf{y}}_n' \end{pmatrix} = \mathbf{P} \begin{pmatrix} \mathbf{y}_1' \\ \mathbf{y}_2' \\ \vdots \\ \mathbf{y}_n' \end{pmatrix},
$$

from which we obtain

$$
\begin{aligned}
\hat{\mathbf{y}}_i' &= \sum_{j=1}^{n} p_{ij}\mathbf{y}_j' \\
&= p_{ii}\mathbf{y}_i' + \sum_{j \neq i} p_{ij}\mathbf{y}_j',
\end{aligned}
$$

where $p_{i1}, p_{i2}, \ldots, p_{in}$ constitute the ith row of $\mathbf{P}$. The coefficient p_{ii} is called the *leverage* of $\mathbf{y}_i'$ in determining $\hat{\mathbf{y}}_i'$. Use of the term *leverage* is borrowed from the Archimedean notion that anything can be moved out of its place with a long enough lever. A point $\mathbf{y}_i$ with large p_{ii} is called a high leverage point because of its potential for influencing regression results.

The residual vectors $\hat{\boldsymbol{\varepsilon}}_i'$ are rows of the residual matrix defined in (7.142):

$$\hat{\boldsymbol{\Xi}} = \begin{pmatrix} \hat{\boldsymbol{\varepsilon}}_1' \\ \hat{\boldsymbol{\varepsilon}}_2' \\ \vdots \\ \hat{\boldsymbol{\varepsilon}}_n' \end{pmatrix}.$$

Many diagnostic influence measures use the following functions of $\hat{\boldsymbol{\varepsilon}}_i$:

$$\tau_i^2 = \frac{1}{1-p_{ii}}\hat{\boldsymbol{\varepsilon}}_i'\mathbf{S}_e^{-1}\hat{\boldsymbol{\varepsilon}}_i, \qquad T_i^2 = \frac{1}{1-p_{ii}}\hat{\boldsymbol{\varepsilon}}_i'\mathbf{S}_{e,-i}^{-1}\hat{\boldsymbol{\varepsilon}}_i,$$

where $\mathbf{S}_e$ is defined in (7.83) as $\mathbf{S}_e = (\mathbf{Y}'\mathbf{Y} - \hat{\mathbf{B}}'\mathbf{X}'\mathbf{Y})/(n-q-1)$ and $\mathbf{S}_{e,-i}$ is $\mathbf{S}_e$ based on $n-1$ observations with the ith observation deleted.

The influence of the ith observation $(\mathbf{y}_i', \mathbf{x}_i')$ on $\hat{\mathbf{B}}$ is given by the Cook type statistic

$$C_i = \frac{1}{q+1}\frac{p_{ii}}{1-p_{ii}}\tau_i^2, \quad i = 1, 2, \ldots, n,$$

where $\mathbf{y}_i'$ is the ith row of $\mathbf{Y}$ and $\mathbf{x}_i'$ is the ith row of $\mathbf{X}$. Large values of C_i are considered significant. The influence of the ith observation on the ith predicted value $\hat{\mathbf{y}}_i'$ is given by the Welsch–Kuh type statistic

$$\text{WK}_i = \frac{p_{ii}}{1-p_{ii}}T_i^2, \quad i = 1, 2, \ldots, n.$$

Large values of WK_i are considered significant. A suggested upper critical value for significance is

$$\frac{q+1}{n} \cdot \frac{p(n-q-2)}{n-q-p-1}F_{\alpha,p,n-q-p-1}.$$

The influence of the ith observation on $\widehat{\text{cov}}(\hat{\boldsymbol{\beta}}_{(1)}), \widehat{\text{cov}}(\hat{\boldsymbol{\beta}}_{(2)}), \ldots, \widehat{\text{cov}}(\hat{\boldsymbol{\beta}}_{(p)})$ is given by

$$\text{CVR}_i = \left(\frac{1}{1-p_{ii}}\right)^p \left[\frac{|\mathbf{S}_{e,-i}|}{|\mathbf{S}_e|}\right]^{q+1}.$$

Either low or high values of CVR_i are considered significant.

The influence of the ith observation on the jth row of $\hat{\mathbf{B}}$ is given by

$$D_{ij} = \frac{T_i^2}{1-p_{ii}} \cdot \frac{w_{ij}^2}{\mathbf{w}_j'\mathbf{w}_j}, \quad i = 1, 2, \ldots, n, \quad j = 1, 2, \ldots, q+1,$$

where w_{ij} is the ith element of $\mathbf{w}_j = (\mathbf{I} - \mathbf{P}_{-j})\mathbf{x}_{(j)}$, in which $\mathbf{P}_{-j}$ is computed without $\mathbf{x}_{(j)}$, the jth column of $\mathbf{X}$. Large values of D_{ij} are considered significant.

Hossain and Naik (1989) gave examples of the application of these diagnostic measures to real data sets.

An outlier was defined at the beginning of this section as an observation whose residual $\hat{\boldsymbol{\varepsilon}}_i$ is unusually distant from the other residuals. Thus an observation $(\mathbf{y}_i', \mathbf{x}_i')$, $i = 1, 2, \ldots, n$, is an outlier if $\hat{\boldsymbol{\varepsilon}}_i'\mathbf{S}_e^{-1}\hat{\boldsymbol{\varepsilon}}_i$ is relatively large. Naik (1989) provided a significance test for outliers in multivariate regression. Schall and Dunne (1987) provided tests of significance for detection of three types of outliers in multivariate regression: additive shift, transformational, and distributional.

7.6.4 Measurement Errors

In some applications, the y's or the x's may be measured with random error; that is, instead of observing $(\mathbf{y}_i', \mathbf{x}_i')$, we observe $(\mathbf{y}_i'^*, \mathbf{x}_i'^*) = (\mathbf{y}_i' + \mathbf{e}_i', \mathbf{x}_i' + \mathbf{f}_i')$, where $\mathbf{e}_i'$ and $\mathbf{f}_i'$ are vectors of measurement errors. These errors can occur in either the fixed-x case or the random-x case (Section 7.7). Gleser (1992) and Fuller (1987, Chapter 4) discuss the estimation of $\mathbf{B}$ in the presence of measurement errors.

7.7 MULTIVARIATE REGRESSION: RANDOM x's

In Sections 7.4–7.6, we assumed that the x's were fixed (constant) in repeated sampling. In many multivariate regression applications, the x's are random variables. We now discuss the random-x case and make comparisons to the fixed-x case. Our development parallels that of Section 7.3.

7.7.1 Multivariate Normal Model for Random x's

We assume $\mathbf{z} = (y_1, \ldots, y_p, x_1, \ldots, x_q)' = (\mathbf{y}', \mathbf{x}')'$ is distributed as $N_{p+q}(\boldsymbol{\mu}_z, \boldsymbol{\Sigma}_z)$, with $\boldsymbol{\mu}_z$ and $\boldsymbol{\Sigma}_z$ partitioned as

$$\boldsymbol{\mu}_z = \begin{pmatrix} \boldsymbol{\mu}_y \\ \boldsymbol{\mu}_x \end{pmatrix}, \qquad \boldsymbol{\Sigma}_z = \begin{pmatrix} \boldsymbol{\Sigma}_{yy} & \boldsymbol{\Sigma}_{yx} \\ \boldsymbol{\Sigma}_{xy} & \boldsymbol{\Sigma}_{xx} \end{pmatrix}. \tag{7.144}$$

From (2.36), the conditional expectation of $\mathbf{y}$ given $\mathbf{x}$ is

$$\begin{aligned} E(\mathbf{y}\,|\,\mathbf{x}) &= \boldsymbol{\mu}_y + \boldsymbol{\Sigma}_{yx}\boldsymbol{\Sigma}_{xx}^{-1}(\mathbf{x} - \boldsymbol{\mu}_x) \\ &= \boldsymbol{\mu}_y - \boldsymbol{\Sigma}_{yx}\boldsymbol{\Sigma}_{xx}^{-1}\boldsymbol{\mu}_x + \boldsymbol{\Sigma}_{yx}\boldsymbol{\Sigma}_{xx}^{-1}\mathbf{x} \\ &= \boldsymbol{\beta}_0 + \mathbf{B}_1'\mathbf{x}, \end{aligned} \tag{7.145}$$

where

$$\boldsymbol{\beta}_0 = \boldsymbol{\mu}_y - \boldsymbol{\Sigma}_{yx}\boldsymbol{\Sigma}_{xx}^{-1}\boldsymbol{\mu}_x, \tag{7.146}$$

$$\mathbf{B}_1 = \boldsymbol{\Sigma}_{xx}^{-1}\boldsymbol{\Sigma}_{xy}. \tag{7.147}$$

From (2.37), the conditional covariance matrix of $\mathbf{y}$ given $\mathbf{x}$ is

$$\mathrm{cov}(\mathbf{y}\,|\,\mathbf{x}) = \boldsymbol{\Sigma}_{yy} - \boldsymbol{\Sigma}_{yx}\boldsymbol{\Sigma}_{xx}^{-1}\boldsymbol{\Sigma}_{xy} = \boldsymbol{\Sigma}. \tag{7.148}$$

We denote $\boldsymbol{\Sigma}_{yy} - \boldsymbol{\Sigma}_{yx}\boldsymbol{\Sigma}_{xx}^{-1}\boldsymbol{\Sigma}_{xy}$ by $\boldsymbol{\Sigma}$, since $\boldsymbol{\Sigma}_{yy} - \boldsymbol{\Sigma}_{yx}\boldsymbol{\Sigma}_{xx}^{-1}\boldsymbol{\Sigma}_{xy}$ does not depend on the random vector $\mathbf{x}$. Thus, as a consequence of the multivariate normality assumption, we have a linear model with constant covariance matrix. As an alternative derivation that is not based on the multivariate normal assumption, it can be shown (see Problem 7.40) that $\boldsymbol{\beta}_0$ and $\mathbf{B}_1$ in (7.146) and (7.147) "minimize" the "mean squared error" matrix $\mathbf{M} = E[(\mathbf{y} - \boldsymbol{\beta}_0 - \mathbf{B}_1'\mathbf{x})(\mathbf{y} - \boldsymbol{\beta}_0 - \mathbf{B}_1'\mathbf{x})']$.

7.7.2 Estimation of $\boldsymbol{\beta}_0$, $\mathbf{B}_1$, and $\boldsymbol{\Sigma}$

For a random sample $(\mathbf{y}_1', \mathbf{x}_1')', (\mathbf{y}_2', \mathbf{x}_2')', \ldots, (\mathbf{y}_n', \mathbf{x}_n')'$, estimators of $\boldsymbol{\beta}_0$, $\mathbf{B}_1$, and $\boldsymbol{\Sigma}$ are given in the following theorem.

Theorem 7.7A. If $\mathbf{z} = (\mathbf{y}', \mathbf{x}')'$ is distributed as $N_{p+q}(\boldsymbol{\mu}_z, \boldsymbol{\Sigma}_z)$, where $\boldsymbol{\mu}_z$ and $\boldsymbol{\Sigma}_z$ are defined in (7.144), then the maximum likelihood estimators for $\boldsymbol{\beta}_0$, $\mathbf{B}_1$, and $\boldsymbol{\Sigma} = \boldsymbol{\Sigma}_{yy} - \boldsymbol{\Sigma}_{yx}\boldsymbol{\Sigma}_{xx}^{-1}\boldsymbol{\Sigma}_{xy}$ are given by

$$\hat{\boldsymbol{\beta}}_0 = \bar{\mathbf{y}} - \mathbf{S}_{yx}\mathbf{S}_{xx}^{-1}\bar{\mathbf{x}}, \tag{7.149}$$

$$\hat{\mathbf{B}}_1 = \mathbf{S}_{xx}^{-1}\mathbf{S}_{xy}, \tag{7.150}$$

$$\frac{n}{n-1}\hat{\boldsymbol{\Sigma}} = \mathbf{S}_{yy} - \mathbf{S}_{yx}\mathbf{S}_{xx}^{-1}\mathbf{S}_{xy}, \tag{7.151}$$

where $n\hat{\boldsymbol{\Sigma}}/(n-1)$ has been corrected for bias.

Proof. By Theorem 2.3A, the maximum likelihood estimators of $\boldsymbol{\mu}_z$ and $\boldsymbol{\Sigma}_z$ in partitioned form are

$$\hat{\boldsymbol{\mu}}_z = \begin{pmatrix} \hat{\boldsymbol{\mu}}_y \\ \hat{\boldsymbol{\mu}}_x \end{pmatrix} = \begin{pmatrix} \bar{\mathbf{y}} \\ \bar{\mathbf{x}} \end{pmatrix},$$

$$\hat{\boldsymbol{\Sigma}}_z = \frac{n-1}{n}\mathbf{S}_z = \frac{n-1}{n}\begin{pmatrix} \mathbf{S}_{yy} & \mathbf{S}_{yx} \\ \mathbf{S}_{xy} & \mathbf{S}_{xx} \end{pmatrix}.$$

By the invariance property (Theorem 2.3B), we substitute these into (7.146)–(7.148) to obtain the maximum likelihood estimators for $\boldsymbol{\beta}_0$, $\mathbf{B}_1$, and $\boldsymbol{\Sigma}$ in (7.149)–(7.151). □

Upon comparing (7.150) with (7.97), we see that $\hat{\mathbf{B}}_1$ in the random-x case is the same as the least squares estimator $\hat{\mathbf{B}}_1$ in the fixed-x case. By (7.100) and (7.107) the estimators in (7.149) and (7.151) are also the same functions of the observations as in the fixed-x case. Of course, even though the estimators in the two cases are the same algebraic functions of the observations, the properties are somewhat different. For example, when $(\mathbf{y}', \mathbf{x}')'$ is multivariate normal, each $\hat{\boldsymbol{\beta}}_{(j)}$ is not multivariate normal as it is in the fixed-x case with normal y's [see Theorem 7.4G(i)].

The estimators $\hat{\boldsymbol{\beta}}_0$ and $\hat{\mathbf{B}}_1$ in (7.149) and (7.150) can be obtained (see Problem 7.41) without the normality assumption by "minimization" of the sample mean squared error matrix $\sum_{i=1}^{n}(\mathbf{y}_i - \hat{\boldsymbol{\beta}}_0 - \hat{\mathbf{B}}_1'\mathbf{x}_i)(\mathbf{y}_i - \hat{\boldsymbol{\beta}}_0 - \hat{\mathbf{B}}_1'\mathbf{x}_i)'/n$.

If the partitioned sample covariance matrix $\mathbf{S}_z$ is expressed in terms of an analogous partitioned correlation matrix $\mathbf{R}$ by (1.52), then we can express $\hat{\mathbf{B}}_1$ in terms of correlation matrices $\mathbf{R}_{xx}$, $\mathbf{R}_{xy}$, and $\mathbf{R}_{yx}$. By (1.52) and (7.60), we have

$$\mathbf{S}_{xx} = \mathbf{D}_x \mathbf{R}_{xx} \mathbf{D}_x, \qquad \mathbf{S}_{xy} = \mathbf{D}_x \mathbf{R}_{xy} \mathbf{D}_y, \tag{7.152}$$

where $\mathbf{D}_x$ and $\mathbf{D}_y$ are diagonal matrices of standard deviations of the x's and y's, respectively; for example, $\mathbf{D}_y = \text{diag}(s_{y_1}, s_{y_2}, \ldots, s_{y_p})$. Substituting (7.152) into (7.150) leads to

$$\hat{\mathbf{B}}_1 = \mathbf{D}_x^{-1} \mathbf{R}_{xx}^{-1} \mathbf{R}_{xy} \mathbf{D}_y. \tag{7.153}$$

We obtain standardized coefficients or beta weights by a direct extension of (7.63):

$$\hat{\mathbf{B}}_1^* = \mathbf{R}_{xx}^{-1} \mathbf{R}_{xy}. \tag{7.154}$$

7.7.3 Tests and Confidence Intervals in the Multivariate Random-x Case

In the multivariate normal model, likelihood ratio tests can be obtained for hypotheses that correspond to those in Section 7.5. These tests turn out to be the same as those based on Wilks' Λ for the fixed-x case in Section 7.5.

A conditional approach suggested by Graybill (1976, pp. 381–385) for the random-x case with one y can be adapted to the multivariate-y case. Since the conditional distribution of $\mathbf{y}_i$ for a given value of $\mathbf{x}_i$ is multivariate normal, a test statistic such as

$$\Lambda = \frac{|\mathbf{Y}'\mathbf{Y} - \hat{\mathbf{B}}'\mathbf{X}'\mathbf{Y}|}{|\mathbf{Y}'\mathbf{Y} - n\bar{\mathbf{y}}\bar{\mathbf{y}}'|}$$

in (7.104) for $H_0 : \mathbf{B}_1 = \mathbf{O}$ is distributed as Wilks' Λ for a given value of the matrix $\mathbf{X}$. The (central) distribution of Λ depends only on p, q, and $n - q - 1$ and does not depend on $\mathbf{X}$. Hence, under H_0, Λ has Wilks' distribution for all values of $\mathbf{X}$. However, when H_0 is false, the noncentral distribution of Λ depends on $\mathbf{X}$, and the test statistic has a different distribution than it has in the fixed-x case. Similar statements can be made about other test statistics.

The confidence intervals of Sections 7.5.4 and 7.5.5 also remain valid for the random-x case, because the conditional distribution of $\mathbf{y}$ for a given $\mathbf{x}$ is multivariate normal. The confidence coefficient (say 95%) for the fixed-x case holds as well for random x's. However, the expected width of the intervals differs in the two cases.

7.8 ADDITIONAL TOPICS

7.8.1 Correlated Response Methods

If each of p responses $y_1, y_2, \ldots, y_p$ depends on $x_1, x_2, \ldots, x_q$, but with different choices of the x's, the model is referred to as a *correlated response model*. Using the notation of Section 7.4.1, we have the models

$$\mathbf{y}_{(j)} = \mathbf{X}_{(j)}\boldsymbol{\beta}_{(j)} + \boldsymbol{\varepsilon}_{(j)}, \quad j = 1, 2, \ldots, p, \tag{7.155}$$

which differ from the models $\mathbf{y}_{(j)} = \mathbf{X}\boldsymbol{\beta}_{(j)} + \boldsymbol{\varepsilon}_{(j)}$ in Section 7.4.1 because in (7.155) each $\mathbf{y}_{(j)}$ has a different $\mathbf{X}_{(j)}$ matrix.

In matrix form, the model in (7.155) becomes

$$(\mathbf{y}_{(1)}, \ldots, \mathbf{y}_{(p)}) = (\mathbf{X}_{(1)}, \ldots, \mathbf{X}_{(p)}) \begin{pmatrix} \boldsymbol{\beta}_{(1)} & \mathbf{0} & \cdots & \mathbf{0} \\ \mathbf{0} & \boldsymbol{\beta}_{(2)} & \cdots & \mathbf{0} \\ \vdots & \vdots & & \vdots \\ \mathbf{0} & \mathbf{0} & \cdots & \boldsymbol{\beta}_{(p)} \end{pmatrix} + (\boldsymbol{\varepsilon}_{(1)}, \ldots, \boldsymbol{\varepsilon}_{(p)})$$

or

$$\mathbf{Y} = \mathbf{XB} + \boldsymbol{\Xi}.$$

This model was introduced by Yates (1939) for bivariate response variables ($p = 2$) and has been discussed by Smith and Choi (1982). The model was referred to as the *multiple design multivariate linear model* by Srivastava (1985) and McDonald (1975) and was called *seemingly unrelated regressions* by Zellner (1962).

By analogy with partitionings in Sections 7.2.4a and 7.2.4b, we partition $\boldsymbol{\beta}_{(j)}$ in two ways:

1. $\boldsymbol{\beta}_{(j)} = \binom{\beta_{0(j)}}{\boldsymbol{\beta}_{1(j)}}, \quad j = 1, 2, \ldots, p;$
2. $\boldsymbol{\beta}_{(j)} = \binom{\boldsymbol{\beta}_{r(j)}}{\boldsymbol{\beta}_{d(j)}}, \quad j = 1, 2, \ldots, p.$

Khuri (1986) provided exact tests for the following hypotheses:

1. $H_0\colon \boldsymbol{\beta}_{(1)} = \boldsymbol{\beta}_{(2)} = \cdots = \boldsymbol{\beta}_{(p)},$
2. $H_0\colon \boldsymbol{\beta}_{1(1)} = \boldsymbol{\beta}_{1(2)} = \cdots = \boldsymbol{\beta}_{1(p)},$
3. $H_0\colon \boldsymbol{\beta}_{d(1)} = \boldsymbol{\beta}_{d(2)} = \cdots = \boldsymbol{\beta}_{d(p)}.$

Percy (1992) considered the predictive aspects of the model.

A related technique is that of *hierarchical linear models* (Kreft et al. 1994, Raudenbush 1993), in which there are models similar to (7.155). These models are then linked together by another model in which the estimated regression coefficients are regressed on second-level predictors.

7.8.2 Categorical Data

Various procedures have been proposed for use when the y's are discrete or categorical, such as logistic regression [Section 6.8; see also McCullagh (1989) and McCullagh and Nelder (1989)] and log-linear models (Bishop et al. 1975). Glonek and McCullagh (1995) extended the logistic model to multivariate response variables. Liang et al. (1992) proposed a model based on marginal expectations of each y. They illustrated their procedure with two data sets and compared it to log-linear models.

7.8.3 Subset Selection

The test on a subset of the x's in Section 7.5.2 can be adapted to a data-driven search for a "best" subset. For a review of this and other procedures, see Rencher (1995, Section 10.7). Bedrick and Tsai (1994) proposed a multivariate selection technique based on Akaike's (1973) information criterion and compared it with other methods.

7.8.4 Other Topics

Bai, Rao, and Wu (1992) developed robust M-estimators (see Section 1.10) for multivariate regression models.

PROBLEMS

7.1 Show that $\sum_{i=1}^{n}(y_i - \mathbf{x}_i'\hat{\boldsymbol{\beta}})^2 = (\mathbf{y} - \mathbf{X}\hat{\boldsymbol{\beta}})'(\mathbf{y} - \mathbf{X}\hat{\boldsymbol{\beta}})$, thus verifying (7.5).

7.2 Show that $(\mathbf{y} - \mathbf{X}\hat{\boldsymbol{\beta}})'(\mathbf{y} - \mathbf{X}\hat{\boldsymbol{\beta}}) = \mathbf{y}'\mathbf{y} - \hat{\boldsymbol{\beta}}'\mathbf{X}'\mathbf{y}$ as in (7.11).

7.3 Show that $E(\text{SSE}) = \sigma^2(n - q - 1)$ as in (7.12).

7.4 Show that $\mathbf{j}'\mathbf{X}_c = \mathbf{0}'$ as in (7.18).

7.5 Show that $\mathbf{X}_c'\mathbf{y}/(n-1) = \mathbf{s}_{yx}$ as in (7.25), even though $\mathbf{y}$ is not corrected for its mean.

7.6 Show that the test of overall regression, $H_0: \boldsymbol{\beta}_1 = \mathbf{0}$, in Section 7.2.4a can be expressed in terms of the full-and-reduced-model approach in Section 7.2.4b.

7.7 Show that $\text{cov}(y, z) = \text{var}(z) = \boldsymbol{\sigma}_{yx}'\boldsymbol{\Sigma}_{xx}^{-1}\boldsymbol{\sigma}_{yx}$ as in (7.64), where $z = \mu_y + \boldsymbol{\sigma}_{yx}'\boldsymbol{\Sigma}_{xx}^{-1}(\mathbf{x} - \boldsymbol{\mu}_x)$.

7.8 Show that $\text{tr}(\mathbf{Y} - \mathbf{X}\mathbf{B}_0)'(\mathbf{Y} - \mathbf{X}\mathbf{B}_0) = \sum_{j=1}^{p}(\mathbf{y}_{(j)} - \mathbf{X}\mathbf{b}_{0(j)})'(\mathbf{y}_{(j)} - \mathbf{X}\mathbf{b}_{0(j)})$ as in (7.72).

7.9 Show that (7.74) follows from (7.73).

7.10 If $\bar{\mathbf{y}}$ is the mean vector of a random sample $\mathbf{y}_1, \mathbf{y}_2, \ldots, \mathbf{y}_n$ from a population with mean vector $\boldsymbol{\mu}$ and covariance matrix $\boldsymbol{\Sigma}$, show that $\bar{\mathbf{y}}$ is the least squares estimator of $\boldsymbol{\mu}$.

7.11 (a) Show that $\text{cov}(\hat{\beta}_{ij}, \hat{\beta}_{kl}) = \sigma_{jl}g_{ik}$ as in (7.80), where $\hat{\beta}_{ij}$ and $\hat{\beta}_{kl}$ are from $\hat{\mathbf{B}}$, σ_{jl} is from $\boldsymbol{\Sigma}$, and g_{ik} is from $\mathbf{G} = (\mathbf{X}'\mathbf{X})^{-1}$.

(b) Show that $\text{cov}(\hat{\boldsymbol{\beta}}_i, \hat{\boldsymbol{\beta}}_k) = g_{ik}\boldsymbol{\Sigma}$ as in (7.81), where $\hat{\boldsymbol{\beta}}_i'$ and $\hat{\boldsymbol{\beta}}_k'$ are rows of $\hat{\mathbf{B}}$.

7.12 In the proof of Theorem 7.4D, show that $\mathbf{A}_2 = (\mathbf{X}'\mathbf{X})^{-1}\mathbf{X}'$ minimizes the diagonal elements of $\mathbf{A}_2\mathbf{A}_2'$.

7.13 Show that $(\mathbf{Y} - \mathbf{X}\hat{\mathbf{B}})'(\mathbf{Y} - \mathbf{X}\hat{\mathbf{B}}) = \mathbf{Y}'\mathbf{Y} - \hat{\mathbf{B}}'\mathbf{X}'\mathbf{Y}$ as in (7.83).

7.14 Show that $E[(\mathbf{y}_{(j)} - \mathbf{X}\hat{\boldsymbol{\beta}}_{(j)})'(\mathbf{y}_{(k)} - \mathbf{X}\hat{\boldsymbol{\beta}}_{(k)})] = (n - q - 1)\sigma_{jk}$ as in (7.85).

7.15 Carry out the following alternative proof of Theorem 7.4E.

(a) Using (7.83), express $\mathbf{E}$ in the form $\mathbf{E} = \mathbf{Y}'[\mathbf{I} - \mathbf{X}(\mathbf{X}'\mathbf{X})^{-1}\mathbf{X}']\mathbf{Y}$.

(b) Use (1.128) in Problem 1.23 to show that $E(\mathbf{S}_e) = \boldsymbol{\Sigma}$, where $\mathbf{S}_e = \mathbf{E}/(n - q - 1)$.

7.16 Show that

$$(\mathbf{Y} - \mathbf{XB})'(\mathbf{Y} - \mathbf{XB}) = (\mathbf{Y} - \mathbf{X}\hat{\mathbf{B}})'(\mathbf{Y} - \mathbf{X}\hat{\mathbf{B}}) + (\hat{\mathbf{B}} - \mathbf{B})'\mathbf{X}'\mathbf{X}(\hat{\mathbf{B}} - \mathbf{B})$$

as in (7.89).

7.17 Show that (7.89) leads to (7.90) in the proof of Theorem 7.4F.

7.18 Differentiate $\ln L(\hat{\mathbf{B}}, \boldsymbol{\Sigma}) = -np \ln\sqrt{2\pi} - (n/2)\ln|\boldsymbol{\Sigma}| - \frac{1}{2}\operatorname{tr}(\boldsymbol{\Sigma}^{-1}\mathbf{E})$ with respect to $\boldsymbol{\Sigma}^{-1}$ to obtain (7.88) in Theorem 7.4F.

7.19 Prove Theorem 7.4G.

7.20 Show that $[(\mathbf{j}, \mathbf{X}_c)'(\mathbf{j}, \mathbf{X}_c)]^{-1}(\mathbf{j}, \mathbf{X}_c)'\mathbf{Y}$ gives the least squares estimators (7.94) and (7.95).

7.21 Show that $\hat{\boldsymbol{\beta}}_0$ in (7.100) can be obtained from (7.99).

7.22 To demonstrate that r_t^2 and r_d^2 in (7.102) are poor measures of association between the y's and the x's, show that $\mathbf{R}^2$ in (7.101) can be written in the equivalent form $\mathbf{R}^2 = \mathbf{S}_{yy}^{-1}\mathbf{S}_{yx}\mathbf{S}_{xx}^{-1}\mathbf{S}_{xy}$. The eigenvalues $r_1^2, \ldots, r_s^2$ of $\mathbf{R}^2$ are called canonical correlations in Chapter 8. Since $0 \leq r_j^2 \leq 1$ for all r_j^2, both $r_t^2 = \operatorname{tr}(\mathbf{R}^2)/p = \sum_{j=1}^p r_j^2/p$ and $r_d^2 = |\mathbf{R}^2| = \prod_{j=1}^p r_j^2$ are small, especially r_d^2.

7.23 Show that the matrix $\mathbf{H} = \hat{\mathbf{B}}'\mathbf{X}'\mathbf{Y} - n\bar{\mathbf{y}}\bar{\mathbf{y}}'$ in (7.103) is adjusted for the intercepts $\beta_{01}, \beta_{02}, \ldots, \beta_{0p}$.

7.24 **(a)** Show that (under H_0) $\mathbf{H} = \hat{\mathbf{B}}'\mathbf{X}'\mathbf{Y} - n\bar{\mathbf{y}}\bar{\mathbf{y}}'$ in (7.103) is distributed as $W_p(q, \boldsymbol{\Sigma})$.

(b) Show that $\mathbf{H} = \hat{\mathbf{B}}'\mathbf{X}'\mathbf{Y} - n\bar{\mathbf{y}}\bar{\mathbf{y}}'$ is independent of $\mathbf{E} = \mathbf{Y}'\mathbf{Y} - \hat{\mathbf{B}}'\mathbf{X}'\mathbf{Y}$.

7.25 Show that under $H_0 : \mathbf{B}_1 = \mathbf{O}$ the maximum likelihood estimator of $\boldsymbol{\Sigma}$ is $\hat{\boldsymbol{\Sigma}}_0 = (\mathbf{Y}'\mathbf{Y} - n\bar{\mathbf{y}}\bar{\mathbf{y}}')/n$ as in (7.106).

7.26 Show that

$$\frac{\hat{\mathbf{B}}'\mathbf{X}'\mathbf{Y} - n\bar{\mathbf{y}}\bar{\mathbf{y}}'}{n - 1} = \mathbf{S}_{yx}\mathbf{S}_{xx}^{-1}\mathbf{S}_{xy},$$

and thus verify (7.107).

7.27 Show that $F(\mathbf{a})$ in (7.111) can be obtained from (7.110).

7.28 Show that $E(\mathbf{H}) = q\boldsymbol{\Sigma} + \mathbf{B}_1'\mathbf{X}_c'\mathbf{X}_c\mathbf{B}_1$ as in (7.116).

7.29 **(a)** Show that $\left|[E(\mathbf{E})]^{-1}E(\mathbf{H}) - \phi\mathbf{I}\right| = 0$ leads to $|\boldsymbol{\Sigma}^{-1}\mathbf{B}_1'\mathbf{X}_c'\mathbf{X}_c\mathbf{B}_1 - [(n - q - 1)\phi - q]\mathbf{I}| = 0$ as in (7.117).

(b) Show that $E(\mathbf{E}^{-1}\mathbf{H}) = [q\mathbf{I} + \boldsymbol{\Sigma}^{-1}\mathbf{B}_1'\mathbf{X}_c'\mathbf{X}_c\mathbf{B}_1]/(n - q - p - 2)$.

7.30 **(a)** Show that (under H_0) $\mathbf{H} = \hat{\mathbf{B}}'\mathbf{X}'\mathbf{Y} - \hat{\mathbf{B}}_r'\mathbf{X}_r'\mathbf{Y}$ in (7.121) is distributed as $W_p(h, \boldsymbol{\Sigma})$.

(b) Show that $\mathbf{H} = \hat{\mathbf{B}}'\mathbf{X}'\mathbf{Y} - \hat{\mathbf{B}}_r'\mathbf{X}_r'\mathbf{Y}$ is independent of $\mathbf{E} = \mathbf{Y}'\mathbf{Y} - \hat{\mathbf{B}}'\mathbf{X}'\mathbf{Y}$.

7.31 Show that Wilks' Λ in (7.122) can be expressed in the form given in (7.123).

7.32 Show that under $H_0 : \mathbf{B}_d = \mathbf{O}$, the maximum likelihood estimator of $\boldsymbol{\Sigma}$ is $\hat{\boldsymbol{\Sigma}}_0 = (\mathbf{Y}'\mathbf{Y} - \hat{\mathbf{B}}_r\mathbf{X}_r'\mathbf{Y})/n$, as in the proof of Theorem 7.5C.

7.33 Prove Theorem 7.5D.

7.34 Show that for the general linear hypothesis $H_0 : \mathbf{CB} = \mathbf{O}$ in Section 7.5.3

$$\mathbf{H} = \hat{\mathbf{B}}'\mathbf{X}'\mathbf{Y} - \hat{\mathbf{B}}_c'\mathbf{X}'\mathbf{Y} = (\mathbf{C}\hat{\mathbf{B}})'\left[\mathbf{C}(\mathbf{X}'\mathbf{X})^{-1}\mathbf{C}'\right]^{-1}\mathbf{C}\hat{\mathbf{B}}$$

as in (7.128), where $\hat{\mathbf{B}}_c$ is given in (7.127).

7.35 Show that $\hat{\mathbf{B}}_c'\mathbf{X}'\mathbf{X}\hat{\mathbf{B}}_c = \hat{\mathbf{B}}_c'\mathbf{X}'\mathbf{Y}$, thus demonstrating directly that the sum of squares and products matrix due to the reduced model is $\hat{\mathbf{B}}_c'\mathbf{X}'\mathbf{Y}$ and that $\mathbf{H} = \hat{\mathbf{B}}'\mathbf{X}'\mathbf{Y} - \hat{\mathbf{B}}_c'\mathbf{X}'\mathbf{Y}$.

7.36 Show that $\mathbf{H} = (\mathbf{C}\hat{\mathbf{B}})'[\mathbf{C}(\mathbf{X}'\mathbf{X})^{-1}\mathbf{C}']^{-1}\mathbf{C}\hat{\mathbf{B}}$ in (7.128) is distributed as a Wishart and that $\mathbf{H}$ and $\mathbf{E} = \mathbf{Y}'\mathbf{Y} - \hat{\mathbf{B}}'\mathbf{X}'\mathbf{Y}$ are independent.

7.37 Show that $\text{var}(\mathbf{a}'\hat{\mathbf{B}}\mathbf{b}) = (\mathbf{b}'\boldsymbol{\Sigma}\mathbf{b})\mathbf{a}'(\mathbf{X}'\mathbf{X})^{-1}\mathbf{a}$, thus verifying (7.133).

7.38 Show that $\mathbf{z}'\mathbf{z} - \hat{\boldsymbol{\beta}}'\mathbf{X}'\mathbf{z} = \mathbf{b}'\mathbf{E}\mathbf{b}$ as in (7.139), where $\mathbf{z} = \mathbf{X}\boldsymbol{\beta} + \boldsymbol{\varepsilon}$ represents the transformed model $\mathbf{Yb} = \mathbf{XBb} + \boldsymbol{\Xi}\mathbf{b}$.

7.39 Show that $(\mathbf{I} - \mathbf{P})\mathbf{Y} = (\mathbf{I} - \mathbf{P})\boldsymbol{\Xi}$ as in Section 7.6.3.

7.40 Show that $\boldsymbol{\beta}_0$ and $\mathbf{B}_1$ in (7.146) and (7.147) "minimize" $\mathbf{M} = E[(\mathbf{y} - \boldsymbol{\beta}_0 - \mathbf{B}_1'\mathbf{x})(\mathbf{y} - \boldsymbol{\beta}_0 - \mathbf{B}_1'\mathbf{x})']$. The matrix $\mathbf{M}$ is minimized in the sense that any other values of $\boldsymbol{\beta}_0$ and $\mathbf{B}_1$ would add a positive semidefinite matrix to $\mathbf{M}$.

7.41 Show that $\hat{\boldsymbol{\beta}}_0$ and $\hat{\mathbf{B}}_1$ in (7.149) and (7.150) minimize the sample mean squared error matrix $\sum_{i=1}^{n}(\mathbf{y}_i - \hat{\boldsymbol{\beta}}_0 - \hat{\mathbf{B}}_1'\mathbf{x}_i)(\mathbf{y}_i - \hat{\boldsymbol{\beta}}_0 - \hat{\mathbf{B}}_1'\mathbf{x}_i)'/n$.

7.42 Show that $\hat{\mathbf{B}}_1$ can be expressed in terms of correlation matrices,

$$\hat{\mathbf{B}}_1 = \mathbf{D}_x^{-1}\mathbf{R}_{xx}^{-1}\mathbf{R}_{xy}\mathbf{D}_y,$$

as in (7.153).

7.43 Consider the biochemical data in Table 4.11 and ignore the grouping variable.

(a) Find $\hat{\mathbf{B}}$ for regression of $(y_1, y_2, \ldots, y_{11})$ on (x_1, x_2) and test the significance of overall regression using all four test statistics in Section 7.5.1.

(b) Test the significance of each x adjusted for the other (Section 7.5.2).

(c) Test H_0: $\boldsymbol{\beta}_1' = \boldsymbol{\beta}_2'$ using an appropriate $\mathbf{C}$ in H_0: $\mathbf{CB} = \mathbf{O}$, where $\boldsymbol{\beta}_1'$ and $\boldsymbol{\beta}_2'$ are the second and third rows of $\mathbf{B}$ (Section 7.5.3).

(d) Test H_0: $2\boldsymbol{\beta}_{(1)} + 2\boldsymbol{\beta}_{(2)} = \boldsymbol{\beta}_{(8)} + \boldsymbol{\beta}_{(9)} + \boldsymbol{\beta}_{(10)} + \boldsymbol{\beta}_{(11)}$ by means of H_0: $\mathbf{CBM} = \mathbf{O}$, where $\boldsymbol{\beta}_{(j)}$ is the jth column of $\mathbf{B}$ (Section 7.5.3).

7.44 Consider the apple data in Table 4.10 and ignore the grouping into four treatments.

(a) Find $\hat{\mathbf{B}}$ for regression of $(y_1, y_2, \ldots, y_6)$ on (x_1, x_2) and test the significance of overall regression using Wilks' Λ.

(b) Find $\hat{\mathbf{B}}$ for regression of (x_1, x_2) on $(y_1, y_2, \ldots, y_6)$ and test the significance of overall regression using Wilks' Λ. Why does Wilks' Λ give the same result as in part (a)?

7.45 Using the diabetes data in Table 1.2, do the following:

(a) Find the least squares estimate $\hat{\mathbf{B}}$ for the regression of (y_1, y_2) on (x_1, x_2, x_3).

(b) Test the significance of overall regression using all four test statistics in Section 7.5.1.

(c) What do the eigenvalues of $\mathbf{E}^{-1}\mathbf{H}$ reveal about the essential rank of $\hat{\mathbf{B}}_1$ and the implications of this rank, such as the relative power of the four tests (Section 7.5.1)?

(d) Test the significance of each of x_1, x_2, and x_3 adjusted for the other two x's (Section 7.5.2).

7.46 Use the glucose data in Table 1.4.

(a) Find $\hat{\mathbf{B}}$ for the regression of (y_1, y_2, y_3) on (x_1, x_2, x_3) and test the significance of overall regression.

(b) What do the eigenvalues of $\mathbf{E}^{-1}\mathbf{H}$ reveal about the essential rank of $\hat{\mathbf{B}}_1$ and the relative power of the four tests (Section 7.5.1)?

(c) Test the significance of each of x_1, x_2, and x_3 adjusted for the other two x's (Section 7.5.2).

7.47 Table 7.1 (Picard and Berk 1990) presents average monthly rainfall (May through September) in inches for nine contiguous counties in west central Kansas. The first three variables are the dependent variables y_1, y_2, y_3, and the last six are the independent variables x_1, $x_2, \ldots, x_6$.

(a) Find the least squares estimate $\hat{\mathbf{B}}$ for the regression of (y_1, y_2, y_3) on $(x_1, x_2, \ldots, x_6)$ and test the significance of overall regression.

(b) Test the significance of (x_5, x_6) adjusted for the other four x's.

(c) Test the significance of (x_4, x_5, x_6) adjusted for the other three x's.

(d) Test the significance of (x_1, x_2, x_3) adjusted for the other three x's.

7.48 For the rainfall data in Table 7.1, test the hypothesis that all six x's act alike in predicting the three y's,

$$H_0\colon \beta_{1j} = \beta_{2j} = \cdots = \beta_{6j}, \qquad j = 1, 2, 3,$$

where

$$\mathbf{B} = \begin{pmatrix} \beta_{01} & \beta_{02} & \beta_{03} \\ \beta_{11} & \beta_{12} & \beta_{13} \\ \vdots & \vdots & \vdots \\ \beta_{61} & \beta_{62} & \beta_{63} \end{pmatrix}.$$

Table 7.1. Average Monthly Rainfall (in inches) for Nine Counties

Observation	y_1	y_2	y_3	x_1	x_2	x_3	x_4	x_5	x_6
1	2.97	1.85	2.36	2.17	2.77	2.42	1.76	1.40	1.70
2	1.88	1.72	1.93	1.25	1.50	1.63	1.50	1.11	1.72
3	2.19	1.94	3.35	2.30	2.64	2.22	2.91	2.60	3.17
4	1.68	.78	1.23	.97	1.66	1.01	1.48	.72	1.39
5	2.40	2.51	2.04	3.19	2.17	2.33	2.38	1.85	2.34
6	3.40	3.55	4.46	3.82	4.06	3.99	3.12	3.56	3.56
7	2.05	2.08	1.67	1.94	2.22	2.55	2.56	1.87	2.16
8	2.11	1.40	1.83	1.94	2.55	1.78	1.78	1.31	2.20
9	2.66	2.36	1.88	2.78	3.18	2.91	1.51	2.48	2.34
10	1.49	2.25	2.34	1.82	2.12	1.90	2.13	2.03	2.37
11	2.45	2.77	3.21	2.57	3.18	2.29	1.83	1.61	2.05
12	2.45	1.70	2.33	2.18	1.95	1.89	2.51	2.97	3.10
13	2.41	2.69	3.22	2.62	3.28	2.85	2.78	2.79	2.89
14	4.10	3.71	3.81	3.30	4.13	4.52	5.06	3.64	3.87
15	3.82	2.98	3.50	3.40	4.28	4.65	2.16	3.23	2.57
16	4.78	4.66	4.45	4.46	4.79	5.37	3.46	4.03	4.18
17	1.25	.95	1.16	.90	1.25	1.57	1.87	1.16	1.09
18	1.88	1.45	1.76	1.33	1.98	1.99	1.65	1.71	1.31
19	1.49	1.36	1.95	2.50	1.50	1.24	1.75	1.38	1.83
20	2.02	2.27	2.28	3.21	2.24	2.66	1.92	2.12	1.41
21	1.36	1.05	1.35	.86	1.39	1.17	1.00	1.29	1.59
22	3.38	2.73	2.92	2.40	4.30	3.84	3.59	3.39	1.85
23	3.16	3.94	2.70	4.30	2.99	3.76	2.71	3.49	4.64
24	2.23	1.94	2.14	2.27	3.52	2.65	2.20	1.92	1.57
25	1.76	1.88	1.71	2.25	2.65	1.16	1.37	1.57	.94
26	3.33	2.84	2.84	2.62	5.44	4.51	3.16	2.32	2.16
27	2.30	2.16	3.07	2.91	3.46	2.94	3.35	2.16	2.82
28	2.59	3.96	2.46	3.02	4.14	3.36	1.81	2.38	1.54
29	2.28	1.90	2.11	1.84	2.75	2.02	3.43	2.11	2.13
30	3.41	3.94	3.89	3.70	3.86	3.30	2.80	3.93	2.74
31	2.56	1.49	2.71	1.44	2.49	1.92	2.63	2.43	2.29
32	2.60	2.92	2.63	3.32	2.78	3.35	2.70	2.28	3.45
33	2.19	2.21	2.40	2.27	3.76	3.83	2.25	1.77	2.08
34	2.48	3.57	2.62	3.18	2.44	2.39	3.63	3.34	3.72
35	2.23	2.63	2.13	2.12	2.55	2.40	1.91	1.84	1.79

CHAPTER 8

Canonical Correlation

8.1 INTRODUCTION

The multiple correlation R, as defined in Sections 7.2.5 and 7.3.3, is a measure of (linear) relationship between one y and several x's. *Canonical correlation* is a measure of (linear) relationship between several y's and several x's.

In a sense, canonical correlation analysis is to multivariate multiple regression analysis as multiple correlation is to multiple regression. The test statistics for H_0: $\mathbf{B}_1 = \mathbf{O}$ and H_0: $\mathbf{B}_r = \mathbf{O}$ in Sections 7.5.1 and 7.5.2 can be expressed in terms of canonical correlations (Rencher 1995, Section 11.6.1). Canonical correlation analysis is also related to MANOVA and discriminant analysis (Rencher 1995, Section 11.6.2) and to the test of independence of two subvectors (Section 8.4.1).

Canonical correlation was proposed by Hotelling (1935, 1936), who applied the technique to a data set in which one set of variables consisted of mental tests and the other set involved physical measurements. The geometry of canonical correlation analysis was considered by Dempster (1966, pp. 98, 176). Estimation has been discussed by Dempster (1966) and Lyttkens (1972). Inference has been considered by Schuenemeyer and Bargmann (1978), Bargmann (1979), and Muirhead and Waternaux (1980). Generalizations of canonical correlation to more than two sets of variables have been given by Kettenring (1971), Gnanadesikan (1977, p. 69), and Van De Geer (1984). Muller (1982) gave linear model and principal component approaches to canonical correlation. A comprehensive survey of canonical correlation analysis was given by Gittens (1985).

8.2 CANONICAL CORRELATIONS AND CANONICAL VARIATES

As in previous chapters, we denote the two subvectors that are to be measured on each sampling unit as $\mathbf{y}$ and $\mathbf{x}$. It is not necessary that both $\mathbf{y}$ and $\mathbf{x}$ be continuous or that both be random vectors. Canonical correlation can be used when either $\mathbf{y}$ or $\mathbf{x}$ is fixed, as, for example, in multivariate regression with $\mathbf{y}$ random and $\mathbf{x}$ fixed (Sections 7.4–7.6).

We assume a random sample of n observation vectors $(y_{i1}, y_{i2}, \ldots, y_{ip}, x_{i1}, x_{i2}, \ldots, x_{iq})'$, $i = 1, 2, \ldots, n$. The overall sample covariance matrix can be partitioned as in (1.59),

$$\mathbf{S} = \begin{pmatrix} \mathbf{S}_{yy} & \mathbf{S}_{yx} \\ \mathbf{S}_{xy} & \mathbf{S}_{xx} \end{pmatrix},$$

where $\mathbf{S}_{yy}$ is the $p \times p$ sample covariance matrix of the y's, $\mathbf{S}_{yx}$ is the $p \times q$ matrix of sample covariances between the y's and the x's, and $\mathbf{S}_{xx}$ is the $q \times q$ sample covariance matrix of the x's.

It can be shown that the squared multiple correlation R^2 in (7.46) or (7.66) is equal to the maximum squared correlation between y and a linear combination of the x's. By extension, we define the *squared canonical correlation* r^2 as the maximum squared correlation between a linear combination of the y's and a linear combination of the x's. Thus we seek coefficient vectors $\mathbf{a}$ and $\mathbf{b}$ for which the sample squared correlation r_{uv}^2 between $u = \mathbf{a}'\mathbf{y}$ and $v = \mathbf{b}'\mathbf{x}$ is maximum. (In seeking the maximum correlation between u and v, we find other dimensions of correlation that we will also designate as canonical correlations.)

By (1.83), the sample correlation between the linear combinations $u = \mathbf{a}'\mathbf{y}$ and $v = \mathbf{b}'\mathbf{x}$ is

$$r_{uv} = \frac{\mathbf{a}'\mathbf{S}_{yx}\mathbf{b}}{\sqrt{(\mathbf{a}'\mathbf{S}_{yy}\mathbf{a})(\mathbf{b}'\mathbf{S}_{xx}\mathbf{b})}}.$$

The *squared canonical correlation* is therefore defined as

$$r^2 = \max_{\mathbf{a},\mathbf{b}} r_{uv}^2,$$

and the linear combinations $u = \mathbf{a}'\mathbf{y}$ and $v = \mathbf{b}'\mathbf{x}$ that provide the maximum squared correlation are called *canonical variates*. The following theorem gives the values of r^2, $\mathbf{a}$, and $\mathbf{b}$.

Theorem 8.2A. The squared canonical correlation $r^2 = \max_{\mathbf{a},\mathbf{b}} r_{uv}^2$ is the first (largest) eigenvalue of either $\mathbf{S}_{yy}^{-1}\mathbf{S}_{yx}\mathbf{S}_{xx}^{-1}\mathbf{S}_{xy}$ or $\mathbf{S}_{xx}^{-1}\mathbf{S}_{xy}\mathbf{S}_{yy}^{-1}\mathbf{S}_{yx}$, and the coefficient vectors $\mathbf{a}$ and $\mathbf{b}$ in the canonical variates $u = \mathbf{a}'\mathbf{y}$ and $v = \mathbf{b}'\mathbf{x}$ are the corresponding eigenvectors.

Proof. Using (A.13.2) and (A.13.3), we differentiate the square of r_{uv},

$$r_{uv}^2 = \frac{(\mathbf{a}'\mathbf{S}_{yx}\mathbf{b})^2}{(\mathbf{a}'\mathbf{S}_{yy}\mathbf{a})(\mathbf{b}'\mathbf{S}_{xx}\mathbf{b})}, \tag{8.1}$$

with respect to $\mathbf{a}$ and with respect to $\mathbf{b}$ and set the results equal to zero. For $\mathbf{a}$ we obtain

$$\frac{\partial r_{uv}^2}{\partial \mathbf{a}} = \frac{(\mathbf{a}'\mathbf{S}_{yy}\mathbf{a})(\mathbf{b}'\mathbf{S}_{xx}\mathbf{b})(2\mathbf{a}'\mathbf{S}_{yx}\mathbf{b})(\mathbf{S}_{yx}\mathbf{b}) - (\mathbf{a}'\mathbf{S}_{yx}\mathbf{b})^2(\mathbf{b}'\mathbf{S}_{xx}\mathbf{b})(2\mathbf{S}_{yy}\mathbf{a})}{(\mathbf{a}'\mathbf{S}_{yy}\mathbf{a})^2(\mathbf{b}'\mathbf{S}_{xx}\mathbf{b})^2} = \mathbf{0}.$$

Multiplying by $(\mathbf{a}'\mathbf{S}_{yy}\mathbf{a})(\mathbf{b}'\mathbf{S}_{xx}\mathbf{b})$ and dividing by 2 yields

$$(\mathbf{a}'\mathbf{S}_{yx}\mathbf{b})\mathbf{S}_{yx}\mathbf{b} - \left[\frac{(\mathbf{a}'\mathbf{S}_{yx}\mathbf{b})^2}{(\mathbf{a}'\mathbf{S}_{yy}\mathbf{a})(\mathbf{b}'\mathbf{S}_{xx}\mathbf{b})}\right](\mathbf{b}'\mathbf{S}_{xx}\mathbf{b})\mathbf{S}_{yy}\mathbf{a} = \mathbf{0}. \tag{8.2}$$

By (8.1), the term in brackets is r_{uv}^2, which we designate as r^2 after differentiation:

$$(\mathbf{a}'\mathbf{S}_{yx}\mathbf{b})\mathbf{S}_{yx}\mathbf{b} - r^2(\mathbf{b}'\mathbf{S}_{xx}\mathbf{b})\mathbf{S}_{yy}\mathbf{a} = \mathbf{0}. \tag{8.3}$$

Similarly, differentiation of (8.1) with respect to $\mathbf{b}$ yields

$$(\mathbf{a}'\mathbf{S}_{yx}\mathbf{b})\mathbf{S}_{xy}\mathbf{a} - r^2(\mathbf{a}'\mathbf{S}_{yy}\mathbf{a})\mathbf{S}_{xx}\mathbf{b} = \mathbf{0}. \tag{8.4}$$

Solving (8.4) for $\mathbf{b}$ in terms of $\mathbf{a}$ gives

$$\mathbf{b} = \frac{\mathbf{a}'\mathbf{S}_{yx}\mathbf{b}}{r^2(\mathbf{a}'\mathbf{S}_{yy}\mathbf{a})}\mathbf{S}_{xx}^{-1}\mathbf{S}_{xy}\mathbf{a}, \tag{8.5}$$

and substituting this into (8.3) yields

$$(\mathbf{S}_{yx}\mathbf{S}_{xx}^{-1}\mathbf{S}_{xy} - r^2\mathbf{S}_{yy})\mathbf{a} = \mathbf{0}. \tag{8.6}$$

Multiplying (8.6) on the left by $\mathbf{S}_{yy}^{-1}$ produces the alternative form

$$(\mathbf{S}_{yy}^{-1}\mathbf{S}_{yx}\mathbf{S}_{xx}^{-1}\mathbf{S}_{xy} - r^2\mathbf{I})\mathbf{a} = \mathbf{0}. \tag{8.7}$$

Similarly, solving (8.3) for $\mathbf{a}$ in terms of $\mathbf{b}$ and substituting into (8.4) yields

$$(\mathbf{S}_{xy}\mathbf{S}_{yy}^{-1}\mathbf{S}_{yx} - r^2\mathbf{S}_{xx})\mathbf{b} = \mathbf{0}, \tag{8.8}$$

$$(\mathbf{S}_{xx}^{-1}\mathbf{S}_{xy}\mathbf{S}_{yy}^{-1}\mathbf{S}_{yx} - r^2\mathbf{I})\mathbf{b} = \mathbf{0}. \tag{8.9}$$

□

The eigenvalue r^2 can be obtained from either of the following:

$$|\mathbf{S}_{yy}^{-1}\mathbf{S}_{yx}\mathbf{S}_{xx}^{-1}\mathbf{S}_{xy} - r^2\mathbf{I}| = 0, \tag{8.10}$$

$$|\mathbf{S}_{xx}^{-1}\mathbf{S}_{xy}\mathbf{S}_{yy}^{-1}\mathbf{S}_{yx} - r^2\mathbf{I}| = 0. \tag{8.11}$$

It is not surprising that (8.10) and (8.11) yield the same eigenvalue r^2, since the two matrices involved are of the form $\mathbf{AB}$ and $\mathbf{BA}$, where $\mathbf{A} = \mathbf{S}_{yy}^{-1}\mathbf{S}_{yx}$ and $\mathbf{B} = \mathbf{S}_{xx}^{-1}\mathbf{S}_{xy}$. By (A.10.9), $\mathbf{AB}$ and $\mathbf{BA}$ have the same eigenvalues but different eigenvectors. Note that the eigenvectors $\mathbf{a}$ and $\mathbf{b}$ from (8.7) and (8.9) differ in size: $\mathbf{a}$ is $p \times 1$ and $\mathbf{b}$ is $q \times 1$.

If $p = 1$, $\mathbf{S}_{yy}^{-1}\mathbf{S}_{yx}\mathbf{S}_{xx}^{-1}\mathbf{S}_{xy}$ in (8.10) becomes $\mathbf{s}_{yx}'\mathbf{S}_{xx}^{-1}\mathbf{s}_{yx}/s_{yy}$, which, by (7.66), is equal to R^2. Thus canonical correlation reduces to multiple correlation when there is only one y.

In addition to the maximum eigenvalue in (8.7) or (8.9), the other (nonzero) eigenvalues are also squared canonical correlations that provide supplemental dimensions of the relationship between **y** and **x**. For simplicity of exposition, the maximum value of r_{uv}^2 was denoted by r^2 in Theorem 8.2A, but the maximum value is actually r_1^2, the first (largest) eigenvalue. Thus there are $s = \min(p, q)$ (nonzero) eigenvalues r_1^2, $r_2^2, \ldots, r_s^2$, with s corresponding eigenvectors $\mathbf{a}_i$ and $\mathbf{b}_i$, $i = 1, 2, \ldots, s$. If $p < q$, then $\mathbf{S}_{xx}^{-1}\mathbf{S}_{xy}\mathbf{S}_{yy}^{-1}\mathbf{S}_{yx}$ is $q \times q$ of rank p, and $q - p$ of the eigenvalues are equal to zero. The corresponding eigenvectors $\mathbf{b}_{p+1}, \ldots, \mathbf{b}_q$ are nonzero, but they are usually ignored. Krzysko (1982) has shown that when $p < q$, prediction of $y_1, y_2, \ldots, y_p$ using all of $u_i = \mathbf{a}_i'\mathbf{y}$, $i = 1, 2, \ldots, q$, is equivalent to their prediction using only $u_1, u_2, \ldots, u_p$, where prediction is in terms of multivariate regression. Similar comments pertain to the case $p > q$.

Thus there are $s = \min(p, q)$ canonical correlations $r_1, r_2, \ldots, r_s$ and s corresponding pairs of canonical variates:

$$\begin{aligned} u_1 &= \mathbf{a}_1'\mathbf{y}, & v_1 &= \mathbf{b}_1'\mathbf{x}, \\ u_2 &= \mathbf{a}_2'\mathbf{y}, & v_2 &= \mathbf{b}_2'\mathbf{x}, \\ &\vdots & &\vdots \\ u_s &= \mathbf{a}_s'\mathbf{y}, & v_s &= \mathbf{b}_s'\mathbf{x}, \end{aligned}$$

or

$$\mathbf{u} = \mathbf{A}\mathbf{y}, \qquad \mathbf{v} = \mathbf{B}\mathbf{x}. \tag{8.12}$$

We now provide an alternative approach to canonical correlation based on a singular value decomposition similar to that given in (1.110) (see also Kshirsagar 1972, pp. 249–255).

Theorem 8.2B. Suppose $p \leq q$ is the rank of $\mathbf{S}_{yx}$. Define the $p \times q$ matrix $\mathbf{C} = \mathbf{S}_{yy}^{-1/2}\mathbf{S}_{yx}\mathbf{S}_{xx}^{-1/2}$, where $\mathbf{S}_{yy}^{-1/2}$ is the inverse of the square root matrix $\mathbf{S}_{yy}^{1/2}$ defined in (A.11.7). Then using a singular value decomposition of **C**, linear combinations $\mathbf{u} = \mathbf{A}\mathbf{y}$ and $\mathbf{v} = \mathbf{B}\mathbf{x}$ can be found whose sample covariance matrix has the simplified form

$$\widehat{\text{cov}}\begin{pmatrix} \mathbf{u} \\ \mathbf{v} \end{pmatrix} = \begin{pmatrix} \mathbf{I} & \mathbf{D} \\ \mathbf{D} & \mathbf{I} \end{pmatrix}, \tag{8.13}$$

where $\mathbf{D} = \text{diag}(r_1, r_2, \ldots, r_p)$ contains the square roots of the nonzero eigenvalues of $\mathbf{CC}'$ or $\mathbf{C}'\mathbf{C}$.

Proof. The matrix **C** has rank equal to p, and the singular value decomposition of **C** can be expressed as

$$\mathbf{C} = \mathbf{PDQ}', \tag{8.14}$$

where the $p \times p$ matrix **P** contains the normalized eigenvectors of $\mathbf{CC}'$ and the $q \times p$ matrix **Q** contains the first p normalized eigenvectors of $\mathbf{C}'\mathbf{C}$.

To establish (8.13), define

$$\mathbf{A} = \mathbf{P}'\mathbf{S}_{yy}^{-1/2} \quad \text{and} \quad \mathbf{B} = \mathbf{Q}'\mathbf{S}_{xx}^{-1/2}. \tag{8.15}$$

Then

$$\mathbf{A}\mathbf{S}_{yy}\mathbf{A}' = \mathbf{P}'\mathbf{S}_{yy}^{-1/2}\mathbf{S}_{yy}\mathbf{S}_{yy}^{-1/2}\mathbf{P} = \mathbf{P}'\mathbf{P} = \mathbf{I}.$$

The columns of $\mathbf{P}$ are orthogonal because they are normalized eigenvectors of the symmetric matrix $\mathbf{C}\mathbf{C}'$ [see (A.11.1)]. Similarly, the columns of $\mathbf{Q}$ are orthogonal, and we obtain $\mathbf{B}\mathbf{S}_{xx}\mathbf{B}' = \mathbf{I}$. By (8.14) and (8.15),

$$\mathbf{A}\mathbf{S}_{yx}\mathbf{B}' = \mathbf{P}'\mathbf{S}_{yy}^{-1/2}\mathbf{S}_{yx}\mathbf{S}_{xx}^{-1/2}\mathbf{Q} = \mathbf{P}'\mathbf{P}\mathbf{D}\mathbf{Q}'\mathbf{Q} = \mathbf{D}.$$

We can now write

$$\widehat{\text{cov}}\begin{pmatrix}\mathbf{u}\\ \mathbf{v}\end{pmatrix} = \widehat{\text{cov}}\begin{pmatrix}\mathbf{Ay}\\ \mathbf{Bx}\end{pmatrix} = \begin{pmatrix}\mathbf{A}\mathbf{S}_{yy}\mathbf{A}' & \mathbf{A}\mathbf{S}_{yx}\mathbf{B}'\\ \mathbf{B}\mathbf{S}_{xy}\mathbf{A}' & \mathbf{B}\mathbf{S}_{xx}\mathbf{B}'\end{pmatrix}$$
$$= \begin{pmatrix}\mathbf{I} & \mathbf{D}\\ \mathbf{D} & \mathbf{I}\end{pmatrix},$$

and (8.13) is established. □

In the following theorem, we establish the connection between the results of Theorem 8.2B and the canonical correlations and variates in Theorem 8.2A and (8.12).

Theorem 8.2C. The eigenvalues $r_1^2, r_2^2, \ldots, r_p^2$ in Theorem 8.2B are squared canonical correlations in (8.10) or (8.11), and the linear functions $\mathbf{u} = \mathbf{Ay}$ and $\mathbf{v} = \mathbf{Bx}$ in Theorem 8.2B are identical to the canonical variates in (8.12).

Proof. By Theorem 8.2B, we have $(\mathbf{C}\mathbf{C}' - r_i^2\mathbf{I})\mathbf{p}_i = \mathbf{0}$ and $(\mathbf{C}'\mathbf{C} - r_i^2\mathbf{I})\mathbf{q}_i = \mathbf{0}$ for $i = 1, 2, \ldots, p$, where $\mathbf{p}_i$ and $\mathbf{q}_i$ represent the ith column of $\mathbf{P}$ and $\mathbf{Q}$, respectively. With $\mathbf{C} = \mathbf{S}_{yy}^{-1/2}\mathbf{S}_{yx}\mathbf{S}_{xx}^{-1/2}$, these become

$$(\mathbf{S}_{yy}^{-1/2}\mathbf{S}_{yx}\mathbf{S}_{xx}^{-1}\mathbf{S}_{xy}\mathbf{S}_{yy}^{-1/2} - r_i^2\mathbf{I})\mathbf{p}_i = \mathbf{0}, \tag{8.16}$$

$$(\mathbf{S}_{xx}^{-1/2}\mathbf{S}_{xy}\mathbf{S}_{yy}^{-1}\mathbf{S}_{yx}\mathbf{S}_{xx}^{-1/2} - r_i^2\mathbf{I})\mathbf{q}_i = \mathbf{0}. \tag{8.17}$$

With $\mathbf{a}_i = \mathbf{S}_{yy}^{-1/2}\mathbf{p}_i$ and $\mathbf{b}_i = \mathbf{S}_{xx}^{-1/2}\mathbf{q}_i$ [see (8.15)], (8.16) and (8.17) are easily shown to be equivalent to

$$(\mathbf{S}_{yy}^{-1}\mathbf{S}_{yx}\mathbf{S}_{xx}^{-1}\mathbf{S}_{xy} - r_i^2\mathbf{I})\mathbf{a}_i = \mathbf{0}, \tag{8.18}$$

$$(\mathbf{S}_{xx}^{-1}\mathbf{S}_{xy}\mathbf{S}_{yy}^{-1}\mathbf{S}_{yx} - r_i^2\mathbf{I})\mathbf{b}_i = \mathbf{0}, \tag{8.19}$$

which are the same as (8.7) and (8.9). □

From Theorem 8.2A, the sample correlation r_1 (positive square root of r_1^2) between u_1 and v_1 is the maximum correlation among all possible linear combinations. The remaining correlations $r_2, r_3, \ldots, r_s$ between u_i and v_i, $i = 2, 3, \ldots, s$, are arranged in decreasing order. The sample covariance structure of the u_i's and v_i's is given in the following theorem.

Theorem 8.2D. All canonical variates are uncorrelated with each other except for the pairs u_i and v_i, $i = 1, 2, \ldots, s$; that is, for $i \neq j$, $\widehat{\text{cov}}(u_i, u_j) = 0$, $\widehat{\text{cov}}(v_i, v_j) = 0$, and $\widehat{\text{cov}}(u_i, v_j) = 0$.

Proof. This follows from (8.13) in Theorem 8.2B. □

Theorem 8.2D provides a simplification in covariance structure of the variables. The pq covariances in $\mathbf{S}_{yx}$ between the p variables in $\mathbf{y}$ and the q variables in $\mathbf{x}$ have been channeled into $s = \min(p, q)$ canonical correlations. The s accompanying canonical variates u_i and v_i are such that the sample correlation between u_1 and v_1 is maximum among all possible linear combinations, the correlation between u_2 and v_2 is greatest among all linear combinations uncorrelated with u_1 and v_1, and so on.

Since the $\mathbf{a}_i$'s and $\mathbf{b}_i$'s are eigenvectors, they are unique only up to multiplication by a scale factor. Two common conventions for scaling are (1) scale so that the canonical variates have unit variance,

$$s_{u_i}^2 = \mathbf{a}_i'\mathbf{S}_{yy}\mathbf{a}_i = s_{v_i}^2 = \mathbf{b}_i'\mathbf{S}_{xx}\mathbf{b}_i = 1, \qquad i = 1, 2, \ldots, s,$$

and (2) scale so that the variances are equal to the squares of the canonical correlations,

$$s_{u_i}^2 = \mathbf{a}_i'\mathbf{S}_{yy}\mathbf{a}_i = s_{v_i}^2 = \mathbf{b}_i'\mathbf{S}_{xx}\mathbf{b}_i = r_i^2, \qquad i = 1, 2, \ldots, s.$$

With either type of scaling it follows from (8.5) and an analogous expression for $\mathbf{a}$ in terms of $\mathbf{b}$ that

$$\mathbf{a}_i = \frac{1}{r_i}\mathbf{S}_{yy}^{-1}\mathbf{S}_{yx}\mathbf{b}_i, \tag{8.20}$$

$$\mathbf{b}_i = \frac{1}{r_i}\mathbf{S}_{xx}^{-1}\mathbf{S}_{xy}\mathbf{a}_i. \tag{8.21}$$

Thus it is not necessary to solve both (8.7) and (8.9).

Significance tests for the canonical correlations are given in Section 8.4. In practice, we also assess the importance of each eigenvalue by its relative size:

$$\frac{r_i^2}{\sum_{j=1}^{s} r_j^2}, \qquad i = 1, 2, \ldots, s. \tag{8.22}$$

Often a few canonical correlations will account for a high percentage of the total linear relationships between the two sets of variables.

8.3 PROPERTIES OF CANONICAL CORRELATIONS AND VARIATES

8.3.1 Properties of Canonical Correlations

We first note that $\mathbf{W} = (n-1)\mathbf{S}$ or $\hat{\boldsymbol{\Sigma}} = (n-1)\mathbf{S}/n$ could be used in place of $\mathbf{S}$ to obtain canonical correlations. If either of these is partitioned and substituted for $\mathbf{S}$ in (8.7) or (8.9), we would obtain the same eigenvectors and eigenvalues.

We next show that the same canonical correlations are obtained from the partitioned correlation matrix

$$\mathbf{R} = \begin{pmatrix} \mathbf{R}_{yy} & \mathbf{R}_{yx} \\ \mathbf{R}_{xy} & \mathbf{R}_{xx} \end{pmatrix}$$

as from $\mathbf{S}$. By (1.52), $\mathbf{S} = \mathbf{DRD}$, where

$$\mathbf{D} = \text{diag}(s_{y_1}, \ldots, s_{y_p}, s_{x_1}, \ldots, s_{x_q}) = \begin{pmatrix} \mathbf{D}_y & \mathbf{O} \\ \mathbf{O} & \mathbf{D}_x \end{pmatrix}.$$

Hence

$$\mathbf{S} = \begin{pmatrix} \mathbf{S}_{yy} & \mathbf{S}_{yx} \\ \mathbf{S}_{xy} & \mathbf{S}_{xx} \end{pmatrix} = \mathbf{DRD} = \begin{pmatrix} \mathbf{D}_y\mathbf{R}_{yy}\mathbf{D}_y & \mathbf{D}_y\mathbf{R}_{yx}\mathbf{D}_x \\ \mathbf{D}_x\mathbf{R}_{xy}\mathbf{D}_y & \mathbf{D}_x\mathbf{R}_{xx}\mathbf{D}_x \end{pmatrix}.$$

Using this form of $\mathbf{S}$ in (8.7), we obtain

$$(\mathbf{D}_y^{-1}\mathbf{R}_{yy}^{-1}\mathbf{R}_{yx}\mathbf{R}_{xx}^{-1}\mathbf{R}_{xy}\mathbf{D}_y - r^2\mathbf{I})\mathbf{a} = \mathbf{0}$$

or

$$(\mathbf{R}_{yy}^{-1}\mathbf{R}_{yx}\mathbf{R}_{xx}^{-1}\mathbf{R}_{xy} - r^2\mathbf{I})\mathbf{c} = \mathbf{0}, \tag{8.23}$$

where

$$\mathbf{c} = \mathbf{D}_y\mathbf{a}. \tag{8.24}$$

Similarly, (8.9) can be written as

$$(\mathbf{R}_{xx}^{-1}\mathbf{R}_{xy}\mathbf{R}_{yy}^{-1}\mathbf{R}_{yx} - r^2\mathbf{I})\mathbf{d} = \mathbf{0}, \tag{8.25}$$

where

$$\mathbf{d} = \mathbf{D}_x\mathbf{b}. \tag{8.26}$$

Thus if we use $\mathbf{R}$ in place of $\mathbf{S}$, the canonical correlations remain the same but the canonical variate coefficient vectors (eigenvectors) differ, as shown in (8.24) and

(8.26). In terms of **R**, the characteristic equations (8.10) and (8.11) become

$$|\mathbf{R}_{yy}^{-1}\mathbf{R}_{yx}\mathbf{R}_{xx}^{-1}\mathbf{R}_{xy} - r^2\mathbf{I}| = 0, \tag{8.27}$$

$$|\mathbf{R}_{xx}^{-1}\mathbf{R}_{xy}\mathbf{R}_{yy}^{-1}\mathbf{R}_{yx} - r^2\mathbf{I}| = 0. \tag{8.28}$$

By (1.53), **R** is the covariance matrix for standardized variables (z-scores). Therefore, **c** and **d** in (8.23) and (8.25) are *standardized coefficient vectors*. This can also be seen by comparing (8.24) and (8.26) with the standardized discriminant function coefficients in (5.10). Thus u and v can be expressed in terms of either standardized or unstandardized variables,

$$u = \mathbf{a}'\mathbf{y} = \mathbf{c}'\mathbf{z}_y, \qquad v = \mathbf{b}'\mathbf{x} = \mathbf{d}'\mathbf{z}_x, \tag{8.29}$$

where $\mathbf{z}_y$ and $\mathbf{z}_x$ are standardized variables (z-scores). To assess the relative contribution of the variables to the canonical variates, it is usually preferable to use the (absolute values of the) standardized coefficients in **c** or **d**, since the coefficients in **a** and **b** involve the same units as the corresponding variables.

As noted above, the same canonical correlations are obtained from **S**, **W**, $\hat{\boldsymbol{\Sigma}}$, and **R**. This invariance property extends to any full-rank linear transformation of the variables, $\mathbf{y}^* = \mathbf{F}\mathbf{y} + \mathbf{f}$ and $\mathbf{x}^* = \mathbf{G}\mathbf{x} + \mathbf{g}$, where **F** and **G** are nonsingular constant matrices and **f** and **g** are constant vectors; that is, the canonical correlations between $\mathbf{y}^*$ and $\mathbf{x}^*$ are equal to those between **y** and **x**.

We now list several properties of canonical correlations, beginning with the invariance feature:

1. Canonical correlations are invariant to any full-rank linear transformations $\mathbf{y}^* = \mathbf{F}\mathbf{y} + \mathbf{f}$ and $\mathbf{x}^* = \mathbf{G}\mathbf{x} + \mathbf{g}$. In particular, the same canonical correlations are obtained from **S**, **W**, $\hat{\boldsymbol{\Sigma}}$, or **R**.
2. A canonical correlation r_i is the ordinary correlation between $u_i = \mathbf{a}_i'\mathbf{y}$ and $v_i = \mathbf{b}_i'\mathbf{x}$. The signs of the coefficients in either $\mathbf{a}_i$ or $\mathbf{b}_i$ could be reversed to change the sign of r_i, but it is customary to take r_i to be positive: $0 \le r_i \le 1$.
3. The first canonical correlation r_1 is greater than (the absolute value of) the largest simple correlation in $\mathbf{R}_{yx}$: $r_1 \ge |r_{ij}|$ for all $i \ne j$.
4. The first squared canonical correlation r_1^2 will increase (with probability 1) if a variable is added to either the y's or the x's (Siotani et al. 1985, p. 544).
5. Any of the canonical correlations r_i, $i = 1, 2, \ldots, s$, is equal to the multiple correlation of u_i with the x's or v_i with the y's: $r_i = R_{u_i|\mathbf{x}} = R_{v_i|\mathbf{y}}$.
6. The first canonical correlation r_1 exceeds the largest multiple correlation of any y_i with the x's or any x_i with the y's: $r_1 \ge \max_i(R_{y_i|\mathbf{x}}, R_{x_i|\mathbf{y}})$. (If $p = 1$ or $q = 1$, then $r_1 = R$.)
7. From property 6 we can deduce that $E(r_1^2) \ge \max(p, q)/(n-1)$ if the true population canonical correlations are all zero (due to $\boldsymbol{\Sigma}_{yx} = \mathbf{O}$), since in this case [see (7.67)] $E(R^2_{y_i|\mathbf{x}}) = q/(n-1)$ and $E(R^2_{x_i|\mathbf{y}}) = p/(n-1)$. If the sample size is small relative to the number of variables $p + q$, this bias can be substantial.

8. If $\mathbf{S}$ is positive definite [see comments following (1.42)], then the first canonical correlation is less than 1.

8.3.2 Properties of Canonical Variates

By (8.29), the same canonical variates are obtained from $\mathbf{R}$ as from $\mathbf{S}$. This invariance property extends to any full-rank linear transformation of the variables.

We now list a few properties of the canonical variates $u_i = \mathbf{a}_i'\mathbf{y}$ and $v_i = \mathbf{b}_i'\mathbf{x}$, beginning with the invariance property:

1. The canonical variates $u_i = \mathbf{a}_i'\mathbf{y}$ and $v_i = \mathbf{b}_i'\mathbf{x}$ are invariant to full-rank linear transformations $\mathbf{y}^* = \mathbf{F}\mathbf{y}$ and $\mathbf{x}^* = \mathbf{G}\mathbf{x}$, where $\mathbf{F}$ and $\mathbf{G}$ are nonsingular constant matrices.
2. The canonical variate coefficient vectors for $\mathbf{y}^* = \mathbf{F}\mathbf{y}$ and $\mathbf{x}^* = \mathbf{G}\mathbf{x}$ are related to those for $\mathbf{y}$ and $\mathbf{x}$ by $\mathbf{a}_i^* = (\mathbf{F}')^{-1}\mathbf{a}_i$ and $\mathbf{b}_i^* = (\mathbf{G}')^{-1}\mathbf{b}_i$.
3. The vectors $\mathbf{a}_i$, $\mathbf{b}_i$, $\mathbf{c}_i$, and $\mathbf{d}_i$ tend to be unstable for small samples. Based on an analysis of eight data sets, Barcikowski and Stevens (1975) recommended a rather large ratio of sample size to number of variables. Weis (1972) and Thorndike and Weiss (1973, 1983) also called for a large ratio of sample size to number of variables.

8.4 TESTS OF SIGNIFICANCE FOR CANONICAL CORRELATIONS

In this section, we assume that $(\mathbf{y}', \mathbf{x}')$ is distributed as a multivariate normal.

8.4.1 Tests of Independence of y and x

If the population covariance matrix $\text{cov}\binom{\mathbf{y}}{\mathbf{x}} = \boldsymbol{\Sigma}$ is partitioned as

$$\boldsymbol{\Sigma} = \begin{pmatrix} \boldsymbol{\Sigma}_{yy} & \boldsymbol{\Sigma}_{yx} \\ \boldsymbol{\Sigma}_{xy} & \boldsymbol{\Sigma}_{xx} \end{pmatrix},$$

then the population counterpart of $\mathbf{S}_{yy}^{-1}\mathbf{S}_{yx}\mathbf{S}_{xx}^{-1}\mathbf{S}_{xy}$ is $\boldsymbol{\Sigma}_{yy}^{-1}\boldsymbol{\Sigma}_{yx}\boldsymbol{\Sigma}_{xx}^{-1}\boldsymbol{\Sigma}_{xy}$, from which we obtain population canonical correlations $\rho_1, \rho_2, \ldots, \rho_s$.

Since $(\mathbf{y}', \mathbf{x}')$ is multivariate normal, $\mathbf{y}$ and $\mathbf{x}$ are independent if and only if $\boldsymbol{\Sigma}_{yx} = \mathbf{O}$. Thus the hypothesis of independence of $\mathbf{y}$ and $\mathbf{x}$ can be expressed as $H_0\colon \boldsymbol{\Sigma}_{yx} = \mathbf{O}$ or

$$H_0\colon \boldsymbol{\Sigma} = \begin{pmatrix} \boldsymbol{\Sigma}_{yy} & \mathbf{O} \\ \mathbf{O} & \boldsymbol{\Sigma}_{xx} \end{pmatrix}. \tag{8.30}$$

If $\boldsymbol{\Sigma}_{yx} = \mathbf{O}$, then $\boldsymbol{\Sigma}_{yy}^{-1}\boldsymbol{\Sigma}_{yx}\boldsymbol{\Sigma}_{xx}^{-1}\boldsymbol{\Sigma}_{xy} = \mathbf{O}$, and the hypothesis of independence is equivalent to the hypothesis that all population canonical correlations are zero:

$$H_0\colon \rho_1 = \rho_2 = \cdots = \rho_s = 0. \tag{8.31}$$

The independence hypothesis H_0: $\boldsymbol{\Sigma}_{yx} = \mathbf{O}$ is also equivalent to the overall regression hypothesis in Section 7.5.1, H_0: $\mathbf{B}_1 = \mathbf{O}$, since by (7.147), $\mathbf{B}_1 = \boldsymbol{\Sigma}_{xx}^{-1}\boldsymbol{\Sigma}_{xy}$. Thus H_0: $\boldsymbol{\Sigma}_{yx} = \mathbf{O}$ can be tested by the Wilks Λ test statistic in (7.109), $\Lambda = |\mathbf{S}|/(|\mathbf{S}_{xx}||\mathbf{S}_{yy}|)$. Critical values for Wilks' Λ are given in Table B.4. In the following theorem, this test is shown to be the likelihood ratio test.

Theorem 8.4A. The likelihood ratio test of the independence hypothesis H_0: $\boldsymbol{\Sigma}_{yx} = \mathbf{O}$ is based on the Wilks Λ test statistic

$$\Lambda = \frac{|\mathbf{S}|}{|\mathbf{S}_{yy}||\mathbf{S}_{xx}|} = \frac{|\mathbf{R}|}{|\mathbf{R}_{yy}||\mathbf{R}_{xx}|}, \tag{8.32}$$

which (under H_0) is distributed as $\Lambda_{p,q,n-1-q}$ or equivalently as $\Lambda_{q,p,n-1-p}$. We reject H_0 if $\Lambda \leq \Lambda_\alpha$.

Proof. Assuming the vector $(\mathbf{y}', \mathbf{x}') = (y_1, \ldots, y_p, x_1, \ldots, x_q)$ has a multivariate normal distribution with covariance matrix

$$\boldsymbol{\Sigma} = \begin{pmatrix} \boldsymbol{\Sigma}_{yy} & \boldsymbol{\Sigma}_{yx} \\ \boldsymbol{\Sigma}_{xy} & \boldsymbol{\Sigma}_{xx} \end{pmatrix},$$

the maximum of the likelihood function under H_1: $\boldsymbol{\Sigma}_{yx} \neq \mathbf{O}$ is easily shown to be [see (3.22)]

$$\max_{H_1} L = \frac{1}{(\sqrt{2\pi})^{n(p+q)}|\hat{\boldsymbol{\Sigma}}|^{n/2}} e^{-n(p+q)/2}, \tag{8.33}$$

where $\hat{\boldsymbol{\Sigma}} = \mathbf{W}/n = (n-1)\mathbf{S}/n$ (Theorem 2.3A). Under H_0, we estimate $\boldsymbol{\Sigma}$ by

$$\hat{\boldsymbol{\Sigma}}_0 = \begin{pmatrix} \hat{\boldsymbol{\Sigma}}_{yy} & \mathbf{O} \\ \mathbf{O} & \hat{\boldsymbol{\Sigma}}_{xx} \end{pmatrix}.$$

Thus by (A.7.8), $|\hat{\boldsymbol{\Sigma}}_0| = |\hat{\boldsymbol{\Sigma}}_{yy}||\hat{\boldsymbol{\Sigma}}_{xx}|$, and

$$\max_{H_0} L = \frac{1}{(\sqrt{2\pi})^{n(p+q)}|\hat{\boldsymbol{\Sigma}}_{yy}|^{n/2}|\hat{\boldsymbol{\Sigma}}_{xx}|^{n/2}} e^{-n(p+q)/2}. \tag{8.34}$$

From (8.33) and (8.34), the likelihood ratio is given by

$$\mathrm{LR} = \left(\frac{|\hat{\boldsymbol{\Sigma}}|}{|\hat{\boldsymbol{\Sigma}}_{yy}||\hat{\boldsymbol{\Sigma}}_{xx}|} \right)^{n/2}, \tag{8.35}$$

from which

$$(\mathrm{LR})^{2/n} = \frac{|\hat{\boldsymbol{\Sigma}}|}{|\hat{\boldsymbol{\Sigma}}_{yy}||\hat{\boldsymbol{\Sigma}}_{xx}|}.$$

The modified likelihood ratio can also be expressed in terms of **S** and **R** as

$$\mathrm{LR}^{2/n} = \frac{|\mathbf{S}|}{|\mathbf{S}_{yy}||\mathbf{S}_{xx}|} \tag{8.36}$$

$$= \frac{|\mathbf{R}|}{|\mathbf{R}_{yy}||\mathbf{R}_{xx}|}, \tag{8.37}$$

where

$$\mathbf{S} = \begin{pmatrix} \mathbf{S}_{yy} & \mathbf{S}_{yx} \\ \mathbf{S}_{xy} & \mathbf{S}_{xx} \end{pmatrix}, \qquad \mathbf{R} = \begin{pmatrix} \mathbf{R}_{yy} & \mathbf{R}_{yx} \\ \mathbf{R}_{xy} & \mathbf{R}_{xx} \end{pmatrix}.$$

To find the distribution of (8.36), we use (A.7.9) for $|\mathbf{S}|$ in (8.36):

$$\mathrm{LR}^{2/n} = \frac{|\mathbf{S}_{xx}||\mathbf{S}_{yy} - \mathbf{S}_{yx}\mathbf{S}_{xx}^{-1}\mathbf{S}_{xy}|}{|\mathbf{S}_{yy}||\mathbf{S}_{xx}|}$$

$$= \frac{|\mathbf{S}_{yy} - \mathbf{S}_{yx}\mathbf{S}_{xx}^{-1}\mathbf{S}_{xy}|}{|\mathbf{S}_{yy}|}. \tag{8.38}$$

The covariance matrix **S** for a single sample is analogous to **E** for the several-group MANOVA situation (with adjustment for degrees of freedom), and (8.38) is equivalent to (4.167). Hence $\mathrm{LR}^{2/n}$ has Wilks' Λ-distribution; that is,

$$\Lambda = \frac{|\mathbf{S}|}{|\mathbf{S}_{yy}||\mathbf{S}_{xx}|} = \frac{|\mathbf{R}|}{|\mathbf{R}_{yy}||\mathbf{R}_{xx}|} \tag{8.39}$$

is distributed as $\Lambda_{p,q,n-1-q}$, and we reject H_0 if $\Lambda \leq \Lambda_\alpha$. [See (7.108) and (7.109), which involved fixed x's.]

By (A.7.8), Λ could alternatively be expressed in the form

$$\Lambda = \frac{|\mathbf{S}_{xx} - \mathbf{S}_{xy}\mathbf{S}_{yy}^{-1}\mathbf{S}_{yx}|}{|\mathbf{S}_{xx}|}, \tag{8.40}$$

from which Λ is distributed as $\Lambda_{q,p,n-1-p}$. This is also required by the symmetry of

$$\frac{|\mathbf{S}|}{|\mathbf{S}_{yy}||\mathbf{S}_{xx}|} = \frac{|\mathbf{S}|}{|\mathbf{S}_{xx}||\mathbf{S}_{yy}|}$$

and is equivalent to property 6 in Section 4.1.2. □

It can be shown (Anderson 1984, pp. 394–396) that the distribution of Λ as given in (8.32) or (8.39) holds if either **y** or **x** has a multivariate normal distribution. Thus, if **y** is multivariate normal, **x** can have any distribution or can be fixed.

We now express Λ in (8.40) in terms of eigenvalues:

$$\begin{aligned}\Lambda &= \frac{|\mathbf{S}_{xx} - \mathbf{S}_{xy}\mathbf{S}_{yy}^{-1}\mathbf{S}_{yx}|}{|\mathbf{S}_{xx}|} \\ &= |\mathbf{S}_{xx}^{-1}||\mathbf{S}_{xx} - \mathbf{S}_{xy}\mathbf{S}_{yy}^{-1}\mathbf{S}_{yx}| \qquad \text{[by (A.7.6)]} \\ &= |\mathbf{S}_{xx}^{-1}(\mathbf{S}_{xx} - \mathbf{S}_{xy}\mathbf{S}_{yy}^{-1}\mathbf{S}_{yx})| \qquad \text{[by (A.7.5)]} \\ &= |\mathbf{I} - \mathbf{S}_{xx}^{-1}\mathbf{S}_{xy}\mathbf{S}_{yy}^{-1}\mathbf{S}_{yx}| \\ &= \prod_{i=1}^{s}(1 - r_i^2) \qquad \text{[by (A.10.8)]},\end{aligned} \tag{8.41}$$

where $s = \min(p, q)$ and the r_i^2 are the nonzero eigenvalues of $\mathbf{S}_{xx}^{-1}\mathbf{S}_{xy}\mathbf{S}_{yy}^{-1}\mathbf{S}_{yx}$. Similarly, from (8.38) we obtain

$$\Lambda = |\mathbf{I} - \mathbf{S}_{yy}^{-1}\mathbf{S}_{yx}\mathbf{S}_{xx}^{-1}\mathbf{S}_{xy}| = \prod_{i=1}^{s}(1 - r_i^2), \tag{8.42}$$

since the eigenvalues of $\mathbf{S}_{yy}^{-1}\mathbf{S}_{yx}\mathbf{S}_{xx}^{-1}\mathbf{S}_{xy}$ are the same as those of $\mathbf{S}_{xx}^{-1}\mathbf{S}_{xy}\mathbf{S}_{yy}^{-1}\mathbf{S}_{yx}$ [see (8.10), (8.11), and the remarks following]. There are $s = \min(p, q)$ nonzero eigenvalues, since s is the rank of both $\mathbf{S}_{yy}^{-1}\mathbf{S}_{yx}\mathbf{S}_{xx}^{-1}\mathbf{S}_{xy}$ and $\mathbf{S}_{xx}^{-1}\mathbf{S}_{xy}\mathbf{S}_{yy}^{-1}\mathbf{S}_{yx}$. For χ^2- and F-approximations for Λ, see (8.48), (8.49), and a remark following (8.49).

The other three multivariate test statistics can also be used to test H_0 in (8.30) or (8.31). From (7.105) and (8.42), we obtain $r_i^2 = \lambda_i/(1 + \lambda_i)$ [see also Rencher (1995, Section 11.6.2)]. Thus Pillai's test statistic in (7.114) becomes

$$V^{(s)} = \sum_{i=1}^{s} r_i^2. \tag{8.43}$$

Table B.6 gives upper percentage points of $V^{(s)}$, based on the parameters

$$s = \min(p, q), \qquad m = \tfrac{1}{2}(|q - p| - 1), \qquad N = \tfrac{1}{2}(n - q - p - 2),$$

as given in (7.113). Two F-approximations for $V^{(s)}$ are given in Section 4.1.4. Using $r_i^2 = \lambda_i/(1 + \lambda_i)$ and (7.115), the Lawley–Hotelling statistic becomes

$$U^{(s)} = \sum_{i=1}^{s} \frac{r_i^2}{1 - r_i^2}. \tag{8.44}$$

Table B.7 gives upper percentage points for $\nu_E U^{(s)}/\nu_H$, where $\nu_H = q$ and $\nu_E = n - q - 1$. Three F-approximations for $U^{(s)}$ are given in Section 4.1.4. By (7.112), Roy's largest root statistic becomes simply

$$\theta = r_1^2. \tag{8.45}$$

Table B.5 gives critical values for θ, based on s, m, and N as defined above for Pillai's test statistic $V^{(s)}$. Note that Roy's test is the union–intersection test (Sections 3.3.8, 3.5.5, 4.1.3) since $\mathbf{y}$ and $\mathbf{x}$ are independent provided that $\mathbf{a}'\mathbf{y}$ and $\mathbf{b}'\mathbf{x}$ are independent for all $\mathbf{a}$ and $\mathbf{b}$. The union–intersection test can therefore be based on the maximum correlation between $\mathbf{a}'\mathbf{y}$ and $\mathbf{b}'\mathbf{x}$.

When H_0: $\rho_1 = \rho_2 = \cdots = \rho_s = 0$ is true, the four test statistics have the same probability α of rejecting H_0. When H_0 is false, the power comparisons are similar to those in Section 7.5.1. If the squared population canonical correlations $\rho_1^2, \rho_2^2, \ldots, \rho_s^2$ are equal or nearly so, $V^{(s)}$, Λ, and $U^{(s)}$ are more powerful than θ. If ρ_1^2 is nonzero and all others are zero, θ is more powerful than the other three. A single nonzero ρ^2 implies rank $(\mathbf{\Sigma}_{yx}) = 1$, which could happen in various ways:

1. A single nonzero row in $\mathbf{\Sigma}_{yx}$ indicates that only one of the y's relates to the x's.
2. A single nonzero column in $\mathbf{\Sigma}_{yx}$ implies that only one of the x's relates to the y's.
3. If all the rows of $\mathbf{\Sigma}_{yx}$ are equal or linear combinations of each other, then the x's behave alike in correlating with the y's. A similar condition on the columns signifies that all the y's correlate alike with the x's.

We can obtain an indication of the dimensionality of the population relationships by examining the sample eigenvalues $r_1^2, r_2^2, \ldots, r_s^2$ as in (8.22), $r_i^2/\sum_{j=1}^{s} r_j^2$, or by use of the tests in Section 8.4.2.

Jensen (1984) has shown that tests for canonical correlations based on normal theory remain valid for distributions that have ellipsoidal symmetry. Thus, for example, we can apply the above four tests to samples from heavy-tailed distributions that have no (population) variance.

8.4.2 Test of Dimension of Relationship between the y's and the x's

If H_0: $\rho_1 = \rho_2 = \cdots = \rho_s = 0$ is rejected, it is natural to ask if all the ρ_i's are nonzero, that is, if all of the canonical correlations are necessary to describe the linear relationships between $\mathbf{y}$ and $\mathbf{x}$. This query can be expressed in the form of a hypothesis:

$$H_{0k}\colon\ \rho_1 \neq 0, \ldots, \rho_{k-1} \neq 0, \quad \rho_k = 0, \ldots, \rho_s = 0. \tag{8.46}$$

If H_{0k} is true for some k, the relationship between $\mathbf{y}$ and $\mathbf{x}$ involves only the first $k-1$ canonical correlations. The hypotheses $H_{01}, \ldots, H_{0k}, \ldots, H_{0s}$ can be tested sequentially to establish a value of k for which all remaining sample canonical correlations can be neglected. If the resulting value of k is small, we have achieved a substantial reduction in the dimensionality of the relationship. This sequence of tests is available in many program packages. Since there are several tests involved, an adjustment to the α-level of each test could be made so as not to exceed the overall α-level (see Section 3.4.5).

Rejection of H_{0k} in (8.46) implies that among $r_k^2, r_{k+1}^2, \ldots, r_s^2$, at least r_k^2 is significantly different from zero. If the sample size is large, there may be more significant values of r_i^2 than we consider to be important. If the ratio $r_i^2/\sum_{j=1}^{s} r_j^2$ is small for some i for which H_{0i} is rejected, we may wish to neglect r_i^2 even though it is statistically significant.

The likelihood ratio for testing H_{0k} is a function of

$$\Lambda_k = \prod_{i=k}^{s}(1 - r_i^2), \tag{8.47}$$

which is distributed as $\Lambda_{p-k+1,q-k+1,n-k-q}$ (Kshirsagar 1972, p. 326). Bartlett's (1938) χ^2-approximation can be applied to Λ_k to obtain

$$\chi^2 = -\left[n - \tfrac{1}{2}(p + q + 3)\right] \ln \Lambda_k \tag{8.48}$$

with $(p - k + 1)(q - k + 1)$ degrees of freedom. Lawley (1959) gave an improved approximation,

$$\chi^2 = -\left[n - k - \frac{1}{2}(p + q + 3) + \sum_{i=1}^{k} \frac{1}{r_i^2}\right] \ln \Lambda_k, \tag{8.49}$$

which (under H_{0k}) is approximately χ^2 with $(p-k+1)(q-k+1)$ degrees of freedom.

The F-approximation for Λ_k in (8.47) is given by

$$F = \frac{1 - \Lambda_k^{1/t}}{\Lambda_k^{1/t}} \frac{\text{df}_2}{\text{df}_1}, \tag{8.50}$$

where

$$\begin{aligned}
\text{df}_1 &= (p - k + 1)(q - k + 1), \\
\text{df}_2 &= wt - \tfrac{1}{2}(p - k + 1)(q - k + 1) + 1, \\
w &= n - \tfrac{1}{2}(p + q + 3), \\
t &= \sqrt{\frac{(p - k + 1)^2(q - k + 1)^2 - 4}{(p - k + 1)^2 + (q - k + 1)^2 - 5}}.
\end{aligned}$$

When $k = 1$, H_{01} in (8.46) is the same as H_0 in (8.31), and Λ_1 in (8.47) is the same as Λ in (8.32) or (8.41). Thus the above χ^2- and F-approximations can be used for the overall hypothesis H_0 by inserting $k = 1$ in (8.48), (8.49), or (8.50).

We note that Λ_k in (8.47) is not the same as Wilks' Λ in (7.122) used to test H_0: $\mathbf{B}_d = \mathbf{O}$, the hypothesis that h of the x's are not significant predictors of the p y's when considered above and beyond the other $q - h$ x's. By (7.123), the test statistic for H_0: $\mathbf{B}_d = \mathbf{O}$ can be expressed as a ratio $\Lambda = \Lambda_f/\Lambda_r$, where $\Lambda_f = \Lambda(x_1, \ldots, x_q)$ is the overall Wilks Λ for the full model with all q of the x's and $\Lambda_r = \Lambda(x_1, \ldots, x_{q-h})$

is the overall Wilks Λ for the $q - h$ x's in the reduced model in which $\mathbf{B}_d = \mathbf{O}$. By (4.14), this ratio can be expressed in terms of eigenvalues of $\mathbf{E}^{-1}\mathbf{H}$:

$$\begin{aligned}\Lambda(x_{q-h+1}, \ldots, x_q \,|\, x_1, \ldots, x_{q-h}) &= \frac{\Lambda(x_1, \ldots, x_q)}{\Lambda(x_1, \ldots, x_{q-h})} \\ &= \frac{\prod_{i=1}^{s} 1/(1+\lambda_i)}{\prod_{i=1}^{t} 1/(1+\lambda_i^*)},\end{aligned} \tag{8.51}$$

where $s = \min(p, q)$, $t = \min(p, q-h)$, the λ_i's are eigenvalues of $\mathbf{E}^{-1}\mathbf{H}$ from the full model, and the λ_i^*'s are eigenvalues of $(\mathbf{E}^*)^{-1}\mathbf{H}^*$ from the reduced model. By (8.41), $\prod_{i=1}^{s} 1/(1+\lambda_i) = \prod_{i=1}^{s}(1 - r_i^2)$, but (8.51) does not reduce to a simpler form; that is,

$$\Lambda(x_{q-h+1}, \ldots, x_q \,|\, x_1, \ldots, x_{q-h}) \neq \prod_{i=t+1}^{s} \frac{1}{1+\lambda_i},$$

because $\lambda_i^* \neq \lambda_i$. Thus (8.51) and (8.47) are not equal.

8.5 VALIDATION

The basic assumptions for canonical correlation analysis can be checked by calculating $u_1 = \mathbf{a}_1'\mathbf{y}$ and $v_1 = \mathbf{b}_1'\mathbf{x}$ for each observation vector $(\mathbf{y}_i', \mathbf{x}_i')$ and plotting the points (u_{1i}, v_{1i}), $i = 1, 2, \ldots, n$. This scatter plot may reveal important characteristics of the sample. The presence of outliers could either induce a synthetic linear relationship or conceal a legitimate one. A curved trend in the plot would indicate nonnormality. We could also plot u_2 versus v_2 or u_3 versus v_3. An alternative method of handling outliers is to use robust versions of $\mathbf{R}$ or $\mathbf{S}$.

As with many other multivariate techniques, canonical correlations are maximized over a particular sample. The results may capitalize on random variation specific to that sample. The chance of spurious relationships increases with the ratio of the number of variables to the sample size, $(p + q)/n$. Two approaches for checking the validity of a canonical correlation analysis are *cross-validation* and the *jackknife*.

Cross-validation has been discussed by Thorndike and Weiss (1973, 1983), Thorndike (1976, 1977, 1978), Wood and Erskine (1976), Huba, Wingard, and Bentler (1980), Lee, McCabe, and Graham (1983), and Gittens (1985, pp. 50–52). In the following brief review of cross-validation we follow Gittens. For simplicity we consider only r_1, $\mathbf{a}_1$, and $\mathbf{b}_1$, but the discussion could be extended to include r_2, $\mathbf{a}_2$, $\mathbf{b}_2$, and so on. Suppose we have obtained r_1, $\mathbf{a}_1$, and $\mathbf{b}_1$ from a sample. If $u_{1i} = \mathbf{a}_1'\mathbf{y}_i$ and $v_{1i} = \mathbf{b}_1'\mathbf{x}_i$, $i = 1, 2, \ldots, n$, are calculated on another sample (using $\mathbf{a}_1$ and $\mathbf{b}_1$ from the first sample), their correlation will ordinarily be smaller than r_1 for the original sample. In cross-validation, we check for unusual instability of this type by splitting the sample into two subsamples, performing the canonical correlation analysis on one subsample, and checking it on the second (holdout) subsample. A ratio of 75–25% has been suggested for the sizes of the two subsamples. The values of $\mathbf{a}_1$ and $\mathbf{b}_1$ from the larger subsample are used in computing $u_{1i} = \mathbf{a}_1'\mathbf{y}_i$ and $v_{1i} = \mathbf{b}_1'\mathbf{x}_i$, $i = 1, 2, \ldots, n$, for the holdout subsample. The correlation between u_{1i} and v_{1i}, $i = 1, 2, \ldots, n$, in the holdout sample is calculated and compared with r_1 for the first subsample. Some decrease is expected, but a large reduction may indicate that the results do not generalize beyond the original sample.

A drawback in the cross-validation procedure is that only part of the data is used to calculate r_1^2, $\mathbf{a}_1$, and $\mathbf{b}_1$. The reduced sample size may, in fact, increase the instability that cross-validation was designed to detect. If the sample size is sufficiently large, we can attempt to avoid this trap by using *double cross-validation*. The sample is split in half and canonical correlation analysis is performed on each subsample. The values for $\mathbf{a}_1$ and $\mathbf{b}_1$ are then applied to the opposite subsample and the correlations are computed. The correlations obtained this way are compared, and if there is good agreement, the generalizability of the results is confirmed. The final canonical analysis would then be based on the total sample.

A jackknife approach has been proposed for a related problem by Dempster (1969, pp. 257–260). The technique can be easily adapted for canonical correlation. We use variables such as u_1 and u_2 so as to focus on stability rather than the amount of relationships. First, the usual analysis is performed on the entire sample and u_{1j} and u_{2j} are plotted. Then one observation is held out and $\mathbf{a}_1$ and $\mathbf{a}_2$ are computed for the remaining $n - 1$ observations. These weights are applied to the deleted observation to estimate u_1 and u_2. This operation is repeated for each of the n observations, and the resulting pairs of u_1 and u_2 are plotted. This plot is compared with the plot of the original (u_1, u_2) pairs. The closeness of the two patterns provides an indication of the stability of the original results applied to a future sample.

Another possible source of instability in the $\mathbf{a}_i$'s and $\mathbf{b}_i$'s is *multicolinearity*. If there are near-linear dependencies among the y's or among the x's, the variances of the a_{ij}'s or b_{ij}'s may be inflated. We can find the squared multiple correlation R^2 of each variable with the others in the same subset of variables (y's or x's) by examining the diagonal elements of $\mathbf{R}_{yy}^{-1}$ or $\mathbf{R}_{xx}^{-1}$, as in (1.102). Large values of R^2 may indicate a multicolinearity problem that could cause instability in the corresponding coefficients in $\mathbf{a}_i$ or $\mathbf{b}_i$. In such a case, it may be possible to obtain improved estimators using a *ridge-type* approach to canonical correlation analysis (Carney 1975, Vinod 1976, Campbell 1980). The *canonical ridge weights* are obtained by extracting eigenvectors of the matrix

$$(\mathbf{S}_{yy} + k_y\mathbf{I})^{-1}\mathbf{S}_{yx}(\mathbf{S}_{xx} + k_x\mathbf{I})^{-1}\mathbf{S}_{xy}$$

or of a transposition of this matrix, as in (8.9). Some guidelines for choosing k_y and k_x can be found in Campbell (1980). This procedure is designed to reduce the variances of a_{ij} and b_{ij} while having minimal impact on r_i^2, u_i, and v_i.

8.6 INTERPRETATION OF CANONICAL VARIATES

We now discuss the contribution of each variable to the canonical variates. Share (1984) and Rencher (1992b) have discussed several aspects of interpretation of canonical variates.

There are three common tools for interpreting canonical variates: (1) standardized coefficients; (2) rotation of the canonical variate coefficients; and (3) the correlation between each variable and the canonical variate. The third of these is the most widely recommended, but we show in Section 8.6.3 that it is the least useful. Likewise, we do not recommend rotation for reasons given in Section 8.6.2. We consider the standardized coefficients as defined in (8.23)–(8.26) to be the most useful for assessing the relative contribution of the variables to canonical variates.

8.6.1 Standardized Coefficients

By (8.29), the first pair of canonical variates u_1 and v_1 can be expressed in terms of standardized coefficient vectors $\mathbf{c}_1$ and $\mathbf{d}_1$ as

$$u_1 = \mathbf{c}_1'\mathbf{z}_y, \qquad v_1 = \mathbf{d}_1'\mathbf{z}_x,$$

where $\mathbf{z}_y$ and $\mathbf{z}_x$ are vectors of standardized variables and $\mathbf{c}_1$ and $\mathbf{d}_1$ are defined in (8.23)–(8.26). The standardized coefficients give the relative contribution of the variables in a multivariate context [see a remark following (8.29)]. Thus in

$$u_1 = c_{11}z_{y1} + c_{12}z_{y2} + \cdots + c_{1p}z_{yp},$$

the coefficients $c_{11}, c_{12}, \ldots, c_{1p}$ reflect the joint contribution of the p y's to the correlation between u_1 and v_1. For example, c_{12} shows the contribution of y_2 in the presence of the other $p - 1$ y's. If some y's are deleted or added, c_{12} will change. This is exactly what c_{12} should do in order to provide a true multivariate measure of the contribution of y_2. For the distinction between assessing the contribution of the variables and interpretation, see the comments at the beginning of Section 5.7.

8.6.2 Rotation of Canonical Variate Coefficients

It is possible to rotate the reference axes of the canonical variate coefficients to obtain more high and low coefficients and fewer intermediate ones (see Section 10.5). Several authors have recommended such a rotation to improve interpretability (Hall 1969, 1977; Cliff and Krus 1976; Krus, Reynolds, and Krus 1976; Skinner 1977, 1978; Huba et al. 1980; Bentler and Huba 1982; Reynolds and Jackosfsky 1981). On the other hand, Share (1984), Cohen (1982), and Rencher (1992b) have questioned the routine use of this technique.

We do not recommend rotation of the canonical variate coefficients because (1) the first canonical correlation is reduced and (2) rotation introduces correlations among canonical variates, such as u_1 and u_2. Thus with rotation two important properties of canonical variates are lost.

8.6.3 Correlations between Variables and Canonical Variates

To assess the contribution of each y_i to the first canonical variate u_1, for example, many writers recommend using the correlation between y_i and u_1. Rencher (1988), however, has shown that this correlation provides no information about the multivariate contribution of y_i to u_1. Instead, such correlations measure only univariate relationships between the y's and the x's.

The vector containing the p (sample) correlations between a canonical variate $u_j = \mathbf{a}_j'\mathbf{y}$ and $y_1, y_2, \ldots, y_p$ is given by $\mathbf{R}_{yy}\mathbf{c}_j$, where $\mathbf{R}_{yy}$ is the correlation matrix for the y's, $\mathbf{c}_j$ is defined in (8.23) and (8.24), and $\mathbf{c}_j$ is scaled so that $\mathbf{c}_j'\mathbf{R}_{yy}\mathbf{c}_j = 1$. Evaluating the vector $\mathbf{R}_{yy}\mathbf{c}_j$ for $j = 1, 2, \ldots, s$, we obtain the ps correlations $r_{y_i u_j}$, $i = 1, 2, \ldots, p$, $j = 1, 2, \ldots, s$. The information provided by the $r_{y_i u_j}$'s is given in the following theorem.

Theorem 8.6A. A weighted sum of the squared correlations $r_{y_i u_1}^2, r_{y_i u_2}^2, \ldots, r_{y_i u_s}^2$ between y_i and $u_1, u_2, \ldots, u_s$ is equal to $R_{y_i|\mathbf{x}}^2$, the squared multiple correlation between y_i and the x's:

$$\sum_{j=1}^{s} r_{y_i u_j}^2 r_j^2 = R_{y_i|\mathbf{x}}^2, \tag{8.52}$$

where r_j^2 is the jth squared canonical correlation. There is an analogous result for the correlations between x_i and $v_1, v_2, \ldots, v_s$.

Proof. The proof of (8.52) is similar to the proof of (5.20) in Theorem 5.7B, so here we present only an outline. We denote the ith row of $\mathbf{R}_{yy}$ and of $\mathbf{R}_{yx}$ by $\mathbf{r}_{yyi}'$ and $\mathbf{r}_{yxi}'$. Then

$$\begin{aligned}\sum_{j=1}^{s} r_{y_i u_j}^2 r_j^2 &= \sum_{j=1}^{s} (\mathbf{r}_{yyi}'\mathbf{c}_j)^2 r_j^2 = \mathbf{r}_{yyi}'\left[\sum_{j=1}^{s} (\mathbf{c}_j\mathbf{c}_j') r_j^2\right]\mathbf{r}_{yyi} \\ &= \mathbf{r}_{yyi}'\mathbf{R}_{yy}^{-1}\mathbf{R}_{yx}\mathbf{R}_{xx}^{-1}\mathbf{R}_{xy}\mathbf{R}_{yy}^{-1}\mathbf{r}_{yyi} \\ &= \mathbf{r}_{yxi}'\mathbf{R}_{xx}^{-1}\mathbf{r}_{yxi} = R_{y_i|\mathbf{x}}^2. \qquad \square\end{aligned}$$

Thus the correlations between y_i and the canonical variates $u_1, u_2, \ldots, u_s$ reflect only the squared multiple correlation $R_{y_i|\mathbf{x}}^2$ for y_i by itself regressed on the x's. The correlations tell us nothing about how the y's contribute jointly to canonical correlation with the x's.

We should not be surprised at this result. Some practitioners reject standardized coefficients because they are "unstable," meaning that they change if some variables are deleted and others added. However, this is what we want the coefficients to do in order to reflect the mutual influence of the variables on each other. In a multivariate setting, we are interested in the joint performance of the particular variables that are available. Only a univariate approach can show the contribution of each variable independent of (ignoring) all other variables. This is precisely what we get with the above correlations. We cannot find a middle ground between the multivariate and univariate realms where we enjoy all the advantages of both.

For additional perspective on these correlations, consider the following analogy to multiple regression relating y to several x's. In order to compare the x's with each other so as to assess their relative contribution to prediction of y, we could examine the standardized regression coefficients (beta weights). We would not evaluate the contribution of x_i to the multiple regression equation by performing a separate simple regression of y on x_i alone. But this corresponds to the use of correlations to interpret canonical variates.

Example 8.6.3. To demonstrate the univariate information provided by correlations between variables and canonical variates, we use a data set described in Timm (1975, p. 314; see diskette for data) that contains seven variables (test scores) observed for 37 students. We wish to relate three achievement and performance tests, SAT, PPVT, and RPMT, to four learning proficiency tests, N, S, NA, and SS. The partitioned correlation matrix **R** is given in Table 8.1.

The canonical correlations are .713, .268, and .174. Since the second and third are not significant, we consider only the first pair of canonical variates u_1 and v_1. In

Table 8.1. Correlation Matrix for Seven Test Scores (n = 37)

	SAT	PPVT	RPMT	N	S	NA	SS
SAT	1.00						
PPVT	.37	1.00					
RPMT	.21	.35	1.00				
N	.19	.44	.35	1.00			
S	.16	.27	.24	.40	1.00		
NA	.26	.67	.34	.65	.65	1.00	
SS	.33	.59	.34	.67	.43	.80	1.00

Table 8.2. Correlations between Each Variable and the First Pair of Canonical Variates, Standardized Coefficients, and R^2

Variable	Standardized Coefficient	Rank	Correlation	Rank	R^2	Rank
y's						
SAT	.001	3	.379	3	.115	3
PPVT	.949	1	.993	1	.502	1
RPMT	.123	2	.460	2	.150	2
x's						
N	.011	4	.653	3	.240	3
S	−.357	2	.399	4	.098	4
NA	1.071	1	.954	1	.463	1
SS	.135	3	.842	2	.378	2

Table 8.2, the correlations between variables and the first pair of canonical variates (u_1 and v_1) are compared with the standardized coefficients and with R^2-values.

As expected from Theorem 8.6A, the correlations rank the variables in the same order as the univariate R^2-values, rather than telling a multivariate story similar to that of the standardized coefficients. Even in the y's, where the rankings are the same, the relative contribution of the three variables provided by the correlations is somewhat different from that given by the standardized coefficients. According to the standardized coefficients, SAT and RPMT are negligible, whereas the correlations would seem to indicate that they are more important.

8.7 REDUNDANCY ANALYSIS

Stewart and Love (1968) and Miller (1969) proposed the *redundancy* as a measure of relationship between two sets of variables. To describe redundancy, we first note that a squared correlation can be interpreted as the proportion of variance each variable accounts for in the other variable [see (7.46) adapted for $q = 1$]. For example, the proportion of variance of y_i explained by the canonical variate u_j is given by $r^2_{y_iu_j}$, as defined in Section 8.6.3. Likewise, r^2_j indicates the proportion of variance of u_j accounted for by v_j.

All the $r^2_{y_iu_j}$'s can be conveniently displayed in the $p \times s$ matrix

$$\mathbf{R}^2_{yu} = \begin{pmatrix} r^2_{y_1u_1} & r^2_{y_1u_2} & \cdots & r^2_{y_1u_s} \\ r^2_{y_2u_1} & r^2_{y_2u_2} & \cdots & r^2_{y_2u_s} \\ \vdots & \vdots & & \vdots \\ r^2_{y_pu_1} & r^2_{y_pu_2} & \cdots & r^2_{y_pu_s} \end{pmatrix}.$$

We assume that the variables $y_1, y_2, \ldots, y_p$ are in standardized form so that the variance of each is 1 and the total variance of the y's is p. Because $u_1, u_2, \ldots, u_s$ are uncorrelated, the squared correlations $r^2_{y_2u_1}, r^2_{y_2u_2}, \ldots, r^2_{y_2u_s}$ in the second row of $\mathbf{R}^2_{yu}$, for example, partition the variance of y_2, and their sum is equal to 1. The same can be said of each row of $\mathbf{R}^2_{yu}$, and the sum of all elements of $\mathbf{R}^2_{yu}$ is thus equal to p.

Consider the second column of $\mathbf{R}^2_{yu}$. The sum of the elements of this column divided by p, $\sum_{i=1}^{p} r^2_{y_iu_2}/p$, is the proportion of total variance in the y's that is associated with u_2. Multiplying by r^2_2 gives the proportion of total variance in the y's that is associated with u_2 and v_2,

$$\text{Rd}(\mathbf{y} \mid v_2) = \sum_{i=1}^{p} \frac{r^2_{y_iu_2} r^2_2}{p}. \tag{8.53}$$

This is called the *redundancy* of the y's given v_2, since the correlation of u_2 with v's other than v_2 is zero. To obtain the *total redundancy* of the y's given the v's, we sum terms illustrated by (8.53)

$$\mathrm{Rd}(\mathbf{y}\,|\,\mathbf{v}) = \sum_{j=1}^{s} \mathrm{Rd}(\mathbf{y}\,|\,v_j) = \sum_{j=1}^{s}\sum_{i=1}^{p} \frac{r_{y_i u_j}^2 r_j^2}{p}, \tag{8.54}$$

where $\mathbf{v} = (v_1, v_2, \ldots, v_s)'$.

Similarily, the redundancy of the x's given u_j is

$$\mathrm{Rd}(\mathbf{x}\,|\,u_j) = \sum_{i=1}^{q} \frac{r_{x_i v_j}^2 r_j^2}{q}, \tag{8.55}$$

and the total redundancy of the x's given the u's is

$$\mathrm{Rd}(\mathbf{x}\,|\,\mathbf{u}) = \sum_{j=1}^{s}\sum_{i=1}^{q} \frac{r_{x_i v_j}^2 r_j^2}{q}. \tag{8.56}$$

Since $r_{y_i u_j}^2 \neq r_{x_i v_j}^2$ and p is typically unequal to q,

$$\mathrm{Rd}(\mathbf{y}\,|\,\mathbf{v}) \neq \mathrm{Rd}(\mathbf{x}\,|\,\mathbf{u}). \tag{8.57}$$

Redundancy analysis has been discussed and extended by Miller and Farr (1971), Miller (1975), Gleason (1976), Nicewander and Wood (1974, 1975), Van Den Wollenberg (1977), Stipak and McDavid (1981), Dawson-Saunders (1982), Muller (1981), and Hollingsworth (1983). It is described in many texts and is widely available in program packages.

However, the total redundancy is not a useful multivariate measure of association. As Stewart and Love (1968) noted (but did not prove), the total redundancy is equal to an average squared multiple correlation.

Theorem 8.7A. Let $R_{y_i|\mathbf{x}}^2$ be the squared multiple correlation of y_i regressed on the x's, and let $R_{x_i|\mathbf{y}}^2$ be the squared multiple correlation of x_i regressed on the y's. Then

$$\mathrm{Rd}(\mathbf{y}\,|\,\mathbf{v}) = \sum_{i=1}^{p} \frac{R_{y_i|\mathbf{x}}^2}{p}, \tag{8.58}$$

$$\mathrm{Rd}(\mathbf{x}\,|\,\mathbf{u}) = \sum_{i=1}^{q} \frac{R_{x_i|\mathbf{y}}^2}{q}. \tag{8.59}$$

Proof. Reversing the order of summation in (8.54) gives

$$\mathrm{Rd}(\mathbf{y}\,|\,\mathbf{v}) = \sum_{i=1}^{p}\sum_{j=1}^{s} \frac{r_{y_i u_j}^2 r_j^2}{p} = \sum_{i=1}^{p} \frac{R_{y_i|\mathbf{x}}^2}{p}$$

by Theorem 8.6A. A similar argument leads to (8.59). □

Note that the form of $\text{Rd}(\mathbf{y} \mid \mathbf{v})$ and $\text{Rd}(\mathbf{x} \mid \mathbf{u})$ in (8.58) and (8.59) further explains the asymmetry exhibited in (8.57).

In (8.58), we see that $\text{Rd}(\mathbf{y} \mid \mathbf{v})$ does not take into account the correlations among the y's and is therefore an average univariate measure of relationship between the y's and the x's. This failure of $\text{Rd}(\mathbf{y} \mid \mathbf{v})$ to be a multivariate measure of association is not surprising, because it is based on correlations between variables and canonical variates. A good measure of association between two sets of variables is provided by r_1^2 itself. Additional measures of association between the y's and the x's are reviewed by Rencher (1995, Section 10.6).

8.8 ADDITIONAL TOPICS

Canonical correlation analysis has been extended to three or more sets of variables by Horst (1961, 1965), Kettenring (1971), and Gnanadesikan (1977). Neuenschwander and Flury (1995) and Goria and Flury (1996) proposed a generalization to several sets of variables based on *common canonical variates* in which the coefficients of the canonical variates are the same across groups but the canonical correlations may vary.

Latour and Styan (1983) and Zinkgraf (1983) considered the use of canonical correlations in the analysis of two-way ANOVA and MANOVA models.

Shi and Taam (1992) proposed an approach to nonlinear canonical correlation analysis. They used the model

$$\mathbf{a}'\mathbf{y} = h(\mathbf{b}'\mathbf{x}) + \varepsilon$$

for some function h and sought values of $\mathbf{a}$ and $\mathbf{b}$ that maximize the correlation between $\mathbf{a}'\mathbf{y}$ and $h(\mathbf{b}'\mathbf{x})$.

Karnel (1991) provided robust estimates of canonical correlations and canonical variates. He also discussed the influence of each observation on the canonical correlation. Romanazzi (1992) discussed the influence of each observation on each canonical correlation and on each pair of canonical variates.

Theil and Chung (1988) considered an information theory approach to assessing the contribution of each variable to the canonical variates.

PROBLEMS

8.1 Verify (8.4) by differentiating (8.1) with respect to $\mathbf{b}$.

8.2 Show that (8.6) can be obtained by solving (8.4) for $\mathbf{b}$ in terms of $\mathbf{a}$ and substituting into (8.3).

8.3 The largest canonical correlation r_1 was defined in Section 8.2 as the maximum correlation between $u = \mathbf{a}'\mathbf{y}$ and $v = \mathbf{b}'\mathbf{x}$. It was shown in Theorem 8.2A that r_1^2 is the largest eigenvalue of $\mathbf{S}_{yy}^{-1}\mathbf{S}_{yx}\mathbf{S}_{xx}^{-1}\mathbf{S}_{xy}$. Show that the same result is obtained if r^2 is defined as the maximum value of $(\mathbf{a}'\mathbf{S}_{yx}\mathbf{b})^2$ subject to the scaling $\mathbf{a}'\mathbf{S}_{yy}\mathbf{a} = 1$ and $\mathbf{b}'\mathbf{S}_{xx}\mathbf{b} = 1$. [Note that r_{uv}^2 in (8.1) is invariant to changes in scale.]

8.4 The maximization in Theorem 8.2A can be carried out in stages, as follows:

(a) Show that if $\mathbf{b}$ is fixed,

$$\max_{\mathbf{a}} r_{uv}^2 = \max_{\mathbf{a}} \frac{(\mathbf{a}'\mathbf{S}_{yx}\mathbf{b})^2}{(\mathbf{a}'\mathbf{S}_{yy}\mathbf{a})(\mathbf{b}'\mathbf{S}_{xx}\mathbf{b})} = \frac{\mathbf{b}'\mathbf{S}_{xx}\mathbf{S}_{yy}^{-1}\mathbf{S}_{yx}\mathbf{b}}{\mathbf{b}'\mathbf{S}_{xx}\mathbf{b}}.$$

(b) Show that

$$\max_{\mathbf{b}} \frac{\mathbf{b}'\mathbf{S}_{xy}\mathbf{S}_{yy}^{-1}\mathbf{S}_{yx}\mathbf{b}}{\mathbf{b}'\mathbf{S}_{xx}\mathbf{b}} = r_1^2,$$

where r_1^2 is the largest eigenvalue of $\mathbf{S}_{xx}^{-1}\mathbf{S}_{xy}\mathbf{S}_{yy}^{-1}\mathbf{S}_{yx}$.

8.5 Show that $\prod_{i=1}^{s} r_i^2 = |\mathbf{S}_{yx}\mathbf{S}_{xx}^{-1}\mathbf{S}_{xy}|/|\mathbf{S}_{yy}|$.

8.6 Show that (8.16) and (8.17) are equivalent to (8.18) and (8.19), respectively.

8.7 As an alternative proof of Theorem 8.2C, show that

$$\mathbf{S}_{yy}^{-1}\mathbf{S}_{yx}\mathbf{S}_{xx}^{-1}\mathbf{S}_{xy}\mathbf{A}' = \mathbf{A}'\mathbf{D}^2,$$

$$\mathbf{S}_{xx}^{-1}\mathbf{S}_{xy}\mathbf{S}_{yy}^{-1}\mathbf{S}_{yx}\mathbf{B}' = \mathbf{B}'\mathbf{D}^2,$$

where $\mathbf{D}^2 = \text{diag}(r_1^2, r_2^2, \ldots, r_p^2)$ contains eigenvalues of $\mathbf{S}_{yy}^{-1}\mathbf{S}_{yx}\mathbf{S}_{xx}^{-1}\mathbf{S}_{xy}$ or $\mathbf{S}_{xx}^{-1}\mathbf{S}_{xy}\mathbf{S}_{yy}^{-1}\mathbf{S}_{yx}$. Thus the rows of $\mathbf{A}$ are eigenvectors of $\mathbf{S}_{yy}^{-1}\mathbf{S}_{yx}\mathbf{S}_{xx}^{-1}\mathbf{S}_{xy}$ and the rows of $\mathbf{B}$ are eigenvectors of $\mathbf{S}_{xx}^{-1}\mathbf{S}_{xy}\mathbf{S}_{yy}^{-1}\mathbf{S}_{yx}$.

8.8 Prove Theorem 8.2D without using Theorem 8.2B.

8.9 Assume that $\mathbf{a}_1'\mathbf{S}_{yy}\mathbf{a}_2 = 0$ and use (8.20) to obtain $\mathbf{b}_1'\mathbf{S}_{xx}\mathbf{b}_2 = 0$ and $\mathbf{a}_1'\mathbf{S}_{yx}\mathbf{b}_2 = 0$.

8.10 Show that with either type of scaling in Section 8.2, (8.20) and (8.21) follow from (8.5) and an analogous expression for $\mathbf{a}$ in terms of $\mathbf{b}$.

8.11 Obtain (8.20) and (8.21) from (A.10.9), which notes that $\mathbf{BA}$ has the same eigenvalues as $\mathbf{AB}$ and that the eigenvectors of $\mathbf{BA}$ are equal to $\mathbf{Bx}_i$, where $\mathbf{x}_i$ is an eigenvector of $\mathbf{AB}$.

8.12 Show that the mean squared error $E(u - v)^2 = E(\mathbf{a}'\mathbf{y} - \mathbf{b}'\mathbf{x})^2$ is minimized by $\mathbf{a} = \boldsymbol{\Sigma}_{yy}^{-1}\boldsymbol{\Sigma}_{yx}\mathbf{b}$ and $\mathbf{b} = \boldsymbol{\Sigma}_{xx}^{-1}\boldsymbol{\Sigma}_{xy}\mathbf{a}$. Note that these are the population analogs of (8.20) and (8.21).

8.13 Verify the expression in (8.23) for canonical correlation in terms of the partitioned correlation matrix.

8.14 Show that the canonical correlations between $\mathbf{y}^* = \mathbf{Fy} + \mathbf{f}$ and $\mathbf{x}^* = \mathbf{Gx} + \mathbf{g}$ are equal to those between $\mathbf{y}$ and $\mathbf{x}$, as in property 1 in Section 8.3.1.

8.15 Prove property 5 in Section 8.3.1.

8.16 Show that if $\mathbf{S}$ is positive definite, then $r_1 < 1$, thus verifying property 8 in Section 8.3.1.

8.17 **(a)** Show that the canonical variates u and v for $\mathbf{y}^* = \mathbf{F}\mathbf{y}$ and $\mathbf{x}^* = \mathbf{G}\mathbf{x}$ are equal to those for $\mathbf{y}$ and $\mathbf{x}$, as in property 1 in Section 8.3.2.

(b) For $\mathbf{y}^* = \mathbf{F}\mathbf{y}$ and $\mathbf{x}^* = \mathbf{G}\mathbf{x}$, show that the canonical variate coefficient vectors are related to those for $\mathbf{y}$ and $\mathbf{x}$ by $\mathbf{a}_i^* = (\mathbf{F}')^{-1}\mathbf{a}_i$ and $\mathbf{b}_i^* = (\mathbf{G}')^{-1}\mathbf{b}_i$, as in property 2 in Section 8.3.2.

8.18 Verify the maxima of the likelihood functions given in (8.33) and (8.34).

8.19 Verify the likelihood ratio in (8.35).

8.20 Show that (8.36) and (8.37) are equal.

8.21 Verify the alternative form of the likelihood ratio statistic in (8.38).

8.22 Show that Λ can be written in the form given in (8.40).

8.23 Show that the vector of sample correlations between $u_j = \mathbf{a}_j'\mathbf{y}$ and $y_1, y_2, \ldots, y_p$ is given by $\mathbf{R}_{yy}\mathbf{c}_j$, where $\mathbf{c}_j$ is scaled so that $\mathbf{c}_j'\mathbf{R}_{yy}\mathbf{c}_j = 1$, as noted in Section 8.6.3.

8.24 Supply the missing steps in the proof of Theorem 8.6A.

8.25 Use the biochemical data of Table 4.11 and ignore the grouping variable.

(a) Find the canonical correlations between $(y_1, y_2, \ldots, y_{11})$ and (x_1, x_2).

(b) Find the standardized coefficients for the canonical variates.

(c) Test the overall significance of the canonical correlations using all four test statistics in Section 8.4.1.

(d) Test the significance of each canonical correlation (Section 8.4.2).

8.26 Use the weight data in Table 4.9.

(a) Find the canonical correlations between (y_1, y_2, y_3) and (x_1, x_2).

(b) Find the standardized coefficients for the canonical variates.

(c) Test the overall significance of the canonical correlations using all four test statistics in Section 8.4.1.

(d) Test the significance of each canonical correlation (Section 8.4.2).

8.27 Use the diabetes data in Table 1.2.

(a) Find the canonical correlations between (y_1, y_2) and (x_1, x_2, x_3).

(b) Find the standardized coefficients for the canonical variates.

(c) Test the overall significance of the canonical correlations using all four test statistics in Section 8.4.1.

(d) Test the significance of each canonical correlation (Section 8.4.2).

8.28 Use the glucose data in Table 1.4.

(a) Find the canonical correlations between (y_1, y_2, y_3) and (x_1, x_2, x_3).

(b) Find the standardized coefficients for the canonical variates.

(c) Test the overall significance of the canonical correlations using all four test statistics in Section 8.4.1.

(d) Test the significance of each canonical correlation (Section 8.4.2).

8.29 Use the glucose data in Table 1.4 and the results of Problem 8.28.

(a) Show that property 3 of Section 8.3.1 holds.

(b) Illustrate property 4 of Section 8.3.1 by leaving out y_1. Then replace y_1 and leave out x_1.

(c) Show that property 5 of Section 8.3.1 holds for u_1 and v_1.

(d) Show that property 6 of Section 8.3.1 holds.

8.30 Using the glucose data in Table 1.4 and the results of Problem 8.28, compare the standardized coefficients and R^2 with the correlations between variables and the first pair of canonical variates u_1 and v_1 as in Example 8.6.3.

8.31 Consider the apple data of Table 4.10 and ignore the grouping into four treatments.

(a) Find the canonical correlations between $(y_1, y_2, \ldots, y_6)$ and (x_1, x_2).

(b) Find the standardized coefficients for the canonical variates.

(c) Test the overall significance of the canonical correlations using all four test statistics in Section 8.4.1.

(d) Test the significance of each canonical correlation (Section 8.4.2).

8.32 Use the rainfall data of Table 7.1.

(a) Find the canonical correlations between (y_1, y_2, y_3) and $(x_1, x_2, \ldots, x_6)$.

(b) Find the standardized coefficients for the canonical variates.

(c) Test the overall significance of the canonical correlations using all four test statistics in Section 8.4.1.

(d) Test the significance of each canonical correlation (Section 8.4.2).

8.33 Use the rainfall data in Table 7.1 and the results of Problem 8.32.

(a) Show that property 3 of Section 8.3.1 holds.

(b) Illustrate property 4 of Section 8.3.1 by leaving out y_1. Then replace y_1 and leave out x_1.

(c) Show that property 5 of Section 8.3.1 holds for u_1 and v_1.

(d) Show that property 6 of Section 8.3.1 holds.

CHAPTER 9

Principal Component Analysis

9.1 INTRODUCTION

In principal component analysis, the p original variables are transformed into linear combinations called *principal components*. This is a dimension reduction technique in which the goal is to find a few principal components that explain a large proportion of the total sample variance of the p variables in $\mathbf{y}$, $\sum_{j=1}^{p} s_j^2 = \text{tr}(\mathbf{S})$, where $\mathbf{S}$ is the sample covariance matrix for a random sample $\mathbf{y}_1, \mathbf{y}_2, \ldots, \mathbf{y}_n$. The transformed variables (principal components) are defined to be orthogonal (and uncorrelated). If the original variables are highly intercorrelated, a few principal components will suffice to reproduce most of the variation as measured by tr($\mathbf{S}$). If the original variables are only slightly intercorrelated, the resulting principal components will largely reflect the original variables, and little parsimony will result.

With an emphasis on modeling the variation in $\mathbf{y}_1, \mathbf{y}_2, \ldots, \mathbf{y}_n$, the principal components can be regarded as dimensions that maximally separate these individual observation vectors. Discriminant functions, on the other hand, are dimensions that maximally separate (the mean vectors of) groups of observations. Principal component analysis is essentially a one-sample procedure (see Section 9.9 for applications to grouped data), whereas discriminant analysis deals with two or more groups.

In other fields, such as physics and engineering, principal components may be called *empirical orthogonal functions* (Jolliffe 1986, page v in the preface) or the *Karhunen–Loeve expansion* (Watanabe 1967). Principal component analysis is sometimes confused with factor analysis. For comments on the distinction, see Section 10.1.

Principal components are often used as input to another analysis, such as regression or MANOVA. In situations in which there are too many variables, the variables can be replaced by a smaller set of principal components.

Principal components were first proposed by Pearson (1901) and further developed by Hotelling (1933). Comprehensive surveys of the field have been given by Jolliffe (1986), Jackson (1991), and Basilevsky (1994). Other reviews are by Rao (1964), Jackson (1980, 1981), Wold et al. (1987), and Dunteman (1989).

9.2 DEFINITION AND PROPERTIES OF PRINCIPAL COMPONENTS

9.2.1 Maximum Variance Property

The first principal component is defined as the linear combination with maximal sample variance among all linear combinations of the variables. For an observation vector **y** in a sample we seek

$$z = a_1 y_1 + a_2 y_2 + \cdots + a_p y_p = \mathbf{a}'\mathbf{y},$$

whose sample variance

$$s_z^2 = \mathbf{a}'\mathbf{S}\mathbf{a}$$

[see (1.73)] is a maximum, where **S** is the sample covariance matrix for a sample $\mathbf{y}_1$, $\mathbf{y}_2, \ldots, \mathbf{y}_n$.

Since s_z^2 has no maximum if the a_j's are allowed to increase without limit, we find the vector **a** that maximizes $\mathbf{a}'\mathbf{S}\mathbf{a}$ relative to the length of **a**. Thus we find the maximum of $\mathbf{a}'\mathbf{S}\mathbf{a}$ as the values of $a_1, a_2, \ldots, a_p$ vary relative to each other; that is, we find the direction of maximum spread of the points in p dimensions. The vector **a** that gives maximum variance is given in the following theorem.

Theorem 9.2A. The vector **a** such that $z = \mathbf{a}'\mathbf{y}$ has maximum sample variance $s_z^2 = \mathbf{a}'\mathbf{S}\mathbf{a}$ (relative to the squared length $\mathbf{a}'\mathbf{a}$) is given by $\mathbf{a}_1$, the first eigenvector of **S**. The maximum value of $\mathbf{a}'\mathbf{S}\mathbf{a}/\mathbf{a}'\mathbf{a}$ is given by λ_1, the first (largest) eigenvalue of **S**.

Proof. Using (A.13.3), we differentiate the ratio

$$\lambda = \frac{\mathbf{a}'\mathbf{S}\mathbf{a}}{\mathbf{a}'\mathbf{a}} \tag{9.1}$$

with respect to **a** and set the result equal to **0** to obtain

$$\frac{\partial \lambda}{\partial \mathbf{a}} = \frac{\mathbf{a}'\mathbf{a}(2\mathbf{S}\mathbf{a}) - \mathbf{a}'\mathbf{S}\mathbf{a}(2\mathbf{a})}{(\mathbf{a}'\mathbf{a})^2} = \mathbf{0}, \quad \text{or} \quad \mathbf{S}\mathbf{a} - \frac{\mathbf{a}'\mathbf{S}\mathbf{a}}{\mathbf{a}'\mathbf{a}}\mathbf{a} = \mathbf{0}.$$

Substitution of $\lambda = \mathbf{a}'\mathbf{S}\mathbf{a}/\mathbf{a}'\mathbf{a}$ from (9.1) yields the expression

$$\mathbf{S}\mathbf{a} - \lambda\mathbf{a} = \mathbf{0}, \quad \text{or} \quad (\mathbf{S} - \lambda\mathbf{I})\mathbf{a} = \mathbf{0}. \tag{9.2}$$

By (A.10.2), λ is an eigenvalue of **S**. Therefore, the maximum value of $\lambda = \mathbf{a}'\mathbf{S}\mathbf{a}/\mathbf{a}'\mathbf{a}$ is given by λ_1, the first (largest) eigenvalue of **S**, and the coefficient vector in $z = \mathbf{a}'\mathbf{y}$ that produces the maximum is the eigenvector $\mathbf{a}_1$ corresponding to λ_1. □

Note that $\mathbf{a}'\mathbf{S}\mathbf{a}/\mathbf{a}'\mathbf{a}$ can be written as

$$\frac{\mathbf{a}'\mathbf{S}\mathbf{a}}{\mathbf{a}'\mathbf{a}} = \mathbf{b}'\mathbf{S}\mathbf{b},$$

where $\mathbf{b} = \mathbf{a}/\sqrt{\mathbf{a}'\mathbf{a}}$ is such that $\mathbf{b}'\mathbf{b} = 1$. Thus, maximizing $\mathbf{a}'\mathbf{S}\mathbf{a}/\mathbf{a}'\mathbf{a}$ is equivalent to maximizing $\mathbf{b}'\mathbf{S}\mathbf{b}$ subject to $\mathbf{b}'\mathbf{b} = 1$, which could be carried out using a Lagrange multiplier to incorporate the constraint.

Since $\mathbf{S}$ has p eigenvectors, we have p principal components:

$$\begin{aligned} z_1 &= \mathbf{a}_1'\mathbf{y}, \\ z_2 &= \mathbf{a}_2'\mathbf{y}, \\ &\vdots \\ z_p &= \mathbf{a}_p'\mathbf{y}. \end{aligned} \tag{9.3}$$

The p principal components in (9.3) can be written in the form

$$\mathbf{z} = \mathbf{A}\mathbf{y}, \tag{9.4}$$

where

$$\mathbf{z} = \begin{pmatrix} z_1 \\ z_2 \\ \vdots \\ z_p \end{pmatrix} \quad \text{and} \quad \mathbf{A} = \begin{pmatrix} \mathbf{a}_1' \\ \mathbf{a}_2' \\ \vdots \\ \mathbf{a}_p' \end{pmatrix}.$$

The z_j's are orthogonal; that is, $\mathbf{a}_j'\mathbf{a}_k = 0$ for all $j \neq k$, because the $\mathbf{a}_j$'s are eigenvectors of the symmetric matrix $\mathbf{S}$.

The last principal component, $z_p = \mathbf{a}_p'\mathbf{y}$, represents the direction of minimum sample variance, and the corresponding minimum value of $\mathbf{a}'\mathbf{S}\mathbf{a}/\mathbf{a}'\mathbf{a}$ is given by λ_p, the smallest eigenvalue. Thus $z_1 = \mathbf{a}_1'\mathbf{y}$ has the largest variance and $z_p = \mathbf{a}_p'\mathbf{y}$ has the smallest. (We assume, as usual, that $\lambda_1 > \lambda_2 > \cdots > \lambda_p$.)

A more precise description of the relationship among the p principal components is given in the following theorem.

Theorem 9.2B. If the eigenvectors $\mathbf{a}_1, \mathbf{a}_2, \ldots, \mathbf{a}_p$ of $\mathbf{S}$ are scaled so that $\mathbf{a}_j'\mathbf{a}_j = 1$, then the first principal component, $z_1 = \mathbf{a}_1'\mathbf{y}$, has the largest sample variance of any linear combination, and the second principal component, $z_2 = \mathbf{a}_2'\mathbf{y}$, has the largest sample variance achievable by any linear combination such that $\mathbf{a}_1'\mathbf{a}_2 = 0$. In general, $z_j = \mathbf{a}_j\mathbf{y}$ has the largest sample variance among all linear combinations orthogonal to $z_1, z_2, \ldots, z_{j-1}$.

Proof. By Theorem 9.2A, the maximum value of $\mathbf{a}'\mathbf{S}\mathbf{a}/\mathbf{a}'\mathbf{a}$ is λ_1, the largest eigenvalue of $\mathbf{S}$, and the vector $\mathbf{a}$ that maximizes $\mathbf{a}'\mathbf{S}\mathbf{a}$ is $\mathbf{a}_1$, the accompanying eigenvector. To show that the second eigenvector $\mathbf{a}_2$ maximizes $\lambda_2 = \mathbf{a}_2'\mathbf{S}\mathbf{a}_2/\mathbf{a}_2'\mathbf{a}_2$ subject to $\mathbf{a}_1'\mathbf{a}_2 = 0$, we use a Lagrange multiplier γ and differentiate with respect to $\mathbf{a}_2$:

$$\frac{\partial}{\partial \mathbf{a}_2}\left(\frac{\mathbf{a}_2'\mathbf{S}\mathbf{a}_2}{\mathbf{a}_2'\mathbf{a}_2} + \gamma \mathbf{a}_1'\mathbf{a}_2\right) = \mathbf{0}.$$

This leads to $\gamma = 0$ and $\mathbf{S}\mathbf{a}_2 = \lambda_2\mathbf{a}_2$ (see Problem 9.2), so that λ_2 is the second eigenvalue of $\mathbf{S}$ and $\mathbf{a}_2$ is the corresponding eigenvector. The analogous properties of $z_3, \ldots, z_p$ follow similarly. □

9.2.2 Principal Components as Projections

As linear combinations of the original variables, principal components represent projections. The first principal component can be regarded as a line onto which the points $\mathbf{y}_1, \mathbf{y}_2, \ldots, \mathbf{y}_n$ are projected orthogonally (perpendicularly). This line is the "best line" in the sense that it is in the direction in which the sum of squared projections is maximum. In addition, the sum of squared orthogonal distances from the points to the line is minimized (Jolliffe 1986, pp. 27–29). The first and second principal components form the "best" fitting plane in a similar sense; the sum of squared orthogonal distances from the points to the plane is minimized. Thus $z_{1i} = \mathbf{a}_1'\mathbf{y}_i$ and $z_{2i} = \mathbf{a}_2'\mathbf{y}_i$ can be evaluated for $i = 1, 2, \ldots, n$ and plotted. The resulting two-dimensional plot (projection) shows the maximum spread in the swarm of points. The plot will show any prominent patterns in the data, such as groupings of points, outliers, or lack of normality.

Since the first principal component is the linear combination with maximum variance and the n observation vectors are projected orthogonally onto this line, the projections $z_{1i} = \mathbf{a}_1'\mathbf{y}_i$, $i = 1, 2, \ldots, n$, can be used to rank the observations. The n observations are more spread out on this dimension than on any other, and a ranking based on all p variables is obtained. Dawkins (1989) illustrated this technique by ranking nations based on their track records for seven events.

Other linear combinations considered previously had similar projection properties. For example, the first two discriminant functions form a plane on which the points can be projected to maximally separate groups.

Additional projection techniques have been proposed. *Projection pursuit*, for example, seeks projections (by iterative methods) that reveal features of the data such as outliers or clusters of points (Friedman and Tukey 1974, Huber 1985, Sibson 1984, Jones and Sibson 1987, Yenyukov 1988, Posse 1990, Nason 1995). Projection pursuit has been compared to principal component analysis by Glover and Hopke (1992).

Caussinus and Ruiz (1990) proposed a projection method for selecting clusters or outliers based on "generalized" principal components involving eigenvalues and eigenvectors of $\mathbf{SM}$, where $\mathbf{S}$ is the sample covariance matrix and $\mathbf{M}$ is a positive definite matrix designed to steer the projection in certain directions. Their method does not require an iterative solution and is thereby computationally simpler than projection pursuit.

9.2.3 Properties of Principal Components

We will now discuss some of the properties of principal components.

1. If the eigenvector $\mathbf{a}_1$ is scaled so that $\mathbf{a}_1'\mathbf{a}_1 = 1$, then from (9.1) and (1.73),

$$\lambda_1 = \frac{\mathbf{a}_1'\mathbf{S}\mathbf{a}_1}{\mathbf{a}_1'\mathbf{a}_1} = \mathbf{a}_1'\mathbf{S}\mathbf{a}_1 = s_{z_1}^2. \tag{9.5}$$

Hence the eigenvalue λ_1 is the sample variance of the first principal component, $z_1 = \mathbf{a}_1'\mathbf{y}$. Similarly, the sample variance of each of the other principal components is equal to its corresponding eigenvalue:

$$s_{z_j}^2 = \mathbf{a}_j'\mathbf{S}\mathbf{a}_j = \lambda_j, \qquad j = 2, 3, \ldots, p. \tag{9.6}$$

2. From (9.6) and (A.10.6), the total sample variance of the components is equal to the total sample variance of the variables:

$$\sum_{j=1}^{p} s_{z_j}^2 = \sum_{j=1}^{p} \lambda_j = \text{tr}(\mathbf{S}). \tag{9.7}$$

Hence in a principal component analysis, we can report "the percent of variance explained" by a particular subset of components; for example,

$$\frac{\sum_{j=1}^{k} \lambda_j}{\sum_{j=1}^{p} \lambda_j} \tag{9.8}$$

gives the proportion of total sample variance accounted for by the first k principal components. This terminology is frequently misapplied to discriminant functions and canonical variates, even though the eigenvalues are not variances of those linear functions, as they are for principal components.

3. In addition to explaining a portion of the total variance, tr(**S**), the first k principal components also indirectly explain the covariances, as can be seen by considering the spectral decomposition of **S** from (A.11.4),

$$\mathbf{S} = \lambda_1\mathbf{a}_1\mathbf{a}_1' + \lambda_2\mathbf{a}_2\mathbf{a}_2' + \cdots + \lambda_p\mathbf{a}_p\mathbf{a}_p'. \tag{9.9}$$

The first k terms of (9.9) provide an approximation to **S**,

$$\tilde{\mathbf{S}} = \lambda_1\mathbf{a}_1\mathbf{a}_1' + \cdots + \lambda_k\mathbf{a}_k\mathbf{a}_k', \tag{9.10}$$

which is the best approximation of rank k in the sense that the sum of squares of the elements of $\mathbf{S} - \tilde{\mathbf{S}}$ is a minimum (Rao 1964).

4. By (A.11.1), the eigenvectors of a symmetric matrix such as **S** are orthogonal: $\mathbf{a}_j'\mathbf{a}_k = 0$ for $j \neq k$. The orthogonality of the principal components also implies that they are uncorrelated (in the sample):

$$s_{z_j z_k} = \mathbf{a}_j'\mathbf{S}\mathbf{a}_k = 0 \quad \text{for } j \neq k. \tag{9.11}$$

5. Principal components are not scale invariant; the eigenvalues and eigenvectors of the original variables do not transform directly to the eigenvalues and eigenvectors of scaled variables. To show this, suppose $\mathbf{y}$ is scaled by transforming to $\mathbf{u} = \mathbf{Cy}$, where $\mathbf{C}$ is nonsingular. Then by (1.77), the sample covariance matrix of $\mathbf{u}$ is $\mathbf{CSC}'$. A principal component of $\mathbf{CSC}'$ has the form $\mathbf{b}_j'\mathbf{u} = \mathbf{b}_j'\mathbf{Cy}$, where $\mathbf{b}_j$ is an eigenvector of $\mathbf{CSC}'$. But $\mathbf{b}_j'\mathbf{Cy}$ does not equal $\mathbf{a}_j'\mathbf{y}$, a principal component of $\mathbf{S}$, because $\mathbf{b}_j = (\mathbf{C}')^{-1}\mathbf{a}_j$ is not an eigenvector of $\mathbf{CSC}'$.

To see that scaling has an effect on the eigenvalues, note that by (9.7)

$$\sum_{j=1}^{p} \lambda_j = \text{tr}(\mathbf{S}) = \sum_{j=1}^{p} s_j^2, \tag{9.12}$$

where s_j^2 is the sample variance of y_j. If an s_j^2 is increased by a scale change, then one or more of the λ_j's will increase.

Due to the lack of scale invariance of principal components, it would be preferable for the variables to be commensurate, that is, similar in measurement scale and in variance. Otherwise, a change in scale of one of the variables, for example, from feet to inches, will change the pattern of both the eigenvalues and eigenvectors. If the variables are not commensurate, we could use the correlation matrix $\mathbf{R}$ instead of $\mathbf{S}$ (see Section 9.4).

6. There are two situations in which a principal component essentially duplicates one of the original variables. If one variable has a much greater variance than the other variables, either naturally or by scaling, then this variable will dominate the first principal component, which will in turn account for most of the total variance, tr($\mathbf{S}$). In this state of affairs, the use of $\mathbf{R}$ would better reflect the mutual effect of the variables (see Section 9.4).

The other situation in which a principal component duplicates a variable is as follows. If one variable, say y_p, is uncorrelated with the other variables, then its variance s_p^2 is one of the eigenvalues of $\mathbf{S}$, and the corresponding eigenvector is of the form $(0, \ldots, 0, 1)'$. Hence y_p itself is the principal component. In general, if all p variables are uncorrelated, the principal components merely reproduce the variables. Thus, if the correlations are all small, there is little to be gained with a principal component analysis.

7. Since there is no inverse or determinant involved in extracting principal components, $\mathbf{S}$ can be singular, in which case some of the eigenvalues will be zero. If the rank of $\mathbf{S}$ is $r < p$, then the total variance $\sum_{j=1}^{p} s_j^2$ can be completely explained by the first r principal components or partially explained by fewer than r principal components.

This property of principal components can be exploited in applications such as regression or MANOVA when the number of variables exceeds the sample size and the test statistic cannot be computed. In such cases, we could replace each observation vector with a small number of principal components and proceed with the test.

8. The first k principal components are the best predictors of the original variables in the following sense. The first principal component is the linear function $z = \mathbf{a}_1'\mathbf{y}$ whose average squared correlation $\sum_{j=1}^{p} r_{z_1 y_j}^2/p$ with each of the y's is maximum. The second principal component $z_2 = \mathbf{a}_2'\mathbf{y}$ is the best linear function for predicting the residuals of the y's after regression on the first principal component, and so on (Schervish 1986, 1987).

9.3 PRINCIPAL COMPONENTS AS A ROTATION OF AXES

If we write $\mathbf{z}_i = \mathbf{A}\mathbf{y}_i$ as in (9.4) for each of $i = 1, 2, \ldots, n$, where the rows of the matrix $\mathbf{A}$ are the p eigenvectors of $\mathbf{S}$, then, by (A.11.1), $\mathbf{A}$ is orthogonal and $\mathbf{z}_i'\mathbf{z}_i = \mathbf{y}_i'\mathbf{A}'\mathbf{A}\mathbf{y}_i = \mathbf{y}_i'\mathbf{y}_i$. Thus distances from the origin are preserved, and principal components represent a rotation of axes. This can be illustrated for $p = 2$ using the constant density elliptical contours of the bivariate normal distribution shown in Figure 2.2, which is repeated as Figure 9.1. The axes are shifted and rotated to become the axes of the ellipse. To plot the axes, we can use the methods of Section 2.1.3, with $\bar{\mathbf{y}}$ and $\mathbf{S}$ substituted for $\boldsymbol{\mu}$ and $\boldsymbol{\Sigma}$. The lengths of the axes are proportional to $\sqrt{\lambda_j}$, and the directions are given by the eigenvectors. (Note that a is a constant of proportionality, not an element of $\mathbf{a}_1$. For the value of a, see Section 2.1.3.) To construct the major axis, we plot a point corresponding to the first eigenvector $\mathbf{a}_1$, using $\bar{\mathbf{y}}$ for the origin, and draw a line through $\bar{\mathbf{y}}$ and $\mathbf{a}_1$. The other axis is similarly

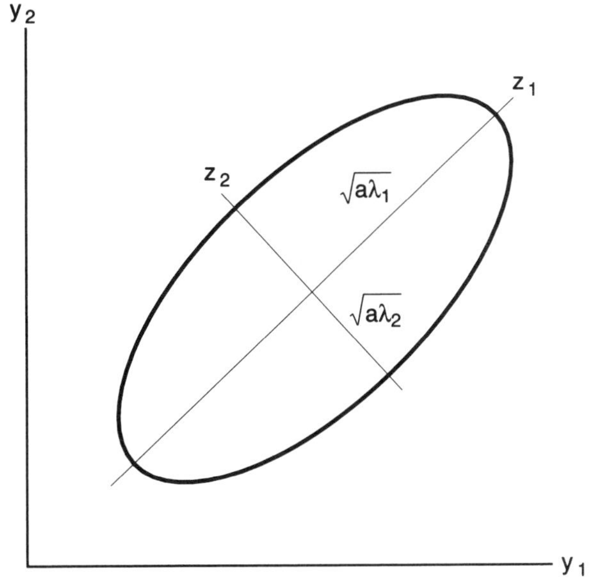

Figure 9.1. Constant density contour for bivariate normal.

established using $\mathbf{a}_2$. The axes will be perpendicular because $\mathbf{a}_1$ and $\mathbf{a}_2$ are orthogonal.

The same axes can be obtained without reference to the elliptical contour by rotating (using $\overline{\mathbf{y}}$ as origin) until the new axes represent uncorrelated dimensions, in which case the major axis will line up with the greatest natural extension (maximum variance) of the swarm of points. Thus we seek an orthogonal transformation $\mathbf{z} = \mathbf{A}\mathbf{y}$ so that the sample covariance matrix of the z's is diagonal:

$$\widehat{\text{cov}}(\mathbf{z}) = \mathbf{A}\mathbf{S}\mathbf{A}' = \begin{pmatrix} s_{z_1}^2 & 0 \\ 0 & s_{z_2}^2 \end{pmatrix}.$$

By (A.11.6), the orthogonal matrix $\mathbf{A}$ that diagonalizes $\mathbf{S}$ is the matrix of normalized eigenvectors of $\mathbf{S}$, and the resulting diagonal elements are eigenvalues of $\mathbf{S}$:

$$s_{z_j}^2 = \lambda_j.$$

The equivalence of principal components to a rotation extends immediately to p dimensions. The rotated axes have the projection properties noted in Section 9.2.2.

9.4 PRINCIPAL COMPONENTS FROM THE CORRELATION MATRIX

The principal components of $\mathbf{S}$ are preferred to those of $\mathbf{R}$ for many purposes (Naik and Khattree 1996), such as for use in further computations (see Section 9.8 for a regression application). However, as suggested in property 5 in Section 9.2.3, we may wish to use $\mathbf{R}$ in situations in which the variances differ widely. In such a case, the principal components from $\mathbf{R}$ will be more interpretable in the sense that all variables can contribute more evenly.

We will now discuss some of the properties of principal components extracted from $\mathbf{R}$.

1. Principal components from $\mathbf{R}$ differ from those obtained from $\mathbf{S}$ because of the lack of scale invariance of components from $\mathbf{S}$ discussed in property 5 of Section 9.2.3. If we express the components from $\mathbf{R}$ in terms of the original variables, they will not be the same as the components from $\mathbf{S}$.

2. If the components of $\mathbf{R}$ are expressed in terms of original variables, they will not be orthogonal. This lack of orthogonality can easily be demonstrated. Let $\mathbf{D}_s^{-1}(\mathbf{y} - \overline{\mathbf{y}})$ be the vector of standardized variables, where $\mathbf{D}_s = \text{diag}(s_1, \ldots, s_p)$ is the diagonal matrix of standard deviations. Then the principal components from $\mathbf{R}$ can be written as

$$z_j^* = \mathbf{b}_j'\mathbf{D}_s^{-1}(\mathbf{y} - \overline{\mathbf{y}}), \qquad j = 1, 2, \ldots, p, \tag{9.13}$$

where $\mathbf{b}_j$ is the jth eigenvector of $\mathbf{R}$. The coefficient vectors of $\mathbf{y}$ in (9.13) are not orthogonal because for $j \neq k$

$$\mathbf{b}_j'\mathbf{D}_s^{-1}(\mathbf{D}_s')^{-1}\mathbf{b}_k \neq 0.$$

The product $\mathbf{b}_j'\mathbf{D}_s^{-1}(\mathbf{D}_s')^{-1}\mathbf{b}_k$ would be zero only if all sample standard deviations s_1, $s_2, \ldots, s_p$ were equal to s, so that $\mathbf{D}_s = s\mathbf{I}$.

3. The percent of variance accounted for by the components of $\mathbf{R}$ [see (9.8)] will not agree with the analogous percent for $\mathbf{S}$. For $\mathbf{R}$ the sum of the eigenvalues is p, since $\sum_{j=1}^{p} \lambda_j = \text{tr}(\mathbf{R}) = p$, and the proportion of variance due to the first k components is

$$\frac{\lambda_1 + \cdots + \lambda_k}{\lambda_1 + \cdots + \lambda_p} = \frac{\lambda_1 + \cdots + \lambda_k}{p}. \tag{9.14}$$

4. Principal components extracted from $\mathbf{R}$ are scale invariant, since a change of scale of the variables does not affect the r_{ij}'s in $\mathbf{R}$.

5. The components from $\mathbf{R}$ do not change if we multiply the off-diagonal elements of $\mathbf{R}$ by any constant k such that $-1 \leq kr_{ij} \leq 1$ for all $i \neq j$, as in

$$\mathbf{R}_k = \begin{pmatrix} 1 & kr_{12} & \cdots & kr_{1p} \\ kr_{21} & 1 & \cdots & kr_{2p} \\ \vdots & \vdots & & \vdots \\ kr_{p1} & kr_{p2} & \cdots & 1 \end{pmatrix}. \tag{9.15}$$

To show this invariance to multiplication by k, we express $\mathbf{R}_k$ in terms of $\mathbf{R}$:

$$\mathbf{R}_k = k\mathbf{R} - (k-1)\mathbf{I}. \tag{9.16}$$

The eigenvalues and eigenvectors of $\mathbf{R}_k$ are defined by

$$(\mathbf{R}_k - \lambda_k \mathbf{I})\mathbf{a}_k = \mathbf{0},$$

which by (9.16) becomes

$$[k\mathbf{R} - (k-1)\mathbf{I} - \lambda_k \mathbf{I}]\,\mathbf{a}_k = \mathbf{0}$$

or

$$\left[\mathbf{R} - \left(\frac{k - 1 + \lambda_k}{k}\right)\mathbf{I}\right]\mathbf{a}_k = \mathbf{0}. \tag{9.17}$$

If we compare (9.17) with $(\mathbf{R} - \lambda\mathbf{I})\mathbf{a} = \mathbf{0}$, we see that the eigenvectors $\mathbf{a}_k$ and $\mathbf{a}$ are the same and that the eigenvalues are related by

$$\lambda = 1 + \frac{\lambda_k - 1}{k}. \tag{9.18}$$

Hence the components of a given $\mathbf{R}$ matrix will serve for other $\mathbf{R}$ matrices of the form given in (9.15), and the statement that a component accounts for a given percent of total sample variance is less meaningful for $\mathbf{R}$ than it is for $\mathbf{S}$.

Example 9.4(a). We illustrate properties 1, 2, and 3 in this section and property 6 in Section 9.2.3. Consider a bivariate case with

$$\mathbf{S} = \begin{pmatrix} 1 & 18 \\ 18 & 400 \end{pmatrix}, \qquad \mathbf{R} = \begin{pmatrix} 1 & .9 \\ .9 & 1 \end{pmatrix}.$$

For $\mathbf{S}$, the eigenvalues and eigenvectors are

$$\lambda_1 = 400.81, \qquad \mathbf{a}_1' = (.045, .998),$$
$$\lambda_2 = .19, \qquad \mathbf{a}_2' = (.998, -.045).$$

Because the variance of y_2 is 400 times as large as the variance of y_1, the first principal component of $\mathbf{S}$, $z_1 = .045y_1 + .998y_2$, accounts for virtually all of the total variance,

$$\frac{\lambda_1}{\lambda_1 + \lambda_2} = \frac{400.81}{401} = .9995,$$

and the coefficient of y_2 in z_1 eclipses that of y_1. This illustrates property 6 in Section 9.2.3.

For $\mathbf{R}$, we have eigenvalues and eigenvectors

$$\lambda_1 = 1.9, \qquad \mathbf{a}_1' = (.707, .707),$$
$$\lambda_2 = .1, \qquad \mathbf{a}_2' = (.707, -.707).$$

In the bivariate case, $\mathbf{a}_1$ and $\mathbf{a}_2$ remain the same for all possible $\mathbf{R}$ matrices, but λ_1 and λ_2 depend on r_{12}. For this example, in which $r_{12} = .9$, the first principal component $z_1 = .707(y_1 - \overline{y}_1)/s_1 + .707(y_2 - \overline{y}_2)/s_2$ accounts for a high proportion of variance,

$$\frac{\lambda_1}{\lambda_1 + \lambda_2} = \frac{1.9}{2} = .95.$$

A comparison of $\lambda_1/(\lambda_1 + \lambda_2) = .95$ for $\mathbf{R}$ with $\lambda_1/(\lambda_1 + \lambda_2) = .9995$ for $\mathbf{S}$ (above) illustrates property 3.

To illustrate properties 1 and 2, we express the components of $\mathbf{R}$ in terms of the original variables:

$$z_1 = .707\left(\frac{y_1 - \overline{y}_1}{1}\right) + .707\left(\frac{y_2 - \overline{y}_2}{20}\right)$$
$$= .707y_1 + .035y_2 - .707\left(\overline{y}_1 + \frac{\overline{y}_2}{20}\right),$$
$$z_2 = .707y_1 - .035y_2 - .707\left(\overline{y}_1 - \frac{\overline{y}_2}{20}\right).$$

These differ from the components of $\mathbf{S}$ and are not orthogonal:

$$.707(.707) + .035(-.035) \neq 0.$$

Example 9.4(b). To illustrate property 5 in this section, consider the following two correlation matrices:

$$\mathbf{R}_1 = \begin{pmatrix} 1 & .9 & .5 \\ .9 & 1 & .1 \\ .5 & .1 & 1 \end{pmatrix}, \qquad \mathbf{R}_2 = \begin{pmatrix} 1 & .0009 & .0005 \\ .0009 & 1 & .0001 \\ .0005 & .0001 & 1 \end{pmatrix}.$$

The components are the same for both $\mathbf{R}_1$ and $\mathbf{R}_2$:

$$\begin{aligned} z_1 &= .69y_1^* + .61y_2^* + .38y_3^*, \\ z_2 &= -.05y_1^* - .48y_2^* + .88y_3^*, \\ z_3 &= -.72y_1^* + .62y_2^* + .30y_3^*, \end{aligned}$$

where y_j^* is the standardized variable $y_j^* = (y_j - \overline{y}_j)/s_j$. The eigenvalues are as follows:

$$\begin{aligned} \mathbf{R}_1 &: 2.074, .915, .011, \\ \mathbf{R}_2 &: 1.0011, .9999, .9990. \end{aligned}$$

Thus z_1 accounts for 69.1% of the variance in $\mathbf{R}_1$ and for 33.4% in $\mathbf{R}_2$, while z_3 accounts for 0.351% of the variance in $\mathbf{R}_1$ and 33.3% in $\mathbf{R}_2$. The first two components of $\mathbf{R}_1$ account for 99.6% of the variance, but the first two components of $\mathbf{R}_2$ account for only 66.7% of the variance. Since the three eigenvalues of $\mathbf{R}_2$ are essentially equal, none of the three components from $\mathbf{R}_2$ can be neglected.

9.5 METHODS FOR DISCARDING COMPONENTS

When carrying out a principal component analysis, the researcher must decide how many components to use to represent the data; the other components will be discarded. The four most common guidelines for discarding components are discussed in Sections 9.5.1–9.5.4; additional methods are described in Section 9.5.5. Fava and Velicer (1992) studied the effects of using too many components. Note that before discarding components, we may wish to examine the "smallest" one for the information it carries (see Section 9.6).

9.5.1 Percent of Variance

The number of components to retain could be based on the percent of variance accounted for, as in (9.8). We would like to achieve a relatively high percentage, say 70–90%. If the reduction in dimensionality is for the purpose of input into some other analysis such as regression or MANOVA, it may be advisable to keep more components than if we are merely aiming for descriptive simplification.

9.5.2 Average Eigenvalue

We could retain those components whose eigenvalues are greater than the average eigenvalue, $\overline{\lambda} = \sum_{j=1}^{p} \lambda_j / p$, which is also the average variance of the variables, since $\sum_j \lambda_j = \text{tr}(\mathbf{S})$. For a correlation matrix, $\overline{\lambda} = 1$.

The average eigenvalue method often works well in practice. When this method errs, it is likely to be on the side of retaining too many components. Cattell and Jaspers (1967), Browne (1968), and Linn (1968) have studied the performance of this criterion in situations where the true dimensionality is known. They found the method to be fairly accurate when the number of variables is ≤ 30 and the variables are rather highly correlated. For larger numbers of variables that are not as highly correlated, the technique tends to overestimate the number of components.

9.5.3 Scree Graph

We could plot the eigenvalues in an attempt to find a visual break between the "large" eigenvalues and the "small" eigenvalues. This plot is called a *scree graph*. The term *scree*, suggested by Cattell (1966), refers to the geological term for the debris at the bottom of a rocky cliff.

An ideal scree graph is shown in Figure 9.2, in which it is easy to distinguish the large eigenvalues from the small ones. The first two eigenvalues form a steep curve; the remaining eigenvalues exhibit a linear trend with small slope. In such a case, it is

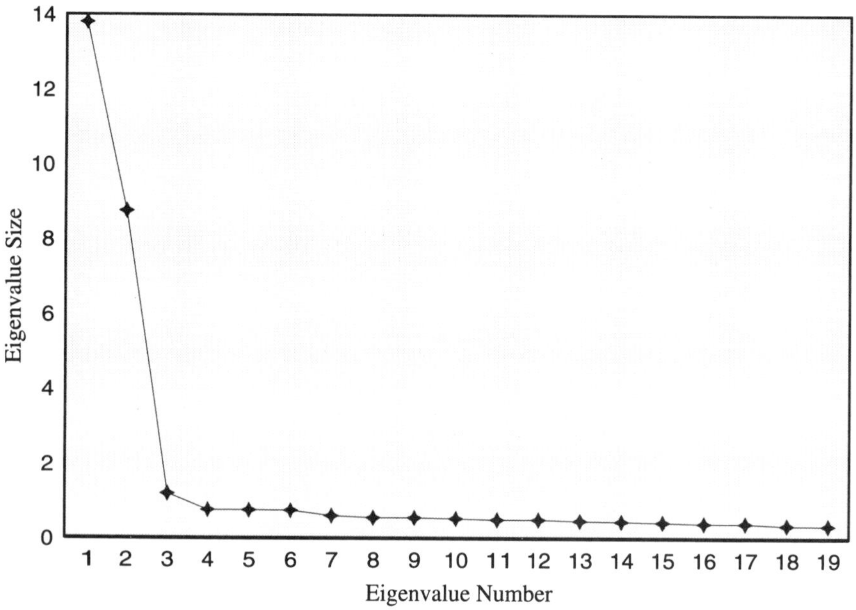

Figure 9.2. Ideal scree graph.

clear that we should delete the components corresponding to the small eigenvalues on the straight line. In practice, this ideal pattern may not appear, and this approach may not be conclusive.

The accuracy of the scree method in choosing the correct number of components has been investigated in several studies. Cattell and Jaspers (1967) found it to give the correct number in 6 out of 8 cases. Linn (1968) found it to be correct in 7 of 10 cases, and Tucker et al. (1969) found it to be accurate in 12 of 18 cases. Hakstian et al. (1982), comparing the average eigenvalue method and the scree method, found both to be accurate when $n > 250$ and the variables are at least moderately intercorrelated. When the correlations were smaller so that more components are needed, both methods were less accurate, the average eigenvalue method performing slightly better than the scree method.

9.5.4 Significance Tests

We could test the significance of the components associated with the larger eigenvalues. The tests assume a random sample from a multivariate normal distribution. (Note that principal components provide valuable descriptive information in many cases where the variables are not normal or are not continuous.)

Before giving a test to determine the number of components to retain, we first discuss some general tests and confidence intervals for the eigenvalues and eigenvectors associated with principal components. Most of these tests are asymptotic (large-sample) results. Generally, the true p-values for small samples will be larger than those shown by these large-sample tests, although in most cases it is not known how far the p-values or confidence levels are distorted for small samples. Most of these tests are applied to **S**; fewer usable results are available for **R**. These results were initially obtained by Girshick (1936, 1939), Anderson (1951, 1963), Bartlett (1954), and Lawley (1956, 1963). For additional details and discussion, see Mardia, Kent, and Bibby (1979, Sections 8.3 and 8.4), Srivastava and Khatri (1979, Sections 9.3–9.7), Anderson (1984, Sections 11.6 and 11.7), and Jolliffe (1986, Section 3.7).

We denote the (population) eigenvalues and eigenvectors of $\boldsymbol{\Sigma}$ by $\gamma_1, \gamma_2, \ldots, \gamma_p$ and $\boldsymbol{\alpha}_1, \boldsymbol{\alpha}_2, \ldots, \boldsymbol{\alpha}_p$, respectively, where $\boldsymbol{\alpha}_j'\boldsymbol{\alpha}_j = 1$. The tests use the sample eigenvalues and eigenvectors $\lambda_1, \lambda_2, \ldots, \lambda_p$ and $\mathbf{a}_1, \mathbf{a}_2, \ldots, \mathbf{a}_p$ from **S**. To test the hypothesis that an eigenvalue has a particular value, H_0: $\gamma_j = \gamma_{j0}$ versus H_1: $\gamma_j \neq \gamma_{j0}$, we compute

$$w = \frac{\lambda_j - \gamma_{j0}}{\gamma_{j0}\sqrt{2/(n-1)}} \tag{9.19}$$

and reject H_0 if $|w| > w_{\alpha/2}$, where $w_{\alpha/2}$ is the upper $100\alpha/2$ percentage point of the standard normal distribution. The corresponding $1 - \alpha$ confidence interval for γ_j is

$$\frac{\lambda_j}{1 + w_{\alpha/2}\sqrt{2/(n-1)}} \leq \gamma_j \leq \frac{\lambda_j}{1 - w_{\alpha/2}\sqrt{2/(n-1)}}. \tag{9.20}$$

To test the hypothesis that an eigenvector of $\mathbf{\Sigma}$ has a particular value, H_0: $\boldsymbol{\alpha}_j = \boldsymbol{\alpha}_{j0}$ versus H_1: $\boldsymbol{\alpha}_j \neq \boldsymbol{\alpha}_{j0}$, we calculate

$$u = (n-1)\boldsymbol{\alpha}'_{j0}(\lambda_j \mathbf{S}^{-1} + \mathbf{S}/\lambda_j - 2\mathbf{I})\boldsymbol{\alpha}_{j0}, \tag{9.21}$$

which has an approximate χ^2-distribution with $p-1$ degrees of freedom. We reject H_0 if $u \geq \chi^2_{\alpha,p-1}$. This could be used, for example, to test the hypothesis that a sample eigenvector does not differ significantly from some simplified form. Schott (1987) gave an improved χ^2-test of H_0: $\boldsymbol{\alpha}_j = \boldsymbol{\alpha}_{j0}$. Schott (1991) provided an approximate χ^2-test and a robust test of the hypothesis that an eigenvector of the population correlation matrix $\mathbf{P}_\rho$ has a specified form.

An asymptotic estimate of the standard deviation of a_{jk}, the kth element of the jth eigenvector $\mathbf{a}_j$, was given by Anderson (1963) as

$$s(a_{jk}) = \left[\frac{1}{n} \lambda_j \sum_{\substack{r=1 \\ r \neq j}}^{p} \frac{\lambda_r}{\lambda_r - \lambda_j} a_{rk}^2 \right]^{1/2}.$$

The values of $s(a_{jk})$ are helpful in judging the significance of the coefficients in the principal components.

We will now discuss a test to determine the number of components to retain. To express the notion that the last k components are not significant descriptors of dimensionality, we use the hypothesis H_{0k}: $\gamma_{p-k+1} = \gamma_{p-k+2} = \cdots = \gamma_p$. Under H_{0k}, the sample eigenvalues would tend to have the straight-line pattern in Figure 9.2. The hypothesis of equality of the last few eigenvalues is more plausible than is a hypothesis that they are zero. Unless one or more variables are linear combinations of others, $\mathbf{\Sigma}$ will ordinarily be nonsingular. Thus all γ_j's are positive, and hypothesizing that $\gamma_{p-k+1} = \gamma_{p-k+2} = \cdots = \gamma_p = 0$ does not represent a realistic simplification.

If H_{0k}: $\gamma_{p-k+1} = \gamma_{p-k+2} = \cdots = \gamma_p$ is true, the basic dimensionality of the swarm of points is $p-k$, since the last k (population) principal components are not uniquely defined and their variances are not maximized in any particular direction. Geometrically, the eigenvectors associated with equal eigenvalues represent axes of a circle, sphere, or hypersphere. For example, if $p = 3$ and $\lambda_2 = \lambda_3$, then the configuration of points is that of a rod with a circular cross section. The first dimension is unique but the other two are arbitrary (apart from being orthogonal).

To test H_{0k}: $\gamma_{p-k+1} = \cdots = \gamma_p$, we use the test statistic

$$u = \left(n - \frac{2p+11}{6}\right)\left(k \ln \overline{\lambda} - \sum_{j=p-k+1}^{p} \ln \lambda_j\right) \tag{9.22}$$

(Bartlett 1951), where

$$\overline{\lambda} = \sum_{j=p-k+1}^{p} \frac{\lambda_j}{k}$$

is the average of the last k sample eigenvalues. The test statistic u is a modification of the likelihood ratio and has an approximate χ^2-distribution when H_{0k} is true. We reject H_{0k} if $u \geq \chi^2_{\alpha,\nu}$, where $\nu = \frac{1}{2}(k-1)(k+2)$. Schott (1988) gave an improved chi-square test of H_{0k}.

We could test $H_{02}, H_{03}, \ldots$ sequentially until H_{0k} is rejected for some value of k. As with similar tests for discriminant functions and canonical variates (Sections 5.5.2 and 8.4.2), this procedure tends to retain one or more "smaller" principal components that are not important descriptors, that is, to reject H_{0k} for a value of k that is too small. A Bonferroni adjustment to preserve the overall α-level would be helpful in this regard.

Before making the test in (9.22), it may be useful to make a preliminary test of complete independence of the variables. This hypothesis can be expressed in terms of the population covariance matrix as

$$H_0\colon \boldsymbol{\Sigma} = \begin{pmatrix} \sigma_{11} & 0 & \cdots & 0 \\ 0 & \sigma_{22} & \cdots & 0 \\ \vdots & \vdots & & \vdots \\ 0 & 0 & \cdots & \sigma_{pp} \end{pmatrix}$$

or $H_0\colon \sigma_{jk} = 0$ for all $j \neq k$. There is no restriction on the σ_{jj}'s. If $\sigma_{jk} = 0$ for all $j \neq k$, the corresponding ρ_{jk}'s are also zero, and an equivalent form of the hypothesis is $H_0\colon \mathbf{P}_\rho = \mathbf{I}$, where $\mathbf{P}_\rho$ is the population correlation matrix. The likelihood ratio statistic (Bartlett 1950) is given by

$$u = \frac{|\mathbf{S}|}{s_{11}s_{22}\cdots s_{pp}} = |\mathbf{R}|, \tag{9.23}$$

and the test statistic

$$u' = -\left[\nu - \frac{2p+5}{6}\right]\ln u \tag{9.24}$$

has an approximate χ^2_f-distribution, where ν is the degrees of freedom of $\mathbf{S}$ or $\mathbf{R}$ and $f = p(p-1)/2$. We reject H_0 if $u' \geq \chi^2_{\alpha,f}$. Exact percentage points for u' for selected values of n and p were given by Mathai and Katiyar (1979).

An alternative test for $H_0\colon \mathbf{P}_\rho = \mathbf{I}$ based on Fisher's (1921) z-transformation of the correlation coefficient r was given by Steiger (1980). In a limited simulation, Wilson and Martin (1983) compared Steiger's test with that of Bartlett in (9.24) and concluded that Steiger's test is more powerful.

Example 9.5.4. We illustrate the test of $H_0\colon \sigma_{jk} = 0,\ j \neq k$, for the probe word data from Table 1.3. The sample correlation matrix $\mathbf{R}$ is given in Example 1.7.1, from which we obtain

$$u = |\mathbf{R}| = .0409, \qquad u' = -\left[n - 1 - \tfrac{1}{6}(2p+5)\right]\ln u = 23.97.$$

Since u' exceeds $\chi^2_{.01,10} = 23.21$, we reject H_0.

9.5.5 Other Methods

Stauffer et al. (1985) gave bootstrap estimators for the standard deviation of any sample eigenvalue λ_j and for the percent of variance remaining after the last component is retained. These estimators can be used to construct tests or confidence intervals. The bootstrap approach is a resampling scheme proposed by Efron (1979). Eastment and Krzanowski (1982) suggested a cross-validatory method for choosing the number of components to retain. Besse and de Falguerolles (1993) gave bootstrap and jackknife estimators of the number of components to retain.

Huang and Tseng (1992) proposed the following method of choosing the number of components to retain: Keep k components if k is the smallest integer such that

$$\frac{\sum_{j=1}^{k} \lambda_j}{\sum_{j=1}^{p} \lambda_j} \geq c,$$

where c is determined so that the probability of retaining the important components is at least some specified level p^*. The sample size n must also be considered.

Gill and Lewbel (1992) proposed a test for the essential rank of $\boldsymbol{\Sigma}$. This rank could be interpreted as the correct number of principal components. The test is based on the Gaussian elimination lower diagonal–upper triangular decomposition (Householder 1964).

9.6 INFORMATION IN THE LAST FEW PRINCIPAL COMPONENTS

Typically, only the first few principal components are considered useful for summarizing data. However, the last few components also carry information that we may wish to examine.

By (9.6), the eigenvalues are variances of the principal components. Hence the last few principal components have the smallest variances. If the variance of a component is close to zero, the component defines a linear relationship among the variables that is nearly constant over the sample. Thus an extremely small eigenvalue may signal a colinearity of which the researcher is unaware. Ramsey (1986) suggests that redundant variables be deleted so that they cannot distort the first few principal components.

Because the last few principal components reflect relationships among the variables, it may be useful to plot the last two or more principal components to aid in the search for outliers that deviate significantly from the correlation structure. Suppose, for example, that y_5 is highly correlated with y_1, y_2, y_3, and y_4, so that $y_5 = \beta_1 y_1 + \cdots + \beta_4 y_4 + \varepsilon$, where the variables have been corrected for their means. Then the linear combination $y_5 - \beta_1 y_1 - \cdots - \beta_4 y_4$ will have a small variance, and one of the smaller principal components will reflect this linear combination. An observation vector that departs from this correlation pattern will likely show up in a plot involving the given principal component.

We now describe a measure for each observation that may be useful to supplement the plots involving the last few principal components. We first show that the observations can be expressed as linear combinations of the eigenvectors. If the eigenvectors of **S** are normalized to length 1, then by (A.9.4),

$$\mathbf{I} = \mathbf{a}_1\mathbf{a}_1' + \mathbf{a}_2\mathbf{a}_2' + \cdots + \mathbf{a}_p\mathbf{a}_p'. \tag{9.25}$$

We multiply (9.25) by the ith observation vector $\mathbf{y}_i$ to obtain

$$\mathbf{I}\mathbf{y}_i = \mathbf{a}_1\mathbf{a}_1'\mathbf{y}_i + \mathbf{a}_2\mathbf{a}_2'\mathbf{y}_i + \cdots + \mathbf{a}_p\mathbf{a}_p'\mathbf{y}_i.$$

Now $\mathbf{a}_1'\mathbf{y}_i = z_{1i}$, the first principal component evaluated for $\mathbf{y}_i$. Similarly, $\mathbf{a}_2'\mathbf{y}_i = z_{2i}$, and so on. Thus

$$\mathbf{y}_i = z_{1i}\mathbf{a}_1 + z_{2i}\mathbf{a}_2 + \cdots + z_{pi}\mathbf{a}_p, \qquad i = 1, 2, \cdots, n. \tag{9.26}$$

We can use the last few terms of (9.26) as an "error" or residual for how well the first components fit $\mathbf{y}_i$. Since the $\mathbf{a}_i$'s are orthogonal, the squared length (see A.2.13) of the residual vector $z_{p-k,i}\mathbf{a}_{p-k} + \cdots + z_{pi}\mathbf{a}_p$ for $\mathbf{y}_i$ is

$$d_i^2 = z_{p-k,i}^2 + \cdots + z_{pi}^2. \tag{9.27}$$

We compute d_i^2 for each of $\mathbf{y}_1, \mathbf{y}_2, \ldots, \mathbf{y}_n$. Then an observation with an unusually large value of d_i^2 will indicate a poor fit by the first $p - k - 1$ components, which may be due to the observation being aberrant with respect to the correlation structure.

Further discussion of the information in the last few principal components can be found in Gnanadesikan (1977, pp. 260–275), Jolliffe (1986, pp. 175–187), and Hawkins and Fatti (1984).

9.7 INTERPRETATION OF PRINCIPAL COMPONENTS

Various approaches for interpretation of principal components have been proposed. In Section 9.4 we discussed the use of **R** instead of **S** in cases where the variables have widely disparate variances. In Section 9.7.1 we present the characteristic pattern of the components from certain patterned covariance and correlation matrices as an additional aid to interpretation. Hypothesis tests for the presence of certain patterns in **S** or **R** are given in Section 9.7.2. In Sections 9.7.3 and 9.7.4, we question the use of two popular approaches to interpretation: further rotation of the principal components and correlations between the variables and the components. Various aspects of interpretation of principal components have been discussed by Rencher (1992b).

9.7.1 Special Patterns in S or R

Some covariance or correlation matrices exhibit characteristic patterns that signal the type of principal components to be expected. In property 6 of Section 9.2.3, for example, two cases were described in which the structure of **S** indicates that a component will reproduce a variable: (1) if one variable has a much larger variance than the other variables, that variable will largely comprise the first component, and (2) when a variable is uncorrelated with all other variables, that variable is itself a principal component.

A pattern of interest for the population covariance matrix is $\boldsymbol{\Sigma} = \sigma^2[(1-\rho)\mathbf{I}+\rho\mathbf{J}]$, or

$$\boldsymbol{\Sigma} = \begin{pmatrix} \sigma^2 & \sigma^2\rho & \cdots & \sigma^2\rho \\ \sigma^2\rho & \sigma^2 & \cdots & \sigma^2\rho \\ \vdots & \vdots & & \vdots \\ \sigma^2\rho & \sigma^2\rho & \cdots & \sigma^2 \end{pmatrix}. \tag{9.28}$$

If ρ is of substantial size, then by (A.11.8), the first eigenvalue of $\boldsymbol{\Sigma}$ is large and the rest are small and equal:

$$\gamma_1 = \sigma^2[1 + (p-1)\rho], \qquad \gamma_j = \sigma^2(1-\rho), \qquad j = 2, 3, \ldots, p. \tag{9.29}$$

By (A.11.10), the normalized eigenvector of $\boldsymbol{\Sigma}$ corresponding to γ_1 is given as

$$\boldsymbol{\alpha}_1' = \left(\frac{1}{\sqrt{p}}, \frac{1}{\sqrt{p}}, \ldots, \frac{1}{\sqrt{p}}\right).$$

Hence the first (population) principal component is proportional to the mean of the p variables; that is, $\boldsymbol{\alpha}_1'\mathbf{y} = \sqrt{p}\sum_{j=1}^{p} y_j/p$. By (A.11.11), the other $p-1$ eigenvectors $\boldsymbol{\alpha}_2, \ldots, \boldsymbol{\alpha}_p$ are orthogonal to $\boldsymbol{\alpha}_1$ (that is, $\boldsymbol{\alpha}_i'\boldsymbol{\alpha}_1 = 0$) and orthogonal to each other.

A pattern for the population correlation matrix related to that in (9.28) for the population covariance matrix is

$$\mathbf{P}_\rho = \begin{pmatrix} 1 & \rho & \cdots & \rho \\ \rho & 1 & \cdots & \rho \\ \vdots & \vdots & & \vdots \\ \rho & \rho & \cdots & 1 \end{pmatrix}, \tag{9.30}$$

in which the correlations are equal but there is no restriction on the variances. The eigenvalues of $\mathbf{P}_\rho$ are given by (A.11.9) as

$$\gamma_1 = 1 + (p-1)\rho, \qquad \gamma_j = 1 - \rho, \qquad j = 2, 3, \ldots, p, \tag{9.31}$$

and the eigenvectors are the same as $\boldsymbol{\alpha}_1, \boldsymbol{\alpha}_2, \ldots, \boldsymbol{\alpha}_p$ described above for $\boldsymbol{\Sigma} = \sigma^2[(1-\rho)\mathbf{I} + \rho\mathbf{J}]$.

If $\mathbf{S}$ resembles (9.28) or $\mathbf{R}$ resembles (9.30), and if the estimate of ρ [see (9.36)] is large, then as suggested in Section 9.5.4, the $p-1$ small components do not have unique direction. Tests for the patterns in (9.28) and (9.30) are given in Section 9.7.2.

Another related pattern for $\mathbf{R}$ or $\mathbf{S}$ is that in which all correlations or covariances are positive but not equal. Meyer (1975) showed that in this situation a lower bound for the first eigenvalue of **R** is

$$\lambda_1 \geq 1 + (p - 1)\bar{r}, \tag{9.32}$$

where $\bar{r}$ is the average of the off-diagonal elements of **R**. Using a few examples and limited simulations, Friedman and Weisberg (1981) demonstrated that λ_1 tends to be close to its lower bound.

If all correlations in **R** or all covariances in **S** are positive, then by the Perron–Frobenius theorem [see (A.11.12)], all elements of the first eigenvector $\mathbf{a}_1$ are positive. The other $p - 1$ eigenvectors are orthogonal to $\mathbf{a}_1$; therefore, $\mathbf{a}_2, \mathbf{a}_3, \ldots, \mathbf{a}_p$ have both positive and negative elements. With all positive elements in $\mathbf{a}_1$, the first component $z_1 = \mathbf{a}_1'\mathbf{y}$ is a weighted average of the variables, sometimes referred to as a *size* component. Because $\mathbf{a}_2, \mathbf{a}_3, \ldots, \mathbf{a}_p$ have both positive and negative elements, $z_2 = \mathbf{a}_2'\mathbf{y}, \ldots, z_p = \mathbf{a}_p'\mathbf{y}$ may be regarded as *shape* components. If we increase all elements of $\mathbf{y}$, then z_1 increases, but an increase in z_2 is typically due to an increase in some elements of $\mathbf{y}$ and a decrease in others. The size and shape characterizations are often applicable when the variables are measurements on an organism (Darroch and Mosimann 1985; Somers 1986, 1989; Bookstein 1989).

As an extension of this pattern of positive covariances or correlations, we may find subgroups of variables with high positive correlations within each subgroup and low correlations with variables outside the subgroup. Then the above pattern will be seen for the principal components that correspond to each subgroup of variables. There will be an overall average or size component for each subgroup and some shape or contrast components. Kotz and Pearn (1984) discussed a pattern of this type and gave an example involving rates of return on common stocks.

Krzanowski (1984a) suggested that a *sensitivity analysis* be carried out to investigate the stability of the principal components. He argued that we can have confidence in the interpretation if the components remain stable when small changes (perturbations) are made in the λ_j's. To illustrate sensitivity analysis, we consider decreasing λ_k by ε and denote by $\mathbf{a}_k^*$ the vector that differs as much as possible from $\mathbf{a}_k$, subject to the restriction that $\text{var}(z_k^*) = \text{var}(\mathbf{a}_k^{*\prime}\mathbf{y}) = \lambda_k - \varepsilon$. Krzanowski (1984a) showed that the angle θ between $\mathbf{a}_k$ and $\mathbf{a}_k^*$ is given by

$$\cos\theta = \left(1 + \frac{\varepsilon}{\lambda_k - \lambda_{k+1}}\right)^{-1/2}. \tag{9.33}$$

Thus if λ_k is close to λ_{k+1}, θ will be large, and the kth principal component $z_k = \mathbf{a}_k'\mathbf{y}$ will be more unstable than if λ_{k+1} is further from λ_k. Similarly, if λ_k is increased by ε, z_k becomes unstable if $\lambda_{k-1} - \lambda_k$ is small. Thus the stability of z_k depends on the distance of λ_k from both λ_{k-1} and λ_{k+1}.

Example 9.7.1(a). Pearce (1965) reported three measurements on 54 apple trees: y_1 = total length of lateral shoots, y_2 = circumference of the trunk, and y_3 = height. In this case, the variances are of different orders of magnitude, and we use the correlation matrix,

$$\mathbf{R} = \begin{pmatrix} 1 & .5792 & .2414 \\ .5792 & 1 & .5816 \\ .2414 & .5816 & 1 \end{pmatrix}.$$

The eigenvalues of $\mathbf{R}$ are 1.94, .76, and .29, with corresponding eigenvectors

$$\mathbf{a}_1 = (.554, .651, .520)',$$
$$\mathbf{a}_2 = (.657, .042, -.753)',$$
$$\mathbf{a}_3 = (.511, -.758, .404)'.$$

If we denote the standardized variables by y_1^*, y_2^*, and y_3^*, where $y_j^* = (y_j - \overline{y}_j)/s_i$, then the first principal component, $z_1 = .554y_1^* + .651y_2^* + .520y_3^*$, is essentially the average of the three variables and is a measure of the overall size of the trees. The second component, $z_2 = .657y_1^* + .042y_2^* - .753y_3^*$, is a shape component that contrasts lateral and height measurements; that is, z_2 essentially measures $y_1^* - y_3^*$, which is large (in absolute value) if y_1^* and y_3^* are far apart and is small if y_1^* and y_3^* are closer together. The third component, $z_3 = .511y_1^* - .758y_2^* + .404y_3^*$, is a comparison of height and lateral measurements versus circumference; thus z_3 depicts another aspect of shape.

Example 9.7.1(b). Constable and Mardia (1992) studied the size and shape of human palms based on various measurements. They carried out principal component analyses separately for males and females on the variables y_1 = proximal extent, y_2 = directional measure, and y_3 = relative locational measure. The correlation matrices for males (M) and females (F) are

$$\mathbf{R}_M = \begin{pmatrix} 1 & .7215 & .5228 \\ .7215 & 1 & .6259 \\ .5228 & .6259 & 1 \end{pmatrix}, \qquad \mathbf{R}_F = \begin{pmatrix} 1 & .7359 & .5737 \\ .7359 & 1 & .6086 \\ .5737 & .6086 & 1 \end{pmatrix}.$$

The eigenvalues for $\mathbf{R}_M$ are 2.25, .49, and .26; and for $\mathbf{R}_F$ they are 2.28, .46, and .26. The eigenvectors for $\mathbf{R}_M$ are

$$\mathbf{a}_{M1} = (.58, .61, .55)', \qquad \mathbf{a}_{M2} = (.57, .18, -.80)', \qquad \mathbf{a}_{M3} = (.58, -.77, .24)'.$$

For $\mathbf{R}_F$, the eigenvectors are

$$\mathbf{a}_{F1} = (.59, .60, .55)', \qquad \mathbf{a}_{F2} = (.46, .31, -.83)', \qquad \mathbf{a}_{F3} = (.67, -.74, .09)'.$$

The results are similar for males and females. In both cases, the first component is a measure of size; the other two components are measures of shape. The second component contrasts y_3 versus y_1 and y_2; the third component represents a comparison of y_1 and y_2.

9.7.2 Testing H_0: $\boldsymbol{\Sigma} = \sigma^2\left[(1-\rho)\mathbf{I} + \rho\mathbf{J}\right]$ and $\mathbf{P}_\rho = (1-\rho)\mathbf{I} + \rho\mathbf{J}$

9.7.2a ***H_0: $\boldsymbol{\Sigma} = \sigma^2[(1-\rho)\mathbf{I} + \rho\mathbf{J}]$: Equal Variances and Equal Covariances***
We now give a test for the hypothesis that the $\boldsymbol{\Sigma}$ has the pattern in (9.28),

$$H_0\colon \boldsymbol{\Sigma} = \begin{pmatrix} \sigma^2 & \sigma^2\rho & \cdots & \sigma^2\rho \\ \sigma^2\rho & \sigma^2 & \cdots & \sigma^2\rho \\ \vdots & \vdots & & \vdots \\ \sigma^2\rho & \sigma^2\rho & \cdots & \sigma^2 \end{pmatrix} \tag{9.34}$$

$$= \sigma^2[(1-\rho)\mathbf{I} + \rho\mathbf{J}]. \tag{9.35}$$

The maximum likelihood estimators (adjusted for bias) of σ^2 and $\sigma^2\rho$ under H_0 are

$$s^2 = \frac{1}{p}\sum_{j=1}^{p} s_{jj} \quad \text{and} \quad s^2 r = \frac{1}{p(p-1)}\sum_{j\neq k} s_{jk}, \tag{9.36}$$

respectively, where s_{jj} and s_{jk} are diagonal and off-diagonal elements of $\mathbf{S}$. From (9.36), the estimator of $\boldsymbol{\Sigma}$ under H_0 is

$$\mathbf{S}_0 = \begin{pmatrix} s^2 & s^2 r & \cdots & s^2 r \\ s^2 r & s^2 & \cdots & s^2 r \\ \vdots & \vdots & & \vdots \\ s^2 r & s^2 r & \cdots & s^2 \end{pmatrix}.$$

As is often the case, the likelihood ratio is a function of the ratio of the determinants of the estimates of $\boldsymbol{\Sigma}$ under H_0 and H_1 (Wilks 1946), and the test statistic based on this ratio is

$$u = (\mathrm{LR})^{2/n} = \frac{|\mathbf{S}|}{|\mathbf{S}_0|}, \tag{9.37}$$

which, by (A.7.7), becomes

$$u = \frac{|\mathbf{S}|}{(s^2)^p(1-r)^{p-1}[1+(p-1)r]}, \tag{9.38}$$

where r, the estimator of ρ, is obtained from (9.36) as $s^2 r/s^2$. Box (1949) showed that

$$u' = -\left[\nu - \frac{p(p+1)^2(2p-3)}{6(p-1)(p^2+p-4)}\right]\ln u \tag{9.39}$$

is approximately distributed as $\chi^2\left[\frac{1}{2}p(p+1) - 2\right]$, where ν is the degrees of freedom of $\mathbf{S}$. We reject H_0 if $u' > \chi^2[\alpha, \frac{1}{2}p(p+1) - 2]$. The reduction of degrees of freedom by 2 is due to estimation of σ^2 and ρ. Nagarsenker (1975) gave the exact distribution of u and provided a table for $p = 4, 5, \ldots, 10$. A nonparametric test was obtained by Choi (1977).

An F-approximation (Box 1950, p. 375) that is more precise for large p or smaller ν is given by

$$F = \frac{-(\delta_2 - \delta_2 c_1 - \delta_1)\nu}{\delta_1 \delta_2} \ln u, \tag{9.40}$$

where

$$c_1 = \frac{p(p+1)^2(2p-3)}{6\nu(p-1)(p^2+p-4)}, \qquad c_2 = \frac{p(p^2-1)(p+2)}{6\nu^2(p^2+p-4)},$$

$$\delta_1 = \tfrac{1}{2}p(p+1) - 2, \qquad \delta_2 = (\delta_1 + 2)/(c_2 - c_1^2).$$

We reject H_0 if $F > F_{\alpha,\delta_1,\delta_2}$.

For an alternative approach to testing H_0 in (9.34), note that by (9.29) all eigenvalues of $\mathbf{\Sigma}$ in (9.34) are equal to $\sigma^2(1-\rho)$ except the first: $\gamma_2 = \gamma_3 = \cdots = \gamma_p = \sigma^2(1-\rho)$. The hypothesis that $\gamma_2 = \gamma_3 = \cdots = \gamma_p$ is equivalent to H_{0k} in Section 9.5.4 with $k = p - 1$. This hypothesis can be tested with the statistic in (9.22).

9.7.2b H_0: $\mathbf{P}_\rho = (1-\rho)\mathbf{I} + \rho\mathbf{J}$: ***Equicorrelation with Unspecified Variances***

A hypothesis of interest for principal components of $\mathbf{R}$ is that the population correlation matrix has the form in (9.30),

$$H_0\colon \mathbf{P}_\rho = \begin{pmatrix} 1 & \rho & \cdots & \rho \\ \rho & 1 & \cdots & \rho \\ \vdots & \vdots & & \vdots \\ \rho & \rho & \cdots & 1 \end{pmatrix}, \tag{9.41}$$

which can also be stated as

$$H_0\colon \rho_{jk} = \rho \text{ for all } j \neq k \text{ versus } H_1\colon \rho_{jk} \neq \rho \text{ for some } j \neq k.$$

There is no restriction on the variances.

The likelihood ratio test is rather intractable, and a heuristic test was proposed by Lawley (1963). The test statistic is

$$u = \frac{\nu}{w^2}\left[\sum_{j<k}(r_{jk} - \bar{r})^2 - q\sum_{k=1}^{p}(\bar{r}_k - \bar{r})^2\right], \tag{9.42}$$

where ν is the number of degrees of freedom for the sample correlation matrix $\mathbf{R} = (r_{jk})$, and

$$\bar{r} = \frac{1}{p(p-1)}\textstyle\sum_{j\neq k} r_{jk}, \qquad w = 1 - \bar{r},$$

$$q = \frac{(p-1)^2(1-w^2)}{p-(p-2)w^2}, \qquad \bar{r}_k = \frac{1}{p-1}\sum_{\substack{j=1 \\ j\neq k}}^{p} r_{jk}.$$

Thus $\bar{r}$ is the overall average and $\bar{r}_k$ is the average of the kth column of $\mathbf{R}$ [excluding the 1 in the (k, k) position]. The statistic u is approximately distributed as $\chi^2\left[\frac{1}{2}(p+1)(p-2)\right]$; we reject H_0 if $u > \chi^2\left[\alpha, \frac{1}{2}(p+1)(p-2)\right]$.

Example 9.7.2. We illustrate the test of H_0: $\rho_{jk} = \rho$ for the probe word data in Table 1.3. The correlation matrix was given in Example 1.7.1 as

$$\mathbf{R} = \begin{pmatrix} 1.00 & .61 & .76 & .58 & .41 \\ .61 & 1.00 & .55 & .75 & .55 \\ .76 & .55 & 1.00 & .61 & .69 \\ .58 & .75 & .61 & 1.00 & .52 \\ .41 & .55 & .69 & .52 & 1.00 \end{pmatrix},$$

from which we obtain

$$\bar{r} = .6026, \qquad w = .3974, \qquad q = 2.9766,$$

$$(\bar{r}_1, \bar{r}_2, \ldots, \bar{r}_5) = (.59, .61, .65, .61, .54).$$

Then

$$u = \frac{10}{(.3974)^2}[.1026 - (2.9766)(.0061361)] = 5.3405.$$

The critical value is $\chi^2_{.05,6} = 12.59$, and we do not reject H_0.

9.7.3 Additional Rotation

We noted in Section 9.3 that principal components represent a rotation of axes so that the new axes represent uncorrelated dimensions that line up with the directions of maximum variance. To simplify the interpretation of principal components, additional rotation is sometimes recommended. The goal is to find new dimensions (linear combinations) in which many of the coefficients are near zero. However, the new rotated "components" are correlated, and they no longer successively account for maximum variance. Therefore, the "rotated principal components" are not principal components as traditionally defined (Jollife 1987).

We now demonstrate these two properties of rotated principal components. If $\mathbf{a}_j$ is the coefficient vector for the jth principal component $z_j = \mathbf{a}_j'\mathbf{y}$, then the corresponding rotated component is $w_j = (\mathbf{C}\mathbf{a}_j)'\mathbf{y} = \mathbf{a}_j'\mathbf{C}'\mathbf{y}$, where $\mathbf{C}$ is orthogonal. By (1.74), the sample covariance of two rotated components is

$$\widehat{\text{cov}}(w_j, w_k) = \mathbf{a}_j'\mathbf{C}'\mathbf{S}\mathbf{C}\mathbf{a}_k,$$

which is not zero because $\mathbf{C}\mathbf{a}_j$ and $\mathbf{C}\mathbf{a}_k$ are not eigenvectors of $\mathbf{S}$ [see (9.11)]. Thus rotated components are correlated.

Since the first principal component is in the direction of maximum variance, any rotation will produce a new dimension with smaller variance. To show that $\widehat{\text{var}}(w_1)$

differs from $\widehat{\text{var}}(z_1)$, note that

$$\widehat{\text{var}}(w_1) = \widehat{\text{var}}(\mathbf{a}_1'\mathbf{C}'\mathbf{y}) = \mathbf{a}_1'\mathbf{C}'\mathbf{S}\mathbf{C}\mathbf{a}_1,$$

which is not equal to $\widehat{\text{var}}(z_1) = \mathbf{a}_1'\mathbf{S}\mathbf{a}_1$.

If we rotate a subset of components, the new variables will account for the same percent of variance as the original subset, but the variance will be distributed differently among the new components. This will be demonstrated for a rotation of the full set of components and for a subset. For the full set, the total variance of the rotated components is

$$\begin{aligned}
\sum_{j=1}^{p} \widehat{\text{var}}(w_j) &= \sum_{j=1}^{p} \mathbf{a}_j'\mathbf{C}'\mathbf{S}\mathbf{C}\mathbf{a}_j && \text{[by (1.73)]} \\
&= \sum_j \text{tr}\left(\mathbf{a}_j'\mathbf{C}'\mathbf{S}\mathbf{C}\mathbf{a}_j\right) \\
&= \sum_j \text{tr}\left(\mathbf{C}'\mathbf{S}\mathbf{C}\mathbf{a}_j\mathbf{a}_j'\right) && \text{[by (A.8.2)]} \\
&= \text{tr}\left(\mathbf{C}'\mathbf{S}\mathbf{C}\sum_j \mathbf{a}_j\mathbf{a}_j'\right) && \text{[by (A.8.1)]} \\
&= \text{tr}\left(\mathbf{C}'\mathbf{S}\mathbf{C}\mathbf{I}\right) && \text{[by (A.9.4)]} \\
&= \sum_{j=1}^{p} \lambda_j && \text{[by (A.10.10)].} \qquad (9.43)
\end{aligned}$$

Thus the total variance of the rotated components is the same as the total variance of the original components.

To rotate k of the components, we express the transformation as $\mathbf{w} = \mathbf{C}\mathbf{z} = \mathbf{C}\mathbf{A}\mathbf{y}$, where $\mathbf{C}$ is an orthogonal matrix that performs the rotation and $\mathbf{z} = \mathbf{A}\mathbf{y}$ is similar to (9.4) with the $k \times p$ matrix $\mathbf{A}$ containing the first k eigenvectors as rows. (Note that $\mathbf{CA}$ transforms the columns of $\mathbf{A}$ rather than the rows as above.) The sample covariance matrix for $\mathbf{w}$ is $\widehat{\text{cov}}(\mathbf{w}) = \mathbf{C}\mathbf{A}\mathbf{S}\mathbf{A}'\mathbf{C}'$. The sum of the variances is given by

$$\begin{aligned}
\sum_{j=1}^{k} \widehat{\text{var}}(w_j) &= \text{tr}(\mathbf{C}\mathbf{A}\mathbf{S}\mathbf{A}'\mathbf{C}') = \text{tr}(\mathbf{C}'\mathbf{C}\mathbf{A}\mathbf{S}\mathbf{A}') \\
&= \text{tr}(\mathbf{A}\mathbf{S}\mathbf{A}') = \sum_{j=1}^{k} \widehat{\text{var}}(z_j),
\end{aligned}$$

since $\widehat{\text{cov}}(\mathbf{z}) = \mathbf{A}\mathbf{S}\mathbf{A}'$.

Thus the sum of the variances of the k rotated components is the same, but the individual rotated components have variances that differ from those of the original components.

Since the rotated components are correlated and lack the maximum variance property, their routine use is questionable. If interpretation of the principal components is not satisfactory, one may wish to try factor analysis, in which rotation does not affect any important properties (see Sections 10.2.3 and 10.5). Rotation of principal components for use in other procedures such as detection of outliers, variable selection, and regression is recommended by Hawkins (1973), Daling and Tamura (1970), Hawkins and Eplett (1982), and Hawkins and Fatti (1984).

Jolliffe (1989) opposes rotation of the larger components but suggests that it may be useful to rotate certain groups of smaller components. By (9.33), if a subset of components has eigenvalues that are nearly equal, then the corresponding eigenvectors are unstable; that is, small changes in any of the λ_j's in this subset may be accompanied by large changes in the associated $\mathbf{a}_j$'s. Jolliffe (1989) therefore recommends that such a subset be rotated to improve interpretability. He provides several examples in which improvement is demonstrated. Jolliffe (1995) notes that the results of the rotation depend on the choice of normalization of the eigenvectors.

9.7.4 Correlations between Variables and Principal Components

For interpretation of a principal component, many authors recommend using the correlations between variables and the component rather than the coefficients in the component. As shown in Sections 5.7.2 and 8.6.3, similar correlations for discriminant functions and canonical variates do not provide multivariate information about how the variables operate in the presence of each other. We now demonstrate an analogous result for principal components (Rencher 1992b).

To obtain an expression for $r_{y_i z_j}$, the correlation between the ith variable and jth component, we begin with the vector of sample covariances between the variables in $\mathbf{y}$ and the jth component $z_j = \mathbf{a}_j'\mathbf{y}$ [see(1.79)]:

$$\begin{aligned} \widehat{\text{cov}}(\mathbf{y}, z_j) &= \widehat{\text{cov}}(\mathbf{Iy}, \mathbf{a}_j'\mathbf{y}) = \mathbf{ISa}_j = \mathbf{Sa}_j \\ &= \lambda_j \mathbf{a}_j, \end{aligned} \tag{9.44}$$

since $\mathbf{a}_j$ is an eigenvector of $\mathbf{S}$. The covariance of y_i and z_j is the ith element of $\lambda_j \mathbf{a}_j$:

$$\widehat{\text{cov}}(y_i, z_j) = \lambda_j a_{ij}. \tag{9.45}$$

The standard deviation of y_i is $\sqrt{s_{ii}}$, the square root of the ith diagonal element of $\mathbf{S}$, and by (9.6), the standard deviation of z_j is $\sqrt{\lambda_j}$. Hence the correlation between the ith variable y_i and the jth component z_j is given by

$$r_{y_i z_j} = \frac{\lambda_j a_{ij}}{\sqrt{s_{ii}}\sqrt{\lambda_j}} = a_{ij}\frac{\sqrt{\lambda_j}}{\sqrt{s_{ii}}}. \tag{9.46}$$

In the following theorem, we show that $r^2_{y_i z_j}$ forms part of the squared multiple correlation of y_i with the principal components.

Theorem 9.7A. Let $r_{y_i z_j}$ be the correlation between the ith variable y_i and the jth principal component z_j. Then

$$r_{y_i z_1}^2 + r_{y_i z_2}^2 + \cdots + r_{y_i z_t}^2 = R_{y_i|z_1,\ldots,z_t}^2, \tag{9.47}$$

where t is the number of components retained and $R_{y_i|z_1,\ldots,z_t}^2$ is the squared multiple correlation of y_i with the z_j's.

Proof. Using (7.66), we can write $R_{y_i|z_1,\ldots,z_t}^2$ in the form

$$R_{y_i|z_1,\ldots,z_t}^2 = \frac{\mathbf{s}_{y_i z}' \mathbf{S}_{zz}^{-1} \mathbf{s}_{y_i z}}{s_{ii}}, \tag{9.48}$$

where $\mathbf{s}_{y_i z}$ is the vector of covariances of y_i with the variables in $\mathbf{z} = (z_1, z_2, \ldots, z_t)'$, $\mathbf{S}_{zz}$ is the covariance matrix of $\mathbf{z}$, and s_{ii} is the variance of y_i. By (9.45), we can write $\mathbf{s}_{y_i z}'$ in the form

$$\mathbf{s}_{y_i z}' = (\lambda_1 a_{i1}, \lambda_2 a_{i2}, \ldots, \lambda_t a_{it}). \tag{9.49}$$

By (9.6), $s_{z_i}^2 = \lambda_i$, and by (9.11), the z_j's are uncorrelated. Hence

$$\mathbf{S}_{zz} = \begin{pmatrix} \lambda_1 & 0 & \cdots & 0 \\ 0 & \lambda_2 & \cdots & 0 \\ \vdots & \vdots & & \vdots \\ 0 & 0 & \cdots & \lambda_t \end{pmatrix}. \tag{9.50}$$

Substituting (9.49) and (9.50) into (9.48) and using (9.46) gives

$$R_{y_i|z_1,\ldots,z_t}^2 = \sum_{j=1}^{t} \frac{\lambda_j a_{ij}^2}{s_{ii}}$$

$$= \sum_{j=1}^{t} r_{y_i z_j}^2. \qquad \square$$

Corollary 1. If all p components are used, then $R_{y_i|z_1,\ldots,z_p}^2 = 1$.

Proof. By (9.46) and (9.47), we have

$$R_{y_i|z_1,\ldots,z_p}^2 = \sum_{j=1}^{p} r_{y_i z_j}^2$$

$$= \frac{1}{s_{ii}} \sum_{j=1}^{p} \lambda_j a_{ij}^2. \tag{9.51}$$

From the spectral decomposition $\mathbf{S} = \sum_{i=1}^{p} \lambda_i \mathbf{a}_i \mathbf{a}_i'$ in (A.11.4), the sum in (9.51) reduces to s_{ii}. Hence $R_{y_i|z_1,\ldots,z_p}^2 = s_{ii}/s_{ii} = 1$. $\square$

By Theorem 9.7A, $r^2_{y_i z_j}$ is part of a partitioning of $R^2_{y_i|z_1,\ldots,z_t}$, which is a univariate measure showing how y_i alone relates to the z's, not what y_i contributes in the presence of the other y's. Thus the correlations $r_{y_i z_j}$ do not provide multivariate information about the joint contribution of the y's to a principal component.

We therefore do not recommend rotation or correlations for interpretation; we simply use the coefficients in $\mathbf{a}_j$, obtained from either $\mathbf{S}$ or $\mathbf{R}$.

9.8 RELATIONSHIP BETWEEN PRINCIPAL COMPONENTS AND REGRESSION

Consider a multiple regression application with one y and several x's in which there are too many x's (the number of x's is close to the sample size n) or the x's contain near-linear dependencies (multicolinearity). In such cases, the $\hat{\beta}_j$'s typically become unstable from sample to sample (have very large variances), and there is a corresponding loss of power in hypothesis tests.

One approach to the problem of multicolinearity is *principal component regression*, in which y is regressed on the principal components of the x's. If we use only the "larger" principal components, the large variances in the $\hat{\beta}_j$'s due to multicolinearity are reduced, but of course we introduce some bias in the new $\hat{\beta}_j$'s. Thus principal component regression is a competitor to *ridge regression* (Hoerl and Kennard 1970a, 1970b), the "lasso" method (Tibshirani 1996), and other biased approaches. If the multicolinearity is severe, we may be able to achieve a substantial reduction in the variances of the $\hat{\beta}_j$'s at the expense of only a small amount of bias.

In biased regression, the goal is to obtain $\hat{\beta}_j$'s that are closer to the β_j's in the expected squared-error sense. This improvement in precision of the estimators of the β_j's may or may not be accompanied by improved prediction of the y's. Better predictors for the actual y's in the sample cannot be obtained, of course, since the least squares estimators already minimize SSE $= \sum_{i=1}^n (y_i - \hat{y}_i)^2$, where $\hat{y} = \hat{\beta}_0 + \hat{\beta}_1 x_1 + \cdots + \hat{\beta}_q x_q$. Similarly, prediction of y for a point $(x_1, x_2, \ldots, x_q)$ that is not in the sample but is in the same region as the sample may not be much improved. However, prediction of y for points $(x_1, x_2, \ldots, x_q)$ outside of this region may be considerably improved since the biased prediction equation may be closer to the true model.

We discuss principal component regression in Section 9.8.1 and a related approach, *latent root regression*, in Section 9.8.2. Principal component regression was first proposed by Hotelling (1957) and Kendall (1957). Marquardt (1970) and Hsuan (1981) have explored the relationship between ridge regression and principal component regression. Baye and Parker (1984) and Nomura and Ohkubo (1985) have shown that additional gains in expected squared error of the estimators can be made by combining ridge regression and principal component regression. Latent root regression was first proposed by Webster, Gunst, and Mason (1974). Coxe (1982) compared latent root regression and principal component regression. Gunst and Mason (1977) compared the same two, along with other biased estimation techniques. Good reviews of the use of principal components in regression have been given by Gunst (1983) and Jolliffe (1986, Chapter 8).

9.8.1 Principal Component Regression

Most expositions of principal component regression use standardized x's (z-scores) so that $\mathbf{X}_s'\mathbf{X}_s$ is proportional to a correlation matrix, where $\mathbf{X}_s$ is the matrix of standardized x-values. However, principal components of the covariance matrix are preferred in many situations, and we will begin with a model that involves centered x's and uses principal components from the covariance matrix of the x's. We will then adapt the procedure for use with correlation matrices.

Consider the centered model (7.13) for multiple regression,

$$y_i = \alpha + \beta_1(x_{i1} - \overline{x}_1) + \cdots + \beta_q(x_{iq} - \overline{x}_q) + \varepsilon_i, \qquad i = 1, 2, \ldots, n.$$

The matrix form of the model for all n observations is given by (7.15) as

$$\mathbf{y} = (\mathbf{j}, \mathbf{X}_c)\begin{pmatrix} \alpha \\ \boldsymbol{\beta}_1 \end{pmatrix} + \boldsymbol{\varepsilon}, \tag{9.52}$$

where $\mathbf{j}$ is a vector of 1s and $\mathbf{X}_c$ is given in (7.16). The estimator for $\boldsymbol{\beta}_1$ is given by (7.20) as

$$\hat{\boldsymbol{\beta}}_1 = (\mathbf{X}_c'\mathbf{X}_c)^{-1}\mathbf{X}_c'\mathbf{y}, \tag{9.53}$$

from which, by an adaption of (7.8), we obtain the covariance matrix of $\hat{\boldsymbol{\beta}}_1$,

$$\text{cov}(\hat{\boldsymbol{\beta}}_1) = \sigma^2(\mathbf{X}_c'\mathbf{X}_c)^{-1}. \tag{9.54}$$

The variance of $\hat{\beta}_j$, $j = 1, 2, \ldots, q$, is the jth diagonal element of (9.54). The total variance $\sum_{j=1}^{q} \text{var}(\hat{\beta}_j)$ can therefore be written as

$$\begin{aligned} \sum_{j=1}^{q} \text{var}(\hat{\beta}_j) &= \sigma^2 \, \text{tr}(\mathbf{X}_c'\mathbf{X}_c)^{-1} \\ &= \sigma^2 \sum_{j=1}^{q} \frac{1}{\lambda_j} \qquad \text{[by (A.10.6)]}, \end{aligned} \tag{9.55}$$

where λ_j is the jth eigenvalue of $\mathbf{X}_c'\mathbf{X}_c$, and by (A.10.5), $1/\lambda_j$ is the corresponding eigenvalue of $(\mathbf{X}_c'\mathbf{X}_c)^{-1}$. If one or more of the λ_j's is small, the total variance of the $\hat{\beta}_j$'s will be large.

A small eigenvalue of $\mathbf{X}_c'\mathbf{X}_c$ that inflates the total variance in (9.55) also induces multicolinearity among the x's. To demonstrate this, suppose that the smallest eigenvalue λ_q is close to zero. Let $\mathbf{a}_q$ be the corresponding eigenvector, normalized so that $\mathbf{a}_q'\mathbf{a}_q = 1$. Then by (A.10.1),

$$\mathbf{X}_c'\mathbf{X}_c\mathbf{a}_q = \lambda_q\mathbf{a}_q.$$

Multiplying on the left by $\mathbf{a}_q'$ gives

$$\mathbf{a}_q'\mathbf{X}_c'\mathbf{X}_c\mathbf{a}_q = \lambda_q\mathbf{a}_q'\mathbf{a}_q = \lambda_q \cong 0,$$

from which it follows that

$$\mathbf{X}_c\mathbf{a}_q \cong \mathbf{0}.$$

By (A.3.4), $\mathbf{X}_c\mathbf{a}_q$ is a linear combination of the columns of $\mathbf{X}_c$, and the expression $\mathbf{X}_c\mathbf{a}_q \cong \mathbf{0}$ therefore represents a near-linear dependency in the columns of $\mathbf{X}_c$.

In the foregoing discussion, we could have used eigenvectors of the covariance matrix of the x's, $\mathbf{S}_{xx} = \mathbf{X}_c'\mathbf{X}_c/(n-1)$, since by (A.10.4) they are the same as those of $\mathbf{X}_c'\mathbf{X}_c$. The eigenvalues of $\mathbf{S}_{xx}$ are obtained from those of $\mathbf{X}_c'\mathbf{X}_c$ by dividing by $n-1$.

The small eigenvalues that inflate the total variance in (9.55) correspond to the last principal components of $\mathbf{X}_c'\mathbf{X}_c$ or of $\mathbf{S}_{xx}$; we can therefore reduce the total variance by replacing $\hat{\boldsymbol{\beta}}_1$ with an estimator based on the first few principal components. We first express $\hat{\boldsymbol{\beta}}_1$ in terms of all q components and then delete the "smaller" components.

Let z_{ij} denote the jth principal component of the ith observation vector $\mathbf{x}_i$:

$$\begin{aligned} z_{ij} &= a_{j1}(x_{i1} - \bar{x}_1) + a_{j2}(x_{i2} - \bar{x}_2) + \cdots + a_{jq}(x_{iq} - \bar{x}_q) \\ &= \mathbf{a}_j'(\mathbf{x}_i - \bar{\mathbf{x}}), \qquad i = 1, 2, \ldots, n, \qquad j = 1, 2, \ldots, q. \end{aligned}$$

Then the ith row $(\mathbf{x}_i - \bar{\mathbf{x}})'$ of $\mathbf{X}_c$ can be transformed to a vector of principal components, $\mathbf{z}_i' = (\mathbf{x}_i - \bar{\mathbf{x}})'\mathbf{A}$, where the $q \times q$ matrix $\mathbf{A} = (\mathbf{a}_1, \mathbf{a}_2, \ldots, \mathbf{a}_q)$ has as columns the q (normalized) eigenvectors of $\mathbf{X}_c'\mathbf{X}_c$ or equivalently of $\mathbf{S}_{xx}$. The $n \times q$ matrix of principal components can therefore be written as

$$\mathbf{Z} = (z_{ij}) = \begin{pmatrix} \mathbf{z}_1' \\ \mathbf{z}_2' \\ \vdots \\ \mathbf{z}_n' \end{pmatrix} = \begin{pmatrix} (\mathbf{x}_1 - \bar{\mathbf{x}})'\mathbf{A} \\ (\mathbf{x}_2 - \bar{\mathbf{x}})'\mathbf{A} \\ \vdots \\ (\mathbf{x}_n - \bar{\mathbf{x}})'\mathbf{A} \end{pmatrix} = \mathbf{X}_c\mathbf{A},$$

and since $\mathbf{A}$ is orthogonal, we can solve for $\mathbf{X}_c$ to obtain $\mathbf{X}_c = \mathbf{Z}\mathbf{A}'$. We then substitute $\mathbf{X}_c = \mathbf{Z}\mathbf{A}'$ in the model (9.52) to express $\mathbf{y}$ as a function of the principal components:

$$\begin{aligned} \mathbf{y} &= \alpha\mathbf{j} + \mathbf{X}_c\boldsymbol{\beta}_1 + \boldsymbol{\varepsilon} = \alpha\mathbf{j} + \mathbf{Z}\mathbf{A}'\boldsymbol{\beta}_1 + \boldsymbol{\varepsilon} \\ &= \alpha\mathbf{j} + \mathbf{Z}\boldsymbol{\delta} + \boldsymbol{\varepsilon}, \end{aligned} \tag{9.56}$$

where $\boldsymbol{\delta} = \mathbf{A}'\boldsymbol{\beta}_1$ is $q \times 1$. Thus the model for each y_i in $\mathbf{y}$ regressed on the principal components z_{ij} is

$$y_i = \alpha + \delta_1 z_{i1} + \cdots + \delta_q z_{iq} + \varepsilon_i, \qquad i = 1, 2, \ldots, n. \tag{9.57}$$

From (7.20) or (9.53), we estimate $\boldsymbol{\delta}$ by

$$\hat{\boldsymbol{\delta}} = (\mathbf{Z}'\mathbf{Z})^{-1}\mathbf{Z}'\mathbf{y}. \tag{9.58}$$

By (7.8), the covariance matrix for $\hat{\boldsymbol{\delta}}$ is

$$\text{cov}(\hat{\boldsymbol{\delta}}) = \sigma^2(\mathbf{Z}'\mathbf{Z})^{-1} = \sigma^2\left[(\mathbf{X}_c\mathbf{A})'\mathbf{X}_c\mathbf{A}\right]^{-1} = \sigma^2\left[\mathbf{A}'\mathbf{X}_c'\mathbf{X}_c\mathbf{A}\right]^{-1}.$$

Since $\mathbf{A}$ is orthogonal with normalized eigenvectors of $\mathbf{X}_c'\mathbf{X}_c$ as columns, it diagonalizes $\mathbf{X}_c'\mathbf{X}_c$, and we have

$$\text{cov}(\hat{\boldsymbol{\delta}}) = \sigma^2\mathbf{D}^{-1} = \sigma^2\begin{pmatrix} \lambda_1 & 0 & \cdots & 0 \\ 0 & \lambda_2 & \cdots & 0 \\ \vdots & \vdots & & \vdots \\ 0 & 0 & \cdots & \lambda_q \end{pmatrix}^{-1},$$

where the λ's are eigenvalues of $\mathbf{X}_c'\mathbf{X}_c$. Hence

$$\text{var}(\hat{\delta}_j) = \frac{\sigma^2}{\lambda_j}, \qquad j = 1, 2, \ldots, q, \tag{9.59}$$

and if λ_j is small, $\text{var}(\hat{\delta}_j)$ will be large. Thus we can reduce the total variance of the $\hat{\delta}_j$'s, $\sum_{j=1}^q \text{var}(\hat{\delta}_j) = \sigma^2\sum_{j=1}^q(1/\lambda_j)$, by deleting some of the smaller components and regressing y on $m < q$ of the larger components $z_1, z_2, \ldots, z_m$.

Since the components are orthogonal, the estimator $\hat{\boldsymbol{\delta}}_m = (\hat{\delta}_1, \hat{\delta}_2, \ldots, \hat{\delta}_m)'$ is the same in the full model and the reduced model. We can therefore obtain $\hat{\boldsymbol{\delta}}_m$ either by taking the first m elements of $\hat{\boldsymbol{\delta}}$ from the full model

$$\hat{\boldsymbol{\delta}} = (\mathbf{Z}'\mathbf{Z})^{-1}\mathbf{Z}'\mathbf{y} = \mathbf{D}^{-1}\mathbf{A}'\mathbf{X}_c'\mathbf{y} \tag{9.60}$$

or by using the reduced model

$$\hat{\boldsymbol{\delta}}_m = \mathbf{D}_m^{-1}\mathbf{A}_m'\mathbf{X}_c'\mathbf{y}, \tag{9.61}$$

where $\mathbf{D}_m$ is a diagonal matrix that contains the first m eigenvalues of $\mathbf{X}_c'\mathbf{X}_c$ and the columns of $\mathbf{A}_m$ consist of the first m eigenvectors.

To obtain regression coefficients for the original x's, we can solve for $\boldsymbol{\beta}_1$ in $\boldsymbol{\delta} = \mathbf{A}'\boldsymbol{\beta}_1$ to obtain $\boldsymbol{\beta}_1 = \mathbf{A}\boldsymbol{\delta}$, since $\mathbf{A}$ is orthogonal. This can be estimated by $\hat{\boldsymbol{\beta}}_1 = \mathbf{A}\hat{\boldsymbol{\delta}}$, which is equivalent to (9.53). In the reduced model based on m components, this becomes

$$\tilde{\boldsymbol{\beta}}_1 = \mathbf{A}_m\hat{\boldsymbol{\delta}}_m = \mathbf{A}_m\mathbf{D}_m^{-1}\mathbf{A}_m'\mathbf{X}_c'\mathbf{y} \tag{9.62}$$

by (9.61). This estimator can also be obtained from the predicted model for an individual observation,

$$\begin{aligned} \hat{y}_i &= \bar{y} + \hat{\delta}_1 z_{i1} + \hat{\delta}_2 z_{i2} + \cdots + \hat{\delta}_m z_{im} \\ &= \bar{y} + \hat{\delta}_1\mathbf{a}_1'(\mathbf{x}_i - \bar{\mathbf{x}}) + \hat{\delta}_2\mathbf{a}_2'(\mathbf{x}_i - \bar{\mathbf{x}}) + \cdots + \hat{\delta}_m\mathbf{a}_m'(\mathbf{x}_i - \bar{\mathbf{x}}) \\ &= \bar{y} + (\hat{\delta}_1\mathbf{a}_1' + \hat{\delta}_2\mathbf{a}_2' + \cdots + \hat{\delta}_m\mathbf{a}_m')(\mathbf{x}_i - \bar{\mathbf{x}}) \\ &= \bar{y} + \tilde{\boldsymbol{\beta}}_1'(\mathbf{x}_i - \bar{\mathbf{x}}), \end{aligned}$$

where, by (A.3.4),

$$\tilde{\boldsymbol{\beta}}_1 = \hat{\delta}_1 \mathbf{a}_1 + \hat{\delta}_2 \mathbf{a}_2 + \cdots + \hat{\delta}_m \mathbf{a}_m = \mathbf{A}_m \hat{\boldsymbol{\delta}}_m.$$

By (1.94), the covariance matrix for $\tilde{\boldsymbol{\beta}}_1 = \mathbf{A}_m \hat{\boldsymbol{\delta}}_m$ is given by

$$\begin{aligned} \text{cov}(\tilde{\boldsymbol{\beta}}_1) &= \mathbf{A}_m \,\text{cov}(\hat{\boldsymbol{\delta}}_m) \mathbf{A}_m' \\ &= \sigma^2 \mathbf{A}_m \mathbf{D}_m^{-1} \mathbf{A}_m', \end{aligned} \tag{9.63}$$

with corresponding estimator

$$\widehat{\text{cov}}(\tilde{\boldsymbol{\beta}}_1) = s^2 \mathbf{A}_m \mathbf{D}_m^{-1} \mathbf{A}_m'. \tag{9.64}$$

The formulation leading to (9.62) and (9.64) is similar for components extracted from the correlation matrix of the x's. Let $\mathbf{X}_s$ indicate the matrix of standardized x-values, in which the x's are centered and divided by their standard deviations ($\mathbf{X}_s$ was called $\mathbf{Z}$ in Section 1.4). Then by (1.53),

$$\frac{\mathbf{X}_s'\mathbf{X}_s}{n-1} = \mathbf{R}_{xx},$$

the sample correlation matrix of the x's. By (A.10.4), the eigenvectors of $\mathbf{R}_{xx}$ and $\mathbf{X}_s'\mathbf{X}_s$ are the same, and the eigenvalues of $\mathbf{R}_{xx}$ are obtained from those of $\mathbf{X}_s'\mathbf{X}_s$ by dividing by $n - 1$.

When the x's are standardized, the regression coefficients become beta weights, defined in (7.63) as $\hat{\boldsymbol{\beta}}_1^* = \mathbf{R}_{xx}^{-1}\mathbf{r}_{yx}$, which is analogous to (9.53) if $\mathbf{X}_c$ is replaced by $\mathbf{X}_s$. In (9.62) and (9.64), $\mathbf{X}_c$ can be similarly replaced by $\mathbf{X}_s$ to obtain

$$\tilde{\boldsymbol{\beta}}_1^* = \mathbf{A}_m \mathbf{D}_m^{-1} \mathbf{A}_m' \mathbf{X}_s' \mathbf{y}, \tag{9.65}$$

$$\widehat{\text{cov}}(\tilde{\boldsymbol{\beta}}_1^*) = s^2 \mathbf{A}_m \mathbf{D}_m^{-1} \mathbf{A}_m', \tag{9.66}$$

where $\mathbf{A}_m$ and $\mathbf{D}_m$ contain the first m eigenvectors and eigenvalues, respectively, of $\mathbf{X}_s'\mathbf{X}_s$.

For simplicity of notation, most of the previous formulation has been in terms of eigenvalues and eigenvectors of $\mathbf{X}_c'\mathbf{X}_c$ or $\mathbf{X}_s'\mathbf{X}_s$. We can easily shift to $\mathbf{S}_{xx}$ or $\mathbf{R}_{xx}$. Since $\mathbf{X}_c'\mathbf{X}_c = (n-1)\mathbf{S}_{xx}$ and $\mathbf{X}_s'\mathbf{X}_s = (n-1)\mathbf{R}_{xx}$, the eigenvectors are unchanged, but the eigenvalue matrix $\mathbf{D}$ should be replaced by $(n-1)\mathbf{D}$ in changing from $\mathbf{X}_c'\mathbf{X}_c$ to $\mathbf{S}_{xx}$ or from $\mathbf{X}_s'\mathbf{X}_s$ to $\mathbf{R}_{xx}$.

In applications, we must decide which components to delete. A deleted component may have a small λ_j but may correlate better with y than another component with a larger λ_j; for example, the component may only account for 3% of $\sum_j \lambda_j$ but may increase R^2 by 12% if included. Thus if one were deleting 10 components, for example, it might be advisable to examine the 10 to see if any of them provide more information about y than some of the components being retained. Keeping components with smaller λ_j will increase the total variance of the estimated coefficients,

Table 9.1. Longley Data

y	x_1	x_2	x_3	x_4	x_5	x_6
1892	83.0	234,289	2356	1590	107,608	1947
1863	88.5	259,426	2325	1456	108,632	1948
1908	88.2	258,054	3682	1616	109,773	1949
1928	89.5	284,599	3351	1650	110,929	1950
2302	96.2	328,975	2099	3099	112,075	1951
2420	98.1	346,999	1932	3594	113,270	1952
2305	99.0	365,385	1870	3547	115,094	1953
2188	100.0	363,112	3578	3350	116,219	1954
2187	101.2	397,469	2904	3048	117,388	1955
2209	104.6	419,180	2822	2857	118,734	1956
2217	108.4	442,769	2936	2798	120,445	1957
2191	110.8	444,546	4681	2637	121,950	1958
2233	112.6	482,704	3813	2552	123,366	1959
2270	114.2	502,601	3931	2514	125,368	1960
2279	115.7	518,173	4806	2572	127,852	1961
2340	116.9	554,894	4007	2827	130,081	1962

but the increase may be offset by the gain in predictability. This trade-off has been discussed by Belinfonte and Coxe (1986), Sutter, Kalivas, and Lang (1992), and Næs and Helland (1993). Jolliffe (1982) cites examples from the literature in which some of the smallest components are important predictors.

Mason and Gunst (1985) discussed a test for significance of the regression coefficient $\hat{\delta}_j$ for each component but found the power of the test to be rather low and did not advocate exclusive reliance on the test to select components. Marx and Smith (1990) proposed an extension of principal component regression to generalized linear regression (see Section 7.1). Boneh and Mendieta (1992) proposed the use of principal component regression as a tool for subset selection in data sets with a high degree of multicolinearity.

Example 9.8.1. To illustrate principal component regression, we use a data set (Table 9.1) by Longley (1967) that has high multicolinearity and has often been used to test regression software for numerical accuracy. This data set has been used to illustrate principal component regression by Hill, Fomby, and Johnson (1977), Karson (1982, pp. 219–223), and Coxe (1975).

The variables are y = number of federal government employees (1000s), x_1 = GNP price deflator (1954 value = 100), x_2 = gross national product ($100,000s), x_3 = unemployed (1000s), x_4 = size of armed forces (1000s), x_5 = population 14 years and over (1000s), x_6 = year. The data were collected for 16 consecutive years, so it is clear that the y's are not independent, but we will ignore this. The means and standard deviations are given below:

Variable	$\overline{x}$	s
x_1	101.7	10.8
x_2	387,698.4	99,394.9
x_3	3,193.3	934.5
x_4	2,606.7	695.9
x_5	117,424.0	6,956.1
x_6	1,954.5	4.8

Because the variances are extremely disparate, we will use principal components of the correlation matrix. The correlations between y and the x's are given by

$$\mathbf{r}'_{yx} = (.7217, .7108, .0525, .8935, .6411, .6733),$$

and the correlation matrix of the six x's is

$$\mathbf{R}_{xx} = \begin{bmatrix} 1 & .9916 & .6206 & .4647 & .9792 & .9911 \\ & 1 & .6043 & .4464 & .9911 & .9953 \\ & & 1 & -.1774 & .6866 & .6683 \\ & & & 1 & .3644 & .4172 \\ & \text{(symmetric)} & & & 1 & .9940 \\ & & & & & 1 \end{bmatrix}.$$

The presence of multicolinearity is indicated by several very high correlations and by the small determinant $|\mathbf{R}_{xx}| = 1.6 \times 10^{-8}$.

The eigenvalues λ_j of $\mathbf{R}_{xx}$ are 4.6034, 1.1753, .2035, .0149, .0025, and .0004. The first three account for 99.7% of the (standardized) total sample variance, and we use them for regression. The first three eigenvectors of $\mathbf{R}_{xx}$ are

$$\begin{aligned} \mathbf{a}_1 &= (.462 \quad .462 \quad .321 \quad .202 \quad .462 \quad .465)', \\ \mathbf{a}_2 &= (.058 \quad .053 \quad -.596 \quad .798 \quad -.046 \quad .001)', \\ \mathbf{a}_3 &= (-.149 \quad -.278 \quad .728 \quad .562 \quad -.196 \quad -.128)', \end{aligned}$$

which become the columns of $\mathbf{A}_m$. We use $(n-1)\lambda_j$ in $\mathbf{D}_m$ because we are working with eigenvalues of $\mathbf{R}_{xx}$ instead of $\mathbf{X}_s'\mathbf{X}_s$. The principal component regression coefficients are given by (9.65), and their estimated variances are given by the diagonal elements of (9.66).

In Table 9.2, we compare the principal component regression coefficients and the standardized least squares regression coefficients [beta weights, $\hat{\boldsymbol{\beta}}_1^* = \mathbf{R}_{xx}^{-1}\mathbf{r}_{yx}$ from (7.63)], along with their corresponding estimated variances. The variances of the $\hat{\beta}_j^*$'s are given by the diagonal elements of $\mathbf{R}_{xx}^{-1}$ multiplied by $\sigma^2/(n-1) = \sigma^2/15$, because $\mathbf{X}_s'\mathbf{X}_s = (n-1)\mathbf{R}_{xx}$. The estimate of σ^2 is obtained from (7.11) as $s^2 = 1572.27$. Since s^2 is rather large, we omit it from the comparison of the

Table 9.2. Comparison of Least Squares and Principal Component Regression Coefficients and Variances

Standardized Variable	Least Squares		Principal Component	
	Coefficient	Var/s^2	Coefficient	Var/s^2
x_1	36.61	9.04	29.02	.01
x_2	1063.98	119.23	25.95	.03
x_3	118.18	2.24	−32.45	.20
x_4	141.23	0.24	109.14	.14
x_5	−179.94	26.61	16.87	.02
x_6	−936.77	50.60	23.38	.01

variances of the regression coefficients. In Table 9.2, the coefficients in the principal component regression differ considerably from those for least squares, and the four large variances are greatly reduced.

9.8.2 Latent Root Regression

Latent root regression uses eigenvalues and eigenvectors from the augmented covariance or correlation matrix, which includes y as well as the x's:

$$\mathbf{S} = \begin{pmatrix} s_{yy} & \mathbf{s}'_{yx} \\ \mathbf{s}_{yx} & \mathbf{S}_{xx} \end{pmatrix} \quad \text{or} \quad \mathbf{R} = \begin{pmatrix} 1 & \mathbf{r}'_{yx} \\ \mathbf{r}_{yx} & \mathbf{R}_{xx} \end{pmatrix}.$$

The principal components based on these eigenvectors are functions of y and the x's. We need not distinguish between $\mathbf{S}_{xx}$ and $\mathbf{X}'_c\mathbf{X}_c$ or between $\mathbf{R}_{xx}$ and $\mathbf{X}'_s\mathbf{X}_s$, as in Section 9.8.1, because the eigenvalues appear in both the numerator and denominator of the expression (9.67) below for the regression coefficients.

We examine the coefficient of y in each of the smaller principal components. If the coefficient of y is small, the principal component is called a *nonpredictive near singularity* and is deleted. If the principal component has a large coefficient for y, it has predictive value and is retained.

We denote the eigenvalues of $\mathbf{S}$ or $\mathbf{R}$ by $\phi_0, \phi_1, \ldots, \phi_q$ and the eigenvectors by $(v_{j0}, \mathbf{v}'_j)$, $j = 0, 1, \ldots, q$, where the coefficient of y is v_{j0} and the coefficient vector of the x's is $\mathbf{v}'_j$. Let M be the subset of the integers $0, 1, \ldots, q$ that correspond to the eigenvectors retained. Then the latent root estimator of $\boldsymbol{\beta}$ is

$$\hat{\boldsymbol{\beta}}_L = \frac{-(n-1)s_y \sum_M (v_{j0}/\phi_j)\mathbf{v}_j}{\sum_M v_{j0}^2/\phi_j}. \tag{9.67}$$

If the eigenvalues and eigenvectors are from $\mathbf{S}$, then $\hat{\boldsymbol{\beta}}_L$ is applied to the centered x's. If the eigenvalues and eigenvectors are from $\mathbf{R}$, the coefficients in $\hat{\boldsymbol{\beta}}_L$ are standardized (beta weights) and are applied to standardized x's (z-scores).

9.9 PRINCIPAL COMPONENT ANALYSIS WITH GROUPED DATA

In this section, we discuss the use of principal components on data consisting of several samples or groups. Principal components have occasionally been used to show separation of groups, but in general they are not the most effective linear combinations for this purpose because they are designed to separate individual observations rather than groups. The directions of the axes of the principal components will ordinarily not be the same as the directions of the axes of the discriminant functions (Chapter 5).

We will discuss four options for principal component analysis of grouped data: (1) ignore the groups, (2) use the pooled within-groups covariance or correlation matrix, (3) define *common principal components* for the groups when pooling is not advisable, or (4) compare the principal components from the groups.

The first option, ignoring the groups and using an overall covariance matrix, would be used, for example, to reduce the number of variables before comparing the groups by means of MANOVA or discriminant analysis.

If we wish to search for size and shape or other aspects of basic dimensionality, we can consider the second option, pooling the covariance matrices of the groups before extracting principal components. In this case, we need to be concerned with homogeneity of covariance matrices before pooling.

If there is reason to believe that the $\boldsymbol{\Sigma}_i$'s for the groups are not equal, we may wish to consider the third option, that of defining *common principal components* for all the groups, as proposed by Flury (1984, 1988) and Krzanowski (1984b). In some cases, the directions of the principal axes are the same for all the groups, so that one set of principal components would suffice for all groups. In Figure 9.3, we see two examples of bivariate population contours that represent covariance matrices

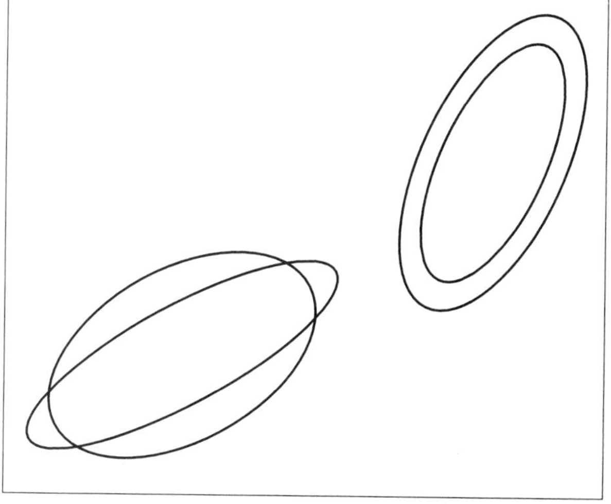

Figure 9.3. Two illustrations of contours with common principal axes.

that are different but have common principal axes. In practice, the sample covariance matrices will not have identical axes, and a compromise is made between the principal component transformations for two or more groups. Lefkovitch (1993) proposed a noniterative approach called consensus principal components. Beaghan and Smith (1996) discussed three-mode principal components and compared the maximum likelihood estimators of the common principal components model with the least squares estimators.

The fourth option involves computing principal components separately for each group and then comparing the results. Krzanowski (1979, 1982) suggested a technique for comparing the principal components from two or more groups. Keramidas, Devlin, and Gnanadesikan (1987) presented a method based on *Q–Q* plots for comparing principal components from several groups. Schott (1991) provided a test for equality of the first m principal components for several groups of data.

9.10 ADDITIONAL TOPICS

In this section, we return to the case of a single sample and discuss various extensions and modifications of principal components. Wold et al. (1987) proposed a method of calculating the *leverage* of each observation, which can be used to indicate the influence of the observation on the principal components. As in Section 7.6.3, we calculate the $n \times n$ projection matrix for the principal components $\mathbf{YA}$ [see (9.4)],

$$\mathbf{P} = \mathbf{YA}(\mathbf{A}'\mathbf{Y}'\mathbf{YA})^{-1}\mathbf{A}'\mathbf{Y}', \tag{9.68}$$

where $\mathbf{Y}$ is the data matrix

$$\mathbf{Y} = \begin{pmatrix} \mathbf{y}_1' \\ \mathbf{y}_2' \\ \vdots \\ \mathbf{y}_n' \end{pmatrix},$$

and $\mathbf{A} = (\mathbf{a}_1, \mathbf{a}_2, \ldots, \mathbf{a}_p)$ has eigenvectors of $\mathbf{S}$ (or $\mathbf{R}$, if $\mathbf{Y}$ is standardized) as columns. Then p_{ii}, the ith diagonal element of $\mathbf{P}$ in (9.68), is the leverage for the ith observation vector $\mathbf{y}_i$. It can be shown that $0 \leq p_{ii} \leq 1$. A high value of p_{ii} may indicate an outlier, which would exert undue influence or leverage on a principal component if it lies far from the axis represented by the component (it has a longer "lever" than other points). Kim and Park (1991) discussed leverage and residuals and also adapted various other measures of influence to principal components, such as the Welch–Kuh distance, Cook's distance, the Andrews–Pregibon statistic, and the Cook–Weisberg statistic (some of these are discussed in Section 7.6.3).

Influence functions for the effect of individual observations on the eigenvalues, eigenvectors, and principal components have been discussed and illustrated by Critchley (1985), Jolliffe (1986), Tanaka (1988), and Pack, Jolliffe, and Morgan (1988). Ramsier (1991) proposed a graphical method for detection of influential observations on the covariance matrix as well as on its eigenvalues and eigenvectors. In a related study, Wang and Nyquist (1991) investigated the effect that removal of an observation has on eigenvalues, eigenvectors, and principal components.

Principal component analysis is sensitive to the presence of outliers. In fact, a principal component could be created by an outlier. We may be able to reduce the effect of outliers by using robust estimates of principal components, obtained by extracting eigenvalues and eigenvectors from a robust covariance matrix (Zhou 1989, Devlin et al., 1981, Gnanadesikan 1977, Chapter 5). Ammann (1989) provided a method for direct robust estimation of principal components. Naga and Antile (1990) warned that robust estimators of the covariance matrix are less stable than ordinary estimators. Xie et al. (1993) discussed a robust approach to principal components based on projection pursuit (see Section 9.3).

Principal component analysis for discrete data, especially data that involve binary variables, has been discussed by Gower (1966, 1967), Cox (1972), and Bloomfield (1974). For principal components on nominal or categorical data, Greenacre (1984) proposed *correspondence analysis*, which can be used to obtain a two-dimensional plot of data from a contingency table. DeLeeuw and Rijckevorsel (1980) and Jolliffe (1986, Section 11.1) describe the connection between correspondence analysis and principal component analysis for nominal data based on a singular value decomposition of a contingency table.

A topic of interest in the psychometric literature is *multidimensional scaling*, in which the pairwise "distances" or similarities among n objects are used to construct a two-dimensional representation showing a spatial pattern of the points. A good review of the various approaches to multidimensional scaling has been given by Mead (1992).

Aitchison (1982, 1983, 1986) and Baxter et al. (1990) have discussed principal components for "compositional data" in which the p variables are proportions such that $\sum_{j=1}^{p} y_j = 1$, arising, for example, from chemical mixtures. Saito, Kariya, and Otsu (1988), DeLeeuw (1982), and Gnanadesikan (1977, Section 2.4.2) discussed nonlinear principal component analysis.

Flury (1990, 1993) defined *principal points* as the k points $\boldsymbol{\xi}_1, \boldsymbol{\xi}_2, \ldots, \boldsymbol{\xi}_k$ that minimize the average distance between points $\mathbf{y}_i$ in the population or sample and the nearest $\boldsymbol{\xi}_i$. The principal points thus typify or represent the distribution or sample. An example was given for $k = 4$.

Hastie and Stuetzle (1989) defined *principal curves* as a generalization of principal components. A principal curve is fitted using the orthogonal or shortest distance to the points. They illustrated the technique with two applications. An alternative definition of principal curves was suggested by Tibshirani (1992).

PROBLEMS

9.1 Show that maximizing $\mathbf{a}'\mathbf{S}\mathbf{a}$ subject to $\mathbf{a}'\mathbf{a} = 1$ gives the same results as maximizing $\mathbf{a}'\mathbf{S}\mathbf{a}/\mathbf{a}'\mathbf{a}$, as noted following Theorem 9.2A.

9.2 Show that

$$\frac{\partial}{\partial \mathbf{a}_2}\left(\frac{\mathbf{a}_2'\mathbf{S}\mathbf{a}_2}{\mathbf{a}_2'\mathbf{a}_2} + \gamma\mathbf{a}_1'\mathbf{a}_2\right) = \mathbf{0}$$

leads to $\gamma = 0$ and $\mathbf{S}\mathbf{a}_2 = \lambda_2\mathbf{a}_2$, as in the proof of Theorem 9.2B.

9.3 As an alternative proof of Theorem 9.2B, use Lagrange multipliers to show that the second eigenvector $\mathbf{a}_2$ maximizes $\mathbf{a}_2'\mathbf{S}\mathbf{a}_2$ subject to $\mathbf{a}_1'\mathbf{a}_2 = 0$ and $\mathbf{a}_2'\mathbf{a}_2 = 1$.

9.4 Prove that the first principal component represents the line on which the sum of squared projections of the points is a maximum. This assertion appears in the first paragraph of Section 9.2.2.

9.5 Show that $s_{z_j}^2 = \lambda_j$ as in (9.6) by writing (9.2) as $\mathbf{S}\mathbf{a}_j = \lambda_j\mathbf{a}_j$ and multiplying by $\mathbf{a}_j'$.

9.6 Show that the orthogonality of $\mathbf{a}_i$ and $\mathbf{a}_j$ for $i \neq j$ leads to $\mathbf{a}_i'\mathbf{S}\mathbf{a}_j = 0$ as in (9.11).

9.7 If $\mathbf{a}$ is an eigenvector of $\mathbf{S}$, show that $\mathbf{b} = (\mathbf{C}')^{-1}\mathbf{a}$ is not an eigenvector of $\mathbf{CSC}'$, where $\mathbf{C}$ is nonsingular. This was noted in property 5 in Section 9.2.3, where it was demonstrated that principal components are not scale invariant.

9.8 Show that if y_p is uncorrelated (in the sample) with the other variables, then its variance s_p^2 is an eigenvalue of $\mathbf{S}$ with corresponding eigenvector $(0, 0, \ldots, 0, 1)'$ (Section 9.2.3, property 6).

9.9 If $\mathbf{z}_i = \mathbf{A}\mathbf{y}_i$, where $\mathbf{A}$ is an orthogonal matrix, show that $\mathbf{z}_i'\mathbf{z}_i = \mathbf{y}_i'\mathbf{y}_i$. Thus principal components represent a rotation of axes, as noted in Section 9.3.

9.10 Obtain (9.18) in the following alternative way.

(a) Show that $\mathbf{R}_k = k\mathbf{R} - (k-1)\mathbf{I}$ as in (9.16).

(b) Show that $(\mathbf{A} - b\mathbf{I})\mathbf{x} = (\lambda - b)\mathbf{x}$, which is an extension of (A.10.8). Use this and (A.10.4) to obtain $\lambda_k = k\lambda - (k-1)$ as an eigenvalue of $k\mathbf{R} - (k-1)\mathbf{I}$; then solve for λ.

9.11 Obtain the confidence interval in (9.20) from $P(-w_{\alpha/2} \leq w \leq w_{\alpha/2}) = 1-\alpha$, where w is given in (9.19).

9.12 Show that the test statistic $u = |\mathbf{S}|/(s_{11}s_{22}\cdots s_{pp})$ in (9.23) is a function of the likelihood ratio.

9.13 Show that if $\mathbf{S} = \mathbf{E}/\nu_E$, where $\mathbf{E}$ is the error matrix from a MANOVA, the test statistic (9.23) can be expressed as

$$u = \frac{|\mathbf{E}|}{e_{11}e_{22}\cdots e_{pp}}.$$

9.14 Show that the squared length of the residual vector $z_{p-k,i}\mathbf{a}_{p-k} + \cdots + z_{pi}\mathbf{a}_p$ is given by $d_i^2 = z_{p-k,i}^2 + \cdots + z_{pi}^2$ as in (9.27).

9.15 It was noted in Section 9.7.1 that the covariance pattern $\mathbf{\Sigma}_0 = \sigma^2[(1-\rho)\mathbf{I}+\rho\mathbf{J}]$ has a size component $z_1 = \boldsymbol{\alpha}_1'\mathbf{y} = (1/\sqrt{p})\mathbf{j}'\mathbf{y}$ and shape components $z_2 = \boldsymbol{\alpha}_2'\mathbf{y}, \ldots, z_p = \boldsymbol{\alpha}_p'\mathbf{y}$, where $\boldsymbol{\alpha}_i'\boldsymbol{\alpha}_1 = 0$, $i = 2, \ldots, p$.

(a) Show that for the case of two groups with a common covariance matrix $\boldsymbol{\Sigma}_0$ the population discriminant function is

$$\boldsymbol{\delta}'\boldsymbol{\Sigma}_0^{-1}\mathbf{y} = \frac{\boldsymbol{\delta}'\mathbf{j}}{\sigma^2 p(1-\rho)}\left[\mathbf{b}'\mathbf{y} + \frac{1-\rho}{1+(p-1)\rho}\mathbf{j}'\mathbf{y}\right],$$

where $\boldsymbol{\delta} = \boldsymbol{\mu}_1 - \boldsymbol{\mu}_2$ and $\mathbf{b} = p\boldsymbol{\delta}/\boldsymbol{\delta}'\mathbf{j} - \mathbf{j}$.

(b) Show that $\mathbf{b}'\mathbf{j} = 0$. Thus $\mathbf{b}'\mathbf{y}$ is a shape term and $\mathbf{j}'\mathbf{y}$ is a size term.

9.16 Show that $\boldsymbol{\Sigma}$ in (9.34) is the same as in (9.35), $\sigma^2[(1-\rho)\mathbf{I} + \rho\mathbf{J}]$.

9.17 Show that s^2 and $s^2 r$ in (9.36) are the maximum likelihood estimators under H_0.

9.18 Suppose $\boldsymbol{\Sigma} = c\mathbf{I} + \mathbf{J}$ is $p \times p$. Show that p is the nonzero eigenvalue of $\mathbf{J}$, that $c + p, c, \ldots, c$ are the eigenvalues of $\boldsymbol{\Sigma}$, and that $z_1 = \sqrt{p}\bar{y}$ is the first principal component of $\boldsymbol{\Sigma}$, where $\bar{y} = \sum_{j=1}^{p} y_j/p$.

9.19 An alternative form of the test statistic in (9.39) is $(1-c)z$, where $z = -\nu \ln u$ and

$$c = \frac{p(p+1)^2(2p-3)}{6\nu(p-1)(p^2+p-4)}.$$

Show that this is equivalent to (9.39).

9.20 Show that $(\mathbf{Z}'\mathbf{Z})^{-1}\mathbf{Z}'\mathbf{y} = \mathbf{D}^{-1}\mathbf{A}'\mathbf{X}_c'\mathbf{y}$ as in (9.60).

9.21 Show that $\hat{\boldsymbol{\beta}}_1 = \mathbf{A}\hat{\boldsymbol{\delta}}$, as given following (9.61), is the same as $\hat{\boldsymbol{\beta}}_1 = (\mathbf{X}_c'\mathbf{X}_c)^{-1}\mathbf{X}_c'\mathbf{y}$ in (9.53).

9.22 Use all nine variables (three y's and six x's) from the rainfall data in Table 7.1.

(a) Test H_0: $\boldsymbol{\Sigma} = \sigma^2[(1-\rho)\mathbf{I} + \rho\mathbf{J}]$ using the χ^2-approximation in (9.39) and the F-approximation in (9.40).

(b) Test H_0: $\rho_{ij} = \rho$ using the approximate χ^2-statistic u in (9.42).

9.23 Use $y_1, y_2, \ldots, y_{11}$ from the biochemical data in Table 4.11 and ignore the grouping variable.

(a) Test H_0: $\boldsymbol{\Sigma} = \sigma^2[(1-\rho)\mathbf{I} + \rho\mathbf{J}]$ using the χ^2-approximation in (9.39) and the F-approximation in (9.40).

(b) Test H_0: $\rho_{jk} = \rho$ using the approximate χ^2-statistic u in (9.42).

9.24 Using $y, x_1, x_2, \ldots, x_6$ from the Longley data in Table 9.1, test H_0: $\rho_{jk} = \rho$ using the approximate χ^2-statistic u in (9.42).

9.25 Carry out a principal component analysis of the probe word data of Table 1.3. Use both **S** and **R**. Show the percent of variance explained for each of the five components. Based on the average eigenvalue or a scree plot, decide how many components to retain. Can you interpret the components of either **S** or **R**? Which do you think is more appropriate for this data set?

9.26 Carry out a principal component analysis on the glucose data of Table 1.4. Use all six variables, $y_1, y_2, y_3, x_1, x_2, x_3$. Use both **S** and **R**. Show the percent of variance explained for each of the six components. Based on the average eigenvalue or a scree plot, decide how many components to retain.

Can you interpret the components of either **S** or **R**? Which do you think is more appropriate for this data set?

9.27 Carry out a principal component analysis on $y_1, y_2, \ldots, y_{11}$ of the biochemical data in Table 4.11. Combine the four groups into a single sample. Use both **S** and **R**. Show the percent of variance explained for each of the 11 components. Based on the average eigenvalue or a scree plot, decide how many components to retain. Can you interpret the components of either **S** or **R**? Which do you think is more appropriate for this data set?

9.28 Carry out a principal component analysis on $y, x_1, x_2, \ldots, x_6$ from the Longley data in Table 9.1.

(a) Use **S** and **R**.

(b) Scale all the variables to three decimal places; for example, divide y by 10, do not change x_1, divide x_2 by 1000. Then use **S** and compare the results with those of part (a).

9.29 Carry out a principal component analysis on $y_1, y_2, \ldots, y_6, x_1, x_2$ from the apple data in Table 4.10. Combine the four groups into a single sample. Use both **S** and **R**. Show the percent of variance explained for each of the eight components. Based on the average eigenvalue or a scree plot, decide how many components to retain. Can you interpret the components of either **S** or **R**? Which do you think is more appropriate for this data set?

9.30 Carry out a principal component analysis on $y_1, y_2, y_3, x_1, x_2, \ldots, x_6$ from the rainfall data of Table 7.1. Use both **S** and **R**. Show the percent of variance explained for each of the nine components. Based on the average eigenvalue or a scree plot, decide how many components to retain. Can you interpret the components of either **S** or **R**? Which do you think is more appropriate for this data set?

9.31 Carry out a principal component regression (Section 9.8.1) on the rainfall data in Table 7.1. Use y_1 as the dependent variable and $y_2, y_3, x_1, x_2, \ldots, x_6$ as the eight independent variables; that is, regress y_1 on the components of $y_2, y_3, x_1, \ldots, x_6$.

CHAPTER 10

Factor Analysis

10.1 INTRODUCTION

In a factor analysis model, the observed variables $y_1, y_2, \ldots, y_p$ are expressed as linear combinations of a few unobserved random variables $f_1, f_2, \ldots, f_m$ $(m < p)$ called *factors*. Factor analysis thus differs from principal component analysis in that principal components are defined as linear combinations of the y's. For additional delineation of the essential distinction between these two techniques, see Gangemi (1986), Sato (1990), Velicer and Jackson (1990), Jolliffe (1992), Widaman (1993), and Rencher (1995, Section 13.8).

The development of factor analysis was initiated by Spearman (1904), who postulated that a single factor, general intelligence, could account for most performance measurements associated with mental endeavor. However, the one-factor model was not adequate for many applications, and it was generalized by Garnett (1919), Thurstone (1931), and others. There is a large body of literature in factor analysis. For a good elementary exposition, see Cooper (1983). Useful monographs are given by Harman (1976), Kim and Mueller (1978), Mulaik (1972), Cureton and D'Agostino (1983), Gorsuch (1983), McDonald (1985), Comrey and Lee (1992), and Basilevsky (1994).

10.2 BASIC FACTOR MODEL

10.2.1 Model and Assumptions

Factor analysis is basically a one-sample procedure; that is, we assume a random sample $\mathbf{y}_1, \mathbf{y}_2, \ldots, \mathbf{y}_n$ from a single population with mean vector $\boldsymbol{\mu}$ and covariance matrix $\boldsymbol{\Sigma}$. The case in which there are groups in the data is discussed in Section 10.8.

Each variable $y_1, y_2, \ldots, y_p$ in the random vector $\mathbf{y}$ is assumed to be a linear function of m factors $f_1, f_2, \ldots, f_m$:

$$
\begin{aligned}
y_1 - \mu_1 &= \lambda_{11} f_1 + \lambda_{12} f_2 + \cdots + \lambda_{1m} f_m + \varepsilon_1 \\
y_2 - \mu_2 &= \lambda_{21} f_1 + \lambda_{22} f_2 + \cdots + \lambda_{2m} f_m + \varepsilon_2 \\
&\vdots \\
y_p - \mu_p &= \lambda_{p1} f_1 + \lambda_{p2} f_2 + \cdots + \lambda_{pm} f_m + \varepsilon_p.
\end{aligned}
\tag{10.1}
$$

If m is substantially smaller than p, the representation in (10.1) may become a useful parsimonious description of the variables. The coefficients λ_{ij}, called *loadings*, can be used in the interpretation of the factors. We interpret f_2, for example, by examining its loadings $\lambda_{12}, \lambda_{22}, \ldots, \lambda_{p2}$ and noting which y's have large loadings on f_2. This subset of y's gives some identification to f_2.

The models in (10.1) for the p variables can be combined in the single matrix expression

$$\mathbf{y} - \boldsymbol{\mu} = \boldsymbol{\Lambda}\mathbf{f} + \boldsymbol{\varepsilon}, \tag{10.2}$$

where $\mathbf{y} = (y_1, y_2, \ldots, y_p)'$, $\boldsymbol{\mu} = (\mu_1, \mu_2, \ldots, \mu_p)'$, $\mathbf{f} = (f_1, f_2, \ldots, f_m)'$, $\boldsymbol{\varepsilon} = (\varepsilon_1, \varepsilon_2, \ldots, \varepsilon_p)'$, and

$$\boldsymbol{\Lambda} = \begin{pmatrix} \lambda_{11} & \lambda_{12} & \cdots & \lambda_{1m} \\ \lambda_{21} & \lambda_{22} & \cdots & \lambda_{2m} \\ \vdots & \vdots & & \vdots \\ \lambda_{p1} & \lambda_{p2} & \cdots & \lambda_{pm} \end{pmatrix}. \tag{10.3}$$

For further simplification, we assume that

$$E(\mathbf{f}) = \mathbf{0}, \tag{10.4}$$

$$\text{cov}(\mathbf{f}) = \mathbf{I}, \tag{10.5}$$

$$E(\boldsymbol{\varepsilon}) = \mathbf{0}, \tag{10.6}$$

$$\text{cov}(\boldsymbol{\varepsilon}) = \boldsymbol{\Psi} = \begin{pmatrix} \psi_1 & 0 & \cdots & 0 \\ 0 & \psi_2 & \cdots & 0 \\ \vdots & \vdots & & \vdots \\ 0 & 0 & \cdots & \psi_p \end{pmatrix}, \tag{10.7}$$

$$\text{cov}(\mathbf{f}, \boldsymbol{\varepsilon}) = \mathbf{O}. \tag{10.8}$$

From the model (10.2) and the assumptions (10.4)–(10.8), we obtain the following covariance structure for $\mathbf{y}$ and $\mathbf{f}$:

$$\text{cov}(\mathbf{y}) = \boldsymbol{\Sigma} = \boldsymbol{\Lambda}\boldsymbol{\Lambda}' + \boldsymbol{\Psi}, \tag{10.9}$$

$$\text{cov}(\mathbf{y}, \mathbf{f}) = \boldsymbol{\Lambda}. \tag{10.10}$$

If standardized variables are used, (10.9) is replaced by a model for the correlation matrix:

$$\mathbf{P}_\rho = \mathbf{\Lambda}\mathbf{\Lambda}' + \mathbf{\Psi}. \tag{10.11}$$

From (10.10) we see that the loadings are covariances of the variables with the factors:

$$\text{cov}(y_i, f_j) = \lambda_{ij}. \tag{10.12}$$

Similarly, in the model (10.11), the loadings represent correlations of the variables with the factors,

$$\text{corr}(y_i, f_j) = \lambda_{ij}. \tag{10.13}$$

The diagonal elements of (10.9) are of the form

$$\begin{aligned}\sigma_{ii} &= \text{var}(y_i) = \lambda_{i1}^2 + \lambda_{i2}^2 + \cdots + \lambda_{im}^2 + \psi_i \\ &= \text{communality} + \text{specific variance},\end{aligned}$$

where

$$\text{Communality} = h_i^2 = \lambda_{i1}^2 + \lambda_{i2}^2 + \cdots + \lambda_{im}^2, \tag{10.14}$$

$$\text{Specific variance} = \psi_i. \tag{10.15}$$

Thus the variance of y_i is partitioned into a part that is common with the other $p - 1$ variables and a part that is unique to y_i.

When the assumptions (10.4)–(10.8) are added to the factor analysis model (10.2), we obtain the simplified covariance structure in (10.9). A critical assumption is (10.7), $\text{cov}(\boldsymbol{\varepsilon}) = \mathbf{\Psi} = \text{diag}(\psi_1, \psi_2, \ldots, \psi_p)$, which specifies that all covariances among the y's are accounted for by the factors; that is, $\mathbf{\Lambda}\mathbf{\Lambda}'$ in (10.9) expresses all the covariances. This clearly fails to hold in fitting the model to some data sets. However, we do not relax assumption (10.7) because it is crucial to estimation of $\mathbf{\Lambda}$.

10.2.2 Scale Invariance of the Model

The basic factor analysis model (10.2) and the covariance partitioning (10.9) retain the same form under a full-rank linear transformation. If $\mathbf{y}$ is transformed to $\mathbf{z} = \mathbf{A}\mathbf{y} + \mathbf{b}$, where $\mathbf{A}$ is nonsingular, the model $\mathbf{y} - \boldsymbol{\mu}_y = \mathbf{\Lambda}_y\mathbf{f} + \boldsymbol{\varepsilon}_y$ in (10.2) becomes

$$\mathbf{A}\mathbf{y} - \mathbf{A}\boldsymbol{\mu}_y = \mathbf{A}\mathbf{\Lambda}_y\mathbf{f} + \mathbf{A}\boldsymbol{\varepsilon}_y$$

or

$$\mathbf{z} - \boldsymbol{\mu}_z = \mathbf{\Lambda}_z\mathbf{f} + \boldsymbol{\varepsilon}_z, \tag{10.16}$$

where $\boldsymbol{\mu}_z = \mathbf{A}\boldsymbol{\mu}_y + \mathbf{b}$ and $\boldsymbol{\Lambda}_z = \mathbf{A}\boldsymbol{\Lambda}_y$. Similarly, the covariance relationship $\boldsymbol{\Sigma}_y = \boldsymbol{\Lambda}_y\boldsymbol{\Lambda}_y' + \boldsymbol{\Psi}_y$ in (10.9) becomes

$$\begin{aligned} \text{cov}(\mathbf{z}) = \boldsymbol{\Sigma}_z = \mathbf{A}\boldsymbol{\Sigma}_y\mathbf{A}' &= \mathbf{A}\boldsymbol{\Lambda}_y\boldsymbol{\Lambda}_y'\mathbf{A}' + \mathbf{A}\boldsymbol{\Psi}_y\mathbf{A}' \\ &= \boldsymbol{\Lambda}_z\boldsymbol{\Lambda}_z' + \boldsymbol{\Psi}_z. \end{aligned} \tag{10.17}$$

Thus a linear transformation of the variables, as, for example, a simple change of scale, carries over to the same transformation of the loadings. One consequence is that the model transforms readily to accommodate standardized variables $\mathbf{z} = \mathbf{D}_\sigma^{-1}(\mathbf{y} - \boldsymbol{\mu}_y)$, where $\mathbf{D}_\sigma = \text{diag}(\sigma_1, \sigma_2, \ldots, \sigma_p)$. In this case, $\boldsymbol{\Sigma}_z$ in (10.17) becomes $\mathbf{P}_\rho$, the population correlation matrix. This invariance property of the model carries over to some estimators of loadings in Section 10.3 but not to others.

10.2.3 Rotation of Factor Loadings in the Model

If the loadings in the model in (10.2) are multiplied by an orthogonal matrix, the covariance structure in (10.9) remains intact. Let $\mathbf{T}$ be an arbitrary $m \times m$ orthogonal matrix such that $\mathbf{T}'\mathbf{T} = \mathbf{T}\mathbf{T}' = \mathbf{I}$ [see (A.9.2) and (A.9.3)]. Then

$$\begin{aligned} \boldsymbol{\Sigma} &= \boldsymbol{\Lambda}\mathbf{T}\mathbf{T}'\boldsymbol{\Lambda}' + \boldsymbol{\Psi} \\ &= \boldsymbol{\Lambda}^*\boldsymbol{\Lambda}^{*\prime} + \boldsymbol{\Psi}, \end{aligned} \tag{10.18}$$

where

$$\boldsymbol{\Lambda}^* = \boldsymbol{\Lambda}\mathbf{T}. \tag{10.19}$$

Since multiplication of a vector by an orthogonal matrix is equivalent to a rotation of axes [see (A.9.5)], the rows of $\boldsymbol{\Lambda}$ are rotated in (10.19) to a new frame of reference. The rotated loadings $\boldsymbol{\Lambda}^*$ correspond to new factors in the basic model:

$$\begin{aligned} \mathbf{y} - \boldsymbol{\mu} &= \boldsymbol{\Lambda}\mathbf{T}\mathbf{T}'\mathbf{f} + \boldsymbol{\varepsilon} \\ &= \boldsymbol{\Lambda}^*\mathbf{f}^* + \boldsymbol{\varepsilon}, \end{aligned} \tag{10.20}$$

where

$$\mathbf{f}^* = \mathbf{T}'\mathbf{f}. \tag{10.21}$$

The rotated factors $\mathbf{f}^*$ satisfy assumptions (10.4), (10.5), and (10.8). The communalities h_i^2 in (10.14) are unaffected by the rotation, since they represent a distance from the origin to the rotated points in the space of loadings.

The estimated loadings in Section 10.3 can likewise be rotated, and this rotation is utilized in Section 10.5 to gain improved interpretation of the factors.

10.3 ESTIMATION OF LOADINGS AND COMMUNALITIES

Many approaches to estimation of the loadings have been proposed. We present four of the most commonly used techniques and briefly describe a few others. In Sections 10.3.1–10.3.3, we discuss three methods that use eigenvectors to factor $\mathbf{S}$ (or $\mathbf{S} - \mathbf{\Psi}$) into $\hat{\mathbf{\Lambda}}\hat{\mathbf{\Lambda}}'$. A maximum likelihood approach is presented in Section 10.3.4. Other methods and some comparisons are briefly covered in Sections 10.3.5 and 10.3.6.

10.3.1 Principal Component Method

In the *principal component method* for estimation of $\mathbf{\Lambda}$ (and in the methods in Sections 10.3.2 and 10.3.3), we use $\mathbf{S}$ in place of $\mathbf{\Sigma}$ in (10.9) and seek an estimator $\hat{\mathbf{\Lambda}}$ such that

$$\mathbf{S} \cong \hat{\mathbf{\Lambda}}\hat{\mathbf{\Lambda}}' + \hat{\mathbf{\Psi}}. \tag{10.22}$$

In the principal component solution for $\hat{\mathbf{\Lambda}}$, we drop $\hat{\mathbf{\Psi}}$ from the right side of (10.22) and take the first m terms of the spectral decomposition (A.11.4) of $\mathbf{S}$,

$$\begin{aligned}\mathbf{S} &\cong \theta_1\mathbf{c}_1\mathbf{c}_1' + \theta_2\mathbf{c}_2\mathbf{c}_2' + \cdots + \theta_m\mathbf{c}_m\mathbf{c}_m' \\ &= \left(\sqrt{\theta_1}\mathbf{c}_1, \sqrt{\theta_2}\mathbf{c}_2, \ldots, \sqrt{\theta_m}\mathbf{c}_m\right)\begin{bmatrix}\sqrt{\theta_1}\mathbf{c}_1' \\ \sqrt{\theta_2}\mathbf{c}_2' \\ \vdots \\ \sqrt{\theta_m}\mathbf{c}_m'\end{bmatrix} \\ &= \hat{\mathbf{\Lambda}}\hat{\mathbf{\Lambda}}',\end{aligned} \tag{10.23}$$

where $\theta_1, \theta_2, \ldots, \theta_m$ are the first m eigenvalues of $\mathbf{S}$ and $\mathbf{c}_1, \mathbf{c}_2, \ldots, \mathbf{c}_m$ are the corresponding normalized eigenvectors ($\mathbf{c}_i'\mathbf{c}_i = 1$). Thus $\hat{\mathbf{\Lambda}}$ has the form

$$\hat{\mathbf{\Lambda}} = \begin{bmatrix}\sqrt{\theta_1}c_{11} & \sqrt{\theta_2}c_{12} & \cdots & \sqrt{\theta_m}c_{1m} \\ \sqrt{\theta_1}c_{21} & \sqrt{\theta_2}c_{22} & \cdots & \sqrt{\theta_m}c_{2m} \\ \vdots & \vdots & & \vdots \\ \sqrt{\theta_1}c_{p1} & \sqrt{\theta_2}c_{p2} & \cdots & \sqrt{\theta_m}c_{pm}\end{bmatrix}, \tag{10.24}$$

so that $\hat{\lambda}_{ij} = \sqrt{\theta_j}c_{ij}$.

This method of estimation is called the *principal component* solution because the columns of $\hat{\mathbf{\Lambda}}$ are multiples of the eigenvectors of $\mathbf{S}$, and the loadings on the jth factor are therefore proportional to coefficients in the jth principal component. However, after rotation of the loadings (Section 10.5), the interpretation of the factors will be different from that of principal components.

The ith diagonal element of $\hat{\mathbf{\Lambda}}\hat{\mathbf{\Lambda}}'$ is the sum of squares of the ith row of $\hat{\mathbf{\Lambda}}$ [see (A.2.23)]. This sum of squares is an estimator of the ith *communality* [see

(10.14)] and is denoted by $\hat{h}_i^2$:

$$\hat{h}_i^2 = \sum_{j=1}^{m} \hat{\lambda}_{ij}^2. \tag{10.25}$$

To complete the partitioning of s_{ii}, we add the estimated specific variance

$$\hat{\psi}_i = s_{ii} - \sum_{j=1}^{m} \hat{\lambda}_{ij}^2 = s_{ii} - \hat{h}_i^2 \tag{10.26}$$

[see (10.15)]. By (10.23), (10.25), and (10.26), we have

$$\mathbf{S} \cong \hat{\mathbf{\Lambda}}\hat{\mathbf{\Lambda}}' + \hat{\mathbf{\Psi}}, \tag{10.27}$$

where

$$\hat{\mathbf{\Psi}} = \text{diag}(\hat{\psi}_1, \hat{\psi}_2, \ldots, \hat{\psi}_p).$$

In (10.27), we have equality for the diagonal elements of **S** but not for the off-diagonal elements. Again, modeling the covariances is the challenge of factor analysis (see the last paragraph of Section 10.2.1).

From (10.25) and (10.27), the total sample variance of the y's can be expressed in terms of the $\hat{\lambda}_{ij}$'s and $\hat{\psi}_i$'s:

$$\begin{aligned} \text{tr}(\mathbf{S}) &= \text{tr}(\hat{\mathbf{\Lambda}}\hat{\mathbf{\Lambda}}') + \text{tr}(\hat{\mathbf{\Psi}}) = \sum_{i=1}^{p} \hat{h}_i^2 + \sum_{i=1}^{p} \hat{\psi}_i \\ &= \sum_{i=1}^{p}\sum_{j=1}^{m} \hat{\lambda}_{ij}^2 + \sum_{i=1}^{p} \hat{\psi}_i. \end{aligned} \tag{10.28}$$

If we reverse the order of summation in (10.28) to obtain

$$\text{tr}(\mathbf{S}) = \sum_{j=1}^{m}\sum_{i=1}^{p} \hat{\lambda}_{ij}^2 + \sum_{i=1}^{p} \hat{\psi}_i, \tag{10.29}$$

we see that the sum of squares of the loadings in the jth column of $\hat{\mathbf{\Lambda}}$, $\sum_{i=1}^{p} \hat{\lambda}_{ij}^2$, is the contribution of the jth factor to the total sample variance. By (10.24), this is equal to the jth eigenvalue, θ_j:

$$\sum_{i=1}^{p} \hat{\lambda}_{ij}^2 = \theta_j. \tag{10.30}$$

Thus the relative size of the jth eigenvalue is the relative contribution of the jth factor to tr(**S**), and the proportion of total sample variance due to the jth factor is $\theta_j/\text{tr}(\mathbf{S})$.

This provides a ranking of the factors for the purpose of choosing m, the number of factors to retain (see Section 10.4).

In practice, we often factor $\mathbf{R}$ instead of $\mathbf{S}$ in (10.23) or (10.24) to obtain the loadings. With the emphasis in factor analysis on modeling the covariances or correlations, the use of $\mathbf{R}$ in place of $\mathbf{S}$ is an appropriate option.

Two properties of the principal component method are that (1) the method does not require that $\mathbf{S}$ or $\mathbf{R}$ be nonsingular and (2) the resulting loadings are not scale invariant. Thus, for example, the loadings obtained from $\mathbf{R}$ are not equal to those from $\mathbf{S}$ divided by $\sqrt{s_{ii}}$.

10.3.2 Principal Factor Method

We can attempt to improve the estimator of $\mathbf{\Lambda}$ obtained from the principal component approach by using an initial estimator of the diagonal elements of $\mathbf{S} - \hat{\mathbf{\Psi}}$ or $\mathbf{R} - \hat{\mathbf{\Psi}}$ for factoring these matrices into $\hat{\mathbf{\Lambda}}\hat{\mathbf{\Lambda}}'$:

$$\mathbf{S} - \hat{\mathbf{\Psi}} \cong \hat{\mathbf{\Lambda}}\hat{\mathbf{\Lambda}}', \tag{10.31}$$

$$\mathbf{R} - \hat{\mathbf{\Psi}} \cong \hat{\mathbf{\Lambda}}\hat{\mathbf{\Lambda}}', \tag{10.32}$$

where $\hat{\mathbf{\Lambda}}$ is $p \times m$. This approach (Thomson 1934) is called the *principal factor* method (it is also known as the *principal axis* method). Approximate equality is used in (10.31) and (10.32) because $\hat{\mathbf{\Lambda}}\hat{\mathbf{\Lambda}}'$ with rank m cannot reproduce $\mathbf{S} - \hat{\mathbf{\Psi}}$ or $\mathbf{R} - \hat{\mathbf{\Psi}}$, whose rank is usually p.

By (10.26) and (10.27), the diagonal elements of $\mathbf{S} - \hat{\mathbf{\Psi}}$ are the communalities

$$\hat{h}_i^2 = s_{ii} - \hat{\psi}_i, \qquad i = 1, 2, \ldots, p. \tag{10.33}$$

Similarly, for $\mathbf{R} - \hat{\mathbf{\Psi}}$ we have diagonal elements

$$\hat{h}_i^2 = 1 - \hat{\psi}_i. \tag{10.34}$$

Note that $\hat{\psi}_i$ and $\hat{h}_i^2$ for $\mathbf{S}$ in (10.33) are different from those for $\mathbf{R}$ in (10.34).

By (10.33), $\hat{h}_i^2$ represents the amount of variation in y_i (the portion of s_{ii}) that is common with the other $p - 1$ y's. A similar statement can be made for $\hat{h}_i^2$ in $\mathbf{R} - \hat{\mathbf{\Psi}}$. Thus a reasonable estimator for a communality (diagonal element) in $\mathbf{R} - \hat{\mathbf{\Psi}}$ is

$$\hat{h}_i^2 = R_i^2 = 1 - \frac{1}{r^{ii}}, \tag{10.35}$$

where r^{ii} is the ith diagonal element of $\mathbf{R}^{-1}$ and, by (1.102), $1 - 1/r^{ii} = R_i^2$, the squared multiple correlation between y_i and the other $p - 1$ variables. For $\mathbf{S} - \hat{\mathbf{\Psi}}$, an analogous initial estimator of communality is

$$\hat{h}_i^2 = s_{ii} - \frac{1}{s^{ii}} = s_{ii}R_i^2 \tag{10.36}$$

(see Problem 10.8), where s^{ii} is the ith diagonal element of $\mathbf{S}^{-1}$. If $\mathbf{R}$ or $\mathbf{S}$ is singular, some other estimator of h_i^2 must be used, such as the square of the largest correlation in the ith row of $\mathbf{R}$.

As is the case with the principal component method, the principal factor method is not scale invariant.

10.3.3 Iterated Principal Factor Method

The principal factor method can easily be iterated by obtaining new communality estimates from the loadings at each stage using (10.25):

$$\hat{h}_i^2 = \sum_{j=1}^{m} \hat{\lambda}_{ij}^2, \qquad i = 1, 2, \ldots, p.$$

The new values of $\hat{h}_1^2, \hat{h}_2^2, \ldots, \hat{h}_p^2$ are used to replace the diagonal elements of $\mathbf{S} - \hat{\boldsymbol{\Psi}}$ or $\mathbf{R} - \hat{\boldsymbol{\Psi}}$, and new estimated loadings are obtained. This two-step procedure is repeated until the communality estimates converge.

For some data sets, the iterated communality estimates do not converge. For other data sets the iterative approach may lead to a communality estimate exceeding s_{ii} when factoring $\mathbf{S}$ or exceeding 1 when factoring $\mathbf{R}$.

10.3.4 Maximum Likelihood Method

Assuming the observations $\mathbf{y}_1, \mathbf{y}_2, \ldots, \mathbf{y}_n$ constitute a random sample from $N_p(\boldsymbol{\mu}, \boldsymbol{\Sigma})$, the log likelihood function is given by (2.49) as

$$\ln L(\boldsymbol{\mu}, \boldsymbol{\Sigma}) = -np \ln \sqrt{2\pi} - \tfrac{1}{2} n \ln |\boldsymbol{\Sigma}| - \tfrac{1}{2}(n-1) \operatorname{tr}(\boldsymbol{\Sigma}^{-1}\mathbf{S}) - \tfrac{1}{2} n (\overline{\mathbf{y}} - \boldsymbol{\mu})' \boldsymbol{\Sigma}^{-1} (\overline{\mathbf{y}} - \boldsymbol{\mu}),$$

where $\overline{\mathbf{y}}$ and $\mathbf{S}$ are the sample mean vector and covariance matrix, respectively. If we substitute the maximum likelihood estimator $\overline{\mathbf{y}}$ for $\boldsymbol{\mu}$ and the assumed structure $\boldsymbol{\Lambda}\boldsymbol{\Lambda}' + \boldsymbol{\Psi}$ for $\boldsymbol{\Sigma}$, the log likelihood function becomes

$$\ln L(\boldsymbol{\Lambda}, \boldsymbol{\Psi}) = -np \ln(\sqrt{2\pi}) - \tfrac{1}{2} n \ln |\boldsymbol{\Lambda}\boldsymbol{\Lambda}' + \boldsymbol{\Psi}| - \tfrac{1}{2}(n-1) \operatorname{tr}\left[(\boldsymbol{\Lambda}\boldsymbol{\Lambda}' + \boldsymbol{\Psi})^{-1}\mathbf{S}\right]. \tag{10.37}$$

Lawley and Maxwell (1971) showed that the maximum likelihood estimators $\hat{\boldsymbol{\Lambda}}$ and $\hat{\boldsymbol{\Psi}}$ satisfy the equations

$$\mathbf{S}\hat{\boldsymbol{\Psi}}\hat{\boldsymbol{\Lambda}} = \hat{\boldsymbol{\Lambda}}(\mathbf{I} + \hat{\boldsymbol{\Lambda}}'\hat{\boldsymbol{\Psi}}^{-1}\hat{\boldsymbol{\Lambda}}), \tag{10.38}$$

$$\hat{\boldsymbol{\Psi}} = \operatorname{diag}(\mathbf{S} - \hat{\boldsymbol{\Lambda}}\hat{\boldsymbol{\Lambda}}'). \tag{10.39}$$

Since $\hat{\boldsymbol{\Lambda}}$ is unique only up to multiplication by an orthogonal matrix (rotation), (10.38) and (10.39) do not have a unique solution. It is customary to impose the computationally convenient uniqueness condition

$$\hat{\boldsymbol{\Lambda}}'\hat{\boldsymbol{\Psi}}^{-1}\hat{\boldsymbol{\Lambda}} \text{ is diagonal} \tag{10.40}$$

(for an additional justification, see Bartholomew 1981).

Unfortunately, (10.38), (10.39), and (10.40) must be solved iteratively (Jöreskog 1967, Lawley 1967, Lawley and Maxwell 1971, Jöreskog and Lawley 1968). In practice, even the latest numerical algorithms sometimes fail to converge. The iterative procedure may also produce a communality estimate greater than s_{ii} when factoring **S** or greater than 1 when factoring **R**. Kosfeld (1987) offered an improved iteration technique.

One advantage of the maximum likelihood method is that the estimators $\hat{\mathbf{\Lambda}}$ and $\hat{\mathbf{\Psi}}$ are scale invariant in the sense of Section 10.2.2. If a variable y_i is replaced by $c_i y_i$, for example, the loadings $\hat{\lambda}_{i1}, \hat{\lambda}_{i2}, \ldots, \hat{\lambda}_{im}$ corresponding to y_i will become $c_i\hat{\lambda}_{i1}, c_i\hat{\lambda}_{i2}, \ldots, c_i\hat{\lambda}_{im}$. Thus, for example, the maximum likelihood loadings extracted from **R** are $1/\sqrt{s_{ii}}$ times those from **S**; that is, $\hat{\mathbf{\Lambda}}_z = \mathbf{D}_s^{-1}\hat{\mathbf{\Lambda}}_y$ and $\hat{\mathbf{\Psi}}_z = \mathbf{D}_s^{-1}\hat{\mathbf{\Psi}}_y\mathbf{D}_s^{-1}$, where z represents a standardized variable and $\mathbf{D}_s = \text{diag}(\sqrt{s_{11}}, \sqrt{s_{22}}, \ldots, \sqrt{s_{pp}})$.

Another advantage of the maximum likelihood approach is that it leads to a significance test for the adequacy of the factor model with m factors. The test is given in Section 10.4.

10.3.5 Other Methods

For completeness, we briefly describe a few other methods of estimating loadings and communalities. See Harman (1976) for detailed descriptions and references.

The *centroid* method is a computationally simplified approximation to the principal factor method. This method was popular before the advent of modern computers, but it is now mainly of historical interest.

Canonical factor analysis seeks factors that are maximally related to the variables in the sense of canonical correlation (Chapter 8).

The minimum residuals or *minres* method seeks to minimize the sum of squared differences between the observed correlations and those predicted by the model.

Image analysis distinguishes between the *image* of a variable and the *anti-image*. The image is the predicted value of the variable from a regression on the other variables, and the anti-image is the error (residual) in the regression.

Alpha factoring seeks factors that have maximum correlation with corresponding factors in the "universe" of variables.

Among the above methods, those that are scale invariant include canonical, image, and alpha.

Kaiser and Derflinger (1990) compared maximum likelihood factor analysis and alpha factor analysis in terms of distributional assumptions, computational complexity, and other properties. They favor the alpha approach. Rubin and Thayer (1982, 1983) discussed the use of the EM algorithm (Section 1.9) in maximum likelihood estimation of the loadings.

10.3.6 Comparison of Methods

The various methods of estimating factor loadings will generally yield different solutions. However, for samples from populations in which the basic factor analysis model (10.1) is valid, most methods yield similar loadings, at least after rotation. Thus if the researcher has data to which a factor analysis model can be successfully fit with large communalities, the choice of technique is not important. To a lesser extent, if the number of variables is large, the various methods will also yield similar results, regardless of the adequacy of fit.

10.4 DETERMINING THE NUMBER OF FACTORS, m

In this section we discuss four criteria for choosing m, the number of factors. These methods are similar to those given in Section 9.5 for deciding how many principal components to retain. The first three methods apply largely to the principal component approach to estimation of $\mathbf{\Lambda}$ (Section 10.3.1).

1. Choose m to be the number of factors required to reach a desired percentage, say 80%, of the total variance tr(**S**) or tr(**R**). By (10.29), (10.30), and a remark following (10.30), the proportion of total sample variance due to the jth factor from **S** is $\theta_j/\text{tr}(\mathbf{S})$, and from **R** the proportion is θ_j/p.

2. Choose m to be the number of eigenvalues greater than the average value of the p eigenvalues. This average is 1 for **R** and $\sum_i \theta_i/p$ for **S**. This approach was first proposed by Guttman (1954). Kaiser (1960, 1974) presented several reasons for using a threshold value of 1 for eigenvalues of **R**. This method is the default in many software packages, and it often works well in practice.

3. Use the *scree test* based on a plot of the eigenvalues of **S** or **R**; that is, plot (i, θ_i), $i = 1, 2, \ldots, p$ (see Figure 9.2). If the curve shows a steep drop followed by a linear trend with a much smaller slope, choose m equal to the number of eigenvalues in the steeper part of the curve. Based on experimentation where the true number of factors was known, Cattell (1966) reported that "the scree invariably began at the kth latent root when k was the true number of factors." Kaiser (1970), however, argued that this "root-staring" approach becomes subjective when there is more than one major bend in the graph.

4. Test the hypothesis $H_0: \mathbf{\Sigma} = \mathbf{\Lambda}\mathbf{\Lambda}' + \mathbf{\Psi}$, where $\mathbf{\Lambda}$ is $p \times m$. The hypothesis test is associated with the maximum likelihood method for estimating $\mathbf{\Lambda}$ and $\mathbf{\Psi}$. [For a test based on the minres method, see Harman (1976, p. 184)]. The likelihood ratio reduces to a ratio of determinants of estimators of $\mathbf{\Sigma}$ under H_1 and under H_0 (Anderson and Rubin 1956),

$$\text{LR} = \left(\frac{|\hat{\mathbf{\Sigma}}|}{|\hat{\mathbf{\Sigma}}_0|} \right)^{n/2},$$

where $\hat{\mathbf{\Sigma}}$ is the unrestricted maximum likelihood estimator under H_1 (see Theorem 2.3A) and $\hat{\mathbf{\Sigma}}_0 = \hat{\mathbf{\Lambda}}\hat{\mathbf{\Lambda}}' + \hat{\mathbf{\Psi}}$ is the estimator under H_0 based on the maximum likelihood estimators $\hat{\mathbf{\Lambda}}$ and $\hat{\mathbf{\Psi}}$. As a test statistic, Bartlett (1951) recommended the following function of LR,

$$\left(n - \frac{2p + 4m + 11}{6} \right) \ln \left(\frac{|\hat{\mathbf{\Lambda}}\hat{\mathbf{\Lambda}}' + \hat{\mathbf{\Psi}}|}{|\mathbf{S}|} \right), \tag{10.41}$$

which is approximately distributed as χ^2_ν when H_0 is true, where $\nu = \frac{1}{2}[(p-m)^2 - p - m]$. If H_0 is rejected, then $\hat{\mathbf{\Lambda}}\hat{\mathbf{\Lambda}}' + \hat{\mathbf{\Psi}}$ is an inadequate fit for $\hat{\mathbf{\Sigma}}$, and a larger value of m should be tried. Geweke and Singleton (1980) showed that the approximate χ^2-distribution is good for fairly small sample sizes.

The degrees of freedom ν of the approximate χ^2-test arises from the following consideration. There are p diagonal elements and $\binom{p}{2}$ unique off-diagonal elements in $\mathbf{\Sigma}$, for a total of $p(p+1)/2$ distinct elements. From this we subtract $pm + p - m(m-1)/2$ for estimation of $\mathbf{\Lambda}$ and $\mathbf{\Psi}$, subject to the constraint imposed by the restriction that $\hat{\mathbf{\Lambda}}'\hat{\mathbf{\Psi}}^{-1}\hat{\mathbf{\Lambda}}$ be diagonal. There are $\binom{m}{2} = m(m-1)/2$ off-diagonal elements of $\hat{\mathbf{\Lambda}}'\hat{\mathbf{\Psi}}^{-1}\hat{\mathbf{\Lambda}}$.

For many data sets, there is not a clear choice for m. However, for a "good" data set (see Section 10.7), the first three methods above will almost always give the same value of m.

Francisco and Finch (1979) used a Monte Carlo approach to compare several methods of choosing m. Yeomans and Golder (1982) examined the performance of method 2, the "eigenvalues greater than 1" rule. Everett (1983) proposed a method of determining m based on a comparison of factor scores (Section 10.6) from two halves of a set of data. Hirsch, Wu, and Tway (1987) compared three methods for determining the number of factors and investigated the effect of outliers on the choice of m. Programs for determining the number of factors have been provided by Coovert and McNelis (1988) and by Bozdogan and Ramirez (1988).

Lambert, Wildt, and Durand (1991) suggested a bootstrap method (Efron 1985, 1987, 1988, Efron and Gong 1983, Efron and Tibshirani 1986) for obtaining confidence intervals for the (rotated) factor loadings. This allows the researcher to compare the loadings to a target value or to each other in a manner somewhat akin to hypothesis testing. Lambert et al. (1991) applied the technique to three data sets and, based on the results, concluded that this method is superior to the traditional rules of thumb.

Some writers distinguish between "exploratory" and "confirmatory" factor analysis. Most of the discussion in this chapter fits in the exploratory category. The hypothesis test in (10.41) falls in the confirmatory class, as do the procrustes and promax methods of rotation to a hypothetical target matrix (Section 10.5.3). For a discussion of the use of factor analysis to test scientific hypotheses, see Eysenck (1986).

10.5 ROTATION OF FACTOR LOADINGS

10.5.1 Introduction

By (10.19), the population loading matrix $\mathbf{\Lambda}$ can be rotated to $\mathbf{\Lambda}^* = \mathbf{\Lambda}\mathbf{T}$, where $\mathbf{T}$ is orthogonal, without destroying any essential features of the model. The estimated loading matrix $\hat{\mathbf{\Lambda}}$ can also be rotated to $\hat{\mathbf{\Lambda}}^* = \hat{\mathbf{\Lambda}}\mathbf{T}$. Since an orthogonal matrix $\mathbf{T}$ has the property $\mathbf{TT}' = \mathbf{I}$ by (A.9.3), the form of the fundamental relationship in (10.27) remains unchanged under rotation:

$$\mathbf{S} \cong \hat{\mathbf{\Lambda}}^* \hat{\mathbf{\Lambda}}^{*\prime} + \hat{\mathbf{\Psi}} = \hat{\mathbf{\Lambda}}\mathbf{TT}'\hat{\mathbf{\Lambda}}' + \hat{\mathbf{\Psi}} = \hat{\mathbf{\Lambda}}\hat{\mathbf{\Lambda}}' + \hat{\mathbf{\Psi}}. \tag{10.42}$$

Each row of $\hat{\mathbf{\Lambda}}^*$ corresponds to a variable [see (10.1)]. Our goal in rotation is to have every point (row of $\hat{\mathbf{\Lambda}}^*$) close to an axis. In this ideal case, each axis would have one or more points close to it, and the variables corresponding to the points would load highly on the factor represented by the axis and have small loadings on the remaining factors. In such a setting, there would be no ambiguity in interpretation of the factors. In the jth column of $\hat{\mathbf{\Lambda}}^*$, we could observe which variables are associated with the jth factor (have high loadings on the jth factor), and the factor would be defined or named accordingly.

We consider two basic types of rotation: *orthogonal* and *oblique*. The rotation $\hat{\mathbf{\Lambda}}^* = \hat{\mathbf{\Lambda}}\mathbf{T}$ used in (10.42) is orthogonal because $\mathbf{T}$ is an orthogonal matrix. In an oblique "rotation" the transformation matrix is not orthogonal, and the transformation is therefore not a rotation in the usual sense. However, the term *oblique rotation* is well established in the literature, and we will conform to this usage. In an oblique rotation, the axes are not perpendicular and therefore may more readily pass through clusters of points. It is hoped that this flexibility will increase the number of high and low loadings, thus enhancing interpretability.

We now discuss orthogonal and oblique rotations and the resulting interpretation. Guertin, Guertin, and Ware (1981) studied the effect of rotating too few or too many factors.

10.5.2 Orthogonal Rotation

10.5.2a Varimax Rotation

Various analytical methods of rotation have been reviewed by Williams (1979). The most widely used is the *varimax* approach (Kaiser 1958, Lawley and Maxwell 1971), which rotates the loadings so that the (sample) variance of the squared loadings in each *column* of $\hat{\mathbf{\Lambda}}^*$ is maximized; that is, the varimax method seeks to find $\hat{\lambda}_{ij}$ such that

$$\sum_{i=1}^{p} \left(\hat{\lambda}_{ij}^2 - \overline{\hat{\lambda}_j^2} \right)^2 \tag{10.43}$$

is maximized for all $j = 1, 2, \ldots, m$, where $\overline{\hat{\lambda}_j^2} = \sum_{i=1}^{p} \hat{\lambda}_{ij}^2/p$. Thus the squared loadings in each column are nudged toward 0 and 1 (when factoring $\mathbf{R}$), and this aids interpretation, as noted in Section 10.5.1. For some refinements, see Kaiser (1958).

10.5.2b Other Orthogonal Rotation Methods

In this section, we briefly describe two other orthogonal rotation methods (Carroll 1953, Saunders 1953, Neuhaus and Wrigley 1954, Ferguson 1954). Harman (1976, pp. 297–299) noted some advantages of the varimax rotation approach over the other orthogonal rotation methods.

The *quartimax* criterion attempts to maximize the variance of the squared loadings in each *row* of $\hat{\mathbf{\Lambda}}$. By an iterative approach,

$$\sum_{j=1}^{m} \left(\hat{\lambda}_{ij}^2 - \overline{\hat{\lambda}_i^2} \right)^2 \tag{10.44}$$

is maximized for all $i = 1, 2, \ldots, p$, where $\overline{\hat{\lambda}_i^2} = \sum_{j=1}^{m} \hat{\lambda}_{ij}^2/m$. This procedure moves the squared loadings in each row toward 0 or 1 (if factoring $\mathbf{R}$) to facilitate interpretation. However, with quartimax rotation there is a greater tendency for variables to have high loadings on more than one factor than is the case with varimax rotation. The quartimax method also has the drawback of occasionally producing a general factor on which all or most of the variables have high loadings. With such a result, the interpretation tends to be the same as that of principal components.

The *equamax* rotation method (Saunders 1962, Kaiser 1974) is an attempted compromise between the varimax and quartimax methods. It seeks to maximize a weighted average of the varimax and quartimax criteria.

10.5.3 Oblique Rotations

In an oblique rotation, the transformation matrix $\mathbf{Q}$ is nonsingular but not orthogonal. Corresponding to (10.21) and (10.19), we have $\mathbf{f}^* = \mathbf{Q}'\mathbf{f}$ and $\hat{\mathbf{\Lambda}}^* = \hat{\mathbf{\Lambda}}\mathbf{Q}$. Since, by (1.87), $\text{cov}(\mathbf{f}^*) = \mathbf{Q}'\mathbf{I}\mathbf{Q} = \mathbf{Q}'\mathbf{Q}$, the new factors in $\mathbf{f}^*$ are correlated.

In the space of transformed loadings, distances and angles are not preserved. Let the rows of $\hat{\mathbf{\Lambda}}^* = \hat{\mathbf{\Lambda}}\mathbf{Q}$ be denoted by $\hat{\boldsymbol{\lambda}}_i^{*\prime}$, $i = 1, 2, \ldots, p$. Then the squared distance of $\hat{\boldsymbol{\lambda}}_i^{*\prime}$ from the origin is

$$\hat{\boldsymbol{\lambda}}_i^{*\prime}\hat{\boldsymbol{\lambda}}_i^* = (\hat{\boldsymbol{\lambda}}_i'\mathbf{Q})(\hat{\boldsymbol{\lambda}}_i'\mathbf{Q})' = \hat{\boldsymbol{\lambda}}_i'\mathbf{Q}\mathbf{Q}'\hat{\boldsymbol{\lambda}}_i \neq \hat{\boldsymbol{\lambda}}_i'\hat{\boldsymbol{\lambda}}_i,$$

since $\mathbf{Q}\mathbf{Q}' \neq \mathbf{I}$. It can be shown in a similar fashion that angles are also altered. Since the factors are correlated, the communalities include cross product terms as well as squares of loadings.

The following oblique rotation methods (among others) have been proposed. *Oblimin* seeks to minimize the covariance of squared loadings between columns. *Oblimax* tries to increase the number of high and low loadings by decreasing those in the middle range. *Quartimin* attempts to minimize the covariances of the structure loadings (defined below). *Covarimin* is somewhat analogous to varimax. *Biquartimin* represents a compromise between the quartimin and covarimin methods. *Orthoblique* begins with orthogonal rotations and then proceeds to an oblique solution. The *procrustes* and *promax* methods use a "target" matrix obtained from a varimax rotation or some hypothetical consideration. For further discussion of these oblique rotations, see Hakstian (1971), Hakstian and Abel (1974), Harman (1976, Chapters 12 and 14), Cureton and D'Agostino (1983, Chapter 9), and McDonald (1985, Section 2.4).

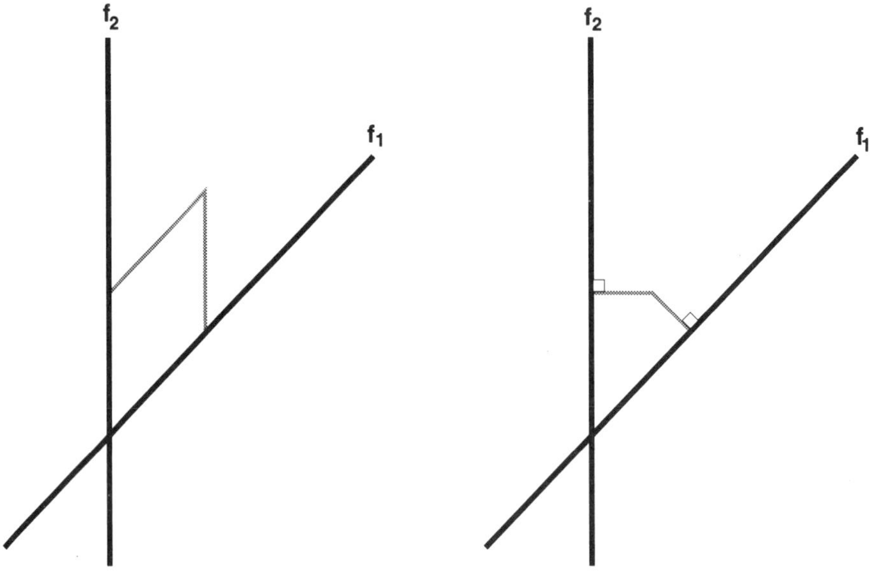

Figure 10.1. Pattern and structure loadings.

After performing an oblique rotation, a program package typically provides a *pattern matrix*, a *structure matrix*, and a matrix of correlations among the oblique factors. The loadings in the pattern matrix are somewhat analogous to regression coefficients, and the loadings in the structure matrix are similar to correlations between the variables and the oblique factors. In an orthogonal rotation, the factor loading matrix $\hat{\mathbf{\Lambda}}^*$ provides both of these features.

Figure 10.1 shows the two different types of projections represented by the pattern and structure loadings. The pattern loadings are obtained by projecting a point (row of $\hat{\mathbf{\Lambda}}^*$) onto an axis along a line parallel to the other axis. The structure loadings arise from a perpendicular projection onto the axes.

For interpretation, the pattern matrix may be more useful for two reasons: (1) the pattern loadings are the coordinates of the point (variable) on the oblique axes and (2) the pattern loadings are the coefficients in the model that relates the variables to the factors.

10.5.4 Interpretation of the Factors

As suggested in Section 10.2.1, to interpret the jth (rotated) factor f_j, we examine the loadings $\hat{\lambda}^*_{1j}, \hat{\lambda}^*_{2j}, \ldots, \hat{\lambda}^*_{pj}$ in the jth column of $\hat{\mathbf{\Lambda}}^*$ and note which y's correspond to the large (in magnitude) loadings. Collectively, these y's give a name or description to f_j.

Many writers recommend that we consider all loadings greater than .3. For most applications, however, a threshold value of .3 is too low and will result in variables that have high loadings on more than one factor. In terms of hypothesis testing, we are making *mp* tests, and the critical value should be increased to avoid inflation of Type I error rates. On the other hand, if m is large, a smaller critical value might be considered since $\hat{h}_i^2 = \sum_{j=1}^m \hat{\lambda}_{ij}^2$ and $\hat{h}_i^2$ is bounded by 1 (when factoring $\mathbf{R}$). Thus m is inversely related to the average squared loading in a row.

Harris (1985, pp. 317–320) recommends using the factor score coefficients (Section 10.6) for interpretation instead of using the factor loadings.

10.6 FACTOR SCORES

As noted in Section 10.1, the underlying factor vectors for the n observations, $\mathbf{f}_1, \mathbf{f}_2, \ldots, \mathbf{f}_n$, are unobserved, where each $\mathbf{f}_i$ is $m \times 1$. In some applications, the researcher is interested in obtaining estimates $\hat{\mathbf{f}}_1, \hat{\mathbf{f}}_2, \ldots, \hat{\mathbf{f}}_n$ to use in place of the n observation vectors $\mathbf{y}_1, \mathbf{y}_2, \ldots, \mathbf{y}_n$. The $\hat{\mathbf{f}}_i$'s are called *factor scores*. Thomson (1951) proposed a regression approach in which the factor scores are obtained as predicted values. For other methods see Harman (1976, Chapter 16).

In regression, a predicted value $\hat{y}$ is an estimator of $E(y|\mathbf{x})$. For multivariate $\mathbf{y}$, we would estimate $E(\mathbf{y}|\mathbf{x})$. In the factor analysis model, we seek to estimate $E(\mathbf{f}|\mathbf{y})$, since $\mathbf{y}$ is observed. If $\mathbf{f}$ and $\mathbf{y}$ are jointly normally distributed, we have, by Theorem 2.2E,

$$E(\mathbf{f}|\mathbf{y}) = \boldsymbol{\mu}_f + \boldsymbol{\Sigma}_{fy}\boldsymbol{\Sigma}_{yy}^{-1}(\mathbf{y} - \boldsymbol{\mu}_y),$$

which can be written as

$$E(\mathbf{f}|\mathbf{y}) = \boldsymbol{\Lambda}'\boldsymbol{\Sigma}_{yy}^{-1}(\mathbf{y} - \boldsymbol{\mu}_y), \tag{10.45}$$

since, by (10.4), $\boldsymbol{\mu}_f = E(\mathbf{f}) = \mathbf{0}$, and by (10.10), $\boldsymbol{\Sigma}_{fy} = \text{cov}(\mathbf{f}, \mathbf{y}) = \boldsymbol{\Lambda}'$.

We can estimate (10.44) by

$$\hat{\mathbf{f}}_i = \hat{\boldsymbol{\Lambda}}'\mathbf{S}^{-1}(\mathbf{y}_i - \overline{\mathbf{y}}), \qquad i = 1, 2, \ldots, n. \tag{10.46}$$

In transposed form, (10.46) becomes

$$\hat{\mathbf{f}}_i' = (\mathbf{y}_i - \overline{\mathbf{y}})'\mathbf{S}^{-1}\hat{\boldsymbol{\Lambda}},$$

and we can write all n factor score vectors as rows in the matrix

$$\hat{\mathbf{F}} = \begin{bmatrix} \hat{\mathbf{f}}_1' \\ \hat{\mathbf{f}}_2' \\ \vdots \\ \hat{\mathbf{f}}_n' \end{bmatrix} = \begin{bmatrix} (\mathbf{y}_1 - \overline{\mathbf{y}})'\mathbf{S}^{-1}\hat{\boldsymbol{\Lambda}} \\ (\mathbf{y}_2 - \overline{\mathbf{y}})'\mathbf{S}^{-1}\hat{\boldsymbol{\Lambda}} \\ \vdots \\ (\mathbf{y}_n - \overline{\mathbf{y}})'\mathbf{S}^{-1}\hat{\boldsymbol{\Lambda}} \end{bmatrix}$$

$$= \begin{bmatrix} (\mathbf{y}_1 - \overline{\mathbf{y}})' \\ (\mathbf{y}_2 - \overline{\mathbf{y}})' \\ \vdots \\ (\mathbf{y}_n - \overline{\mathbf{y}})' \end{bmatrix} \mathbf{S}^{-1}\hat{\boldsymbol{\Lambda}}$$

$$= \mathbf{Y}_c\mathbf{S}^{-1}\hat{\mathbf{\Lambda}}, \tag{10.47}$$

where $\mathbf{Y}_c$ is the $n \times p$ matrix of centered y values [see (1.43)].

If we use $\mathbf{R}$ instead of $\mathbf{S}$, (10.47) becomes

$$\hat{\mathbf{F}} = \mathbf{Y}_s\mathbf{R}^{-1}\hat{\mathbf{\Lambda}}, \tag{10.48}$$

where $\mathbf{Y}_s$ is the observed matrix of standardized variables and $\hat{\mathbf{\Lambda}}$ is the matrix of loadings obtained from $\mathbf{R}$.

We would usually be interested in factor scores for the rotated factors rather than for the original factors. They can be obtained by substituting rotated loadings $\hat{\mathbf{\Lambda}}^*$ for $\hat{\mathbf{\Lambda}}$ in (10.47) or (10.48).

Note that $\mathbf{S}$ or $\mathbf{R}$ must be nonsingular in (10.47) or (10.48). Hofmann (1978) provided a generalized regression method of finding factor scores when $\mathbf{R}$ is singular.

10.7 APPLICABILITY OF THE FACTOR ANALYSIS MODEL

The model in (10.9) implies $\mathbf{\Sigma} - \mathbf{\Psi} = \mathbf{\Lambda}\mathbf{\Lambda}'$, where $\mathbf{\Lambda}\mathbf{\Lambda}'$ is of rank m. Many population covariance matrices do not come close to this pattern unless m is close to p. If we attempt to impose a small value of m, the model will not successfully fit data from such a population.

In the cases where the factor analysis model provides a satisfactory fit to a set of data, we should be cautious in interpreting the factors until we can check the procedure on another sample. If the same factors emerge in another sample from the same population, then we can be more confident that the factors are actually present.

Since the factor analysis model $\mathbf{\Sigma} = \mathbf{\Lambda}\mathbf{\Lambda}' + \mathbf{\Psi}$ does not successfully fit all data sets, its applications and results are sometimes criticized. However, we are more likely to recognize when the factor analysis model does not fit the data than we are with other techniques. In such cases, the estimates show two readily discernible problems:

1. the number of factors m is unclear and
2. interpretation of the factors is unsatisfactory.

The problem sometimes lies with the data rather than the model. Even if the model holds in the population, we may not be able to successfully estimate the parameters in a particular sample. Francis (1973, 1974) found it difficult in some cases to retrieve known factors in a simulation study. Chalmer and Schneider (1981), who compared various methods for factor extraction, rotation, and choosing m, found that in small samples the true factor structure did not always emerge.

Since a basic goal of factor analysis is to model the covariance structure, the variables must be correlated. We could make a preliminary check for the significance of the sample correlations by testing $H_0 : \mathbf{P}_\rho = \mathbf{I}$ using (9.24). Some writers have advocated an examination of $\mathbf{R}^{-1}$ for additional clues as to the suitability of $\mathbf{R}$ for factoring. Guttman (1953) and Kaiser (1963) suggested that $\mathbf{R}^{-1}$ should be close to a diagonal matrix in order to successfully fit a factor analysis model to the data. A similar recommendation based on psychometric theory was made by Dziuban and Shirkey (1974). However, this criterion does not appear to be a consistent predictor of success in practice.

Lawley and Maxwell (1973) and Basilevsky (1981) discussed applications of factor analysis to regression with measurement errors. Riccia and Shapiro (1983) demonstrated a connection between factor analysis and discriminant analysis for several groups.

10.8 FACTOR ANALYSIS AND GROUPED DATA

Up to this point we have assumed that the data constituted a single sample from a homogeneous population with mean $\boldsymbol{\mu}$ and covariance matrix $\boldsymbol{\Sigma}$. We now discuss some possible applications of factor analysis when there are groups in the data, such as in a one-way MANOVA setting. An extension could be made for data from a two-way or higher order design. We consider three approaches:

1. If a comparison of means is of interest, we could combine the groups into a single sample, then carry out a factor analysis to obtain factor scores, followed by a MANOVA to compare the means of the factor scores for the groups.
2. If the grouping is incidental to our purposes, so that we wish only to eliminate it as a source of variation, we can do a MANOVA first followed by factor analysis on $\mathbf{S} = \mathbf{E}/\nu_E$.
3. A third alternative is to carry out a factor analysis on each group separately and then compare the results. Harman (1976, Sections 15.1–15.4), discussed several types of comparisons: (a) two different solutions from the same sample, (b) different samples but the same variables, (c) different variables but the same samples, and (d) rotation of two different factor loading matrices to obtain solutions that match as closely as possible. Everett (1983) and Everett and Entrekin (1980) discussed a measure of comparability based on cross-correlation coefficients between factor scores from two halves of a set of data. Wiseman (1993) provided an algorithm for the computation of similarity coefficients that compare factor loading matrices from two samples.

Example 10.8. Brown, Strong, and Rencher (1973) manipulated the speaking rate of 4 voices without altering the pitch. This produced five rates for each voice: the normal rate, two levels of increased rate, and two levels of reduced rate. The resulting 20 voices were played to several judges, who rated them on 15 paired-opposite adjectives (variables) over a 14-point Likert scale. The judges' ratings for each voice were averaged.

All four methods of estimating $\boldsymbol{\Lambda}$ discussed in Section 10.3 produced almost identical results (after rotation) for two factors that were called "competence" and "benevolence." The two factors accounted for 92% of tr($\mathbf{R}$), and the same two factors emerged in other studies with different voices and different judges.

The two factor scores were computed for each of the 20 voices so as to determine the effect of the rate manipulations on the two factors; that is, does perception of competence and benevolence change when the speaking rate is increased or decreased?

The factor scores for five rates of speaking on four voices constitute a one-way MANOVA with five groups and four replications. It is of interest to see if the two factor scores (dependent variables) differ significantly on the five speaking rates. The resulting Wilks' Λ-value was $\Lambda = .048$. When this is compared with the critical value $\Lambda_{.01,2,4,15} = .365$, we reject the hypothesis and conclude that the perceived competence and benevolence are significantly altered by speeding up or slowing down the speaking rate.

In a similar study, Brown, Strong, and Rencher (1974) simultaneously manipulated speaking rate, pitch, and intonation. The factor scores were analyzed in a three-way MANOVA.

10.9 ADDITIONAL TOPICS

Browne (1987) discussed robustness to nonnormality of factor estimators and tests. De Ligny et al. (1981) discussed the use of factor analysis with missing data. Tanaka (1983) proposed four methods of variable selection in factor analysis. Amemiya (1993) and Amemiya and Yalcin (1995) proposed the use of nonlinear models in factor analysis and gave methods for fitting such models.

Mislevy (1986) discussed factor analysis of dichotomous data. Knol and Berger (1991) showed that for dichotomous data factor analysis performed better than the theoretically correct item response model.

In some cases, the data can be organized in a three-way table. For example, several variables could be measured on each subject and then measured again at other time periods. To handle such data, *three-mode factor analysis* was proposed by Tucker (1966). Three-mode factor analysis produces three two-way factor-loading matrices and one three-way matrix relating the three modes. For additional discussion, see Kroonenberg (1983), Zeng and Hopke (1990), and Pham and Mocks (1992).

Regression based on factor analysis was proposed by Horst (1941) in an attempt to increase predictive validity. Properties of the method have been discussed by Lawley and Maxwell (1973) and by Browne (1988). Let

$$\mathbf{S} = \begin{pmatrix} s_{yy} & \mathbf{s}'_{yx} \\ \mathbf{s}_{yx} & \mathbf{S}_{xx} \end{pmatrix}$$

be factored as in (10.22), $\mathbf{S}_F \cong \hat{\mathbf{\Lambda}}\hat{\mathbf{\Lambda}}' + \hat{\mathbf{\Psi}}$, where the subscript F indicates the factor analytic approximation to $\mathbf{S}$. Then the regression expression $\hat{\boldsymbol{\beta}}_1 = \mathbf{S}_{xx}^{-1}\mathbf{s}_{yx}$ in (7.26) becomes

$$\hat{\boldsymbol{\beta}}_F = \mathbf{S}_{xx,F}^{-1}\mathbf{s}_{yx,F},$$

where $\mathbf{S}_F$ is partitioned as

$$\mathbf{S}_F = \begin{pmatrix} s_{yy,F} & \mathbf{s}'_{yx,F} \\ \mathbf{s}_{yx,F} & \mathbf{S}_{xx,F} \end{pmatrix}.$$

The detection of influential observations in factor analysis has been discussed by Tanaka and Odaka (1989). The influence functions show the effect of individual observations on $\hat{\mathbf{\Lambda}}$ and $\hat{\mathbf{\Psi}}$. The methods were illustrated by examples.

PROBLEMS

10.1 Show that the covariance structure $\mathbf{\Sigma} = \mathbf{\Lambda}\mathbf{\Lambda}' + \mathbf{\Psi}$ in (10.8) follows from the model (10.2) with assumptions (10.4)–(10.8).

10.2 Show that the covariance relationship $\text{cov}(\mathbf{y}, \mathbf{f}) = \mathbf{\Lambda}$ in (10.10) follows from the model (10.2) with assumptions (10.4)–(10.8).

10.3 Show that $\mathbf{z} - \boldsymbol{\mu}_z = \mathbf{\Lambda}_z\mathbf{f} + \boldsymbol{\varepsilon}_z$ in (10.16) results from the transformation $\mathbf{z} = \mathbf{A}\mathbf{y} + \mathbf{b}$ applied to $\mathbf{y} - \boldsymbol{\mu}_y = \mathbf{\Lambda}_y\mathbf{f} + \boldsymbol{\varepsilon}_y$.

10.4 Show that for standardized variables $\mathbf{z} = \mathbf{D}_\sigma^{-1}(\mathbf{y} - \boldsymbol{\mu}_y)$, as in Section 10.2.2, the transformed loadings have the form $\mathbf{\Lambda}_z = \mathbf{D}_\sigma^{-1}\mathbf{\Lambda}_y$, where $\mathbf{D}_\sigma = \text{diag}(\sigma_1, \sigma_2, \ldots, \sigma_p)$. Thus $\hat{\lambda}_{ij,z} = \hat{\lambda}_{ij,y}/\sigma_i$.

10.5 Show that the rotated factors $\mathbf{f}^* = \mathbf{T}'\mathbf{f}$ given in (10.21) satisfy the assumptions (10.4), (10.5), and (10.8).

10.6 Show that $\text{tr}(\hat{\mathbf{\Lambda}}\hat{\mathbf{\Lambda}}') = \sum_{i=1}^{p}\sum_{j=1}^{m}\hat{\lambda}_{ij}^2$ as in (10.28).

10.7 Show that for the principal component method of obtaining $\hat{\mathbf{\Lambda}}$ the sum of squares of each column of $\hat{\mathbf{\Lambda}}$ is equal to the corresponding eigenvalue of $\mathbf{S}$; that is, $\sum_{i=1}^{p}\hat{\lambda}_{ij}^2 = \theta_j$ as in (10.30).

10.8 Show that $s_{ii} - 1/s^{ii} = s_{ii}R_i^2$ as in (10.36), where s^{ii} is the ith diagonal element of $\mathbf{S}^{-1}$ and R_i^2 is the squared multiple correlation between y_i and the other $p - 1$ variables.

10.9 Show that the maximum likelihood method of estimating loadings (Section 10.3.4) is scale invariant.

Use the correlation matrix $\mathbf{R}$ in problems 10.10–10.15.

10.10 Carry out a factor analysis of the Longley data in Table 9.1 using all seven variables, $y, x_1, \ldots, x_6$.

- **(a)** Obtain principal component loadings and varimax rotations for two factors. Show the variance accounted for and convert to percent. Find the communalities. What interpretation is suggested by the rotated loadings?

(b) Show a scree graph of the eigenvalues and compare with the other three methods of choosing m in Section 10.4. Do all four methods agree on $m = 2$?

(c) Obtain principal factor loadings, varimax rotations, and communalities for two factors.

(d) Obtain maximum likelihood loadings, varimax rotations, and communalities for two factors.

(e) Compare the rotated loadings and communalities in parts (a), (c), and (d).

10.11 Carry out a factor analysis on the biochemical data of Table 4.11 using all 13 variables, $y_1, \ldots, y_{11}, x_1, x_2$.

(a) Obtain principal component loadings and varimax rotations for four factors. Show the variance accounted for and convert to percent. Find the communalities.

(b) Did the varimax rotation lead to the ideal pattern discussed in Section 10.5.1, in which every variable loads highly on one factor, with small loadings on the remaining factors? Try an oblique rotation to improve the pattern of the loadings.

(c) Show a scree graph of the eigenvalues and compare with the other three methods of choosing m in Section 10.4. Do all four methods agree on $m = 4$?

(d) Obtain principal factor loadings, varimax rotations, and communalities for four factors.

(e) Obtain maximum likelihood loadings, varimax rotations, and communalities for four factors.

(f) Compare the rotated loadings and communalities in parts (a), (b), (d), and (e).

10.12 Use the probe word data in Table 1.3.

(a) Obtain loadings for two factors by the principal component method and carry out a varimax rotation.

(b) Notice the near duplication of loadings for y_2 and y_4 in both original and rotated form. Is there any indication in the correlation matrix as to why this is so?

(c) Is the rotation satisfactory? Try an oblique rotation.

10.13 Carry out a factor analysis on the eight variables $y_1, \ldots, y_6, x_1, x_2$ of the apple data in Table 4.10. Combine the four groups into a single sample.

10.14 Carry out a factor analysis on the rainfall data in Table 7.1. Use all nine variables, $y_1, y_2, y_3, x_1, \ldots, x_6$.

10.15 Patrons of a genealogical library were asked to fill out a questionnaire that included the following seven questions.

1. Do you feel that the library staff has the necessary training and skills to help you?
2. Do you know how to use our library?

3. Do you feel comfortable asking for help from the library staff?
4. Does the library staff refer you to the appropriate areas in the library when necessary?
5. Have your research efforts been successful?
6. Is the library staff helpful?
7. Are our physical facilities adequate?

The responses, $y_1, y_2, \ldots, y_7$, were given on a seven-point semantic differential (Likert) scale ranging from 1 to 7. Partial data are given in Table 10.1.

Table 10.1. Ratings of Library Services

y_1	y_2	y_3	y_4	y_5	y_6	y_7
4	4	4	4	4	4	7
3	4	5	4	6	5	6
4	4	6	3	5	4	7
6	5	2	4	5	5	5
5	5	5	4	4	3	7
4	6	6	7	5	5	6
5	5	5	5	6	6	7
5	7	7	6	7	6	3
4	6	6	6	6	6	6
7	5	3	5	6	6	6
5	4	6	6	7	6	6
4	5	7	5	6	7	7
6	5	5	6	6	6	6
5	6	6	6	6	6	7
6	3	4	6	2	6	7
6	4	7	6	4	6	6
6	7	7	6	7	7	5
5	7	7	6	7	7	7
7	4	5	5	4	7	7
6	6	7	6	7	6	7
6	4	7	7	6	6	7
6	6	6	7	6	7	7
6	4	7	6	4	7	7
5	4	7	7	4	7	7
7	5	6	6	6	7	7
6	3	7	7	5	7	7
6	3	7	7	7	7	7
7	2	7	7	3	6	7
7	5	7	7	7	7	6
7	2	7	6	2	7	7

(a) From an examination of the eigenvalues greater than 1, the scree plot, and the percentages, is there a clear choice of m?

(b) Extract two factors by the principal component method and carry out a varimax rotation.

(c) Is the rotation an improvement? Try an oblique rotation.

(d) Extract three factors by the principal component method and carry out a varimax rotation.

APPENDIX A

Review of Matrix Algebra

In this appendix we review the elements of matrix algebra used throughout the book. With the exception of a few derivations that seemed instructive, most of the results are given without proof. For the proofs, see any general monograph on matrix theory or one of the specialized matrix texts oriented to statistics, such as Graybill (1969), Searle (1982), or Healy (1986). For numerical illustrations and exercises (with answers), see Rencher (1995, Chapter 2).

A.1 INTRODUCTION

A.1.1 Basic Definitions

A *matrix* is a rectangular or square array of numbers or variables denoted by an uppercase boldface letter:

$$\mathbf{A} = \begin{pmatrix} a_{11} & a_{12} & \cdots & a_{1p} \\ a_{21} & a_{22} & \cdots & a_{2p} \\ \vdots & \vdots & & \vdots \\ a_{n1} & a_{n2} & \cdots & a_{np} \end{pmatrix} = (a_{ij}). \tag{A.1.1}$$

The first subscript in a_{ij} represents the row, and the second subscript is the column designation. We consider only matrices whose elements a_{ij} are real. If a matrix $\mathbf{A}$ such as (A.1.1) has n rows and p columns, we say that $\mathbf{A}$ is $n \times p$ or that the *size* of $\mathbf{A}$ is $n \times p$.

We assume that the reader is familiar with matrix equality and basic operations such as transpose, addition, and multiplication of matrices.

A column *vector* has the form

$$\mathbf{x} = \begin{pmatrix} x_1 \\ x_2 \\ \vdots \\ x_p \end{pmatrix}. \tag{A.1.2}$$

Row vectors are expressed as transposed column vectors:

$$\mathbf{x}' = (x_1, x_2, \ldots, x_p). \tag{A.1.3}$$

Vectors are represented by lowercase boldface letters. In the context of vectors and matrices, a single real number is called a *scalar*. Scalars are usually denoted by lowercase nonbolded letters—for example, $a = 12.59$.

A.1.2 Matrices with Special Patterns

The elements $a_{11}, a_{22}, \ldots, a_{pp}$ constitute the *diagonal* of a $p \times p$ matrix $\mathbf{A}$. A matrix with zeros in all off-diagonal positions is called a *diagonal matrix*:

$$\mathbf{D} = \begin{pmatrix} a_{11} & 0 & \cdots & 0 \\ 0 & a_{22} & \cdots & 0 \\ \vdots & \vdots & & \vdots \\ 0 & 0 & \cdots & a_{pp} \end{pmatrix} \tag{A.1.4}$$

$$= \text{diag}(a_{11}, a_{22}, \ldots, a_{pp}). \tag{A.1.5}$$

Some of the a_{ii}'s could be zero. A diagonal matrix, denoted by diag($\mathbf{A}$), can be obtained from a square matrix $\mathbf{A}$ by replacing off-diagonal elements by zeros.

An *identity* matrix is a diagonal matrix with a 1 in each diagonal position:

$$\mathbf{I} = \begin{pmatrix} 1 & 0 & \cdots & 0 \\ 0 & 1 & \cdots & 0 \\ \vdots & \vdots & & \vdots \\ 0 & 0 & \cdots & 1 \end{pmatrix}. \tag{A.1.6}$$

An *upper triangular matrix* is a square matrix whose elements below the diagonal are all zeros. Similarly, a *lower triangular matrix* has zeros above the diagonal.

A vector of 1s is denoted by $\mathbf{j}$:

$$\mathbf{j} = \begin{pmatrix} 1 \\ 1 \\ \vdots \\ 1 \end{pmatrix}. \tag{A.1.7}$$

The product $\mathbf{jj}'$ is a square matrix of 1s that is often denoted by $\mathbf{J}$:

$$\mathbf{J} = \begin{pmatrix} 1 & 1 & \cdots & 1 \\ 1 & 1 & \cdots & 1 \\ \vdots & \vdots & & \vdots \\ 1 & 1 & \cdots & 1 \end{pmatrix} = \mathbf{jj}'. \tag{A.1.8}$$

If $\mathbf{j}$ is $n \times 1$, we have

$$\mathbf{j}'\mathbf{j} = n, \tag{A.1.9}$$

$$\mathbf{J}^2 = \mathbf{j}\mathbf{j}'\mathbf{j}\mathbf{j}' = n\mathbf{j}\mathbf{j}' = n\mathbf{J}. \tag{A.1.10}$$

A vector of zeros is denoted by $\mathbf{0}$, and a matrix (square or rectangular) of zeros is denoted by $\mathbf{O}$.

A.2 PROPERTIES OF MATRIX ADDITION AND MULTIPLICATION

We first review some basic results involving addition and multiplication.

$$\mathbf{A} + \mathbf{B} = \mathbf{B} + \mathbf{A}, \tag{A.2.1}$$

$$(\mathbf{A} \pm \mathbf{B})' = \mathbf{A}' \pm \mathbf{B}', \tag{A.2.2}$$

$$(\mathbf{x} \pm \mathbf{y})' = \mathbf{x}' \pm \mathbf{y}'. \tag{A.2.3}$$

In a few special cases $\mathbf{AB}$ equals $\mathbf{BA}$, but in general,

$$\mathbf{AB} \neq \mathbf{BA}. \tag{A.2.4}$$

Other properties are as follows:

$$\mathbf{A}(\mathbf{B} \pm \mathbf{C}) = \mathbf{AB} \pm \mathbf{AC}, \tag{A.2.5}$$

$$(\mathbf{A} \pm \mathbf{B})\mathbf{C} = \mathbf{AC} \pm \mathbf{BC}, \tag{A.2.6}$$

$$(\mathbf{A} - \mathbf{B})(\mathbf{C} - \mathbf{D}) = \mathbf{AC} - \mathbf{BC} - \mathbf{AD} + \mathbf{BD}, \tag{A.2.7}$$

$$(\mathbf{AB})' = \mathbf{B}'\mathbf{A}'. \tag{A.2.8}$$

In (A.2.8), $\mathbf{A}$ and $\mathbf{B}$ need not be square.

Products involving a matrix $\mathbf{A}$ and vectors $\mathbf{a}$, $\mathbf{b}$, and $\mathbf{c}$ take the following form: $\mathbf{Ab}$ is a column vector, $\mathbf{c}'\mathbf{A}$ is a row vector, $\mathbf{c}'\mathbf{Ab}$ is a scalar, and $\mathbf{a}'\mathbf{b}$ is a scalar. Since $\mathbf{c}'\mathbf{Ab}$ and $\mathbf{a}'\mathbf{b}$ are scalars, they are equal to their transposes: $\mathbf{c}'\mathbf{Ab} = \mathbf{b}'\mathbf{A}'\mathbf{c}$ and $\mathbf{a}'\mathbf{b} = \mathbf{b}'\mathbf{a}$.

In the product $\mathbf{a}'\mathbf{b}$, both $\mathbf{a}$ and $\mathbf{b}$ must be the same size,

$$\mathbf{a}'\mathbf{b} = a_1b_1 + a_2b_2 + \cdots + a_nb_n, \tag{A.2.9}$$

but in the product $\mathbf{ab}'$, they can be of different sizes:

$$\mathbf{ab}' = \begin{pmatrix} a_1 \\ a_2 \\ \vdots \\ a_n \end{pmatrix} (b_1 \; b_2 \; \cdots \; b_p)$$

$$= \begin{pmatrix} a_1b_1 & a_1b_2 & \cdots & a_1b_p \\ a_2b_1 & a_2b_2 & \cdots & a_2b_p \\ \vdots & \vdots & & \vdots \\ a_nb_1 & a_nb_2 & \cdots & a_nb_p \end{pmatrix}. \tag{A.2.10}$$

There are two products of a vector by itself:

$$\mathbf{a}'\mathbf{a} = a_1^2 + a_2^2 + \cdots + a_n^2, \tag{A.2.11}$$

$$\mathbf{aa}' = \begin{pmatrix} a_1^2 & a_1a_2 & \cdots & a_1a_n \\ a_2a_1 & a_2^2 & \cdots & a_2a_n \\ \vdots & \vdots & & \vdots \\ a_na_1 & a_na_2 & \cdots & a_n^2 \end{pmatrix}. \tag{A.2.12}$$

The square root of $\mathbf{a}'\mathbf{a}$ is the *distance* from the origin to the point $\mathbf{a}$ and is also known as the *length* of $\mathbf{a}$:

$$\text{Length of } \mathbf{a} = \sqrt{\mathbf{a}'\mathbf{a}} = \sqrt{\sum_{i=1}^{n} a_i^2}. \tag{A.2.13}$$

A vector whose length is equal to 1 is said to be *normalized*.

The vector $\mathbf{j}$ can be used to sum the elements of vectors and of the rows or columns of matrices; for example,

$$\mathbf{a}'\mathbf{j} = \mathbf{j}'\mathbf{a} = \sum_{i=1}^{n} a_i, \tag{A.2.14}$$

$$\mathbf{Aj} = \begin{pmatrix} \sum_j a_{1j} \\ \sum_j a_{2j} \\ \vdots \\ \sum_j a_{nj} \end{pmatrix}. \tag{A.2.15}$$

Because $\mathbf{a}'\mathbf{b} = \mathbf{b}'\mathbf{a}$, we can write $(\mathbf{a}'\mathbf{b})^2$ in the form

$$(\mathbf{a}'\mathbf{b})^2 = \mathbf{a}'(\mathbf{bb}')\mathbf{a}. \tag{A.2.16}$$

We now list some factoring results, obtained as extensions of (A.2.5), (A.2.6), and (A.2.16). If each of $\mathbf{a}$ and $\mathbf{x}_1, \mathbf{x}_2, \ldots, \mathbf{x}_n$ is $p \times 1$ and $\mathbf{A}$ is $p \times p$, then

$$\sum_{i=1}^{n} \mathbf{a}'\mathbf{x}_i = \mathbf{a}' \sum_{i=1}^{n} \mathbf{x}_i, \tag{A.2.17}$$

$$\sum_{i=1}^{n} \mathbf{Ax}_i = \mathbf{A} \sum_{i=1}^{n} \mathbf{x}_i, \tag{A.2.18}$$

$$\sum_{i=1}^{n} (\mathbf{a}'\mathbf{x}_i)^2 = \mathbf{a}' \left(\sum_{i=1}^{n} \mathbf{x}_i\mathbf{x}_i' \right) \mathbf{a}, \tag{A.2.19}$$

$$\sum_{i=1}^{n} \mathbf{Ax}_i(\mathbf{Ax}_i)' = \mathbf{A} \left(\sum_{i=1}^{n} \mathbf{x}_i\mathbf{x}_i' \right) \mathbf{A}'. \tag{A.2.20}$$

We can express a matrix in terms of its row vectors or in terms of its column vectors. For example,

$$\mathbf{A} = \begin{pmatrix} \mathbf{a}_1' \\ \mathbf{a}_2' \\ \mathbf{a}_3' \end{pmatrix}, \qquad \mathbf{B} = (\mathbf{b}_1, \mathbf{b}_2, \mathbf{b}_3, \mathbf{b}_4). \tag{A.2.21}$$

If a matrix $\mathbf{A}$ is expressed in terms of both row vectors and column vectors, we use the notation $\mathbf{a}_i'$ for rows and $\mathbf{a}_{(j)}$ for columns. To illustrate, we can write a 3×4 matrix $\mathbf{A}$ as

$$\begin{aligned} \mathbf{A} &= \begin{pmatrix} a_{11} & a_{12} & a_{13} & a_{14} \\ a_{21} & a_{22} & a_{23} & a_{24} \\ a_{31} & a_{32} & a_{33} & a_{34} \end{pmatrix} \\ &= \begin{pmatrix} \mathbf{a}_1' \\ \mathbf{a}_2' \\ \mathbf{a}_3' \end{pmatrix} = (\mathbf{a}_{(1)}, \mathbf{a}_{(2)}, \mathbf{a}_{(3)}, \mathbf{a}_{(4)}), \end{aligned} \tag{A.2.22}$$

where

$$\mathbf{a}_i' = (a_{i1}, a_{i2}, a_{i3}, a_{i4}),$$

$$\mathbf{a}_{(j)} = \begin{pmatrix} a_{1j} \\ a_{2j} \\ a_{3j} \end{pmatrix}.$$

If a matrix $\mathbf{A}$ is expressed in terms of its rows as

$$\mathbf{A} = \begin{pmatrix} \mathbf{a}_1' \\ \mathbf{a}_2' \\ \vdots \\ \mathbf{a}_n' \end{pmatrix},$$

then the elements of $\mathbf{AA}'$ are products of these rows:

$$\mathbf{AA}' = \begin{pmatrix} \mathbf{a}_1'\mathbf{a}_1 & \mathbf{a}_1'\mathbf{a}_2 & \cdots & \mathbf{a}_1'\mathbf{a}_n \\ \mathbf{a}_2'\mathbf{a}_1 & \mathbf{a}_2'\mathbf{a}_2 & \cdots & \mathbf{a}_2'\mathbf{a}_n \\ \vdots & \vdots & & \vdots \\ \mathbf{a}_n'\mathbf{a}_1 & \mathbf{a}_n'\mathbf{a}_2 & \cdots & \mathbf{a}_n'\mathbf{a}_n \end{pmatrix}. \tag{A.2.23}$$

On the other hand, when $\mathbf{A}$ is written in terms of its columns,

$$\mathbf{A} = (\mathbf{a}_{(1)}, \mathbf{a}_{(2)}, \ldots, \mathbf{a}_{(p)}),$$

then the elements of $\mathbf{A}'\mathbf{A}$ are products of these columns:

$$\mathbf{A}'\mathbf{A} = \begin{pmatrix} \mathbf{a}_{(1)}'\mathbf{a}_{(1)} & \mathbf{a}_{(1)}'\mathbf{a}_{(2)} & \cdots & \mathbf{a}_{(1)}'\mathbf{a}_{(p)} \\ \mathbf{a}_{(2)}'\mathbf{a}_{(1)} & \mathbf{a}_{(2)}'\mathbf{a}_{(2)} & \cdots & \mathbf{a}_{(2)}'\mathbf{a}_{(p)} \\ \vdots & \vdots & & \vdots \\ \mathbf{a}_{(p)}'\mathbf{a}_{(1)} & \mathbf{a}_{(p)}'\mathbf{a}_{(2)} & \cdots & \mathbf{a}_{(p)}'\mathbf{a}_{(p)} \end{pmatrix}. \tag{A.2.24}$$

Matrix products can be expressed in terms of the rows or columns of one of the matrices:

$$\mathbf{AB} = \begin{pmatrix} \mathbf{a}_1' \\ \mathbf{a}_2' \\ \vdots \\ \mathbf{a}_n' \end{pmatrix} \mathbf{B} = \begin{pmatrix} \mathbf{a}_1'\mathbf{B} \\ \mathbf{a}_2'\mathbf{B} \\ \vdots \\ \mathbf{a}_n'\mathbf{B} \end{pmatrix}, \tag{A.2.25}$$

$$\mathbf{CA} = \mathbf{C}(\mathbf{a}_{(1)}, \mathbf{a}_{(2)}, \ldots, \mathbf{a}_{(p)}) = (\mathbf{Ca}_{(1)}, \mathbf{Ca}_{(2)}, \ldots, \mathbf{Ca}_{(p)}). \tag{A.2.26}$$

Multiplication of a matrix $\mathbf{A}$ on the left (right) by a diagonal matrix $\mathbf{D}$ has the effect of multiplying the rows (columns) of $\mathbf{A}$ by the diagonal elements of $\mathbf{D}$. To illustrate, suppose $\mathbf{A}$ is 3×4, $\mathbf{D}_1 = \text{diag}(d_1, d_2, d_3)$, and $\mathbf{D}_2 = \text{diag}(d_1, d_2, d_3, d_4)$. Then by (A.2.22), $\mathbf{A}$ can be expressed as

$$\mathbf{A} = \begin{pmatrix} a_{11} & a_{12} & a_{13} & a_{14} \\ a_{21} & a_{22} & a_{23} & a_{24} \\ a_{31} & a_{32} & a_{33} & a_{34} \end{pmatrix} = \begin{pmatrix} \mathbf{a}_1' \\ \mathbf{a}_2' \\ \mathbf{a}_3' \end{pmatrix} = (\mathbf{a}_{(1)}, \mathbf{a}_{(2)}, \mathbf{a}_{(3)}, \mathbf{a}_{(4)}),$$

and we have

$$\mathbf{D}_1\mathbf{A} = \begin{pmatrix} d_1\mathbf{a}_1' \\ d_2\mathbf{a}_2' \\ d_3\mathbf{a}_3' \end{pmatrix}, \tag{A.2.27}$$

$$\mathbf{AD}_2 = (d_1\mathbf{a}_{(1)}, d_2\mathbf{a}_{(2)}, d_3\mathbf{a}_{(3)}, d_4\mathbf{a}_{(4)}). \tag{A.2.28}$$

If $\mathbf{S}$ is a symmetric matrix and $\mathbf{a}$ is a vector, the second-order sum of squares and products

$$\mathbf{a}'\mathbf{Sa} = \sum_i a_i^2 s_{ii} + \sum_{i \neq j} a_i a_j s_{ij} \tag{A.2.29}$$

is called a *quadratic form*. Similarly, for a matrix $\mathbf{S}$ (not necessarily symmetric) and vectors $\mathbf{a}$ and $\mathbf{b}$,

$$\mathbf{a}'\mathbf{Sb} = \sum_{i,j} a_i b_j s_{ij} \tag{A.2.30}$$

is called a *bilinear form*. Both of these forms are scalars.

A.3 PARTITIONED MATRICES

A matrix can be partitioned into submatrices, for example,

$$\mathbf{A} = \begin{pmatrix} \mathbf{A}_{11} & \mathbf{A}_{12} \\ \mathbf{A}_{21} & \mathbf{A}_{22} \end{pmatrix}.$$

A product of two appropriately partitioned matrices can be found by multiplying the submatrices, for example,

$$\mathbf{AB} = \begin{pmatrix} \mathbf{A}_{11} & \mathbf{A}_{12} \\ \mathbf{A}_{21} & \mathbf{A}_{22} \end{pmatrix} \begin{pmatrix} \mathbf{B}_{11} & \mathbf{B}_{12} \\ \mathbf{B}_{21} & \mathbf{B}_{22} \end{pmatrix}$$

$$= \begin{pmatrix} \mathbf{A}_{11}\mathbf{B}_{11} + \mathbf{A}_{12}\mathbf{B}_{21} & \mathbf{A}_{11}\mathbf{B}_{12} + \mathbf{A}_{12}\mathbf{B}_{22} \\ \mathbf{A}_{21}\mathbf{B}_{11} + \mathbf{A}_{22}\mathbf{B}_{21} & \mathbf{A}_{21}\mathbf{B}_{12} + \mathbf{A}_{22}\mathbf{B}_{22} \end{pmatrix}. \tag{A.3.1}$$

The following are special cases of (A.3.1) involving a matrix $\mathbf{A}$ and a vector $\mathbf{b}$:

1. the product $\mathbf{Ab}$ in partitioned form,

$$\mathbf{Ab} = (\mathbf{A}_1, \mathbf{A}_2)\begin{pmatrix} \mathbf{b}_1 \\ \mathbf{b}_2 \end{pmatrix} = \mathbf{A}_1\mathbf{b}_1 + \mathbf{A}_2\mathbf{b}_2; \tag{A.3.2}$$

2. a subvector $\mathbf{b}_2$ expressed in terms of the full vector $\mathbf{b}$,

$$\mathbf{Ab} = (\mathbf{O}, \mathbf{I})\begin{pmatrix} \mathbf{b}_1 \\ \mathbf{b}_2 \end{pmatrix} = \mathbf{O}\mathbf{b}_1 + \mathbf{I}\mathbf{b}_2 = \mathbf{b}_2; \tag{A.3.3}$$

3. the product $\mathbf{Ab}$ as a linear combination of the columns of $\mathbf{A}$,

$$\mathbf{Ab} = (\mathbf{a}_1, \mathbf{a}_2, \ldots, \mathbf{a}_p)\begin{pmatrix} b_1 \\ b_2 \\ \vdots \\ b_p \end{pmatrix} = b_1\mathbf{a}_1 + b_2\mathbf{a}_2 + \cdots + b_p\mathbf{a}_p. \tag{A.3.4}$$

If only one of two matrices is partitioned, the results are similar to (A.2.25) and (A.2.26); for example,

$$\mathbf{Ab} = \begin{pmatrix} \mathbf{A}_1\mathbf{b} \\ \mathbf{A}_2\mathbf{b} \end{pmatrix}, \tag{A.3.5}$$

$$\mathbf{AB} = \begin{pmatrix} \mathbf{A}_1\mathbf{B} \\ \mathbf{A}_2\mathbf{B} \end{pmatrix}. \tag{A.3.6}$$

The transpose of the partitioned matrix $\mathbf{A} = (\mathbf{A}_1, \mathbf{A}_2)$ is given by

$$\mathbf{A}' = (\mathbf{A}_1, \mathbf{A}_2)' = \begin{pmatrix} \mathbf{A}_1' \\ \mathbf{A}_2' \end{pmatrix}. \tag{A.3.7}$$

If $\mathbf{A}$ is written in terms of rows,

$$\mathbf{A} = \begin{pmatrix} \mathbf{a}_1' \\ \mathbf{a}_2' \\ \vdots \\ \mathbf{a}_n' \end{pmatrix},$$

then by (A.3.1) and (A.3.7),

$$\mathbf{A}'\mathbf{A} = (\mathbf{a}_1, \mathbf{a}_2, \ldots, \mathbf{a}_n)\begin{pmatrix} \mathbf{a}_1' \\ \mathbf{a}_2' \\ \vdots \\ \mathbf{a}_n' \end{pmatrix} = \sum_{i=1}^{n} \mathbf{a}_i\mathbf{a}_i'. \tag{A.3.8}$$

This is in contrast to $\mathbf{A}'\mathbf{A}$ in terms of columns as given in (A.2.24).

A.4 RANK OF MATRICES

The vectors $\mathbf{a}_1, \mathbf{a}_2, \ldots, \mathbf{a}_n$ are said to be *linearly dependent* if there exist constants c_1, $c_2, \ldots, c_n$ (not all zero) such that

$$c_1\mathbf{a}_1 + c_2\mathbf{a}_2 + \cdots + c_n\mathbf{a}_n = \mathbf{0}. \tag{A.4.1}$$

Otherwise the vectors $\mathbf{a}_1, \mathbf{a}_2, \ldots, \mathbf{a}_n$ are *linearly independent.*

The *rank* of any square or rectangular matrix $\mathbf{A}$ is defined as

$$\begin{aligned} \text{rank}(\mathbf{A}) &= \text{number of linearly independent rows of } \mathbf{A} \\ &= \text{number of linearly independent columns of } \mathbf{A}. \end{aligned} \tag{A.4.2}$$

The number of linearly independent rows of a matrix is equal to the number of linearly independent columns.

The maximum possible rank of the $n \times p$ matrix $\mathbf{A}$ is $\min(n, p)$. If $\text{rank}(\mathbf{A}) = \min(n, p)$, $\mathbf{A}$ is said to be of *full rank.* If $n \neq p$, either the columns or the rows are linearly dependent. This property of rectangular matrices (or of non-full-rank square matrices) leads to possibilities such as the following:

$$\mathbf{Ac} = \mathbf{0}, \text{ where } \mathbf{A} \neq \mathbf{O} \text{ and } \mathbf{c} \neq \mathbf{0}; \tag{A.4.3}$$

$$\mathbf{AB} = \mathbf{CB}, \text{ where } \mathbf{A} \neq \mathbf{C}. \tag{A.4.4}$$

Note, however, the following exception:

$$\text{If } \mathbf{Ax} = \mathbf{Bx} \text{ holds for all possible values of } \mathbf{x}, \text{ then } \mathbf{A} = \mathbf{B}. \tag{A.4.5}$$

This can be proved using $\mathbf{x}_1 = (1, 0, \ldots, 0)'$, $\mathbf{x}_2 = (0, 1, 0, \ldots, 0)', \ldots$, $\mathbf{x}_n = (0, \ldots, 0, 1)'$. Thus if $\mathbf{Ax}_1 = \mathbf{Bx}_1$, then the first column of $\mathbf{A}$ equals the first column of $\mathbf{B}$; if $\mathbf{Ax}_2 = \mathbf{Bx}_2$, then the second column of $\mathbf{A}$ equals the second column of $\mathbf{B}$; and so on.

Any matrix $\mathbf{A}$ has the following property:

$$\text{rank}(\mathbf{A}'\mathbf{A}) = \text{rank}(\mathbf{A}). \tag{A.4.6}$$

A.5 INVERSE MATRICES

A full-rank square matrix $\mathbf{A}$ is said to be *nonsingular*, in which case $\mathbf{A}$ has a unique *inverse* $\mathbf{A}^{-1}$ such that

$$\mathbf{A}\mathbf{A}^{-1} = \mathbf{A}^{-1}\mathbf{A} = \mathbf{I}. \tag{A.5.1}$$

If $\mathbf{A}$ is square and is not full rank, then $\mathbf{A}$ does not have an inverse and is said to be *singular*. The following are some properties of nonsingular matrices:

1. If $\mathbf{A}$ is nonsingular, $\text{rank}(\mathbf{AB}) = \text{rank}(\mathbf{B})$. (A.5.2)
2. If $\mathbf{A}$ and $\mathbf{B}$ are nonsingular, $(\mathbf{AB})^{-1} = \mathbf{B}^{-1}\mathbf{A}^{-1}$. (A.5.3)
3. If $\mathbf{A}$ is nonsingular, $\mathbf{AB} = \mathbf{AC}$ implies $\mathbf{B} = \mathbf{C}$. (A.5.4)

If $\mathbf{A}$ is $n \times p$ of rank $p < n$ (full column rank), then, by (A.4.6), $\mathbf{A}'\mathbf{A}$ is $p \times p$ of rank p and is nonsingular.

Suppose $\mathbf{A}$ is symmetric and nonsingular and is partitioned in the form

$$\mathbf{A} = \begin{pmatrix} \mathbf{A}_{11} & \mathbf{a}_{12} \\ \mathbf{a}_{12}' & a_{22} \end{pmatrix},$$

in which $\mathbf{A}_{11}$ is square, a_{22} is a scalar, and $\mathbf{a}_{12}$ is a vector. Then, if $\mathbf{A}_{11}$ is nonsingular, $\mathbf{A}^{-1}$ is given by

$$\mathbf{A}^{-1} = \frac{1}{b}\begin{pmatrix} b\mathbf{A}_{11}^{-1} + \mathbf{A}_{11}^{-1}\mathbf{a}_{12}\mathbf{a}_{12}'\mathbf{A}_{11}^{-1} & -\mathbf{A}_{11}^{-1}\mathbf{a}_{12} \\ -\mathbf{a}_{12}'\mathbf{A}_{11}^{-1} & 1 \end{pmatrix}, \tag{A.5.5}$$

where $b = a_{22} - \mathbf{a}_{12}'\mathbf{A}_{11}^{-1}\mathbf{a}_{12}$. The determinant of a matrix partitioned in this form is given in (A.7.11).

For a more general partitioning of a nonsingular matrix,

$$\mathbf{A} = \begin{pmatrix} \mathbf{A}_{11} & \mathbf{A}_{12} \\ \mathbf{A}_{21} & \mathbf{A}_{22} \end{pmatrix},$$

the inverse is given by

$$\mathbf{A}^{-1} = \begin{pmatrix} \mathbf{A}_{11}^{-1} + \mathbf{A}_{11}^{-1}\mathbf{A}_{12}\mathbf{B}^{-1}\mathbf{A}_{21}\mathbf{A}_{11}^{-1} & -\mathbf{A}_{11}^{-1}\mathbf{A}_{12}\mathbf{B}^{-1} \\ -\mathbf{B}^{-1}\mathbf{A}_{21}\mathbf{A}_{11}^{-1} & \mathbf{B}^{-1} \end{pmatrix}, \tag{A.5.6}$$

provided that $\mathbf{A}_{11}^{-1}$ and $\mathbf{B}^{-1}$ exist, where $\mathbf{B} = \mathbf{A}_{22} - \mathbf{A}_{21}\mathbf{A}_{11}^{-1}\mathbf{A}_{12}$. The determinant of a matrix $\mathbf{A}$ partitioned in this form is given in (A.7.8) and (A.7.9).

The following $p \times p$ patterned matrix is of interest in Section 9.7.2:

$$\boldsymbol{\Sigma}_0 = \begin{pmatrix} \sigma^2 & \sigma^2\rho & \cdots & \sigma^2\rho \\ \sigma^2\rho & \sigma^2 & \cdots & \sigma^2\rho \\ \vdots & \vdots & & \vdots \\ \sigma^2\rho & \sigma^2\rho & \cdots & \sigma^2 \end{pmatrix} = \sigma^2 \begin{pmatrix} 1 & \rho & \cdots & \rho \\ \rho & 1 & \cdots & \rho \\ \vdots & \vdots & & \vdots \\ \rho & \rho & \cdots & 1 \end{pmatrix} = \sigma^2[(1-\rho)\mathbf{I} + \rho\mathbf{J}].$$

The inverse of $\mathbf{\Sigma}_0$ is given by

$$\mathbf{\Sigma}_0^{-1} = \frac{1}{\sigma^2(1-\rho)}\left[\mathbf{I} - \frac{\rho}{1+(p-1)\rho}\mathbf{J}\right]. \qquad \text{(A.5.7)}$$

The determinant of $\mathbf{\Sigma}_0$ is given in (A.7.7), and the eigenvalues of an analogous matrix are given in (A.11.8).

A.6 POSITIVE DEFINITE AND POSITIVE SEMIDEFINITE MATRICES

A symmetric matrix $\mathbf{A}$ is defined to be *positive definite* if $\mathbf{x}'\mathbf{A}\mathbf{x} > 0$ for all $\mathbf{x} \neq \mathbf{0}$. Similarly, if $\mathbf{x}'\mathbf{A}\mathbf{x} \geq 0$ for all $\mathbf{x} \neq \mathbf{0}$, then $\mathbf{A}$ is said to be *positive semidefinite*. The square of a real number $b \neq 0$ is positive. There is an analogous matrix property:

If $\mathbf{B}$ is $n \times p$ of rank $p < n$, then $\mathbf{B}'\mathbf{B}$ is positive definite. (A.6.1)

If $\mathbf{B}$ is not full rank, then $\mathbf{B}'\mathbf{B}$ is positive semidefinite. To prove (A.6.1), note that because $\mathbf{B}$ is full rank, $\mathbf{B}\mathbf{x} \neq \mathbf{0}$ for $\mathbf{x} \neq \mathbf{0}$ [see (A.3.4) and the definition of linear independence following (A.4.1)]. Therefore $\mathbf{x}'\mathbf{B}'\mathbf{B}\mathbf{x} = (\mathbf{B}\mathbf{x})'(\mathbf{B}\mathbf{x}) = \mathbf{z}'\mathbf{z} = \sum_{i=1}^n z_i^2 > 0$.

We now give some properties of positive definite and positive semidefinite matrices.

1. If $\mathbf{A}$ is positive definite, then there exists at least one nonsingular $\mathbf{T}$ such that $\mathbf{A} = \mathbf{T}'\mathbf{T}$. (A.6.2)
2. If $\mathbf{A}$ is positive definite, then the diagonal elements a_{ii} are positive. (A.6.3)
3. If $\mathbf{A}$ is positive semidefinite, then $a_{ii} \geq 0$. (A.6.4)

One approach for factoring $\mathbf{A}$ in (A.6.2) is the *Cholesky decomposition*. If $\mathbf{A} = (a_{ij})$ and $\mathbf{T} = (t_{ij})$ are $n \times n$, then $\mathbf{T}$ is defined as follows:

$$t_{11} = \sqrt{a_{11}}, \qquad t_{1j} = a_{1j}/t_{11}, \qquad 2 \leq j \leq n;$$

$$t_{ii} = \sqrt{a_{ii} - \sum_{k=1}^{i-1} t_{ki}^2}, \qquad 2 \leq i \leq n;$$

$$t_{ij} = \left(a_{ij} - \sum_{k=1}^{i-1} t_{ki}t_{kj}\right)\Big/ t_{ii} \qquad 2 \leq i < j \leq n;$$

$$t_{ij} = 0, \qquad 1 \leq j < i \leq n.$$

The matrix $\mathbf{T}$ obtained in this way is upper triangular.

To prove (A.6.3), let $\mathbf{x}_i' = (0, \ldots, 0, 1, 0, \ldots, 0)$ where the 1 is in the ith position. Then, by the definition of a positive definite matrix, $\mathbf{x}_i'\mathbf{A}\mathbf{x}_i = a_{ii} > 0$ for all i.

A.7 DETERMINANTS

The *determinant* of an $n \times n$ matrix $\mathbf{A}$ is a scalar denoted by $|\mathbf{A}|$, which is defined as the sum of $n!$ specified products involving the elements of $\mathbf{A}$. Properties of determinants include the following:

1. If $\mathbf{D} = \text{diag}(d_1, d_2, \ldots, d_n)$, then $|\mathbf{D}| = \Pi_{i=1}^{n} d_i$. (A.7.1)
2. If c is scalar, $|c\mathbf{A}| = c^n |\mathbf{A}|$. (A.7.2)
3. If $\mathbf{A}$ is singular, $|\mathbf{A}| = 0$. If $\mathbf{A}$ is nonsingular, $|\mathbf{A}| \neq 0$. (A.7.3)
4. If $\mathbf{A}$ is positive definite, $|\mathbf{A}| > 0$. (A.7.4)
5. If $\mathbf{A}$ and $\mathbf{B}$ are $n \times n$, $|\mathbf{AB}| = |\mathbf{A}||\mathbf{B}|$. (A.7.5)
6. If $\mathbf{A}$ is nonsingular, $|\mathbf{A}^{-1}| = 1/|\mathbf{A}|$. (A.7.6)

The result in (A.7.6) is easily demonstrated by taking the determinant of both sides of $\mathbf{AA}^{-1} = \mathbf{I}$ and using (A.7.5).

The determinant of the $p \times p$ matrix $\mathbf{\Sigma}_0 = \sigma^2[(1 - \rho)\mathbf{I} + \rho\mathbf{J}]$ defined in Section A.5 is given by

$$|\mathbf{\Sigma}_0| = (\sigma^2)^p (1 - \rho)^{p-1} [1 + (p - 1)\rho]. \tag{A.7.7}$$

If $\mathbf{A}$ is partitioned as

$$\mathbf{A} = \begin{pmatrix} \mathbf{A}_{11} & \mathbf{A}_{12} \\ \mathbf{A}_{21} & \mathbf{A}_{22} \end{pmatrix},$$

where $\mathbf{A}_{11}$ and $\mathbf{A}_{22}$ are square and nonsingular, then the determinant of $\mathbf{A}$ can be found as

$$\begin{vmatrix} \mathbf{A}_{11} & \mathbf{A}_{12} \\ \mathbf{A}_{21} & \mathbf{A}_{22} \end{vmatrix} = |\mathbf{A}_{11}||\mathbf{A}_{22} - \mathbf{A}_{21}\mathbf{A}_{11}^{-1}\mathbf{A}_{12}| \tag{A.7.8}$$

$$= |\mathbf{A}_{22}||\mathbf{A}_{11} - \mathbf{A}_{12}\mathbf{A}_{22}^{-1}\mathbf{A}_{21}|. \tag{A.7.9}$$

The inverse of a matrix partitioned in this form is given in (A.5.6).

From (A.7.8) and (A.7.9), we obtain two special cases. If $\mathbf{A}_{11} = \mathbf{B}$, $\mathbf{A}_{12} = \mathbf{c}$, $\mathbf{A}_{21} = -\mathbf{c}'$, and $\mathbf{A}_{22} = 1$, then equating the right-hand sides of (A.7.8) and (A.7.9) gives

$$|\mathbf{B} + \mathbf{cc}'| = |\mathbf{B}|(1 + \mathbf{c}'\mathbf{B}^{-1}\mathbf{c}). \tag{A.7.10}$$

If $\mathbf{A}$ is symmetric with $\mathbf{A}_{22} = a_{22}$, $\mathbf{A}_{12} = \mathbf{a}_{12}$, and $\mathbf{A}_{21} = \mathbf{a}_{12}'$, then (A.7.8) becomes

$$\begin{vmatrix} \mathbf{A}_{11} & \mathbf{a}_{12} \\ \mathbf{a}_{12}' & a_{22} \end{vmatrix} = |\mathbf{A}_{11}|(a_{22} - \mathbf{a}'_{12}\mathbf{A}_{11}^{-1}\mathbf{a}_{12}). \tag{A.7.11}$$

The inverse of a matrix in this form is given in (A.5.5).

A.8 TRACE OF A MATRIX

The *trace* of an $n \times n$ matrix $\mathbf{A}$ is defined as $\text{tr}(\mathbf{A}) = \sum_{i=1}^{n} a_{ii}$. The trace has the following properties:

1. $\text{tr}(\mathbf{A} + \mathbf{B}) = \text{tr}(\mathbf{A}) + \text{tr}(\mathbf{B})$. (A.8.1)
2. $\text{tr}(\mathbf{AB}) = \text{tr}(\mathbf{BA})$. (A.8.2)
3. If $\mathbf{A} = \begin{pmatrix} \mathbf{a}_1' \\ \vdots \\ \mathbf{a}_n' \end{pmatrix}$, then $\text{tr}(\mathbf{AA}') = \sum_{i=1}^{n} \mathbf{a}_i'\mathbf{a}_i$. (A.8.3)
4. If $\mathbf{A} = (\mathbf{a}_{(1)}, \ldots, \mathbf{a}_{(n)})$, then $\text{tr}(\mathbf{A}'\mathbf{A}) = \sum_{i=1}^{n} \mathbf{a}_{(i)}'\mathbf{a}_{(i)}$. (A.8.4)

In (A.8.1), $\mathbf{A}$ and $\mathbf{B}$ must be square and of the same size, whereas (A.8.2) holds for any matrices $\mathbf{A}$ and $\mathbf{B}$ such that $\mathbf{AB}$ and $\mathbf{BA}$ are both defined. The results in (A.8.3) and (A.8.4) can easily be proved using (A.2.23) and (A.2.24).

A.9 ORTHOGONAL VECTORS AND MATRICES

Two $n \times 1$ vectors $\mathbf{a}$ and $\mathbf{b}$ are said to be *orthogonal* if

$$\mathbf{a}'\mathbf{b} = a_1 b_1 + a_2 b_2 + \cdots + a_n b_n = 0. \tag{A.9.1}$$

A square matrix $\mathbf{C}$ is defined to be an *orthogonal matrix* if

$$\mathbf{C}'\mathbf{C} = \mathbf{I}. \tag{A.9.2}$$

If $\mathbf{C}$ is orthogonal and satisfies (A.9.2), then it is also true that

$$\mathbf{CC}' = \mathbf{I}. \tag{A.9.3}$$

By (A.9.2), the columns of $\mathbf{C} = (\mathbf{c}_1, \mathbf{c}_2, \ldots, \mathbf{c}_n)$ are normalized (length equal 1) and mutually orthogonal ($\mathbf{c}_i'\mathbf{c}_j = 0$ for $i \neq j$). By (A.9.3), the rows of $\mathbf{C}$ are also normalized and mutually orthogonal.

Letting $\mathbf{A} = \mathbf{C}'$ in (A.3.8), we obtain

$$\mathbf{I} = \mathbf{CC}' = \mathbf{c}_1\mathbf{c}_1' + \mathbf{c}_2\mathbf{c}_2' + \cdots + \mathbf{c}_n\mathbf{c}_n', \tag{A.9.4}$$

where $\mathbf{c}_i$ is the ith column of the orthogonal matrix $\mathbf{C}$.

Multiplication of every vector in a set of vectors by an orthogonal matrix has the effect of rotating the axes; that is, if $\mathbf{x}$ is transformed to $\mathbf{z} = \mathbf{Cx}$, where $\mathbf{C}$ is orthogonal, the distance from the origin to $\mathbf{z}$ is the same as the distance to $\mathbf{x}$:

$$\mathbf{z}'\mathbf{z} = (\mathbf{Cx})'(\mathbf{Cx}) = \mathbf{x}'\mathbf{C}'\mathbf{Cx} = \mathbf{x}'\mathbf{x}. \tag{A.9.5}$$

A.10 EIGENVALUES AND EIGENVECTORS

An $n \times n$ matrix $\mathbf{A}$ has n *eigenvalues* $\lambda_1, \lambda_2, \ldots, \lambda_n$ and accompanying nonzero *eigenvectors* $\mathbf{x}_1, \mathbf{x}_2, \ldots, \mathbf{x}_n$. Each λ and the corresponding $\mathbf{x}$ satisfy

$$\mathbf{Ax} = \lambda\mathbf{x}. \tag{A.10.1}$$

If we write (A.10.1) as

$$(\mathbf{A} - \lambda\mathbf{I})\mathbf{x} = \mathbf{0}, \tag{A.10.2}$$

we see by (A.4.3) that $\mathbf{A} - \lambda\mathbf{I}$ must be singular [otherwise $\mathbf{x} = \mathbf{0}$ is the only solution to (A.10.2)]. Hence, by (A.7.3),

$$|\mathbf{A} - \lambda\mathbf{I}| = 0, \tag{A.10.3}$$

which is called the *characteristic equation*. An eigenvector $\mathbf{x}$ is unique only up to multiplication by a scalar; each $\mathbf{x}$ is often scaled so that $\mathbf{x}'\mathbf{x} = 1$.

We now list some properties of eigenvalues and eigenvectors:

1. If c is a scalar, $c\mathbf{A}$ has eigenvalues $c\lambda_1, \ldots, c\lambda_n$, but the eigenvectors are the same as those of $\mathbf{A}$:
$$(c\mathbf{A} - c\lambda\mathbf{I})\mathbf{x} = \mathbf{0}. \tag{A.10.4}$$
2. The inverse, $\mathbf{A}^{-1}$, has eigenvalues $1/\lambda_1, 1/\lambda_2, \ldots, 1/\lambda_n$, but the eigenvectors are the same as those for $\mathbf{A}$:
$$\left(\mathbf{A}^{-1} - \frac{1}{\lambda}\mathbf{I}\right)\mathbf{x} = \mathbf{0}. \tag{A.10.5}$$
3. For any $n \times n$ matrix $\mathbf{A}$, $\text{tr}(\mathbf{A}) = \sum_{i=1}^{n} \lambda_i$. (A.10.6)
4. For any $n \times n$ matrix $\mathbf{A}$, $|\mathbf{A}| = \prod_{i=1}^{n} \lambda_i$. (A.10.7)
5. For any $n \times n$ matrix $\mathbf{A}$, $|\mathbf{I} + \mathbf{A}| = \prod_{i=1}^{n} (1 + \lambda_i)$, (A.10.8)
where $\lambda_1, \lambda_2, \ldots, \lambda_n$ are the eigenvalues of $\mathbf{A}$. The eigenvalues of $\mathbf{I} + \mathbf{A}$ are of the form $1 + \lambda_i$, since $(\mathbf{I} + \mathbf{A})\mathbf{x} = (1 + \lambda)\mathbf{x}$. (This pattern for the eigenvalues does not generalize to $\mathbf{A} + \mathbf{B}$ for arbitrary $n \times n$ matrices $\mathbf{A}$ and $\mathbf{B}$.)
6. If $\mathbf{A}$ and $\mathbf{B}$ are $n \times n$ or if $\mathbf{A}$ is $n \times p$ and $\mathbf{B}$ is $p \times n$, then the (nonzero) eigenvalues of $\mathbf{AB}$ are the same as those of $\mathbf{BA}$. If $\mathbf{x}$ is an eigenvector of $\mathbf{AB}$, then $\mathbf{Bx}$ is an eigenvector of $\mathbf{BA}$. (A.10.9)
7. If $\mathbf{A}$ is $n \times n$ and $\mathbf{C}$ is an orthogonal matrix of the same size, then $\mathbf{C}'\mathbf{AC}$ has the same eigenvalues as $\mathbf{A}$. (A.10.10)

A.11 EIGENSTRUCTURE OF SYMMETRIC AND POSITIVE DEFINITE MATRICES

We now give some useful properties involving the eigenvalues $\lambda_1, \lambda_2, \ldots, \lambda_n$ and eigenvectors $\mathbf{x}_1, \mathbf{x}_2, \ldots, \mathbf{x}_n$ of symmetric and positive definite matrices:

1. The eigenvectors $\mathbf{x}_1, \mathbf{x}_2, \ldots, \mathbf{x}_n$ of an $n \times n$ symmetric matrix $\mathbf{A}$ are mutually orthogonal; that is, $\mathbf{x}_i'\mathbf{x}_j = 0$ for $i \neq j$. (A.11.1)
2. For a positive definite matrix $\mathbf{A}$, $\lambda_i > 0$ for all i. (A.11.2)
3. For a positive semidefinite matrix $\mathbf{A}$, $\lambda_i \geq 0$ for all i; the number of eigenvalues for which $\lambda_i > 0$ is given by the rank of $\mathbf{A}$. (A.11.3)
4. If $\mathbf{A}$ is symmetric with eigenvalues $\lambda_1, \lambda_2, \ldots, \lambda_n$ and normalized eigenvectors $\mathbf{x}_1, \mathbf{x}_2, \ldots, \mathbf{x}_n$, then $\mathbf{A}$ can be expressed as

$$\mathbf{A} = \sum_{i=1}^{n} \lambda_i \mathbf{x}_i \mathbf{x}_i' \tag{A.11.4}$$

$$= \mathbf{CDC}', \tag{A.11.5}$$

where $\mathbf{D} = \text{diag}(\lambda_1, \lambda_2, \ldots, \lambda_n)$ and $\mathbf{C} = (\mathbf{x}_1, \mathbf{x}_2, \ldots, \mathbf{x}_n)$. By (A.11.1), $\mathbf{C}$ is orthogonal. The results in (A.11.4) and (A.11.5), known as the *spectral decomposition* of $\mathbf{A}$, can be proved as follows. By (A.9.4),

$$\mathbf{I} = \sum_{i=1}^{n} \mathbf{x}_i \mathbf{x}_i',$$

which, when multiplied by $\mathbf{A}$, gives

$$\begin{aligned}
\mathbf{A} &= \mathbf{A} \sum_{i=1}^{n} \mathbf{x}_i \mathbf{x}_i' = \sum_{i=1}^{n} \mathbf{A}\mathbf{x}_i \mathbf{x}_i' \\
&= \sum_{i=1}^{n} \lambda_i \mathbf{x}_i \mathbf{x}_i' && \text{[by (A.10.1)]} \\
&= (\lambda_1 \mathbf{x}_1, \lambda_2 \mathbf{x}_2, \ldots, \lambda_n \mathbf{x}_n) \begin{pmatrix} \mathbf{x}_1' \\ \mathbf{x}_2' \\ \vdots \\ \mathbf{x}_n' \end{pmatrix} && \text{[by (A.3.8)]} \\
&= \mathbf{CDC}' && \text{[by (A.2.28)]},
\end{aligned}$$

where

$$\mathbf{D} = \begin{pmatrix} \lambda_1 & 0 & \cdots & 0 \\ 0 & \lambda_2 & \cdots & 0 \\ \vdots & \vdots & & \vdots \\ 0 & 0 & \cdots & \lambda_n \end{pmatrix}.$$

5. A symmetric matrix $\mathbf{A}$ can be *diagonalized* by an orthogonal matrix $\mathbf{C}$ that contains normalized eigenvectors of $\mathbf{A}$ as columns:

$$\mathbf{C}'\mathbf{A}\mathbf{C} = \mathbf{D}. \tag{A.11.6}$$

This follows from (A.11.5) since $\mathbf{C}$ is orthogonal.

6. If $\mathbf{A}$ is positive definite, then, by (A.11.2), the eigenvalues are positive, and we can obtain a *square root matrix* by a modification of (A.11.4) or (A.11.5):

$$\mathbf{A}^{1/2} = \sum_{i=1}^{n} \sqrt{\lambda_i}\mathbf{x}_i\mathbf{x}_i' = \mathbf{C}\mathbf{D}^{1/2}\mathbf{C}', \tag{A.11.7}$$

where

$$\mathbf{D}^{1/2} = \begin{pmatrix} \sqrt{\lambda_1} & 0 & \cdots & 0 \\ 0 & \sqrt{\lambda_2} & \cdots & 0 \\ \vdots & \vdots & & \vdots \\ 0 & 0 & \cdots & \sqrt{\lambda_n} \end{pmatrix}.$$

The matrix $\mathbf{A}^{1/2}$ has the property

$$\mathbf{A}^{1/2}\mathbf{A}^{1/2} = (\mathbf{A}^{1/2})^2 = \mathbf{A}.$$

7. The $p \times p$ patterned matrix

$$\mathbf{S}_0 = \begin{pmatrix} s^2 & s^2r & \cdots & s^2r \\ s^2r & s^2 & \cdots & s^2r \\ \vdots & \vdots & & \vdots \\ s^2r & s^2r & \cdots & s^2 \end{pmatrix} = s^2 \begin{pmatrix} 1 & r & \cdots & r \\ r & 1 & \cdots & r \\ \vdots & \vdots & & \vdots \\ r & r & \cdots & 1 \end{pmatrix} = s^2\mathbf{R}_0$$

is the sample analogue of $\mathbf{\Sigma}_0$ in Section A.5. The eigenvalues of $\mathbf{S}_0$ are

$$\lambda_1 = s^2\left[1 + (p-1)r\right], \qquad \lambda_i = s^2(1-r), \qquad i = 2, \ldots, p. \tag{A.11.8}$$

Note that we can use these values of $\lambda_1, \lambda_2, \ldots, \lambda_p$ and the relationship in (A.10.7) to obtain $|\mathbf{S}_0|$ given in (A.7.7). By (A.10.4), the eigenvalues of $\mathbf{R}_0$ are

$$\lambda_1 = 1 + (p-1)r, \qquad \lambda_i = 1 - r, \qquad i = 2, \ldots, p. \tag{A.11.9}$$

For both $\mathbf{S}_0$ and $\mathbf{R}_0$ [see (A.10.4)], the first eigenvector is

$$\mathbf{x}_1 = \left(\frac{1}{\sqrt{p}}, \frac{1}{\sqrt{p}}, \ldots, \frac{1}{\sqrt{p}}\right)' = \frac{1}{\sqrt{p}}\mathbf{j}. \tag{A.11.10}$$

Since the remaining eigenvectors $\mathbf{x}_2, \ldots, \mathbf{x}_p$ are orthogonal to $\mathbf{x}_1$,

$$\mathbf{x}_i'\mathbf{x}_1 = 0, \qquad i = 2, 3, \ldots, p, \tag{A.11.11}$$

each of the eigenvectors $\mathbf{x}_2, \ldots, \mathbf{x}_p$ has both positive and negative elements. We can prove that the first eigenvector of $\mathbf{R}_0$ is proportional to $\mathbf{j}$ as follows. The matrix $\mathbf{R}_0$ can be expressed as $\mathbf{R}_0 = (1-r)\mathbf{I} + r\mathbf{J}$. Then for $\lambda_1 = 1 + (p-1)r$, (A.10.1) becomes

$$\begin{aligned}
\mathbf{R}_0\mathbf{x}_1 &= \lambda_1\mathbf{x}_1, \\
[(1-r)\mathbf{I} + r\mathbf{J}]\mathbf{x}_1 &= [1 + (p-1)r]\mathbf{x}_1, \\
(1-r)\mathbf{x}_1 + r\mathbf{J}\mathbf{x}_1 &= \mathbf{x}_1 + (p-1)r\mathbf{x}_1, \\
(1-r)\mathbf{x}_1 + r\mathbf{j}\mathbf{j}'\mathbf{x}_1 &= \mathbf{x}_1 + pr\mathbf{x}_1 - r\mathbf{x}_1 \quad \text{[by (A.1.8)]} \\
&= (1-r)\mathbf{x}_1 + pr\mathbf{x}_1, \\
r\mathbf{j}\mathbf{j}'\mathbf{x}_1 &= pr\mathbf{x}_1.
\end{aligned}$$

If we substitute $\mathbf{x}_1 = \mathbf{j}$, we obtain $pr\mathbf{j} = pr\mathbf{j}$, since $\mathbf{j}'\mathbf{j} = p$.

8. The pattern of eigenvectors in (A.11.10) and (A.11.11) can be extended to the following more general result known as the *Perron–Frobenius theorem* (See Bellman 1960, p. 278). If the positive definite matrix $\mathbf{A}$ has all $a_{ij} > 0$, then all elements of the first eigenvector are positive. (A.11.12)

A.12 IDEMPOTENT MATRICES

A matrix $\mathbf{A}$ is defined to be *idempotent* if $\mathbf{A}^2 = \mathbf{A}$. In this book, all idempotent matrices are symmetric. The eigenvalues of idempotent matrices have the following property:

If the $n \times n$ idempotent matrix $\mathbf{A}$ has rank r, then $\mathbf{A}$ has r eigenvalues equal to 1 and $n - r$ eigenvalues equal to 0. (A.12.1)

A.13 DIFFERENTIATION

In this section, we consider differentiation of certain scalar functions with respect to vectors and with respect to matrices. We begin by defining differentiation with respect to a vector.

Let $u = f(x_1, x_2, \ldots, x_p)$ be a scalar function of the variables $x_1, x_2, \ldots, x_p$, and suppose we wish to differentiate u with respect to each x_i. It is notationally convenient to express $\partial u/\partial x_1, \partial u/\partial x_2, \ldots, \partial u/\partial x_p$ in vector form as

$$\frac{\partial u}{\partial \mathbf{x}} = \begin{pmatrix} \dfrac{\partial u}{\partial x_1} \\ \dfrac{\partial u}{\partial x_2} \\ \vdots \\ \dfrac{\partial u}{\partial x_p} \end{pmatrix}. \tag{A.13.1}$$

Let $u = \mathbf{a}'\mathbf{x} = \mathbf{x}'\mathbf{a}$. Then

$$\frac{\partial(\mathbf{a}'\mathbf{x})}{\partial \mathbf{x}} = \frac{\partial(\mathbf{x}'\mathbf{a})}{\partial \mathbf{x}} = \mathbf{a}. \tag{A.13.2}$$

This can be proved by noting that

$$\frac{\partial(\mathbf{a}'\mathbf{x})}{\partial x_i} = \frac{\partial(a_1x_1 + a_2x_2 + \cdots + a_px_p)}{\partial x_i} = a_i,$$

and therefore by (A.13.1),

$$\frac{\partial(\mathbf{a}'\mathbf{x})}{\partial \mathbf{x}} = \begin{pmatrix} a_1 \\ a_2 \\ \vdots \\ a_p \end{pmatrix} = \mathbf{a}.$$

Let $u = \mathbf{x}'\mathbf{A}\mathbf{x}$, where $\mathbf{A}$ is symmetric. Then

$$\frac{\partial(\mathbf{x}'\mathbf{A}\mathbf{x})}{\partial \mathbf{x}} = 2\mathbf{A}\mathbf{x}. \tag{A.13.3}$$

We demonstrate (A.13.3) for the case in which $\mathbf{A}$ is 3×3. The illustration could be generalized to any size $\mathbf{A}$. Let

$$\mathbf{x} = \begin{pmatrix} x_1 \\ x_2 \\ x_3 \end{pmatrix} \quad \text{and} \quad \mathbf{A} = \begin{pmatrix} a_{11} & a_{12} & a_{13} \\ a_{12} & a_{22} & a_{23} \\ a_{13} & a_{23} & a_{33} \end{pmatrix} = \begin{pmatrix} \mathbf{a}_1' \\ \mathbf{a}_2' \\ \mathbf{a}_3' \end{pmatrix}.$$

Then $\mathbf{x}'\mathbf{A}\mathbf{x} = x_1^2a_{11} + 2x_1x_2a_{12} + 2x_1x_3a_{13} + x_2^2a_{22} + 2x_2x_3a_{23} + x_3^2a_{33}$, and

$$\frac{\partial(\mathbf{x}'\mathbf{A}\mathbf{x})}{\partial x_1} = 2x_1a_{11} + 2x_2a_{12} + 2x_3a_{13} = 2\mathbf{a}_1'\mathbf{x},$$

$$\frac{\partial(\mathbf{x}'\mathbf{A}\mathbf{x})}{\partial x_2} = 2x_1a_{12} + 2x_2a_{22} + 2x_3a_{23} = 2\mathbf{a}_2'\mathbf{x},$$

$$\frac{\partial(\mathbf{x}'\mathbf{A}\mathbf{x})}{\partial x_3} = 2x_1a_{13} + 2x_2a_{23} + 2x_3a_{33} = 2\mathbf{a}_3'\mathbf{x}.$$

Thus by (A.13.1),

$$\frac{\partial(\mathbf{x}'\mathbf{A}\mathbf{x})}{\partial \mathbf{x}} = \begin{pmatrix} \dfrac{\partial(\mathbf{x}'\mathbf{A}\mathbf{x})}{\partial x_1} \\ \dfrac{\partial(\mathbf{x}'\mathbf{A}\mathbf{x})}{\partial x_2} \\ \dfrac{\partial(\mathbf{x}'\mathbf{A}\mathbf{x})}{\partial x_3} \end{pmatrix} = 2\begin{pmatrix} \mathbf{a}_1'\mathbf{x} \\ \mathbf{a}_2'\mathbf{x} \\ \mathbf{a}_3'\mathbf{x} \end{pmatrix} = 2\mathbf{A}\mathbf{x}.$$

If $u = f(\mathbf{A}) = f(a_{11}, a_{12}, \ldots, a_{pp})$, where $\mathbf{A}$ is a symmetric matrix, we define

$$\frac{\partial u}{\partial \mathbf{A}} = \left(\frac{\partial u}{\partial a_{ij}}\right) = \begin{pmatrix} \frac{\partial u}{\partial a_{11}} & \frac{\partial u}{\partial a_{12}} & \cdots & \frac{\partial u}{\partial a_{1p}} \\ \frac{\partial u}{\partial a_{21}} & \frac{\partial u}{\partial a_{22}} & \cdots & \frac{\partial u}{\partial a_{2p}} \\ \vdots & \vdots & & \vdots \\ \frac{\partial u}{\partial a_{p1}} & \frac{\partial u}{\partial a_{p2}} & \cdots & \frac{\partial u}{\partial a_{pp}} \end{pmatrix}. \qquad \text{(A.13.4)}$$

Two functions of interest are $\text{tr}(\mathbf{AB})$ and $\ln|\mathbf{A}|$. For $\text{tr}(\mathbf{AB})$, where $\mathbf{A}$ is symmetric, we have

$$\frac{\partial \, \text{tr}(\mathbf{AB})}{\partial \mathbf{A}} = \mathbf{B} + \mathbf{B}' - \text{diag}(\mathbf{B}). \qquad \text{(A.13.5)}$$

For $\ln|\mathbf{A}|$, where $\mathbf{A}$ is symmetric and nonsingular, the derivative is given by

$$\frac{\partial \ln|\mathbf{A}|}{\partial \mathbf{A}} = 2\mathbf{A}^{-1} - \text{diag}(\mathbf{A}^{-1}). \qquad \text{(A.13.6)}$$

APPENDIX B

Tables

Table B.1. Upper Percentage Points of Hotelling's T^2-Distribution

Degrees of Freedom, ν	Number of Variables, p									
	1	2	3	4	5	6	7	8	9	10
					$\alpha = 0.05$					
2	18.513									
3	10.128	57.000								
4	7.709	25.472	114.986							
5	6.608	17.361	46.383	192.468						
6	5.987	13.887	29.661	72.937	289.446					
7	5.591	12.001	22.720	44.718	105.157	405.920				
8	5.318	10.828	19.028	33.230	62.561	143.050	541.890			
9	5.117	10.033	16.766	27.202	45.453	83.202	186.622	697.356		
10	4.965	9.459	15.248	23.545	36.561	59.403	106.649	235.873	872.317	
11	4.844	9.026	14.163	21.108	31.205	47.123	75.088	132.903	290.806	1066.774
12	4.747	8.689	13.350	19.376	27.656	39.764	58.893	92.512	161.967	351.421
13	4.667	8.418	12.719	18.086	25.145	34.911	49.232	71.878	111.676	193.842
14	4.600	8.197	12.216	17.089	23.281	31.488	42.881	59.612	86.079	132.582
15	4.543	8.012	11.806	16.296	21.845	28.955	38.415	51.572	70.907	101.499
16	4.494	7.856	11.465	15.651	20.706	27.008	35.117	45.932	60.986	83.121
17	4.451	7.722	11.177	15.117	19.782	25.467	32.588	41.775	54.041	71.127
18	4.414	7.606	10.931	14.667	19.017	24.219	30.590	38.592	48.930	62.746
19	4.381	7.504	10.719	14.283	18.375	23.189	28.975	36.082	45.023	56.587
20	4.351	7.415	10.533	13.952	17.828	22.324	27.642	34.054	41.946	51.884
21	4.325	7.335	10.370	13.663	17.356	21.588	26.525	32.384	39.463	48.184
22	4.301	7.264	10.225	13.409	16.945	20.954	25.576	30.985	37.419	45.202
23	4.279	7.200	10.095	13.184	16.585	20.403	24.759	29.798	35.709	42.750
24	4.260	7.142	9.979	12.983	16.265	19.920	24.049	28.777	34.258	40.699
25	4.242	7.089	9.874	12.803	15.981	19.492	23.427	27.891	33.013	38.961
26	4.225	7.041	9.779	12.641	15.726	19.112	22.878	27.114	31.932	37.469
27	4.210	6.997	9.692	12.493	15.496	18.770	22.388	26.428	30.985	36.176
28	4.196	6.957	9.612	12.359	15.287	18.463	21.950	25.818	30.149	35.043
29	4.183	6.919	9.539	12.236	15.097	18.184	21.555	25.272	29.407	34.044
30	4.171	6.885	9.471	12.123	14.924	17.931	21.198	24.781	28.742	33.156
35	4.121	6.744	9.200	11.674	14.240	16.944	19.823	22.913	26.252	29.881
40	4.085	6.642	9.005	11.356	13.762	16.264	18.890	21.668	24.624	27.783
45	4.057	6.564	8.859	11.118	13.409	15.767	18.217	20.781	23.477	26.326
50	4.034	6.503	8.744	10.934	13.138	15.388	17.709	20.117	22.627	25.256
55	4.016	6.454	8.652	10.787	12.923	15.090	17.311	19.600	21.972	24.437
60	4.001	6.413	8.577	10.668	12.748	14.850	16.992	19.188	21.451	23.790
70	3.978	6.350	8.460	10.484	12.482	14.485	16.510	18.571	20.676	22.834
80	3.960	6.303	8.375	10.350	12.289	14.222	16.165	18.130	20.127	22.162
90	3.947	6.267	8.309	10.248	12.142	14.022	15.905	17.801	19.718	21.663
100	3.936	6.239	8.257	10.167	12.027	13.867	15.702	17.544	19.401	21.279
110	3.927	6.216	8.215	10.102	11.934	13.741	15.540	17.340	19.149	20.973
120	3.920	6.196	8.181	10.048	11.858	13.639	15.407	17.172	18.943	20.725
150	3.904	6.155	8.105	9.931	11.693	13.417	15.121	16.814	18.504	20.196
200	3.888	6.113	8.031	9.817	11.531	13.202	14.845	16.469	18.083	19.692
400	3.865	6.052	7.922	9.650	11.297	12.890	14.447	15.975	17.484	18.976
1000	3.851	6.015	7.857	9.552	11.160	12.710	14.217	15.692	17.141	18.570
∞	3.841	5.991	7.815	9.488	11.070	12.592	14.067	15.507	16.919	18.307

(*continued*)

Table B.1. ***(continued)***

Degrees of Freedom, ν	Number of Variables, p									
	1	2	3	4	5	6	7	8	9	10
					$\alpha = 0.01$					
2	98.503									
3	34.116	297.000								
4	21.198	82.177	594.997							
5	16.258	45.000	147.283	992.494						
6	13.745	31.857	75.125	229.679	1489.489					
7	12.246	25.491	50.652	111.839	329.433	2085.984				
8	11.259	21.821	39.118	72.908	155.219	446.571	2781.978			
9	10.561	19.460	32.598	54.890	98.703	205.293	581.106	3577.472		
10	10.044	17.826	28.466	44.838	72.882	128.067	262.076	733.045	4472.464	
11	9.646	16.631	25.637	38.533	58.618	93.127	161.015	325.576	902.392	5466.956
12	9.330	15.722	23.588	34.251	49.739	73.969	115.640	197.555	395.797	1089.149
13	9.074	15.008	22.041	31.171	43.745	62.114	90.907	140.429	237.692	472.742
14	8.862	14.433	20.834	28.857	39.454	54.150	75.676	109.441	167.499	281.428
15	8.683	13.960	19.867	27.060	36.246	48.472	65.483	90.433	129.576	196.853
16	8.531	13.566	19.076	25.626	33.762	44.240	58.241	77.755	106.391	151.316
17	8.400	13.231	18.418	24.458	31.788	40.975	52.858	68.771	90.969	123.554
18	8.285	12.943	17.861	23.487	30.182	38.385	48.715	62.109	80.067	105.131
19	8.185	12.694	17.385	22.670	28.852	36.283	45.435	56.992	71.999	92.134
20	8.096	12.476	16.973	21.972	27.734	34.546	42.779	52.948	65.813	82.532
21	8.017	12.283	16.613	21.369	26.781	33.088	40.587	49.679	60.932	75.181
22	7.945	12.111	16.296	20.843	25.959	31.847	38.750	46.986	56.991	69.389
23	7.881	11.958	16.015	20.381	25.244	30.779	37.188	44.730	53.748	64.719
24	7.823	11.820	15.763	19.972	24.616	29.850	35.846	42.816	51.036	60.879
25	7.770	11.695	15.538	19.606	24.060	29.036	34.680	41.171	48.736	57.671
26	7.721	11.581	15.334	19.279	23.565	28.316	33.659	39.745	46.762	54.953
27	7.677	11.478	15.149	18.983	23.121	27.675	32.756	38.496	45.051	52.622
28	7.636	11.383	14.980	18.715	22.721	27.101	31.954	37.393	43.554	50.604
29	7.598	11.295	14.825	18.471	22.359	26.584	31.236	36.414	42.234	48.839
30	7.562	11.215	14.683	18.247	22.029	26.116	30.589	35.538	41.062	47.283
35	7.419	10.890	14.117	17.366	20.743	24.314	28.135	32.259	36.743	41.651
40	7.314	10.655	13.715	16.750	19.858	23.094	26.502	30.120	33.984	38.135
45	7.234	10.478	13.414	16.295	19.211	22.214	25.340	28.617	32.073	35.737
50	7.171	10.340	13.181	15.945	18.718	21.550	24.470	27.504	30.673	33.998
55	7.119	10.228	12.995	15.667	18.331	21.030	23.795	26.647	29.603	32.682
60	7.077	10.137	12.843	15.442	18.018	20.613	23.257	25.967	28.760	31.650
70	7.011	9.996	12.611	15.098	17.543	19.986	22.451	24.957	27.515	30.139
80	6.963	9.892	12.440	14.849	17.201	19.536	21.877	24.242	26.642	29.085
90	6.925	9.813	12.310	14.660	16.942	19.197	21.448	23.710	25.995	28.310
100	6.895	9.750	12.208	14.511	16.740	18.934	21.115	23.299	25.496	27.714
110	6.871	9.699	12.125	14.391	16.577	18.722	20.849	22.972	25.101	27.243
120	6.851	9.657	12.057	14.292	16.444	18.549	20.632	22.705	24.779	26.862
150	6.807	9.565	11.909	14.079	16.156	18.178	20.167	22.137	24.096	26.054
200	6.763	9.474	11.764	13.871	15.877	17.819	19.720	21.592	23.446	25.287
400	6.699	9.341	11.551	13.569	15.473	17.303	19.080	20.818	22.525	24.209
1000	6.660	9.262	11.426	13.392	15.239	17.006	18.743	20.376	22.003	23.600
∞	6.635	9.210	11.345	13.277	15.086	16.812	18.475	20.090	21.666	23.209

Table B.2. Bonferroni t-values, $t_{\alpha/2k,\nu}$, $\alpha = 0.05$

	k									
	1	2	3	4	5	6	7	8	9	10
	$100\alpha/k$									
ν	5.0000	2.5000	1.6667	1.2500	1.0000	.8333	.7143	.6250	.5556	.5000
2	4.3027	6.2053	7.6488	8.8602	9.9248	10.8859	11.7687	12.5897	13.3604	14.0890
3	3.1824	4.1765	4.8567	5.3919	5.8409	6.2315	6.5797	6.8952	7.1849	7.4533
4	2.7764	3.4954	3.9608	4.3147	4.6041	4.8510	5.0675	5.2611	5.4366	5.5976
5	2.5706	3.1634	3.5341	3.8100	4.0321	4.2193	4.3818	4.5257	4.6553	4.7733
6	2.4469	2.9687	3.2875	3.5212	3.7074	2.8630	3.9971	4.1152	4.2209	4.3168
7	2.3646	2.8412	3.1276	3.3353	3.4995	3.6358	3.7527	3.8552	3.9467	4.0293
8	2.3060	2.7515	3.0158	3.2060	3.3554	3.4789	3.5844	3.6766	3.7586	3.8325
9	2.2622	2.6850	2.9333	3.1109	3.2498	3.3642	3.4616	3.5465	3.6219	3.6897
10	2.2281	2.6338	2.8701	3.0382	3.1693	3.2768	3.3682	3.4477	3.5182	3.5814
11	2.2010	2.5931	2.8200	2.9809	3.1058	3.2081	3.2949	3.3702	3.4368	3.4966
12	2.1788	2.5600	2.7795	2.9345	3.0545	3.1527	3.2357	3.3078	3.3714	3.4284
13	2.1604	2.5326	2.7459	2.8961	3.0123	3.1070	3.1871	3.2565	3.3177	3.3725
14	2.1448	2.5096	2.7178	2.8640	2.9768	3.0688	3.1464	3.2135	3.2727	3.3257
15	2.1314	2.4899	2.6937	2.8366	2.9467	3.0363	3.1118	3.1771	3.2346	3.2860
16	2.1199	2.4729	2.6730	2.8131	2.9208	3.0083	3.0821	3.1458	3.2019	3.2520
17	2.1098	2.4581	2.6550	2.7925	2.8982	2.9840	3.0563	3.1186	3.1735	3.2224
18	2.1009	2.4450	2.6391	2.7745	2.8784	2.9627	3.0336	3.0948	3.1486	3.1966
19	2.0930	2.4334	2.6251	2.7586	2.8609	2.9439	3.0136	3.0738	3.1266	3.1737
20	2.0860	2.4231	2.6126	2.7444	2.8453	2.9271	2.9958	3.0550	3.1070	3.1534
21	2.0796	2.4138	2.6013	2.7316	2.8314	2.9121	2.9799	3.0382	3.0895	3.1352
22	2.0739	2.4055	2.5912	2.7201	2.8188	2.8985	2.9655	3.0231	3.0737	3.1188
23	2.0687	2.3979	2.5820	2.7097	2.8073	2.8863	2.9525	3.0095	3.0595	3.1040
24	2.0639	2.3909	2.5736	2.7002	2.7969	2.8751	2.9406	2.9970	3.0465	3.0905
25	2.0595	2.3846	2.5660	2.6916	2.7874	2.8649	2.9298	2.9856	3.0346	3.0782
26	2.0555	2.3788	2.5589	2.6836	2.7787	2.8555	2.9199	2.9752	3.0237	3.0669
27	2.0518	2.3734	2.5525	2.6763	2.7707	2.8469	2.9107	2.9656	3.0137	3.0565
28	2.0484	2.3685	2.5465	2.6695	2.7633	2.8389	2.9023	2.9567	3.0045	3.0469
29	2.0452	2.3638	2.5409	2.6632	2.7564	2.8316	2.8945	2.9485	2.9959	3.0380
30	2.0423	2.3596	2.5357	2.6574	2.7500	2.8247	2.8872	2.9409	2.9880	3.0298
35	2.0301	2.3420	2.5145	2.6334	2.7238	2.7966	2.8575	2.9097	2.9554	2.9960
40	2.0211	2.3289	2.4989	2.6157	2.7045	2.7759	2.8355	2.8867	2.9314	2.9712
45	2.0141	2.3189	2.4868	2.6021	2.6896	2.7599	2.8187	2.8690	2.9130	2.9521
50	2.0086	2.3109	2.4772	2.5913	2.6778	2.7473	2.8053	2.8550	2.8984	2.9370
55	2.0040	2.3044	2.4694	2.5825	2.6682	2.7370	2.7944	2.8436	2.8866	2.9247
60	2.0003	2.2990	2.4630	2.5752	2.6603	2.7286	2.7855	2.8342	2.8768	2.9146
70	1.9944	2.2906	2.4529	2.5639	2.6479	2.7153	2.7715	2.8195	2.8615	2.8987
80	1.9901	2.2844	2.4454	2.5554	2.6387	2.7054	2.7610	2.8086	2.8502	2.8870
90	1.9867	2.2795	2.4395	2.5489	2.6316	2.6978	2.7530	2.8002	2.8414	2.8779
100	1.9840	2.2757	2.4349	2.5437	2.6259	2.6918	2.7466	2.7935	2.8344	2.8707
110	1.9818	2.2725	2.4311	2.5394	2.6213	2.6868	2.7414	2.7880	2.8287	2.8648
120	1.9799	2.2699	2.4280	2.5359	2.6174	2.6827	2.7370	2.7835	2.8240	2.8599
250	1.9695	2.2550	2.4102	2.5159	2.5956	2.6594	2.7124	2.7577	2.7972	2.8322
500	1.9647	2.2482	2.4021	2.5068	2.5857	2.6488	2.7012	2.7460	2.7850	2.8195
1000	1.9623	2.2448	2.3980	2.5022	2.5808	2.6435	2.6957	2.7402	2.7790	2.8133
∞	1.9600	2.2414	2.3940	2.4977	2.5758	2.6383	2.6901	2.7344	2.7729	2.8070

(*continued*)

Table B.2. (*continued*)

	k								
	11	12	13	14	15	16	17	18	19
	$100\alpha/k$								
ν	.4545	.4167	.3846	.3571	.3333	.3125	.2941	.2778	.2632
2	14.7818	15.4435	16.0780	16.6883	17.2772	17.8466	18.3984	18.9341	19.4551
3	7.7041	7.9398	8.1625	8.3738	8.5752	8.7676	8.9521	9.1294	9.3001
4	5.7465	5.8853	6.0154	6.1380	6.2541	6.3643	6.4693	6.5697	6.6659
5	4.8819	4.9825	5.0764	5.1644	5.2474	5.3259	5.4005	5.4715	5.5393
6	4.4047	4.4858	4.5612	4.6317	4.6979	4.7604	4.8196	4.8759	4.9295
7	4.1048	4.1743	4.2388	4.2989	4.3553	4.4084	4.4586	4.5062	4.5514
8	3.8999	3.9618	4.0191	4.0724	4.1224	4.1693	4.2137	4.2556	4.2955
9	3.7513	3.8079	3.8602	3.9088	3.9542	3.9969	4.0371	4.0752	4.1114
10	3.6388	3.6915	3.7401	3.7852	3.8273	3.8669	3.9041	3.9394	3.9728
11	3.5508	3.6004	3.6462	3.6887	3.7283	3.7654	3.8004	3.8335	3.8648
12	3.4801	3.5274	3.5709	3.6112	3.6489	3.6842	3.7173	3.7487	3.7783
13	3.4221	3.4674	3.5091	3.5478	3.5838	3.6176	3.6493	3.6793	3.7076
14	3.3736	3.4173	3.4576	3.4949	3.5296	3.5621	3.5926	3.6214	3.6487
15	3.3325	3.3749	3.4139	3.4501	3.4837	3.5151	3.5447	3.5725	3.5989
16	3.2973	3.3386	3.3765	3.4116	3.4443	3.4749	3.5036	3.5306	3.5562
17	3.2667	3.3070	3.3440	3.3783	3.4102	3.4400	3.4680	3.4944	3.5193
18	3.2399	3.2794	3.3156	3.3492	3.3804	3.4095	3.4369	3.4626	3.4870
19	3.2163	3.2550	3.2906	3.3235	3.3540	3.3826	3.4094	3.4347	3.4585
20	3.1952	3.2333	3.2683	3.3006	3.3306	3.3587	3.3850	3.4098	3.4332
21	3.1764	3.2139	3.2483	3.2802	3.3097	3.3373	3.3632	3.3876	3.4106
22	3.1595	3.1965	3.2304	3.2618	3.2909	3.3181	3.3436	3.3676	3.3903
23	3.1441	3.1807	3.2142	3.2451	3.2739	3.3007	3.3259	3.3495	3.3719
24	3.1302	3.1663	3.1994	3.2300	3.2584	3.2849	3.3097	3.3331	3.3552
25	3.1175	3.1532	3.1859	3.2162	3.2443	3.2705	3.2950	3.3181	3.3400
26	3.1058	3.1412	3.1736	3.2035	3.2313	3.2572	3.2815	3.3044	3.3260
27	3.0951	3.1301	3.1622	3.1919	3.2194	3.2451	3.2691	3.2918	3.3132
28	3.0852	3.1199	3.1517	3.1811	3.2084	3.2339	3.2577	3.2801	3.3013
29	3.0760	3.1105	3.1420	3.1712	3.1982	3.2235	3.2471	3.2694	3.2904
30	3.0675	3.1017	3.1330	3.1620	3.1888	3.2138	3.2373	3.2594	3.2802
35	3.0326	3.0658	3.0962	3.1242	3.1502	3.1744	3.1971	3.2185	3.2386
40	3.0069	3.0393	3.0690	3.0964	3.1218	3.1455	3.1676	3.1884	3.2081
45	2.9872	3.0191	3.0482	3.0751	3.1000	3.1232	3.1450	3.1654	3.1846
50	2.9716	3.0030	3.0318	3.0582	3.0828	3.1057	3.1271	3.1472	3.1661
55	2.9589	2.9900	3.0184	3.0446	3.0688	3.0914	3.1125	3.1324	3.1511
60	2.9485	2.9792	3.0074	3.0333	3.0573	3.0796	3.1005	3.1202	3.1387
70	2.9321	2.9624	2.9901	3.0156	3.0393	3.0613	3.0818	3.1012	3.1194
80	2.9200	2.9500	2.9773	3.0026	3.0259	3.0476	3.0679	3.0870	3.1050
90	2.9106	2.9403	2.9675	2.9924	3.0156	3.0371	3.0572	3.0761	3.0939
100	2.9032	2.9327	2.9596	2.9844	3.0073	3.0287	3.0487	3.0674	3.0851
110	2.8971	2.9264	2.9532	2.9778	3.0007	3.0219	3.0417	3.0604	3.0779
120	2.8921	2.9212	2.9479	2.9724	2.9951	3.0162	3.0360	3.0545	3.0720
250	2.8635	2.8919	2.9178	2.9416	2.9637	2.9842	3.0034	3.0213	3.0383
500	2.8505	2.8785	2.9041	2.9276	2.9494	2.9696	2.9885	3.0063	3.0230
1000	2.8440	2.8719	2.8973	2.9207	2.9423	2.9624	2.9812	2.9988	3.0154
∞	2.8376	2.8653	2.8905	2.9137	2.9352	2.9552	2.9738	2.9913	3.0078

Table B.3. $\beta = 1-$ Power of the F-test

ν_2	ϕ=.5	1.0	1.2	1.4	1.6	1.8	2.0	2.2	2.6	3.0
					$\nu_1 = 1$					
2	.9271	.8617	.8256	.7847	.7402	.6927	.6432	.5926	.4915	.3950
4	.9141	.8048	.7415	.6694	.5910	.5095	.4284	.3509	.2169	.1198
6	.9077	.7768	.7010	.6153	.5238	.4315	.3431	.2629	.1374	.0611
8	.9040	.7610	.6784	.5858	.4883	.3916	.3015	.2223	.1054	.0413
10	.9017	.7510	.6642	.5675	.4666	.3680	.2775	.1997	.0890	.0322
12	.9000	.7440	.6544	.5551	.4521	.3524	.2620	.1854	.0793	.0272
14	.8988	.7390	.6474	.5462	.4418	.3414	.2513	.1756	.0728	.0240
16	.8979	.7351	.6420	.5394	.4341	.3333	.2433	.1685	.0683	.0219
18	.8972	.7321	.6379	.5342	.4281	.3270	.2373	.1631	.0649	.0203
20	.8966	.7297	.6345	.5300	.4233	.3220	.2325	.1589	.0623	.0192
22	.8961	.7277	.6317	.5265	.4194	.3180	.2287	.1555	.0603	.0183
24	.8957	.7260	.6294	.5236	.4161	.3146	.2255	.1527	.0586	.0175
26	.8954	.7246	.6274	.5212	.4134	.3118	.2228	.1504	.0573	.0169
28	.8951	.7233	.6258	.5192	.4111	.3094	.2206	.1485	.0561	.0165
30	.8948	.7223	.6243	.5173	.4090	.3073	.2186	.1468	.0551	.0160
40	.8939	.7185	.6192	.5110	.4020	.3001	.2119	.1410	.0518	.0147
60	.8930	.7147	.6140	.5047	.3949	.2930	.2053	.1354	.0487	.0134
120	.8920	.7108	.6087	.4983	.3879	.2859	.1998	.1300	.0457	.0123
∞	.8910	.7070	.6036	.4920	.3810	.2791	.1926	.1248	.0430	.0112
					$\nu_1 = 2$					
2	.9324	.8814	.8527	.8201	.7840	.7451	.7038	.6608	.5722	.4837
4	.9201	.8239	.7657	.6976	.6219	.5414	.4598	.3804	.2400	.1353
6	.9129	.7891	.7135	.6257	.5303	.4330	.3396	.2554	.1264	.0520
8	.9083	.7672	.6810	.5821	.4769	.3729	.2773	.1955	.0821	.0273
10	.9052	.7523	.6592	.5536	.4430	.3361	.2408	.1624	.0609	.0175
12	.9030	.7417	.6438	.5336	.4197	.3115	.2173	.1419	.0490	.0126
14	.9013	.7337	.6323	.5189	.4028	.2941	.2010	.1281	.0416	.0099
16	.9000	.7274	.6234	.5077	.3901	.2812	.1892	.1183	.0367	.0082
18	.8989	.7225	.6164	.4988	.3802	.2713	.1802	.1110	.0331	.0071
20	.8980	.7184	.6107	.4917	.3723	.2634	.1732	.1054	.0305	.0063
22	.8973	.7150	.6059	.4858	.3658	.2570	.1675	.1009	.0285	.0057
24	.8967	.7122	.6019	.4808	.3603	.2517	.1629	.0973	.0269	.0052
26	.8961	.7097	.5985	.4767	.3558	.2472	.1590	.0943	.0256	.0048
28	.8957	.7076	.5956	.4730	.3518	.2434	.1558	.0918	.0245	.0045
30	.8953	.7058	.5930	.4699	.3484	.2401	.1530	.0896	.0236	.0043
40	.8938	.6992	.5839	.4588	.3365	.2288	.1434	.0824	.0207	.0035
60	.8923	.6924	.5746	.4476	.3247	.2177	.1341	.0756	.0181	.0029
120	.8908	.6855	.5651	.4364	.3129	.2069	.1253	.0692	.0157	.0024
∞	.8892	.6785	.5556	.4251	.3013	.1963	.1168	.0632	.0137	.0019
					$\nu_1 = 3$					
2	.9342	.8882	.8623	.8327	.7998	.7640	.7260	.6861	.6030	.5187
4	.9221	.8302	.7735	.7064	.6311	.5505	.4683	.3880	.2453	.1384
6	.9144	.7909	.7134	.6226	.5235	.4225	.3264	.2407	.1132	.0435
8	.9092	.7643	.6733	.5683	.4570	.3482	.2504	.1694	.0639	.0184
10	.9056	.7454	.6453	.5314	.4134	.3019	.2059	.1307	.0419	.0098
12	.9028	.7314	.6249	.5050	.3831	.2709	.1776	.1074	.0305	.0061
14	.9007	.7207	.6093	.4853	.3611	.2490	.1583	.0922	.0238	.0042
16	.8990	.7122	.5972	.4701	.3443	.2328	.1444	.0817	.0196	.0031
18	.8976	.7054	.5874	.4581	.3313	.2204	.1340	.0740	.0167	.0025
20	.8965	.6997	.5794	.4483	.3208	.2106	.1259	.0682	.0146	.0020
22	.8955	.6950	.5728	.4402	.3122	.2026	.1195	.0637	.0131	.0017
24	.8947	.6909	.5671	.4333	.3051	.1961	.1143	.0601	.0119	.0015
26	.8940	.6875	.5623	.4275	.2990	.1907	.1100	.0571	.0110	.0013
28	.8934	.6845	.5581	.4225	.2938	.1860	.1064	.0547	.0103	.0012
30	.8928	.6818	.5544	.4182	.2894	.1820	.1033	.0526	.0097	.0011
40	.8909	.6723	.5414	.4028	.2738	.1684	.0930	.0458	.0078	.0008
60	.8888	.6624	.5279	.3872	.2583	.1552	.0833	.0397	.0062	.0006
120	.8866	.6522	.5142	.3716	.2431	.1425	.0743	.0342	.0049	.0004
∞	.8843	.6415	.5000	.3557	.2280	.1304	.0659	.0293	.0038	.0003

(continued)

Table B.3. ***(continued)***

ν_2	$\phi = .5$	1.0	1.2	1.4	1.6	1.8	2.0	2.2	2.6	3.0
					$\nu_1 = 4$					
2	.9351	.8917	.8672	.8391	.8079	.7738	.7375	.6993	.6193	.5374
4	.9232	.8332	.7771	.7103	.6350	.5542	.4714	.3905	.2466	.1389
6	.9151	.7906	.7112	.6178	.5158	.4122	.8143	.2282	.1030	.0375
8	.9094	.7602	.6649	.5549	.4389	.3271	.2286	.1493	.0515	.0132
10	.9052	.7378	.6315	.5110	.3876	.2736	.1788	.1076	.0301	.0059
12	.9020	.7208	.6066	.4791	.3516	.2380	.1475	.0833	.0199	.0032
14	.8995	.7076	.5875	.4550	.3253	.2129	.1266	.0680	.0143	.0019
16	.8975	.6970	.5723	.4363	.3054	.1945	.1118	.0577	.0109	.0013
18	.8958	.6883	.5600	.4214	.2898	.1804	.1009	.0503	.0087	.0009
20	.8945	.6811	.5498	.4092	.2774	.1695	.0926	.0449	.0073	.0007
22	.8933	.6750	.5413	.3991	.2672	.1607	.0861	.0408	.0062	.0006
24	.8923	.6698	.5341	.3907	.2587	.1535	.0808	.0376	.0054	.0005
26	.8914	.6653	.5279	.3834	.2516	.1475	.0765	.0349	.0049	.0004
28	.8906	.6614	.5225	.3772	.2455	.1424	.0730	.0328	.0044	.0003
30	.8899	.6579	.5178	.3718	.2402	.1381	.0700	.0311	.0040	.0003
40	.8874	.6454	.5009	.3526	.2219	.1234	.0601	.0254	.0029	.0002
60	.8848	.6322	.4833	.3332	.2040	.1095	.0511	.0206	.0021	.0001
120	.8819	.6183	.4652	.3136	.1865	.0965	.0431	.0164	.0015	.0001
∞	.8789	.6038	.4466	.2940	.1695	.0844	.0360	.0130	.0011	.0000
					$\nu_1 = 5$					
2	.9356	.8939	.8702	.8431	.8128	.7798	.7445	.7074	.6293	.5490
4	.9238	.8349	.7791	.7124	.6369	.5558	.4727	.3914	.2467	.1386
6	.9154	.7897	.7087	.6131	.5088	.4033	.3044	.2181	.0952	.0333
8	.9093	.7561	.6573	.5432	.4237	.3099	.2115	.1342	.0430	.0100
10	.9048	.7308	.6193	.4933	.3660	.2506	.1579	.0909	.0227	.0038
12	.9012	.7111	.5904	.4566	.3254	.2118	.1249	.0665	.0136	.0018
14	.8983	.6956	.5679	.4287	.2957	.1845	.1033	.0516	.0090	.0010
16	.8960	.6829	.5499	.4069	.2732	.1646	.0883	.0418	.0064	.0006
18	.8941	.6725	.5352	.3895	.2557	.1496	.0774	.0351	.0048	.0004
20	.8924	.6638	.5231	.3753	.2417	.1380	.0693	.0303	.0038	.0003
22	.8910	.6564	.5129	.3635	.2303	.1288	.0630	.0267	.0031	.0002
24	.8898	.6501	.5042	.3536	.2209	.1213	.0580	.0240	.0026	.0001
26	.8888	.6445	.4967	.3451	.2129	.1151	.0540	.0218	.0022	.0001
28	.8878	.6397	.4901	.3379	.2062	.1099	.0507	.0201	.0019	.0001
30	.8870	.6354	.4844	.3315	.2003	.1055	.0479	.0186	.0017	.0001
40	.8840	.6198	.4638	.3091	.1803	.0908	.0390	.0142	.0011	.0000
60	.8807	.6033	.4423	.2864	.1609	.0772	.0313	.0106	.0007	.0000
120	.8771	.5857	.4201	.2638	.1422	.0649	.0247	.0078	.0004	.0000
∞	.8733	.5671	.3971	.2412	.1245	.0538	.0192	.0056	.0003	.0000
					$\nu_1 = 6$					
2	.9360	.8953	.8722	.8457	.8161	.7839	.7493	.7129	.6361	.5569
4	.9242	.8361	.7800	.7136	.6380	.5567	.4733	.3916	.2464	.1381
6	.9156	.7887	.7063	.6090	.5028	.3959	.2962	.2100	.0893	.0301
8	.9092	.7525	.6500	.5332	.4109	.2958	.1978	.1225	.0369	.0080
10	.9042	.7245	.6086	.4782	.3480	.2325	.1417	.0784	.0177	.0026
12	.9003	.7024	.5761	.4373	.3036	.1908	.1077	.0544	.0097	.0011
14	.8972	.6847	.5506	.4061	.2711	.1619	.0859	.0401	.0059	.0005
16	.8946	.6702	.5301	.3816	.2465	.1412	.0710	.0312	.0039	.0003
18	.8924	.6582	.5132	.3621	.2275	.1257	.0605	.0252	.0028	.0002
20	.8905	.6480	.4992	.3461	.2124	.1139	.0528	.0210	.0021	.0001
22	.8889	.6394	.4874	.3328	.2001	.1045	.0469	.0179	.0016	.0001
24	.8875	.6319	.4773	.3216	.1900	.0970	.0423	.0157	.0013	.0001
26	.8863	.6253	.4686	.3121	.1815	.0908	.0387	.0139	.0011	.0000
28	.8852	.6196	.4610	.3039	.1744	.0857	.0357	.0125	.0009	.0000
30	.8843	.6145	.4543	.2968	.1682	.0814	.0333	.0114	.0008	.0000
40	.8807	.5960	.4302	.2717	.1471	.0672	.0256	.0081	.0004	.0000
60	.8768	.5760	.4050	.2464	.1270	.0545	.0193	.0055	.0002	.0000
120	.8724	.5547	.3789	.2214	.1082	.0434	.0141	.0037	.0001	.0000
∞	.8677	.5319	.3520	.1967	.0907	.0339	.0101	.0024	.0000	.0000

(continued)

Table B.3. (*continued*)

ν_2	$\phi = .5$	1.0	1.2	1.4	1.6	1.8	2.0	2.2	2.6	3.0
					$\nu_1 = 7$					
2	.9363	.8963	.8736	.8476	.8185	.7868	.7527	.7168	.6410	.5627
4	.9245	.8368	.7811	.7144	.6387	.5571	.4735	.3916	.2460	.1376
6	.9157	.7878	.7042	.6054	.4978	.3897	.2895	.2035	.0846	.0278
8	.9090	.7492	.6449	.5247	.4002	.2841	.1868	.1133	.0323	.0065
10	.9038	.7189	.5992	.4652	.3328	.2174	.1288	.0689	.0143	.0019
12	.8996	.6947	.5636	.4207	.2852	.1738	.0944	.0454	.0072	.0007
14	.8961	.6750	.5353	.3866	.2505	.1439	.0726	.0320	.0041	.0003
16	.8933	.6588	.5125	.3598	.2243	.1226	.0502	.0238	.0025	.0001
18	.8908	.6452	.4936	.3383	.2041	.1070	.0482	.0185	.0017	.0001
20	.8888	.6336	.4779	.3208	.1882	.0951	.0409	.0149	.0012	.0000
22	.8870	.6238	.4646	.3062	.1753	.0858	.0355	.0123	.0009	.0000
24	.8854	.6152	.4532	.2940	.1647	.0785	.0314	.0105	.0007	.0000
26	.8840	.6077	.4433	.2836	.1559	.0725	.0282	.0091	.0005	.0000
28	.8828	.6011	.4347	.2747	.1485	.0676	.0256	.0080	.0004	.0000
30	.8817	.5952	.4272	.2669	.1421	.0634	.0234	.0071	.0003	.0000
40	.8776	.5737	.3998	.2396	.1206	.0501	.0170	.0046	.0002	.0000
60	.8730	.5504	.3713	.2124	.1005	.0387	.0119	.0029	.0001	.0000
120	.8679	.5253	.3417	.1857	.0821	.0290	.0080	.0017	.0000	.0000
∞	.8622	.4983	.3112	.1597	.0656	.0211	.0052	.0010	.0000	.0000
					$\nu_1 = 8$					
2	.9365	.8971	.8747	.8490	.8203	.7889	.7553	.7198	.6448	.5671
4	.9274	.8374	.7817	.7149	.6891	.5574	.4735	.3914	.2456	.1371
6	.9158	.7869	.7024	.6023	.4935	.3845	.2839	.1981	.0809	.0259
8	.9088	.7464	.6398	.5173	.3910	.2744	.1777	.1059	.0289	.0055
10	.9033	.7140	.5910	.4540	.3200	.2049	.1184	.0615	.0118	.0014
12	.8989	.6878	.5526	.4063	.2697	.1598	.0838	.0387	.0055	.0005
14	.8951	.6663	.5218	.3696	.2330	.1292	.0624	.0260	.0029	.0002
16	.8920	.6484	.4968	.3407	.2056	.1077	.0485	.0186	.0017	.0001
18	.8894	.6334	.4761	.3176	.1846	.0921	.0390	.0139	.0010	.0000
20	.8871	.6205	.4588	.2988	.1680	.0804	.0323	.0108	.0007	.0000
22	.8851	.6095	.4441	.2832	.1548	.0713	.0274	.0087	.0005	.0000
24	.8834	.5999	.4315	.2700	.1439	.0642	.0237	.0072	.0003	.0000
26	.8819	.5915	.4206	.2589	.1349	.0585	.0208	.0060	.0003	.0000
28	.8805	.5840	.4111	.2493	.1274	.0538	.0186	.0052	.0002	.0000
30	.8793	.5774	.4027	.2410	.1209	.0499	.0168	.0045	.0002	.0000
40	.8746	.5530	.3725	.2120	.0995	.0377	.0114	.0027	.0001	.0000
60	.8694	.5264	.3408	.1834	.0798	.0275	.0074	.0015	.0000	.0000
120	.8635	.4975	.3081	.1556	.0623	.0193	.0046	.0008	.0000	.0000
∞	.8568	.4663	.2745	.1292	.0472	.0130	.0027	.0004	.0000	.0000
					$\nu_1 = 9$					
2	.9366	.8977	.8756	.8501	.8217	.7906	.7573	.7221	.6477	.5705
4	.9249	.8378	.7821	.7153	.6394	.5575	.4735	.3912	.2452	.1366
6	.9158	.7861	.7007	.5996	.4898	.3800	.2792	.1936	.0778	.0245
8	.9087	.7439	.6354	.5109	.3832	.2661	.1702	.0998	.0262	.0048
10	.9029	.7096	.5838	.4442	.3089	.1944	.1099	.0555	.0100	.0011
12	.8982	.6816	.5428	.3937	.2564	.1481	.0753	.0334	.0043	.0003
14	.8943	.6584	.5097	.3547	.2182	.1171	.0543	.0216	.0021	.0001
16	.8909	.6390	.4827	.3241	.1898	.0956	.0409	.0148	.0011	.0000
18	.8881	.6226	.4604	.2996	.1681	.0801	.0320	.0107	.0007	.0000
20	.8856	.6086	.4416	.2796	.1511	.0686	.0259	.0080	.0004	.0000
22	.8835	.5964	.4257	.2630	.1376	.0599	.0214	.0062	.0003	.0000
24	.8816	.5858	.4120	.2492	.1266	.0531	.0181	.0050	.0002	.0000
26	.8799	.5765	.4002	.2374	.1176	.0477	.0156	.0041	.0001	.0000
28	.8784	.5683	.3898	.2274	.1110	.0433	.0137	.0034	.0001	.0000
30	.8770	.5609	.3807	.2186	.1036	.0397	.0121	.0029	.0001	.0000
40	.8718	.5337	.3477	.1883	.0825	.0286	.0077	.0016	.0000	.0000
60	.8660	.5038	.3133	.1587	.0636	.0197	.0046	.0008	.0000	.0000
120	.8592	.4713	.2778	.1304	.0473	.0129	.0026	.0004	.0000	.0000
∞	.8514	.4361	.2417	.1041	.0337	.0080	.0014	.0002	.0000	.0000

(*continued*)

Table B.3. ***(continued)***

ν_2	$\phi = .5$	1.0	1.2	1.4	1.6	1.8	2.0	2.2	2.6	3.0
					$\nu_1 = 10$					
2	.9368	.8981	.8762	.8510	.8228	.7920	.7589	.7240	.6500	.5732
4	.9250	.8381	.7825	.7156	.6395	.5575	.4734	.3910	.2448	.1362
6	.9158	.7854	.6992	.5972	.4865	.3762	.2751	.1898	.0752	.0233
8	.9085	.7417	.6315	.5053	.3764	.2591	.1638	.0948	.0241	.0042
10	.9026	.7057	.5774	.4357	.2993	.1854	.1028	.0508	.0086	.0009
12	.8976	.6761	.5340	.3826	.2448	.1383	.0683	.0293	.0035	.0002
14	.8935	.6514	.4990	.3417	.2055	.1070	.0478	.0182	.0016	.0001
16	.8899	.6305	.4702	.3094	.1763	.0856	.0350	.0120	.0008	.0000
18	.8869	.6128	.4463	.2836	.1541	.0704	.0267	.0083	.0004	.0000
20	.8843	.5976	.4262	.2627	.1368	.0592	.0210	.0061	.0003	.0000
22	.8819	.5844	.4091	.2454	.1232	.0508	.0170	.0046	.0002	.0000
24	.8799	.5729	.3944	.2309	.1122	.0443	.0141	.0035	.0001	.0000
26	.8780	.5627	.3817	.2186	.1031	.0393	.0119	.0028	.0001	.0000
28	.8764	.5537	.3705	.2082	.0956	.0352	.0102	.0023	.0001	.0000
30	.8749	.5456	.3607	.1991	.0893	.0319	.0089	.0019	.0000	.0000
40	.8692	.5157	.3253	.1678	.0687	.0218	.0053	.0010	.0000	.0000
60	.8627	.4827	.2884	.1377	.0508	.0142	.0029	.0004	.0000	.0000
120	.8551	.4467	.2506	.1094	.0359	.0086	.0015	.0002	.0000	.0000
∞	.8462	.4047	.2124	.0836	.0240	.0049	.0007	.0000	.0000	.0000
					$\nu_1 = 11$					
2	.9369	.8989	.8772	.8524	.8245	.7941	.7614	.7268	.6536	.5774
4	.9252	.8385	.7829	.7159	.6397	.5575	.4731	.3905	.2441	.1355
6	.9159	.7842	.6968	.5934	.4813	.3700	.2686	.1837	.0713	.0214
8	.9082	.7379	.6250	.4960	.3653	.2476	.1536	.0869	.0208	.0034
10	.9019	.6991	.5666	.4213	.2836	.1711	.0917	.0435	.0067	.0006
12	.8966	.6666	.5192	.3641	.2260	.1227	.0577	.0234	.0024	.0001
14	.8921	.6391	.4805	.3197	.1847	.0913	.0382	.0135	.0010	.0000
16	.8882	.6157	.4485	.2849	.1544	.0703	.0265	.0083	.0004	.0000
18	.8848	.5956	.4219	.2571	.1316	.0557	.0192	.0053	.0002	.0000
20	.8818	.5793	.3994	.2345	.1142	.0452	.0144	.0036	.0001	.0000
22	.8792	.5631	.3803	.2160	.1005	.0376	.0111	.0026	.0001	.0000
24	.8768	.5498	.3638	.2006	.0896	.0318	.0088	.0019	.0000	.0000
26	.8747	.5381	.3496	.1876	.0808	.0274	.0072	.0014	.0000	.0000
28	.8728	.5276	.3371	.1766	.0736	.0239	.0059	.0011	.0000	.0000
30	.8710	.5182	.3261	.1671	.0676	.0211	.0050	.0009	.0000	.0000
40	.8643	.4831	.2865	.1347	.0485	.0131	.0026	.0004	.0000	.0000
60	.8565	.4443	.2456	.1044	.0329	.0075	.0012	.0001	.0000	.0000
120	.8472	.4016	.2042	.0770	.0207	.0039	.0005	.0000	.0000	.0000
∞	.8359	.3548	.1632	.0535	.0120	.0018	.0002	.0000	.0000	.0000

Table B.4. Lower Critical Values of Wilks Λ, α = 0.05

$$\Lambda = \frac{|\mathbf{E}|}{|\mathbf{E}+\mathbf{H}|} = \prod_{i=1}^{s} \frac{1}{1+\lambda_i}, \text{ where } \lambda_1, \lambda_2, \ldots, \lambda_s \text{ are eigenvalues of } \mathbf{E}^{-1}\mathbf{H}.$$

Reject H_0 if $\Lambda \leq$ table value.

	ν_H											
ν_E	1	2	3	4	5	6	7	8	9	10	11	12
						$p = 1$						
1	6.16[a]	2.50[a]	1.54[a]	1.11[a]	.868[a]	.712[a]	.603[a]	.523[a]	.462[a]	.413[a]	.374[a]	.341[a]
2	.098	.050	.034	.025	.020	.017	.015	.013	.011	.010	9.28[a]	8.51[a]
3	.229	.136	.097	.076	.062	.053	.046	.041	.036	.033	.030	.028
4	.342	.224	.168	.135	.113	.098	.086	.076	.069	.063	.058	.053
5	.431	.302	.236	.194	.165	.144	.128	.115	.104	.096	.088	.082
6	.501	.368	.296	.249	.215	.189	.169	.153	.140	.129	.119	.111
7	.556	.425	.349	.298	.261	.232	.209	.190	.175	.161	.150	.140
8	.601	.473	.396	.343	.303	.271	.246	.225	.208	.193	.180	.169
9	.638	.514	.437	.382	.341	.308	.281	.258	.239	.223	.209	.196
10	.668	.549	.473	.418	.376	.341	.313	.289	.269	.251	.236	.222
11	.694	.580	.505	.450	.407	.372	.343	.318	.297	.278	.262	.247
12	.717	.607	.534	.479	.436	.400	.370	.345	.323	.304	.286	.271
13	.736	.631	.560	.506	.462	.426	.396	.370	.347	.327	.310	.294
14	.753	.652	.583	.529	.486	.450	.420	.393	.370	.350	.332	.315
15	.768	.671	.603	.551	.508	.473	.442	.415	.392	.371	.352	.336
16	.781	.688	.622	.571	.529	.493	.462	.436	.412	.391	.372	.355
17	.792	.703	.639	.589	.548	.512	.482	.455	.431	.410	.390	.373
18	.803	.717	.655	.606	.565	.530	.499	.473	.449	.427	.408	.390
19	.813	.730	.669	.621	.581	.546	.516	.490	.466	.444	.425	.407
20	.821	.741	.683	.636	.596	.562	.532	.505	.482	.460	.440	.423
21	.829	.752	.695	.649	.610	.576	.547	.520	.497	.475	.455	.437
22	.836	.762	.706	.661	.623	.590	.561	.534	.511	.489	.470	.452
23	.843	.771	.717	.673	.635	.603	.574	.548	.524	.503	.483	.465
24	.849	.779	.727	.684	.647	.615	.586	.560	.537	.516	.496	.478
25	.855	.787	.736	.694	.658	.626	.598	.572	.549	.528	.508	.490
26	.860	.794	.744	.703	.668	.637	.609	.583	.560	.539	.520	.502
27	.865	.801	.752	.712	.677	.647	.619	.594	.571	.551	.531	.513
28	.870	.807	.760	.721	.686	.656	.629	.604	.582	.561	.542	.524
29	.874	.813	.767	.729	.695	.665	.638	.614	.592	.571	.552	.535
30	.878	.819	.774	.736	.703	.674	.647	.623	.601	.581	.562	.544
40	.907	.861	.824	.793	.766	.741	.718	.696	.677	.658	.641	.625
60	.938	.905	.879	.856	.835	.816	.798	.781	.766	.751	.736	.723
80	.953	.928	.907	.889	.873	.858	.843	.829	.816	.804	.792	.780
100	.962	.942	.925	.910	.897	.884	.872	.860	.849	.838	.828	.818
120	.968	.951	.937	.925	.913	.902	.891	.882	.872	.863	.854	.845
140	.973	.958	.946	.935	.925	.915	.906	.897	.889	.881	.873	.865
170	.978	.965	.955	.946	.937	.929	.922	.914	.907	.900	.893	.887
200	.981	.970	.962	.954	.947	.940	.933	.926	.920	.914	.908	.902
240	.984	.975	.968	.961	.955	.949	.944	.938	.933	.928	.923	.918
320	.988	.981	.976	.971	.966	.962	.957	.953	.949	.945	.941	.937
440	.991	.986	.982	.979	.975	.972	.969	.966	.963	.960	.957	.954
600	.994	.990	.987	.984	.982	.979	.977	.975	.972	.970	.968	.966
800	.995	.993	.990	.988	.986	.984	.983	.981	.979	.977	.976	.974
1000	.996	.994	.992	.991	.989	.988	.986	.985	.983	.982	.981	.979

[a]Multiply entry by 10^{-3}.

(continued)

Table B.4. *(continued)*

ν_E	ν_H 1	2	3	4	5	6	7	8	9	10	11	12
						$p = 2$						
1	.000	.000	.000	.000	.000	.000	.000	.000	.000	.000	.000	.000
2	2.50[a]	.641[a]	.287[a]	.162[a]	.104[a]	.072[a]	.053[a]	.041[a]	.032[a]	.026[a]	.022[a]	.018[a]
3	.050	.018	9.53[a]	5.84[a]	3.95[a]	2.85[a]	2.15[a]	1.68[a]	1.35[a]	1.11[a]	.928[a]	.787[a]
4	.136	.062	.036	.023	.017	.012	9.56[a]	7.62[a]	6.21[a]	5.17[a]	4.36[a]	3.73[a]
5	.224	.117	.074	.051	.037	.028	.023	.018	.015	.013	.011	.009
6	.302	.175	.116	.084	.063	.049	.040	.033	.027	.023	.020	.017
7	.368	.230	.160	.119	.092	.074	.060	.050	.042	.036	.032	.028
8	.426	.280	.203	.155	.122	.099	.082	.069	.059	.051	.045	.040
9	.473	.326	.243	.190	.153	.126	.106	.090	.078	.068	.060	.053
10	.514	.367	.281	.223	.183	.152	.129	.111	.097	.085	.075	.067
11	.549	.404	.316	.255	.212	.179	.153	.133	.116	.102	.091	.082
12	.580	.437	.348	.286	.240	.204	.176	.154	.136	.120	.108	.097
13	.607	.467	.378	.314	.266	.229	.199	.175	.155	.138	.124	.112
14	.631	.495	.405	.340	.291	.252	.221	.195	.174	.156	.141	.128
15	.652	.519	.431	.365	.315	.275	.242	.215	.193	.174	.157	.143
16	.671	.542	.454	.389	.337	.296	.263	.235	.211	.191	.174	.159
17	.688	.562	.476	.410	.359	.317	.282	.254	.229	.208	.190	.174
18	.703	.581	.496	.431	.379	.337	.301	.272	.246	.225	.206	.189
19	.717	.598	.515	.450	.398	.355	.320	.289	.263	.241	.221	.204
20	.730	.614	.532	.468	.416	.373	.337	.306	.279	.256	.236	.218
21	.741	.629	.548	.485	.433	.390	.354	.322	.295	.271	.251	.232
22	.752	.643	.564	.501	.449	.406	.370	.338	.310	.286	.265	.246
23	.762	.656	.578	.516	.465	.422	.385	.353	.325	.300	.279	.259
24	.771	.668	.591	.530	.479	.436	.399	.367	.339	.314	.292	.272
25	.779	.679	.604	.544	.493	.450	.413	.381	.353	.328	.305	.285
26	.787	.689	.616	.556	.506	.464	.427	.395	.366	.341	.318	.297
27	.794	.699	.627	.568	.519	.477	.440	.407	.379	.353	.330	.309
28	.801	.708	.638	.580	.531	.489	.452	.420	.391	.365	.342	.321
29	.807	.717	.648	.591	.542	.501	.464	.432	.403	.377	.354	.332
30	.813	.725	.657	.601	.553	.512	.475	.443	.414	.388	.365	.344
40	.858	.786	.730	.682	.640	.602	.568	.537	.509	.484	.460	.439
60	.903	.853	.811	.774	.741	.710	.682	.656	.632	.609	.588	.568
80	.927	.888	.854	.825	.798	.772	.749	.727	.706	.686	.667	.649
100	.941	.909	.882	.857	.834	.813	.793	.774	.755	.738	.721	.705
120	.951	.924	.900	.879	.860	.841	.823	.807	.791	.775	.760	.746
140	.958	.934	.914	.895	.878	.862	.846	.831	.817	.803	.790	.777
170	.965	.946	.929	.913	.898	.885	.871	.859	.846	.834	.823	.812
200	.970	.954	.939	.926	.913	.901	.889	.878	.867	.857	.847	.837
240	.975	.961	.949	.938	.927	.917	.907	.897	.888	.879	.870	.862
320	.981	.971	.962	.953	.945	.937	.929	.922	.914	.907	.901	.894
440	.986	.979	.972	.965	.959	.953	.948	.942	.937	.932	.926	.921
600	.990	.984	.979	.975	.970	.966	.961	.957	.953	.949	.945	.942
800	.993	.988	.984	.981	.977	.974	.971	.968	.965	.962	.959	.956
1000	.994	.991	.987	.985	.982	.979	.977	.974	.972	.969	.967	.964

[a]Multiply entry by 10^{-3}.

(continued)

Table B.4. (*continued*)

ν_E	ν_H 1	2	3	4	5	6	7	8	9	10	11	12
						$p = 3$						
1	.000	.000	.000	.000	.000	.000	.000	.000	.000	.000	.000	.000
2	.000	.000	.000	.000	.000	.001[a]	.002[a]	.004[a]	.005[a]	.008[a]	.010[a]	.013[a]
3	1.70[a]	.354[a]	.179[a]	.127[a]	.105[a]	.095[a]	.091[a]	.090[a]	.091[a]	.092[a]	.095[a]	.098[a]
4	.034	.010	.004	.002	.001	.001	.809[a]	.659[a]	.562[a]	.496[a]	.449[a]	.416[a]
5	.097	.036	.018	.010	6.36[a]	4.37[a]	3.20[a]	2.46[a]	1.97[a]	1.64[a]	1.40[a]	1.22[a]
6	.168	.074	.040	.024	.016	.011	.008	.006	.004	3.94[a]	3.28[a]	2.79[a]
7	.236	.116	.068	.043	.029	.021	.016	.012	9.49[a]	7.67[a]	6.35[a]	5.35[a]
8	.296	.160	.099	.066	.046	.034	.026	.020	.016	.013	.011	9.00[a]
9	.349	.203	.131	.091	.066	.049	.038	.030	.024	.020	.016	.014
10	.396	.243	.164	.117	.086	.066	.052	.041	.034	.028	.023	.020
11	.437	.281	.196	.143	.108	.084	.067	.054	.044	.037	.031	.026
12	.473	.316	.226	.169	.130	.103	.083	.067	.056	.047	.040	.034
13	.505	.348	.255	.194	.152	.122	.099	.082	.068	.058	.049	.042
14	.534	.378	.283	.219	.174	.141	.116	.096	.081	.069	.059	.051
15	.560	.405	.309	.243	.195	.160	.133	.111	.095	.081	.070	.061
16	.583	.431	.334	.266	.216	.179	.149	.127	.108	.093	.081	.071
17	.603	.454	.357	.288	.236	.197	.166	.142	.122	.106	.092	.081
18	.622	.476	.379	.309	.256	.215	.183	.157	.136	.118	.104	.092
19	.639	.496	.399	.329	.275	.233	.199	.172	.149	.131	.115	.102
20	.655	.515	.419	.348	.293	.250	.215	.187	.163	.144	.127	.113
21	.669	.532	.437	.366	.310	.266	.230	.201	.177	.156	.139	.124
22	.683	.548	.454	.383	.327	.282	.246	.215	.190	.169	.150	.135
23	.695	.564	.470	.399	.343	.298	.260	.229	.203	.181	.162	.146
24	.706	.578	.486	.415	.359	.313	.275	.243	.216	.193	.173	.156
25	.717	.591	.500	.430	.374	.327	.289	.256	.229	.205	.185	.167
26	.727	.604	.514	.444	.388	.341	.302	.269	.241	.217	.196	.178
27	.736	.616	.527	.458	.401	.355	.315	.282	.253	.229	.207	.188
28	.744	.627	.540	.471	.415	.368	.328	.294	.265	.240	.218	.199
29	.752	.638	.552	.483	.427	.380	.340	.306	.277	.251	.229	.209
30	.760	.648	.563	.495	.439	.392	.352	.318	.288	.262	.239	.219
40	.816	.724	.651	.591	.539	.494	.454	.419	.387	.359	.334	.311
60	.875	.808	.752	.704	.661	.623	.587	.555	.526	.498	.473	.449
80	.905	.853	.808	.769	.733	.700	.670	.641	.615	.590	.566	.544
100	.924	.881	.844	.810	.780	.751	.725	.700	.676	.654	.632	.612
120	.936	.900	.868	.839	.813	.788	.764	.742	.721	.700	.681	.663
140	.945	.913	.886	.861	.837	.815	.794	.774	.755	.736	.719	.702
170	.955	.928	.905	.884	.864	.845	.827	.809	.792	.776	.761	.746
200	.961	.939	.919	.900	.883	.866	.850	.835	.820	.806	.792	.779
240	.968	.949	.932	.916	.901	.887	.873	.860	.848	.835	.823	.811
320	.976	.961	.948	.936	.925	.914	.903	.893	.883	.873	.864	.854
440	.982	.972	.962	.953	.945	.937	.929	.921	.913	.906	.899	.891
600	.987	.979	.972	.966	.959	.953	.947	.941	.936	.930	.924	.919
800	.990	.984	.979	.974	.969	.965	.960	.956	.951	.947	.943	.939
1000	.992	.987	.983	.979	.975	.972	.968	.964	.961	.957	.954	.950

[a]Multiply entry by 10^{-3}.

(continued)

Table B.4. ***(continued)***

ν_E	ν_H 1	2	3	4	5	6	7	8	9	10	11	12
						$p = 4$						
1	.000	.000	.000	.000	.000	.000	.000	.000	.000	.000	.000	.000
2	.000	.000	.000	.000	.000	.000	.000	.000	.000	.000	.000	.000
3	.000	.000	.000	.000	.000	.001[a]	.001[a]	.001[a]	.002[a]	.002[a]	.002[a]	.003[a]
4	1.38[a]	.292[a]	.127[a]	.075[a]	.052[a]	.040[a]	.033[a]	.029[a]	.026[a]	.025[a]	.023[a]	.022[a]
5	.026	6.09[a]	2.31[a]	1.13[a]	.647[a]	.416[a]	.292[a]	.218[a]	.172[a]	.141[a]	.120[a]	.105[a]
6	.076	.024	.010	5.07[a]	2.90[a]	1.82[a]	1.22[a]	.872[a]	.652[a]	.508[a]	.409[a]	.338[a]
7	.135	.051	.024	.013	7.74[a]	4.94[a]	3.34[a]	2.36[a]	1.74[a]	1.33[a]	1.05[a]	.848[a]
8	.194	.084	.043	.025	.015	.010	6.98[a]	4.99[a]	3.70[a]	2.82[a]	2.21[a]	1.77[a]
9	.249	.119	.066	.040	.026	.017	.012	8.91[a]	6.66[a]	5.11[a]	4.01[a]	3.21[a]
10	.298	.155	.091	.057	.038	.027	.019	.014	.011	8.29[a]	6.54[a]	5.25[a]
11	.343	.190	.117	.077	.053	.037	.027	.021	.016	.012	9.84[a]	7.95[a]
12	.382	.223	.143	.097	.068	.049	.037	.028	.022	.017	.014	.011
13	.418	.255	.169	.117	.085	.063	.047	.037	.029	.023	.019	.015
14	.450	.286	.194	.138	.102	.077	.059	.046	.037	.030	.024	.020
15	.479	.314	.219	.159	.119	.091	.071	.056	.045	.037	.030	.025
16	.506	.340	.243	.180	.136	.106	.083	.067	.054	.044	.037	.031
17	.529	.365	.266	.200	.154	.121	.096	.078	.064	.053	.044	.037
18	.551	.389	.288	.219	.171	.136	.109	.089	.074	.061	.051	.044
19	.571	.410	.309	.239	.188	.151	.123	.101	.084	.070	.059	.051
20	.589	.431	.329	.257	.205	.166	.136	.113	.094	.079	.068	.058
21	.606	.450	.348	.275	.221	.181	.149	.124	.105	.089	.076	.065
22	.621	.468	.366	.292	.237	.195	.162	.136	.115	.098	.085	.073
23	.636	.485	.383	.309	.253	.210	.175	.148	.126	.108	.093	.081
24	.649	.501	.399	.325	.268	.224	.188	.160	.137	.118	.102	.089
25	.661	.516	.415	.340	.283	.237	.201	.172	.148	.128	.111	.097
26	.673	.530	.430	.355	.297	.251	.214	.183	.158	.138	.120	.106
27	.684	.544	.444	.369	.311	.264	.226	.195	.169	.147	.129	.114
28	.694	.556	.458	.383	.324	.277	.238	.206	.180	.157	.138	.122
29	.703	.568	.471	.396	.337	.289	.250	.217	.190	.167	.147	.131
30	.712	.580	.483	.409	.349	.301	.261	.228	.200	.177	.157	.139
40	.779	.668	.583	.513	.455	.406	.364	.327	.295	.267	.243	.221
60	.849	.767	.700	.643	.592	.547	.507	.471	.438	.409	.382	.357
80	.885	.821	.766	.718	.675	.636	.600	.567	.536	.508	.482	.457
100	.908	.854	.809	.768	.730	.696	.664	.634	.606	.580	.555	.532
120	.923	.877	.838	.802	.770	.739	.711	.684	.658	.634	.611	.590
140	.934	.894	.860	.828	.799	.772	.746	.721	.698	.676	.655	.635
170	.945	.912	.883	.856	.831	.808	.785	.764	.743	.724	.705	.687
200	.953	.925	.900	.876	.855	.834	.814	.795	.777	.759	.742	.726
240	.961	.937	.916	.896	.877	.859	.842	.826	.810	.795	.780	.765
320	.971	.952	.936	.921	.907	.893	.879	.866	.854	.841	.829	.818
440	.979	.965	.953	.942	.931	.921	.911	.901	.891	.882	.872	.863
600	.984	.974	.966	.957	.949	.941	.934	.926	.919	.912	.905	.898
800	.988	.981	.974	.968	.961	.956	.950	.944	.938	.933	.927	.922
1000	.991	.985	.979	.974	.969	.964	.960	.955	.950	.946	.941	.937

[a]Multiply entry by 10^{-3}.

(continued)

Table B.4. ***(continued)***

ν_E	ν_H 1	2	3	4	5	6	7	8	9	10	11	12
						$p = 5$						
1	.000	.000	.000	.000	.000	.000	.000	.000	.000	.000	.000	.000
2	.000	.000	.000	.000	.000	.000	.000	.000	.000	.000	.000	.000
3	.000	.000	.000	.000	.000	.000	.000	.000	.000	.000	.000	.000
4	.000	.000	.000	.000	.001[a]	.001[a]	.001[a]	.001[a]	.001[a]	.001[a]	.001[a]	.001[a]
5	1.60[a]	.291[a]	.105[a]	.052[a]	.031[a]	.021[a]	.015[a]	.012[a]	.010[a]	.008[a]	.007[a]	.007[a]
6	.021	4.39[a]	1.48[a]	.647[a]	.335[a]	.197[a]	.126[a]	.087[a]	.064[a]	.049[a]	.039[a]	.032[a]
7	.063	.017	6.36[a]	2.90[a]	1.51[a]	.872[a]	.544[a]	.361[a]	.253[a]	.185[a]	.141[a]	.110[a]
8	.114	.037	.016	7.74[a]	4.21[a]	2.48[a]	1.56[a]	1.03[a]	.716[a]	.516[a]	.385[a]	.296[a]
9	.165	.063	.029	.015	8.79[a]	5.35[a]	3.43[a]	2.30[a]	1.61[a]	1.16[a]	.861[a]	.657[a]
10	.215	.092	.046	.026	.015	9.64[a]	6.34[a]	4.34[a]	3.06[a]	2.22[a]	1.66[a]	1.27[a]
11	.261	.122	.066	.038	.024	.015	.010	7.22[a]	5.17[a]	3.80[a]	2.86[a]	2.19[a]
12	.303	.153	.086	.053	.034	.022	.015	.011	7.99[a]	5.95[a]	4.51[a]	3.49[a]
13	.341	.183	.108	.068	.045	.031	.022	.016	.012	8.68[a]	6.66[a]	5.19[a]
14	.376	.212	.130	.085	.057	.040	.029	.021	.016	.012	9.31[a]	7.32[a]
15	.407	.239	.152	.102	.070	.050	.037	.027	.021	.016	.012	9.88[a]
16	.436	.266	.174	.119	.084	.061	.045	.034	.026	.020	.016	.013
17	.462	.291	.195	.136	.098	.072	.054	.042	.032	.025	.020	.016
18	.486	.315	.216	.154	.113	.084	.064	.050	.039	.031	.025	.020
19	.508	.337	.236	.171	.127	.096	.074	.058	.046	.037	.030	.024
20	.529	.359	.256	.188	.142	.109	.085	.067	.053	.043	.035	.029
21	.548	.379	.275	.205	.156	.121	.095	.076	.061	.050	.041	.034
22	.565	.398	.293	.221	.171	.134	.106	.085	.069	.057	.047	.039
23	.581	.416	.310	.237	.185	.146	.117	.095	.077	.064	.053	.044
24	.596	.433	.327	.253	.199	.159	.128	.104	.086	.071	.060	.050
25	.610	.449	.343	.268	.213	.171	.139	.114	.094	.079	.066	.056
26	.623	.465	.359	.283	.226	.183	.150	.124	.103	.087	.073	.062
27	.635	.479	.374	.297	.239	.195	.161	.134	.112	.094	.080	.068
28	.647	.493	.388	.311	.252	.207	.172	.143	.121	.102	.087	.075
29	.658	.506	.401	.324	.265	.219	.182	.153	.130	.110	.094	.081
30	.668	.519	.415	.337	.277	.230	.193	.163	.138	.118	.102	.088
40	.744	.617	.522	.446	.384	.333	.291	.255	.224	.198	.176	.156
60	.825	.729	.652	.587	.531	.482	.438	.400	.366	.336	.308	.284
80	.867	.791	.727	.672	.623	.578	.538	.502	.469	.438	.410	.385
100	.893	.830	.776	.728	.685	.645	.609	.576	.544	.516	.489	.464
120	.910	.856	.810	.768	.730	.694	.661	.631	.602	.575	.549	.525
140	.923	.876	.835	.798	.763	.731	.701	.673	.647	.621	.598	.575
170	.936	.897	.862	.830	.801	.773	.747	.722	.698	.675	.654	.633
200	.945	.912	.882	.854	.828	.803	.780	.758	.736	.716	.696	.677
240	.954	.926	.900	.877	.855	.833	.813	.793	.775	.757	.739	.722
300	.966	.944	.925	.906	.889	.872	.856	.841	.825	.811	.797	.783
440	.975	.959	.945	.931	.918	.905	.893	.881	.870	.858	.847	.836
600	.982	.970	.959	.949	.939	.930	.920	.911	.903	.894	.885	.877
800	.986	.977	.969	.961	.954	.947	.940	.933	.926	.919	.913	.906
1000	.989	.982	.975	.969	.963	.957	.951	.946	.940	.935	.929	.924

[a]Multiply entry by 10^{-3}.

(continued)

Table B.4. ***(continued)***

ν_E	ν_H 1	2	3	4	5	6	7	8	9	10	11	12
						$p = 6$						
1	.000	.000	.000	.000	.000	.000	.000	.000	.000	.000	.000	.000
2	.000	.000	.000	.000	.000	.000	.000	.000	.000	.000	.000	.000
3	.000	.000	.000	.000	.000	.000	.000	.000	.000	.000	.000	.000
4	.000	.000	.000	.000	.000	.000	.000	.000	.000	.000	.000	.000
5	.007[a]	.002[a]	.001[a]	.001[a]	.001[a]	.000	.000	.000	.000	.000	.000	.000
6	2.04[a]	.315[a]	.095[a]	.040[a]	.021[a]	.012[a]	.008[a]	.006[a]	.004[a]	.003[a]	.003[a]	.002[a]
7	.019	3.48[a]	1.05[a]	.416[a]	.197[a]	.106[a]	.063[a]	.040[a]	.027[a]	.020[a]	.015[a]	.011[a]
8	.054	.013	4.37[a]	1.82[a]	.872[a]	.465[a]	.270[a]	.168[a]	.111[a]	.076[a]	.055[a]	.041[a]
9	.098	.029	.011	4.94[a]	2.48[a]	1.36[a]	.798[a]	.497[a]	.325[a]	.222[a]	.157[a]	.115[a]
10	.144	.050	.021	.010	5.35[a]	3.04[a]	1.83[a]	1.16[a]	.762[a]	.521[a]	.369[a]	.269[a]
11	.189	.074	.034	.017	9.64[a]	5.67[a]	3.51[a]	2.26[a]	1.51[a]	1.05[a]	.744[a]	.543[a]
12	.232	.099	.049	.027	.015	9.35[a]	5.94[a]	3.92[a]	2.66[a]	1.86[a]	1.34[a]	.983[a]
13	.271	.126	.066	.037	.022	.014	9.17[a]	6.17[a]	4.27[a]	3.03[a]	2.20[a]	1.63[a]
14	.308	.152	.084	.049	.031	.020	.013	9.07[a]	6.38[a]	4.59[a]	3.37[a]	2.52[a]
15	.341	.179	.103	.063	.040	.026	.018	.013	9.00[a]	6.57[a]	4.88[a]	3.68[a]
16	.372	.204	.122	.077	.050	.034	.024	.017	.012	8.97[a]	6.74[a]	5.14[a]
17	.400	.229	.141	.091	.061	.042	.030	.021	.016	.012	8.97[a]	6.90[a]
18	.426	.252	.160	.106	.072	.051	.037	.027	.020	.015	.012	8.97[a]
19	.450	.275	.179	.121	.084	.060	.044	.033	.025	.019	.015	.011
20	.473	.296	.197	.136	.096	.070	.052	.039	.030	.023	.018	.014
21	.493	.317	.215	.151	.109	.080	.060	.045	.035	.027	.021	.017
22	.512	.337	.233	.166	.121	.090	.068	.052	.041	.032	.025	.020
23	.530	.355	.250	.181	.134	.101	.077	.060	.047	.037	.030	.024
24	.546	.373	.266	.195	.146	.111	.086	.067	.053	.042	.034	.028
25	.562	.390	.282	.210	.159	.122	.095	.075	.060	.048	.039	.032
26	.576	.406	.298	.224	.171	.133	.104	.083	.066	.054	.044	.036
27	.590	.422	.313	.237	.183	.143	.113	.091	.073	.060	.049	.040
28	.603	.436	.327	.251	.195	.154	.123	.099	.080	.066	.054	.045
29	.615	.450	.341	.264	.207	.165	.132	.107	.088	.072	.060	.050
30	.626	.464	.355	.277	.219	.175	.142	.116	.095	.079	.066	.055
40	.711	.570	.467	.387	.324	.273	.232	.198	.170	.147	.127	.110
60	.802	.693	.608	.536	.476	.424	.379	.340	.305	.275	.249	.225
80	.849	.762	.690	.629	.574	.526	.483	.445	.410	.378	.350	.324
100	.878	.806	.745	.691	.642	.599	.559	.523	.489	.458	.430	.404
120	.898	.836	.783	.735	.692	.652	.616	.582	.551	.521	.494	.468
140	.912	.858	.811	.769	.730	.694	.660	.629	.599	.572	.546	.521
170	.927	.882	.842	.806	.772	.740	.710	.682	.656	.630	.607	.584
200	.938	.899	.864	.832	.803	.774	.748	.722	.698	.675	.653	.632
240	.948	.915	.886	.858	.833	.808	.785	.763	.741	.721	.701	.682
320	.961	.936	.913	.892	.872	.852	.834	.816	.799	.782	.766	.750
440	.972	.953	.936	.920	.905	.890	.876	.862	.849	.836	.823	.811
600	.979	.965	.953	.941	.930	.918	.908	.897	.887	.877	.867	.857
800	.984	.974	.964	.955	.947	.938	.930	.922	.914	.906	.898	.891
1000	.987	.979	.971	.964	.957	.950	.944	.937	.930	.924	.918	.912

[a]Multiply entry by 10^{-3}.

(continued)

Table B.4. ***(continued)***

ν_E	ν_H 1	2	3	4	5	6	7	8	9	10	11	12
						$p = 7$						
1	.000	.000	.000	.000	.000	.000	.000	.000	.000	.000	.000	.000
2	.000	.000	.000	.000	.000	.000	.000	.000	.000	.000	.000	.000
3	.000	.000	.000	.000	.000	.000	.000	.000	.000	.000	.000	.000
4	.000	.000	.000	.000	.000	.000	.000	.000	.000	.000	.000	.000
5	.000	.000	.000	.000	.000	.000	.000	.000	.000	.000	.000	.000
6	.043[a]	.006[a]	.002[a]	.001[a]	.001[a]	.000	.000	.000	.000	.000	.000	.000
7	2.62[a]	.350[a]	.091[a]	.033[a]	.015[a]	.008[a]	.005[a]	.003[a]	.002[a]	.002[a]	.001[a]	.001[a]
8	.018	2.95[a]	.809[a]	.292[a]	.126[a]	.063[a]	.034[a]	.020[a]	.013[a]	.009[a]	.006[a]	.005[a]
9	.048	.010	3.20[a]	1.22[a]	.543[a]	.270[a]	.147[a]	.086[a]	.053[a]	.035[a]	.024[a]	.017[a]
10	.087	.023	8.07[a]	3.34[a]	1.56[a]	.798[a]	.440[a]	.259[a]	.160[a]	.104[a]	.070[a]	.049[a]
11	.128	.040	.016	6.97[a]	3.43[a]	1.83[a]	1.04[a]	.619[a]	.387[a]	.252[a]	.170[a]	.119[a]
12	.170	.060	.026	.012	6.34[a]	3.51[a]	2.05[a]	1.25[a]	.796[a]	.525[a]	.357[a]	.249[a]
13	.209	.083	.038	.019	.010	5.94[a]	3.57[a]	2.23[a]	1.45[a]	.967[a]	.665[a]	.468[a]
14	.246	.106	.052	.027	.015	9.17[a]	5.67[a]	3.63[a]	2.40[a]	1.62[a]	1.13[a]	.804[a]
15	.281	.129	.067	.037	.022	.013	8.37[a]	5.48[a]	3.68[a]	2.54[a]	1.79[a]	1.28[a]
16	.313	.153	.083	.047	.029	.018	.012	7.80[a]	5.34[a]	3.73[a]	2.66[a]	1.94[a]
17	.343	.176	.099	.059	.037	.024	.016	.011	7.38[a]	5.24[a]	3.78[a]	2.78[a]
18	.370	.199	.116	.071	.045	.030	.020	.014	9.81[a]	7.06[a]	5.16[a]	3.83[a]
19	.396	.221	.133	.083	.054	.037	.025	.018	.013	9.20[a]	6.80[a]	5.10[a]
20	.420	.242	.149	.096	.064	.044	.031	.022	.016	.012	8.72[a]	6.60[a]
21	.442	.263	.166	.109	.074	.052	.037	.026	.019	.014	.011	8.34[a]
22	.462	.283	.183	.123	.085	.060	.043	.031	.023	.018	.013	.010
23	.482	.301	.199	.136	.095	.068	.050	.037	.028	.021	.016	.013
24	.499	.320	.215	.149	.106	.077	.057	.042	.032	.025	.019	.015
25	.516	.337	.230	.162	.117	.086	.064	.048	.037	.029	.022	.018
26	.532	.354	.246	.175	.128	.095	.071	.055	.042	.033	.026	.020
27	.547	.370	.260	.188	.139	.104	.079	.061	.047	.037	.029	.024
28	.561	.385	.275	.201	.150	.113	.087	.068	.053	.042	.033	.027
29	.574	.399	.289	.214	.161	.123	.095	.074	.059	.047	.037	.030
30	.586	.413	.302	.226	.172	.132	.103	.081	.064	.052	.042	.034
40	.679	.526	.417	.335	.273	.224	.185	.154	.128	.108	.091	.077
60	.779	.660	.566	.490	.426	.373	.327	.288	.254	.225	.200	.178
80	.832	.735	.656	.588	.530	.479	.434	.394	.358	.326	.298	.272
100	.864	.783	.715	.656	.603	.556	.513	.475	.439	.408	.378	.352
120	.886	.817	.757	.704	.657	.613	.574	.537	.504	.473	.444	.418
140	.902	.841	.788	.741	.698	.658	.621	.587	.556	.526	.498	.472
170	.919	.868	.823	.782	.744	.709	.676	.645	.616	.589	.563	.539
200	.931	.887	.848	.812	.778	.747	.717	.689	.662	.637	.613	.590
240	.942	.905	.871	.841	.812	.784	.758	.733	.709	.687	.665	.644
320	.957	.928	.902	.878	.855	.833	.812	.792	.773	.754	.736	.719
440	.968	.947	.928	.910	.893	.876	.860	.844	.829	.814	.800	.786
600	.977	.961	.947	.933	.920	.908	.895	.883	.872	.860	.849	.838
800	.982	.971	.960	.950	.940	.930	.920	.911	.902	.893	.884	.876
1000	.986	.977	.968	.959	.951	.943	.936	.928	.921	.914	.906	.899

[a]Multiply entry by 10^{-3}.

(continued)

Table B.4. (*continued*)

ν_E	ν_H 1	2	3	4	5	6	7	8	9	10	11	12
						$p = 8$						
1	.000	.000	.000	.000	.000	.000	.000	.000	.000	.000	.000	.000
2	.000	.000	.000	.000	.000	.000	.000	.000	.000	.000	.000	.000
3	.000	.000	.000	.000	.000	.000	.000	.000	.000	.000	.000	.000
4	.000	.000	.000	.000	.000	.000	.000	.000	.000	.000	.000	.000
5	.000	.000	.000	.000	.000	.000	.000	.000	.000	.000	.000	.000
6	.000	.000	.000	.000	.000	.000	.000	.000	.000	.000	.000	.000
7	.138[a]	.015[a]	.004[a]	.001[a]	.001[a]	.000	.000	.000	.000	.000	.000	.000
8	3.30[a]	.393[a]	.090[a]	.029[a]	.012[a]	.006[a]	.003[a]	.002[a]	.001[a]	.001[a]	.001[a]	.000
9	.017	2.63[a]	.659[a]	.218[a]	.087[a]	.040[a]	.020[a]	.011[a]	.007[a]	.004[a]	.003[a]	.002[a]
10	.044	8.63[a]	2.46[a]	.872[a]	.361[a]	.168[a]	.086[a]	.047[a]	.028[a]	.017[a]	.011[a]	.008[a]
11	.078	.019	6.15[a]	2.36[a]	1.03[a]	.497[a]	.259[a]	.144[a]	.085[a]	.052[a]	.034[a]	.023[a]
12	.116	.033	.012	4.99[a]	2.30[a]	1.16[a]	.619[a]	.351[a]	.209[a]	.130[a]	.084[a]	.056[a]
13	.154	.051	.020	8.91[a]	4.34[a]	2.26[a]	1.25[a]	.727[a]	.441[a]	.278[a]	.181[a]	.122[a]
14	.190	.070	.030	.014	7.22[a]	3.92[a]	2.23[a]	1.33[a]	.824[a]	.527[a]	.347[a]	.235[a]
15	.225	.090	.041	.021	.011	6.17[a]	3.63[a]	2.22[a]	1.40[a]	.910[a]	.608[a]	.416[a]
16	.258	.111	.054	.028	.016	9.06[a]	5.48[a]	3.42[a]	2.20[a]	1.46[a]	.987[a]	.683[a]
17	.289	.133	.067	.037	.021	.013	7.80[a]	4.98[a]	3.27[a]	2.20[a]	1.51[a]	1.06[a]
18	.318	.154	.082	.046	.027	.017	.011	6.92[a]	4.62[a]	3.15[a]	2.19[a]	1.56[a]
19	.345	.175	.096	.056	.034	.021	.014	9.23[a]	6.26[a]	4.34[a]	3.06[a]	2.19[a]
20	.370	.195	.111	.067	.042	.027	.018	.012	8.22[a]	5.77[a]	4.12[a]	2.99[a]
21	.393	.215	.127	.078	.050	.033	.022	.015	.010	7.46[a]	5.39[a]	3.95[a]
22	.415	.235	.142	.089	.058	.039	.026	.018	.013	9.40[a]	6.86[a]	5.08[a]
23	.436	.254	.157	.101	.067	.045	.031	.022	.016	.012	8.56[a]	6.39[a]
24	.455	.272	.172	.113	.076	.052	.037	.026	.019	.014	.010	7.88[a]
25	.473	.289	.187	.124	.085	.060	.042	.031	.023	.017	.013	9.56[a]
26	.490	.306	.201	.136	.095	.067	.048	.035	.026	.020	.015	.011
27	.505	.322	.215	.148	.104	.075	.055	.040	.030	.023	.017	.013
28	.520	.338	.229	.160	.114	.083	.061	.045	.034	.026	.020	.016
29	.534	.353	.243	.172	.124	.091	.068	.051	.039	.030	.023	.018
30	.548	.367	.256	.183	.134	.099	.074	.056	.043	.034	.026	.021
40	.649	.485	.372	.290	.229	.182	.146	.118	.096	.079	.065	.054
60	.758	.627	.527	.447	.381	.327	.282	.244	.212	.184	.161	.141
80	.815	.709	.623	.551	.489	.435	.389	.348	.313	.281	.253	.229
100	.851	.761	.687	.622	.566	.516	.471	.431	.395	.362	.333	.306
120	.875	.798	.732	.675	.623	.577	.535	.496	.461	.429	.399	.372
140	.892	.825	.767	.715	.667	.625	.585	.549	.515	.484	.455	.428
170	.911	.854	.804	.759	.717	.679	.644	.610	.579	.550	.523	.497
200	.924	.875	.831	.791	.755	.720	.688	.657	.629	.602	.576	.551
240	.936	.895	.858	.823	.791	.761	.732	.705	.679	.655	.631	.609
320	.952	.920	.891	.865	.839	.815	.792	.770	.748	.728	.708	.689
440	.965	.942	.920	.900	.880	.862	.844	.827	.810	.794	.778	.762
600	.974	.957	.941	.926	.911	.897	.883	.870	.857	.844	.831	.819
800	.981	.968	.955	.944	.933	.922	.911	.901	.890	.880	.871	.861
1000	.985	.974	.964	.955	.946	.937	.928	.920	.911	.903	.895	.887

[a]Multiply entry by 10^{-3}.

Table B.5. Upper Critical Values for Roy's Test, $\alpha = .05$

Roy's test statistic is given by

$$\theta = \lambda_1/(1 + \lambda_1),$$

where λ_1 is the largest eigenvalue of $\mathbf{E}^{-1}\mathbf{H}$. The parameters are

$$s = \min(\nu_H, p),$$

$$m = \tfrac{1}{2}(|\nu_H - p| - 1),$$

$$N = \tfrac{1}{2}(\nu_E - p - 1).$$

Reject H_0 if $\theta >$ table value.

	m												
N	−.5	0	1	2	3	4	5	6	7	8	9	10	15
							s = 1						
−.5	.994	.997	.999	.999	.999	1.000	1.000	1.000	1.000	1.000	1.000	1.000	1.000
0	.902	.950	.975	.983	.987	.990	.991	.993	.994	.994	.995	.995	.997
1	.658	.776	.865	.902	.924	.937	.947	.954	.959	.963	.967	.970	.979
3	.399	.527	.657	.729	.775	.807	.831	.850	.865	.877	.877	.896	.925
5	.283	.393	.521	.600	.655	.696	.729	.755	.776	.794	.809	.822	.868
10	.164	.238	.339	.410	.466	.511	.548	.580	.608	.632	.653	.672	.742
15	.115	.171	.250	.310	.359	.401	.437	.468	.496	.521	.544	.564	.643
20	.088	.133	.198	.249	.292	.330	.363	.392	.419	.443	.465	.485	.567
25	.072	.109	.164	.208	.246	.280	.310	.337	.362	.385	.406	.425	.506
30	.061	.092	.140	.179	.213	.243	.270	.295	.318	.340	.360	.379	.457
40	.046	.070	.108	.139	.167	.192	.215	.236	.257	.275	.293	.310	.382
50	.037	.057	.088	.114	.137	.159	.179	.197	.215	.231	.247	.262	.328
100	.019	.029	.046	.060	.073	.085	.097	.108	.118	.129	.138	.148	.192
500	.004	.006	.009	.012	.015	.018	.021	.023	.026	.028	.031	.033	.044
1000	.002	.003	.005	.006	.008	.009	.010	.012	.013	.014	.015	.017	.023
							s = 2						
1	.858	.894	.930	.947	.958	.965	.970	.973	.976	.979	.981	.982	.987
3	.638	.702	.776	.820	.849	.870	.885	.898	.908	.916	.922	.928	.948
5	.498	.565	.651	.706	.746	.776	.799	.818	.834	.847	.858	.868	.901
10	.318	.374	.455	.514	.561	.598	.629	.656	.679	.698	.716	.732	.789
15	.232	.278	.348	.402	.446	.483	.515	.542	.567	.589	.609	.627	.696
20	.183	.221	.281	.329	.369	.404	.434	.461	.486	.508	.528	.546	.620
25	.151	.184	.236	.278	.314	.346	.375	.401	.424	.446	.465	.484	.558
30	.129	.157	.203	.241	.274	.303	.330	.354	.376	.396	.416	.433	.507
40	.099	.122	.159	.190	.218	.243	.266	.287	.306	.325	.342	.359	.428
50	.081	.099	.130	.157	.180	.202	.222	.241	.259	.275	.291	.306	.370
100	.042	.052	.069	.084	.097	.110	.122	.134	.145	.155	.166	.176	.220
500	.009	.011	.014	.018	.021	.024	.027	.029	.032	.035	.037	.040	.052
1000	.004	.005	.007	.009	.010	.012	.013	.015	.016	.018	.019	.020	.027

(continued)

Table B.5. *(continued)*

	m												
N	−.5	0	1	2	3	4	5	6	7	8	9	10	15
							$s = 3$						
1	.922	.938	.956	.965	.972	.976	.979	.982	.984	.985	.986	.987	.991
3	.756	.792	.839	.868	.887	.902	.913	.922	.930	.936	.941	.945	.960
5	.625	.669	.729	.770	.800	.822	.840	.855	.867	.877	.886	.894	.920
10	.429	.472	.537	.586	.625	.656	.683	.705	.725	.741	.756	.770	.819
15	.324	.362	.422	.469	.508	.541	.569	.594	.616	.635	.653	.669	.730
20	.260	.293	.346	.390	.427	.458	.486	.511	.533	.554	.572	.589	.656
25	.218	.246	.294	.333	.367	.397	.424	.448	.470	.490	.508	.525	.594
30	.187	.212	.255	.291	.322	.350	.375	.398	.419	.439	.457	.473	.543
40	.146	.166	.201	.232	.259	.283	.305	.326	.345	.363	.379	.395	.462
50	.119	.136	.167	.192	.216	.237	.257	.275	.292	.309	.324	.339	.402
100	.063	.072	.089	.104	.118	.131	.143	.155	.166	.177	.187	.197	.242
500	.013	.015	.019	.022	.026	.029	.031	.034	.037	.040	.043	.045	.058
1000	.007	.008	.010	.011	.013	.015	.016	.018	.019	.020	.022	.023	.030
							$s = 4$						
1	.950	.959	.969	.975	.979	.982	.985	.986	.988	.989	.990	.991	.993
3	.824	.846	.877	.898	.912	.923	.931	.938	.944	.948	.952	.956	.967
5	.708	.739	.782	.813	.836	.854	.868	.880	.889	.898	.905	.911	.933
10	.513	.547	.601	.641	.674	.700	.723	.742	.759	.773	.786	.798	.840
15	.399	.431	.482	.523	.558	.587	.612	.634	.654	.671	.687	.701	.756
20	.326	.354	.402	.441	.474	.503	.529	.552	.572	.591	.607	.623	.684
25	.275	.301	.344	.380	.412	.440	.464	.487	.507	.526	.543	.559	.624
30	.238	.261	.301	.334	.364	.390	.414	.435	.455	.474	.491	.507	.572
40	.188	.207	.240	.269	.294	.318	.339	.359	.377	.395	.411	.426	.490
50	.155	.171	.199	.224	.247	.268	.287	.305	.322	.338	.353	.367	.428
100	.082	.091	.108	.123	.137	.150	.162	.174	.185	.196	.206	.216	.261
500	.017	.020	.023	.027	.030	.033	.036	.039	.042	.045	.048	.050	.063
1000	.009	.010	.012	.014	.015	.017	.018	.020	.021	.023	.024	.026	.032
							$s = 5$						
1	.966	.971	.977	.981	.984	.986	.988	.989	.990	.991	.992	.992	.994
3	.866	.881	.903	.918	.929	.937	.944	.949	.953	.957	.960	.963	.972
5	.766	.788	.821	.845	.863	.877	.888	.898	.906	.913	.918	.924	.942
10	.580	.607	.651	.685	.713	.735	.755	.771	.786	.799	.810	.820	.857
15	.462	.488	.533	.569	.599	.625	.648	.667	.685	.701	.715	.728	.777
20	.383	.407	.449	.485	.515	.542	.565	.586	.604	.621	.637	.651	.708
25	.326	.349	.388	.422	.451	.477	.500	.521	.540	.557	.573	.588	.648
30	.284	.305	.341	.373	.400	.425	.448	.468	.487	.504	.520	.535	.597
40	.226	.243	.275	.302	.327	.349	.370	.389	.406	.423	.438	.453	.514
50	.187	.202	.230	.254	.276	.296	.315	.332	.348	.364	.378	.392	.451
100	.101	.110	.126	.141	.155	.168	.180	.191	.203	.213	.224	.233	.278
500	.022	.024	.027	.031	.034	.038	.041	.044	.047	.049	.052	.055	.068
1000	.011	.012	.014	.016	.017	.019	.021	.022	.024	.025	.027	.028	.035
							$s = 6$						
1	.975	.978	.983	.985	.988	.989	.990	.991	.992	.993	.993	.944	.995
3	.895	.906	.922	.933	.941	.948	.953	.957	.961	.964	.966	.968	.976
5	.808	.825	.850	.869	.883	.895	.904	.912	.918	.924	.929	.934	.949
10	.633	.655	.692	.721	.744	.764	.781	.795	.808	.819	.829	.838	.871
15	.514	.537	.576	.608	.635	.658	.678	.696	.711	.726	.739	.750	.795
20	.432	.454	.491	.523	.551	.575	.596	.615	.632	.648	.662	.676	.728
25	.372	.392	.428	.458	.485	.509	.531	.550	.568	.584	.599	.613	.669
30	.326	.345	.378	.407	.433	.457	.478	.497	.514	.531	.546	.560	.618
40	.261	.278	.307	.333	.356	.378	.397	.416	.432	.448	.463	.477	.536
50	.218	.232	.258	.281	.302	.322	.340	.357	.372	.387	.401	.414	.472
100	.119	.128	.144	.158	.172	.184	.197	.208	.219	.230	.240	.250	.294
500	.026	.028	.031	.035	.039	.042	.045	.048	.051	.054	.057	.059	.073
1000	.013	.014	.016	.018	.020	.021	.023	.024	.026	.028	.029	.031	.037

(continued)

Table B.5. ***(continued)***

N	m −.5	0	1	2	3	4	5	6	7	8	9	10	15
							$s = 7$						
1	.981	.983	.986	.988	.990	.991	.992	.993	.993	.994	.994	.995	.996
3	.915	.923	.935	.944	.950	.956	.960	.963	.966	.969	.971	.973	.979
5	.840	.852	.872	.887	.899	.908	.917	.923	.929	.934	.938	.941	.955
10	.677	.695	.726	.750	.771	.788	.802	.815	.826	.836	.845	.853	.882
15	.560	.579	.613	.641	.665	.686	.704	.720	.734	.747	.759	.769	.810
20	.475	.494	.528	.557	.582	.604	.624	.641	.657	.671	.685	.697	.745
25	.412	.431	.463	.491	.516	.538	.558	.576	.593	.608	.622	.635	.688
30	.364	.381	.412	.439	.463	.485	.505	.523	.540	.555	.569	.583	.638
40	.294	.309	.337	.362	.384	.404	.423	.440	.456	.471	.485	.499	.555
50	.247	.260	.285	.307	.327	.346	.363	.379	.395	.409	.422	.435	.491
100	.136	.145	.160	.175	.188	.200	.212	.224	.235	.245	.255	.265	.309
500	.030	.032	.036	.039	.042	.046	.049	.052	.055	.058	.061	.064	.077
1000	.015	.016	.018	.020	.022	.023	.025	.027	.028	.030	.031	.033	.040
							$s = 8$						
1	.985	.986	.989	.990	.992	.992	.993	.994	.994	.995	.995	.995	.997
3	.930	.936	.945	.952	.958	.962	.965	.968	.971	.973	.974	.976	.982
5	.864	.874	.890	.902	.912	.920	.927	.932	.937	.941	.945	.948	.959
10	.713	.728	.754	.775	.793	.808	.821	.832	.842	.851	.859	.865	.892
15	.598	.615	.645	.670	.692	.710	.727	.741	.754	.766	.776	.786	.824
20	.513	.531	.561	.587	.610	.630	.648	.664	.679	.692	.705	.716	.761
25	.449	.466	.495	.521	.544	.565	.583	.600	.616	.630	.643	.655	.705
30	.398	.414	.443	.468	.491	.511	.530	.547	.563	.577	.590	.603	.655
40	.325	.339	.365	.388	.409	.428	.446	.463	.478	.493	.506	.519	.573
50	.274	.286	.310	.331	.351	.368	.385	.401	.416	.429	.442	.455	.508
100	.153	.161	.176	.190	.203	.216	.228	.239	.250	.260	.270	.279	.323
500	.034	.036	.040	.043	.047	.050	.053	.056	.059	.062	.065	.068	.081
1000	.017	.018	.020	.022	.024	.025	.027	.029	.030	.032	.033	.035	.042
							$s = 9$						
1	.988	.989	.991	.992	.993	.994	.994	.995	.995	.995	.996	.996	.997
3	.942	.946	.953	.959	.963	.967	.970	.972	.974	.976	.977	.979	.984
5	.883	.891	.904	.914	.922	.929	.935	.939	.944	.947	.950	.953	.963
10	.743	.756	.778	.797	.812	.825	.837	.847	.855	.863	.870	.876	.901
15	.632	.647	.674	.696	.715	.732	.747	.760	.771	.782	.792	.801	.835
20	.548	.563	.591	.614	.635	.654	.670	.685	.698	.711	.722	.733	.775
25	.482	.497	.525	.549	.570	.589	.606	.622	.636	.650	.662	.673	.720
30	.430	.445	.471	.495	.516	.535	.552	.569	.583	.597	.610	.622	.671
40	.353	.366	.391	.413	.433	.451	.468	.484	.499	.512	.525	.538	.590
50	.299	.311	.333	.354	.373	.390	.406	.421	.435	.448	.461	.473	.524
100	.169	.177	.192	.206	.219	.231	.242	.253	.264	.274	.284	.293	.336
500	.038	.040	.043	.047	.051	.054	.057	.060	.063	.066	.069	.072	.086
1000	.019	.020	.022	.024	.026	.028	.029	.031	.032	.034	.036	.037	.044
							$s = 10$						
1	.990	.991	.992	.993	.994	.994	.995	.995	.996	.996	.996	.997	.997
3	.950	.954	.960	.964	.968	.971	.973	.975	.977	.978	.980	.981	.985
5	.898	.905	.916	.924	.931	.937	.941	.946	.949	.952	.955	.958	.967
10	.769	.780	.799	.815	.829	.840	.851	.859	.867	.874	.880	.886	.908
15	.662	.675	.699	.719	.736	.751	.764	.776	.787	.797	.806	.814	.846
20	.578	.592	.617	.639	.658	.675	.690	.704	.716	.728	.738	.748	.787
25	.512	.526	.551	.573	.593	.611	.627	.642	.655	.667	.679	.690	.734
30	.459	.473	.497	.519	.539	.557	.573	.589	.603	.615	.627	.639	.686
40	.380	.392	.415	.436	.455	.473	.489	.504	.518	.531	.543	.555	.605
50	.323	.335	.356	.375	.393	.410	.425	.440	.453	.466	.478	.490	.540
100	.185	.193	.207	.220	.233	.245	.256	.267	.278	.287	.297	.306	.348
500	.042	.044	.047	.051	.054	.058	.061	.064	.067	.070	.073	.076	.090
1000	.021	.022	.024	.026	.028	.030	.031	.033	.034	.036	.038	.039	.047

Table B.6. Upper Critical Values of Pillai's Statistic $V^{(s)}$, $\alpha = .05$

$V^{(s)} = \sum_{i=1}^{s} \frac{\lambda_i}{1+\lambda_i}$, where $\lambda_1, \lambda_2, \ldots, \lambda_s$ are eigenvalues of $\mathbf{E}^{-1}\mathbf{H}$. Reject H_0 if $V^{(s)}$ > table value. The parameters s, m, and N are defined in Table B.5.

	N													
m	0	1	2	3	4	5	6	7	8	9	10	15	20	25
							$s = 2$							
0	1.536	1.232	1.031	0.890	0.782	0.698	0.629	0.573	0.526	0.485	0.451	0.333	0.263	0.218
1	1.706	1.452	1.258	1.109	0.991	0.896	0.817	0.751	0.694	0.646	0. 604	0.455	0.364	0.304
2	1.784	1.573	1.397	1.254	1.137	1.039	0.956	0.886	0.825	0.772	0. 725	0.556	0.451	0.379
3	1.829	1.649	1.492	1.358	1.245	1.149	1.065	0.993	0.930	0.875	0. 825	0.643	0.526	0.445
4	1.859	1.703	1.560	1.436	1.329	1.235	1.153	1.081	1.018	0.961	0. 910	0.719	1.594	0.506
5	1.880	1.742	1.613	1.497	1.395	1.305	1.226	1.155	1.091	1.034	0. 983	0.786	0.655	0.561
6	1.895	1.772	1.654	1.546	1.450	1.364	1.286	1.217	1.154	1.098	1. 046	0.846	0.710	0.612
7	1.907	1.796	1.687	1.586	1.495	1.413	1.338	1.270	1.209	1.153	1. 102	0.901	0.761	0.658
8	1.917	1.815	1.714	1.620	1.534	1.455	1.383	1.317	1.257	1.202	1. 151	0.950	0.808	0.702
9	1.924	1.831	1.737	1.649	1.567	1.491	1.422	1.358	1.299	1.245	1. 195	0.995	0.851	0.743
10	1.931	1.844	1.757	1.673	1.595	1.523	1.456	1.394	1.337	1.284	1 .235	1.036	0.891	0.781
15	1.951	1.888	1.822	1.758	1.695	1.636	1.580	1.527	1.477	1.430	1 .386			
20	1.963	1.913	1.860	1.807	1.756	1.706	1.658	1.612	1.568	1.527	1 .487			
25	1.969	1.929	1.885	1.840	1.796	1.753	1.711	1.671	1.632	1.595	1 .559			
							$s = 3$							
0	2.037	1.710	1.473	1.294	1.153	1.040	0.947	0.869	0.803	0.746	0.697	0.524	0.420	0.350
1	2.297	1.988	1.751	1.564	1.412	1.287	1.183	1.094	1.017	0.950	0.892	0.682	0.552	0.453
2	2.447	2.168	1.943	1.759	1.606	1.477	1.367	1.273	1.190	1.117	1.053	0.818	0.668	0.565
3	2.544	2.294	2.084	1.907	1.757	1.628	1.517	1.420	1.334	1.258	1.190	0.937	0.772	0.656
4	2.612	2.386	2.191	2.023	1.878	1.752	1.641	1.543	1.456	1.378	1.308	1.042	0.866	0.740
5	2.662	2.457	2.276	2.117	1.978	1.854	1.745	1.648	1.561	1.482	1.411	1.137	0.952	0.818
6	2.701	2.514	2.345	2.194	2.061	1.941	1.835	1.739	1.652	1.573	1.502	1.222	1.030	0.890
7	2.732	2.559	2.402	2.259	2.131	2.016	1.912	1.818	1.732	1.654	1.582	1.300	1.103	0.957
8	2.757	2.597	2.449	2.314	2.192	2.081	1.979	1.887	1.803	1.726	1.655	1.371	1.170	1.020
9	2.777	2.629	2.490	2.362	2.244	2.137	2.039	1.949	1.866	1.790	1.720	1.436	1.23	
10	2.795	2.656	2.525	2.403	2.291	2.187	2.092	2.004	1.923	1.848	1.779	1.496	1.3	
15	2.853	2.748	2.646	2.549	2.457	2.370	2.288	2.211	21.39	2.071	2.007			
20	2.885	2.802	2.718	2.637	2.560	2.485	2.414	2.347	2.283	2.222	2.163			
25	2.906	2.836	2.766	2.697	2.630	2.565	2.503	2.443	2.385					

(continued)

Table B.6. ***(continued)***

	N												
m	0	1	2	3	4	5	6	7	8	9	10	15	20	25
							$s = 4$							
0	2.549	2.194	1.926	1.717	1.548	1.410	1.294	1.196	1.112	1.038	0.974	0.744	0.602	
1	2.852	2.510	2.241	2.023	1.844	1.693	1.566	1.456	1.360	1.277	1.203	0.932	0.761	
2	3.052	2.733	2.472	2.256	2.074	1.919	1.786	1.670	1.567	1.477	1.396	1.097	0.903	
3	3.193	2.898	2.650	2.440	2.260	2.104	1.969	1.849	1.743	1.649	1.564	1.243	1.032	
4	3.298	3.025	2.791	2.589	2.413	2.259	2.123	2.002	1.895	1.798	1.710	1.375	1.149	
5	3.378	3.126	2.905	2.711	2.541	2.390	2.255	2.135	2.027	1.929	1.840	1.494		
6	3.442	3.208	2.999	2.814	2.649	2.502	2.370	2.251	2.143	2.044	1.955	1.602		
7	3.494	3.276	3.079	2.902	2.743	2.600	2.470	2.353	2.246	2.148	2.058	1.70		
8	3.537	3.333	3.146	2.977	2.824	2.685	2.559	2.444	2.338	2.241	2.151	1.8		
9	3.574	3.382	3.205	3.043	2.896	2.761	2.638	2.525	2.421	2.325	2.236			
10	3.605	3.424	3.256	3.101	2.959	2.829	2.708	2.598	2.496	2.401	2.313			
15	3.710	3.570	3.436	3.310	3.191	3.079	2.974	2.876	2.783	2.696	2.615			
20	3.771	3.657	3.546	3.440	3.338	3.241	3.149							
							$s = 5$							
0	3.055	2.681	2.389	2.155	1.962	1.801	1.664	1.547	1.445	1.356	1.277			
1	3.390	3.025	2.731	2.488	2.285	2.122	1.964	1.835	1.722	1.622	1.533			
2	3.628	3.281	2.993	2.751	2.545	2.367	2.213	2.077	1.957	1.850	1.754			
3	3.805	3.478	3.201	2.964	2.759	2.580	2.423	2.284	2.160	2.048	1.948			
4	3.941	3.635	3.370	3.140	2.938	2.761	2.604	2.463	2.337	2.222	2.119			
5	4.050	3.762	3.510	3.288	3.091	2.916	2.760	2.619	2.492	2.377	2.271			
6	4.138	3.868	3.627	3.414	3.223	3.052	2.897	2.758	2.630	2.514	2.408			
7	4.212	3.957	3.728	3.522	3.337	3.170	3.018	2.880						
8	4.274	4.033	3.815	3.617	3.438	3.275	3.126							
9	4.327	4.099	3.890	3.700	3.527	3.369								
10	4.372	4.156	3.957	3.774	3.607	3.45								

(continued)

Table B.6. *(continued)*

	N													
m	0	1	2	3	4	5	6	7	8	9	10	15	20	25
							$s = 6$							
0	3.559	3.171	2.859	2.604	2.390	2.209	2.053	1.918	1.799	1.694	1.601			
1	3.917	3.535	3.221	2.958	2.734	2.542	2.375	2.229	2.099	1.984	1.881			
2	4.185	3.817	3.508	3.245	3.018	2.821	2.647	2.494	2.358	2.235	2.125			
3	4.391	4.041	3.741	3.482	3.256	3.057	2.881	2.724	2.583	2.456	2.341			
4	4.556	4.223	3.934	3.681	3.458	3.260	3.084	2.925	2.782	2.652	2.534			
5	4.690	4.375	4.097	3.851	3.633	3.438	3.262	3.103	2.959	2.827	2.706			
6	4.802	4.502	4.236	3.998	3.785									
7	4.896	4.611	4.356	4.126	3.919									
8	4.976	4.706	4.461	4.239										
9	5.045	4.788	4.553											
10	5.106	4.860	4.635											

Table B.7. Upper Critical Values for the Lawley–Hotelling Statistic, $\alpha = .05$

The test statistic is $\nu_E U^{(s)}/\nu_H$, where $U^{(s)}$ is the Lawley–Hotelling statistic. Reject H_0 if $\nu_E U^{(s)}/\nu_H >$ table value.

	ν_H												
ν_E	2	3	4	5	6	8	10	12	15	20	25	40	60
							$p = 2$						
2	9.8591[a]	10.659[a]	11.098[a]	11.373[a]	11.562[a]	11.952[a]	11.804[a]	12.052[a]	12.153[a]	12.254[a]	12.316[a]	12.409[a]	12.461[a]
3	58.428	58.915	59.161	59.308	59.407	59.531	59.606	59.655	59.705	59.755	59.785	59.830	59.855
4	23.999	23.312	22.918	22.663	22.484	22.250	22.104	22.003	21.901	21.797	21.733	21.636	21.582
5	15.639	14.864	14.422	14.135	13.934	13.670	13.504	13.391	13.275	13.156	13.083	12.972	12.909
6	12.175	11.411	10.975	10.691	10.491	10.228	10.063	9.9489	9.8320	9.7118	9.6381	9.5251	9.4610
7	10.334	9.5937	9.1694	8.8927	8.6975	8.4396	8.2765	8.16399	8.0480	7.9285	7.8549	7.7417	7.6773
8	9.2069	8.4881	8.0752	7.8054	7.6145	7.3614	7.2008	7.0896	6.9748	6.8560	6.7826	6.6694	6.6048
10	7.9095	7.2243	6.8294	6.5702	6.3860	6.1405	5.9837	5.8745	5.7612	5.6433	5.5701	5.4564	5.3910
12	7.1902	6.5284	6.1461	5.8942	5.7147	5.4744	5.3200	5.2122	5.0997	4.9820	4.9085	4.7938	4.7274
14	6.7350	6.0902	4.7168	5.4703	5.2941	5.0574	4.9048	4.7977	4.6856	4.5678	4.4939	4.3780	4.3105
16	6.4217	5.7895	5.4230	5.1804	5.0067	4.7727	4.6213	4.5147	4.4028	4.2846	4.2102	4.0930	4.0243
18	6.1932	5.5708	5.2095	4.9700	4.7982	4.5663	4.4157	4.3094	4.1976	4.0791	4.0042	3.8855	3.8158
20	6.0192	5.4046	5.0475	4.8105	4.6402	4.4099	4.2600	4.1539	4.0420	3.9231	3.8477	3.7278	3.6569
25	5.7244	5.1237	4.7741	2.5415	4.3740	4.1465	3.9977	3.8919	3.7798	3.6598	3.5832	3.4605	3.3868
30	5.5401	4.9487	4.6040	4.3743	4.2086	3.9829	3.8347	3.7291	3.6166	3.4957	3.4181	3.2926	3.2168
35	5.4140	4.8291	4.8880	4.2604	4.0959	3.8715	3.7237	3.6181	3.5054	3.3836	3.3051	3.1774	3.1000
40	5.3224	4.7424	4.4039	4.1778	4.0143	3.7908	3.6433	3.5377	3.4247	3.3022	3.2230	3.0933	3.0140
50	5.1981	4.6249	4.2900	4.0661	3.9039	3.6817	3.5346	3.4289	3.3154	3.1919	3.1115	2.9787	2.8965
60	5.1178	4.5490	4.2166	3.9941	3.8328	3.6114	3.4646	3.3588	3.2450	3.1206	3.0392	2.9041	2.8196
70	5.0616	4.4960	4.1653	3.9439	3.7831	3.5624	3.4157	3.3099	3.1957	3.0706	2.9886	2.8516	2.7652
80	5.0200	4.4569	4.1275	3.9068	3.7465	3.5262	3.3796	3.2737	3.1594	3.0338	2.9512	2.8126	2.7247
100	4.9628	4.4030	4.0754	3.8557	3.6961	3.4764	3.3300	3.2240	3.1093	2.9829	2.8994	2.7586	2.6683
200	4.8514	4.2982	3.9742	3.7567	3.5983	3.3798	3.2336	3.1275	3.0120	2.8838	2.7984	2.6520	2.5559
∞	4.7442	4.1973	3.8769	3.6614	3.5044	3.2870	3.1410	3.0346	2.9182	2.7879	2.7002	2.5470	2.4428

[a]Multiply entry by 100.

(continued)

Table B.7. ***(continued)***

ν_E	ν_H 3	4	5	6	8	10	12	15	20	25	40	60
						$p = 3$						
3	25.930[a]	26.996[a]	27.665[a]	28.125[a]	28.712[a]	29.073[a]	29.316[a]	29.561[a]	29.809[a]	29.959[a]	30.19[a]	30.31[a]
4	1.1880[a]	1.1929[a]	1.1959[a]	1.1978[a]	1.2003[a]	1.2018[a]	1.2028[a]	1.2038[a]	1.2048[a]	1.2054[a]	1.2063[a]	1.2068[a]
5	42.474	41.764	1.305	40.983	40.562	40.300	40.120	39.937	39.750	39.635	39.462	39.366
6	25.456	24.715	24.235	23.899	23.458	23.182	22.992	22.799	22.600	22.479	22.294	22.190
7	18.752	18.056	17.605	17.288	16.870	16.608	16.427	16.241	16.051	15.934	15.755	15.653
8	15.308	14.657	14.233	13.934	13.540	13.290	13.118	12.941	12.758	12.646	12.473	12.375
10	11.893	11.306	10.921	10.649	10.287	10.057	9.8974	9.7320	9.5603	9.4541	9.2897	9.1955
12	10.229	9.6825	9.3234	9.0680	8.7271	8.5088	8.3566	8.1982	8.0330	7.9301	7.7700	7.6777
14	9.2550	8.7356	8.3935	8.1495	7.8225	7.6122	7.4649	7.3110	7.1497	7.0488	6.8908	6.7991
16	8.6180	8.1183	7.7884	7.5526	7.2355	7.0307	6.8868	6.7360	6.5772	6.4774	6.3204	6.2287
18	8.1701	7.6851	7.3644	7.1347	6.8251	6.6244	6.4830	6.3343	6.1771	6.0780	5.9212	5.8292
20	7.8384	7.3649	7.0513	6.8263	6.5224	6.3249	6.1853	6.0383	5.8822	5.7834	5.6266	5.5341
25	7.2943	6.8407	6.5394	6.3227	6.0287	5.8365	5.7001	5.5555	5.4010	5.3025	5.1446	5.0503
30	6.9654	6.5245	6.2311	6.0196	5.7319	5.5431	5.4085	5.2654	5.1116	5.0129	4.8535	4.7575
35	6.7453	6.3132	6.0253	5.8175	5.5341	5.3476	5.2143	5.0720	4.9185	4.8195	4.6586	4.5608
40	6.5877	6.1621	5.8783	5.6732	5.3929	5.2081	5.0757	4.9340	4.7806	4.6813	4.5189	4.4195
50	6.3773	5.9606	5.6823	5.4809	5.2050	5.0224	4.8911	4.7502	4.5967	4.4968	4.3319	4.2297
60	6.2433	5.8324	5.5577	5.3587	5.0856	4.9044	4.7739	4.6334	4.4798	4.3793	4.2123	4.1078
70	6.1504	5.7436	5.4715	5.2742	5.0031	4.8229	4.6929	4.5526	4.3988	4.2979	4.1292	4.0227
80	6.0823	5.6786	5.4084	5.2122	4.9426	4.7632	4.6336	4.4935	4.3395	4.2381	4.0680	3.9600
100	5.9891	5.5896	5.3220	5.1276	4.8601	4.6817	4.5525	4.4126	4.2583	4.1563	3.9840	3.8734
200	5.8099	5.4186	5.1562	4.9653	4.7017	4.5252	4.3970	4.2574	4.1023	3.9988	3.8212	3.7042
∞	5.6397	5.2565	4.9992	4.8116	4.5519	4.3773	4.2499	4.1104	3.9541	3.8487	3.6642	3.5384

[a]Multiply entry by 100.

(*continued*)

Table B.7. (*continued*)

ν_E	ν_H 4	5	6	8	10	12	15	20	25	40	60
						$p = 4$					
4	49.964[a]	51.204[a]	52.054[a]	53.142[a]	53.808[a]	54 .258[a]	54.71[a]	55.17[a]	55.46[a]		
5	1.9964[a]	2.0013[a]	2.0046[a]	2.0087[a]	2.0112[a]	2.0128[a]	2.0145[a]	2.0171[a]	2.0171[a]	2.019[a]	
6	65.715	64.999	64.497	63.841	63.432	63.151	62.866	62.573	62.396	62.13	
7	37.343	36.629	36.129	35.474	35.064	34.782	34.495	34.200	34.019	33.75	
8	26.516	25.868	25.413	24.814	24.437	24.178	23.912	23.639	23.471	23.214	23.072
10	17.875	17.326	16.938	16.424	16.098	15.872	15.640	15.399	15.250	15.021	14.891
12	14.338	13.848	13.500	13.037	12.741	12.535	12.321	12.099	11.961	11.747	11.624
14	12.455	12.002	11.680	11.248	10.972	10.778	10.577	10.366	10.234	10.029	9.9103
16	11.295	10.868	10.563	10.154	9.8904	9.7054	9.5119	9.3085	9.1810	8.9808	8.8644
18	10.512	10.104	9.8121	9.4190	9.1647	8.9857	8.7978	8.5996	8.4748	8.2778	8.1626
20	9.9500	9.5560	9.2736	8.8926	8.6453	8.4708	8.2871	8.0926	7.9696	7.7748	7.6601
25	9.0585	8.6884	8.4223	8.0616	7.8261	7.6590	7.4821	7.2933	7.1730	6.9805	6.8659
30	8.5377	8.1825	7.9265	7.5784	7.3502	7.1876	7.0147	6.8291	6.7101	6.5181	6.4026
35	8.1968	7.8517	7.6026	7.2631	7.0397	6.8801	6.7099	6.5262	6.4079	6.2156	6.0989
40	7.9566	7.6188	7.3746	7.0413	6.8214	6.6640	6.4955	6.3131	6.1952	6.0023	5.8844
50	7.6404	7.3125	7.0751	6.7501	6.5350	6.3804	6.2143	6.0334	5.9157	5.7214	5.6011
60	7.4417	7.1202	6.8872	6.5676	6.3555	6.2027	6.0381	5.8581	5.7403	5.5446	5.4222
70	7.3054	6.9884	6.7584	6.4426	6.2325	6.0809	5.9173	5.7378	5.6200	5.4230	5.2987
80	7.2061	6.8924	6.6646	6.3515	6.1430	5.9924	5.8294	5.6503	5.5323	5.3343	5.2084
100	7.0711	6.7619	6.5372	6.2279	6.0215	5.8721	5.7101	5.5313	5.4131	5.2133	5.0849
200	6.8143	6.5139	6.2952	5.9933	5.7910	5.6439	5.4836	5.3053	5.1863	4.9819	4.8471
∞	6.5741	6.2821	6.0692	5.7743	5.5758	5.4309	5.2721	5.0940	4.9737	4.7629	4.6190

[a]Multiply entry by 100.

(*continued*)

Table B.7. ***(continued)***

	ν_H									
ν_E	5	6	8	10	12	15	20	25	40	60
					$p = 5$					
5	81.991[a]	83.352[a]	85.093[a]	86.160[a]	86.88[a]					
6	3.0093[a]	3.0142[a]	3.0204[a]	3.0241[a]	3.0266[a]	3.0291[a]	3.032[a]			
7	93.762	93.042	92.102	91.515	91.113	90.705	90.29	90.04		
8	51.339	50.646	49.739	49.170	48.780	48.382	47.973	47.723	47.35	
10	27.667	27.115	26.387	25.927	25.610	25.284	24.947	24.740	24.422	
12	20.169	19.701	19.079	18.683	18.409	18.124	17.830	17.647	17.365	17.20
14	16.643	16.224	15.666	15.309	15.059	14.800	14.530	14.361	14.100	13.95
16	14.624	14.239	13.722	13.389	13.157	12.914	12.659	12.499	12.250	12.105
18	13.326	12.963	12.476	12.161	11.939	11.708	11.463	11.310	11.068	10.928
20	12.424	12.078	11.612	11.310	11.097	10.874	10.637	10.488	10.252	10.113
25	11.046	10.728	10.297	10.016	9.8168	9.6061	9.3814	9.2386	9.0102	8.8745
30	10.270	9.9689	9.5592	9.2907	9.0995	8.8964	8.6785	8.5389	8.3141	8.1790
35	9.7739	9.4836	9.0879	8.8277	8.6419	8.4437	8.2301	8.0926	7.8693	7.7339
40	9.4292	9.1469	8.7613	8.5070	8.3250	8.1303	7.9195	7.7833	7.5607	7.4247
50	8.9825	8.7107	8.3385	8.0921	7.9150	7.7248	7.5177	7.3829	7.1605	7.0229
60	8.7057	8.4406	8.0769	7.8355	7.6615	7.4741	7.2692	7.1351	6.9124	6.7730
70	8.5174	8.2570	7.8991	7.6612	7.4894	7.3039	7.1004	6.9667	6.7434	6.6024
80	8.3811	8.1241	7.7705	7.5351	7.3648	7.1807	6.9782	6.8448	6.6208	6.4785
100	8.1969	7.9446	7.5969	7.3649	7.1968	7.0145	6.8133	6.6801	6.4550	6.3103
200	7.8505	7.6070	7.2706	7.0451	6.8811	6.7023	6.5032	6.3702	6.1416	5.9908
∞	7.5305	7.2955	6.9698	6.7505	6.5902	6.4144	6.2171	6.0838	5.8499	5.6899

[a]Multiply entry by 100.

(*continued*)

Table B.7. (*continued*)

ν_E	ν_H 6	8	10	12	15	20	25	30	35
					$p = 6$				
10	45.722	44.677	44.019	43.567	43.103	42.626	42.334	42.136	41.993
12	28.959	28.121	27.590	27.223	26.843	26.451	26.209	26.044	25.925
14	22.321	21.600	21.141	20.821	20.489	20.144	19.929	19.783	19.677
16	18.858	18.210	17.795	17.505	17.202	16.886	16.688	16.553	16.455
18	16.755	16.157	15.772	15.501	15.218	14.921	14.735	14.607	14.513
20	15.351	14.788	14.424	14.168	13.899	13.615	13.436	13.313	13.223
25	13.293	12.786	12.456	12.222	11.975	11.711	11.544	11.428	11.343
30	12.180	11.705	11.395	11.173	10.939	10.687	10.526	10.414	10.331
35	11.484	11.031	10.733	10.520	10.293	10.049	9.8921	9.7820	9.7003
40	11.009	10.571	10.282	10.075	9.8535	9.6142	9.4596	9.3508	9.2699
50	10.402	9.9832	9.7060	9.5067	9.2927	9.0598	8.9082	8.8009	8.7207
60	10.031	9.6246	9.3547	9.1602	8.9507	8.7215	8.5717	8.4651	8.3851
70	9.7813	9.3830	9.1182	8.9269	8.7204	8.4938	8.3450	8.2388	8.1589
80	9.6014	9.2093	8.9480	8.7591	8.5548	8.3300	8.1819	8.0759	7.9959
100	9.3598	8.9760	8.7197	8.5340	8.3326	8.1102	7.9629	7.8572	7.7771
200	8.9099	8.5419	8.2950	8.1153	7.9193	7.7011	7.5552	7.4494	7.3685
∞	8.4997	8.1463	7.9082	7.7340	7.5430	7.3284	7.1832	7.0768	6.9945

Table B.8. Test for Equal Covariance Matrices, $\alpha = .05$

	k								
ν	2	3	4	5	6	7	8	9	10
					$p = 2$				
3	12.18	18.70	24.55	30.09	35.45	40.68	45.81	50.87	55.86
4	10.70	16.65	22.00	27.07	31.97	36.75	41.45	46.07	50.64
5	9.97	15.63	20.73	25.57	30.23	34.79	39.26	43.67	48.02
6	9.53	15.02	19.97	24.66	29.19	33.61	37.95	42.22	46.45
7	9.24	14.62	19.46	24.05	28.49	32.83	37.08	41.26	45.40
8	9.04	14.33	19.10	23.62	27.99	32.26	36.44	40.57	44.64
9	8.88	14.11	18.83	23.30	27.62	31.84	35.98	40.05	44.08
10	8.76	13.94	18.61	23.05	27.33	31.51	35.61	39.65	43.64
11	8.67	13.81	18.44	22.85	27.10	31.25	35.32	39.33	43.29
12	8.59	13.70	18.30	22.68	26.90	31.03	35.08	39.07	43.00
13	8.52	13.60	18.19	22.54	26.75	30.85	34.87	38.84	42.76
14	8.47	13.53	18.10	22.42	26.61	30.70	34.71	38.66	42.56
15	8.42	13.46	18.01	22.33	26.50	30.57	34.57	38.50	42.38
16	8.38	13.40	17.94	22.24	26.40	30.45	34.43	38.36	42.23
17	8.35	13.35	17.87	22.17	26.31	30.35	34.32	38.24	42.10
18	8.32	13.30	17.82	22.10	26.23	30.27	34.23	38.13	41.99
19	8.28	13.26	17.77	22.04	26.16	30.19	34.14	38.04	41.88
20	8.26	13.23	17.72	21.98	26.10	30.12	34.07	37.95	41.79
25	8.17	13.10	17.55	21.79	25.87	29.86	33.78	37.63	41.44
30	8.11	13.01	17.44	21.65	25.72	29.69	33.59	37.42	41.21

	k								
ν	2	3	4	5	6	7	8	9	10
					$p = 3$				
4	22.41	35.00	46.58	57.68	68.50	79.11	89.60	99.94	110.21
5	19.19	30.52	40.95	50.95	60.69	70.26	79.69	89.03	98.27
6	17.57	28.24	38.06	47.49	56.67	65.69	74.58	83.39	92.09
7	16.59	26.84	36.29	45.37	54.20	62.89	71.44	79.90	88.30
8	15.93	25.90	35.10	43.93	52.54	60.99	69.32	77.57	85.73
9	15.46	25.22	34.24	42.90	51.33	59.62	67.78	75.86	83.87
10	15.11	24.71	33.59	42.11	50.42	58.57	66.62	74.58	82.46
11	14.83	24.31	33.08	41.50	49.71	57.76	65.71	73.57	81.36
12	14.61	23.99	32.67	41.00	49.13	57.11	64.97	72.75	80.45
13	14.43	23.73	32.33	40.60	48.65	56.56	64.36	72.09	79.72
14	14.28	23.50	32.05	40.26	48.26	56.11	63.86	71.53	79.11
15	14.15	23.32	31.81	39.97	47.92	55.73	63.43	71.05	78.60
16	14.04	23.16	31.60	39.72	47.63	55.40	63.06	70.64	78.14
17	13.94	23.02	31.43	39.50	47.38	55.11	62.73	70.27	77.76
18	13.86	22.89	31.26	39.31	47.16	54.86	62.45	69.97	77.41
19	13.79	22.78	31.13	39.15	46.96	54.64	62.21	69.69	77.11
20	13.72	22.69	31.01	39.00	46.79	54.44	61.98	69.45	76.84
25	13.48	22.33	30.55	38.44	46.15	53.70	61.16	68.54	75.84
30	13.32	22.10	30.25	38.09	45.73	53.22	60.62	67.94	75.18

(continued)

Table B.8. ***(continued)***

ν	k = 2	3	4	5	6	7	8	9	10
					$p = 4$				
5	35.39	56.10	75.36	93.97	112.17	130.11	147.81	165.39	182.80
6	30.06	48.62	65.90	82.60	98.93	115.03	130.94	146.69	162.34
7	27.31	44.69	60.89	76.56	91.88	106.98	121.90	136.71	151.39
8	25.61	42.24	57.77	72.77	87.46	101.94	116.23	130.43	144.50
9	24.45	40.57	55.62	70.17	84.42	98.46	112.32	126.08	139.74
10	23.62	39.34	54.04	68.26	82.19	95.90	109.46	122.91	136.24
11	22.98	38.41	52.84	66.81	80.48	93.95	107.27	120.46	133.57
12	22.48	37.67	51.90	65.66	79.14	92.41	105.54	118.55	131.45
13	22.08	37.08	51.13	64.73	78.04	91.15	104.12	116.98	129.74
14	21.75	36.59	50.50	63.95	77.13	90.12	102.97	115.69	128.32
15	21.47	36.17	49.97	63.30	76.37	89.26	101.99	114.59	127.14
16	21.24	35.82	49.51	62.76	75.73	88.51	101.14	113.67	126.10
17	21.03	35.52	49.12	62.28	75.16	87.87	100.42	112.87	125.22
18	20.86	35.26	48.78	61.86	74.68	87.31	99.80	112.17	124.46
19	20.70	35.02	48.47	61.50	74.25	86.82	99.25	111.56	123.79
20	20.56	34.82	48.21	61.17	73.87	86.38	98.75	111.02	123.18
25	20.06	34.06	47.23	59.98	72.47	84.78	96.95	109.01	120.99
30	19.74	33.59	46.61	59.21	71.58	83.74	95.79	107.71	119.57

ν	k = 2	3	4	5	6	7	8	9	10
					$p = 5$				
6	51.11	81.99	110.92	138.98	166.54	193.71	220.66	247.37	273.88
7	43.40	71.06	97.03	122.22	146.95	171.34	195.49	219.47	243.30
8	39.29	65.15	89.45	113.03	136.18	159.04	181.65	204.14	226.48
9	36.71	61.39	84.62	107.17	129.30	151.17	172.80	194.27	215.64
10	34.93	58.78	81.25	103.06	124.48	145.64	166.56	187.37	208.02
11	33.62	56.85	78.75	100.02	120.92	141.54	161.98	182.24	202.37
12	32.62	55.37	76.83	97.68	118.15	138.38	158.38	178.23	198.03
13	31.83	54.19	75.30	95.82	115.96	135.86	155.54	175.10	194.51
14	31.19	53.23	74.05	94.29	114.16	133.80	153.21	172.49	191.68
15	30.66	52.44	73.01	93.02	112.66	132.07	151.29	170.36	189.38
16	30.22	51.76	72.14	91.94	111.41	130.61	149.66	166.53	187.32
17	29.83	51.19	71.39	91.03	110.34	129.38	148.25	166.99	185.61
18	29.51	50.69	70.74	90.23	109.39	128.29	147.03	165.65	184.10
19	29.22	50.26	70.17	89.54	108.57	127.36	145.97	164.45	182.81
20	28.97	49.88	69.67	88.93	107.85	126.52	145.02	163.38	181.65
25	28.05	48.48	67.86	86.70	105.21	123.51	141.62	159.60	177.49
30	27.48	47.61	66.71	85.29	103.56	121.60	139.47	157.22	174.87

Note: Table contains upper percentage points for

$$-2\ln M = \nu\left(k\ln|\mathbf{S}| - \sum_{i=1}^{k}\ln|\mathbf{S}_i|\right)$$

for k samples, each with ν degrees of freedom. Reject H_0: $\boldsymbol{\Sigma}_1 = \boldsymbol{\Sigma}_2 = \cdots = \boldsymbol{\Sigma}_k$ if $-2\ln M >$ table value.

APPENDIX C

Answers and Hints to Selected Problems

Chapter 1

1.1 Use (1.3) and (1.4).

1.2 Expand $(y - \mu)^2$ and use (1.3) and (1.4).

1.3 Expand $(y_i - \overline{y})^2$ and sum each of the three terms.

1.4 The proof can be outlined as follows:

$$\begin{aligned} E(\mathbf{y}'\mathbf{A}\mathbf{y}) &= E[\text{tr}(\mathbf{A}\mathbf{y}\mathbf{y}')] = \text{tr}[\mathbf{A}E(\mathbf{y}\mathbf{y}')] \\ &= \text{tr}[\mathbf{A}\boldsymbol{\Sigma} + \mathbf{A}\boldsymbol{\mu}\boldsymbol{\mu}'] = \text{tr}(\mathbf{A}\boldsymbol{\Sigma}) + \boldsymbol{\mu}'\mathbf{A}\boldsymbol{\mu}. \end{aligned}$$

1.5 Write $(n - 1)s^2$ as a quadratic form $\mathbf{y}'\mathbf{A}\mathbf{y}$, where $\mathbf{y}' = (y_1, y_2, \ldots, y_n)$ and $\mathbf{A} = [\mathbf{I} - (1/n)\mathbf{J}]$. This can be done using $(n - 1)s^2 = \sum_i y_i^2 - n\overline{y}^2$ in (1.11), with $\sum_i y_i^2 = \mathbf{y}'\mathbf{y}$ and $\overline{y} = (1/n)\mathbf{y}'\mathbf{j}$. Now apply (1.125) in Problem 1.4 to find $E[\mathbf{y}'(\mathbf{I} - (1/n)\mathbf{J})\mathbf{y}]$. Note that $E(\mathbf{y}) = \mu\mathbf{j}$ and $\text{cov}(\mathbf{y}) = \sigma^2\mathbf{I}$.

1.6 $\text{var}(ay) = E(ay - a\mu)^2 = E[a^2(y - \mu)^2]$.

1.8 By an extension of (1.2),

$$\begin{aligned} E(xy) &= \int\int xyf(x, y)\,dx\,dy = \int\int xyg(x)h(y)\,dx\,dy \\ &= \int xg(x)\,dx \int yh(y)\,dy = E(x)E(y). \end{aligned}$$

1.9 Use (1.16) and (1.18).

1.11 This can be proved in a manner analogous to the approach used in the proof of (1.125), which is outlined in the hint for Problem 1.4.

1.12 Write $(n - 1)s_{xy}$ in the form $(n - 1)s_{xy} = \mathbf{x}'[\mathbf{I} - (1/n)\mathbf{J}]\mathbf{y}$ and apply (1.126) in Problem 1.11. Note that $\boldsymbol{\Sigma}_{yx} = \sigma_{xy}\mathbf{I}$.

1.14 $\sum_{i=1}^{n} \mathbf{y}_i = \sum_i (y_{i1}, y_{i2}, \ldots, y_{ip})' = (\sum_i y_{i1}, \ldots, \sum_i y_{ip})'$.

1.15 Use (A.2.15).

1.16 $\mathbf{y}_i - \overline{\mathbf{y}} = (y_{i1} - \overline{y}_1, y_{i2} - \overline{y}_2, \ldots, y_{ip} - \overline{y}_p)'$. Then in $\sum_{i=1}^n (\mathbf{y}_i - \overline{\mathbf{y}})(\mathbf{y}_i - \overline{\mathbf{y}})'$, the (1, 2) element, for example, is $\sum_{i=1}^n (y_{i1} - \overline{y}_1)(y_{i2} - \overline{y}_2) = (n-1)s_{12}$.

1.17 $\sum_{i=1}^n (\mathbf{y}_i - \overline{\mathbf{y}})(\mathbf{y}_i - \overline{\mathbf{y}})' = \sum_i \mathbf{y}_i\mathbf{y}_i' - \sum_i \mathbf{y}_i\overline{\mathbf{y}}' - \sum_i \overline{\mathbf{y}}\mathbf{y}_i' + n\overline{\mathbf{y}}\,\overline{\mathbf{y}}'$. Use $n\overline{\mathbf{y}} = \sum_i \mathbf{y}_i$.

1.18 Use (A.1.10).

1.19 Rank$(\mathbf{Y}_c)$ = p, the number of columns, as long as there are sufficient rows. Show that $\mathbf{j}'\mathbf{Y}_c = \mathbf{0}'$. Thus each column sums to 0, and we need $n - 1 > p$.

1.20 **(a)** Add and subtract $\overline{\mathbf{y}}$ in $\mathbf{S}_a$.

(b) Use (A.7.10), with $\sqrt{n/(n-1)}(\overline{\mathbf{y}} - \mathbf{a})$.

(c) The result follows from part (a).

1.21 $E(\mathbf{y} - \boldsymbol{\mu})(\mathbf{y} - \boldsymbol{\mu})' = E(\mathbf{y}\mathbf{y}' - \mathbf{y}\boldsymbol{\mu}' - \boldsymbol{\mu}\mathbf{y}' + \boldsymbol{\mu}\boldsymbol{\mu}') = E(\mathbf{y}\mathbf{y}') - [E(\mathbf{y})]\boldsymbol{\mu}' - \boldsymbol{\mu}E(\mathbf{y}') + \boldsymbol{\mu}\boldsymbol{\mu}'$ by extensions of (1.3) and (1.4) [see also (1.86) and (1.88)–(1.90)].

1.22 **(a)** Show that $\text{cov}(\overline{\mathbf{y}}) = (1/n)\boldsymbol{\Sigma}$. By (1.46), $E(\mathbf{y}\mathbf{y}') = \boldsymbol{\Sigma} + \boldsymbol{\mu}\boldsymbol{\mu}'$. Use (1.38) and (1.31) to obtain the result.

(b) To establish (1.127), begin with $\sum_{i=1}^n (\mathbf{y}_i - \boldsymbol{\mu})(\mathbf{y}_i - \boldsymbol{\mu})'$ and add and subtract $\overline{\mathbf{y}}$; that is, expand $\sum_{i=1}^n (\mathbf{y}_i - \overline{\mathbf{y}} + \overline{\mathbf{y}} - \boldsymbol{\mu})(\mathbf{y}_i - \overline{\mathbf{y}} + \overline{\mathbf{y}} - \boldsymbol{\mu})'$ in terms of $\mathbf{y}_i - \overline{\mathbf{y}}$ and $\overline{\mathbf{y}} - \boldsymbol{\mu}$. Then show that $E[\sum_{i=1}^n (\mathbf{y}_i - \boldsymbol{\mu})(\mathbf{y}_i - \boldsymbol{\mu})'] = n\boldsymbol{\Sigma}$ and that $E[(\overline{\mathbf{y}} - \boldsymbol{\mu})(\overline{\mathbf{y}} - \boldsymbol{\mu})'] = \boldsymbol{\Sigma}/n$.

1.23 By analogy with (A.2.29), we can write

$$
\begin{aligned}
E(\mathbf{Y}'\mathbf{A}\mathbf{Y}) &= E\left[\sum_{i=1}^n a_{ii}\mathbf{y}_i\mathbf{y}_i' + \sum_{i \neq j} a_{ij}\mathbf{y}_i\mathbf{y}_j'\right] \\
&= \sum_i a_{ii}(\boldsymbol{\Sigma} + \boldsymbol{\mu}_i\boldsymbol{\mu}_i') + \sum_{i \neq j} a_{ij}\boldsymbol{\mu}_i\boldsymbol{\mu}_j' \qquad \text{[by (1.46)]} \\
&= \sum_i a_{ii}\boldsymbol{\Sigma} + \sum_i a_{ii}\boldsymbol{\mu}_i\boldsymbol{\mu}_i' + \sum_{i \neq j} a_{ij}\boldsymbol{\mu}_i\boldsymbol{\mu}_j'.
\end{aligned}
$$

1.24 By (1.40), $\mathbf{S} = \mathbf{Y}'[\mathbf{I} - (1/n)\mathbf{J}]\mathbf{Y}/(n-1)$. Show that $E(\mathbf{Y}) = \mathbf{j}\boldsymbol{\mu}'$ and $\text{tr}[\mathbf{I} - (1/n)\mathbf{J}] = n - 1$.

1.27 Use Problem 1.26(a) to show that $\mathbf{Z}$ can be written as $\mathbf{Z} = [\mathbf{I} - (1/n)\mathbf{J}]\mathbf{Y}\mathbf{D}_s^{-1} = \mathbf{Y}_c\mathbf{D}_s^{-1}$. Then use (1.42).

1.28 Use (1.76) and (1.77).

1.29 **(a)** $\mathbf{S}$ in (1.59) is defined as

$$
\mathbf{S} = \frac{1}{n-1}\sum_{i=1}^n \left[\begin{pmatrix}\mathbf{y}_i \\ \mathbf{x}_i\end{pmatrix} - \begin{pmatrix}\overline{\mathbf{y}} \\ \overline{\mathbf{x}}\end{pmatrix}\right]\left[\begin{pmatrix}\mathbf{y}_i \\ \mathbf{x}_i\end{pmatrix} - \begin{pmatrix}\overline{\mathbf{y}} \\ \overline{\mathbf{x}}\end{pmatrix}\right]'.
$$

(b) Use the method that was used to obtain (1.40).

1.32 $s_{zw} = \sum_{i=1}^n (z_i - \overline{z})(w_i - \overline{w})/(n-1) = \sum_i (\mathbf{a}'\mathbf{y}_i - \mathbf{a}'\overline{\mathbf{y}})(\mathbf{b}'\mathbf{y}_i - \mathbf{b}'\overline{\mathbf{y}})/(n-1)$. Follow the steps in the proof of (1.73).

1.33 $\overline{\mathbf{z}} = \sum_{i=1}^n \mathbf{z}_i/n = \sum_i \mathbf{A}\mathbf{y}_i/n = \mathbf{A}\sum_i \mathbf{y}_i/n = \mathbf{A}\overline{\mathbf{y}}$. $\mathbf{S}_z = \sum_{i=1}^n (\mathbf{z}_i - \overline{\mathbf{z}})(\mathbf{z}_i - \overline{\mathbf{z}})'/(n-1) = \sum_i (\mathbf{A}\mathbf{y}_i - \mathbf{A}\overline{\mathbf{y}})(\mathbf{A}\mathbf{y}_i - \mathbf{A}\overline{\mathbf{y}})'/(n-1) = \mathbf{A}[\sum_i (\mathbf{y}_i - \overline{\mathbf{y}})(\mathbf{y}_i - \overline{\mathbf{y}})'/(n-1)]\mathbf{A}'$.

1.34 Show that $\bar{\mathbf{z}} = \mathbf{A}\bar{\mathbf{y}} + \mathbf{b}$ and substitute this and $\mathbf{z}_i = \mathbf{A}\mathbf{y}_i + \mathbf{b}$ into $\sum_{i=1}^n (\mathbf{z}_i - \bar{\mathbf{z}})(\mathbf{z}_i - \bar{\mathbf{z}})'$.

1.35 By Problem 1.29(a), $\widehat{\text{cov}}(\mathbf{Ay}, \mathbf{By}) = \sum_{i=1}^n (\mathbf{A}\mathbf{y}_i - \mathbf{A}\bar{\mathbf{y}})(\mathbf{B}\mathbf{y}_i - \mathbf{B}\bar{\mathbf{y}})'/(n-1)$. Factor out $\mathbf{A}$ and $\mathbf{B}'$.

1.37 $\widehat{\text{cov}}(\mathbf{a}'\mathbf{y}, \mathbf{b}'\mathbf{x}) = \sum_{i=1}^n (\mathbf{a}'\mathbf{y}_i - \mathbf{a}'\bar{\mathbf{y}})(\mathbf{b}'\mathbf{x}_i - \mathbf{b}'\bar{\mathbf{x}})'/(n-1)$. Factor out $\mathbf{a}'$ and $\mathbf{b}$ and use Problem 1.29(a).

1.38 Use Problem 1.29(a) twice.

1.39 Use the definition $\sigma_z^2 = E(z - \mu_z)^2$ with $z = \mathbf{a}'\mathbf{y}$ and $\mu_z = \mathbf{a}'\boldsymbol{\mu}$, where $\boldsymbol{\mu} = E(\mathbf{y})$. Then

$$\begin{aligned}\sigma_z^2 &= E(z - \mu_z)^2 = E(\mathbf{a}'\mathbf{y} - \mathbf{a}'\boldsymbol{\mu})^2 = E(\mathbf{a}'\mathbf{y} - \mathbf{a}'\boldsymbol{\mu})(\mathbf{a}'\mathbf{y} - \mathbf{a}'\boldsymbol{\mu}) \\ &= E(\mathbf{a}'\mathbf{y} - \mathbf{a}'\boldsymbol{\mu})(\mathbf{y}'\mathbf{a} - \boldsymbol{\mu}'\mathbf{a}) = E[\mathbf{a}'(\mathbf{y} - \boldsymbol{\mu})(\mathbf{y} - \boldsymbol{\mu})'\mathbf{a}].\end{aligned}$$

Now apply (1.89).

1.40 By (1.15), $\text{cov}(\mathbf{a}'\mathbf{y}, \mathbf{b}'\mathbf{y}) = E[(\mathbf{a}'\mathbf{y} - \mathbf{a}'\boldsymbol{\mu})(\mathbf{b}'\mathbf{y} - \mathbf{b}'\boldsymbol{\mu})]$. Show that this becomes $E[\mathbf{a}'(\mathbf{y} - \boldsymbol{\mu})(\mathbf{y} - \boldsymbol{\mu})'\mathbf{b}]$.

1.41 Designate the rows of $\mathbf{A}$ as $\mathbf{a}_1', \mathbf{a}_2', \ldots, \mathbf{a}_q'$. Then

$$E(\mathbf{Ay}) = E\left[\begin{pmatrix}\mathbf{a}_1' \\ \vdots \\ \mathbf{a}_q'\end{pmatrix}\mathbf{y}\right] = E\begin{pmatrix}\mathbf{a}_1'\mathbf{y} \\ \vdots \\ \mathbf{a}_q'\mathbf{y}\end{pmatrix} = \begin{pmatrix}\mathbf{a}_1'E(\mathbf{y}) \\ \vdots \\ \mathbf{a}_q'E(\mathbf{y})\end{pmatrix}.$$

1.43 Show that $E(\mathbf{Ay} + \mathbf{b}) = \mathbf{A}\boldsymbol{\mu} + \mathbf{b}$. Then $\text{cov}(\mathbf{Ay} + \mathbf{b}) = E[(\mathbf{Ay} + \mathbf{b} - \mathbf{A}\boldsymbol{\mu} - \mathbf{b})(\mathbf{Ay} + \mathbf{b} - \mathbf{A}\boldsymbol{\mu} - \mathbf{b})']$.

1.45 By (1.15), $\text{cov}(\mathbf{a}'\mathbf{y}, \mathbf{b}'\mathbf{x}) = E[(\mathbf{a}'\mathbf{y} - \mathbf{a}'\boldsymbol{\mu}_y)(\mathbf{b}'\mathbf{x} - \mathbf{b}'\boldsymbol{\mu}_x)]$. Show that this becomes $E[\mathbf{a}'(\mathbf{y} - \boldsymbol{\mu}_y)(\mathbf{x} - \boldsymbol{\mu}_x)'\mathbf{b}] = \mathbf{a}'E[(\mathbf{y} - \boldsymbol{\mu}_y)(\mathbf{x} - \boldsymbol{\mu}_x)']\mathbf{b}$. Use (1.65).

1.47 Show that $\sum_{i=1}^p \lambda_i^2 = \sum_{i=1}^p \sum_{j=1}^p r_{ij}^2$ and that the single nonzero eigenvalue of $\mathbf{J}$ is p.

1.48 Adapt (A.5.5) to obtain

$$\begin{pmatrix} a & \mathbf{b}' \\ \mathbf{b} & \mathbf{C}\end{pmatrix}^{-1} = \begin{pmatrix} 1/d & -\mathbf{b}'\mathbf{C}^{-1}/d \\ -\mathbf{C}^{-1}\mathbf{b}/d & \mathbf{C}^{-1} + \mathbf{C}^{-1}\mathbf{b}\mathbf{b}'\mathbf{C}^{-1}/d\end{pmatrix},$$

where $\mathbf{C}$ is symmetric and $d = a - \mathbf{b}'\mathbf{C}^{-1}\mathbf{b}$. Apply this formula to $\mathbf{R}$ partitioned in the form

$$\mathbf{R} = \begin{pmatrix} 1 & \mathbf{r}_1' \\ \mathbf{r}_1 & \mathbf{R}_1\end{pmatrix},$$

where $\mathbf{r}_1' = (r_{12}, r_{13}, \ldots, r_{1p})$ and

$$\mathbf{R}_1 = \begin{pmatrix} 1 & r_{23} & r_{24} & \cdots & r_{2p} \\ r_{32} & 1 & r_{34} & \cdots & r_{3p} \\ \vdots & \vdots & \vdots & & \vdots \\ r_{p2} & r_{p3} & r_{p4} & \cdots & 1\end{pmatrix}$$

is the matrix of correlations among $x_2, x_3, \ldots, x_p$. From the above formula for the inverse, we have

$$\begin{aligned} r^{11} &= \frac{1}{d} = \frac{1}{a - \mathbf{b}'\mathbf{C}^{-1}\mathbf{b}} \\ &= \frac{1}{1 - \mathbf{r}_1'\mathbf{R}_1^{-1}\mathbf{r}_1} = \frac{1}{1 - R_1^2}. \end{aligned}$$

To show that $R_1^2 = \mathbf{r}_1'\mathbf{R}_1^{-1}\mathbf{r}_1$, substitute (7.60) and (7.61) into (7.66).

1.51 Use the square root matrix defined in (A.11.7) to rewrite (1.105) as

$$\begin{aligned} \Delta^2 &= (\bar{\mathbf{y}} - \boldsymbol{\mu})'\boldsymbol{\Sigma}^{-1}(\bar{\mathbf{y}} - \boldsymbol{\mu}) = (\bar{\mathbf{y}} - \boldsymbol{\mu})'(\boldsymbol{\Sigma}^{1/2}\boldsymbol{\Sigma}^{1/2})^{-1}(\bar{\mathbf{y}} - \boldsymbol{\mu}) \\ &= \left[(\boldsymbol{\Sigma}^{1/2})^{-1}(\bar{\mathbf{y}} - \boldsymbol{\mu})\right]' \left[(\boldsymbol{\Sigma}^{1/2})^{-1}(\bar{\mathbf{y}} - \boldsymbol{\mu})\right] = \mathbf{z}'\mathbf{z}, \text{ say.} \end{aligned}$$

Now show that $\text{cov}(\mathbf{z}) = \mathbf{I}/n$ (see Problem 1.22).

1.52 $r_{xy} = .889$, $r_{xy}^* = .895$.

1.53

$$\bar{\mathbf{y}} = \begin{pmatrix} 36.09 \\ 25.55 \\ 34.09 \\ 27.27 \\ 30.73 \end{pmatrix}, \qquad \mathbf{S} = \begin{pmatrix} 65.09 & 33.65 & 47.59 & 36.77 & 25.43 \\ 33.65 & 46.07 & 28.95 & 40.34 & 28.36 \\ 47.59 & 28.95 & 60.69 & 37.37 & 41.13 \\ 36.77 & 40.34 & 37.37 & 62.82 & 31.68 \\ 25.43 & 28.36 & 41.13 & 31.68 & 58.22 \end{pmatrix}.$$

1.54 $\bar{z} = 67.273$, $s_z^2 = 825.42$.

1.55 **(a)** $s_{zw} = -230.1$, **(b)** $r_{zw} = -.4308$.

1.56 Use (1.75) to obtain $r_{zw} = .7744$.

1.57 **(a)**

$$\bar{\mathbf{z}} = \begin{pmatrix} 153.7 \\ 134.0 \\ -94.4 \end{pmatrix}, \quad \mathbf{S}_z = \begin{pmatrix} 995.4 & 780.2 & -476.1 \\ 780.2 & 778.8 & -457.7 \\ -476.1 & -457.7 & 629.7 \end{pmatrix}.$$

1.58 $\mathbf{S}_{zw} = \begin{pmatrix} -423.82 & 86.89 \\ -421.20 & -15.30 \\ 35.51 & 1.35 \end{pmatrix}.$

1.59

$$\begin{pmatrix} \bar{\mathbf{y}} \\ \hline \bar{\mathbf{x}} \end{pmatrix} = \begin{pmatrix} .8988 \\ 90.44 \\ \hline 339.24 \\ 172.24 \\ 88.04 \end{pmatrix}, \qquad \mathbf{S} = \left(\begin{array}{cc|ccc} .0128 & .0464 & .2740 & -.1997 & 1.069 \\ .0464 & 42.51 & 8.973 & 17.43 & -16.81 \\ \hline .2740 & 8.973 & 1122.4 & 512.7 & -16.80 \\ -.1997 & 17.43 & 512.7 & 1853.2 & 305.0 \\ 1.069 & -16.81 & -16.80 & 305.0 & 1129.5 \end{array}\right).$$

1.60 **(a)**

$$\mathbf{R} = \begin{pmatrix} 1.000 & .063 & .072 & -.041 & .281 \\ .063 & 1.000 & .041 & .062 & -.077 \\ .072 & .041 & 1.000 & .355 & -.015 \\ -.041 & .062 & .355 & 1.000 & .211 \\ .281 & -.077 & -.015 & .211 & 1.000 \end{pmatrix}$$

The eigenvalues of $\mathbf{R}$ are 1.451, 1.213, 1.028, .818, .489.

(b) $g = .165, q_1 = .056, q_2 = .131, q_3 = .149, q_4 = .156, q_5 = .598, q_6 = .127.$

1.61 $s_u^2 = 382.05, s_v^2 = 6360.17, s_{uv} = 131.58, r_{uv} = .0844.$

1.62 **(a)** $\mathbf{S}_u = \begin{pmatrix} 42.61 & -42.43 \\ -42.43 & 42.37 \end{pmatrix}, \quad \mathbf{S}_v = \begin{pmatrix} 7365.9 & -3813.6 \\ -3813.6 & 6360.2 \end{pmatrix}.$

(b) $\widehat{\text{cov}}(\mathbf{u}, \mathbf{v}) = \begin{pmatrix} 57.96 & -40.96 \\ -64.15 & 46.18 \end{pmatrix}.$

1.63

$$\begin{pmatrix} \overline{\mathbf{y}} \\ \overline{\mathbf{x}} \end{pmatrix} = \begin{pmatrix} 68.7 \\ 69.7 \\ 73.5 \\ \hline 106.8 \\ 100.6 \\ 115.8 \end{pmatrix},$$

$$\mathbf{S} = \left(\begin{array}{ccc|ccc} 61.06 & -5.09 & -5.58 & 28.24 & -6.81 & 37.99 \\ -5.09 & 41.48 & -2.47 & -41.18 & 66.14 & -5.43 \\ -5.58 & -2.47 & 64.37 & 9.50 & -27.11 & 17.84 \\ \hline 28.24 & -41.18 & 9.50 & 876.9 & 268.3 & 143.4 \\ -6.81 & 66.14 & -27.11 & 268.3 & 621.6 & -.0316 \\ 37.99 & -5.43 & 17.84 & 143.4 & -.0316 & 293.0 \end{array}\right).$$

1.64 $s_u^2 = 817.4, s_v^2 = 8680.4, s_{uv} = -648.2, r_{uv} = -.2433.$

1.65 **(a)** $\mathbf{S}_u = \begin{pmatrix} 140.62 & 95.19 \\ 95.19 & 502.33 \end{pmatrix}, \quad \mathbf{S}_v = \begin{pmatrix} 9575.7 & -5621.6 & 788.9 \\ -5621.6 & 4443.1 & 2144.1 \\ 788.9 & 2144.1 & 6749.2 \end{pmatrix}.$

(b) $\widehat{\text{cov}}(\mathbf{u}, \mathbf{v}) = \begin{pmatrix} -78.92 & -10.59 & -53.16 \\ 26.45 & 341.5 & 923.6 \end{pmatrix}.$

Chapter 2

2.1 Use (A.7.6) to show that $|\boldsymbol{\Sigma}^{-1/2}| = 1/|\boldsymbol{\Sigma}^{1/2}|$. Apply (A.7.5) to $\boldsymbol{\Sigma} = \boldsymbol{\Sigma}^{1/2}\boldsymbol{\Sigma}^{1/2}$ to show that $|\boldsymbol{\Sigma}^{1/2}| = |\boldsymbol{\Sigma}|^{1/2}$.

2.2 $a = -2\ln[(\sqrt{2\pi})^p|\boldsymbol{\Sigma}|^{1/2}c]$.

2.3 See (A.2.29).

2.4 The moment generating function for $\mathbf{y}-\boldsymbol{\mu}$ is given by $M_{\mathbf{y}-\boldsymbol{\mu}}(\mathbf{t}) = E\left(e^{\mathbf{t}'(\mathbf{y}-\boldsymbol{\mu})}\right) = E\left(e^{\mathbf{t}'\mathbf{y}-\mathbf{t}'\boldsymbol{\mu}}\right) = e^{-\mathbf{t}'\boldsymbol{\mu}}e^{\mathbf{t}'\boldsymbol{\mu}+\mathbf{t}'\boldsymbol{\Sigma}\mathbf{t}/2} = e^{\mathbf{t}'\boldsymbol{\Sigma}\mathbf{t}/2}$. Differentiate this with respect to t_i, t_j, and t_k and set the t's equal to 0 to obtain the desired result.

2.5 Differentiate $M_{\mathbf{y}-\boldsymbol{\mu}}(t) = e^{\mathbf{t}'\boldsymbol{\Sigma}\mathbf{t}/2}$ with respect to t_i, t_j, t_k, and t_l and set the t's equal to 0.

2.7 **(a)** -2, **(b)** 4, **(c)** 0, **(d)** 0, **(e)** 0, **(f)** 20, **(g)** 32, **(h)** 108.

2.8 **(a)** -6, **(b)** 2, **(c)** 0, **(d)** 0, **(e)** 0, **(f)** -69, **(g)** 87, **(h)** 177, **(i)** 40, **(j)** 243.

2.9 Multiply out the third term on the right side in terms of $\mathbf{y}-\boldsymbol{\mu}$ and $\boldsymbol{\Sigma}\mathbf{t}$.

2.10 $E\left(e^{\mathbf{t}'\mathbf{A}\mathbf{y}}\right) = E\left(e^{(\mathbf{A}'\mathbf{t})'\mathbf{y}}\right)$. Now use Theorem 2.2A with $\mathbf{A}'\mathbf{t}$ in place of $\mathbf{t}$.

2.11 Use (A.5.2), (A.6.1), and (A.6.2).

2.12 Show that $E[\mathbf{A}^{-1}(\mathbf{y}-\boldsymbol{\mu})] = \mathbf{0}$ and $\text{cov}[\mathbf{A}^{-1}(\mathbf{y}-\boldsymbol{\mu})] = \mathbf{I}$. Then use Theorem 2.2B(ii).

2.14 See (A.3.3).

2.16 $\mathbf{y}+\mathbf{x} = (\mathbf{I},\mathbf{I})\binom{\mathbf{y}}{\mathbf{x}}$, where each $\mathbf{I}$ is $p \times p$. Apply Theorem 2.2B(ii).

2.17 $\text{cov}(\mathbf{y}-\mathbf{B}\mathbf{x},\mathbf{x}) = \text{cov}[(\mathbf{I},-\mathbf{B})\binom{\mathbf{y}}{\mathbf{x}}, (\mathbf{O},\mathbf{I})\binom{\mathbf{y}}{\mathbf{x}}]$. Use (1.95).

2.19 Write $g(\mathbf{y},\mathbf{x})$ in terms of $\binom{\boldsymbol{\mu}_y}{\boldsymbol{\mu}_x}$ and

$$\boldsymbol{\Sigma} = \begin{pmatrix} \boldsymbol{\Sigma}_{yy} & \boldsymbol{\Sigma}_{yx} \\ \boldsymbol{\Sigma}_{xy} & \boldsymbol{\Sigma}_{xx} \end{pmatrix}.$$

For $|\boldsymbol{\Sigma}|$ and $\boldsymbol{\Sigma}^{-1}$, see (A.7.9) and (A.5.7). After canceling $h(\mathbf{x})$ in (2.38), show that $f(\mathbf{y}|\mathbf{x})$ can be written in the form

$$f(\mathbf{y}|\mathbf{x}) = \frac{1}{(2\pi)^{p/2}|\boldsymbol{\Sigma}_{y\cdot x}|^{1/2}} e^{-(\mathbf{y}-\boldsymbol{\mu}_{y\cdot x})'\boldsymbol{\Sigma}_{y\cdot x}^{-1}(\mathbf{y}-\boldsymbol{\mu}_{y\cdot x})/2},$$

where $\boldsymbol{\mu}_{y\cdot x} = \boldsymbol{\mu}_y + \boldsymbol{\Sigma}_{yx}\boldsymbol{\Sigma}_{xx}^{-1}(\mathbf{x}-\boldsymbol{\mu}_x)$ and $\boldsymbol{\Sigma}_{y\cdot x} = \boldsymbol{\Sigma}_{yy} - \boldsymbol{\Sigma}_{yx}\boldsymbol{\Sigma}_{xx}^{-1}\boldsymbol{\Sigma}_{xy}$.

2.21 Use (A.8.1) and (A.8.2).

2.23 Use (A.8.1) and (A.8.2).

2.24 Expand $(\bar{\mathbf{y}}-\boldsymbol{\mu})'\boldsymbol{\Sigma}^{-1}(\bar{\mathbf{y}}-\boldsymbol{\mu})$ to obtain four terms. Differentiate these using (A.13.2) and (A.13.3) to obtain $\mathbf{0}-\boldsymbol{\Sigma}^{-1}\mathbf{y}-\boldsymbol{\Sigma}^{-1}\mathbf{y}+2\boldsymbol{\Sigma}^{-1}\boldsymbol{\mu}$.

2.27 **(a)** Show that $\text{cov}\left(\mathbf{j}'\mathbf{y}_{(i)},\mathbf{A}\mathbf{y}_{(j)}\right) = \mathbf{j}'\text{cov}\left(\mathbf{y}_{(i)},\mathbf{y}_{(j)}\right)\mathbf{A}'$ and that $\text{cov}\left(\mathbf{y}_{(i)},\mathbf{y}_{(j)}\right) = \sigma_{ij}\mathbf{I}$, where σ_{ij} is from $\boldsymbol{\Sigma}$. Show that $\mathbf{j}'\mathbf{A}' = \mathbf{0}'$.

(b) Using (1.65), show that $\text{cov}(\mathbf{y}_i-\bar{\mathbf{y}},\bar{\mathbf{y}}) = E[(\mathbf{y}_i-\bar{\mathbf{y}})\bar{\mathbf{y}}'] - E(\mathbf{y}_i-\bar{\mathbf{y}})E(\bar{\mathbf{y}}')$, which equals $E(\mathbf{y}_i\bar{\mathbf{y}}')-E(n\bar{\mathbf{y}}\bar{\mathbf{y}}')$ since $E(\mathbf{y}_i-\bar{\mathbf{y}}) = \mathbf{0}$. Show that $E(\mathbf{y}_i\bar{\mathbf{y}}') = E[\mathbf{y}_i(1/n)\sum_{j=1}^n \mathbf{y}_j']$ $= (1/n)(n\boldsymbol{\mu}\boldsymbol{\mu}'+\boldsymbol{\Sigma})$ and that $E(n\bar{\mathbf{y}}\bar{\mathbf{y}}') = (1/n)\boldsymbol{\Sigma}+\boldsymbol{\mu}\boldsymbol{\mu}'$.

2.28 **(a)**

$$\mathbf{y} = \begin{pmatrix} \mathbf{y}_1 \\ \mathbf{y}_2 \\ \vdots \\ \mathbf{y}_n \end{pmatrix}, \quad \boldsymbol{\mu}_y = \begin{pmatrix} \boldsymbol{\mu} \\ \boldsymbol{\mu} \\ \vdots \\ \boldsymbol{\mu} \end{pmatrix}, \quad \boldsymbol{\Sigma}_y = \begin{pmatrix} \boldsymbol{\Sigma} & \mathbf{O} & \cdots & \mathbf{O} \\ \mathbf{O} & \boldsymbol{\Sigma} & \cdots & \mathbf{O} \\ \vdots & \vdots & & \vdots \\ \mathbf{O} & \mathbf{O} & \cdots & \boldsymbol{\Sigma} \end{pmatrix}.$$

Then $\mathbf{A}\mathbf{y}$ is $N_p(\mathbf{A}\boldsymbol{\mu}_y, \mathbf{A}\boldsymbol{\Sigma}_y\mathbf{A}')$. Show that $\mathbf{A}\boldsymbol{\mu}_y = \sum_{i=1}^n a_i\boldsymbol{\mu}$ and $\mathbf{A}\boldsymbol{\Sigma}_y\mathbf{A}' = \sum_{i=1}^n a_i^2\boldsymbol{\Sigma}$.

(b) $M_{\sum_i a_i \mathbf{y}_i}(\mathbf{t}) = E(e^{\mathbf{t}'\sum_i a_i \mathbf{y}_i}) = E(e^{\sum_i a_i \mathbf{t}'\mathbf{y}_i}) = E\left[\prod_{i=1}^n e^{a_i \mathbf{t}'\mathbf{y}_i}\right]$. Using the independence of the $\mathbf{y}_i$'s, show that this is equal to $e^{\mathbf{t}'(\sum_i a_i \boldsymbol{\mu}) + \mathbf{t}'(\sum_i a_i^2 \boldsymbol{\Sigma})\mathbf{t}/2}$.

2.29 Define $\mathbf{B} = (b_1\mathbf{I}, b_2\mathbf{I}, \ldots, b_n\mathbf{I})$ and use (1.95).

2.30 **(a)** By (A.3.4), $\mathbf{Z}'\mathbf{x}_i = \sum_{k=1}^n x_{ik}\mathbf{z}_k$, where $\mathbf{x}_i = (x_{i1}, x_{i2}, \ldots, x_{in})'$ and $\mathbf{z}_k$ is $N_p(\mathbf{0}, \boldsymbol{\Sigma})$. By (2.64), $\sum_{k=1}^n x_{ik}\mathbf{z}_k$ is $N_p(\sum_{k=1}^n x_{ik}\mathbf{0}, \sum_{k=1}^n x_{ik}^2 \boldsymbol{\Sigma}) = N_p(\mathbf{0}, \boldsymbol{\Sigma})$, since $\mathbf{x}_i'\mathbf{x}_i = 1$.

(b)

$$\begin{aligned} \text{cov}(\mathbf{Z}'\mathbf{x}_i, \mathbf{Z}'\mathbf{x}_j) &= \text{cov}\left(\sum_{k=1}^n x_{ik}\mathbf{z}_k, \sum_{k=1}^n x_{jk}\mathbf{z}_k\right) \\ &= \mathbf{O} \end{aligned}$$

by the result of Problem 2.29, since $\mathbf{x}_i'\mathbf{x}_j = 0$ (eigenvectors of a symmetric matrix).

2.31 By Problem 2.29,

$$\begin{aligned} \text{cov}(\mathbf{Z}'\mathbf{x}_i, \mathbf{Z}'\mathbf{u}_j) &= \text{cov}\left(\sum_{k=1}^n x_{ik}\mathbf{z}_k, \sum_{k=1}^n u_{ik}\mathbf{z}_k\right) \\ &= \sum_{k=1}^n x_{ik}u_{jk}\boldsymbol{\Sigma}. \end{aligned}$$

Show that $\sum_{k=1}^n x_{ik}u_{jk} = 0$ follows from $\mathbf{AB} = \mathbf{O}$ by expressing $\mathbf{A}$ and $\mathbf{B}$ in terms of their spectral decompositions (A.11.4).

2.32 By (1.40), $(n-1)\mathbf{S} = \mathbf{Y}'[\mathbf{I} - (1/n)\mathbf{J}]\mathbf{Y}$. Show that $\mathbf{I} - (1/n)\mathbf{J}$ is idempotent and use Theorem 2.3G.

2.33 **(a)** $\bar{\mathbf{y}}$ is $N_8(\boldsymbol{\mu}, \boldsymbol{\Sigma}/25)$.

(b) $24\mathbf{S}$ is $W_8(24, \boldsymbol{\Sigma})$.

(c) $24\mathbf{CSC}'$ is $W_3(24, \mathbf{C\Sigma C}')$.

2.34 $\mathbf{W}_1$ and $\mathbf{W}_2$ can be written in the form $\mathbf{W}_1 = \sum_{i=1}^{\nu_1}(\mathbf{y}_{1i} - \boldsymbol{\mu}_1)(\mathbf{y}_{1i} - \boldsymbol{\mu}_1)'$, $\mathbf{W}_2 = \sum_{i=1}^{\nu_2}(\mathbf{y}_{2i} - \boldsymbol{\mu}_2)(\mathbf{y}_{2i} - \boldsymbol{\mu}_2)'$, where $\mathbf{y}_{11}, \mathbf{y}_{12}, \ldots, \mathbf{y}_{1\nu_1}$ is a random sample from $N_p(\boldsymbol{\mu}_1, \boldsymbol{\Sigma})$; $\mathbf{y}_{21}, \mathbf{y}_{22}, \ldots, \mathbf{y}_{2\nu_2}$ is a random sample from $N_p(\boldsymbol{\mu}_2, \boldsymbol{\Sigma})$; and all the $\mathbf{y}_{1i}$'s are independent of the $\mathbf{y}_{2i}$'s. Then $\mathbf{W}_1 + \mathbf{W}_2 = \sum_{i=1}^{\nu_1}(\mathbf{y}_{1i} - \boldsymbol{\mu}_1)(\mathbf{y}_{1i} - \boldsymbol{\mu}_1)' + \sum_{i=1}^{\nu_2}(\mathbf{y}_{2i} - \boldsymbol{\mu}_2)(\mathbf{y}_{2i} - \boldsymbol{\mu}_2)'$, and the result follows by the definition in (2.59).

2.35 By the definition in (2.59), $\mathbf{W}$ can be written in the form $\mathbf{W} = \sum_{i=1}^{\nu}(\mathbf{y}_i - \boldsymbol{\mu})(\mathbf{y}_i - \boldsymbol{\mu})'$, where $\mathbf{y}_i$ is $N_p(\boldsymbol{\mu}, \boldsymbol{\Sigma})$. Now by Theorem 2.2B(ii), $\mathbf{v}_i = \mathbf{C}\mathbf{y}_i$ is $N_q(\mathbf{C}\boldsymbol{\mu}, \mathbf{C\Sigma C}')$. Then $\sum_{i=1}^{\nu}(\mathbf{v}_i - \mathbf{C}\boldsymbol{\mu})(\mathbf{v}_i - \mathbf{C}\boldsymbol{\mu})'$ is $\mathbf{W}_q(\nu, \mathbf{C\Sigma C}')$. Show that $\sum_{i=1}^{\nu}(\mathbf{v}_i - \mathbf{C}\boldsymbol{\mu})(\mathbf{v}_i - \mathbf{C}\boldsymbol{\mu})' = \mathbf{CWC}'$.

2.36 By Theorem 2.2F, $d^2 = (\mathbf{y} - \boldsymbol{\mu})'\boldsymbol{\Sigma}^{-1}(\mathbf{y} - \boldsymbol{\mu})$ is distributed as $\chi^2(p)$. Therefore, $E(d^2) = p$.

2.37 **(a)** Use the sum over the sample, $\sum_{i=1}^n u_{ii} = \sum_{i=1}^n (\mathbf{y}_i - \bar{\mathbf{y}})'\mathbf{S}^{-1}(\mathbf{y}_i - \bar{\mathbf{y}})$, which is equal to $\sum_i \text{tr}(\mathbf{y}_i - \bar{\mathbf{y}})'\mathbf{S}^{-1}(\mathbf{y}_i - \bar{\mathbf{y}}) = \sum_i \text{tr}[\mathbf{S}^{-1}(\mathbf{y}_i - \bar{\mathbf{y}})(\mathbf{y}_i - \bar{\mathbf{y}})'] = \text{tr}[\sum_i \mathbf{S}^{-1}(\mathbf{y}_i - \bar{\mathbf{y}})(\mathbf{y}_i - \bar{\mathbf{y}})'] = \text{tr}[\mathbf{S}^{-1}\sum_i(\mathbf{y}_i - \bar{\mathbf{y}})(\mathbf{y}_i - \bar{\mathbf{y}})'] = (n-1)\,\text{tr}[\mathbf{S}^{-1}\mathbf{S}] = (n-1)p$. Then $E[\sum_i(\mathbf{y}_i - \bar{\mathbf{y}})'\mathbf{S}^{-1}(\mathbf{y}_i - \bar{\mathbf{y}})] = nE(\mathbf{y}_i - \bar{\mathbf{y}})'\mathbf{S}^{-1}(\mathbf{y}_i - \bar{\mathbf{y}})$.

(b) Show that $\sum_{i=1}^{n} u_{ij} = 0$. In the hint to part (a), we have $\sum_{i=1}^{n} u_{ii} = (n-1)p$. Since $\mathbf{y}_1, \mathbf{y}_2, \ldots, \mathbf{y}_n$ arise from a random sample, $E(u_{ii}) = \alpha$, say, for all i, and $E(u_{ij}) = \beta$, say, for all $i \neq j$. Then $E(\sum_{i=1}^{n} u_{ij}) = (n-1)\beta + \alpha = 0$, and $E(\sum_{i=1}^{n} u_{ii}) = n\alpha = (n-1)p$.

2.38 Use Theorem 2.3K.

2.39 By Theorem 2.3K, $\mathbf{a}'\mathbf{W}\mathbf{a}$ is $W_1(\nu, \mathbf{a}\boldsymbol{\Sigma}\mathbf{a})$. By the definition in (2.59), a Wishart random variable with $p = 1$ must be divided by the variance in order to have a χ^2-distribution.

2.40 Use (1.45) and (2.19).

Chapter 3

3.1 The likelihood function is given by (2.45). Use (2.46) and (2.47) to show that $\text{LR} = e^{-Z^2/2}$. Thus a small value of LR corresponds to a large value of Z^2.

3.2 **(a)** $\bar{z} = \sum_{i=1}^{n} z_i/n = \sum_i (ay_i + b)/n = a\sum_i y_i/n + nb/n$.

(b) $s_z^2 = \sum_{i=1}^{n}(z_i - \bar{z})^2/(n-1) = \sum_i (ay_i + b - a\bar{y} - b)^2/(n-1) = \sum_i [a(y_i - \bar{y})]^2/(n-1)$.

3.3 Differentiate $\ln L$ with respect to μ and with respect to σ^2 and set the results equal to 0.

3.4 Differentiate $\ln L(\mu_0, \sigma^2)$ with respect to σ^2.

3.6 Use (2.56) in Theorem 2.3K, with $a_1 = \sqrt{n}$ and $a_2 = \cdots = a_n = 0$. Use Theorem 2.3I and Theorem 2.3D.

3.7 Using $\sqrt{n}\mathbf{c}$ in place of $\mathbf{c}$, show that (A.7.10) can be written in the form

$$|\mathbf{B} + n\mathbf{c}\mathbf{c}'| = |\mathbf{B}|(1 + n\mathbf{c}'\mathbf{B}^{-1}\mathbf{c}),$$

so that

$$1 + n\mathbf{c}'\mathbf{B}^{-1}\mathbf{c} = \frac{|\mathbf{B} + n\mathbf{c}\mathbf{c}'|}{|\mathbf{B}|}. \qquad (*)$$

Then write T^2 as

$$T^2 = n(\bar{\mathbf{y}} - \boldsymbol{\mu}_0)'\mathbf{S}^{-1}(\bar{\mathbf{y}} - \boldsymbol{\mu}_0) = 1 + n(\bar{\mathbf{y}} - \boldsymbol{\mu}_0)'\mathbf{S}^{-1}(\bar{\mathbf{y}} - \boldsymbol{\mu}_0) - 1$$

and apply $(*)$ to the first two terms on the right.

3.8 In partitioned form, the sample covariance matrix after adding x is

$$\mathbf{S} = \begin{pmatrix} \mathbf{S}_{yy} & \mathbf{s}_{xy} \\ \mathbf{s}'_{xy} & s_x^2 \end{pmatrix}.$$

By (A.5.5), the inverse of $\mathbf{S}$ can be written in the form

$$\mathbf{S}^{-1} = \frac{1}{s_{x\cdot y}^2}\begin{pmatrix} s_{x\cdot y}^2\mathbf{S}_{yy}^{-1} + \mathbf{S}_{yy}^{-1}\mathbf{s}_{xy}\mathbf{s}'_{xy}\mathbf{S}_{yy}^{-1} & -\mathbf{S}_{yy}^{-1}\mathbf{s}_{xy} \\ -\mathbf{s}'_{xy}\mathbf{S}_{yy}^{-1} & 1 \end{pmatrix},$$

where $s_{x\cdot y}^2 = s_x^2 - \mathbf{s}_{xy}'\mathbf{S}_{yy}^{-1}\mathbf{s}_{xy}$. With the addition of x to the p variables in $\mathbf{y}$, T^2 becomes

$$\begin{aligned} T_{y,x}^2 &= n\begin{pmatrix} \overline{\mathbf{y}} - \boldsymbol{\mu}_{0y} \\ \overline{x} - \mu_{0x} \end{pmatrix}' \mathbf{S}^{-1} \begin{pmatrix} \overline{\mathbf{y}} - \boldsymbol{\mu}_{0y} \\ \overline{x} - \mu_{0x} \end{pmatrix} \\ &= n(\overline{\mathbf{y}} - \boldsymbol{\mu}_{0y})'\mathbf{S}_{yy}^{-1}(\overline{\mathbf{y}} - \boldsymbol{\mu}_{0y}) + \frac{n}{s_{x\cdot y}^2}\left[\mathbf{s}_{xy}'\mathbf{S}_{yy}^{-1}(\overline{\mathbf{y}} - \boldsymbol{\mu}_{0y}) - (\overline{x} - \mu_{0x})\right]^2. \end{aligned}$$

3.9 From Problem 1.34, we have $\overline{\mathbf{z}} = \mathbf{A}\overline{\mathbf{y}} + \mathbf{b}$ and $\mathbf{S}_z = \mathbf{ASA}'$. Show that $\boldsymbol{\mu}_{0z} = \mathbf{A}\boldsymbol{\mu}_y + \mathbf{b}$. Substitute these in $T_z^2 = n(\overline{\mathbf{z}} - \boldsymbol{\mu}_{0z})'\mathbf{S}_z^{-1}(\overline{\mathbf{z}} - \boldsymbol{\mu}_{0z})$.

3.10 Show that the exponent in the likelihood function $L(\boldsymbol{\mu}_0, \boldsymbol{\Sigma})$ in the proof of Theorem 3.3D can be written as $-\frac{1}{2}\,\mathrm{tr}(\boldsymbol{\Sigma}^{-1}\mathbf{W}_0)$, where $\mathbf{W}_0 = \sum_{i=1}^n (\mathbf{y}_i - \boldsymbol{\mu}_0)(\mathbf{y}_i - \boldsymbol{\mu}_0)'$. Then follow the steps leading to (2.52) and (2.53) in the proof of Theorem 2.3A.

3.11 To see that the exponent reduces to $-np/2$, we use (A.8.1) and (A.8.2):

$$\begin{aligned} \sum_{i=1}^n (\mathbf{y}_i - \boldsymbol{\mu}_0)'\hat{\boldsymbol{\Sigma}}_0^{-1}(\mathbf{y}_i - \boldsymbol{\mu}_0) &= \sum_{i=1}^n \mathrm{tr}\left[(\mathbf{y}_i - \boldsymbol{\mu}_0)'\hat{\boldsymbol{\Sigma}}_0^{-1}(\mathbf{y}_i - \boldsymbol{\mu}_0)\right] \\ &= \sum_{i=1}^n \mathrm{tr}\left[\hat{\boldsymbol{\Sigma}}_0^{-1}(\mathbf{y}_i - \boldsymbol{\mu}_0)(\mathbf{y}_i - \boldsymbol{\mu}_0)'\right] \\ &= \mathrm{tr}\left[\hat{\boldsymbol{\Sigma}}_0^{-1}\sum_{i=1}^n (\mathbf{y}_i - \boldsymbol{\mu}_0)(\mathbf{y}_i - \boldsymbol{\mu}_0)'\right] \\ &= \mathrm{tr}\left[\hat{\boldsymbol{\Sigma}}_0^{-1} n\hat{\boldsymbol{\Sigma}}_0\right] = \mathrm{tr}(n\mathbf{I}) = np. \end{aligned}$$

3.12 See (*) in the hint to Problem 3.7.

3.14 Apply Theorem 3.3E with $\mathbf{S}_{\mathrm{pl}}$ in place of $\mathbf{S}$.

3.16 Let $\mathbf{C} = \mathbf{c}' = (1, -1)$ and let $\mathbf{y}_i = \binom{y_i}{x_i}$. Then $d_i = y_i - x_i = \mathbf{c}'\mathbf{y}_i$, and the hypothesis is H_0: $\mathbf{c}'\boldsymbol{\mu} = \mu_y - \mu_x = 0$. By (3.51), $T^2 = n(\mathbf{c}'\overline{\mathbf{y}})'(\mathbf{c}'\mathbf{Sc})^{-1}(\mathbf{c}'\overline{\mathbf{y}})$, where $\overline{\mathbf{y}}' = (\overline{y}, \overline{x})$ and

$$\mathbf{S} = \begin{pmatrix} s_y^2 & s_{yx} \\ s_{yx} & s_x^2 \end{pmatrix}.$$

Show that $\mathbf{c}'\overline{\mathbf{y}} = \overline{y} - \overline{x} = \overline{d}$ and $\mathbf{c}'\mathbf{Sc} = s_y^2 - 2s_{yx} + s_x^2 = s_d^2$. Then $T^2 = t^2 = n\overline{d}^2/s_d^2$.

3.17 As in Section 3.4.6, define the nonsingular matrix $\mathbf{A} = \binom{\mathbf{j}'}{\mathbf{C}}$, where $\mathbf{Cj} = \mathbf{0}$, and define $\mathbf{B} = (\mathbf{S}^{-1/2}\mathbf{j}, \mathbf{S}^{1/2}\mathbf{C}')$, which is nonsingular, since $(\mathbf{S}^{1/2}\mathbf{C}')'\mathbf{S}^{-1/2}\mathbf{j} = \mathbf{0}$. Substitute into the identity $\mathbf{I} = \mathbf{B}(\mathbf{B}'\mathbf{B})^{-1}\mathbf{B}'$ to obtain

$$\mathbf{I} = \mathbf{S}^{-1/2}\mathbf{G}\mathbf{S}^{-1/2} + \mathbf{S}^{1/2}\mathbf{C}'(\mathbf{CSC}')^{-1}\mathbf{C}\mathbf{S}^{1/2},$$

where $\mathbf{G} = \mathbf{jj}'/\mathbf{j}'\mathbf{S}^{-1}\mathbf{j}$ (note that $\mathbf{B}'\mathbf{B}$ simplifies to a block diagonal form because $\mathbf{Cj} = \mathbf{0}$). Multiply this on the left and on the right by $\mathbf{S}^{-1/2}$ to obtain $\mathbf{S}^{-1} = \mathbf{S}^{-1}\mathbf{G}\mathbf{S}^{-1} + \mathbf{C}'(\mathbf{CSC}')^{-1}\mathbf{C}$. Multiply on the left by $\overline{\mathbf{y}}'$ and on the right by $\overline{\mathbf{y}}$.

3.18 See (3.4) and the comments that follow (3.4).

3.19 Adapt the hint for Problem 3.24 to a univariate setting.

3.20 $E(\mathbf{S}_1) = E(\mathbf{S}_2) = \mathbf{\Sigma}$.

3.22 See the hint for Problem 3.7.

3.23 See the hint for Problem 3.8. Use $\mathbf{S} = \begin{pmatrix} \mathbf{S}_{\text{pl}} & \mathbf{s}_{yx} \\ \mathbf{s}'_{yx} & s_x^2 \end{pmatrix}$.

3.24 Assuming $\mathbf{\Sigma}_1 = \mathbf{\Sigma}_2 = \mathbf{\Sigma}$, the likelihood function is

$$L(\boldsymbol{\mu}_1, \boldsymbol{\mu}_2, \mathbf{\Sigma}) = \prod_{i=1}^{n_1} f(\mathbf{y}_{1i}; \boldsymbol{\mu}_1, \mathbf{\Sigma}) \prod_{i=1}^{n_2} f(\mathbf{y}_{2i}; \boldsymbol{\mu}_2, \mathbf{\Sigma})$$

$$= \frac{1}{(\sqrt{2\pi})^{(n_1+n_2)p} |\mathbf{\Sigma}|^{(n_1+n_2)/2}} e^{-\sum_{i=1}^{n_1} (\mathbf{y}_{1i}-\boldsymbol{\mu}_1)' \mathbf{\Sigma}^{-1} (\mathbf{y}_{1i}-\boldsymbol{\mu}_1)/2}$$

$$\times e^{-\sum_{i=1}^{n_2} (\mathbf{y}_{2i}-\boldsymbol{\mu}_2)' \mathbf{\Sigma}^{-1} (\mathbf{y}_{2i}-\boldsymbol{\mu}_2)/2}.$$

Under H_1, L is maximized (see Theorem 2.3A) by

$$\hat{\boldsymbol{\mu}}_1 = \overline{\mathbf{y}}_1, \qquad \hat{\boldsymbol{\mu}}_2 = \overline{\mathbf{y}}_2, \qquad \hat{\mathbf{\Sigma}} = (\mathbf{W}_1 + \mathbf{W}_2)/(n_1 + n_2),$$

where $\mathbf{W}_1 = \sum_{i=1}^{n_1} (\mathbf{y}_{1i} - \overline{\mathbf{y}}_1)(\mathbf{y}_{1i} - \overline{\mathbf{y}}_1)'$ and $\mathbf{W}_2 = \sum_{i=1}^{n_2} (\mathbf{y}_{2i} - \overline{\mathbf{y}}_2)(\mathbf{y}_{2i} - \overline{\mathbf{y}}_2)'$. Show that

$$\max_{H_1} L = \frac{(n_1 + n_2)^{(n_1+n_2)p/2}}{(\sqrt{2\pi})^{(n_1+n_2)p} |\mathbf{W}_1 + \mathbf{W}_2|^{(n_1+n_2)/2}} e^{-(n_1+n_2)p/2}.$$

Under H_0: $\boldsymbol{\mu}_1 = \boldsymbol{\mu}_2 = \boldsymbol{\mu}$, L is maximized by

$$\hat{\boldsymbol{\mu}} = \frac{n_1 \overline{\mathbf{y}}_1 + n_2 \overline{\mathbf{y}}_2}{n_1 + n_2},$$

$$\hat{\mathbf{\Sigma}} = \left[\sum_{i=1}^{n_1} (\mathbf{y}_{1i} - \hat{\boldsymbol{\mu}})(\mathbf{y}_{1i} - \hat{\boldsymbol{\mu}})' + \sum_{i=1}^{n_2} (\mathbf{y}_{2i} - \hat{\boldsymbol{\mu}})(\mathbf{y}_{2i} - \hat{\boldsymbol{\mu}})' \right] / (n_1 + n_2).$$

Show that this can be written [see (2.47)] as

$$\hat{\mathbf{\Sigma}} = \left[\mathbf{W}_1 + \mathbf{W}_2 + \frac{n_1 n_2}{n_1 + n_2} (\overline{\mathbf{y}}_1 - \overline{\mathbf{y}}_2)(\overline{\mathbf{y}}_1 - \overline{\mathbf{y}}_2)' \right] / (n_1 + n_2)$$

and that

$$\max_{H_0} L = \frac{(n_1 + n_2)^{(n_1+n_2)p/2}}{(\sqrt{2\pi})^{(n_1+n_2)p} |\mathbf{W}_1 + \mathbf{W}_2 + c(\overline{\mathbf{y}}_1 - \overline{\mathbf{y}}_2)(\overline{\mathbf{y}}_1 - \overline{\mathbf{y}}_2)'|^{(n_1+n_2)/2}} e^{-(n_1+n_2)p/2},$$

where $c = n_1 n_2/(n_1 + n_2)$. The likelihood ratio is then given by

$$\text{LR} = \frac{\max_{H_0} L}{\max_{H_1} L} = \frac{|\mathbf{W}_1 + \mathbf{W}_2|^{(n_1+n_2)/2}}{|\mathbf{W}_1 + \mathbf{W}_2 + c(\overline{\mathbf{y}}_1 - \overline{\mathbf{y}}_2)(\overline{\mathbf{y}}_1 - \overline{\mathbf{y}}_2)'|^{(n_1+n_2)/2}}.$$

Using (A.7.10), show that this can be written as

$$\mathrm{LR} = \frac{|\mathbf{S}_{\mathrm{pl}}|^{(n_1+n_2)/2}}{|\mathbf{S}_{\mathrm{pl}}|^{(n_1+n_2)/2}\left[1 + c(\bar{\mathbf{y}}_1 - \bar{\mathbf{y}}_2)'\mathbf{S}_{\mathrm{pl}}^{-1}(\bar{\mathbf{y}}_1 - \bar{\mathbf{y}}_2)/(n_1+n_2-2)\right]^{(n_1+n_2)/2}}$$

$$= \left[\frac{1}{1 + T^2/(n_1+n_2-2)}\right]^{(n_1+n_2)/2}.$$

3.25 See the proof of Theorem 3.3E.

3.29 Substitute $n_1 = n_2 = n$ in both (3.57) and (3.81).

3.30 With $N = n_1 + n_2$, we have $n_1 n_2/(n_1 + n_2) = n_1(N - n_1)/N$. Differentiate this with respect to n_1 and set the result equal to 0.

3.31 Write Δ_{p+q}^2 in (3.89) in partitioned form,

$$\Delta_{p+q}^2 = \boldsymbol{\delta}'\boldsymbol{\Sigma}^{-1}\boldsymbol{\delta} = (\boldsymbol{\delta}_y', \boldsymbol{\delta}_x')\begin{pmatrix}\boldsymbol{\Sigma}_{yy} & \boldsymbol{\Sigma}_{yx}\\ \boldsymbol{\Sigma}_{xy} & \boldsymbol{\Sigma}_{xx}\end{pmatrix}^{-1}\begin{pmatrix}\boldsymbol{\delta}_y\\ \boldsymbol{\delta}_x\end{pmatrix}.$$

Use (A.5.6) to show that this can be written as follows:

$$\Delta_{p+q}^2 = \boldsymbol{\delta}_{x\cdot y}'\boldsymbol{\Sigma}_{x\cdot y}^{-1}\boldsymbol{\delta}_{x\cdot y} + \boldsymbol{\delta}_y'\boldsymbol{\Sigma}_{yy}^{-1}\boldsymbol{\delta}_y,$$

where $\boldsymbol{\delta}_{x\cdot y} = \boldsymbol{\delta}_x - \boldsymbol{\Sigma}_{xy}\boldsymbol{\Sigma}_{yy}^{-1}\boldsymbol{\delta}_y$ and $\boldsymbol{\Sigma}_{x\cdot y} = \boldsymbol{\Sigma}_{xx} - \boldsymbol{\Sigma}_{xy}\boldsymbol{\Sigma}_{yy}^{-1}\boldsymbol{\Sigma}_{yx}$. By (3.90), $\Delta_p^2 = \boldsymbol{\delta}_y'\boldsymbol{\Sigma}_{yy}^{-1}\boldsymbol{\delta}_y$. Thus H_0: $\Delta_{p+q}^2 = \Delta_p^2$ implies $\boldsymbol{\delta}_{x\cdot y} = \mathbf{0}$ since $\boldsymbol{\Sigma}_{x\cdot y}^{-1}$ is positive definite.

3.32 Express (3.91) in terms of partitioned $\boldsymbol{\Sigma}$ and $\boldsymbol{\delta}$:

$$\boldsymbol{\alpha} = \begin{pmatrix}\boldsymbol{\alpha}_y\\ \boldsymbol{\alpha}_x\end{pmatrix} = \begin{pmatrix}\boldsymbol{\Sigma}_{yy} & \boldsymbol{\Sigma}_{yx}\\ \boldsymbol{\Sigma}_{xy} & \boldsymbol{\Sigma}_{xx}\end{pmatrix}^{-1}\begin{pmatrix}\boldsymbol{\delta}_y\\ \boldsymbol{\delta}_x\end{pmatrix}.$$

Use (A.5.6) to solve for $\boldsymbol{\alpha}_x$ and obtain the result.

3.33 This can be done as in the decomposition of Δ_{p+q}^2 in Problem 3.31 by use of (A.5.6). For an alternative approach, define

$$\mathbf{A} = \begin{pmatrix}\mathbf{I} & \mathbf{O}\\ -\mathbf{S}_{yx}\mathbf{S}_{xx}^{-1} & \mathbf{I}\end{pmatrix}, \text{ where } \mathbf{S}_{\mathrm{pl}} = \begin{pmatrix}\mathbf{S}_{yy} & \mathbf{S}_{yx}\\ \mathbf{S}_{xy} & \mathbf{S}_{xx}\end{pmatrix}.$$

Since $\mathbf{A}$ is nonsingular, we have $\hat{\boldsymbol{\delta}}'\mathbf{S}_{\mathrm{pl}}^{-1}\hat{\boldsymbol{\delta}} = (\mathbf{A}\hat{\boldsymbol{\delta}})'(\mathbf{A}\mathbf{S}_{\mathrm{pl}}\mathbf{A}')^{-1}\mathbf{A}\hat{\boldsymbol{\delta}}$, and expanding the right side leads to the result.

3.34 The following form of H_0 is implied by (3.94) and leads to the same likelihood ratio:

$$H_0\colon\ \boldsymbol{\delta}_x = \mathbf{0}, \text{ with } \boldsymbol{\delta}_y = \mathbf{0} \text{ given.}$$

Thus H_1 becomes H_1: $\boldsymbol{\delta}_x \neq \mathbf{0}$, with $\boldsymbol{\delta}_y = \mathbf{0}$. Define

$$\mathbf{W}_{\mathrm{pl}} = \mathbf{W}_1 + \mathbf{W}_2 = \begin{pmatrix}\mathbf{W}_{yy} & \mathbf{W}_{yx}\\ \mathbf{W}_{xy} & \mathbf{W}_{xx}\end{pmatrix}.$$

Then $|\mathbf{W}_{\text{pl}}| = |\mathbf{W}_{yy}||\mathbf{W}_{xx} - \mathbf{W}_{xy}\mathbf{W}_{yy}^{-1}\mathbf{W}_{yx}|$, and we substitute $\mathbf{W}_{yy} + c\hat{\boldsymbol{\delta}}_y\hat{\boldsymbol{\delta}}_y'$ for $\mathbf{W}_{yy}$ in order to incorporate $\boldsymbol{\delta}_y = \mathbf{0}$ for H_1 [see (2.47) with $\boldsymbol{\mu} = \mathbf{0}$], where $c = n_1n_2/(n_1 + n_2)$. Thus

$$\max_{H_1} L = \frac{(n_1 + n_2)^{(n_1+n_2)(p+q)/2}}{(\sqrt{2\pi})^{(n_1+n_2)(p+q)}\left(|\mathbf{W}_{yy} + c\hat{\boldsymbol{\delta}}_y\hat{\boldsymbol{\delta}}_y'||\mathbf{W}_{x\cdot y}|\right)^{(n_1+n_2)/2}} e^{-(n_1+n_2)(p+q)/2},$$

where $\mathbf{W}_{x\cdot y} = \mathbf{W}_{xx} - \mathbf{W}_{xy}\mathbf{W}_{yy}^{-1}\mathbf{W}_{yx}$. [For a more detailed development, see Siotani, Hayakawa, and Fujikoshi (1985, p. 216).] In H_0, we have $\boldsymbol{\delta}_y = \mathbf{0}$ and $\boldsymbol{\delta}_x = \mathbf{0}$, so that

$$\max_{H_0} L = \frac{(n_1 + n_2)^{(n_1+n_2)(p+q)/2}}{(\sqrt{2\pi})^{(n_1+n_2)(p+q)}|\mathbf{W}_{\text{pl}} + c\hat{\boldsymbol{\delta}}\hat{\boldsymbol{\delta}}'|^{(n_1+n_2)/2}} e^{-(n_1+n_2)(p+q)/2}.$$

Then

$$\text{LR} = \frac{|\mathbf{W}_{yy} + c\hat{\boldsymbol{\delta}}_y\hat{\boldsymbol{\delta}}_y'|^{(n_1+n_2)/2}|\mathbf{W}_{x\cdot y}|^{(n_1+n_2)/2}}{|\mathbf{W}_{\text{pl}} + c\hat{\boldsymbol{\delta}}\hat{\boldsymbol{\delta}}'|^{(n_1+n_2)/2}}.$$

Show that, by (A.7.10), this can be written in the form

$$\text{LR} = \frac{(1 + c\hat{\boldsymbol{\delta}}_y'\mathbf{S}_{yy}^{-1}\hat{\boldsymbol{\delta}}_y)^{(n_1+n_2)/2}}{(1 + c\hat{\boldsymbol{\delta}}'\mathbf{S}_{\text{pl}}^{-1}\hat{\boldsymbol{\delta}}_y)^{(n_1+n_2)/2}} = \frac{(1 + T_p^2/\nu)^{(n_1+n_2)/2}}{(1 + T_{p+q}^2/\nu)^{(n_1+n_2)/2}},$$

where $\nu = n_1 + n_2 - 2$. Show that

$$\frac{1}{(\text{LR})^{2/(n_1+n_2)}} - 1 = \frac{T_{p+q}^2 - T_p^2}{\nu + T_p^2}.$$

3.37

$$\bar{\mathbf{y}} = \begin{pmatrix} 7.6 \\ 5.6 \end{pmatrix}, \quad \mathbf{S}_y = \begin{pmatrix} 9.3 & 1.8 \\ 1.8 & 3.3 \end{pmatrix}, \quad T_y^2 = .8634, \quad \bar{\mathbf{z}} = \begin{pmatrix} 10.2 \\ 9.0 \end{pmatrix},$$

$$\mathbf{S}_z = \begin{pmatrix} 16.2 & 6.0 \\ 6.0 & 9.0 \end{pmatrix}, \quad T_z^2 = .8634$$

3.38 (a) $T^2 = 85.3327$.

	(b) Simultaneous		(c) Bonferroni		(d) Test
Mean	Lower	Upper	Lower	Upper	t_i
μ_1	21.38	50.80	28.38	43.80	2.504
μ_2	13.17	37.92	19.06	32.03	.267
μ_3	19.89	48.29	26.65	41.54	−2.516
μ_4	12.82	41.72	19.70	34.85	.951
μ_5	16.82	44.64	23.44	38.02	.316

For the Bonferroni tests in part (d), we reject if $|t_i| > t_{\alpha/2p,n-1} = t_{.05/10,10} = 3.1693$. There are no rejections.

(e) Since $T^2 = 85.3327 > T^2_{.05,5,10} = 36.561$, we reject H_0: $\boldsymbol{\mu} = \boldsymbol{\mu}_0$, and we can test H_{0i}: $\mu_i = \mu_{0i}$, $i = 1, 2, \ldots, 5$, using $t_{.025,10} = 2.228$. With this critical value, we reject H_{01}: $\mu_1 = 30$ and H_{03}: $\mu_3 = 40$, since $t_1 = 2.504 > 2.228$ and $|t_3| = 2.516 > 2.228$.

3.39 **(a)** We use a layout similar to that in Example 3.3.5.

x	$\hat{t}_x$	t_x	R^2	$T^2_{y,x} - T^2_y$
y_1	−1.990	2.504	.683	63.721
y_2	2.332	.267	.655	12.367
y_3	1.728	−2.516	.764	76.161
y_4	−.545	.951	.616	5.831
y_5	−3.027	.316	.588	27.131

The effect of y_1 and of y_3 is enhanced in each case because $\hat{t}_x$ and t_x are opposite in sign. The effect of y_5 is due almost entirely to the large (absolute) value of $\hat{t}_x$.

(b) Using (3.102) and (3.103), we obtain $T^2(y_4, y_5|y_1, y_2, y_3) = 3.7327$ and $F = 1.5997$.

(c) By (3.100) adapted for the one-sample case, $t^2(y_1|y_2, y_3, y_4, y_5) = 12.095$, $t^2(y_2|y_1, y_3, y_4, y_5) = .8944$, $t^2(y_3|y_1, y_2, y_4, y_5) = 23.835$, $t^2(y_4|y_1, y_2, y_3, y_5) = .3909$, $t^2(y_5|y_1, y_2, y_3, y_4) = 2.3868$.

3.40 **(a)** Using (3.51), where $\mathbf{C}$ is a full-rank contrast matrix as in Section 3.4.6, we obtain $T^2 = 30.286$.

(b) Using (3.51) with $\mathbf{C} = (\frac{1}{2}, \frac{1}{2}, -\frac{1}{3}, -\frac{1}{3}, -\frac{1}{3})$, we obtain $t^2 = .00802$.

(c) $t^2 = 15.734$, **(d)**$T^2 = 28.524$.

3.41 **(a)** $T^2 = 133.487$.

(b) $\mathbf{a} = (.345, -.130, -.106, -.143)'$.

(c) By (3.100), $T^2(y_3, y_4|y_1, y_2) = 12.521$.

(d) By (3.100), $t^2(y_1|y_2, y_3, y_4) = 35.934$. $t^2(y_2|y_1, y_3, y_4) = 5.799$. $t^2(y_3|y_1, y_2, y_4) = 1.775$, $t^2(y_4|y_1, y_2, y_3) = 8.259$.

3.42

x	$\hat{t}_x$	t_x	R^2	$T^2_{y,x} - T^2_y$
y_1	−2.896	3.888	.475	87.601
y_2	−.682	−3.865	.592	24.843
y_3	−3.350	−5.691	.352	8.459
y_4	.120	−5.043	.200	33.320

The large effect of y_1 is due to a large t_x and the opposite sign of $\hat{t}_x$. The effect of y_4 is due to a large (absolute) value of t_x. Even though y_3 has the largest (absolute) value of t_x, its effect on T^2 is the least because $\hat{t}_x$ is fairly close to t_x.

3.43 $T^{*2} = 135.288$.

(a) James: $A = 1.054$, $B = .0141$, the critical value is $(\chi^2_{.05,4})(A + B\chi^2_{.05,4}) = 11.268$.

(b) Yao: $\nu = 35.433$.

(c) Johansen: $D = .27713$, $C = 4.4355$, test statistic $= 136.155$, $\nu = 28.87$.

(d) Nel and van der Merwe: $\nu = 36.323$.

(e) Kim: $b = .1052$, $c = 3.8595$, $\nu = 35.433$, $\nu - p - 1 = 32.433$, $F = 27.243$, p-value $= 7.6 \times 10^{-10}$.

3.44

	(a) Simultaneous		(b) Bonferroni	
Parameter	Lower	Upper	Lower	Upper
$\mu_{11} - \mu_{21}$	1.887	27.960	4.845	25.002
$\mu_{12} - \mu_{22}$	−44.614	−2.881	−39.879	−7.615
$\mu_{13} - \mu_{23}$	−31.666	−7.997	−28.981	−10.682
$\mu_{14} - \mu_{24}$	−38.997	−7.608	−35.436	−11.169

3.45 **(a)** $T^2 = 132.686$, **(b)** $T^2 = 107.054$.

3.46 **(a)** $T^2 = 142.122$.

(b) By (3.102), $T^2(d_1|d_2, d_3) = .3416$, $T^2(d_2|d_1, d_3) = 1.627$, $T^2(d_3|d_1, d_2) = 22.858$.

(c) $t_1 = -5.731$, $t_2 = -5.998$, $t_3 = -10.547$.

(d) $\mathbf{a} = (-.0157, -.0420, -.1232)'$.

3.47 For $n = 11$, power $= .804$; for $n = 14$, power $= .905096$.

3.48 For $n = 14$, power $= .816$; for $n = 17$, power $= .905076$.

3.49 $T^2 = 117.138$ (based on ranks), $v = 28.878$, which exceeds $\chi^2_{.05,4} = 9.488$.

Chapter 4

4.1 Multiply (4.2) and (4.4) to obtain four terms and substitute $\bar{\mathbf{y}}_{i.} = \mathbf{y}_{i.}/n$ and $\bar{\mathbf{y}}_{..} = \mathbf{y}_{..}/kn$.

4.3 Using the forms of $\mathbf{H}$ and $\mathbf{E}$ in (4.3) and (4.5), we obtain $\mathbf{E} + \mathbf{H} = \sum_{ij} \mathbf{y}_{ij}\mathbf{y}_{ij}' - \mathbf{y}_{..}\mathbf{y}_{..}'$, which leads to the desired result.

4.6 By (4.3), $\mathbf{H} = \sum_{i=1}^k (1/n)\mathbf{y}_{i.}\mathbf{y}_{i.}' - (1/kn)\mathbf{y}_{..}\mathbf{y}_{..}'$. Show that $\mathbf{y}_{i.} = \mathbf{Y}_i'\mathbf{j}_n$, where $\mathbf{j}_n$ is an n-vector of 1s. Then $\mathbf{y}_{i.}\mathbf{y}_{i.}' = \mathbf{Y}_i'\mathbf{j}_n\mathbf{j}_n'\mathbf{Y}_i = \mathbf{Y}_i'\mathbf{J}_n\mathbf{Y}_i$. Similarly, $\mathbf{y}_{..} = \mathbf{Y}'\mathbf{j}_{kn}$, and $\mathbf{y}_{..}\mathbf{y}_{..}' = \mathbf{Y}'\mathbf{J}_{kn}\mathbf{Y}$. Show that $\sum_{i=1}^k \mathbf{y}_{i.}\mathbf{y}_{i.}' = \sum_{i=1}^k \mathbf{Y}_i'\mathbf{J}_n\mathbf{Y}_i = \mathbf{Y}'\mathbf{G}\mathbf{Y}$.

4.7 $\mathbf{A}^2 = (1/n^2)\mathbf{G}^2 + (1/k^2n^2)\mathbf{J}_{kn}^2 - (1/kn^2)\mathbf{G}\mathbf{J}_{kn} - (1/kn^2)\mathbf{J}_{kn}\mathbf{G}$. Now $\mathbf{G}^2 = n\mathbf{G}$, since $\mathbf{J}_n^2 = \mathbf{j}_n\mathbf{j}_n'\mathbf{j}_n\mathbf{j}_n' = n\mathbf{j}_n\mathbf{j}_n' = n\mathbf{J}_n$. Show that $(1/k^2n^2)\mathbf{J}_{kn}^2 = (1/kn^2)\mathbf{G}\mathbf{J}_{kn} = (1/kn^2)\mathbf{J}_{kn}\mathbf{G} = (1/kn)\mathbf{J}_{kn}$.

4.8 Show that $\mathbf{H}_z = \mathbf{F}\mathbf{H}_y\mathbf{F}'$ and $\mathbf{E}_z = \mathbf{F}\mathbf{E}_y\mathbf{F}'$.

4.9 Use (A.7.5) and (A.10.8):

$$\Lambda = \frac{|\mathbf{E}|}{|\mathbf{E} + \mathbf{H}|} = \frac{|\mathbf{E}^{-1}||\mathbf{E}|}{|\mathbf{E}^{-1}||\mathbf{E} + \mathbf{H}|} = \frac{|\mathbf{E}^{-1}\mathbf{E}|}{|\mathbf{E}^{-1}(\mathbf{E} + \mathbf{H})|}$$

$$= \frac{1}{|\mathbf{I} + \mathbf{E}^{-1}\mathbf{H}|}$$

$$= \frac{1}{\prod_{i=1}^{p}(1+\lambda_i)} = \prod_{i=1}^{p} \frac{1}{1+\lambda_i}.$$

4.10 See (A.10.9).

4.11 **(c)** See (4.20) and a remark following (A.10.3).

4.14 **(a)** Replace $\mathbf{E}$ by $\mathbf{U}'\mathbf{U}$ in (4.20) and multiply on the left by $(\mathbf{U}')^{-1}$ to obtain $[(\mathbf{U}')^{-1}\mathbf{H} - \lambda\mathbf{U}]\mathbf{a} = \mathbf{0}$. Insert $\mathbf{U}^{-1}\mathbf{U}$ preceding $\mathbf{a}$ to obtain $[(\mathbf{U}')^{-1}\mathbf{H}\mathbf{U}^{-1} - \lambda\mathbf{I}]\mathbf{U}\mathbf{a} = \mathbf{0}$. **(b)** For the relationship between $(\mathbf{U}')^{-1}\mathbf{H}\mathbf{a}$ and $\mathbf{U}\mathbf{a}$, consider $[(\mathbf{U}')^{-1}\mathbf{H} - \lambda\mathbf{U}]\mathbf{a} = \mathbf{0}$ in part (a) above and property 1 in Section A.10.

4.15 Show that $(2N+s+1)/(2m+s+1)$ in F_1 can be written as $(\nu_E - p + s)/[|\nu_H - p| + \min(\nu_H, p)]$ and that the denominator of F_1 is equal to $d = \max(p, \nu_H)$.

4.16 Using an adaptation of (2.47), write $\mathbf{H}$ as

$$\mathbf{H} = n\sum_{i=1}^{k}(\overline{\mathbf{y}}_{i.} - \boldsymbol{\mu})(\overline{\mathbf{y}}_{i.} - \boldsymbol{\mu})' - kn(\overline{\mathbf{y}}_{..} - \boldsymbol{\mu})(\overline{\mathbf{y}}_{..} - \boldsymbol{\mu})',$$

where $\boldsymbol{\mu}$ is the common value of $\boldsymbol{\mu}_1, \boldsymbol{\mu}_2, \ldots, \boldsymbol{\mu}_k$ under H_0. Then show that

$$\operatorname{tr}(\mathbf{E}^{-1}\mathbf{H}) = \frac{n}{\nu_E}\sum_{i=1}^{k}(\overline{\mathbf{y}}_{i.} - \boldsymbol{\mu})'\mathbf{S}_{\text{pl}}^{-1}(\overline{\mathbf{y}}_{i.} - \boldsymbol{\mu}) - \frac{kn}{\nu_E}(\overline{\mathbf{y}}_{..} - \boldsymbol{\mu})'\mathbf{S}_{\text{pl}}^{-1}(\overline{\mathbf{y}}_{..} - \boldsymbol{\mu}),$$

where $\mathbf{S}_{\text{pl}} = \mathbf{E}/\nu_E$. Write the terms on the right side in terms of T^2-statistics.

4.17 In the hint for Problem 4.8, we have $\mathbf{H}_z = \mathbf{F}\mathbf{H}\mathbf{F}'$ and $\mathbf{E}_z = \mathbf{F}\mathbf{E}\mathbf{F}'$ for the transformation $\mathbf{z}_{ij} = \mathbf{F}\mathbf{y}_{ij} + \mathbf{g}$, where $\mathbf{F}$ is nonsingular. To show that $|(\mathbf{F}\mathbf{E}\mathbf{F}')^{-1}(\mathbf{F}\mathbf{H}\mathbf{F}') - \lambda\mathbf{I}| = |\mathbf{E}^{-1}\mathbf{H} - \lambda\mathbf{I}|$, replace $\mathbf{I}$ by $(\mathbf{F}')^{-1}\mathbf{F}'$ to obtain $|(\mathbf{F}')^{-1}\mathbf{E}^{-1}\mathbf{H}\mathbf{F}' - \lambda(\mathbf{F}')^{-1}\mathbf{F}'| = |(\mathbf{F}')^{-1}[\mathbf{E}^{-1}\mathbf{H} - \lambda\mathbf{I}]\mathbf{F}'|$. Now use (A.7.5).

4.18 We have a random sample of size n_i from each of $N_p(\boldsymbol{\mu}_i, \boldsymbol{\Sigma}_i)$, $i = 1, 2, \ldots, k$. By (2.48), the likelihood function is

$$L(\boldsymbol{\mu}_1, \ldots, \boldsymbol{\mu}_k, \boldsymbol{\Sigma}_1, \ldots, \boldsymbol{\Sigma}_k) = \prod_{i=1}^{k} L(\boldsymbol{\mu}_i, \boldsymbol{\Sigma}_i)$$

$$= \frac{1}{(\sqrt{2\pi})^{Np}} \prod_{i=1}^{k} \frac{1}{|\boldsymbol{\Sigma}_i|^{n_i/2}} e^{-\frac{1}{2}\operatorname{tr}\boldsymbol{\Sigma}_i^{-1}[\mathbf{W}_i + n_i(\overline{\mathbf{y}}_i - \boldsymbol{\mu}_i)(\overline{\mathbf{y}}_i - \boldsymbol{\mu}_i)']},$$

where $N = \sum_{i=1}^{k} n_i$ and $\mathbf{W}_i$ is defined below (4.42). Show that with the maximum likelihood estimators $\hat{\boldsymbol{\mu}}_i = \overline{\mathbf{y}}_i$ and $\hat{\boldsymbol{\Sigma}}_i = \mathbf{W}_i/n_i$, we obtain

$$\max_{H_1} L = \frac{1}{(\sqrt{2\pi})^{Np}} \prod_{i=1}^{k} \frac{1}{|\hat{\boldsymbol{\Sigma}}_i|^{n_i/2}} e^{-pn_i/2}.$$

Show that under H_0 we have

$$\max_{H_0} L = \frac{1}{(\sqrt{2\pi})^{Np}|\hat{\boldsymbol{\Sigma}}_{\text{pl}}|^{N/2}} e^{-Np/2},$$

where $\hat{\boldsymbol{\Sigma}}_{\text{pl}} = (\mathbf{W}_1 + \mathbf{W}_2 + \cdots + \mathbf{W}_k)/N$. From these, we obtain LR in (4.42).

4.19 If $\nu_i < p$, what is $|\mathbf{S}_i|$?

4.22 Show that $\mathbf{z} = (\sum_{i=1}^k c_i^2/n)^{-1/2}\hat{\boldsymbol{\delta}}$ is $N_p(\mathbf{0}, \boldsymbol{\Sigma})$ when H_0: $\boldsymbol{\delta} = \mathbf{0}$ is true. By property 1 in Section 4.1.2, $\mathbf{E}$ is $W_p(\nu_E, \boldsymbol{\Sigma})$. Argue that $\mathbf{z}$ and $\mathbf{E}$ are independent. Thus $\mathbf{z}'(\mathbf{E}/\nu_E)^{-1}\mathbf{z}$ satisfies (3.16) and its attendant assumptions. Show that $\mathbf{z}'(\mathbf{E}/\nu_E)^{-1}\mathbf{z} = T^2$ in (4.64).

4.24 See Section 3.4.4.

4.26 See the hint for Problem 4.22.

4.27 See (A.3.1).

4.29 By (1.94), $\text{cov}(\hat{\boldsymbol{\mu}}) = \mathbf{A}\,\text{cov}(\mathbf{y})\mathbf{A}'$, where $\text{cov}(\mathbf{y}) = \sigma^2\mathbf{I}$ and $\mathbf{A} = (\mathbf{W}'\mathbf{W})^{-1}\mathbf{W}'$ from (4.94).

4.30 Substitute $\hat{\boldsymbol{\mu}} = (\mathbf{W}'\mathbf{W})^{-1}\mathbf{W}'\mathbf{y}$ from (4.94) into $\text{SSE} = (\mathbf{y} - \mathbf{W}\hat{\boldsymbol{\mu}})'(\mathbf{y} - \mathbf{W}\hat{\boldsymbol{\mu}})$ in (4.96) to obtain $\text{SSE} = \mathbf{y}'[\mathbf{I} - \mathbf{W}(\mathbf{W}'\mathbf{W})^{-1}\mathbf{W}']\mathbf{y}$ in (4.97).

4.33 $\mathbf{a}'(\mathbf{W}'\mathbf{W})^{-1}\mathbf{a} = 3.833$.

4.34 $\mathbf{B}(\mathbf{W}'\mathbf{W})^{-1}\mathbf{B}' = \frac{1}{3}\begin{pmatrix} 25 & -1 \\ -1 & 7 \end{pmatrix}$.

4.36 Use (1.93).

4.37 To demonstrate the equivalence of H_{01}, H_{02}, and H_{03} in Section 4.9.1, first use (A.5.6) to show that

$$\text{tr}(\boldsymbol{\Sigma}^{-1}\boldsymbol{\Omega}) = \text{tr}(\boldsymbol{\Sigma}_{yy}^{-1}\boldsymbol{\Omega}_{yy}) + \text{tr}[\boldsymbol{\Sigma}_{x\cdot y}^{-1}(-\mathbf{B}, \mathbf{I})\boldsymbol{\Omega}(-\mathbf{B}, \mathbf{I})'],$$

where $\boldsymbol{\Sigma}_{x\cdot y} = \boldsymbol{\Sigma}_{xx} - \boldsymbol{\Sigma}_{xy}\boldsymbol{\Sigma}_{yy}^{-1}\boldsymbol{\Sigma}_{yx}$, $\mathbf{B} = \boldsymbol{\Sigma}_{xy}\boldsymbol{\Sigma}_{yy}^{-1}$, and $\mathbf{I}$ is $q \times q$. By (A.2.28) and (A.3.8), $\boldsymbol{\Omega} = \sum_{i=1}^k n_i(\boldsymbol{\mu}_i - \boldsymbol{\mu})(\boldsymbol{\mu}_i - \boldsymbol{\mu})'$ can be factored into $\boldsymbol{\Omega} = \mathbf{UDU}'$, where $\mathbf{U} = (\boldsymbol{\mu}_1 - \boldsymbol{\mu}, \ldots, \boldsymbol{\mu}_k - \boldsymbol{\mu})$ and $\mathbf{D} = \text{diag}(n_1, n_2, \ldots, n_k)$. Thus $(-\mathbf{B}, \mathbf{I})\boldsymbol{\Omega}(-\mathbf{B}, \mathbf{I})' = (-\mathbf{B}, \mathbf{I})\mathbf{UDU}'(-\mathbf{B}, \mathbf{I})'$, and it follows that $(-\mathbf{B}, \mathbf{I})\mathbf{U} = \mathbf{O}$ (which is equivalent to H_0: $\boldsymbol{\mu}_1 = \boldsymbol{\mu}_2 = \cdots = \boldsymbol{\mu}_k$) implies both H_{02} and H_{03}. Now suppose H_{01} holds. Let $\gamma_1, \ldots, \gamma_s$ be the eigenvalues of $\boldsymbol{\Sigma}^{-1}\boldsymbol{\Omega}$. Show that $\boldsymbol{\Omega}\boldsymbol{\alpha} = \gamma\boldsymbol{\Sigma}\boldsymbol{\alpha}$ reduces to $\boldsymbol{\Omega}_{yy}\boldsymbol{\alpha}_y = \gamma\boldsymbol{\Sigma}_{yy}\boldsymbol{\alpha}_y$ under H_{01} and that this leads to H_{03}.

4.39 See the proof of Theorem 4.9A.

4.40

	Source	Λ	$V^{(s)}$	$U^{(s)}$	θ	Significant
(a)	Treatments	.0316	1.928	7.645	.822	yes
(c)	Contrasts					
	1 vs. 2, 3, 4	.2471	.753	3.047	.753	yes
	2 vs. 3, 4	.5231	.477	.912	.477	yes
	3 vs. 4	.2157	.784	3.635	.784	yes

(b) The eigenvalues of $\mathbf{E}^{-1}\mathbf{H}$ are 4.6165, 2.3476, and .6813. The essential dimensionality of the mean vectors is 2.

(d)

Variable	y_1	y_2	y_3	y_4	y_5	y_6
F	13.50	14.13	6.25	23.89	8.86	3.05

All are significant at the .05 level.

4.41 **(a)** By (4.134), $\Lambda(y_1, y_2 | y_3, y_4, y_5, y_6) = \Lambda(y_1, \ldots, y_6)/\Lambda(y_3, \ldots, y_6) = .03163/.06212 = .5092$.

(b) By (4.135), $\Lambda(y_1 | y_2, \ldots, y_6) = \Lambda(y_1, \ldots, y_6)/\Lambda(y_2, \ldots, y_6) = .03163/.04650 = .6802$.

$$\begin{aligned}
\Lambda(y_2 | y_1, y_3, \ldots, y_6) &= .03163/.03399 = .9308, \\
\Lambda(y_3 | y_1, y_2, y_4, y_5, y_6) &= .03163/.05173 = .6115, \\
\Lambda(y_4 | y_1, y_2, y_3, y_5, y_6) &= .03163/.1127 = .2808, \\
\Lambda(y_5 | y_1, \ldots, y_4, y_6) &= .03163/.05435 = .5821, \\
\Lambda(y_6 | y_1, \ldots, y_5) &= .03163/.05062 = .6250.
\end{aligned}$$

(c) The values of F_x are given above in Problem 4.40(d). The values of $\Lambda_{y,x}$ and Λ_y are given above in part (b) of this problem.

Variable	y_1	y_2	y_3	y_4	y_5	y_6
R_e^2	.7310	.7151	.8271	.7761	.5646	.7922
R_{e+h}^2	.8088	.8547	.8117	.7281	.5699	.7355

We illustrate (4.39),

$$\Lambda_{y,x} = \Lambda_y \frac{1 - R_e^2}{(1 + cF_x)(1 - R_{e+h}^2)}, \quad \text{where } c = \frac{\nu_H}{\nu_E} = \frac{3}{38} = .07895.$$

For y_1, F_x is relatively large, but $R_{e+h}^2 > R_e^2$, and Λ reduces from .04650 to .03163:

$$\begin{aligned}
\Lambda_{y,1} = .03163 &= (.04650)\frac{1 - .7310}{[1 + (.07895)(13.50)](1 - .8088)} \\
&= (.04650)\frac{.269}{(2.0658)(.1912)}.
\end{aligned}$$

For y_4, F_x is large and $R_e^2 > R_{e+h}^2$, and therefore Λ reduces from .1127 to .03163:

$$\begin{aligned}
\Lambda_{y,4} = .03163 &= (.1127)\frac{1 - .7761}{[1 + (.07895)(23.89)](1 - .7281)} \\
&= (.1127)\frac{.2239}{(2.8861)(.2719)}.
\end{aligned}$$

For y_6, F_x is relatively small, but $R_e^2 > R_{e+h}^2$, and Λ reduces from .05062 to .03163:

$$\begin{aligned}
\Lambda_{y,6} = .03163 &= (.05062)\frac{1 - .7922}{[1 + (.07895)(3.05)](1 - .7355)} \\
&= (.05062)\frac{.2078}{(1.2408)(.2645)}.
\end{aligned}$$

(d) $c_1 = .2796$, $c_2 = .09778$, $a_1 = 63$, $a_2 = 3321.057$, $b_1 = .01113$, $b_2 = .0002167$, $u = 78.481$, $F = 1.213$.

4.42

Source	Λ	$V^{(s)}$	$U^{(s)}$	θ	Significant
Treatments	.0154	2.160	10.350	.847	yes
Blocks	.4049	.739	1.129	.407	no
Interaction	.1658	1.471	2.249	.460	no

4.43 By (4.165), $\Lambda_1 = .01407/.1526 = .09222$. By (4.167), $\Lambda_2 = .2528$. By (4.170), $\Lambda_3 = .3104$.

4.44 For treatments,

$$\Lambda_1 = \frac{|\mathbf{E}|/|\mathbf{E}+\mathbf{H}|}{|\mathbf{E}_{xx}|/|\mathbf{E}_{xx}+\mathbf{H}_{xx}|} = \frac{8.1129 \times 10^{39}/1.385 \times 10^{42}}{1.9994 \times 10^{7}/1.9303 \times 10^{8}} = .05655,$$

$$\Lambda_2 = \frac{4.0576 \times 10^{32}}{3.0251 \times 10^{33}} = .13413 \quad \text{(same as for blocks and interaction)},$$

$$\Lambda_3 = \frac{9.2659 \times 10^{31}}{4.0576 \times 10^{32}} = .22836.$$

For blocks,

$$\Lambda_1 = \frac{8.1129 \times 10^{39}/2.1489 \times 10^{40}}{1.9994 \times 10^{7}/2.4852 \times 10^{7}} = .46926,$$

$$\Lambda_2 = .13413 \quad \text{(same as for treatments and interaction)},$$

$$\Lambda_3 = \frac{5.2692 \times 10^{31}}{4.0576 \times 10^{32}} = .12986.$$

For interaction,

$$\Lambda_1 = \frac{8.1129 \times 10^{39}/1.8489 \times 10^{41}}{1.9994 \times 10^{7}/3.5708 \times 10^{7}} = .07836,$$

$$\Lambda_2 = .13413 \quad \text{(same as for treatments and blocks)}.$$

Λ_3 was not computed for interaction. Some of the error matrices were nearly singular because of small sample sizes.

4.45 By (4.165), $\Lambda_1 = .05226/.6595 = .07922$. By (4.167), $\Lambda_2 = .1323$. By (4.170), $\Lambda_3 = .04904$.

4.46

	Source	Λ	$V^{(s)}$	$U^{(s)}$	θ	Significant
(a)	Group	.0689	1.460	6.702	.870	yes
(c)	Contrasts					
	1, 2 vs. 3, 4	.1625	.837	5.153	.837	yes
	1 vs. 2	.4818	.518	1.076	.518	yes
	3 vs. 4	.5603	.440	.785	.440	yes

(b) The eigenvalues of $\mathbf{E}^{-1}\mathbf{H}$ are 5.6805, .8412, and .1806. The essential dimensionality of the mean vectors is 2.

(d)

Variable	y_1	y_2	y_3	y_4	y_5	y_6
F	.55	12.40	4.61	6.09	3.22	5.47

Variable	y_7	y_8	y_9	y_{10}	y_{11}
F	4.52	1.61	2.55	1.11	2.43

The F-values for y_2, y_3, y_4, y_5, y_6, and y_7 are significant at the .05 level.

(e) By (4.134), $\Lambda(y_1, \ldots, y_5 | y_6, \ldots, y_{11}) = .06886/.4201 = .1639$.

4.47 $c_1 = .1591$, $c_2 = .02837$, $a_1 = 6$, $a_2 = 2605.4$, $u = 27.66$, $F = 4.597$.

Chapter 5

5.1 First show that the ratio (5.1) is unchanged if $\mathbf{a}$ is replaced by $c\mathbf{a}$. Thus $\mathbf{a}$ is unique only up to multiplication by a scalar. Now differentiate (5.1) with respect to $\mathbf{a}$, set the result equal to $\mathbf{0}$, and show that it can be written in the form

$$\left[\frac{1}{\lambda}(\overline{\mathbf{y}}_1 - \overline{\mathbf{y}}_2)'\mathbf{a}\right] \mathbf{S}_{\text{pl}}^{-1}(\overline{\mathbf{y}}_1 - \overline{\mathbf{y}}_2) = \mathbf{a},$$

where λ is the maximum value of the original ratio in (5.1). Since $(\overline{\mathbf{y}}_1 - \overline{\mathbf{y}}_2)'\mathbf{a}$ is a scalar, $\mathbf{a}$ is proportional to $\mathbf{S}_{\text{pl}}^{-1}(\overline{\mathbf{y}}_1 - \overline{\mathbf{y}}_2)$.

5.2 The coefficient vector is given by $\mathbf{b} = \mathbf{S}_u^{-1}(\overline{\mathbf{u}}_1 - \overline{\mathbf{u}}_2)$, which is easily shown to be equal to $(\mathbf{A}\mathbf{S}_{\text{pl}}\mathbf{A}')^{-1}(\mathbf{A}\overline{\mathbf{y}}_1 - \mathbf{A}\overline{\mathbf{y}}_2)$. Then $\mathbf{b}'\mathbf{u}$ readily reduces to $\mathbf{a}'\mathbf{y}$.

5.3 See Section 4.1.3.

5.4 Multiply $(\mathbf{H} - \lambda_1\mathbf{E})\mathbf{a}_1 = \mathbf{0}$ by $\mathbf{a}_2'$.

5.5 This follows from the result in Problem 5.4.

5.6 Show that for $\mathbf{u} = \mathbf{A}\mathbf{y}$, we have $\mathbf{H}_u = \mathbf{A}\mathbf{H}\mathbf{A}'$ and $\mathbf{E}_u = \mathbf{A}\mathbf{E}\mathbf{A}'$. Then show that the eigenvalues of $\mathbf{E}_u^{-1}\mathbf{H}_u$ are the same as those of $\mathbf{E}^{-1}\mathbf{H}$ and that the ith eigenvector of $\mathbf{E}_u^{-1}\mathbf{H}_u$ can be expressed as $\mathbf{b}_i = (\mathbf{A}^{-1})'\mathbf{a}_i$, where $\mathbf{a}_i$ is the ith eigenvector of $\mathbf{E}^{-1}\mathbf{H}$. It now follows immediately that $z = \mathbf{b}_i'\mathbf{u} = \mathbf{a}_i'\mathbf{y}$.

5.7 Denote an observation vector by $(\mathbf{y}', u)'$ and let the mean vectors and pooled sample covariance matrix be partitioned correspondingly as

$$\begin{pmatrix} \overline{\mathbf{y}}_1 \\ \overline{u}_1 \end{pmatrix}, \quad \begin{pmatrix} \overline{\mathbf{y}}_2 \\ \overline{u}_2 \end{pmatrix}, \quad \mathbf{S}_{\text{pl}} = \begin{pmatrix} \mathbf{S}_{yy} & \mathbf{s}_{uy} \\ \mathbf{s}_{uy}' & s_u^2 \end{pmatrix}.$$

The coefficients in the discriminant function $\mathbf{a}_y'\mathbf{y} + a_u u$ are given by (5.2) as

$$\begin{pmatrix} \mathbf{a}_y \\ a_u \end{pmatrix} = \mathbf{S}_{\text{pl}}^{-1}\begin{pmatrix} \overline{\mathbf{y}}_1 - \overline{\mathbf{y}}_2 \\ \overline{u}_1 - \overline{u}_2 \end{pmatrix}.$$

Using (A.5.6), show that this becomes

$$\begin{pmatrix} \mathbf{a}_y \\ a_u \end{pmatrix} = \frac{1}{s_{u\cdot y}^2}\begin{pmatrix} \mathbf{G}\mathbf{S}_{yy}^{-1}(\overline{\mathbf{y}}_1 - \overline{\mathbf{y}}_2) \\ \overline{u}_1 - \overline{u}_2 - \mathbf{s}_{uy}'\mathbf{S}_{yy}^{-1}(\overline{\mathbf{y}}_1 - \overline{\mathbf{y}}_2) \end{pmatrix},$$

where $s_{u\cdot y}^2 = s_u^2 - \mathbf{s}_{uy}'\mathbf{S}_{yy}^{-1}\mathbf{s}_{uy}$ and $\mathbf{G} = (s_{u\cdot y}^2 - \overline{u}_1 + \overline{u}_2)\mathbf{I} + \mathbf{S}_{yy}^{-1}\mathbf{s}_{uy}\mathbf{s}_{uy}'$. Then, with the definitions of $\hat{\boldsymbol{\beta}}$ and R^2 in Theorem 5.4A, we have

$$a_u = \frac{\overline{u}_1 - \overline{u}_2 - \hat{\boldsymbol{\beta}}'(\overline{\mathbf{y}}_1 - \overline{\mathbf{y}}_2)}{s_u^2(1 - R^2)},$$

which, with (5.10), leads to (5.11).

5.8 The invariance of sample discriminant functions in (5.3) applies to population discriminant functions as well. Let $\mathbf{u} = \mathbf{A}\mathbf{y}$, where $\mathbf{A}$ is nonsingular. Then $\boldsymbol{\beta}'\mathbf{u} = \boldsymbol{\alpha}'\mathbf{y}$, where $\boldsymbol{\beta} = \boldsymbol{\Sigma}_u^{-1}(\boldsymbol{\mu}_{u_1} - \boldsymbol{\mu}_{u_2})$, $\boldsymbol{\Sigma}_u = \mathbf{A}\boldsymbol{\Sigma}\mathbf{A}'$, and $\boldsymbol{\mu}_{u_i} = \mathbf{A}E(\overline{\mathbf{y}}_i)$, $i = 1, 2$. Now let $\mathbf{A} = (\boldsymbol{\alpha}_0, \boldsymbol{\alpha}_2, \ldots, \boldsymbol{\alpha}_p)'$, so that H_0 in (5.13) is equivalent to H_0: $\boldsymbol{\beta}_2 = \mathbf{0}$, where $\boldsymbol{\beta} = (\beta_1, \boldsymbol{\beta}_2')'$, since $\beta_1 u_1 = \beta_1\boldsymbol{\alpha}_0'\mathbf{y}$. Show that (3.98) and (3.99) lead to (5.15) and (5.14).

5.10 The vector of correlations between $\mathbf{y}$ and z can be obtained as the vector of covariances of the standardized forms of $\mathbf{y}$ and z, say $\mathbf{y}_s$ and z_s. Then by (1.53), $\mathbf{S}_{y_s} = \mathbf{R}$, and by (1.79),

$$\begin{aligned}\mathbf{r}_{yz} &= \widehat{\text{cov}}(\mathbf{y}_s, z_s) = \widehat{\text{cov}}(\mathbf{I}\mathbf{y}_s, \mathbf{a}^{*\prime}\mathbf{y}_s) \\ &= \mathbf{I}\mathbf{S}_{y_s}\mathbf{a}^* = \mathbf{R}\mathbf{a}^*.\end{aligned}$$

5.11 Use (A.2.19).

5.12 See the hint to Problem 4.14.

5.13 Use (A.2.20).

5.15 Show that $\widehat{\text{cov}}(\mathbf{A}\mathbf{y}) = \mathbf{A}\mathbf{S}\mathbf{A}' = \begin{pmatrix} \mathbf{a}_1'\mathbf{S}\mathbf{a}_1 & \mathbf{a}_1'\mathbf{S}\mathbf{a}_2 \\ \mathbf{a}_2'\mathbf{S}\mathbf{a}_1 & \mathbf{a}_2'\mathbf{S}\mathbf{a}_2 \end{pmatrix}$.

5.16 Use (A.8.2).

5.17 See (A.10.8).

5.18 **(a)** $\mathbf{a} = (.345, -.130, -.106, -.143)'$.

(b) $\mathbf{a}^* = (4.137, -2.501, -1.158, -2.068)'$.

(c) $F(y_1|y_2, y_3, y_4) = 35.934$, $F(y_2|y_1, y_3, y_4) = 5.799$, $F(y_3|y_1, y_2, y_4) = 1.775$, $F(y_4|y_1, y_2, y_3) = 8.259$.

5.19 **(a)**

Variable	y_1	y_2	y_3	y_4
D_u	1.246	−1.238	−1.823	−1.615
$\hat{D}_u$	−.928	−.219	−1.073	.038
R^2	.474	.592	.352	.200

(b) Using (5.15), $F = .4789$, p-value $= .6991$

5.20 **(a)** By (5.2), $\mathbf{a} = (.00292, .01992, -.05509, .00010, -.05149, .04541)'$.

(b) By (5.10), $\mathbf{a}* = (2.3170, 3.3839, -5.0447, .0772, -1.9609, 1.5356)'$.

(c) By (3.100), we have

$$F(y_1|y_2, y_3, y_4, y_5, y_6) = 1.260, F(y_2|y_1, y_3, y_4, y_5, y_6) = 2.732,$$

$$F(y_3|y_1, y_2, y_4, y_5, y_6) = 10.128, F(y_4|y_1, y_2, y_3, y_5, y_6) = .00110,$$

$$F(y_5|y_1, y_2, y_3, y_4, y_6) = 1.357, F(y_6|y_1, y_2, y_3, y_4, y_5) = .559.$$

5.21 **(a)**, **(d)**, and **(e)** Eigenvectors $\mathbf{a}_i$, standardized coefficients $\mathbf{a}_i^*$, and the partial F for each variable adjusted for the other five variables:

Variable	$\mathbf{a}_1$	$\mathbf{a}_2$	$\mathbf{a}_3$	$\mathbf{a}_1^*$	$\mathbf{a}_2^*$	$\mathbf{a}_3^*$	Partial F
y_1	.0006	.0011	−.0001	.619	1.113	−.113	5.171
y_2	−.0006	.0018	−.0014	−.157	.483	−.382	.818
y_3	−.0071	−.0054	.0147	−.866	−.663	1.806	6.988
y_4	.0021	−.0006	.0011	1.806	−.537	.900	28.177
y_5	.0140	−.0220	.0238	.507	−.797	.865	7.898
y_6	−.0199	−.0026	−.0489	−.742	−.098	−1.822	6.601

The standardized coefficients for the first discriminant function rank the variables in the order y_4, y_3, y_6, y_1, y_5, y_2. The partial F's rank them in the order y_4, y_5, y_3, y_6, y_1, y_2.

(b)

i	λ_i	$\lambda_i / \sum_{j=1}^{3} \lambda_j$
1	4.6165	.6038
2	2.3476	.3071
3	.6813	.0891

(c)

Test	Λ	Approximate F	p-Value for F
1	.0316	12.461	$< .0001$
2	.1777	9.332	$< .0001$
3	.5948	5.962	.0009

5.22 **(a)**, **(d)**, and **(e)** Eigenvectors $\mathbf{a}_i$, standardized coefficients $\mathbf{a}_i^*$, and the partial F for each variable adjusted for the other 10 variables:

Variable	$\mathbf{a}_1$	$\mathbf{a}_2$	$\mathbf{a}_3$	$\mathbf{a}_1^*$	$\mathbf{a}_2^*$	$\mathbf{a}_3^*$	Partial F
y_1	−.728	−.951	−1.149	−.290	−.378	−.457	1.412
y_2	2.252	.533	.391	1.792	.424	.311	31.522
y_3	.049	−.046	.116	.173	−.164	.410	.451
y_4	−.927	.707	−.778	−.853	.651	−.716	1.956
y_5	−4.009	6.815	1.171	−.361	.613	.105	2.430
y_6	1.348	−.479	2.227	.857	−.305	1.416	1.066
y_7	−1.224	.237	−.568	−1.083	.210	−.503	2.956
y_8	.198	−.308	.017	.378	−.590	.033	1.451
y_9	.064	.115	−.350	.094	.171	−.519	.484
y_{10}	−.096	−.052	.007	−.561	−.302	.042	1.878
y_{11}	−6.021	−3.860	7.466	−.294	−.188	.364	.524

(b)

i	λ_i	$\lambda_i / \sum_{j=1}^{3} \lambda_j$
1	5.6805	.8475
2	.8412	.1255
3	.1806	.0270

(c)

Test	Λ	Approximate F	p-Value for F
1	.0689	4.127	< .0001
2	.4600	1.518	.1061
3	.8470	.662	.7361

5.23

SUMMARY TABLE

Step	Variable Entered	Variable Removed	Overall Λ	Partial Λ	Partial F	p-Value
1	y_4		.3465	.3465	23.886	< .0001
2	y_2		.1409	.4065	18.009	< .0001
3	y_3		.1001	.7104	4.892	.0059
4	y_5		.0734	.7337	4.233	.0118
5	y_6		.0465	.6334	6.560	.0013
6	y_1		.0316	.6802	5.171	.0049
7		y_2	.0340	.9308	.818	.4930

The overall Λ values are all significant. After removing y_2, there remain five variables.

5.24

SUMMARY TABLE

Step	Variable Entered	Overall Λ	Partial Λ	Partial F	p-Value
1	y_2	.5243	.5243	12.399	< .0001
2	y_7	.1969	.3755	22.170	< .0001
3	y_{10}	.1549	.7869	3.520	.0237
4	y_4	.1315	.8489	2.254	.0977

5.25 (a)

EIGENVALUES

Treatment	Block	Interaction
5.5242	.6874	.8520
3.7215	.3841	.5410
1.1042	.0576	.4150
0	0	.2481
0	0	.1676
0	0	.0249

DISCRIMINANT FUNCTION COEFFICIENTS

Treatments (Eigenvectors of $\mathbf{E}^{-1}\mathbf{H}_T$)			Blocks (Eigenvectors of $\mathbf{E}^{-1}\mathbf{H}_B$)		
$\mathbf{a}_1$	$\mathbf{a}_2$	$\mathbf{a}_3$	$\mathbf{a}_1$	$\mathbf{a}_2$	$\mathbf{a}_3$
.084	.240	−.023	.169	−.064	−.257
−.235	.406	.126	.199	−.091	.551
−.913	−1.562	−2.498	.549	.439	.446
.413	−.106	−.180	.007	−.225	.207
4.095	−4.173	−5.877	3.498	−5.591	5.906
−4.314	1.369	9.111	−1.986	7.993	−.245

DISCRIMINANT FUNCTION COEFFICIENTS

Interaction (Eigenvectors of $\mathbf{E}^{-1}\mathbf{H}_{TB}$)		
$\mathbf{a}_1$	$\mathbf{a}_2$	$\mathbf{a}_3$
.128	.165	.156
.268	.351	−1.25
−1.436	−.164	1.124
−.291	.083	.081
−7.449	5.047	−.754
7.794	−2.334	−.481

(b) **STANDARDIZED COEFFICIENTS**

Treatments			Blocks		
$\mathbf{a}_1^*$	$\mathbf{a}_2^*$	$\mathbf{a}_3^*$	$\mathbf{a}_1^*$	$\mathbf{a}_2^*$	$\mathbf{a}_3^*$
.364	1.042	−.101	.733	−.276	−1.119
−.300	.517	.161	.253	−.116	.702
−.521	−.891	−1.425	.313	.250	.254
1.892	−.486	−.826	.031	−1.032	.947
.742	−.756	−1.065	.634	−1.013	1.070
−.782	.248	1.652	−.360	1.449	−.044

STANDARDIZED COEFFICIENTS

Interaction		
$\mathbf{a}_1^*$	$\mathbf{a}_2^*$	$\mathbf{a}_3^*$
.554	.716	.677
.341	.448	−1.601
−.819	−.093	.641
−1.332	.378	.371
−1.350	.915	−.137
1.413	−.423	−.087

Chapter 6

6.1 Let R_1 be the set of $\mathbf{y}$ values that we classify as belonging to G_1, and let R_2 be all other $\mathbf{y}$ values (which we classify into G_2). The total probability of misclassification is given by (6.27) as

$$\begin{aligned}
&\text{Total probability of misclassification}\\
&\quad = p_1 P(\text{classify as } G_2|G_1) + p_2 P(\text{classify as } G_1|G_2)\\
&\quad = p_1 \int_{R_2} f(\mathbf{y}|G_1)\,d\mathbf{y} + p_2 \int_{R_1} f(\mathbf{y}|G_2)\,d\mathbf{y}\\
&\quad = p_1 \left[1 - \int_{R_1} f(\mathbf{y}|G_1)\,d\mathbf{y}\right] + p_2 \int_{R_1} f(\mathbf{y}|G_2)\,d\mathbf{y}\\
&\quad = p_1 + \int_{R_1} \left[p_2 f(\mathbf{y}|G_2) - p_1 f(\mathbf{y}|G_1)\right]\,d\mathbf{y}.
\end{aligned}$$

Since $p_1, p_2, f(\mathbf{y}|G_1)$, and $f(\mathbf{y}|G_2)$ are positive, this will be minimized by letting R_1 be the set of $\mathbf{y}$-values for which $p_2 f(\mathbf{y}|G_2) - p_1 f(\mathbf{y}|G_1) \le 0$. This choice of R_1 will maximize the quantity subtracted from p_1. For a formal justification of this assertion, see the Neyman–Pearson Lemma (Hogg and Craig 1995, p. 397).

6.2 Write the exponent in (6.6) and in (6.5) in terms of $\mathbf{y}'\boldsymbol{\Sigma}^{-1}\mathbf{y}$, $\mathbf{y}'\boldsymbol{\Sigma}^{-1}\boldsymbol{\mu}_i$, and so on.

6.3 Expand $D_1^2 < D_2^2$ and (6.8) into terms of the form $\mathbf{y}'\mathbf{S}_{\text{pl}}^{-1}\mathbf{y}$, $\mathbf{y}'\mathbf{S}_{\text{pl}}^{-1}\overline{\mathbf{y}}_i$, and so on.

6.5 Substitute $N_p(\boldsymbol{\mu}_i, \boldsymbol{\Sigma})$ from (2.11) for $f(\mathbf{y}|G_i)$ in (6.21), and use estimators of $\boldsymbol{\mu}_i$ and $\boldsymbol{\Sigma}$.

6.6 $P(\text{classify as } G_2|G_1) = P[\boldsymbol{\alpha}'\mathbf{y} \le \frac{1}{2}(\boldsymbol{\mu}_1 - \boldsymbol{\mu}_2)'\boldsymbol{\Sigma}^{-1}(\boldsymbol{\mu}_1 - \boldsymbol{\mu}_2) + \ln(p_1/p_2)]$. Now subtract $\boldsymbol{\alpha}'\boldsymbol{\mu}_1$ from both sides of the inequality and divide by Δ.

6.7 In the proof of Theorem 3.3B, it was noted that $\mathbf{z}'\boldsymbol{\Sigma}^{-1}\mathbf{z}/\mathbf{z}'\mathbf{W}^{-1}\mathbf{z}$ is distributed as $\chi^2_{\nu-p+1}$, where $\mathbf{z}$ is $N_p(\mathbf{0}, \boldsymbol{\Sigma})$, $\mathbf{W}$ is $W_p(\nu, \boldsymbol{\Sigma})$, and $\mathbf{z}$ and $\mathbf{W}$ are independent. This also holds if $\mathbf{z}$ is constant, say $\mathbf{z} = \mathbf{a}$. Thus $\mathbf{a}'\boldsymbol{\Sigma}^{-1}\mathbf{a}/\mathbf{a}'\mathbf{W}^{-1}\mathbf{a}$ is $\chi^2_{\nu-p+1}$ (Seber 1984, p. 29). Now if u is χ^2_q, then $E(1/u) = 1/(q-2)$. Hence $E(\mathbf{a}'\mathbf{W}^{-1}\mathbf{a}) = \mathbf{a}'\boldsymbol{\Sigma}^{-1}\mathbf{a}/(\nu - p - 1)$. Let $\mathbf{a}' = (0, \ldots, 0, 1, 0, \ldots, 0)$, with a 1 in the ith position.

Then $E(w^{ii}) = \sigma^{ii}/(\nu - p - 1)$, where w^{ii} and σ^{ii} are the ith diagonal elements of $\mathbf{W}^{-1}$ and $\boldsymbol{\Sigma}^{-1}$, respectively. Letting $\mathbf{a}' = (1, 1, 0, \ldots, 0)$, for example, gives $E(w^{11} + 2w^{12} + w^{22}) = (\sigma^{11} + 2\sigma^{12} + \sigma^{22})/(\nu - p - 1)$, which leads to $E(w^{12}) = \sigma^{12}/(\nu - p - 1)$.

6.8 $\binom{p}{2} = p!/2!(p-2)!$.

6.9 Let n represent either n_1 or n_2. Then the probability that a given observation is not included is $[(n-1)/n]^n$. For $n = 5$, this value is .328. For $n = 10$ and 25, the values are .349 and .360. As n increases, this value approaches $e^{-1} = .368$.

6.10 See (6.6).

6.11 Divide the numerator and denominator of (6.16) by $f(\mathbf{y}|G_2)$ and use (6.43).

6.12 Divide the numerator and denominator of (6.16) by $f(\mathbf{y}|G_1)$ and use (6.43).

6.13 **(a)** Assign $\mathbf{y}$ to G_1 if $.345y_1 - .130y_2 - .106y_3 - .143y_4 > -15.805$.

LINEAR CLASSIFICATION

Actual Group	Number of Observations	Predicted Group 1	Predicted Group 2
1	19	19	0
2	20	1	19

Apparent error rate = $1/39 = .0256$.

(b) $\frac{1}{2}\ln(|\mathbf{S}_2|/|\mathbf{S}_1|) - \frac{1}{2}(\bar{\mathbf{y}}_1'\mathbf{S}_1^{-1}\bar{\mathbf{y}}_1 - \bar{\mathbf{y}}_2'\mathbf{S}_2^{-1}\bar{\mathbf{y}}_2) = 37.418$,

$\bar{\mathbf{y}}_1'\mathbf{S}_1^{-1} - \bar{\mathbf{y}}_2'\mathbf{S}_2^{-1} = (-1.345, .263, 1.263, -.299)$,

$$\tfrac{1}{2}(\mathbf{S}_1^{-1} - \mathbf{S}_2^{-1}) = (.001)\begin{pmatrix} -2.639 & 1.290 & -3.231 & -1.354 \\ 1.290 & -.553 & .892 & -.076 \\ -3.231 & .892 & 4.835 & 1.921 \\ -1.354 & -.076 & 1.921 & -.382 \end{pmatrix}$$

Using resubstitution based on quadratic classification, we obtain the same classification table as in part (a). The apparent error rate is therefore the same: .0256.

6.14 **(a)** Assign $\mathbf{y}$ to G_1 if $4.872y_1 + .0296y_2 + .1187y_3 > 39.499$.

LINEAR CLASSIFICATION

Actual Group	Number of Observations	Predicted Group 1	Predicted Group 2
1	13	12	1
2	10	0	10

Apparent error rate = $1/23 = .0435$.

(b) $\frac{1}{2}\ln(|\mathbf{S}_2|/|\mathbf{S}_1|) - \frac{1}{2}(\bar{\mathbf{y}}_1'\mathbf{S}_1^{-1}\bar{\mathbf{y}}_1 - \bar{\mathbf{y}}_2'\mathbf{S}_2^{-1}\bar{\mathbf{y}}_2) = 74.886$,

$\bar{\mathbf{y}}_1'\mathbf{S}_1^{-1} - \bar{\mathbf{y}}_2'\mathbf{S}_2^{-1} = (-23.942, -.759, -.126)$,

$$\tfrac{1}{2}(\mathbf{S}_1^{-1} - \mathbf{S}_2^{-1} = \begin{pmatrix} -2.1102 & -.0401 & .0172 \\ -.0401 & -.0066 & .0060 \\ .0172 & .0060 & -.0214 \end{pmatrix}.$$

Using resubstitution based on quadratic classification, we obtain the same classification table as in part (a). The apparent error rate is therefore the same: .0435.

(c) Using p_1 and p_2 proportional to sample sizes, we obtain the same classification table as in part (a).

6.15 **(a)** Assign $\mathbf{y}$ to G_1 if $(-.00292, -.0199, .0551, -.000095, .0515, -.0454)\mathbf{y} > -8.693$.

LINEAR CLASSIFICATION

Actual Group	Number of Observations	Predicted Group 1	Predicted Group 2
1	10	10	0
2	11	0	11

Apparent error rate $= 0/21 = 0$.

(b) Using (6.36), plug-in error rate $= .0301$.

(c) Using (6.39), modified plug-in error rate $= .0677$.

(d) Assuming multivariate normality, we obtain the following rule from (6.8) and (6.43). Assign $\mathbf{y}$ to G_1 if $(-.00292, -.0199, .0551, -.000095, .0515, -.0454)\mathbf{y} + 8.693 > 0$.

6.16 **(a)**

SUMMARY TABLE

Step	Variable Entered	Overall Λ	Partial Λ	Partial F	p-Value
1	y_2	.4486	.4486	23.358	$< .0001$
2	y_3	.3006	.6701	8.863	.0081
3	y_4	.2464	.8198	3.737	.0700

(b) Assign $\mathbf{y}$ to G_1 if $-.0250y_2 + .0406y_3 - .00224y_4 > -32.262$.

LINEAR CLASSIFICATION

Actual Group	Number of Observations	Predicted Group 1	Predicted Group 2
1	10	10	0
2	11	0	11

This is the same classification result obtained in Problem 6.15(a) using all six variables.

6.17 **(a)** With $p_1 = p_2 = p_3 = p_4$, $\ln p_i$ can be deleted to obtain $L_i(\mathbf{y}) = \bar{\mathbf{y}}_i'\mathbf{S}_{\text{pl}}^{-1}\mathbf{y} - \frac{1}{2}\bar{\mathbf{y}}_i'\mathbf{S}_{\text{pl}}^{-1}\bar{\mathbf{y}}_i = \mathbf{c}_i'\mathbf{y} + c_{0i}$. The vectors $\binom{c_{0i}}{\mathbf{c}_i}$, $i = 1, 2, 3, 4$, are as follows:

Group 1	Group 2	Group 3	Group 4
−213.967	−207.153	−262.691	−193.905
−.019	−.015	−.014	−.014
.035	.040	.035	.042
.079	.027	.033	.053
.028	.030	.039	.028
.693	.633	.737	.622
−.108	−.075	−.222	−.152

(b)

LINEAR CLASSIFICATION

Actual Group	Number of Observations	Predicted Group			
		1	2	3	4
1	10	9	1	0	0
2	11	0	10	0	1
3	11	0	0	11	0
4	10	0	0	0	10

Apparent error rate $= 2/42 = .0476$.

(c) The expression following (6.23) can be written as $Q_i(\mathbf{y}) = a_i + \mathbf{b}_i'\mathbf{y} - \frac{1}{2}\mathbf{y}'\mathbf{S}_i^{-1}\mathbf{y}$. The vectors $(a_i, \mathbf{b}_i')$ are as follows:

i	$(a_i, \mathbf{b}_i')$
1	$(-400.8, .0154, .0682, .229, .0248, .545, .119)$
2	$(-302.2, -.0135, .0832, -.0609, .0609, .868, -.659)$
3	$(-3491.2, -.301, 1.331, -3.417, .107, 5.241, 15.60)$
4	$(-256.0, -.0222, .00877, .346, .0676, 1.125, -1.633)$

(d)

QUADRATIC CLASSIFICATION

Actual Group	Number of Observations	Predicted Group			
		1	2	3	4
1	10	10	0	0	0
2	11	0	11	0	0
3	11	0	0	11	0
4	10	0	0	0	10

Apparent error rate $= 0/42 = 0$.

6.18 **(a)** As in the answer to Problem 6.17(a), we can write $L_i(\mathbf{y}) = \mathbf{c}_i'\mathbf{y} + c_{0i}$. The vectors $\binom{c_{0i}}{\mathbf{c}_i}$, $i = 1, 2, 3, 4$, are as follows:

Group 1	Group 2	Group 3	Group 4
−155.453	−153.786	−156.730	−148.553
39.369	41.224	44.581	42.499
6.955	1.749	−5.045	−6.210
2.409	2.417	2.180	2.056
−5.249	−4.816	−1.648	.499
32.721	33.490	42.051	63.018
7.470	7.012	1.161	−.027
−4.830	−3.028	.893	2.427
4.435	4.426	3.915	2.994
−.090	−.651	−.516	−.555
−.682	−.404	−.127	−.127
102.040	125.643	136.538	139.674

(b) **LINEAR CLASSIFICATION**

Actual Group	Number of Observations	Predicted Group 1	2	3	4
1	12	10	2	0	0
2	14	4	9	1	0
3	11	0	1	9	1
4	8	0	0	1	7

Apparent error rate = $10/45 = .222$.

(c) **HOLDOUT METHOD**

Actual Group	Number of Observations	Predicted Group 1	2	3	4
1	12	8	4	0	0
2	14	4	8	1	1
3	11	0	2	6	3
4	8	0	0	4	4

Apparent error rate = $19/45 = .422$.

6.19 **(a)** $L_i(\mathbf{y}) = \mathbf{c}_i'\mathbf{y} + c_{0i}$, where $\mathbf{y} = (y_2, y_4, y_7, y_{10})'$. The vectors $\binom{c_{0i}}{\mathbf{c}_i}$, $i = 1, 2, 3, 4$, are as follows:

Group 1	Group 2	Group 3	Group 4
−20.468	−9.301	−4.037	−8.223
12.988	8.215	2.923	1.779
−1.952	−1.377	−.112	1.595
−3.356	−1.704	.923	2.218
−.451	−.177	−.031	−.130

(b)

LINEAR CLASSIFICATION

Actual Group	Number of Observations	Predicted Group 1	2	3	4
1	12	10	2	0	0
2	14	2	11	0	1
3	11	0	1	5	5
4	8	0	0	3	5

Apparent error rate $= 14/45 = .311$.

6.20 **(a)** The variables selected in problem Problem 5.22 were y_1, y_3, y_4, y_5, y_6.

LINEAR CLASSIFICATION

Actual Group	Number of Observations	Predicted Group 1	2	3	4
1	10	9	1	0	0
2	11	0	10	0	1
3	11	0	0	11	0
4	10	0	0	0	10

Apparent error rate $= 2/42 = .0476$.

Chapter 7

7.1 By (A.2.9), $\hat{\beta}_0 + \hat{\beta}_1 x_{i1} + \cdots + \hat{\beta}_q x_{iq} = \mathbf{x}_i'\hat{\boldsymbol{\beta}}$. By (A.2.11) and (A.2.25), $\sum_{i=1}^n (y_i - \mathbf{x}_i'\hat{\boldsymbol{\beta}})^2 = (\mathbf{y} - \mathbf{X}\hat{\boldsymbol{\beta}})'(\mathbf{y} - \mathbf{X}\hat{\boldsymbol{\beta}})$.

7.2 $(\mathbf{y} - \mathbf{X}\hat{\boldsymbol{\beta}})'(\mathbf{y} - \mathbf{X}\hat{\boldsymbol{\beta}}) = \mathbf{y}'\mathbf{y} - \mathbf{y}'\mathbf{X}\hat{\boldsymbol{\beta}} - \hat{\boldsymbol{\beta}}'\mathbf{X}'\mathbf{y} + \hat{\boldsymbol{\beta}}'\mathbf{X}'\mathbf{X}\hat{\boldsymbol{\beta}}$. Use (7.6) to show that $\hat{\boldsymbol{\beta}}'\mathbf{X}'\mathbf{X}\hat{\boldsymbol{\beta}} = \hat{\boldsymbol{\beta}}'\mathbf{X}\mathbf{y}$.

7.3 Using (7.11) we can write SSE in the form $\text{SSE} = \mathbf{y}'\mathbf{y} - \hat{\boldsymbol{\beta}}'\mathbf{X}'\mathbf{y} = \mathbf{y}'\mathbf{y} - \mathbf{y}'\mathbf{X}(\mathbf{X}'\mathbf{X})^{-1}\mathbf{X}'\mathbf{y} = \mathbf{y}'[\mathbf{I} - \mathbf{X}(\mathbf{X}'\mathbf{X})^{-1}\mathbf{X}']\mathbf{y}$, so that SSE is a quadratic form in $\mathbf{y}$. Then use (1.123) in Problem 1.4.

7.4 Use $\mathbf{J} = \mathbf{j}\mathbf{j}'$ in $\mathbf{j}'\mathbf{X}_c = \mathbf{j}'[\mathbf{I} - (1/n)\mathbf{J}]\mathbf{X}_1$.

7.5 The second element of $\mathbf{s}_{yx}$, for example, is $s_{y2} = \sum_{i=1}^n (x_{i2} - \overline{x}_2)(y_i - \overline{y})/(n - 1) = \left[\sum_i (x_{i2} - \overline{x}_2)y_i - \sum_i (x_{i2} - \overline{x}_2)\overline{y}\right]/(n-1) = \sum_i (x_{i2} - \overline{x}_2)y_i/(n-1)$, which is the second element of $\mathbf{X}_c'\mathbf{y}/(n-1)$.

7.6 Partition $\mathbf{X}$ in the form $\mathbf{X} = (\mathbf{j}, \mathbf{X}_1)$ to conform to the partitioning $\boldsymbol{\beta}' = (\beta_0, \boldsymbol{\beta}_1')$. Then corresponding to (7.30), we have

$$\mathbf{y} = (\mathbf{j}, \mathbf{X}_1)\begin{pmatrix} \beta_0 \\ \boldsymbol{\beta}_1 \end{pmatrix} + \boldsymbol{\varepsilon} = \beta_0\mathbf{j} + \mathbf{X}_1\boldsymbol{\beta}_1 + \boldsymbol{\varepsilon}.$$

The reduced model under H_0: $\boldsymbol{\beta}_1 = \mathbf{0}$ is $\mathbf{y} = \beta_0\mathbf{j} + \boldsymbol{\varepsilon}$. Show that for this reduced model, $\text{SS}(\beta_0) = n\overline{y}^2$, and then (7.32) leads to (7.28).

7.7 Express y and z in terms of $\binom{y}{\mathbf{x}}$ as follows: $y = (1, 0, \ldots, 0)\binom{y}{\mathbf{x}} = \mathbf{a}'\binom{y}{\mathbf{x}}$, $z = (0, \boldsymbol{\sigma}_{yx}'\boldsymbol{\Sigma}_{xx})\binom{y}{\mathbf{x}}$ + constant $= \mathbf{b}'\binom{y}{\mathbf{x}}$ + constant. Then use (1.87) and (1.91) with $\boldsymbol{\Sigma}$ partitioned as in (7.48).

7.9 Multiply in terms of $\mathbf{y}_{(j)} - \mathbf{X}\hat{\boldsymbol{\beta}}_{(j)}$ and $\mathbf{X}\hat{\boldsymbol{\beta}}_{(j)} - \mathbf{X}\mathbf{b}_{0(j)}$.

7.10 The model $\mathbf{Y} = \mathbf{X}\mathbf{B} + \boldsymbol{\Xi}$ in (7.68) becomes $\mathbf{Y} = \mathbf{j}\boldsymbol{\mu}' + \boldsymbol{\Xi}$, and $\hat{\mathbf{B}} = (\mathbf{X}'\mathbf{X})^{-1}\mathbf{X}'\mathbf{Y}$ becomes $\hat{\boldsymbol{\mu}}' = (\mathbf{j}'\mathbf{j})^{-1}\mathbf{j}'\mathbf{Y}$.

7.11 **(a)** (7.80) follows from Theorem 7.4C. **(b)** (7.81) follows from part (a).

7.12

$$
\begin{aligned}
\mathbf{A}_2\mathbf{A}_2' &= \left[\mathbf{A}_2 - (\mathbf{X}'\mathbf{X})^{-1}\mathbf{X}' + (\mathbf{X}'\mathbf{X})^{-1}\mathbf{X}'\right]\left[\mathbf{A}_2 - (\mathbf{X}'\mathbf{X})^{-1}\mathbf{X}' + (\mathbf{X}'\mathbf{X})^{-1}\mathbf{X}'\right]' \\
&= \left[\mathbf{A}_2 - (\mathbf{X}'\mathbf{X})^{-1}\mathbf{X}'\right]\left[\mathbf{A}_2 - (\mathbf{X}'\mathbf{X})^{-1}\mathbf{X}'\right]' + (\mathbf{X}'\mathbf{X})^{-1}.
\end{aligned}
$$

The other two terms vanish because they are of the form $\mathbf{A}_2\mathbf{X}(\mathbf{X}'\mathbf{X})^{-1} - (\mathbf{X}'\mathbf{X})^{-1}$, which is $\mathbf{O}$ because $\mathbf{A}_2\mathbf{X} = \mathbf{I}$. The term $[\mathbf{A}_2 - (\mathbf{X}'\mathbf{X})^{-1}\mathbf{X}'][\mathbf{A}_2 - (\mathbf{X}'\mathbf{X})^{-1}\mathbf{X}']'$ is a positive semidefinite matrix, and by (A.6.4), the diagonal elements are greater than or equal to zero. We therefore choose $\mathbf{A}_2 = (\mathbf{X}'\mathbf{X})^{-1}\mathbf{X}'$, which is compatible with $\mathbf{A}_2\mathbf{X} = \mathbf{I}$.

7.13 $(\mathbf{Y} - \mathbf{X}\hat{\mathbf{B}})'(\mathbf{Y} - \mathbf{X}\hat{\mathbf{B}}) = \mathbf{Y}'\mathbf{Y} - \mathbf{Y}'\mathbf{X}\hat{\mathbf{B}} - \hat{\mathbf{B}}'\mathbf{X}'\mathbf{Y} + \hat{\mathbf{B}}'\mathbf{X}'\mathbf{X}\hat{\mathbf{B}}$. Use (7.70) to show that $\hat{\mathbf{B}}'\mathbf{X}'\mathbf{X}\hat{\mathbf{B}} = \mathbf{Y}'\mathbf{X}\hat{\mathbf{B}}$.

7.14 Show that $(\mathbf{y}_{(j)} - \mathbf{X}\hat{\boldsymbol{\beta}}_{(j)})'(\mathbf{y}_{(k)} - \mathbf{X}\hat{\boldsymbol{\beta}}_{(k)}) = \mathbf{y}_{(j)}'[\mathbf{I} - \mathbf{X}(\mathbf{X}'\mathbf{X})^{-1}\mathbf{X}']\mathbf{y}_{(k)}$ and use (1.126) in Problem 1.10. Note that $\text{cov}(\mathbf{y}_{(j)}, \mathbf{y}_{(k)}) = \sigma_{jk}\mathbf{I}$, as used in the proof of Theorem 7.4C.

7.15 **(b)** $\text{tr}(\mathbf{A}) = \text{tr}[\mathbf{I} - \mathbf{X}(\mathbf{X}'\mathbf{X})^{-1}\mathbf{X}'] = \text{tr}(\mathbf{I}) - \text{tr}[\mathbf{X}'\mathbf{X}(\mathbf{X}'\mathbf{X})^{-1}]$ by (A.8.1) and (A.8.2).

7.16 Adding and subtracting $\mathbf{X}\hat{\mathbf{B}}$, we obtain $(\mathbf{Y} - \mathbf{X}\mathbf{B})'(\mathbf{Y} - \mathbf{X}\mathbf{B}) = (\mathbf{Y} - \mathbf{X}\hat{\mathbf{B}} + \mathbf{X}\hat{\mathbf{B}} - \mathbf{X}\mathbf{B})'(\mathbf{Y} - \mathbf{X}\hat{\mathbf{B}} + \mathbf{X}\hat{\mathbf{B}} - \mathbf{X}\mathbf{B})$. Expanding the right side in terms of $\mathbf{Y} - \mathbf{X}\hat{\mathbf{B}}$ and $\mathbf{X}\hat{\mathbf{B}} - \mathbf{X}\mathbf{B}$ gives four terms, two of which vanish for $\hat{\mathbf{B}} = (\mathbf{X}'\mathbf{X})^{-1}\mathbf{X}'\mathbf{Y}$.

7.17 Use (A.8.2).

7.19 **(i)** Use Theorem 2.2B.

(ii) Express $\mathbf{E}$ in the form $\mathbf{E} = \mathbf{Y}'[\mathbf{I} - \mathbf{X}(\mathbf{X}'\mathbf{X})^{-1}\mathbf{X}']\mathbf{Y}$ as in Problem 7.15(a). Show that $\mathbf{I} - \mathbf{X}(\mathbf{X}'\mathbf{X})^{-1}\mathbf{X}'$ is idempotent and use Theorem 2.3G.

(iii) Adapt the hint to Problem 2.27. $\hat{\mathbf{B}} = (\mathbf{X}'\mathbf{X})^{-1}\mathbf{X}'\mathbf{Y} = \mathbf{C}\mathbf{Y} = \mathbf{C}(\mathbf{y}_{(1)}, \mathbf{y}_{(2)}, \ldots, \mathbf{y}_{(p)}) = (\mathbf{C}\mathbf{y}_{(1)}, \ldots, \mathbf{C}\mathbf{y}_{(p)})$. $\mathbf{E} = \mathbf{Y}'\mathbf{A}\mathbf{Y} = \mathbf{Y}'\mathbf{A}'\mathbf{A}\mathbf{Y}$, where $\mathbf{A} = \mathbf{I} - \mathbf{X}(\mathbf{X}'\mathbf{X})^{-1}\mathbf{X}'$, as in part (ii) above. $\hat{\mathbf{B}}$ and $\mathbf{Y}'\mathbf{A}\mathbf{Y}$ are independent if $\text{cov}(\mathbf{C}\mathbf{y}_{(i)}, \mathbf{A}\mathbf{y}_{(j)}) = \mathbf{O}$ for all i and j, where $\mathbf{A}\mathbf{Y} = (\mathbf{A}\mathbf{y}_{(1)}, \ldots, \mathbf{A}\mathbf{y}_{(p)})$. Show that $\text{cov}(\mathbf{C}\mathbf{y}_{(i)}, \mathbf{A}\mathbf{y}_{(j)}) = \sigma_{ij}\mathbf{C}\mathbf{A}' = \sigma_{ij}\mathbf{O}$, where σ_{ij} is from $\boldsymbol{\Sigma}$.

(iv) Note that (7.90) has the form of (2.54); that is, $L(\mathbf{B}, \boldsymbol{\Sigma})$ depends on $\mathbf{Y}$ only through $\hat{\mathbf{B}}$ and $\hat{\boldsymbol{\Sigma}} = \mathbf{E}/n$.

(v) This follows from (iv) and the Rao–Blackwell Theorem (see Hogg and Craig 1995, pp. 326–332).

7.20 Show that

$$
[(\mathbf{j}, \mathbf{X}_c)'(\mathbf{j}, \mathbf{X}_c)]^{-1}(\mathbf{j}, \mathbf{X}_c)'\mathbf{Y} = \begin{pmatrix} n & \mathbf{0}' \\ \mathbf{0} & \mathbf{X}_c'\mathbf{X}_c \end{pmatrix}^{-1} \begin{pmatrix} n(\bar{y}_1, \bar{y}_2, \ldots, \bar{y}_p) \\ \mathbf{X}_c'\mathbf{Y} \end{pmatrix}.
$$

7.21 Use (7.94), (7.96), and (7.97).

7.22 See (7.107) and Problem 7.26.

7.23 The reduced model under H_0: $\mathbf{B}_1 = \mathbf{O}$ becomes $\mathbf{Y} = \mathbf{j}\boldsymbol{\beta}_0' + \boldsymbol{\Xi}$. Show that for this reduced model, $\mathbf{H}$ becomes $\mathbf{H}_{\text{red}} = n\bar{\mathbf{y}}\bar{\mathbf{y}}'$.

7.24 (a) and (b) $\mathbf{H} = \hat{\mathbf{B}}'\mathbf{X}'\mathbf{Y} - n\bar{\mathbf{y}}\bar{\mathbf{y}}'$ is a special case of $\mathbf{H} = \hat{\mathbf{B}}'\mathbf{X}'\mathbf{Y} - \hat{\mathbf{B}}_r'\mathbf{X}_r'\mathbf{Y}$ in (7.121), where $\mathbf{X}_r = \mathbf{j}$ and $\hat{\mathbf{B}}_r = \hat{\boldsymbol{\beta}}_0'$. Show that $\hat{\boldsymbol{\beta}}_0 = \bar{\mathbf{y}}$ and $\hat{\mathbf{B}}_r'\mathbf{X}_r'\mathbf{Y} = n\bar{\mathbf{y}}\bar{\mathbf{y}}'$. The distribution of $\mathbf{H}$ and the independence of $\mathbf{E}$ and $\mathbf{H}$ follow from the results of Problem 7.30.

7.25 Under H_0: $\mathbf{B}_1 = \mathbf{O}$, the model becomes $\mathbf{Y} = \mathbf{j}\boldsymbol{\beta}_0' + \boldsymbol{\Xi}$ or $\mathbf{y}_i = \boldsymbol{\beta}_0$, $i = 1, 2, \ldots, n$. Then $L(\boldsymbol{\beta}_0', \boldsymbol{\Sigma}) = [1/(\sqrt{2\pi})^{np}|\boldsymbol{\Sigma}|^{n/2}]e^{-\frac{1}{2}\sum_i(\mathbf{y}_i-\boldsymbol{\beta}_0)'\boldsymbol{\Sigma}^{-1}(\mathbf{y}_i-\boldsymbol{\beta}_0)}$. The result follows from Theorem 2.3A and (1.38) and (1.39).

7.26 See Sections 7.2.3 and 7.4.6 and the hint to Problem 7.28.

7.27 Show that $\hat{\boldsymbol{\beta}}'\mathbf{X}'\mathbf{z} = \mathbf{a}'\hat{\mathbf{B}}'\mathbf{X}'\mathbf{Y}\mathbf{a}$, where $\hat{\boldsymbol{\beta}} = (\mathbf{X}'\mathbf{X})^{-1}\mathbf{X}'\mathbf{z} = (\mathbf{X}'\mathbf{X})^{-1}\mathbf{X}'\mathbf{Y}\mathbf{a} = \hat{\mathbf{B}}\mathbf{a}$. Using $\bar{z} = \mathbf{j}'\mathbf{z}/n$ and (1.29), show that $n\bar{z}^2 = n\mathbf{a}'\bar{\mathbf{y}}\bar{\mathbf{y}}'\mathbf{a}$.

7.28 To show that $\mathbf{H} = \hat{\mathbf{B}}'\mathbf{X}'\mathbf{Y} - n\bar{\mathbf{y}}\bar{\mathbf{y}}' = \hat{\mathbf{B}}_1'\mathbf{X}_c'\mathbf{Y}$, write $\hat{\mathbf{B}}'\mathbf{X}'\mathbf{Y}$ as follows:

$$\hat{\mathbf{B}}'\mathbf{X}'\mathbf{Y} = \begin{pmatrix} \hat{\boldsymbol{\beta}}_0' \\ \hat{\mathbf{B}}_1 \end{pmatrix}' (\mathbf{j}, \mathbf{X}_1)'\mathbf{Y} = (\hat{\boldsymbol{\beta}}_0, \hat{\mathbf{B}}_1') \begin{pmatrix} \mathbf{j}' \\ \mathbf{X}_1' \end{pmatrix} \mathbf{Y}$$

$$= (\hat{\boldsymbol{\beta}}_0, \hat{\mathbf{B}}_1') \begin{pmatrix} n\bar{\mathbf{y}}' \\ \mathbf{X}_1'\mathbf{Y} \end{pmatrix} = n\hat{\boldsymbol{\beta}}_0\bar{\mathbf{y}}' + \hat{\mathbf{B}}_1'\mathbf{X}_1'\mathbf{Y}.$$

With $\hat{\boldsymbol{\beta}}_0 = \bar{\mathbf{y}} - \hat{\mathbf{B}}_1'\bar{\mathbf{x}}$ from (7.100) and $\mathbf{X}_c = [\mathbf{I} - (1/n)\mathbf{J}]\mathbf{X}_1$ from (7.22), this becomes

$$\hat{\mathbf{B}}'\mathbf{X}'\mathbf{Y} = n(\bar{\mathbf{y}} - \hat{\mathbf{B}}_1'\bar{\mathbf{x}})\bar{\mathbf{y}}' + \hat{\mathbf{B}}_1'[\mathbf{X}_c' + (1/n)\mathbf{X}_1'\mathbf{J}]\mathbf{Y}$$

$$= n\bar{\mathbf{y}}\bar{\mathbf{y}}' - n\hat{\mathbf{B}}_1'\bar{\mathbf{x}}\bar{\mathbf{y}}' + \hat{\mathbf{B}}_1'\mathbf{X}_c'\mathbf{Y} + (1/n)\hat{\mathbf{B}}_1'\mathbf{X}_1'\mathbf{J}\mathbf{Y}.$$

The last term can be written as

$$\frac{1}{n}\hat{\mathbf{B}}_1'\mathbf{X}_1'\mathbf{J}\mathbf{Y} = \frac{1}{n}\hat{\mathbf{B}}_1'\mathbf{X}_1'\mathbf{j}\mathbf{j}'\mathbf{Y} = \frac{1}{n}\hat{\mathbf{B}}_1'n^2\bar{\mathbf{x}}\bar{\mathbf{y}}',$$

so that $\hat{\mathbf{B}}'\mathbf{X}'\mathbf{Y} = n\bar{\mathbf{y}}\bar{\mathbf{y}}' + \hat{\mathbf{B}}_1'\mathbf{X}_c'\mathbf{Y}$. Show that $E(\mathbf{Y}) = \mathbf{j}\boldsymbol{\alpha}' + \mathbf{X}_c\mathbf{B}_1$ [see (7.92)] and use (1.128).

7.29 (a) Since $E(\mathbf{E}) = (n-q-1)\boldsymbol{\Sigma}$ and $E(\mathbf{H}) = q\boldsymbol{\Sigma} + \mathbf{B}_1'\mathbf{X}_c'\mathbf{X}_c\mathbf{B}_1$, we have

$$|[E(\mathbf{E})]^{-1}E(\mathbf{H}) - \phi\mathbf{I}| = \left| \frac{\boldsymbol{\Sigma}^{-1}\mathbf{B}_1'\mathbf{X}_c'\mathbf{X}_c\mathbf{B}_1}{n-q-1} - \frac{(n-q-1)\phi - q}{n-q-1}\mathbf{I} \right| = 0.$$

(b) Since $\mathbf{E}$ and $\mathbf{H}$ are independent, $E(\mathbf{E}^{-1}\mathbf{H}) = E(\mathbf{E}^{-1})E(\mathbf{H})$. Show that $E(\mathbf{E}^{-1}) = \boldsymbol{\Sigma}^{-1}/(n-q-p-2)$ (see Problem 6.7).

7.30 (a) $\mathbf{E} = \mathbf{Y}'[\mathbf{I} - \mathbf{X}(\mathbf{X}'\mathbf{X})^{-1}\mathbf{X}']\mathbf{Y} = \mathbf{Y}'(\mathbf{I} - \mathbf{A}_1)\mathbf{Y}$ and

$$\mathbf{H} = \mathbf{Y}'\mathbf{X}(\mathbf{X}'\mathbf{X})^{-1}\mathbf{X}'\mathbf{Y} - \mathbf{Y}'\mathbf{X}_r(\mathbf{X}_r'\mathbf{X}_r)^{-1}\mathbf{X}_r'\mathbf{Y}$$

$$= \mathbf{Y}'[\mathbf{X}(\mathbf{X}'\mathbf{X})^{-1}\mathbf{X}' - \mathbf{X}_r(\mathbf{X}_r'\mathbf{X}_r)^{-1}\mathbf{X}_r']\mathbf{Y}$$

$$= \mathbf{Y}'(\mathbf{A}_1 - \mathbf{A}_2)\mathbf{Y}.$$

The matrix $\mathbf{I} - \mathbf{A}_1$ was shown to be idempotent in Problem 7.19. To demonstrate that $\mathbf{A}_1 - \mathbf{A}_2$ is idempotent, first show that $(\mathbf{I} - \mathbf{A}_1)\mathbf{X} = \mathbf{O}$, from which $\mathbf{X} = \mathbf{A}_1\mathbf{X}$. By writing $\mathbf{X} = (\mathbf{X}_r, \mathbf{X}_d)$, show that $\mathbf{X}_r = \mathbf{A}_1\mathbf{X}_r$. Then $\mathbf{A}_1\mathbf{A}_2 = \mathbf{A}_1\mathbf{X}_r(\mathbf{X}_r'\mathbf{X}_r)^{-1}\mathbf{X}_r' = \mathbf{X}_r(\mathbf{X}_r'\mathbf{X}_r)^{-1}\mathbf{X}_r' = \mathbf{A}_2$. Show that $\mathbf{A}_2\mathbf{A}_1 = \mathbf{A}_2$, that $\mathbf{A}_2^2 = \mathbf{A}_2$ and that $(\mathbf{A}_1 - \mathbf{A}_2)^2 = \mathbf{A}_1 - \mathbf{A}_2$.

To show that $\text{rank}(\mathbf{A}_1 - \mathbf{A}_2) = h$, note that, by (A.10.6), $\text{tr}(\mathbf{A}_1 - \mathbf{A}_2) = \sum_{i=1}^{n} \lambda_i$, which is the rank of $\mathbf{A}_1 - \mathbf{A}_2$ by (A.13.1). Show that $\text{tr}(\mathbf{A}_1 - \mathbf{A}_2) = h$. Now by Theorem 2.3G, $\mathbf{H}$ is $W_p(h, \boldsymbol{\Sigma})$.

(b) Show that $(\mathbf{I} - \mathbf{A}_1)(\mathbf{A}_1 - \mathbf{A}_2) = \mathbf{O}$. Then by Theorem 2.3H, $\mathbf{E}$ and $\mathbf{H}$ are independent.

7.31 Divide the numerator and denominator by $|\mathbf{Y}'\mathbf{Y} - n\bar{\mathbf{y}}\bar{\mathbf{y}}'|$.

7.33 Follow the steps in the proof of Theorem 7.5B.

7.34 Substitute $\hat{\mathbf{B}}_c$ in (7.127) into $\mathbf{H} = \hat{\mathbf{B}}'\mathbf{X}'\mathbf{Y} - \hat{\mathbf{B}}_c'\mathbf{X}'\mathbf{Y}$.

7.35 By (7.127)

$$
\begin{aligned}
\hat{\mathbf{B}}_c'\mathbf{X}'\mathbf{X}\hat{\mathbf{B}}_c &= \hat{\mathbf{B}}_c'\mathbf{X}'\mathbf{X}\left\{\hat{\mathbf{B}} - (\mathbf{X}'\mathbf{X})^{-1}\mathbf{C}'[\mathbf{C}(\mathbf{X}'\mathbf{X})^{-1}\mathbf{C}']^{-1}\mathbf{C}\hat{\mathbf{B}}\right\} \\
&= \hat{\mathbf{B}}_c'\mathbf{X}'\mathbf{X}\hat{\mathbf{B}} - \hat{\mathbf{B}}_c'\mathbf{C}'\left[\mathbf{C}(\mathbf{X}'\mathbf{X})^{-1}\mathbf{C}'\right]^{-1}\mathbf{C}\hat{\mathbf{B}} \\
&= \hat{\mathbf{B}}_c'\mathbf{X}'\mathbf{Y} - \hat{\mathbf{B}}_c'\mathbf{C}'\left[\mathbf{C}(\mathbf{X}'\mathbf{X})^{-1}\mathbf{C}'\right]^{-1}\mathbf{C}\hat{\mathbf{B}}.
\end{aligned}
$$

Show that the second term on the right vanishes by substituting $\hat{\mathbf{B}}_c$ from (7.127).

7.36 Express $\mathbf{H} = (\mathbf{C}\hat{\mathbf{B}})'[\mathbf{C}(\mathbf{X}'\mathbf{X})^{-1}\mathbf{C}']^{-1}\mathbf{C}\hat{\mathbf{B}}$ in terms of $\mathbf{Y}$ by substituting $\hat{\mathbf{B}} = (\mathbf{X}'\mathbf{X})^{-1}\mathbf{X}'\mathbf{Y}$ to obtain $\mathbf{H} = \mathbf{Y}'\mathbf{D}\mathbf{Y}$. Show that $\mathbf{D}$ is idempotent. Note that the rows of $\mathbf{Y}$ do not have means equal to $\mathbf{0}'$ as assumed in Theorem 2.3G. Thus $\mathbf{H} = \mathbf{Y}'\mathbf{D}\mathbf{Y}$ has a noncentral Wishart distribution.

To show that $\mathbf{H}$ and $\mathbf{E}$ are independent, express $\mathbf{E}$ as $\mathbf{E} = \mathbf{Y}'(\mathbf{I} - \mathbf{A}_1)\mathbf{Y}$ and show that $\mathbf{D}(\mathbf{I} - \mathbf{A}_1) = \mathbf{O}$. Then by Theorem 2.3H, $\mathbf{H}$ and $\mathbf{E}$ are independent.

7.37 $\text{var}(\mathbf{a}'\hat{\mathbf{B}}\mathbf{b}) = \text{var}[\mathbf{a}'(\hat{\mathbf{B}}\mathbf{b})] = \mathbf{a}'\,\text{cov}(\hat{\mathbf{B}}\mathbf{b})\mathbf{a} = \mathbf{a}'\,\text{cov}\begin{pmatrix} \hat{\boldsymbol{\beta}}_0'\mathbf{b} \\ \vdots \\ \hat{\boldsymbol{\beta}}_q'\mathbf{b} \end{pmatrix}\mathbf{a}$ [by (A.2.25)],

where $\hat{\boldsymbol{\beta}}_i'$ is the ith row of $\hat{\mathbf{B}}$, $i = 0, 1, \ldots, q$. Using Problem 7.33(b), show that the iith element of $\text{cov}\begin{pmatrix} \hat{\boldsymbol{\beta}}_0'\mathbf{b} \\ \vdots \\ \hat{\boldsymbol{\beta}}_q'\mathbf{b} \end{pmatrix}$ is $\mathbf{b}'\,\text{cov}(\hat{\boldsymbol{\beta}}_i)\mathbf{b} = \mathbf{b}'g_{ii}\boldsymbol{\Sigma}\mathbf{b}$ and that the ijth element is $\mathbf{b}'\,\text{cov}(\hat{\boldsymbol{\beta}}_i, \hat{\boldsymbol{\beta}}_j)\mathbf{b} = \mathbf{b}'g_{ij}\boldsymbol{\Sigma}\mathbf{b}$, where g_{ij} is from $(\mathbf{X}'\mathbf{X})^{-1}$.

7.38 See the hint for Problem 7.33.

7.39 Substitute $\mathbf{Y} = \mathbf{X}\mathbf{B} + \boldsymbol{\Xi}$ and $\mathbf{P} = \mathbf{X}(\mathbf{X}'\mathbf{X})^{-1}\mathbf{X}'$.

7.40 Add and subtract $\boldsymbol{\mu}_y$ and $\boldsymbol{\mu}_x$ and show that $\mathbf{M}$ can be written in the form $\mathbf{M} = \boldsymbol{\Sigma}_{yy} - \boldsymbol{\Sigma}_{yx}\boldsymbol{\Sigma}_{xx}^{-1}\boldsymbol{\Sigma}_{xy} + (\boldsymbol{\beta}_0 - \boldsymbol{\mu}_y + \mathbf{B}_1'\boldsymbol{\mu}_x)(\boldsymbol{\beta}_0 - \boldsymbol{\mu}_y + \mathbf{B}_1'\boldsymbol{\mu}_x)' + (\mathbf{B}_1' - \boldsymbol{\Sigma}_{yx}\boldsymbol{\Sigma}_{xx}^{-1})\boldsymbol{\Sigma}_{xx}(\mathbf{B}_1' -$

$\boldsymbol{\Sigma}_{yx}\boldsymbol{\Sigma}_{xx}^{-1})'$. The last two terms are positive semidefinite matrices that vanish for $\boldsymbol{\beta}_0 = \boldsymbol{\mu}_y - \mathbf{B}_1'\boldsymbol{\mu}_x$ and $\mathbf{B}_1 = \boldsymbol{\Sigma}_{xx}^{-1}\boldsymbol{\Sigma}_{xy}$.

7.41 Follow the steps in the hint for Problem 7.40, using $\overline{\mathbf{y}}$, $\overline{\mathbf{x}}$, $\mathbf{S}_{xx}$ and $\mathbf{S}_{yx}$ in place of $\boldsymbol{\mu}_y$, $\boldsymbol{\mu}_x$, $\boldsymbol{\Sigma}_{xx}$, and $\boldsymbol{\Sigma}_{yx}$.

7.43 **(a)** $$\hat{\mathbf{B}} = \begin{pmatrix} 5.624 & 4.588 & 13.009 & 1.987 & -.059 & 2.271 & 2.491 & -.497 & 6.044 & .207 & .257 \\ -.001 & -.007 & .016 & -.005 & .000 & -.003 & -.005 & .002 & -.007 & -.006 & .000 \\ -.001 & .001 & -.063 & .048 & .008 & .016 & .048 & .195 & -.042 & .344 & -.002 \end{pmatrix}.$$

$\Lambda = .115$, $V^{(s)} = 1.208$, $U^{(s)} = 4.892$, $\theta = .809$.

(b) By (7.122) or (7.123), $\Lambda(x_1|x_2) = .407$, $F = 4.241$, $p = .0007$. Similarly, $\Lambda(x_2|x_1) = .394$, $F = 4.473$, $p = .0004$.

(c) $\mathbf{C}$ consists of the single row $(0, 1, -1)$. $\Lambda = .372$, $F = 4.918$, $p = .0002$.

(d) $\mathbf{M}$ consists of the single column $(2, 2, 0, 0, 0, 0, 0, -1, -1, -1, -1)'$. $\Lambda = .788$, $F = 6.624$, $p = .0032$.

7.44 **(a)** $$\hat{\mathbf{B}} = \begin{pmatrix} 3966.406 & 1763.246 & 888.148 & 4491.706 & 308.711 & 299.590 \\ 2.044 & 1.088 & -.222 & 57.252 & -1.007 & 1.039 \\ 26.044 & 10.457 & 3.429 & 11.428 & -1.301 & .285 \end{pmatrix}.$$

$\Lambda = .0867$.

(b) $$\hat{\mathbf{B}} = \begin{pmatrix} 38.078 & 31.170 \\ -.004 & -.005 \\ -.012 & .036 \\ -.017 & .022 \\ .009 & -.006 \\ -.125 & -.275 \\ .111 & .079 \end{pmatrix}.$$

$\Lambda = .0867$.

7.45 **(a)** $$\hat{\mathbf{B}} = \begin{pmatrix} .6264 & 83.2425 \\ .0009 & .0287 \\ -.0010 & -.0127 \\ .0015 & -.0044 \end{pmatrix}.$$

(b) $\Lambda = .724$, $V^{(s)} = .280$, $U^{(s)} = .375$, $\theta = .264$.

(c) $\lambda_1 = .3594$, $\lambda_2 = .0160$. The essential rank of $\hat{\mathbf{B}}_1$ is 1, and θ is more powerful than $U^{(s)}$, Λ, and $V^{(s)}$.

(d) $\Lambda(x_1|x_2, x_3) = .931$, $\Lambda(x_2|x_1, x_3) = .887$, $\Lambda(x_3|x_1, x_2) = .762$.

7.46 **(a)** $$\hat{\mathbf{B}} = \begin{pmatrix} 54.870 & 65.679 & 58.106 \\ .054 & -.048 & .018 \\ -.024 & .163 & .012 \\ .107 & -.036 & .125 \end{pmatrix}, \Lambda = .6650.$$

(b) $\lambda_1 = .3159$, $\lambda_2 = .1385$, $\lambda_3 = .0037$. The essential rank of $\hat{\mathbf{B}}_1$ is 2, and $V^{(s)}$, Λ, and $U^{(s)}$ are more powerful than θ.

(c) $\Lambda(x_1|x_2, x_3) = .942$, $\Lambda(x_2|x_1, x_3) = .847$, $\Lambda(x_3|x_1, x_2) = .829$.

7.47 **(a)**
$$\hat{\mathbf{B}} = \begin{pmatrix} .3957 & -.1483 & .1287 \\ .0005 & .5353 & .1017 \\ .1867 & .1183 & .3856 \\ .2928 & .1067 & -.1190 \\ -.0129 & -.0428 & .0406 \\ .2594 & .2905 & .3087 \\ .0868 & .0120 & .2291 \end{pmatrix}.$$

(b) $\Lambda = .0461$.

(c) $\Lambda(x_5, x_6 | x_1, x_2, x_3, x_4) = .6712$.

7.48 H_0 can be expressed in the form H_0: $\mathbf{CB} = \mathbf{O}$, where

$$\mathbf{C} = \begin{pmatrix} 0 & 1 & -1 & 0 & 0 & 0 & 0 \\ 0 & 0 & 1 & -1 & 0 & 0 & 0 \\ 0 & 0 & 0 & 1 & -1 & 0 & 0 \\ 0 & 0 & 0 & 0 & 1 & -1 & 0 \\ 0 & 0 & 0 & 0 & 0 & 1 & -1 \end{pmatrix}.$$

$$\Lambda = .4841.$$

See Section 7.5.3 for a test statistic.

Chapter 8

8.1 The steps are similar to those leading to (8.3). Note that $\mathbf{S}'_{yx} = \mathbf{S}_{xy}$.

8.3 Use Lagrange multipliers γ_1 and γ_2. Differentiate $F = (\mathbf{a}'\mathbf{S}_{yx}\mathbf{b})^2 + \gamma_1(\mathbf{a}'\mathbf{S}_{yy}\mathbf{a} - 1) + \gamma_2(\mathbf{b}'\mathbf{S}_{xx}\mathbf{b} - 1)$ with respect to each of $\mathbf{a}$, $\mathbf{b}$, γ_1, and γ_2 and set the results equal to zero. Multiply $\partial F/\partial \mathbf{a}$ by $\mathbf{a}'$ and $\partial F/\partial \mathbf{b}$ by $\mathbf{b}'$ to obtain $\gamma_1 = \gamma_2 = (\mathbf{a}'\mathbf{S}_{yx}\mathbf{b})^2 = r^2$ (by definition). Solve for $\mathbf{b}$ in terms of $\mathbf{a}$ to obtain (8.6). Thus $r^2 = r_1^2$.

8.4 **(a)** Differentiate with respect to $\mathbf{a}$ and set the result equal to $\mathbf{0}$ to obtain $\mathbf{a} = (\mathbf{a}'\mathbf{S}_{yy}\mathbf{a}/\mathbf{a}'\mathbf{S}_{yx}\mathbf{b})\mathbf{S}_{yy}^{-1}\mathbf{S}_{yx}\mathbf{b}$. Substitute this into r_{uv}^2 expressed as

$$r_{uv}^2 = (\mathbf{a}'\mathbf{S}_{yx}\mathbf{b})(\mathbf{a}'\mathbf{S}_{yx}\mathbf{b})/(\mathbf{a}'\mathbf{S}_{yy}\mathbf{a})(\mathbf{b}'\mathbf{S}_{xx}\mathbf{b}).$$

(b) Use Theorem 4.1C.

8.5 Use (8.10) and (A.10.7).

8.6 Multiply (8.16) on the left by $\mathbf{S}_{yy}^{-1/2}$. Replace $\mathbf{p}_i$ by $\mathbf{S}_{yy}^{1/2}\mathbf{S}_{yy}^{-1/2}\mathbf{p}_i$ to obtain $(\mathbf{S}_{yy}^{-1}\mathbf{S}_{yx}\mathbf{S}_{xx}^{-1}\mathbf{S}_{xy} - r_i^2\mathbf{I})\mathbf{S}_{yy}^{-1/2}\mathbf{p}_i = \mathbf{0}$. By (8.15), $\mathbf{S}_{yy}^{-1/2}\mathbf{p}_i = \mathbf{a}_i$, where $\mathbf{a}'_i$ is the ith row of $\mathbf{A}$.

8.7 Substitute $\mathbf{A}' = \mathbf{S}_{yy}^{-1/2}\mathbf{P}$ into $\mathbf{S}_{yy}^{-1}\mathbf{S}_{yx}\mathbf{S}_{xx}^{-1}\mathbf{S}_{xy}\mathbf{A}'$ and use (8.14) and its transpose to show that $\mathbf{S}_{yy}^{-1}\mathbf{S}_{yx}\mathbf{S}_{xx}^{-1}\mathbf{S}_{xy}\mathbf{A}'$ can be written as

$$\begin{aligned} \mathbf{S}_{yy}^{-1}\mathbf{S}_{yx}\mathbf{S}_{xx}^{-1}\mathbf{S}_{xy}\mathbf{A}' &= \mathbf{S}_{yy}^{-1/2}\mathbf{PDQ}'\mathbf{QDP}'\mathbf{P} \\ &= \mathbf{S}_{yy}^{-1/2}\mathbf{PD}^2. \end{aligned}$$

8.8 To simplify notation, use $i = 1$ and $j = 2$. To show that $\widehat{\text{cov}}(u_1, u_2) = 0$, use (8.6) to write

$$(\mathbf{S}_{yx}\mathbf{S}_{xx}^{-1}\mathbf{S}_{xy} - r_1^2\mathbf{S}_{yy})\mathbf{a}_1 = \mathbf{0}, \tag{1}$$

$$(\mathbf{S}_{yx}\mathbf{S}_{xx}^{-1}\mathbf{S}_{xy} - r_2^2\mathbf{S}_{yy})\mathbf{a}_2 = \mathbf{0}. \tag{2}$$

Show that multiplying (1) by $\mathbf{a}_2'$, (2) by $\mathbf{a}_1'$, and then subtracting give

$$(r_1^2 - r_2^2)\mathbf{a}_1'\mathbf{S}_{yy}\mathbf{a}_2 = 0.$$

Since $r_1^2 \neq r_2^2$ with probability 1, it follows that $\mathbf{a}_1'\mathbf{S}_{yy}\mathbf{a}_2 = 0$, and therefore, by (1.76), the (sample) correlation between u_1 and u_2 is 0.

To show that $\widehat{\text{cov}}(v_1, v_2) = 0$, write (8.8) in terms of r_1^2, r_2^2, $\mathbf{b}_1$, and $\mathbf{b}_2$:

$$(\mathbf{S}_{xy}\mathbf{S}_{yy}^{-1}\mathbf{S}_{yx} - r_1^2\mathbf{S}_{xx})\mathbf{b}_1 = \mathbf{0}, \tag{3}$$

$$(\mathbf{S}_{xy}\mathbf{S}_{yy}^{-1}\mathbf{S}_{yx} - r_2^2\mathbf{S}_{xx})\mathbf{b}_2 = \mathbf{0}. \tag{4}$$

Show that multiplying (3) and (4) by $\mathbf{b}_2'$ and $\mathbf{b}_1'$, respectively, leads to $\mathbf{b}_1'\mathbf{S}_{xx}\mathbf{b}_2 = 0$, from which $\widehat{\text{cov}}(v_1, v_2) = 0$.

For u_1 and v_2, show that multiplying (1) by $\mathbf{b}_2'\mathbf{S}_{xy}\mathbf{S}_{yy}^{-1}$ and (4) by $\mathbf{a}_1'\mathbf{S}_{yx}\mathbf{S}_{xx}^{-1}$ gives $\mathbf{a}_1'\mathbf{S}_{yx}\mathbf{b}_2 = 0$. By (1.84), the correlation between u_1 and v_2 is 0. The correlation between u_2 and v_1 is similarly shown to be 0.

8.10 From (8.3), $\mathbf{a} = (\mathbf{a}'\mathbf{S}_{yx}\mathbf{b}/r^2\mathbf{b}'\mathbf{S}_{xx}\mathbf{b})\mathbf{S}_{yy}^{-1}\mathbf{S}_{yx}\mathbf{b}$.

8.11 If $\mathbf{a}$ is an eigenvector of $\mathbf{AB} = \mathbf{S}_{yy}^{-1}\mathbf{S}_{yx}\mathbf{S}_{xx}^{-1}\mathbf{S}_{xy}$, then $\mathbf{Ba} = \mathbf{S}_{xx}^{-1}\mathbf{S}_{xy}\mathbf{a}$ is an eigenvector of $\mathbf{BA} = \mathbf{S}_{xx}^{-1}\mathbf{S}_{xy}\mathbf{S}_{yy}^{-1}\mathbf{S}_{yx}$.

8.12 To simplify notation, assume that $E(\mathbf{y}', \mathbf{x}') = \mathbf{0}'$. Then

$$\begin{aligned} E(\mathbf{a}'\mathbf{y} - \mathbf{b}'\mathbf{x})^2 &= E(\mathbf{a}'\mathbf{y} - \mathbf{b}'\mathbf{x})(\mathbf{y}'\mathbf{a} - \mathbf{x}'\mathbf{b}) \\ &= E(\mathbf{a}'\mathbf{y}\mathbf{y}'\mathbf{a} - \mathbf{a}'\mathbf{y}\mathbf{x}'\mathbf{b} - \mathbf{b}'\mathbf{x}\mathbf{y}'\mathbf{a} + \mathbf{b}'\mathbf{x}\mathbf{x}'\mathbf{b}) \\ &= \mathbf{a}'\boldsymbol{\Sigma}_{yy}\mathbf{a} - 2\mathbf{a}'\boldsymbol{\Sigma}_{yx}\mathbf{b} + \mathbf{b}'\boldsymbol{\Sigma}_{xx}\mathbf{b}. \end{aligned}$$

Differentiating with respect to $\mathbf{a}$ and with respect to $\mathbf{b}$ leads to the result.

8.13 Multiply the expression preceding (8.23) by $\mathbf{D}_y$ (on the left) and factor out $\mathbf{D}_y$ on the right.

8.14 By Problem 1.28, $\widehat{\text{cov}}(\mathbf{y}^*) = \mathbf{S}_{y^*y^*} = \mathbf{F}\mathbf{S}_{yy}\mathbf{F}'$ and $\widehat{\text{cov}}(\mathbf{x}^*) = \mathbf{S}_{x^*x^*} = \mathbf{G}\mathbf{S}_{xx}\mathbf{G}'$. By (1.84), $\widehat{\text{cov}}(\mathbf{y}^*, \mathbf{x}^*) = \mathbf{S}_{y^*x^*} = \mathbf{F}\mathbf{S}_{yx}\mathbf{G}'$. Solve these for $\mathbf{S}_{yy}$, $\mathbf{S}_{xx}$, $\mathbf{S}_{yx}$, and $\mathbf{S}_{xy}$ and substitute into (8.7) to obtain $[\mathbf{F}'\mathbf{S}_{y^*y^*}^{-1}\mathbf{S}_{y^*x^*}\mathbf{S}_{x^*x^*}^{-1}\mathbf{S}_{x^*y^*}(\mathbf{F}')^{-1} - r^2\mathbf{I}]\mathbf{a} = \mathbf{0}$. Multiply (on the left) by $(\mathbf{F}')^{-1}$ and factor it out on the right.

8.15 By (7.66), $R^2_{u_i|\mathbf{x}} = \mathbf{s}_{u_ix}'\mathbf{S}_{xx}^{-1}\mathbf{s}_{u_ix}/s_{u_i}^2$. Using (1.79), show that $\mathbf{s}_{u_ix} = \mathbf{S}_{xy}\mathbf{a}_i$. By (1.72), $s_{u_i}^2 = \mathbf{a}_i'\mathbf{S}_{yy}\mathbf{a}_i$. Then $R^2_{u_i|\mathbf{x}} = \mathbf{a}_i'\mathbf{S}_{yx}\mathbf{S}_{xx}^{-1}\mathbf{S}_{xy}\mathbf{a}_i/\mathbf{a}_i'\mathbf{S}_{yy}\mathbf{a}_i$. Use (8.6) to show that this is equal to r_i^2.

8.16 r_1 cannot be greater than 1 because it is a correlation. To show that it cannot equal 1, factor $|\mathbf{S}|$ using (A.7.4):

$$\begin{aligned}|\mathbf{S}| &= |\mathbf{S}_{xx}||\mathbf{S}_{yy} - \mathbf{S}_{yx}\mathbf{S}_{xx}^{-1}\mathbf{S}_{xy}| \\ &= |\mathbf{S}_{xx}||\mathbf{S}_{yy}||\mathbf{S}_{yy}^{-1}||\mathbf{S}_{yy} - \mathbf{S}_{yx}\mathbf{S}_{xx}^{-1}\mathbf{S}_{xy}| \\ &= |\mathbf{S}_{xx}||\mathbf{S}_{yy}||\mathbf{I} - \mathbf{S}_{yy}^{-1}\mathbf{S}_{yx}\mathbf{S}_{xx}^{-1}\mathbf{S}_{xy}|.\end{aligned}$$

If $r_1 = 1$, then $|\mathbf{S}| = 0$ (Why?), which contradicts $|\mathbf{S}| > 0$.

8.17 **(a)** $u^* = \mathbf{a}^{*\prime}\mathbf{y}^* = [(\mathbf{F}')^{-1}\mathbf{a}]'\mathbf{F}\mathbf{y}$, where $\mathbf{y}^* = \mathbf{F}\mathbf{y}$.

(b) See the hint for Problem 8.14.

8.18 $\max_{H_1} L$ in (8.33) follows from (3.22). Under H_0, $\boldsymbol{\mu} = (\boldsymbol{\mu}_y', \boldsymbol{\mu}_x')'$ is estimated by $\hat{\boldsymbol{\mu}} = (\bar{\mathbf{y}}', \bar{\mathbf{x}}')'$, and $\boldsymbol{\Sigma}$ and $\hat{\boldsymbol{\Sigma}}$ become

$$\boldsymbol{\Sigma}_0 = \begin{pmatrix} \boldsymbol{\Sigma}_{yy} & \mathbf{O} \\ \mathbf{O} & \boldsymbol{\Sigma}_{xx} \end{pmatrix} \text{ and } \hat{\boldsymbol{\Sigma}}_0 = \begin{pmatrix} \hat{\boldsymbol{\Sigma}}_{yy} & \mathbf{O} \\ \mathbf{O} & \hat{\boldsymbol{\Sigma}}_{xx} \end{pmatrix}.$$

Using (2.48), we obtain

$$\max_{H_0} L = \frac{1}{\left(\sqrt{2\pi}\right)^{np} |\hat{\boldsymbol{\Sigma}}_{yy}|^{n/2} |\hat{\boldsymbol{\Sigma}}_{xx}|^{n/2}} e^{-\operatorname{tr}\hat{\boldsymbol{\Sigma}}_0^{-1}(n\hat{\boldsymbol{\Sigma}}_0)/2},$$

and (8.34) follows.

8.19 Verify (8.33) and (8.34) and take the ratio.

8.20 Using the definitions and results at the beginning of Section 8.3.1, we obtain

$$\frac{|\mathbf{S}|}{|\mathbf{S}_{yy}||\mathbf{S}_{xx}|} = \frac{|\mathbf{D}\mathbf{R}\mathbf{D}|}{|\mathbf{D}_y\mathbf{R}_{yy}\mathbf{D}_y||\mathbf{D}_x\mathbf{R}_{xx}\mathbf{D}_x|} = \frac{|\mathbf{D}||\mathbf{R}||\mathbf{D}|}{|\mathbf{D}_x||\mathbf{R}_{xx}||\mathbf{D}_x||\mathbf{D}_y||\mathbf{R}_{yy}||\mathbf{D}_y|}.$$

8.23 See (5.23) and Problem 5.10.

8.24 See the proof of Theorem 5.7B.

8.25 **(a)** $r_1 = .8993$, $r_2 = .6318$.

(b)

	$\mathbf{c}_1$	$\mathbf{c}_2$
y_1	.041	.089
y_2	.135	−.686
y_3	−.017	.218
y_4	.410	.085
y_5	.144	.400
y_6	−.273	−.071
y_7	.457	.036
y_8	.414	.466
y_9	.196	−.042
y_{10}	.116	.350
y_{11}	.076	−.159

	$\mathbf{d}_1$	$\mathbf{d}_2$
x_1	−.557	1.042
x_2	.585	1.026

(c) $\Lambda = .115$, $V^{(s)} = 1.208$, $U^{(s)} = 4.892$, $\theta = .809$.

(d)

k	Λ	Approximate F	p-Value
1	.115	5.671	< .0001
2	.601	2.192	.0441

8.26 **(a)** $r_1 = .7609$, $r_2 = .1213$.

(b)

	$\mathbf{c}_1$	$\mathbf{c}_2$
y_1	−.562	.050
y_2	.062	1.233
y_3	−.654	.742

	$\mathbf{d}_1$	$\mathbf{d}_2$
x_1	.947	.990
x_2	−.076	1.368

(c) $\Lambda = .415$, $V^{(s)} = .594$, $U^{(s)} = 1.390$, $\theta = .579$.

(d)

k	Λ	Approximate F	p-Value
1	.415	4.973	.0004
2	.985	.209	.8124

8.27 **(a)** $r_1 = .2156$, $r_2 = .1195$.

(b)

	$\mathbf{c}_1$	$\mathbf{c}_2$
y_1	1.015	−.043
y_2	.223	.992

	$\mathbf{d}_1$	$\mathbf{d}_2$
x_1	−.434	.567
x_2	.719	.766
x_3	.382	−.696

(c) $\Lambda = .940$, $V^{(s)} = .061$, $U^{(s)} = .063$, $\theta = .0465$.

(d)

k	Λ	Approximate F	p-Value
1	.940	.210	.972
2	.986	.152	.860

8.28 **(a)** $r_1 = .5909$, $r_2 = .3090$, $r_3 = .0526$.

(b)

	$\mathbf{c}_1$	$\mathbf{c}_2$	$\mathbf{c}_3$
y_1	−.132	.928	.375
y_2	.940	.318	−.170
y_3	−.241	.424	−.879

	$\mathbf{d}_1$	$\mathbf{d}_2$	$\mathbf{d}_3$
x_1	−.753	−.161	.823
x_2	.992	.192	.384
x_3	.018	1.026	−.219

(c) $\Lambda = .587$, $V^{(s)} = .447$, $U^{(s)} = .645$, $\theta = .349$.

(d)

k	Λ	Approximate F	p-Value
1	.587	.930	.512
2	.902	.397	.809
3	.997	.044	.836

8.29 **(a)** $\mathbf{R}_{yx} = \begin{pmatrix} .122 & -.035 & .284 \\ -.216 & .412 & -.049 \\ .040 & -.136 & .130 \end{pmatrix}.$

The largest correlation in $\mathbf{R}_{yx}$ is $.412 < r_1 = .5909$.

(b) From Problem 8.28, we have $r_1^2 = .3491$ for all three y's and all three x's. Leaving out y_1 (using y_2, y_3, x_1, x_2, x_3), we obtain $r_1^2 = .3446$. Leaving out x_1 (using y_1, y_2, y_3, x_2, x_3), we obtain $r_1^2 = .1973$.

(c) $R^2_{u_1|\mathbf{x}} = .3491$, $R^2_{v_1|\mathbf{y}} = .3491$, $r_1^2 = .3491$.

(d) $r_1^2 = .3491$ exceeds each of the following: $R^2_{y_1|\mathbf{x}} = .086$, $R^2_{y_2|\mathbf{x}} = .329$, $R^2_{y_3|\mathbf{x}} = .039$, $R^2_{x_1|\mathbf{y}} = .058$, $R^2_{x_2|\mathbf{y}} = .183$, $R^2_{x_3|\mathbf{y}} = .105$.

8.31 **(a)** $r_1 = .8728$, $r_2 = .7975$.

(b)

	$\mathbf{c}_1$	$\mathbf{c}_2$
y_1	$-.427$	$-.308$
y_2	$-.211$	.766
y_3	$-.140$	.189
y_4	.804	$-.591$
y_5	$-.430$	$-.621$
y_6	.302	.132

	$\mathbf{d}_1$	$\mathbf{d}_2$
x_1	.993	$-.121$
x_2	.102	.995

(c) $\Lambda = .087$, $V^{(s)} = 1.398$, $U^{(s)} = 4.946$, $\theta = .762$.

(d)

k	Λ	Approximate F	p-Value
1	.087	13.578	$< .0001$
2	.364	12.228	$< .0001$

8.32 **(a)** $r_1 = .9636$, $r_2 = .4937$, $r_3 = .3826$.

(b)

	$\mathbf{c}_1$	$\mathbf{c}_2$	$\mathbf{c}_3$
y_1	.390	1.565	-1.088
y_2	.456	-1.549	$-.668$
y_3	.229	$-.013$	1.837

	$\mathbf{d}_1$	$\mathbf{d}_2$	$\mathbf{d}_3$
x_1	.269	-1.609	$-.364$
x_2	.272	.356	-1.373
x_3	.185	.906	-2.123
x_4	$-.014$	.074	.301
x_5	.318	.062	.273
x_6	.104	.272	.890

(c) $\Lambda = .0461$, $V^{(s)} = 1.319$, $U^{(s)} = 13.483$, $\theta = .929$.

(d)

k	Λ	Approximate F	p-Value
1	.046	8.089	$< .0001$
2	.646	1.321	.2433
3	.854	1.201	.3325

8.33 **(a)** $\mathbf{R}_{yx} = \begin{pmatrix} .748 & .818 & .885 & .646 & .813 & .637 \\ .883 & .720 & .806 & .571 & .820 & .676 \\ .746 & .734 & .759 & .668 & .821 & .693 \end{pmatrix}$

The largest correlation in $\mathbf{R}_{yx}$ is $.885 < r_1 = .9636$.

(b) From Problem 8.32, we have $r_1^2 = .9285$ based on all three y's and all six x's. Leaving out y_1 (using $y_2, y_3, x_1, \ldots, x_6$), we obtain $r_1^2 = .8999$. Leaving out x_1 (using $y_1, y_2, y_3, x_2, \ldots, x_6$) we obtain $r_1^2 = .9146$.

(c) $R^2_{u_1|\mathbf{x}} = .9285$, $R^2_{v_1|\mathbf{y}} = .9285$, $r_1^2 = .9285$.

(d) $r_1^2 = .9285$ exceeds each of the following: $R^2_{y_1|\mathbf{x}} = .846$, $R^2_{y_2|\mathbf{x}} = .848$, $R^2_{y_3|\mathbf{x}} = .778$, $R^2_{x_1|\mathbf{y}} = .791$, $R^2_{x_2|\mathbf{y}} = .688$, $R^2_{x_3|\mathbf{y}} = .812$, $R^2_{x_4|\mathbf{y}} = .476$, $R^2_{x_5|\mathbf{y}} = .777$, $R^2_{x_6|\mathbf{y}} = .531$.

Chapter 9

9.1 Using a Lagrange multiplier γ, we first differentiate with respect to $\mathbf{a}$:

$$\frac{\partial}{\partial \mathbf{a}}\left[\mathbf{a}'\mathbf{S}\mathbf{a} + \gamma(1 - \mathbf{a}'\mathbf{a})\right] = \mathbf{0},$$

$$2\mathbf{S}\mathbf{a} - \gamma 2\mathbf{a} = \mathbf{0},$$

from which we obtain $\gamma = \lambda_1$, the first eigenvalue. Differentiating with respect to γ yields the constraint $\mathbf{a}'\mathbf{a} = 1$. Thus the maximum of $\mathbf{a}'\mathbf{S}\mathbf{a}$ is given by λ_1, the first eigenvalue, and the value of $\mathbf{a}$ that produces the maximum is $\mathbf{a}_1$, the associated eigenvector, subject to $\mathbf{a}_1'\mathbf{a}_1 = 1$.

9.2 After differentiation with respect to $\mathbf{a}_2$, we obtain

$$\mathbf{S}\mathbf{a}_2 - \lambda_2\mathbf{a}_2 + \tfrac{1}{2}\gamma(\mathbf{a}_2'\mathbf{a}_2)\mathbf{a}_1 = \mathbf{0}.$$

Multiplying by $\mathbf{a}_1'$ gives $\gamma = -2\mathbf{a}_1'\mathbf{S}\mathbf{a}_2$, since $\mathbf{a}_1'\mathbf{a}_2 = 0$ and $\mathbf{a}_2'\mathbf{a}_2 = \mathbf{a}_1'\mathbf{a}_1 = 1$. To show that $\gamma = -2\mathbf{a}_1'\mathbf{S}\mathbf{a}_2 = 0$, multiply $\mathbf{S}\mathbf{a}_1 = \lambda_1\mathbf{a}_1$ by $\mathbf{a}_2'$ (see Problem 9.6 below).

9.3 We seek to maximize

$$\mathbf{a}_2'\mathbf{S}\mathbf{a}_2 + \gamma(1 - \mathbf{a}_2'\mathbf{a}_2) + \theta\mathbf{a}_1'\mathbf{a}_2,$$

where γ and θ are Lagrange multipliers. Differentiate with respect to $\mathbf{a}_2$ to obtain

$$2\mathbf{S}\mathbf{a}_2 - \gamma\mathbf{a}_2 + \theta\mathbf{a}_1 = \mathbf{0}.$$

Show that multiplication on the left by $\mathbf{a}_1'$ leads to $\theta = 0$.

9.4 The sum of squared projections onto the line is given by

$$\sum_{i=1}^{n} z_{1i}^2 = \sum_{i=1}^{n} [\mathbf{a}_1'(\mathbf{y}_i - \overline{\mathbf{y}})]^2,$$

where we have used the centered form. Show that this is equal to $(n-1)\mathbf{a}_1'\mathbf{S}\mathbf{a}_1$, which equals $(n-1)\lambda_1$, where λ_1 is the largest eigenvalue of $\mathbf{S}$.

9.5 Use (9.2) for $\mathbf{a}_j$ and multiply by $\mathbf{a}_j'$.

9.6 Use (9.2) for $\mathbf{a}_j$ and multiply by $\mathbf{a}_i'$ or use $\mathbf{A}'\mathbf{S}\mathbf{A} = \mathbf{D}_\lambda$, where $\mathbf{A}$ is defined in (9.4).

9.7 Multiply $(\mathbf{S} - \lambda\mathbf{I})\mathbf{a} = \mathbf{0}$ by $\mathbf{C}$ and insert $\mathbf{C}'(\mathbf{C}')^{-1}$ to obtain

$$(\mathbf{CSC}' - \lambda\mathbf{CC}')(\mathbf{C}')^{-1}\mathbf{a} = \mathbf{0}.$$

Thus λ is not an eigenvalue of $\mathbf{CSC}'$ and $(\mathbf{C}')^{-1}\mathbf{a}$ is not an eigenvector. Multiply by $(\mathbf{CC}')^{-1}$ to show that λ and $(\mathbf{C}')^{-1}\mathbf{a}$ correspond to $(\mathbf{CC}')^{-1}\mathbf{CSC}'$.

9.8 If y_p is uncorrelated with $y_1, \ldots, y_{p-1}$, then $\mathbf{S}\mathbf{a}_p = \lambda_p\mathbf{a}_p$ for the pth eigenvalue and eigenvector becomes

$$\begin{pmatrix} \mathbf{S}_{11} & \mathbf{0} \\ \mathbf{0}' & s_p^2 \end{pmatrix} \begin{pmatrix} \mathbf{a}_{p-1} \\ a_{pp} \end{pmatrix} = \begin{pmatrix} \lambda_p\mathbf{a}_{p-1} \\ \lambda_p a_{pp} \end{pmatrix},$$

where $\mathbf{a}_{p-1}$ contains the first $p-1$ elements of $\mathbf{a}_p$ and a_{pp} is the pth element. From $s_p^2 a_{pp} = \lambda_p a_{pp}$, we obtain $\lambda_p = s_p^2$. Show that $\mathbf{S}_{11}\mathbf{a}_{p-1} = \lambda_p\mathbf{a}_{p-1}$ leads to $\mathbf{a}_{p-1} = \mathbf{0}$.

9.9 $\mathbf{z}_i'\mathbf{z}_i = \mathbf{y}_i'\mathbf{A}'\mathbf{A}\mathbf{y}_i = \mathbf{y}_i'\mathbf{y}_i$, since $\mathbf{A}'\mathbf{A} = \mathbf{I}$.

9.11 $P(-w_{\alpha/2} \le w \le w_{\alpha/2}) = P\{-w_{\alpha/2} \le [(\lambda_i - \gamma_i)/\gamma_i\sqrt{2/(n-2)}] \le w_{\alpha/2}\}$. Solve this for γ_i to obtain (9.20).

9.12 Under H_1, we obtain [see (3.22) or (8.33)]

$$\max_{H_1} L = \frac{1}{(\sqrt{2\pi})^{np}|\hat{\mathbf{\Sigma}}|^{n/2}} e^{-np/2}.$$

Under H_0, $\mathbf{\Sigma}_0 = \text{diag}(\sigma_{11}, \sigma_{22}, \ldots, \sigma_{pp})$ and $\hat{\mathbf{\Sigma}}_0 = \text{diag}(\hat{\sigma}_{11}, \hat{\sigma}_{22}, \ldots, \hat{\sigma}_{pp})$, where $\hat{\sigma}_{ii} = \sum_{k=1}^n (y_{ki} - \bar{y}_i)^2/n$ is the ith diagonal element of $\mathbf{W}/n$ [see (1.33) and Theorem 2.3A]. Since $\hat{\boldsymbol{\mu}} = \bar{\mathbf{y}}$, we obtain from (2.48)

$$\max_{H_0} L = \frac{1}{(\sqrt{2\pi})^{np}|\hat{\mathbf{\Sigma}}_0|^{n/2}} e^{-\operatorname{tr}\hat{\mathbf{\Sigma}}_0^{-1}(n\hat{\mathbf{\Sigma}}_0)/2} = \frac{1}{(\sqrt{2\pi})^{np}\prod_{i=1}^n \hat{\sigma}_{ii}} e^{-np/2}.$$

9.13 Substitute $\mathbf{E}/\nu_E$ for $\mathbf{S}$ and e_{ii}/ν_E for s_{ii}.

9.14 $(z_{p-k,i}\mathbf{a}_{p-k} + \cdots + z_{pi}\mathbf{a}_p)'(z_{p-k,i}\mathbf{a}_{p-k} + \cdots + z_{pi}\mathbf{a}_p) = \sum_{j=p-k}^p z_{ji}^2\mathbf{a}_j'\mathbf{a}_j + \sum_{j\ne r} z_{ji}z_{ri}\mathbf{a}_j'\mathbf{a}_r$. The result follows from $\mathbf{a}_j'\mathbf{a}_j = 1$ and $\mathbf{a}_j'\mathbf{a}_r = 0$, $j \ne r$.

9.15 (a) Use (A.5.7).

9.17 Using (2.48), (A.5.7), and (A.7.7), show that the likelihood function that corresponds to $\mathbf{\Sigma}_0 = \sigma^2[(1-\rho)\mathbf{I} + \rho\mathbf{J}]$ and $\hat{\mathbf{u}} = \bar{\mathbf{y}}$ can be written as

$$\begin{aligned} L(\bar{\mathbf{y}}, \mathbf{\Sigma}_0) &= \frac{1}{(\sqrt{2\pi})^{np}(\lambda_1\lambda_2^{p-1})^{n/2}} e^{-\operatorname{tr}(\mathbf{W} - \rho\sigma^2\mathbf{JW}/\lambda_1)/2\lambda_2} \\ &= \frac{1}{(\sqrt{2\pi})^{np}(\lambda_1\lambda_2^{p-1})^{n/2}} e^{-\sum_{i=1}^p s_{ii}/2\lambda_2 + \rho\sigma^2\sum_{ij} s_{ij}/2\lambda_1\lambda_2}, \end{aligned}$$

where $\lambda_1 = \sigma^2[1 + (p-1)\rho]$ and $\lambda_2 = \sigma^2(1-\rho)$. Differentiate $\ln L$ with respect to σ^2 and $\sigma^2\rho = t$, say.

9.18 For $(\mathbf{J} - \lambda\mathbf{I})\mathbf{x} = (\mathbf{jj}' - \lambda\mathbf{I})\mathbf{x} = \mathbf{0}$, a solution is given by $\lambda = p$ and $\mathbf{x} = \mathbf{j}$. The other eigenvalues of $\mathbf{J}$ are 0 (Why?), and by property 5 in Section A.10, the eigenvalues of $c\mathbf{I} + \mathbf{J}$ are $c + p, c, \ldots, c$. Show that $(1/\sqrt{p})\mathbf{j}$ is a (normalized) solution to $[c\mathbf{I} + \mathbf{J} - (c + p)\mathbf{I}]\mathbf{x}_1 = \mathbf{0}$.

9.20 $(\mathbf{Z}'\mathbf{Z})^{-1}\mathbf{Z}'\mathbf{y} = (\mathbf{A}'\mathbf{X}_c'\mathbf{X}_c\mathbf{A})^{-1}\mathbf{A}'\mathbf{X}_c'\mathbf{y} = \mathbf{D}^{-1}\mathbf{A}'\mathbf{X}_c'\mathbf{y}$.

9.21 $\hat{\boldsymbol{\beta}}_1 = \mathbf{A}\hat{\boldsymbol{\delta}} = \mathbf{A}\mathbf{D}^{-1}\mathbf{A}'\mathbf{X}_c'\mathbf{y} = \mathbf{A}\mathbf{A}^{-1}(\mathbf{X}_c'\mathbf{X}_c)^{-1}(\mathbf{A}')^{-1}\mathbf{A}'\mathbf{X}_c'\mathbf{y}$.

9.22 **(a)** χ^2-approximation: $s^2 = .843$, $s^2 r = .597$, $r = .708$, $\nu = 34$, $u = .03998$, $u' = 98.927$, .05 critical value $= 59.304$. F-approximation: $c_1 = .0962$, $c_2 = .0133$, $\delta_1 = .0962$, $\delta_2 = 11{,}178.3$, $F = 2.291$, .05 critical value $= 1.381$.

(b) $\bar{r} = .716$, $q = 6.976$, $u = 105.06$, .05 critical value $= 49.802$.

9.23 **(a)** $u' = 1185.76$, .05 critical value $= 83.675$; $F = 18.456$, .05 critical value $= 1.308$.

(b) $u = 270.382$, .05 critical value $= 72.153$.

9.24 $u = 131.039$, .05 critical value $= 31.410$.

9.25 **R** is given in Example 1.7; **S** is given in the answer to Problem 1.53. The eigenvalues of **S** and **R** are as follows:

S			R		
λ_i	$\lambda_i/\sum_j \lambda_j$	Cumulative	λ_i	$\lambda_i/\sum_j \lambda_j$	Cumulative
200.4	.684	.684	3.42	.683	.683
36.1	.123	.807	.61	.123	.806
34.1	.116	.924	.57	.114	.921
15.0	.051	.975	.27	.054	.975
7.4	.025	1.000	.13	.025	1.000

The first three eigenvectors of **S** and **R** are as follows:

S			R		
$\mathbf{a}_1$	$\mathbf{a}_2$	$\mathbf{a}_3$	$\mathbf{a}_1$	$\mathbf{a}_2$	$\mathbf{a}_3$
.47	−.58	−.42	.44	−.20	−.68
.39	−.11	.45	.45	−.43	.35
.49	.10	−.48	.47	.37	−.38
.47	−.12	.62	.45	−.39	.33
.41	.80	−.09	.41	.70	.41

The variances in **S** are 65.1, 46.1, 60.7, 62.8, and 58.2. These are nearly identical. The covariances in **S** are also similar in magnitude, as are the correlations in **R**. Consequently, the percent of variance explained by the eigenvalues of **S** and **R** are essentially the same. The interpretation (based on the second eigenvector) of the second principal component from **S** is slightly different from that of the second one from **R**.

9.26 The covariance matrix **S** is given in the answer to Problem 1.63. The variances on the diagonal of **S** are 61.1, 41.5, 64.4, 876.9, 621.6, and 293.0.

$$\mathbf{R} = \begin{pmatrix} 1.000 & -.101 & -.089 & .122 & -.035 & .284 \\ -.101 & 1.000 & -.048 & -.216 & .412 & -.049 \\ -.089 & -.048 & 1.000 & .040 & -.136 & .130 \\ .122 & -.216 & .040 & 1.000 & .363 & .283 \\ -.035 & .412 & -.136 & .363 & 1.000 & -.001 \\ .284 & -.049 & .130 & .283 & -.001 & 1.000 \end{pmatrix}.$$

The eigenvalues of **S** and **R** are as follows:

S			R		
λ_i	$\lambda_i/\sum_j \lambda_j$	Cumulative	λ_i	$\lambda_i/\sum_j \lambda_j$	Cumulative
1066.37	.544	.544	1.556	.259	.259
496.87	.254	.798	1.483	.247	.506
251.84	.129	.927	1.063	.177	.684
67.94	.035	.961	.965	.161	.844
48.82	.025	.986	.629	.105	.949
26.64	.014	1.000	.304	.051	1.000

The relative sizes of the eigenvalues of **S** would suggest retention of either two or three components. The first three components from **S** account for 93% of the variance; it requires five components from **R** to reach a similar percentage. The first three eigenvectors from **S** and the first five from **R** are as follows:

	S			R				
	$\mathbf{a}_1$	$\mathbf{a}_2$	$\mathbf{a}_3$	$\mathbf{a}_1$	$\mathbf{a}_2$	$\mathbf{a}_3$	$\mathbf{a}_4$	$\mathbf{a}_5$
y_1	.026	−.07	.13	.45	−.09	−.52	.40	.59
y_2	−.002	.16	.10	−.27	.57	.07	.57	−.03
y_3	−.003	−.07	.04	.10	−.24	.82	.28	.43
y_4	.845	−.43	−.30	.60	.24	.19	−.49	.07
y_5	.510	.82	.20	.13	.74	.09	−.12	.18
y_6	.158	−.33	.92	.58	−.04	.10	.43	−.66

In **S** the first three components are heavily influenced by the last three variables since the variances of these variables are much larger than the variances of the first three variables. In the components of **R**, all six variables contribute significantly.

9.27 The variances on the diagonal of **S** are .154, 1.13, 15.5, 1.14, .0093, .527, .972, 3.81, 2.43, 34.1, and .0026. By extension of property 6 in Section 9.2.2, we would expect y_3 and y_{10} to determine the first two principal components of **S** because of the large variances of y_3 and y_{10}. Similarly, the next two largest variances are those of y_8 and y_9. These variables could be expected to influence the third and fourth components. This pattern does indeed emerge in the eigenvectors of **S** shown below.

The first component of **S** represents a contrast or shape component comparing y_3 and y_{10}. The second component contrasts y_3 and y_{10} versus all other variables. The third and fourth components are dominated by y_8 and y_9, respectively.

In **R**, the pattern is different from that of **S**. All variables contribute to the first few components.

The first two components of **S** account for 86% of the variance, and the first four account for 97%. On the other hand, five components of **R** are required to reach 84%. The eigenvalues of **S** and **R** are as follows.

S			**R**		
λ_i	$\lambda_i/\sum_j \lambda_j$	Cumulative	λ_i	$\lambda_i/\sum_j \lambda_j$	Cumulative
37.234	.62255	.62255	4.37	.397	.397
14.172	.23695	.85950	1.73	.157	.555
3.886	.06497	.92448	1.37	.125	.680
2.724	.04555	.97003	1.06	.096	.776
.874	.01462	.98464	.65	.059	.835
.562	.00940	.99404	.49	.045	.880
.181	.00302	.99707	.46	.042	.922
.124	.00208	.99914	.36	.032	.954
.044	.00073	.99988	.29	.026	.980
.006	.00011	.99998	.17	.015	.995
.001	.00002	1.00000	.05	.005	1.000

Some eigenvectors of **S** and **R** are as follows:

S

	$\mathbf{a}_1$	$\mathbf{a}_2$	$\mathbf{a}_3$	$\mathbf{a}_4$
y_1	.017	−.005	−.042	.063
y_2	.050	−.058	.159	.396
y_3	−.313	.909	−.104	.163
y_4	.110	−.115	−.043	.121
y_5	.001	−.006	.018	−.007
y_6	.075	−.078	−.095	.100
y_7	.091	−.127	.038	.178
y_8	.014	.103	.963	.086
y_9	−.008	−.103	−.137	.865
y_{10}	.934	.337	−.049	−.001
y_{11}	−.002	−.005	−.012	.003

R

	$\mathbf{a}_1$	$\mathbf{a}_2$	$\mathbf{a}_3$	$\mathbf{a}_4$	$\mathbf{a}_5$
y_1	.199	−.276	.301	.395	.768
y_2	.235	.208	.564	−.054	−.287
y_3	−.347	−.126	.146	.214	−.094
y_4	.431	.075	−.108	.064	.014
y_5	.080	.536	−.252	−.342	.434
y_6	.452	−.052	−.137	−.004	−.042
y_7	.428	.163	.028	−.060	.013
y_8	−.091	.626	.261	.203	−.024
y_9	.159	−.190	.590	−.403	−.027
y_{10}	.267	.112	−.083	.658	−.294
y_{11}	.317	−.320	−.229	−.182	−.202

9.28 (a) The variances on the diagonal of **S** are 30,428; 116; 9.88×10^9; 873,223; 484,304; 4.84×10^7; and 23. The eigenvalues of **S** and **R** are as follows:

S			R		
λ_i	$\lambda_i/\sum_j \lambda_j$	Cumulative	λ_i	$\lambda_i/\sum_j \lambda_j$	Cumulative
9.93×10^9	.99982	.99982	5.1411	.7344	.7344
1.51×10^6	.00015	.99997	1.5878	.2268	.9613
1.87×10^5	.00002	.99999	.2056	.0294	.9906
1.16×10^5	.00001	1.00000	.0487	.0070	.9976
2137.14	.00000	1.00000	.0147	.0021	.9997
.85	.00000	1.00000	.0019	.0003	1.0000
.01	.00000	1.00000	.0002	.0000	1.0000

Due to the large variance of y_2, the first eigenvalue of **S** accounts for 99.98% of the variance, and y_2 completely dominates the first component, as seen in the first eigenvector of **S** below. The first three eigenvectors of **S** and **R** are as follows:

	S			R		
	$\mathbf{a}_1$	$\mathbf{a}_2$	$\mathbf{a}_3$	$\mathbf{a}_1$	$\mathbf{a}_2$	$\mathbf{a}_3$
y	.00124	−.0789	.1383	.343	.477	.092
x_1	.00011	−.0001	.0006	.437	−.058	−.164
x_2	.99758	−.0520	−.0267	.435	−.064	−.293
x_3	.00567	.5586	.2391	.265	−.573	.755
x_4	.00312	−.3949	.9049	.240	.634	.496
x_5	.06920	.7233	.3226	.430	−.147	−.205
x_6	.00005	.0003	.0004	.435	−.110	−.145

(b) After scaling the variables, the variances on the diagonal of **S** are 304, 116, 9879, 8732, 4843, 48, and .227. In this case, there are three large variances, as opposed to a single dominating variance in part (a), and the first three principal components reflect this. The eigenvalues of the scaled **S** are as follows:

S (scaled data)

λ_i	$\lambda_i / \sum_j \lambda_j$	Cumulative
15429.3	.644932	.64493
7265.8	.303705	.94864
1205.8	.050402	.99904
21.3	.000890	.99993
1.6	.000066	.99999
.1	.000005	1.00000
.0	.000000	1.00000

The first three eigenvectors of **S** are as follows:

S (scaled data)

	$\mathbf{a}_1$	$\mathbf{a}_2$	$\mathbf{a}_3$
y	.078	.161	.038
x_1	.083	.032	−.042
x_2	.758	.297	−.562
x_3	.619	−.586	.522
x_4	.165	.736	.638
x_5	.054	.012	−.035
x_6	.004	.001	−.002

9.29 The variances of y_1 and y_4 are almost identical: 1,960,077.8 and 1,960,769.9. The other variances range from 309.7 to 139,018.6. The first two components of **S** account for 98.9% of the variance. The first component of **S** is largely a weighted average of y_1 and y_4, and the second component represents a contrast comparing y_1 and y_4. These can aptly be described as size and shape components. A different pattern emerges in the components of **R**.

S			**R**		
λ_i	$\lambda_i / \sum_j \lambda_j$	Cumulative	λ_i	$\lambda_i / \sum_j \lambda_j$	Cumulative
3129355	.76603	.76603	4.430	.554	.554
912008	.22325	.98928	1.662	.208	.761
36849	.00902	.99830	1.067	.133	.895
5150	.00126	.99956	.347	.043	.938
1137	.00028	.99984	.180	.023	.961
412	.00010	.99994	.134	.017	.977
177	.00004	.99998	.103	.013	.990
71	.00002	1.00000	.077	.010	1.000

	S		R		
	$\mathbf{a}_1$	$\mathbf{a}_2$	$\mathbf{a}_1$	$\mathbf{a}_2$	$\mathbf{a}_3$
y_1	.7028	.6724	.407	−.175	.321
y_2	.1692	.1338	.428	−.241	.019
y_3	.0588	.0637	.401	−.262	.248
y_4	.6880	−.7250	.360	.450	.034
y_5	−.0143	.0040	−.346	−.002	.591
y_6	.0171	−.0059	.343	.297	.428
x_1	.0041	−.0135	.175	.664	−.263
x_2	.0050	.0061	.305	−.332	−.482

9.30 The variances on the diagonal of **S** are .624, .902, .693, .827, 1.059, 1.203, .702, .759, and .814. These are close to the same size, and the covariances are all positive and similar in size. The pattern is close to that in (9.28): $\boldsymbol{\Sigma} = \sigma^2[(1-\rho)\mathbf{I} + \rho\mathbf{J}]$. As expected from (9.29), there is one large eigenvalue, and the others are relatively small. The first component is a weighted average of the variables, with the weights nearly equal. The remaining components are contrasts in the variables, or shape components. The eigenvalues and eigenvectors of **S** and **R** are as follows:

S			R		
λ_i	$\lambda_i/\sum_j \lambda_j$	Cumulative	λ_i	$\lambda_i/\sum_j \lambda_j$	Cumulative
5.714	.753	.753	6.761	.751	.751
.735	.097	.850	.809	.090	.841
.431	.057	.907	.552	.061	.902
.185	.024	.932	.240	.027	.929
.164	.022	.953	.199	.022	.951
.126	.017	.970	.169	.019	.970
.085	.011	.981	.106	.012	.982
.075	.010	.991	.095	.011	.992
.068	.009	1.000	.068	.008	1.000

	S			R		
	$\mathbf{a}_1$	$\mathbf{a}_2$	$\mathbf{a}_3$	$\mathbf{a}_1$	$\mathbf{a}_2$	$\mathbf{a}_3$
y_1	.31	−.12	.07	.35	−.19	.10
y_2	.36	.04	−.41	.35	−.06	−.38
y_3	.31	.06	.09	.35	.02	.07
x_1	.33	.13	−.55	.34	.03	−.55
x_2	.36	−.59	.17	.32	−.56	.23
x_3	.42	−.35	.04	.35	−.33	.07
x_4	.26	.27	.70	.29	.36	.69
x_5	.33	.24	.05	.35	.21	−.01
x_6	.29	.61	.01	.30	.60	−.13

9.31 The means and standard deviations of the eight independent variables are as follows:

	y_2	y_3	x_1	x_2	x_3	x_4	x_5	x_6
Mean	2.40	2.53	2.49	2.91	2.70	2.42	2.28	2.36
St.Dev.	.95	.83	.91	1.03	1.10	.84	.87	.90

The variances are close together, and we will use principal components of the covariance matrix $\mathbf{S}_{xx}$. The eigenvalues of $\mathbf{S}_{xx}$ are 5.192, .726, .429, .185, .156, .116, .084, and .071. The first three account for 91.2% of the total sample variance, and we use them for regression. The first three (normalized) eigenvectors of $\mathbf{S}_{xx}$ are

$$\mathbf{a}_1 = (.382, .326, .351, .378, .442, .274, .346, .304)',$$

$$\mathbf{a}_2 = (.012, .050, .112, -.607, -.370, .262, .226, .600)',$$

$$\mathbf{a}_3 = (-.403, .095, -.538, .195, .058, .703, .051, .007)',$$

which become the columns of $\mathbf{A}_m$ in (9.61) or (9.62). From (9.61), we obtain $\hat{\boldsymbol{\delta}}_m = (.3140, -.1125, .0654)'$, and by (9.62), we obtain $\tilde{\boldsymbol{\beta}}_1 = \mathbf{A}_m\hat{\boldsymbol{\delta}}_m$, which is compared with $\hat{\boldsymbol{\beta}}_1$ in the table below.

COMPARISON OF $\hat{\beta}_1$ FROM LEAST SQUARES AND $\tilde{\beta}_1$ FROM PRINCIPAL COMPONENT REGRESSION AND THE VARIANCES

Variable	$\hat{\boldsymbol{\beta}}_1$	$\widehat{\text{var}}(\hat{\beta}_{1j})/s^2$	$\tilde{\boldsymbol{\beta}}_1$	$\widehat{\text{var}}(\tilde{\beta}_{1j})/s^2$
y_2	$-.016$	.215	.092	.407
y_3	.158	.191	.103	.045
x_1	$-.007$	.202	.062	.716
x_2	.128	.196	.200	.624
x_3	.313	.301	.184	.234
x_4	$-.020$	.119	.103	1.262
x_5	.215	.214	.086	.099
x_6	.051	.158	.028	.514

The principal component regression does not appear to be an improvement over least squares regression. Only one-half of the $\tilde{\beta}_{1j}$'s are smaller than the $\hat{\beta}_{1j}$'s, and only three of the eight estimated variances of the $\tilde{\beta}_{1j}$'s are smaller. The rainfall data do not exhibit high multicolinearity (the smallest eigenvalue is .0715) and were therefore not a good candidate for a biased regression approach.

Chapter 10

10.1 By (10.8), (1.94), (10.5), and (10.7), we obtain $\text{cov}(\boldsymbol{\Lambda}\mathbf{f} + \boldsymbol{\varepsilon}) = \text{cov}(\boldsymbol{\Lambda}\mathbf{f}) + \text{cov}(\boldsymbol{\varepsilon}) = \boldsymbol{\Lambda}\mathbf{I}\boldsymbol{\Lambda}' + \boldsymbol{\Psi}$.

10.2 By (1.65), $\text{cov}(\mathbf{y}, \mathbf{f}) = \text{cov}(\mathbf{\Lambda f} + \boldsymbol{\varepsilon}, \mathbf{f}) = E[\mathbf{\Lambda f} + \boldsymbol{\varepsilon} - E(\mathbf{\Lambda f} + \boldsymbol{\varepsilon})][\mathbf{f} - E(\mathbf{f})]' = E(\mathbf{\Lambda ff}') + E(\boldsymbol{\varepsilon}\mathbf{f}')$. Show that this equals $\mathbf{\Lambda}$.

10.3 Substitute $\mathbf{y} = \mathbf{A}^{-1}(\mathbf{z} - \mathbf{b})$ and $\boldsymbol{\mu}_y = \mathbf{A}^{-1}(\boldsymbol{\mu}_z - \mathbf{b})$ for $\mathbf{y}$ and $\boldsymbol{\mu}_y$ in $\mathbf{y} - \boldsymbol{\mu}_y = \mathbf{\Lambda}_y\mathbf{f} + \boldsymbol{\varepsilon}_y$.

10.4 $\mathbf{A\Lambda}_y = \mathbf{D}_\sigma^{-1}\mathbf{\Lambda}_y$.

10.5 $E(\mathbf{f}^*) = E(\mathbf{T}'\mathbf{f}) = \mathbf{T}'E(\mathbf{f}) = \mathbf{T}'\mathbf{0} = \mathbf{0}$. To show that (10.5) holds for $\mathbf{f}^*$, use (1.94). For (10.8), use (1.98).

10.8 $R_i^2 = 1 - 1/r^{ii}$, as noted following (10.35). By (1.51), $\mathbf{R} = \mathbf{D}^{-1}\mathbf{S}\mathbf{D}^{-1}$, where $\mathbf{D} = \text{diag}(\sqrt{s_{11}}, \sqrt{s_{22}}, \ldots, \sqrt{s_{pp}})$ and $\mathbf{R}^{-1} = \mathbf{D}\mathbf{S}^{-1}\mathbf{D}$, from which $r^{ii} = s_{ii}s^{ii}$.

10.9 We need to show that $\mathbf{z} = \mathbf{Dy}$ implies $\hat{\mathbf{\Lambda}}_z = \mathbf{D}\hat{\mathbf{\Lambda}}_y$, where $\mathbf{D}$ is a nonsingular diagonal matrix. First show that $\mathbf{S}_y = \hat{\mathbf{\Lambda}}_y\hat{\mathbf{\Lambda}}_y' + \hat{\mathbf{\Psi}}_y$ becomes $\mathbf{S}_z = \hat{\mathbf{\Lambda}}_z\hat{\mathbf{\Lambda}}_z' + \hat{\mathbf{\Psi}}_z$, where $\mathbf{S}_z = \mathbf{D}\mathbf{S}_y\mathbf{D}$, $\hat{\mathbf{\Lambda}}_z = \mathbf{D}\hat{\mathbf{\Lambda}}_y$, and $\hat{\mathbf{\Psi}}_z = \mathbf{D}\hat{\mathbf{\Psi}}_y\mathbf{D}$. Then use these relationships in the defining equations for the maximum likelihood estimators (10.38), (10.39), and (10.40) to show that there is no essential change.

10.10 **(a)**

Variables	Principal Component Loadings $\hat{\lambda}_{1j}$	Principal Component Loadings $\hat{\lambda}_{2j}$	Varimax Rotated Loadings $\hat{\lambda}_{1j}$	Varimax Rotated Loadings $\hat{\lambda}_{2j}$	Communalities $\hat{h}_i^2$
y	.778	.601	.380	**.907**	.967
x_1	.990	−.073	**.896**	.426	.985
x_2	.987	−.081	**.898**	.418	.981
x_3	.600	−.723	**.879**	−.332	.882
x_4	.545	.799	.079	**.964**	.936
x_5	.975	−.186	**.939**	.321	.985
x_6	.986	−.139	**.926**	.367	.993
Variance accounted for	5.141	1.588	4.273	2.456	6.729
Proportion of total variance	.734	.227	.610	.351	.961
Cumulative proportion	.734	.961	.610	.961	.961

The boldface rotated loadings indicate that the first factor is associated with variables x_1, x_2, x_3, x_5, and x_6; the second factor is associated with y and x_4. This groups the number of government employees and the size of the armed forces into factor 2 and the other variables into factor 1, a partitioning that may be informative to the researcher.

(b) Method 1: Two factors account for 96.1% of the total variance.

Method 2: The eigenvalues are 5.14, 1.59, .21, .049, .015, .0019, and .0002, two of which exceed 1.

Method 3: The scree test conclusively indicates two factors.

Method 4: The test of the hypothesis that two factors provide an adequate fit [see (10.41)] yielded (approximate) $\chi^2 = 36.371$, with 8 df, which has a p-value less than .0001. This indicates that more than two factors are needed, as opposed to the other three methods, which agree on $m = 2$.

(c) The initial communality estimates are squared multiple correlations, as in (10.35). The final communality estimates are the same for principal factor loadings and rotated loadings.

	Principal Factor Loadings		Varimax Rotated Loadings		Initial Communalities	Final Communalities
Variables	$\hat{\lambda}_{i1}$	$\hat{\lambda}_{i2}$	$\hat{\lambda}_{i1}$	$\hat{\lambda}_{i2}$	$\hat{h}_i^2$	$\hat{h}_i^2$
y	.774	.598	.376	**.903**	.969	.956
x_1	.989	−.069	**.894**	.430	.993	.984
x_2	.988	−.077	**.897**	.423	.999	.983
x_3	.599	−.718	**.876**	−.327	.979	.874
x_4	.540	.788	.079	**.952**	.960	.913
x_5	.976	−.183	**.938**	.324	.998	.986
x_6	.987	−.137	**.925**	.371	.999	.994
Variance	5.132	1.558	4.254	2.435		6.689
Proportion	.733	.223	.608	.348		.956
Cumulative	.733	.956	.608	.956		.956

(d)

	Maximum Likelihood Loadings		Varimax Rotated Loadings		Communalities
Variables	$\hat{\lambda}_{i1}$	$\hat{\lambda}_{i2}$	$\hat{\lambda}_{i1}$	$\hat{\lambda}_{i2}$	$\hat{h}_i^2$
y	.713	−.691	.363	**.925**	.986
x_1	.994	−.019	**.897**	.429	.988
x_2	.998	.001	**.909**	.412	.996
x_3	.629	.580	**.813**	−.267	.732
x_4	.456	−.819	.076	**.935**	.879
x_5	.991	.097	**.943**	.322	.992
x_6	.998	.054	**.930**	.364	.998
Variance	5.074	1.498	4.182	2.390	6.572

(e) The pattern of the rotated loadings and communalities is essentially the same in parts (a), (c), and (d).

10.11 **(a)**

	Principal Component Loadings				Varimax Rotated Loadings				Communalities
Variables	$\hat{\lambda}_{i1}$	$\hat{\lambda}_{i2}$	$\hat{\lambda}_{i3}$	$\hat{\lambda}_{i4}$	$\hat{\lambda}_{i1}$	$\hat{\lambda}_{i2}$	$\hat{\lambda}_{i3}$	$\hat{\lambda}_{i4}$	$\hat{h}_i^2$
y_1	.38	−.33	.33	.49	.19	−.07	.23	**.71**	.598
y_2	.58	.24	.57	−.05	.24	.34	**.72**	.16	.719
y_3	−.71	.05	.22	.26	**−.76**	−.04	−.15	.13	.623
y_4	.89	−.09	−.21	.06	**.88**	.14	.11	.20	.846
y_5	.25	.57	−.42	−.41	.35	.46	−.12	**−.61**	.728
y_6	.89	−.28	−.21	.00	**.93**	−.05	.13	.21	.921
y_7	.92	.04	−.06	−.07	**.84**	.23	.30	.10	.857
y_8	−.01	.90	.14	.17	−.24	**.89**	.07	−.11	.862
y_9	.35	−.18	.72	−.31	.08	−.17	**.85**	.07	.760
y_{10}	.55	.04	−.23	.65	.49	.35	−.23	**.60**	.777
y_{11}	.56	−.58	−.19	−.18	**.70**	−.46	.08	.09	.718
x_1	−.80	−.20	−.32	.12	−.56	−.34	**−.61**	−.08	.799
x_2	.63	.62	−.06	.14	.45	**.76**	.13	.03	.800
Variance	5.25	2.19	1.48	1.09	4.44	2.26	1.90	1.41	10.01

(b) For some of the variables, the varimax rotation did not achieve the ideal pattern of loading highly on only one factor, for example, y_5, y_{10}, y_{11}, x_1, and x_2. An oblique rotation follows:

	Orthoblique Pattern Loadings			
Variables	$\hat{\lambda}_{i1}$	$\hat{\lambda}_{i2}$	$\hat{\lambda}_{i3}$	$\hat{\lambda}_{i4}$
y_1	.01	**.70**	.15	.10
y_2	.35	.12	.17	**.63**
y_3	.15	.26	**−.79**	−.05
y_4	−.03	.06	**.93**	−.08
y_5	.27	**−.68**	.44	−.15
y_6	−.23	.06	**.96**	−.04
y_7	.07	−.04	**.87**	.13
y_8	**.92**	−.06	−.17	.00
y_9	−.13	.04	−.07	**.88**
y_{10}	.32	.55	**.60**	−.47
y_{11}	−.60	−.03	**.68**	.03
x_1	−.25	.01	**−.53**	−.48
x_2	**.66**	−.03	.54	.06

In this case the oblique rotation did not improve the pattern of the loadings for any of the five variables noted above. Also, the interpretation of the four factors changed somewhat in the oblique rotation, as indicated by the boldface loadings. The correlations among the four oblique factors are as follows:

$$\begin{pmatrix} 1.000 & -.171 & .125 & .081 \\ -.171 & 1.000 & .184 & .188 \\ .125 & .184 & 1.000 & .342 \\ .081 & .188 & .342 & 1.000 \end{pmatrix}.$$

Most of the correlations are small; this indicates that the factors are basically orthogonal.

(c) Method 1: Four factors account for 77% of the total variance.

Method 2: The eigenvalues are 5.25, 2.19, 1.48, 1.09, .68, .52, .47, .40, .31, .22, .18, .16, and .05. The first four exceed 1.

Method 3: The scree test indicates four factors.

Method 4: The test of the hypothesis that four factors provide an adequate fit [see (10.41)] yielded (approximate) $\chi^2 = 17.699$, with 32 df, which does not lead to rejection of the hypothesis. The conclusion is that four factors are adequate. Thus all four methods are in general agreement that $m = 4$.

(d) The initial communality estimates are squared multiple correlations, as in (10.35). The final communality estimates are the same for principal factor loadings and rotated loadings.

	Principal Factor Loadings				Varimax Rotated Loadings				Initial Commu-nalities	Final Commu-nalities
Variables	$\hat{\lambda}_{i1}$	$\hat{\lambda}_{i2}$	$\hat{\lambda}_{i3}$	$\hat{\lambda}_{i4}$	$\hat{\lambda}_{i1}$	$\hat{\lambda}_{i2}$	$\hat{\lambda}_{i3}$	$\hat{\lambda}_{i4}$	$\hat{h}_i^2$	$\hat{h}_i^2$
y_1	.33	−.21	−.21	.28	.16	−.19	.28	**.37**	.23	.27
y_2	.54	.24	−.49	.01	.17	.24	**.68**	.16	.54	.58
y_3	−.66	.06	−.13	.21	**−.67**	−.13	−.18	−.04	.49	.50
y_4	.89	−.10	.20	.07	**.81**	.14	.18	.38	.87	.85
y_5	.23	.43	.34	−.35	.30	**.55**	−.13	−.25	.41	.47
y_6	.91	−.31	.18	.02	**.89**	−.05	.19	.34	.93	.95
y_7	.92	.04	.03	−.09	**.77**	.26	.37	.23	.83	.85
y_8	−.01	.85	−.03	.10	−.30	**.80**	.09	.06	.68	.74
y_9	.31	−.12	−.57	−.12	.11	−.16	**.65**	−.03	.42	.46
y_{10}	.52	.04	.20	.48	.33	.16	−.01	**.63**	.50	.54
y_{11}	.53	−.54	.08	−.12	**.66**	−.37	.12	.09	.61	.59
x_1	−.78	−.21	.33	.09	−.46	−.31	**−.65**	−.16	.71	.76
x_2	.61	.59	.12	.08	.34	**.72**	.20	.27	.72	.74
Variance	4.97	1.82	.99	.54	3.62	1.93	1.71	1.05		8.31

(e)

	Maximum Likelihood Loadings				Varimax Rotated Loadings				Communalities
Variables	$\hat{\lambda}_{i1}$	$\hat{\lambda}_{i2}$	$\hat{\lambda}_{i3}$	$\hat{\lambda}_{i4}$	$\hat{\lambda}_{i1}$	$\hat{\lambda}_{i2}$	$\hat{\lambda}_{i3}$	$\hat{\lambda}_{i4}$	$\hat{h}_i^2$
y_1	.33	−.07	−.23	.29	.16	−.15	.25	**.37**	.25
y_2	.38	.47	−.50	.07	.16	.24	**.71**	.16	.62
y_3	−.64	−.12	.04	.22	**−.63**	−.09	−.24	.00	.47
y_4	.92	.10	.08	.04	**.83**	.13	.20	.34	.86
y_5	.19	.40	.40	−.48	.38	**.53**	−.15	−.38	.58
y_6	.99	−.08	.03	.01	**.91**	−.07	.19	.34	.98
y_7	.86	.29	−.07	−.09	**.77**	.24	.39	.19	.84
y_8	−.22	.83	.17	.10	−.27	**.84**	.10	.01	.78
y_9	.27	.06	−.56	−.06	.13	−.16	**.59**	.01	.39
y_{10}	.53	.14	.21	.48	.36	.21	−.02	**.64**	.58
y_{11}	.66	−.37	−.13	−.11	**.64**	−.39	.15	.12	.60
x_1	−.63	−.49	.41	.05	−.45	−.30	**−.71**	−.13	.81
x_2	.46	.71	.23	.04	.39	**.74**	.17	.19	.76
Variance	4.69	2.10	1.09	.64	3.69	2.00	1.78	1.05	8.52

10.12 **(a)** and **(c)**

	Principal Component Loadings		Varimax Rotated Loadings		Orthoblique Pattern Loadings	
Variables	$\hat{\lambda}_{i1}$	$\hat{\lambda}_{i2}$	$\hat{\lambda}_{i1}$	$\hat{\lambda}_{i2}$	$\hat{\lambda}_{i1}$	$\hat{\lambda}_{i2}$
y_1	.817	−.157	**.732**	.395	**.737**	.131
y_2	.838	−.336	**.861**	.271	**.963**	−.092
y_3	.874	.288	.494	**.776**	.248	**.734**
y_4	.838	−.308	**.844**	.292	**.931**	−.057
y_5	.762	.547	.244	**.905**	−.134	**1.023**
Variance	3.416	.614	2.294	1.736		
Proportion	.683	.123	.459	.347		

(b) The loadings for y_2 and y_4 are nearly identical in all three sets of loadings. In **R** (below), we see that the correlations of y_2 and y_4 with the other variables are very similar:

$$\mathbf{R} = \begin{pmatrix} 1.000 & .614 & .757 & .575 & .413 \\ .614 & 1.000 & .547 & .750 & .548 \\ .757 & .547 & 1.000 & .605 & .692 \\ .575 & .750 & .605 & 1.000 & .524 \\ .413 & .548 & .692 & .524 & 1.000 \end{pmatrix}.$$

(c) The oblique rotation improved the pattern of the loadings, in that it generally produced higher and lower loadings, but the interpretation is the same as for the varimax rotated loadings. The correlation between the two oblique factors is .680, suggesting the possibility of using only one factor instead of two. This is also suggested by the large gap between the first two eigenvalues, 3.416 and .614.

10.13 The eigenvalues are 4.430, 1.662, 1.067, .347, .180, .134, .103, and .077. The three eigenvalues greater than 1 account for 89.5% of the total variance.

	Principal Component Loadings			Varimax Rotated Loadings			Communalities
Variables	$\hat{\lambda}_{i1}$	$\hat{\lambda}_{i2}$	$\hat{\lambda}_{i3}$	$\hat{\lambda}_{i1}$	$\hat{\lambda}_{i2}$	$\hat{\lambda}_{i3}$	$\hat{h}_i^2$
y_1	.856	−.225	.332	**.904**	.118	.249	.893
y_2	.901	−.311	.020	**.775**	.093	.548	.910
y_3	.844	−.337	.257	**.881**	.020	.339	.892
y_4	.758	.581	.035	.430	**.836**	.167	.913
y_5	-.727	−.003	.610	−.180	−.372	−**.855**	.901
y_6	.722	.383	.442	**.702**	.599	−.108	.863
x_1	.368	.857	−.272	.123	**.956**	.120	.943
x_2	.643	−.428	−.498	.303	−.064	**.865**	.844
Variance	4.430	1.662	1.067	3.012	2.137	2.010	7.159
Proportion	.554	.208	.133	.376	.267	.251	.895
Cumulative	.554	.762	.895	.376	.644	.895	.895

10.14 The eigenvalues are 6.76, .81, .55, .24, .20, .17, .11, .09, and .07. The first accounts for 75% of the variance, the first two for 84%, and the first three for 90%. We could extract one, two, or three factors. We show all three posibilities. (It is not necessary to rotate one factor.)

VARIMAX ROTATED LOADINGS $\hat{\lambda}_{ij}$ AND COMMUNALITIES $\hat{h}_i^2$ FOR ONE, TWO, AND THREE FACTORS

	One Factor		Two Factors			Three Factors			
Variables	$\hat{\lambda}_{i1}$	$\hat{h}_i^2$	$\hat{\lambda}_{i1}$	$\hat{\lambda}_{i2}$	$\hat{h}_i^2$	$\hat{\lambda}_{i1}$	$\hat{\lambda}_{i2}$	$\hat{\lambda}_{i3}$	$\hat{h}_i^2$
y_1	.92	.85	**.80**	.50	.88	**.73**	.45	.38	.89
y_2	.91	.82	**.70**	.58	.83	.56	**.73**	.22	.90
y_3	.90	.81	**.65**	.62	.81	**.57**	.51	.47	.81
x_1	.87	.76	**.62**	.61	.76	.45	**.83**	.17	.93
x_2	.82	.67	**.94**	.19	.92	**.93**	.21	.20	.95
x_3	.91	.82	**.87**	.39	.91	**.81**	.41	.29	.91
x_4	.76	.58	.34	**.75**	.68	.35	.18	**.88**	.94
x_5	.91	.84	.55	**.76**	.87	.45	**.61**	.54	.87
x_6	.78	.61	.21	**.93**	.91	.08	**.71**	.63	.92
Variance	6.76	6.76	4.03	3.54	7.57	3.25	2.82	2.05	8.12
Proportion	1.00	1.00	.448	.393	.841	.361	.314	.228	.902
Cumulative	1.00	1.00	.448	.841	.841	.361	.674	.902	.902

10.15 **(a)** The eigenvalues are 2.456, 2.088, .978, .597, .324, .286, and .272. The first two account for 64.9% of the total variance, and the first three account for 78.9%. From these and the scree plot, we could choose m to be either 2 or 3.
(b) and (c)

	Principal Component Loadings		Varimax Rotated Loadings		Communalities	Orthoblique Pattern Loadings	
Variables	$\hat{\lambda}_{i1}$	$\hat{\lambda}_{i2}$	$\hat{\lambda}_{i1}$	$\hat{\lambda}_{i2}$	$\hat{h}_i^2$	$\hat{\lambda}_{i1}$	$\hat{\lambda}_{i2}$
y_1	**.608**	−.287	**.562**	−.368	.452	−.354	**.565**
y_2	−.162	**.873**	−.040	**.888**	.789	**.886**	−.045
y_3	**.699**	.208	**.721**	.109	.532	.128	**.721**
y_4	**.876**	.129	**.885**	.006	.784	.029	**.885**
y_5	.074	**.866**	.193	**.847**	.755	**.852**	.188
y_6	**.872**	.138	**.882**	.016	.779	.039	**.882**
y_7	.196	**−.643**	.105	**−.664**	.453	**−.662**	.109
Variance	2.456	2.088	2.449	2.095	4.544		
Proportion	.351	.298	.350	.299	.649		
Cumulative	.351	.649	.350	.649	.649		

In this case, we have the unusual result that neither of the rotations produced an improvement in the pattern of the original principal component loadings. In fact, the correlation between the two oblique factors is only −.0198, indicating that the natural factors are orthogonal.

The first factor is associated with variables y_1, y_3, y_4, and y_6; the second factor is associated with variables y_2, y_5, and y_7. This groups all questions regarding the library staff into the first factor and all questions about personal use of the library into the second factor. This partitioning is a meaningful representation of the two dimensions with which a person would respond to the questionnaire.

(d)

	Principal Component Loadings			Varimax Rotated Loadings			Communalities
Variables	$\hat{\lambda}_{i1}$	$\hat{\lambda}_{i2}$	$\hat{\lambda}_{i3}$	$\hat{\lambda}_{i1}$	$\hat{\lambda}_{i2}$	$\hat{\lambda}_{i3}$	$\hat{h}_i^2$
y_1	.608	−.287	**−.654**	.128	−.165	**.915**	.880
y_2	−.162	**.873**	−.039	.039	**.865**	−.204	.791
y_3	**.699**	.208	.542	**.901**	−.013	−.117	.826
y_4	**.876**	.129	.021	**.777**	.030	.423	.784
y_5	.074	**.866**	.068	.288	**.806**	−.166	.760
y_6	**.872**	.138	−.122	**.706**	.077	.538	.793
y_7	.196	**−.643**	.484	.259	**−.764**	−.188	.687
Variance	2.456	2.088	.978	2.082	2.015	1.424	5.521
Proportion	.351	.298	.140	.297	.288	.203	.789
Cumulative	.351	.649	.789	.297	.585	.789	.789

Again, the rotation did not change the essential pattern of the loadings. However, the rotation produced a more satisfactory appearance of high and low loadings.

The interpretation is the same as for the two factors in part (b), except that variable y_1 is moved to a third factor associated with the skills and training of the staff, a plausible third dimension.

APPENDIX D

About the Diskette

The Diskette that accompanies the book contains the following:

1. All the data sets, including those referred to but not shown in the text.
2. SAS command files for all the examples in the text.
3. SAS command files for all the numerical problems in the text.
4. An annotated example introducing SAS IML (denoted by *example.sas*).

The command files include illustrations for the use of the following SAS Procedures: IML, CANCORR, CANDISC, DISCRIM, FACTOR, GLM, PRINCOMP, REG, STEPDISC. The command files can be adapted for analyzing the reader's data sets.

The files can be located under the main program directory in three subdirectories called DATA, EXAMPLES, and PROGRAMS. The examples and programs files are further divided into chapter directories.

Hardware and Software Requirements

The disk can be installed on any PC-compatible computer. The data sets are provided in ASCII text format. The SAS command files can be run on any standard platform that supports SAS.

Installing the Diskette Files

To install the files included on the disk, you will need approximately 150 KB of free disk space on your hard drive. To load the files you may move the MULTSTAT directory to your hard drive by copying the directory or dragging it to your desktop. If you are using a version of Windows on your system, you can also install the files by double-clicking on the INSTALL.EXE to run a WINZIP installation program for the files. The default installation directory is MULTSTAT.

Technical Assistance and Customer Information

If you have a damaged diskette, you can reach Wiley's technical support group at 212-850-6753 or by e-mail at: techhelp@wiley.com

Further information about Wiley products can be obtained at the company's web site at: http://www.wiley.com

Bibliography

Ahmed, S. W. and Lachenbruch, P. A. (1977). Discriminant analysis when scale contamination is present in the initial sample. In J. V. Ryzin (ed.), *Classification and Clustering*, pp. 331–353. New York: Academic Press.

Aitchison, J. and Begg, C. B. (1976). Statistical diagnosis when basic cases are not classified with certainty. *Biometrika* **63**, 1–12.

Aitchison, J., Habbema, J. D. F., and Kay, J. W. (1977). A critical comparison of two methods of statistical discrimination. *Applied Statistics* **26**, 15–25.

Aitchison, J. A. (1982). The statistical analysis of compositional data (with discussion). *Journal of the Royal Statistical Society, Series B* **44**, 139–177.

Aitchison, J. A. (1983). Principal component analysis of compositional data (with discussion). *Biometrika* **70**, 57–65.

Aitchison, J. A. (1986). *The Statistical Analysis of Compositional Data*. London: Chapman & Hall.

Akaike, H. (1973). Information theory and an extension of the maximum likelihood principle. In B. Petrov and F. Csaki (eds.), *Second International Symposium on Information Theory*, pp. 267–281. Budapest: Akademia Kiado.

Albert, A. and Anderson, J. A. (1981). Probit and logistic discriminant functions. *Communications in Statistics—Series A, Theory and Methods* **10**, 641–657.

Albert, A. and Lesaffre, E. (1986). Multiple group logistic discrimination. *Computers and Mathematics with Applications (Ser. A)* **12**, 209–224.

Algina, J. and Oshima, T. C. (1990). Robustness of the independent samples Hotelling's T^2 to variance-covariance heteroscedasticity when sample sizes are unequal in small ratios. *Psychological Bulletin* **108**, 308–313.

Algina, J., Oshima, T. C., and Tang, K. L. (1991). Robustness of Yao's, James', and Johansen's tests under variance-covariance heteroscedasticity and nonnormality. *Journal of Educational Statistics* **16**, 125–139.

Algina, J. and Tang, K. L. (1988). Type I error rates for Yao's and James' tests of equality of mean vectors under variance-covariance heteroscedasticity. *Journal of Educational Statistics* **13**, 281–290.

Allison, M. J., Zappasodi, P., and Lurie, M. B. (1962). The correlation of a biphasic metabolic response with a biphasic response in resistance to tuberculosis in rabbits. *Journal of Experimental Medicine* **115**, 881–890.

Alt, F. and Smith, N. (1988). *Multivariate Process Control*, Chap. 7, pp. 333–351. Amsterdam: North Holland Publishing Co.

Amemiya, Y. (1993). Instrumental variable estimation for nonlinear factor analysis. In C. M. Cuadras and C. R. Rao (eds.), *Multivariate Analysis: Future Directions*, Vol. 2, pp. 113–129. Amsterdam: North-Holland Publishing Co.

Amemiya, Y. and Yalcin, I. (1995). Model fitting procedures for nonlinear factor analysis using the errors-in-variables parameterization. In M. Berkane (ed.), *Latent Variable Modeling and Application to Causality*. Springer-Verlag Lecture Note Series. New York: Springer-Verlag.

Ammann, L. P. (1989). Robust principal components. *Communications in Statistics: Simulation and Computation* **18**, 857–874.

Anderson, J. A. (1972). Separate sample logistic discrimination. *Biometrika* **59**, 19–35.

Anderson, J. A. (1975). Quadratic logistic discrimination. *Biometrika* **62**, 149–154.

Anderson, J. A. (1982). Logistic discrimination. In P. R. Krishnaiah and L. N. Kanal (eds.), *Handbook of Statistics*, Vol. 2. Amsterdam: North-Holland Publishing Co.

Anderson, T. W. (1951). The asymptotic distribution of certain characteristic roots and vectors. In *Proceedings of the Second Berkeley Symposium on Mathematical Statistics and Probability*, pp. 103–130. Berkeley, CA: University of California Press.

Anderson, T. W. (1963). Asymptotic theory for principal component analysis. *Annals of Mathematical Statistics* **34**, 122–148.

Anderson, T. W. (1984). *Introduction to Multivariate Statistical Analysis* (2nd edition). New York: Wiley.

Anderson, T. W. and Fang, K. T. (1990). Statistical inference in elliptically contoured and related distributions. In K. T. Fang and T. W. Anderson (eds.), *Inference in Multivariate Elliptically Contoured Distributions based on Maximum Likelihood*, pp. 201–216. New York: Allerton Press.

Anderson, T. W. and Rubin, H. (1956). Statistical inference in factor analysis. In J. Neyman (Ed.), *Proceedings of the Third Berkeley Symposium on Mathematical Statistics and Probability* **5**, 111–150. Berkeley, CA: University of California Press.

Andrews, D. F. and Herzberg, A. M. (1985). *Data*. New York: Springer-Verlag.

Andrews, D. F., Gnanadesikan, R., and Warner, J. L. (1971). Transformation of multivariate data. *Biometrics* **27**, 825–840.

Andrews, D. F., Bicknel, P. J., Hampel, F. R., Huber, P. J., Rogers, W. H., and Tukey, J. W. (1972). *Robust Estimates of Location-Survey and Advances*. Princeton, NJ: Princeton University Press.

Ashikaga, T. and Chang, P. C. (1981). Robustness of Fisher's linear discriminant function under two-component mixed normal models. *Journal of the American Statistical Association* **76**, 676–680.

Back, J., Gray, H. L., and Woodward, W. A. (1994). A bootstrap generalized likelihood ratio test in discriminant analysis. Presented at the Annual Meeting of the American Statistical Association, Toronto, Canada, August 1994.

Bai, Z. D., Rao, C. R., and Wu, Y. (1992). M-estimation of multivariate linear regression parameters under a convex discrepancy function. *Statistica Sinica* **2**, 237–254.

Bailey, B. J. R. (1977). Tables of the Bonferroni *t* statistic. *Journal of the American Statistical Association* **72**, 469–479.

Barcikowski, R. and Stevens, J. P. (1975). A Monte Carlo study of the stability of canonical correlations, canonical weights and canonical variate-variable correlations. *Multivariate Behavioral Research* **10**, 353–364.

Bargmann, R. E. (1979). Structural analysis of singular matrices using union-intersection statistics. In L. Orlici, C. R. Rao, and W. M. Stiteler (eds.), *Multivariate Methods in Ecological Works*, pp. 1–9. Fairland, MD: International Cooperative Publishing House.

Barrett, B. E. and Ling, R. F. (1992). General classes of influence measures for multivariate regression. *Journal of the American Statistical Association* **87**, 184–191.

Bartholomew, D. J. (1981). Posterior analysis of the factor model. *British Journal of Mathematical and Statistical Psychology* **34**, 93–99.

Bartlett, M. S. (1938). Further aspects of the theory of multiple regression. *Proceedings of the Cambridge Philosophical Society* **34**, 33–40.

Bartlett, M. S. (1947). Multivariate analysis. *Journal of the Royal Statistical Society* **9** (Suppl.), 176–197.

Bartlett, M. S. (1950). Tests of significance in factor analysis. *British Journal of Psychology* **3**, 77–85.

Bartlett, M. S. (1951). The effect of standardization on an approximation in factor analysis. *Biometrika* **38**, 337–344.

Bartlett, M. S. (1954). A note on multiplying factors for various chi-squared approximations. *Journal of the Royal Statistical Society, Series B* **16**, 296–298.

Barton, C. N. and Cramer, E. C. (1989). Hypothesis tests in multivariate linear models with randomly missing data. *Communications in Statistics—Series B, Simulation and Computation* **18**, 875–895.

Basilevsky, A. (1981). Factor analysis regression. *La Revue Canadienne de Statistique* **9**(1), 109–117.

Basilevsky, A. (1994). *Statistical Factor Analysis and Related Methods*. New York: Wiley.

Baxter, M. J., Cool, H. E. M., and Heyworth, M. P. (1990). Principal component and correspondence analysis of compositional data: Some similarities. *Journal of Applied Statistics* **17**, 229–235.

Baye, M. R. and Parker, D. F. (1984). Combining ridge and principal component regression: A money demand illustration. *Communications in Statistics* **13**(2), 197–205.

Beaghan, M. and Smith, E. P. (1996). Least squares and maximum likelihood approaches to common principal components. Presented at the Annual Meeting of the American Statistical Association, Chicago, IL, August 1996.

Beall, G. (1945). Approximate methods in calculating discriminant functions. *Psychometrika* **10**(3), 205–217.

Beauchamp, J. J. and Robson, D. S. (1986). Transformation considerations in discriminant analysis. *Communications in Statistics: Simulation and Computation* **15**(1), 147–179.

Beauchamp, J. J., Folkert, J. E., and Robson, D. S. (1980). A note on the effect of logarithmic transformation on the probability of misclassification. *Communications in Statistics—Series A, Theory and Methods* **9**, 777–794.

Beckman, R. J. and Cook, R. D. (1983). Outliers (with comments). *Technometrics* **25**, 119–163.

Bedrick, E. J. and Tsai, C. (1994). Model selection for multivariate regression in small samples. *Biometrics* **50**, 226–231.

Behrens, W. U. (1929). Ein Beitrag zur Fehlerberechnung bei wenigen Beobachtungen. *Landwirtsch Jahrbucher* **68**, 607–837.

Belinfonte, A. and Coxe, K. (1986). Principal components regression-selection rules and application. In *Proceedings of the Section on Business and Economic Statistics*. Annual Meeting of the American Statistical Association, Chicago, IL, August 18–21, 1986.

Bellman, R. (1960). *Introduction to Matrix Analysis*. New York: McGraw-Hill.

Belsley, D., Kuh, E., and Welsch, R. (1980). *Regression Diagnostics: Identifying Data and Sources of Collinearity*. New York: Wiley.

Bennett, B. M. (1951). Note on a solution of the generalized Behrens-Fisher problem. *Annals of the Institute of Statistical Mathematics* **2**, 87–90.

Bentler, P. M. and Huba, G. J. (1982). Symmetric and asymmetric rotations in canonical correlation analysis: New methods with drug variable examples. In N. Hirschberg and L. G. Humphreys (eds.), *Multivariate Applications in the Social Sciences*, pp. 21–46. Hillsdale, New Jersey: Lawrence Erlbaum.

Berk, K. N. (1978). Comparing subset regression procedures. *Technometrics* **20**, 1–6.

Berk, K. N. (1980). Forward and backward stepping in variable selection. *Journal of Statistical Computation and Simulation* **10**, 177–185.

Berk, K. N. and Booth, D. E. (1993). Seeing a curve in multiple regression. Presented at the Annual Meeting of the American Statistical Association, San Francisco, CA, August, 1993.

Besse, P. and de Falguerolles, A. (1993). Application of resampling methods to the choice of dimension in principal component analysis. *Computer Intensive Methods in Statistics*, 167–176.

Betz, M. A. (1987). An approximation for the Hotelling-Lawley trace in the noncentral case. *Communications in Statistics: Theory and Methods* **16**, 3169–3183.

Bhattacharajee, S. K. and Dunsmore, I. R. (1991). The influence of variables in a logistic model. *Biometrika* **78**, 851–856.

Bickel, P. J. (1964). On some alternative estimates for shift in the p-variate one-sample problem. *Annals of Mathematical Statistics* **35**, 1079–1090.

Bingham, C. and Feinberg, S. E. (1982). Textbook analysis of covariance—is it correct?. *Biometrics* **38**, 747–753.

Biscay, R., Valdes, P., and Pascual, R. (1990). Modified Fisher's linear discriminant function with reduction of dimensionality. *Journal of Statistical Computation and Simulation* **36**, 1–8.

Bishop, Y., Fienbar, S., and Holland, P. (1975). *Discrete Multivariate Analysis: Theory and Practice*. Cambridge, MA: Massachusetts Institute of Technology Press.

Blackwell, D. (1947). Conditional expectation and unbiased sequential estimation. *Annals of Mathematical Statistics* **18**, 105–110.

Bloomfield, P. (1974). Linear transformations for multivariate binary data. *Biometrics* **30**, 609–617.

Boik, R. J. (1988). The mixed model for multivariate repeated measures: Validity conditions and an approximate test. *Psychometrika* **53**, 469–486.

Boneh, S. and Mendieta, G. R. (1992). Regression modelling using principal components. In *Proceedings of the Kansas State University Conference on Applied Statistics*, pp. 222–232. Wichita, KS: Wichita State University Press.

Bonferroni, C. E. (1936). Il calcolo delle assicurazioni su gruppi di teste. *Studii in Onore del Professor S. O. Carboni*. Rome.

Bookstein, F. L. (1989). Size and shape: A comment on semantics. *Systematic Zoology* **38**, 173–180.

Box, G. E. P. (1949). A general distribution theory for a class of likelihood criteria. *Biometrika* **36**, 317–346.

Box, G. E. P. (1950). Problems in the analysis of growth and linear curves. *Biometrika* **7**, 362–389.

Box, G. E. P. and Cox, D. R. (1964). An analysis of transformations. *Journal of the Royal Statistical Society, Series B* **26**, 211–252.

Boyles, R. A. (1996). Multivariate process analysis with lattice data. *Technometrics* **38**(1), 37–49.

Bozdogan, H. and Ramirez, D. E. (1988). FACAIC: Model selection algorithm for the orthogonal factor model using AIC and CAIC. *Psychometrika* **53**, 407–415.

Breiman, L., Friedman, J. H., Olshen, R. A., Stone, C. J., and Gordon, A. D. (1984). Review of "classification of regression trees." *Biometrics* **40**, 874–874.

Bristol, D. (1993). *P*-value adjustments for subgroup analyses. Technical Report 93-002, Department of Biostatistics, Schering-Plough Research Institute, Kenilworth, NJ.

Broadbent, K. L. (1993). A comparison of six Bonferroni procedures. Master's thesis, Department of Statistics, Brigham Young University.

Broffitt, J. D. (1982). Nonparametric classification. In P. R. Krishnaiah and L. N. Kanal (eds.), *Handbook of Statistics*, Vol. 2, pp. 139–168. Amsterdam: North-Holland Publishing Co.

Bronson, R. (1975). *Linear Algebra*. New York: Academic Press.

Brown, B. L., Strong, W. J., and Rencher, A. C. (1973). Perceptions of personality from speech: Effects of manipulations of acoustical parameters. *Journal of the Acoustical Society of America* **54**, 29–35.

Brown, B. L., Strong, W. J., and Rencher, A. C. (1974). Fifty-four voices from two: The effects of simultaneous manipulations of rate, mean fundamental frequency, and variance of fundamental frequency on ratings of personality from speech. *Journal of the Acoustical Society of America* **55**(2), 313–318.

Brown, J. D. and Beerstecher, E. (1951). Metabolic patterns of underweight and overweight individuals. University of Texas Publication, Biochemical Institute Studies IV, No. 5109.

Browne, M. W. (1968). A comparison of factor analytic techniques. *Psychometrika* **33**, 267–334.

Browne, M. W. (1987). Robustness of statistical inference in factor analysis and related models. *Biometrika* **74**(2), 375–384.

Browne, M. W. (1988). Properties of the maximum likelihood solution in factor analysis regression. *Psychometrika* **53**, 585–590.

Bryce, G. R. (1975). The one-way model. *American Statistician* **29**, 69–70.

Bryce, G. R. and Maynes, D. D. (1979). Generation of multivariate data sets. Technical Report SD-015-R, Brigham Young University, Department of Statistics.

Bryce, G. R., Carter, M. W., and Reader, M. W. (1976). Nonsingular and singular transformations in the fixed model. Presented at the Annual Meeting of the American Statistical Association, Boston, MA, August 1976.

Bryce, G. R., Scott, D. T., and Carter, M. W. (1980). Estimation and hypothesis testing in linear models—A reparameterization approach. *Communications in Statistics—Series A, Theory and Methods* **9**, 131–150.

Buck, S. F. A. (1960). A method of estimation of missing values in multivariate data suitable for use with an electronic computer. *Journal of the Royal Statistical Society, Series B* **22**, 302–307.

Buja, A. and Tukey, P. A. (eds.). (1991). *Computing and Graphics in Statistics*. New York: Springer-Verlag.

Bull, S. B. and Donner, A. (1987a). Derivation of large sample efficiency of multinomial logistic regression compared to multiple group discriminant analysis. In I. B. MacNeill and G. J. Umphrey (eds.), *Proceedings of the Symposia in Statistics and Festschrift in Honour of V. M. Joshi (Advances in the Statistical Sciences-Biostatistics)*, Vol. 5, pp. 177–197. Boston: D. Reidel.

Bull, S. B. and Donner, A. (1987b). The efficiency of multinomial logistic regression compared with multiple group discriminant analysis. *Journal of the American Statistical Association* **82**, 1118–1122.

Butler, R. W. and Huzurbazar, S. (1992). Saddlepoint approximations for the Bartlett-Nanda-Pillai trace statistic in multivariate analysis. *Biometrika* **79**, 705–715.

Buturovic, L. J. (1993). Improving k-nearest neighbor density and error estimates. *Pattern Recognition* **26**(4), 611–616.

Campbell, N. A. (1978). The influence function as an aid in outlier detection in discriminant analysis. *Applied Statistics* **27**, 251–258.

Campbell, N. A. (1980). Shrunken estimators in discriminant and canonical variate analysis. *Applied Statistics* **29**, 5–14.

Campbell, N. A. (1981). Graphical comparison of covariance matrices. *Australian Journal of Statistics* **23**, 21–37.

Campbell, N. A. (1982). Robust procedures in multivariate analysis II. Robust canonical variate analysis. *Applied Statistics* **31**, 1–8.

Carmer, S. G. and Swanson, M. R. (1973). An evaluation of ten pairwise multiple comparison procedures by Monte Carlo methods. *Journal of the American Statistical Association* **68**, 66–74.

Carney, E. J. (1975). Ridge estimates for canonical analysis. In J. W. Frane (ed.), *Computer Science and Statistics: Proceedings of the 8th Symposium on the Interface*, pp. 252–256. Health Sciences Computing Facility, Univ. of California, Los Angeles, CA.

Caroni, C. (1987). Residuals and influence in the multivariate linear model. *Statistician* **36**, 365–370.

Carpenter, W. T., Jr., Strauss, J. S., and Bartko, J. J. (1973). Flexible system for diagnosis of schizophrenia: Report from the WHO international pilot study of schizophrenia. *Science* **182**, 1275–1278.

Carroll, J. B. (1953). An analytic solution for approximating simple structure in factor analysis. *Psychometrika* **18**, 23–38.

Carter, E. M. (1989). Power function studies. *American Journal of Mathematical and Management Sciences* **9**, 51–67.

Cattell, R. B. (1966). The Scree test for the number of factors. *Multivariate Behavioral Research* **1**, 245–276.

Cattell, R. B. and Jaspers, J. A. (1967). A general plasmode for factor analytic exercises and research. *Multivariate Behavioral Research Monographs* **67**, 1–212.

Caussinus, H. and Ruiz, A. (1990). Interesting projections of multidimensional data by means of generalized principal component analyses. *Compstat* **9**, 121–126.

Celeux, G. and Mkhadri, A. (1992). Discrete regularized discriminant analysis. *Statistics and Computing* **2**, 143–151.

Chalmer, B. J. and Schneider, W. D. (1981). A simulation study of factor analysis using small samples. In *Proceedings of the Section on Statistical Computing*. Annual Meeting of the American Statistical Association, Detroit, MI, August 10–13, 1981.

Chatterjee, S. and Price, B. (1977). *Regression Analysis by Example*. New York: Wiley.

Cherkaoui, O. and Cleroux, R. (1991). Comparative study of six classification methods for mixtures of variables. *Computing Science and Statistics: Proceedings of the 24th Symposium* **23**, 233–236.

Chernick, M. R. and Murthy, V. K. (1985). Properties of bootstrap samples. *American Journal of Mathematical and Management Sciences* **5**, 161–170.

Chernick, M. R., Murthy, V. K., and Nealy, C. D. (1985). Application of bootstrap and other resampling techniques: Evaluation of classifier performance. *Pattern Recognition Letter* **3**, 167–178.

Chinganda, E. F. and Subrahmaniam, K. (1979). Robustness of the linear discriminant function to nonnormality: Johnson's system. *Journal of Statistical Planning and Inference* **3**, 69–77.

Choi, S. C. (1977). Tests of equality of dependent correlation coefficients. *Biometrika* **64**, 645–647.

Christensen, R. R. (1989). Lack of fit tests based on near or exact replicates. *Annals of Statistics* **17**, 673–683.

Christensen, W. F. and Rencher, A. C. (1997). A comparison of type I error rates and power levels for seven solutions to the multivariate Behrens-Fisher problem. *Communications in Statistics—Simulation and Computation* **26**, Issue 4.

Clayton, J. K., Anderson, J. A., and McNicol, G. P. (1976). Preoperative prediction of postoperative deep vein thrombosis. *British Medical Journal* **2**, 910–912.

Cliff, N. and Krus, D. J. (1976). Interpretation of canonical analysis: Rotated vs. unrotated solutions. *Psychometrika* **41**, 35–42.

Cohen, J. (1982). Set correlation as a general multivariate data-analysis method. *Multivariate Behavioral Research* **17**, 301–341.

Comrey, A. L. and Lee, H. B. (1992). *A First Course in Factor Analysis* (2nd edition). Hillsdale, NJ: Lawrence Erlbaum.

Constable, P. D. L. and Mardia, K. V. (1992). Size and shape analysis of palmar interdigital areas. *Journal of Applied Statistics* **19**, 285–292.

Constanza, M. C. and Afifi, A. A. (1980). Comparison of stopping rules in forward stepwise discriminant analysis. *Journal of the American Statistical Association* **74**, 777–785.

Cook, R. D. (1977). Detection of influential observations in linear regression. *Technometrics* **19**, 15–18.

Cook, R. D. (1979). Influential observations in linear regression. *Journal of the American Statistical Association* **74**, 169–174.

Cook, R. D. and Weisberg, S. (1982). *Residuals and Influence in Regression*. New York: Chapman & Hall.

Cook, R. D. (1996). Added-variable plots and curvature in linear regression. *Technometrics* **38**(3), 275–278.

Cooper, J. C. B. (1983). Factor analysis: An overview. *American Statistician* **37**, 141–147.

Coovert, M. D. and McNelis, K. (1988). Determining the number of common factors in factor analysis: A review and program. *Educational and Pyschological Measurement* **48**, 687–692.

Cordeiro, G. M. and McCullagh, P. (1991). Bias correction in generalized linear models. *Journal of the Royal Statistical Society* **3**, 629–643.

Cornfield, J. (1962). Joint dependence of risk of coronary heart disease on serum cholesterol and systolic blood pressure: A discriminant function analysis. *Proceedings of the Federal American Society of Experimental Biology* **21**(2), 58–61.

Cox, D. R. (1966). Some procedures associated with the logistic qualitative response curve. In F. N. David (ed.), *Research Papers in Statistics: Festschrift for J. Neyman*. New York: Wiley.

Cox, D. R. (1972). The analysis of multivariate binary data. *Applied Statistics* **21**, 113–120.

Cox, G. M. and Martin, W. P. (1937). Use of a discriminant function for differentiating soils with different azotobacter populations. Journal Paper No. J451 of the Iowa Agricultural Experiment Station, Ames, IA.

Cox, T. F. and Ferry, G. (1991). Robust logistic discrimination. *Biometrika* **78**, 841–849.

Coxe, K. (1975). Do principal components solve multicolinearity? the Longley data revisited. In *Proceedings of the Section on Statistical Computing*, pp. 90–95. Annual Meeting of the American Statistical Association.

Coxe, K. (1982). Selection rules for principal components regression: Comparison with latent root regression. In *Proceedings of the Section on Business and Economics*, pp. 222–227. Annual Meeting of the American Statistical Association, Cincinnati, OH, August 16–19, 1982.

Crawley, D. R. (1979). Logistic discrimination as an alternative to Fisher's linear discrimination function. *New Zealand Statistician* **14**(2), 21–25.

Critchley, F. (1985). Influence in principal components analysis. *Biometrika* **72**, 627–636.

Critchley, F. and Vitiello, C. (1991). The influence of observations on misclassification probability estimates in linear discriminant analysis. *Biometrika* **78**, 677–690.

Cureton, E. E. and D'Agostino, R. B. (1983). *Factor Analysis: An Applied Approach*. Hillsdale, NJ: Lawrence Erlbaum.

Daling, J. R. and Tamura, H. (1970). Use of orthogonal factors for selection of variables in a regression equation—an illustration. *Applied Statistics* **19**, 260–268.

Darroch, J. N. and Mosimann, J. E. (1985). Canonical and principal components of shape. *Biometrika* **72**, 241–252.

Das Gupta, S. (1968). Some aspects of discrimination function coefficients. *Sankhya (Series A)* **30**, 387–400.

Davis, A. W. (1970a). Exact distributions of Hotelling's generalized T_0^2-test. *Biometrika* **57**, 187–191.

Davis, A. W. (1970b). Further applications of a differential equation for Hotelling's generalized T_0^2-test. *Annals of the Institute of Statistical Mathematics* **22**, 77–87.

Davis, A. W. (1980a). Further tabulation of Hotelling's generalized T_0^2-test. *Communications in Statistics—Series B, Simulation and Computation* **9**, 321–336.

Davis, A. W. (1980b). On the effects of moderate multivariate nonnormality on Wilks' likelihood ratio criterion. *Biometrika* **67**, 419–427.

Davis, A. W. (1982). On the effects of moderate multivariate nonnormality on Roy's largest root test. *Journal of the American Statistical Association* **77**, 896–900.

Davison, A. C. and Hall, P. (1992). On the bias and variability of bootstrap and cross-validation estimates of error rate in discrimination problems. *Biometrika* **79**, 279–284.

Dawkins, B. (1989). Multivariate analysis of national track records. *American Statistician* **43**, 110–115.

Dawson-Saunders, B. (1982). Correcting for bias in the canonical redundancy statistic. *Educational and Psychological Measurement* **42**, 131–143.

Day, N. E. and Kerridge, D. F. (1967). A general maximum likelihood discriminant. *Biometrics* **23**, 313–323.

Delaney, N. J. (1987). Classification using robust multivariate estimation with financial applications. In *Proceedings of the Business and Economic Statistics Section*. Annual Meeting of the American Statistical Association, San Francisco, CA, August 17–20, 1987.

DeLeeuw, J. (1982). Nonlinear principal component analysis. In H. Caussinus, P. Eltinger, and R. Tomassone (eds.), *Compstat*, Vol. 82, pp. 77–86. Wien: Physica-Verlag.

DeLeeuw, J. and Rijckevorsel, J. V. (1980). Homals and princals: Some generalizations of principal components analysis. In E. Diday, L. Lebart, J. P. Pages, and R. Tomassone (eds.), *Data Analysis and Informatics*, pp. 231–242. Amsterdam: North-Holland.

De Ligny, C. L., Nieuwdorp, G. H. E., Brederode, W. K., and Hammers, W. E. (1981). An application of factor analysis with missing data. *Technometrics* **23**(1), 91–95.

Dempster, A. P. (1966). Estimation in multivariate analysis. In P. R. Krishnaiah (ed.), *Multivariate Analysis*, pp. 315–334. New York: Academic Press.

Dempster, A. P. (1969). *Elements of Continuous Multivariate Analysis*. Reading, MA: Addison-Wesley.

Dempster, A. P., Laird, N. M., and Rubin, D. B. (1977). Maximum likelihood from incomplete data via the EM algorithm. *Journal of the Royal Statistical Society, Series B* **39**, 1–38.

Devlin, S. J., Gnanadesikan, R., and Kettenring, J. R. (1981). Robust estimation of dispersion matrices and principal components. *Journal of the American Statistical Association* **76**, 354–362.

Diehr, G. and Hoflin, D. R. (1974). Approximating the distribution of the sample R^2 in best subset regressions. *Technometrics* **16**, 317–321.

DiPillo, P. J. (1976). The application of bias to discriminant analysis. *Communications in Statistics—Series A, Theory and Methods* **5**(9), 834–844.

DiPillo, P. J. (1979). Biased discriminant analysis: Evaluation of the optimum probability of misclassification. *Communications in Statistics—Series A, Theory and Methods* **A8**(14), 1447–1457.

Dobson, A. J. (1990). *An Introduction to Generalized Linear Models*. New York: Chapman & Hall.

Dorveo, A. S. S. (1993). Comparison of two asymptotic techniques for estimating error rates in discriminant analysis. *Communications in Statistics—Series B, Simulation and Computation* **22**(2), 461–470.

Draper, N. R. and Smith, H. (1981). *Applied Regression Analysis* (2nd edition). New York: Wiley.

Dubuisson, B. and Masson, M. (1993). A statistical decision rule with incomplete knowledge about classes. *Pattern Recognition* **26**(1), 155–165.

Dunnett, C. W. and Sobel, M. (1954). A bivariate generalization of Student's t-distribution with tables for certain special cases. *Biometrika* **41**, 153–169.

Dunteman, G. H. (1989). *Principal Components Analysis*. Newbury Park, CA: SAGE Publications.

Dyer, R. H. and Ansher, A. F. (1977). Analysis of German wines by gas chromatography and atomic absorption. *Journal of Food Science* **42**, 534–536.

Dziuban, C. D. and Shirkey, E. C. (1974). When is a correlation matrix appropriate for factor analysis. *Psychological Bulletin* **81**, 358–361.

Eastment, H. T. and Krzanowski, W. J. (1982). Cross-validatory choice of the number of components from a principal component analysis. *Technometrics* **24**, 73–77.

Efron, B. (1975). The efficiency of logistic regression compared to normal discriminant analysis. *Journal of the American Statistical Association* **70**, 892–898.

Efron, B. (1979). Bootstrap methods: Another look at the jackknife. *Annals of Statistics* **7**, 1–26.

Efron, B. (1981). Nonparametric standard errors and confidence intervals. *Canadian Journal of Statistics* **9**, 139–172.

Efron, B. (1982). *The Jackknife, the Bootstrap, and Other Resampling Plans*. Philadelphia: Society for Industrial and Applied Mathematics.

Efron, B. (1983). Estimating the error rate of a prediction rule: Improvement on cross-validation. *Journal of the American Statistical Association* **78**, 316–331.

Efron, B. (1985). Bootstrap confidence intervals for a class of parametric problems. *Biometrika* **72**, 45–58.

Efron, B. (1987). Better bootstrap confidence intervals. *Journal of the American Statistical Association* **82**, 171–185.

Efron, B. (1988). Bootstrap confidence intervals: Good or bad? *Psychological Bulletin* **104**, 293–296.

Efron, B. and Gong, G. (1983). A leisurely look at the bootstrap, the jackknife and cross-validation. *American Statistician* **37**, 36–48.

Efron, B. and Tibshirani, R. (1986). Bootstrap methods for standard errors, confidence intervals and other measures of statistical accuracy. *Statistical Science* **1**, 54–75.

Efron, B. and Tibshirani, R. (1993). *An Introduction to the Bootstrap*. London: Chapman & Hall.

Eisenbis, R. A., Gilbert, G. C., and Avery, R. B. (1973). Investigating the relative importance of individual variables and variable subsets in discriminant analysis. *Communications in Statistics* **2**, 205–219.

Espahbodi, P. (1991). Identification of problem banks and binary choice models. *Journal of Banking and Finance* **15**, 53–71.

Everett, J. E. (1983). Factor comparability as a means of determining the number of factors and their rotation. *Multivariate Behavioral Research* **18**, 197–218.

Everett, J. E. and Entrekin, L. V. (1980). Factor comparability and the advantages of multiple-group factor analysis. *Multivariate Behavioral Research* **15**, 165–180.

Everitt, B. S. (1979). A Monte Carlo investigation of the robustness of Hotelling's one and two-sample T^2 statistic. *Journal of the American Statistical Association* **74**, 48–51.

Everitt, B. S. (1994). Exploring multivariate data graphically: A brief review with examples. *Journal of Applied Statistics* **21**(3), 63–93.

Eysenck, H. J. (1986). The why and wherefore of factor analysis. In R. J. Brook, G. C. Arnold, T. H. Hassard, and R. M. Pringle (eds.), *The Fascination of Statistics*. New York: Dekker.

Fang, K. and Zhang, Y. (1990). *Generalized Multivariate Analysis*. New York: Springer-Verlag.

Farmer, J. H. and Freund, R. J. (1975). Variable selection in the multivariate analysis of variance (MANOVA). *Communications in Statistics* **4**, 87–98.

Farrar, D. E. and Glauber, R. R. (1967). Multicollinearity in regression analysis (the problem revisited). *Review of Economic Statistics* **49**, 92–107.

Farver, T. B. and Dunn, O. J. (1979). Stepwise variable selection in classification problems. *Biometrical Journal* **21**, 145–153.

Fatti, L. P. (1983). The random-effects model in discriminant analysis. *Journal of the American Statistical Association* **78**, 679–687.

Fatti, L. P. and Hawkins, D. M. (1986). Variable selection in heteroscedastic discriminant analysis. *Journal of the American Statistical Association* **81**, 494–500.

Fava, J. L. and Velicer, W. F. (1992). The effects of overextraction on factor and component analysis. *Multivariate Behavioral Research* **27**(3), 387–415.

Feiveson, A. H. (1983). Classification by thresholding. *IEEE Transactions on Pattern Analysis and Machine Intelligence* **1**(PAMI-5), 48–54.

Feldmann, U. (1993). Partial distributional canonical discriminant analysis. *Biometrical Journal* **35**(4), 427–443.

Ferguson, G. A. (1954). The concept of parsimony in factor analysis. *Psychometrika* **19**, 281–290.

Finney, D. J. (1971). *Probit Analysis* (3rd edition). Cambridge: Cambridge University Press.

Fisher, R. A. (1921). On the probable error of a coefficient of correlation deduced from a small sample. *Metron* **1**, 1–32.

Fisher, R. A. (1936). The use of multiple measurement in taxonomic problems. *Annals of Eugenics* **7**, 179–188.

Flack, V. F. and Chang, P. C. (1987). Frequency of selecting noise variables in subset regression analysis: A simulation study. *American Statistician* **41**, 84–86.

Flury, B. (1983). Some relations between the comparison of covariance matrices and the principal component analysis. *Computational Statistics and Data Analysis* **1**, 97–109.

Flury, B. (1984). Common principal components in *k* groups. *Journal of the American Statistical Association* **79**, 892–898.

Flury, B. (1988). *Common Principal Components and Related Multivariate Models*. New York: Wiley.

Flury, B. (1990). Principal points. *Biometrika* **77**, 33–41.

Flury, B. (1993). Estimation of principal points. *Applied Statistics* **42**, 139–151.

Flury, B. W. (1989). Understanding partial statistics and redundancy of variables in regression and discriminant analysis. *American Statistician* **43**(1), 27–31.

Francis, I. (1973). Factor analysis: Its purpose, practice, and packaged progams. In *Proceedings of the Section on Statistical Computing*. Annual Meeting of the American Statistical Association, New York, August 1973.

Francis, I. (1974). Factor analysis: Fact or fabrication. *Mathematical Chronicle* **3**, 9–44.

Francisco, C. A. and Finch, M. D. (1979). A comparison of methods used for determining the number of factors to retain in factor analysis. In *Proceedings of the Section on the Statistical Computing*. Annual Meeting of the American Statistical Association, Washington, DC.

Freed, N. and Glover, F. A. (1981a). A linear programming approach to the discriminant problem. *Decision Sciences* **12**, 68–74.

Freed, N. and Glover, F. A. (1981b). Simple but powerful goal programming models for discriminant problems. *European Journal of Operations Research* **7**, 44–60.

Freed, N. and Glover, F. A. (1982). Linear programming and statistical discrimination pages—The LP side. *Decision Sciences* **13**, 172–175.

Frets, G. P. (1921). Heredity of head form in man. *Genetica* **3**, 193–384.

Friedman, J. H. (1989). Regularized discriminant analysis. *Journal of the American Statistical Association* **84**, 165–175.

Friedman, J. H. and Tukey, J. W. (1974). A projection pursuit algorithm for exploratory data analysis. *IEEE Transactions on Component Parts* **9**, 881–890.

Friedman, S. and Weisberg, H. F. (1981). Interpreting the first eigenvalue of a correlation matrix. *Educational and Psychological Measurement* **41**, 11–21.

Frost, T. (1994). A comparison of four test statistics in multivariate analysis of variance. Master's thesis, Brigham Young University, Department of Statistics.

Fukunaga, K. and Kessell, D. (1973). Nonparametric Bayes error estimation using unclassified samples. *IEEE Transactions on Information Theory* **4**, 434–440.

Fuller, W. A. (1987). *Measurement Error Models*. New York: Wiley.

Fung, W. K. (1992). Some diagnostic measures in discriminant analysis. *Statistics and Probability Letters* **13**, 279–285.

Fung, W. K. (1995). Diagnostics in linear discriminant analysis. *Journal of the American Statistical Association* **90**, 952–953.

Furnival, G. M. (1971). All possible regressions with less computation. *Technometrics* **13**, 403–408.

Furnival, G. M. and Wilson, R. W. (1974). Regression by leaps and bounds. *Technometrics* **16**, 499–511.

Gabriel, K. R. (1968). Simultaneous test procedures in multivariate analysis of variance. *Biometrika* **55**, 489–504.

Gabriel, K. R. (1978). Least squares approximation of matrices by additive and multiplicative models. *Journal of the Royal Statistical Society* **40**, 186–196.

Ganeshanandam, S. and Krzanowski, W. J. (1989). On selecting variables and assessing their perfomance in linear discriminant analysis. *Australian Journal of Statistics* **31**, 433–447.

Ganeshanandam, S. and Krzanowski, W. J. (1990). Error-rate estimation in two-group discriminant analysis using the linear discriminant function. *Journal of Statistical Computation and Simulation* **36**, 157–175.

Gangemi, G. (1986). Epistemological reasons for preferring component analysis to factor analysis. *Quality and Quantity* **20**, 75–84.

Garnett, J. C. M. (1919). On certain independent factors in mental measurement. *Proceedings of the Royal Society of London* **96**, 91–111.

Geweke, J. F. and Singleton, K. J. (1980). Interpreting the likelihood ratio statistic in factor models when sample size is small. *Journal of the American Statistical Association* **75**, 133–137.

Gilbert, E. S. (1968). On discrimination using qualitative variables. *Journal of the American Statistical Association* **63**, 1399–1412.

Gill, L. and Lewbel, A. (1992). Testing the rank and definiteness of estimated matrices with applications to factor, state-space and ARMA models. *Journal of the American Statistical Association* **87**, 766–776.

Gill, P. E. and Murray, W. (1972). Quasi-Newton methods for unconstrained optimization. *Journal of the Institute of Mathematics and Its Applications* **9**, 91–108.

Girshick, M. A. (1936). Principal components. *Journal of the American Statistical Association* **31**, 519–528.

Girshick, M. A. (1939). On the sampling theory of roots of determinantal equations. *Annals of Mathematical Statistics* **10**, 203–224.

Gittens, R. (1985). *Canonical Analysis: A Review with Applications in Ecology*. New York: Springer.

Gleason, T. C. (1976). On redundancy in canonical analysis. *Psychological Bulletin* **83**, 1004–1006.

Gleason, T. C. and Staelin, R. (1975). A proposal for handling missing data. *Psychometrika* **40**, 229–252.

Gleser, L. J. (1992). The importance of assessing measurement reliability in multivariate regression. *Journal of the American Statistical Association* **87**, 696–707.

Glick, N. (1978). Additive estimators for probabilities of correct classification. *Pattern Recognition* **10**, 211–222.

Glonek, G. F. V. and McCullagh, P. (1995). Multivariate logistic models. *Journal of the Royal Statistical Society, Series B* **57**(3), 533–546.

Glover, D. M. and Hopke, P. K. (1992). Exploration of multivariate chemical data by projection pursuit. *Chemometrics and Intelligent Laboratory Systems* **16**, 45–59.

Gnanadesikan, R. (1977). *Methods for Statistical Data Analysis of Multivariate Observations*. New York: Wiley.

Gnanadesikan, R. and Kettenring, J. R. (1972). Robust estimates, residuals and outlier detection with multiresponse data. *Biometrics* **28**, 81–124.

Gnanadesikan, R. and Lee, E. T. (1970). Graphical techniques for internal comparisons amongst equal degree of freedom groupings in multiresponse experiments. *Biometrika* **57**, 229–337.

Gnanadesikan, R., Laub, E., Snyder, M., and Yao, Y. (1965). Efficiency comparisons of certain multivariate analysis of variance test procedures. *Annals of Mathematical Statistics* **36**, 356–357 (Abstract).

Gong, G. (1986). Cross-validation, the jackknife and the bootstrap: Excess error estimation in forward logistic regression. *Journal of the American Statistical Association* **81**, 108–113.

Goria, M. N. and Flury, B. D. (1996). Common canonical variates in k independent groups. *Journal of the American Statistical Association* **91**(436), 1735–1742.

Gorsuch, R. L. (1983). *Factor Analysis* (2nd edition). Hillsdale, NJ: Lawrence Erlbaum.

Gower, J. C. (1966). Some distance properties of latent root and vector methods used in multivariate analysis. *Biometrika* **53**, 325–338.

Gower, J. C. (1967). Multivariate analysis and multidimensional geometry. *Statistician* **17**, 13–28.

Grablowsky, M. and Talley, W. (1981). Probit and discriminant functions for classifying credit applicants: A comparison. *Journal of Economics and Business* **3**, 254–261.

Granville, V. and Rasson, J. P. (1995). Multivariate discriminant analysis and maximum penalized likelihood density estimation. *Journal of the Royal Statistical Society, Series B* **57**(3), 501–517.

Graybill, F. A. (1969). *Introduction to Matrices with Applications in Statistics*. Belmont, CA: Wadsworth Publishing Company.

Graybill, F. A. (1976). *Theory and Application of the Linear Model.* North Scituate, MA: Duxbury Press.

Green, B. F. (1979). The two kinds of linear discriminant functions and their relationship. *Journal of Educational Statistics* **4**, 247–263.

Green, J. R. (1971). Testing departure from a regression without using replication. *Technometrics* **13**, 609–615.

Greenacre, M. J. (1984). *Theory and Applications of Correspondence Analysis.* London: Academic Press.

Greene, T. and Rayens, W. S. (1989). Partially pooled covariance matrix estimation in discriminant analysis. *Communications in Statistics: Theory and Methods* **18**, 3679–3702.

Griffith, R. A., Young, J. C., and Mason, R. L. (1996). Behavior of the Hotelling T^2 statistics when sampling from non-normal multivariate distributions. Presented at the Annual Meeting of the American Statistical Association, Chicago, IL, August 1996.

Guertin, A. S., Guertin, W. H., and Ware, W. B. (1981). Distortion as a function of the number of factors rotated under varying levels of common variance and error. *Educational and Psychological Measurement* **41**, 1–9.

Gunst, R. F. (1983). Regression analysis with multicollinearity predictor variables: Definition, detection and effects. *Communications in Statistics—Series A, Theory and Methods* **12**, 2217–2260.

Gunst, R. F. and Mason, R. L. (1977). Biased estimation in regression: An evaluation using mean squared error. *Journal of the American Statistical Association* **72**, 616–628.

Guttman, I. (1982). *Linear Models: An Introduction.* New York: Wiley.

Guttman, L. (1953). Image theory for the structure of quantitative variates. *Psychometrika* **18**, 277–296.

Guttman, L. (1954). Some necessary conditions for common factor analysis. *Psychometrika* **19**, 149–161.

Habbema, J. D. F. and Hermans, J. (1977). Selection of variables in discriminant analysis by F-statistic and error rate. *Technometrics* **19**, 487–493.

Haitovsky, Y. (1969). Multicollinearity in regression analysis: Comment. *Review of Economics and Statistics* **50**, 486–489.

Hakstian, A. R. (1971). A comparative evaluation of several prominent methods of oblique factor transformation. *Psychometrika* **36**, 175–193.

Hakstian, A. R. and Abel, R. A. (1974). A further comparison of oblique factor transformation methods. *Psychometrika* **39**, 429–444.

Hakstian, A. R., Roed, J. C., and Lind, J. C. (1979). Two-sample T^2 procedure and the assumption of homogeneous covariance matrices. *Pyschological Bulletin* **86**, 1255–1263.

Hakstian, A. R., Rogers, W. D., and Cattell, R. B. (1982). The behavior of numbers of factors rules with simulated data. *Multivariate Behavioral Research* **17**, 193–219.

Hall, C. E. (1969). Rotation of canonical variates in multivariate analysis of variance. *Journal of Experimental Education* **38**, 31–38.

Hall, C. E. (1977). Some elements of the design of multivariate experiments. *Journal of Experimental Education* **45**, 26–37.

Hamilton, D. (1987). Sometimes $R^2 > r^2_{yx_1} + r^2_{yx_2}$: Correlated variables are not always redundant. *American Statistician* **41**(2), 129–132.

Hamilton, D. C. and Lesperance, M. L. (1995). A comparison of methods for univariate and multivariate acceptance sampling by variables. *Technometrics* **37**(3), 329–339.

Han, C. P. and Huang, M. H. (1987). Effect of significance level for testing equality of covariance matrices in discriminant analysis. In *Proceedings of the Section on Statistical Computing*. Annual Meeting of the American Statistical Association, San Francisco, CA, August 17–20.

Hand, D. J. (1981). *Discriminant Analysis*. New York: Wiley.

Hand, D. J. (1992). Statistical methods in diagnosis. *Statistical Methods in Medical Research* **1**, 49–67.

Hand, D. J. (1994). Assessing classification rules. *Journal of Applied Statistics* **21**(3), 3–16.

Harman, H. H. (1976). *Modern Factor Analysis* (3rd revised edition). Chicago: The University of Chicago Press.

Harrell, F. E. and Lee, K. L. (1985). A comparison of discriminant analysis and logistic regression under multivariate normality. In P. K. Sen (ed.), *Biostatistics: Statistics in Biomedical, Public Health and Environmental Sciences*, pp. 333–343. Amsterdam: Elsevier Science Publishers (North-Holland Publishing Company).

Harris, R. J. (1985). *A Primer of Multivariate Statistics* (2nd edition). New York: Academic Press.

Hastie, T. and Stuetzle, W. (1989). Principal curves. *Journal of the American Statistical Association* **84**, 502–516.

Hastie, T. and Tibshirani, R. (1996). Discriminant analysis by Gaussian mixtures. *Journal of the Royal Statistical Society, Series B* **58**(1), 155–176.

Hawkins, D. M. (1973). On the investigation of alternative regressions by principal component analysis. *Applied Statistics* **22**, 275–286.

Hawkins, D. M. (1976). The subset problem in multivariate analysis of variance. *Journal of the Royal Statistical Society, Series B* **38**, 132–139.

Hawkins, D. M. (1981). A new test for multivariate normality and homoscedasticity. *Technometrics* **23**, 105–110.

Hawkins, D. M. (1993). Regression adjustment for variables in multivariate quality control. *Journal of Quality Technology* **25**, 170–182.

Hawkins, D. M. (1996). Multivariate quality control and regression-adjusted variables. Presented at the Annual Meeting of the American Statistical Association, Chicago, IL, August 1996.

Hawkins, D. M. and Eplett, W. J. R. (1982). The Cholesky factorization of the inverse correlation or covariance matrix in multiple regression. *Technometrics* **24**, 191–198.

Hawkins, D. M. and Fatti, L. P. (1984). Exploring multivariate data using the minor principal components. *American Statistician* **33**, 325–338.

Healy, J. D. (1987). A note on multivariate CUSUM procedures. *Technometrics* **29**(4), 409–412.

Healy, J. D. (1986). Adaptive CUSUM procedures. ASQC Technical Conference Transactions. American Society for Quality Control, 161 West Wisconsin Ave., Milwaukee, WI 53203.

Hecker, R. and Wegener, H. (1978). The valuation of classification rates in stepwise discriminant analysis. *Biometrical Journal* **20**, 713–727.

Heiberger, R. M. (1977). Regression with the pairwise-present covariance matrix: A dangerous practice. Technical Report 19, The Wharton School, University of Pennsylvania, Department of Statistics.

Heitjan, D. F. and Basu, S. (1996). Distinguishing "missing at random" and "missing completely at random." *American Statistician* **50**(3), 207–213.

Helland, I. S. (1987). On the interpretation and use of R^2 in regression analysis. *Biometrics* **43**, 61–69.

Hemel, J. B., van der Voet, H., Hindriks, F. R., and van der Slik, W. (1987). Stepwise deletion: A technique for missing-data handling in multivariate analysis. *Analytica Chimica Acta* **193**, 255–268.

Hendrix, L. J., Carter, M. W., and Scott, D. T. (1982). Covariance analysis with heterogeneity of slopes in fixed models. *Biometrics* **38**, 641–650.

Heo, T. Y. (1987). The comparison of eigensystem techniques for measuring multicolinearity in multivariate normal data. Master's thesis, Brigham Young University, Department of Statistics.

Hermans, J. and Habbema, J. D. F. (1976). Manual for the ALLOC discriminant analysis programs. Technical Report, University of Leiden, Department of Statistics.

Hermans, J., Habbema, J. D. F., and Schafer, J. R. (1982). The ALLOC80 package for discriminant analysis. *Statistics Software Newsletter* **8**(1), 15–20.

Higbee, K. T. (1994). Variably regularized discriminant analysis. Presented at the Annual Meeting of the American Statistical Association, Toronto, Canada, August 1994.

Higbee, K. T., Redgate, T., Miller, N., Anderson, D., and Anderson, K. (1996). Issues in selecting a classification method. In *Proceedings of the Section on Statistical Computing*. Annual Meeting of the American Statistical Association, Chicago, IL, August 1996.

Hill, R. C., Fomby, T. B., and Johnson, S. R. (1977). Component selection norms for principal components regression. *Communications in Statistics—Series A, Theory and Methods* **6**, 309–334.

Hills, M. (1966). Allocation rules and their error rates. *Journal of the Royal Statistical Society* **28**, 1–20.

Hirsch, R. F., Wu, G. L., and Tway, P. C. (1987). Reliability of factor analysis in the presence of random noise or outlying data. *Chemometrics and Intelligent Laboratory Systems* **1**, 265–272.

Hirst, D. (1996). Error-rate estimation in multiple-group linear discriminant analysis. *Technometrics* **38**(4), 389–399.

Hochberg, Y. (1988). A sharper Bonferroni procedure for multiple tests of significance. *Biometrika* **75**, 800–802.

Hocking, R. R. (1976). The analysis and selection of variables in linear regression. *Biometrics* **32**, 1–51.

Hocking, R. R. (1985). *The Analysis of Linear Models*. Monterey, CA: Brooks/Cole.

Hocking, R. R. (1996). *Methods and Applications of Linear Models*. New York: Wiley.

Hocking, R. R. and Speed, F. M. (1975). A full rank analysis of some linear model problems. *Journal of the American Statistical Association* **70**, 706–712.

Hoerl, A. E. and Kennard, R. W. (1970a). Ridge regression: Biased estimation for nonorthogonal problems. *Technometrics* **12**, 55–67.

Hoerl, A. E. and Kennard, R. W. (1970b). Ridge regression: Applications to nonorthogonal problems. *Technometrics* **12**, 69–82.

Hofmann, R. J. (1978). A generalized regression estimate formula for factor scores: Its application to singular data. *Journal of General Psychology* **99**, 141–147.

Hogg, R. V. (1979). Statistical robustness: One view of its use in applications today. *American Statistician* **33**, 108–115.

Hogg, R. V. and Craig, A. T. (1995). *Introduction to Mathematical Statistics* (5th edition). Englewood Cliffs, NJ: Prentice-Hall.

Holland, B. (1991). On the application of three modified Bonferroni procedures to pairwise multiple comparisons in balanced repeated measures designs. *Computational Statistics Quarterly* **3**, 219–231.

Holland, B. and Copenhaver, M. D. (1987). An improved sequentially rejective Bonferroni test procedure. *Biometrics* **43**, 417–423.

Hollingsworth, H. H. (1983). On the bias of the canonical correlation redundancy statistic. In *Proceedings of the Section on Social Sciences*. Annual Meeting of the American Statistical Association, Toronto, Canada, 1983.

Holloway, L. N. and Dunn, O. J. (1967). The robustness of Hotelling's T^2. *Journal of the American Statistical Association* **62**, 124–136.

Holm, S. (1979). A simple sequentially rejective multiple test procedure. *Scandinavian Journal of Statistics* **6**, 65–70.

Hommel, G. (1988). A stagewise rejective multiple test procedure based on a modified Bonferroni test. *Biometrika* **75**, 383–386.

Hopkins, J. W. and Clay, P. P. F. (1963). Some empirical distributions of bivariate t^2 and homoscedasticity criterion m under unequal variance and leptokurtosis. *Journal of the American Statistical Association* **58**, 1048–1053.

Horst, P. (1941). Approximating a multiple correlation system by one of lower rank as a basis for deriving more stable prediction weights. In P. Horst (ed.), *The Prediction of Personal Adjustment*. New York: Social Science Research Council.

Horst, P. (1961). Relations among m sets of measures. *Psychometrika* **26**, 129–149.

Horst, P. (1965). *Factorial Analysis of Data Matrices*. New York: Holt, Rinehart and Winston.

Hosmer, D., Jovanovic, B., and Lemeshow, S. (1989). Best subsets logistic regression. *Biometrics* **45**, 1265–1270.

Hossain, A. and Naik, D. N. (1989). Detection of influential observations in multivariate regression. *Journal of Applied Statistics* **16**, 25–37.

Hotelling, H. (1931). The generalization of student's ratio. *Annals of Mathematical Statistics* **2**, 360–378.

Hotelling, H. (1933). Analysis of a complex of statistical variables into principal components. *Journal of Education Psychology* **24**, 417–441.

Hotelling, H. (1935). The most predictable criterion. *Journal of Education Psychology* **26**, 139–142.

Hotelling, H. (1936). Relations between two sets of variates. *Biometrika* **28**, 321–327.

Hotelling, H. (1951). A generalized t^2 test and measure of multivariate dispersion. *Proceedings of the Second Berkeley Symposium on Mathematical Statistics and Probability* **1**, 23–41.

Hotelling, H. (1957). The relations of the newer multivariate statistical methods to factor analysis. *British Journal of Psychology* **10**, 69–79.

Householder, A. S. (1964). *The Theory of Matrices in Numerical Analysis*. New York: Blaisdell.

Householder, A. S. and Young, G. (1938). Matrix approximation and latent roots. *American Mathematics Monthly* **45**, 165–171.

Houshmand, A. A. (1993). Misclassification probabilities for quadratic discriminant functions. *Communications in Statistics—Series B, Simulation and Computation* **22**(1), 81–98.

Hsuan, F. C. (1981). Ridge regression from a principal component point of view. *Communications in Statistics—Series A, Theory and Methods* **10**, 1981–1995.

Hu, M., Fisher, D. M., Booth, D. E., and Hung, M. S. (1988). Robust discriminant analysis in marketing research: Methods and applications. *Journal of the Industrial Mathematics Society* **38**, 181–196.

Huang, D. and Tseng, S. (1992). A decision procedure for determining the number of components in principal component analysis. *Journal of Statistical Planning and Inference* **30**, 63–71.

Huba, G. J., Wingard, J. A., and Bentler, P. M. (1980). Longitudinal analysis of the role of peer support, adult models and peer subcultures in beginning adolescent substance use: An application of setwise canonical correlation methods. *Multivariate Behavioral Research* **15**, 259–279.

Huber, P. J. (1964). Robust estimation of a location parameter. *Annals of Mathematical Statistics* **35**, 73–101.

Huber, P. J. (1981). *Robust Statistics*. New York: Wiley.

Huber, P. J. (1985). Projection pursuit. *Annals of Statistics* **13**, 435–525.

Huberty, C. J. (1975). Discriminant analysis. *Review of Educational Research* **45**, 543–598.

Huberty, C. J. (1984). Issues in the use and interpretation of discriminant analysis. *Psychological Bulletin* **95**, 156–171.

Huberty, C. J. (1994). *Applied Discriminant Analysis*. New York: Wiley.

Huberty, C. J. (1996). Some issues and problems in discriminant analysis. Presented at the Annual Meeting of the American Statistical Association, Chicago, IL, August 1996.

Huberty, C. J. and Curry, A. R. (1978). Linear versus quadratic multivariate classification. *Multivariate Behavioral Research* **13**, 237–245.

Huberty, C. J. and Smith, D. U. (1976). Variable contribution in discriminant analysis. Presented at the Annual Meeting of the American Educational Research Association. San Francisco, CA.

Huberty, C. J., Wisenbaker, J. M., and Smith, J. C. (1987). Assessing predictive accuracy in discriminant analysis. *Multivariate Behavioral Research* **22**, 307–329.

Hummel, T. J. and Sligo, J. (1971). Empirical comparison of univariate and multivariate analysis of variance procedures. *Psychological Bulletin* **76**, 49–57.

Hwang, H. L. and Paulson, A. S. (1986). Some methods for the multivariate two-sample location problem. Unpublished Manuscript, Oak Ridge National Laboratory, Oak Ridge, Tennessee.

Imrey, P. B., Koch, G. G., and Stokes, M. E. (1981). Categorical data analysis: Some reflections on the log linear model and logistic regression. Part I: Historical and methodological overview. *International Statistics Review* **49**, 265–283.

Imrey, P. B., Koch, G. G., and Stokes, M. E. (1982). Categorical data analysis: Some reflections on the log linear model and logistic regression. Part II: Data analysis. *International Statistics Review* **50**, 35–63.

Ito, K. (1969). On the effect of heteroscedasticity and nonnormality upon some multivariate test procedures. In P. R. Krishnaiah (ed.), *Multivariate Analysis*, Vol. 2, pp. 87–120. New York: Academic Press.

Ito, K. and Schull, W. J. (1964). On the robustness of the T_0^2 test in multivariate analysis of variance when variance-covariance matrices are not equal. *Biometrika* **51**, 71–82.

Ito, P. K. (1980). Robustness of ANOVA and MANOVA test procedures. In P. R. Krishnaiah (ed.), *Handbook of Statistics*, Vol. 1, pp. 199–236. Amsterdam: North-Holland Publishing Company.

Jackson, J. E. (1980). Principal components and factor analysis: Part I—Principal components. *Journal of Quality Technology* **12**, 201–213.

Jackson, J. E. (1981). Principal components and factor analysis: Part II—Additional topics related to principal components. *Journal of Quality Technology* **13**, 46–58.

Jackson, J. E. (1985). Multivariate quality control. *Communications in Statistics: Theory and Methods* **14**(11), 2657–2688.

Jackson, J. E. (1991). *A User's Guide to Principal Components*. New York: Wiley.

Jambu, M. (1991). *Exploratory and Multivariate Data Analysis*. Boston: Harcourt Brace Jovanovich.

James, G. S. (1954). Tests of linear hypotheses in univariate and multivariate analysis when the ratios of the population variances are unknown. *Biometrika* **41**, 19–43.

James, M. (1985). *Classification Algorithms*. New York: Wiley.

Jan, S. and Randles, R. H. (1994). A multivariate signed sum test for the one-sample location problem. *Journal of Nonparametric Statistics* **4**, 49–63.

Jennrich, R. L. (1977). Stepwise discriminant analysis. In K. Enslein, A. Ralston, and H. S. Wilf (eds.), *Statistical Methods for Digital Computers*, Vol. 3, pp. 76–95. New York: Wiley.

Jensen, D. R. (1972). Some simultaneous multivariate procedures using Hotelling's T^2 statistics. *Biometrics* **28**, 39–53.

Jensen, D. R. (1984). The structure of ellipsoidal distributions: I. Canonical analysis. *Biometrika* **26**, 779–787.

Jensen, D. R. (1985). Multivariate distributions. In S. Kotz and N. L. Johnson (eds.), *Encyclopedia of Statistical Sciences*, Vol. 6, pp. 43–55. New York: Wiley.

Joglekar, G., Schuenemeyer, J. H., and LaRiccia, V. (1989). Lack of fit when replicates are not available. *American Statistician* **43**, 135–143.

Johansen, S. (1980). The Welch-James approximation to the distribution of the residual sum of squares in a weighted linear regression. *Biometrika* **67**, 85–92.

Johnson, R. A. and Wichern, D. W. (1992). *Applied Multivariate Statistical Analysis* (3rd edition). Englewood Cliffs, NJ: Prentice-Hall.

Johnson, W. (1985). Influence measures for logistic regression: Another point of view. *Biometrika* **72**, 59–65.

Johnson, W. (1987). The detection of influential observation for allocation, separation and the determination of probabilities in a Bayesian framework. *Journal of Business and Economic Statistics* **5**, 369–381.

Jolliffe, I. T. (1982). A note on the use of principal components in regression. *Applied Statistics* **31**, 300–303.

Jolliffe, I. T. (1986). *Principal Component Analysis*. New York: Springer.

Jollife, I. T. (1987). Rotation of principal components: Some comments. *Journal of Climatology* **7**, 507–510.

Jolliffe, I. T. (1989). Rotation of ill-defined principal components. *Applied Statistics* **38**, 139–147.

Jolliffe, I. T. (1992). Principal component analysis and exploratory factor analysis. *Statistical Methods in Medical Research* **1**, 69–95.

Jolliffe, I. T. (1995). Rotation of principal components: Choice of normalization constraints. *Journal of Applied Statistics* **22**(1), 29–35.

Jones, M. C. and Sibson, R. (1987). What is projection pursuit? *Journal of the Royal Statistical Society* **150**, 1–36.

Jöreskog, K. G. (1967). Some contributions to maximum likelihood factor analysis. *Psychometrika* **32**, 443–482.

Jöreskog, K. G. and Lawley, D. N. (1968). New methods in maximum likelihood factor analysis. *British Journal of Mathematical and Statistical Psychology* **21**, 85–96.

Kaiser, H. F. (1958). The varimax criterion for analytic rotation in factor analysis. *Psychometrika* **23**, 187–200.

Kaiser, H. F. (1960). The application of electronic computers to factor analysis. *Educational and Psychological Measurement* **20**, 141–151.

Kaiser, H. F. (1963). Image analysis. In C. W. Harris (ed.), *Problems in Measuring Change*. University of Wisconsin Press.

Kaiser, H. F. (1970). A second generation little jiffy. *Psychometrika* **35**, 401–415.

Kaiser, H. F. (1974). An index of factorial simplicity. *Psychometrika* **39**, 31–36.

Kaiser, H. F. and Derflinger, G. (1990). Some contrasts between maximum likelihood factor analysis and alpha factor analysis. *Applied Psychological Measurement* **14**, 29–32.

Kariya, T. and Cohen, A. (1992). On the invariance structure of the one-sided testing problem for a multivariate normal mean. *Statistica Sinica* **2**, 221–236. (Correction p. 619).

Karnel, G. (1991). Robust canonical correlation and correspondence analysis. *The Frontiers of Statistical Scientific Theory and Industrial Applications* **2**, 335–354.

Karson, M. J. (1982). *Multivariate Statistical Methods*. Ames, IA: The Iowa State University Press.

Kendall, M. G. (1957). *A Course in Multivariate Analysis*. London: Griffin.

Keramidas, E. M., Devlin, S. J., and Gnanadesikan, R. (1987). A graphical procedure for comparing the principal components of several covariance matrices. *Communications in Statistics—Series B, Simulation and Computation* **16**, 161–191.

Kettenring, J. R. (1971). Canonical analysis of several sets of variables. *Biometrika* **58**, 433–541.

Khuri, A. I. (1985). A test for lack of fit of a linear multiresponse model. *Technometrics* **27**, 213–218.

Khuri, A. I. (1986). Exact tests for the comparison of correlated response models with an unknown dispersion matrix. *Technometrics* **28**, 347–357.

Khuri, A. I., Mathew, T., and Nel, D. G. (1994). A test to determine closeness of multivariate Satterthwaite's approximation. *Journal of Multivariate Analysis* **51**, 201–209.

Kim, J. O. and Mueller, C. W. (1978). *Factor Analysis: Statistical Methods and Practical Issues*. Beverly Hills: Sage Publications.

Kim, S. (1992). A practical solution to the multivariate Behrens-Fisher problem. *Biometrika* **79**, 171–176.

Kim, S. and Park, S. H. (1991). On sensitivity analysis in principal component regression. *Journal of the Korean Statistical Society* **20**, 177–190.

Kleinbaum, D. G. (1994). *Logistic Regression*. New York: Springer-Verlag.

Klemm, R. and Gust, C. (1982). A comparison of procedures for the classification of multivariate observations. Presented at Proceedings of the SAS User's Group. P.O. Box 8000, Cary, NC 27511.

Knol, D. L. and Berger, M. P. F. (1991). Empirical comparison between factor analysis and multidimensional item response models. *Multivariate Behavioral Research* **26**, 457–477.

Koehler, G. J. and Erenguc, S. S. (1990). Minimizing misclassifications in linear discriminant analysis. *Decision Sciences* **21**, 63–85.

Korin, B. P. (1972). Some comments on the homoscedasticity criterion *m* and the multivariate analysis of variance tests T^2, *W* and *R*. *Biometrika* **59**, 215–216.

Kosfeld, R. (1987). Efficient iteration procedures for maximum likelihood factor analysis. *Statistische Hefte* **28**, 301–315.

Kotz, S. and Pearn, W. L. (1984). Eigenvalue-eigenvector analysis for a class of patterned correlation matrices with an application. *Statistics and Probability Letters* **2**, 119–125.

Kourti, T. and MacGregor, J. F. (1996). Multivariate SPC methods for process and product monitoring. *Journal of Quality Technology* **28**(4), 409–428.

Kramer, C. Y. and Jensen, D. R. (1969a). Fundamentals of multivariate analysis, Part I. Inference about means. *Journal of Quality Technology* **1**(2), 120–133.

Kramer, C. Y. and Jensen, D. R. (1969b). Fundamentals of multivariate analysis, part III. Analysis of variance for one-way classifications. *Journal of Quality Technology* **1**(4), 264–276.

Kreft, I. G. G., de Leeuw, J., and van der Leeden, R. (1994). Review of five multilevel analysis programs: BMDP-5V, GENMOD, HLM, ML3, VARCL. *American Statistician* **48**(4), 324–335.

Krishnaiah, P. R. (1982). Selection of variables in discriminant analysis. In P. R. Krishnaiah and L. N. Kanal (eds.), *Handbook of Statistics*, Vol. 2, pp. 883–892. Amsterdam: North-Holland Publishing.

Kroonenberg, P. M. (1983). *Three-Mode Principal Component Analysis: Theory and Applications*. Leiden: DSWO Press.

Krus, D. J., Reynolds, T. J., and Krus, P. H. (1976). Rotation in canonical variate analysis. *Educational and Psychological Measurement* **36**, 725–730.

Krusinska, E. (1988). Robust methods in discriminant analysis. *Rivista di Statistica Applicata* **21**, 239–253.

Krusinska, E. and Liebhart, J. (1989). Some further remarks on robust selection of variables in discriminant analysis. *Biometrical Journal* **31**, 227–233.

Krusinska, E. and Liebhart, J. (1990a). On the Wilks' Λ statistic as a tool for model selection in mixed variable discriminant analysis. *Zastosowania Matematyki Applicationes Mathematicae* **20**, 551–560.

Krusinska, E. and Liebhart, J. (1990b). Robust discriminant functions in assisting medical diagnosis. Application to the chronic obturative lung disease data. *Biometrical Journal* **32**, 915–929.

Krzanowski, W. J. (1977). The performance of Fisher's linear discriminant function under non-optimal conditions. *Technometrics* **19**, 191–200.

Krzanowski, W. J. (1979). Some linear transformations for mixtures of binary and continuous variables, with particular reference to linear discriminant analysis. *Biometrika* **66**, 33–39.

Krzanowski, W. J. (1980). Mixtures of continuous and categorical variables in discriminant analysis. *Biometrics* **36**, 493–499.

Krzanowski, W. J. (1982). Between-groups comparison of principal components—Some sampling results. *Journal of Statistical Computation and Simulation* **15**, 141–154.

Krzanowski, W. J. (1983). Stepwise location model choice in mixed variable discrimination. *Applied Statistics* **32**, 260–266.

Krzanowski, W. J. (1984a). Principal component analysis in the presence of group structure. *Applied Statistics* **33**, 164–168.

Krzanowski, W. J. (1984b). Sensitivity of principal components. *Journal of the Royal Statistical Society, Series B* **46**, 558–563.

Krzanowski, W. J. (1989). On confidence regions in canonical variate analysis. *Biometrika* **76**, 107–116.

Krzanowski, W. J. (1993). The location model for mixtures of categorical and continuous variables. *Journal of Classification* **10**, 25–49.

Krzysko, M. (1982). Canonical analysis. *Biometrical Journal* **24**, 211–228.

Kshirsagar, A. M. (1972). *Multivariate Analysis*. New York: M. Dekker.

Kshirsagar, A. M. and Arseven, E. (1975). A note on the equivalency of two discrimination procedures. *American Statistician* **29**, 38–39.

Kshirsagar, A. M., Kocherlakota, S., and Kocherlakota, K. (1990). Classification procedures using principal component analysis and stepwise discriminant functions. *Communications in Statistics: Theory and Methods* **19**, 91–109.

Kwan, W. O. and Kowalski, B. R. (1978). Classification of wines by applying pattern recognition to chemical composition data. *Journal of Food Science* **43**, 1320–1323.

Lachenbruch, P. A. (1965). Estimation of error rates in discriminant analysis. Ph.D. thesis, University of California at Los Angeles, Department of Statistics.

Lachenbruch, P. A. (1966). Discriminant analysis when the initial samples are misclassified. *Technometrics* **8**, 657–662.

Lachenbruch, P. A. (1967). An almost unbiased method of obtaining confidence intervals for the probability of misclassification in discriminant analysis. *Biometrics* **23**, 639–645.

Lachenbruch, P. A. (1968). On expected probabilities of misclassification in discriminant analysis, necessary sample size and a relation with the multiple correlation coefficient. *Biometrics* **24**, 823–834.

Lachenbruch, P. A. (1974). Discriminant analysis when the initial samples are misclassified: II: Non-random misclassification models. *Technometrics* **16**, 419–424.

Lachenbruch, P. A. (1975). *Discriminant Analysis*. New York: Hafner Press.

Lachenbruch, P. A. (1979). Note on initial misclassification effects on the quadratic discrimination function. *Technometrics* **21**, 129–132.

Lachenbruch, P. A. and Goldstein, M. (1979). Discriminant analysis. *Biometrics* **35**, 69–85.

Lachenburch, P. A. and Mickey, M. R. (1968). Estimation of error rates in discriminant analysis. *Technometrics* **10**, 1–11.

Lachenbruch, P. A., Sneeringer, C., and Revo, L. T. (1973). Robustness of the linear and quadratic discriminant function to certain types of non-normality. *Communications in Statistics* **1**, 39–57.

Lambert, Z. V., Wildt, A. R., and Durand, R. M. (1991). Approximating confidence intervals for factor loadings. *Multivariate Behavioral Research* **26**, 421–434.

Larsen, W. A. and McCleary, S. A. (1972). The use of partial residual plots in regression analysis. *Technometrics* **14**, 781–790.

Larson, S. F. (1980). A general procedure for robust discrimination. Master's thesis, Brigham Young University, Department of Statistics.

Latour, D. and Styan, G. P. H. (1983). Canonical correlations in the two-way layout. Proceedings of the First Tampere Sem. Linear Models.

Lauter, J. (1978). Sample size requirements for the T^2 test of MANOVA (tables for one-way classification). *Biometrical Journal* **20**, 389–406.

Lawley, D. N. (1938). A generalization of Fisher's z-test. *Biometrika* **30**, 180–187.

Lawley, D. N. (1956). Tests of significance for the latent roots of covariance and correlation matrices. *Biometrika* **43**, 128–136.

Lawley, D. N. (1959). Tests of significance in canonical analysis. *Biometrika* **46**, 59–66.

Lawley, D. N. (1963). On testing a set of correlation coefficients for equality. *Annals of Mathematical Statistics* **34**, 149–151.

Lawley, D. N. (1967). Some new results in maximum likelihood factor analysis. *Proceedings of the Royal Society of Education* **67**, 256–264.

Lawley, D. N. and Maxwell, A. E. (1971). *Factor Analysis as a Statistical Method* (2nd edition). London: Butterworth.

Lawley, D. N. and Maxwell, A. E. (1973). Regression and factor analysis. *Biometrika* **60**, 331–338.

Le Cessie, S. and Van Houwelingen, J. C. (1992). Ridge estimators in logistic regression. *Applied Statistics* **41**, 191–201.

Lee, C. K. and Ord, J. K. (1990). Discriminant analysis using least absolute deviations. *Decision Sciences* **21**, 86–96.

Lee, J. C., Chang, T. C., and Krishnaiah, P. R. (1977), Approximations to the distributions of the likelihood ratio statistics for testing certain structures on the covariance matrices of real multivariate normal populations, In P. R. Krishnaiah (ed.), *Multivariate Analysis*, Vol. 4, pp. 105–118, Amsterdam: North-Holland.

Lee, R., McCabe, D. J., and Graham, W. K. (1983). Multivariate relationships between job characteristics and job satisfaction in the public sector: A triple cross-validation study. *Multivariate Behavioral Research* **18**, 47–62.

Lee, Y. S. (1971). Asymptotic formulas for the distribution of a multivariate test statistic: Power comparisons of certain multivariate tests. *Biometrika* **58**, 647–651.

Lefkovitch, L. P. (1993). Consensus principal components. *Biometrical Journal* **35**(5), 567–580.

Lesaffre, E., Willems, J. L., and Albert, A. (1989). Estimation of error rate in multiple group logistic discrimination. The approximate leaving-out method. *Communications in Statistics: Theory and Methods* **18**, 2989–3007.

Lesaffre, E., Albert, A., Van Der Meulen, E. C., and Willems, J. L. (1991). Some new development in multiple group logistic discrimination. *Frontiers of Statistical Scientific Theory and Industrial Applications* **2**, 217–249.

Levy, M. S. and Neill, J. W. (1990). Testing for lack of fit in linear multiresponse models based on exact or near replicates. *Communications in Statistics—Series A, Theory and Methods* **19**, 1987–2002.

Li, K. H., Raghunathan, T. E., and Rubin, D. B. (1991). Large-sample significance levels from multiply imputed data using moment-based statistics and an F reference distribution. *Journal of the American Statistical Association* **86**, 1065–1073.

Li, K. H., Meng, X. L., Raghunathan, T. E., and Rubin, D. B. (1991). Significance levels from repeated p-values with multiply-imputed data. *Statistica Sinica* **1**, 65–92.

Liang, K., Zeger, S. L., and Qaqish, B. (1992). Multivariate regression analysis for categorical data. *Journal of the Royal Statistical Society* **54**, 3–40.

Linn, R. L. (1968). A Monte Carlo approach to the number of factors problem. *Psychometrika* **33**, 37–71.

Little, R. J. A. (1988). A test of missing completely at random for multivariate data with missing values. *Journal of the American Statistical Association* **83**, 1198–1202.

Little, R. J. A. (1992). Regression with missing *X*'s: A review. *Journal of the American Statistical Association* **87**, 1227–1237.

Little, R. J. A. and Rubin, D. B. (1987). *Statistical Analysis with Missing Data*. New York: Wiley.

Little, R. J. A. and Schenker, N. (1992). Missing data. In G. Arminger, C. C. Clogg, and M. E. Sobel (eds.), *Handbook of Statistical Modeling in the Behavioral and Social Sciences*. New York: Plenum.

Little, R. J. A. and Smith, P. J. (1987). Editing and imputation for quantitative survey data. *Journal of the American Statistical Association* **82**, 58–68.

Liu, C. and Rubin, D. B. (1994). The ECME algorithm: A simple extension of EM and ECM with faster monotone convergence. *Biometrika* **81**, 633–648.

Liu, R. Y. (1995). Control charts for multivariate processes. *Journal of the American Statistical Association* **90**(432), 1380–1387.

Loh, W. (1993). On linear discriminant analysis with adaptive ridge classification rules. Presented at the Annual Meeting of the American Statistical Association, San Francisco, CA, August 8–12, 1993.

Longley, J. W. (1967). An appraisal of least squares programs for the electronic computer from the point of view of the user. *Journal of the American Statistical Association* **62**, 819–841.

Looney, S. W. (1988). A statistical technique for comparing the accuracies of several classifiers. *Pattern Recognition Letters* **8**, 5–9.

Lopuhaä, H. P. (1989). On the relation between S-estimators and M-estimators of multivariate location and covariance. *Annals of Statistics* **17**, 1662–1683.

Lowry, C. A., Woodall, W. H., Champ, C. W., and Rigdon, S. E. (1992). A multivariate exponentially weighted moving average control chart. *Technometrics* **34**(1), 46–53.

Lubischew, A. A. (1962). On the use of discriminant functions in taxonomy. *Biometrics* **18**, 455–477.

Ludwig, O., Gottlieb, R., and Lienert, G. A. (1986). Tables of Bonferroni-limits for simultaneous F-tests. *Biometrical Journal* **28**, 548–576.

Lunneborg, C. E. and Abbott, R. D. (1983). *Elementary Multivariate Analysis for the Behavioral Sciences*. Amsterdam: North-Holland Publishing Company.

Lyttkens, E. (1972). Regression aspects of canonical correlation. *Journal of Multivariate Analysis* **2**, 418–439.

Mahalanobis, P. C. (1936). On the generalized distance in statistics. *Proceedings of the National Institute of Science of India* **12**, 49–55.

Manly, R. F. J. and Rayner, J. C. W. (1987). The comparison of sample covariance matrices using likelihood ratio tests. *Biometrika* **74**, 841–847.

Marden, J. L. (1982). Minimal complete classes of tests of hypotheses with multivariate one-sided alternatives. *Annals of Statistics* **10**, 962–970.

Mardia, K. V. (1971). The effect of nonnormality on some multivariate tests and robustness to nonnormality in the linear model. *Biometrika* **58**, 105–121.

Mardia, K. V., Kent, J. T., and Bibby, J. M. (1979). *Multivariate Analysis*. New York: Academic Press.

Marks, S. and Dunn, O. J. (1974). Discrimination functions when covariance matrices are unequal. *Journal of the American Statistical Association* **69**, 555–559.

Maronna, R. A. (1976). Robust M-estimators of multivariate location and scatter. *Annals of Statistics* **4**, 51–67.

Marquardt, D. W. (1970). Generalized inverses, ridge regression, biased linear estimation and nonlinear estimation. *Technometrics* **12**, 591–612.

Marshall, R. J. and Chisholm, E. M. (1985). Hypothesis testing in the polychotomous logistic model with an application to detecting gastrointestinal cancer. *Statistics in Medicine* **4**, 337–344.

Marx, B. D. and Smith, E. P. (1990). Principal component estimation for generalized linear regression. *Biometrika* **77**, 23–31.

Mason, R. L. and Gunst, R. F. (1985). Selecting principal components in regression. *Statistics & Probability Letters* **3**, 299–301.

Mason, R. L., Gunst, R. F., and Webster, J. T. (1975). Regression analysis and the problem of multicollinearity. *Communications in Statistics* **4**, 277–292.

Mason, R. L., Tracy, N. D., and Young, J. C. (1995). Decomposition of T^2 for multivariate control chart interpretation. *Journal of Quality Technology* **27**(2), 157–158.

Mathai, A. M. and Katiyar, R. S. (1979). Exact percentage points for testing independence. *Biometrika* **66**, 353–356.

McCabe, G. P. (1975). Computations for variable selection in discriminant analysis. *Technometrics* **17**, 103–109.

McCullagh, P. and Nelder, J. A. (1989). *Generalized Linear Models* (2nd edition). New York: Chapman & Hall.

McCulloch, R. E. (1986). Some remarks on allocatory and separatory linear discrimination. *Journal of Statistical Planning and Inference* **14**, 323–330.

McDonald, L. (1975). Tests for the general linear hypothesis under the multiple design multivariate linear model. *Annals of Statistics* **3**, 461–466.

McDonald, R. P. (1985). *Factor Analysis and Related Methods*. Hillsdale, NJ: Lawrence Erlbaum.

McHenry, C. E. (1978). Computation of a best subset in multivariate analysis. *Applied Statistics* **27**, 291–296.

McKay, R. J. (1977). Variable selection in multivariate regression: An application of simultaneous test procedures. *Journal of the Royal Statistical Society, Series B* **39**, 371–380.

McKay, R. J. and Campbell, N. A. (1982a). Variable selection techniques in discriminant analysis: I. Description. *British Journal of Mathematical and Statistical Psychology* **35**, 1–29.

McKay, R. J. and Campbell, N. A. (1982b). Variable selection techniques in discriminant analysis: II. Allocation. *British Journal of Mathematical and Statistical Psychology* **35**, 30–41.

McKeon, J. J. (1974). F approximations to the distribution of Hotelling's T_0^2. *Biometrika* **61**, 381–383.

McLachlan, G. J. (1972). Asymptotic results for discriminant analysis when the initial samples are misclassified. *Technometrics* **14**, 414–422.

McLachlan, G. J. (1974). An asymptotic unbiased technique for estimating the error rates in discriminant analysis. *Biometrics* **30**, 239–249.

McLachlan, G. J. (1975). Confidence intervals for the conditional probability of misallocation in discriminant analysis. *Biometrics* **31**, 161–167.

McLachlan, G. J. (1976). The bias of the apparent error rate in discriminant analysis. *Biometrika* **63**, 239–244.

McLachlan, G. J. (1980a). The efficiency of Efron's "Bootstrap" approach applied to error rate estimation in discriminant analysis. *Journal of Statistical Computation and Simulation* **11**, 273–279.

McLachlan, G. J. (1980b). On the relationship between the F-test and the overall error rate for variable selection in two-group discriminant analysis. *Biometrics* **11**, 273–279.

McLachlan, G. J. (1992). *Discriminant Analysis and Statistical Pattern Recognition*. New York: Wiley-Interscience.

Mead, A. (1992). Review of the development of multidimensional scaling methods. *Statistician* **41**, 27–39.

Mehrotra, D. V. (1995). Robust elementwise estimation of a dispersion matrix. *Biometrics* **51**, 1344–1351.

Meng, X. and Rubin, D. B. (1992). Performing likelihood ratio tests with multiply-imputed data sets. *Biometrika* **79**(1), 103–111.

Meyer, E. P. (1975). A measure of the average intercorrelation. *Educational and Psychological Measurement* **35**, 67–72.

Michaelis, J. (1973). Simulation experiments with multiple group linear and quadratic discriminant analysis. In T. Cacoullos (ed.), *Discriminant Analysis and Applications*. New York: Academic Press.

Mihalisin, T., Schwegler, J., and Timlin, J. (1992). A new multivariate plotting method. In *Proceedings of the Section on Statistical Graphics*, pp. 69–74. Annual Meeting of the American Statistical Association, Boston, MA, August 9–13.

Miller, F. R., Neill, J. W., and Scherfey, B. W. (1996). Maximin clusters for near replicate regression lack of fit tests. Presented at the Annual Meeting of the American Statistical Association, Chicago, IL, August 1996.

Miller, J. K. (1969). The development and application of bi-multivariate correlations: A measure of statistical association. Ed.D. thesis, Faculty of Education Studies, State University of New York at Buffalo.

Miller, J. K. (1975). The sampling distribution and a test for the significance of the bimultivariate redundancy statistic: A Monte Carlo study. *Multivariate Behavioral Research* **10**, 233–244.

Miller, J. K. and Farr, S. D. (1971). Bimultivariate redundancy: A comprehensive measure of interbattery relationship. *Multivariate Behavioral Research* **6**, 313–324.

Miller, R. G. (1962). Selecting predictors for multiple discriminant analysis. *Meteorological Monographs* **4**, 11–14.

Mislevy, R. J. (1986). Recent developments in the factor analysis of categorical variables. *Journal of Educational Statistics* **11**, 3–31.

Mood, A. M. (1941). On the joint distribution of the median in samples from the multivariate normal population. *Annals of Mathematical Statistics* **12**, 268–278.

Moore, D. H. (1973). Evaluation of five discrimination procedures for binary variables. *Journal of the American Statistical Association* **68**, 399–404.

Morrison, D. F. (1972). The analysis of a single sample of repeated measurements. *Biometrics* **28**, 55–71.

Morrison, D. F. (1983). *Applied Linear Statistical Methods*. Englewood Cliffs, NJ: Prentice-Hall.

Mosteller, F. and Tukey, J. W. (1977). *Data Analysis and Regression*. Reading, MA: Addison-Wesley.

Mucciardi, A. N. and Gose, E. E. (1971). A comparison of seven techniques for choosing subsets of pattern recognition properties. *IEEE Transactions on Computers* **C-20**, 1023–1031.

Mudholkar, G. S. and Srivastava, D. K. (1996a). Trimmed T^2: A robust analog of Hotelling's T^2. Presented at the Annual Meeting of the American Statistical Association, Chicago, IL, August 1996.

Mudholkar, G. S. and Srivastava, D. K. (1996b). Robust analogs of Hotelling's two-sample T^2. Presented at the Annual Meeting of the American Statistical Association, Chicago, IL, August 1996.

Mueller, R. O. and Cozad, J. B. (1993). Standardized discriminant coefficients: A rejoinder. *Journal of Educational Statistics* **18**(1), 108–114.

Muirhead, R. J. and Waternaux, C. M. (1980). Asymptotic distributions in canonical correlation analysis and other multivariate procedures for nonnormal populations. *Biometrika* **67**, 31–43.

Mulaik, S. A. (1972). *The Foundations of Factor Analysis*. New York: McGraw-Hill.

Muller, K. E. (1981). Relationships between redundancy analysis, canonical correlation and multivariate regression. *Psychometrika* **46**, 139–142.

Muller, K. E. (1982). Understanding canonical correlation through the general linear model and principal components. *American Statistician* **36**, 342–354.

Muller, K. E. and Peterson, B. L. (1984). Practical methods for computing power in testing the multivariate general linear hypothesis. *Computational Statistics and Data Analysis* **2**, 143–158.

Muller, K. E., LaVange, L. M., Ramey, S. L., and Ramey, C. T. (1992). Power calculations for general linear multivariate models including repeated measures applications. *Journal of the American Statistical Association* **87**, 1209–1226.

Murray, G. D. (1977). A cautionary note on selection of variables in discriminant analysis. *Applied Statistics* **26**, 246–250.

Murty, B. R. and Federer, W. T. (1991). Missing observations in multivariate analysis. *Journal of the Indian Society of Agricultural Statistics* **43**, 107–126.

Myers, R. H. (1990). *Classical and Modern Regression with Applications* (2nd edition). Boston: Duxbury Press.

Nachtsheim, C. J. and Johnson, M. E. (1988). A new family of multivariate distributions with applications to Monte Carlo studies. *Journal of the American Statistical Association* **83**, 984–989.

Næs, T. and Helland, I. S. (1993). Relevant components in regression. *Scandinavian Journal of Statistics* **20**, 239–250.

Naga, R. A. and Antile, G. (1990). Stability of robust and non-robust principal components analysis. *Computational Statistics and Data Analysis* **10**, 169–174.

Nagarsenker, B. N. (1975). Percentage points of Wilks' L_{vc} criterion. *Communications in Statistics* **4**, 629–641.

Nagarsenker, B. N. (1979). Noncentral distributions of Wilks' statistic for tests of three hypotheses. *Sankhya: The Indian Journal of Statistics* **41**, 67–81.

Naik, D. N. (1989). Detection of outliers in the multivariate linear regression model. *Communications in Statistics—Series A, Theory and Methods* **18**, 2225–2232.

Naik, D. N. and Khattree, R. (1996). Revisiting Olympic track records: Some practical considerations in the principal component analysis. *American Statistician* **50**(2), 140–144.

Naik, D. N. and Rao, S. S. (1994). Analysis of multivariate repeated measurements. Annual Meeting of the American Statistical Association, Toronto, Canada, August 1994.

Nason, G. (1995). Three-dimensional projection pursuit. *Journal of Applied Statistics* **44**(4), 411–430.

Nath, R. and Pavur, R. (1985). A new statistic in the one-way multivariate analysis of variance. *Computational Statistics and Data Analysis* **22**, 297–315.

Nel, D. G. and van der Merwe, C. A. (1986). A solution to the multivariate Behrens-Fisher problem. *Communications in Statistics—Series A, Theory and Methods* **15**, 3719–3735.

Nel, D. G., van der Merwe, C. A., and Moser, B. K. (1990). The exact distributions of the univariate and multivariate Behrens-Fisher statistics with a comparison of several solutions in the univariate case. *Communications in Statistics—Series A, Theory and Methods* **19**, 279–298.

Nelder, J. A. (1974). Letter to the editor, *Journal of the Royal Statistical Society* (Series C), **23**, 232.

Neter, J., Kutner, M. H., Nachtsheim, C. J., and Wasserman, W. (1996). *Applied Linear Statistical Models* (4th edition). Homewood, IL: Richard D. Irwin.

Neuenschwander, B. E. and Flury, B. D. (1995). Common canonical variates. *Biometrika* **82**(3), 553–60.

Neuhaus, J. O. and Wrigley, C. (1954). The quartimax method: An analytical approach to orthogonal simple structure. *British Journal of Statistical Psychology* **7**, 81–91.

Nicewander, W. A. and Wood, D. A. (1974). Comments on "A general canonical correlation index." *Psychological Bulletin* **81**, 92–94.

Nicewander, W. A. and Wood, D. A. (1975). On the mathematical bases of the general canonical correlation index: Rejoinder to Miller. *Psychological Bulletin* **82**, 210–212.

Nomikos, P. and MacGregor, J. F. (1995). Multivariate SPC charts for monitoring batch processes. *Technometrics* **37**(1), 41–59.

Nomura, M. and Ohkubo, T. (1985). A note on combining ridge and principal component regression. *Communications in Statistics: Theory and Methods* **14**(10), 2489–2493.

Novak, T. P. and Cramer, E. M. (1987). Graphical representation of MANOVA. In *Proceedings of the Section on Statistical Graphics*. Annual Meeting of the American Statistical Association, San Francisco, CA, August 17–20, 1987.

O'Brien, P. C. (1992). Robust procedures for testing equality of covariance matrices. *Biometrics* **48**, 819–827.

O'Brien, P. C., and Muller, K. E. (1992), A Unified Approach to Statistical Power for *t* Tests to Multivariate Models. In L. K. Edwards (ed.), *Applied Analysis of Variance in Behavioral Sciences*. New York: Marcel Dekker.

O'Brien, P. N., Parente, F. J., and Schmitt, C. J. (1982). A Monte Carlo study on the robustness of four MANOVA criterion tests. *Journal of Statistics and Computer Simulation* **15**, 183–192.

O'Gorman, T. W. and Woolson, R. F. (1991). Variable selection to discriminate between two groups: Stepwise logistic regression or stepwise discriminant analysis? *American Statistician* **45**, 187–193.

Okamoto, M. (1963). An asymptotic expansion for the distribution of the linear discriminant function. *Annals of Mathematical Statistics* **39**, 1358–1359.

Olkin, I. and Pratt, J. W. (1958). Unbiased estimation of certain correlation coefficients. *Annals of Mathematical Statistics* **29**, 201–211.

Olson, C. L. (1974). Comparative robustness of six tests in multivariate analysis of variance. *Journal of the American Statistical Association* **69**, 894–908.

Olson, C. L. (1975). A Monte Carlo investigation of the robustness of multivariate analysis of variance. *Psychological Bulletin* **86**, 1350–1352.

Olson, C. L. (1976). On choosing a test statistic in multivariate analysis of variance. *Psychological Bulletin* **83**, 579–586.

Olson, C. L. (1979). Practical considerations in choosing a MANOVA test statistic, a rejoinder to Stevens. *Psychological Bulletin* **86**, 1350–1352.

Orchard, T. and Woodbury, M. A. (1972). A missing information principle: Theory and applications. In *Sixth Berkeley Symposium on Mathematical Statistics and Probability*, pp. 697–715. Berkeley, CA: University of California Press.

O'Sullivan, J. B. and Mahan, C. M. (1966). Glucose tolerance test: Variability in pregnant and non-pregnant women. *American Journal of Clinical Nutrition* **19**, 345–351.

Overall, J. E. and Spiegel, D. K. (1969). Concerning least squares analysis of experimental data. *Psychological Bulletin* **72**, 311–322.

Pack, P., Jolliffe, I. T., and Morgan, B. J. T. (1988). Influential observations in principal component analysis: A case study. *Journal of Applied Statistics* **15**, 39–52.

Page, J. T. (1985). Error-rate estimation in discriminant analysis. *Technometrics* **27**, 189–198.

Patuwo, E., Hu, M. Y., and Hung, M. S. (1993). Two-group classification using neural networks. *Decision Sciences* **24**(4), 825–845.

Pearce, S. C. (1965). The measurement of a living organism. *Biometrie-Praximetrie* **6**, 143–152.

Pearson, E. S. (1969). Some comments on the accuracy of Box's approximations to the distribution of *M*. *Biometrika* **56**, 219–220.

Pearson, E. S. and Hartley, H. O. (eds.). (1972). *Biometrika Tables for Statisticians*, Vol. 2. Cambridge: Cambridge University Press.

Pearson, K. (1901). On lines and planes of closest fit to systems of points in space. *Philos. Magazine* **2**, 559–572.

Pederzoli, G. and Rathie, P. N. (1983). The exact distribution of Bartlett's criterion for testing the equality of covariance matrices. *Metron* **41**(3), 83–89.

Peele, L. C. and Ryan, T. P. (1979). The merits of some new ridge regression estimators. Presented at the Annual Meeting of the American Statistical Association Washington, DC, August 1979.

Percy, D. F. (1992). Blocked arteries and multivariate regression. *Biometrics* **48**, 683–693.

Percy, D. F. (1993). Prediction for generalized linear models. *Journal of Applied Statistics* **20**, 285–291.

Perlman, M. D. (1969). One-sided testing problems in multivariate analysis. *Annals of Mathematical Statistics* **40**, 549–567.

Perlman, M. D. (1980). Unbiasedness of multivariate tests: Recent results. *Multivariate Analysis* **5**, 413–432.

Pham, T. D. and Mocks, J. (1992). Beyond principal component analysis: A trilinear decomposition model and least squares estimation. *Pyschometrika* **57**, 203–215.

Picard, R. R. and Berk, K. N. (1990). Data splitting. *American Statistician* **44**(2), 140–147.

Pignatiello, Jr., J. J. and Runger, G. C. (1990). Comparisons of multivariate CUSUM charts. *Journal of Quality Technology* **3**, 173–186.

Pillai, K. C. S. (1954). On some distribution problems in multivariate analysis. Mimeograph Series No. 88, Institute of Statistics, University of North Carolina, Chapel Hill.

Pillai, K. C. S. (1955). Some new test criteria in multivariate analysis. *Annals of Mathematical Statistics* **26**, 117–121.

Pillai, K. C. S. (1956a). On the distribution of the largest or the smallest root of a matrix in multivariate analysis. *Biometrika* **43**, 122–127.

Pillai, K. C. S. (1956b). Some results useful in multivariate analysis. *Annals of Mathematical Statistics* **27**, 1106–13.

Pillai, K. C. S. (1960). *Statistical Tables for Tests of Multivariate Hypotheses*. Manila: University of the Philippines, Statistical Center.

Pillai, K. C. S. (1964). On the distribution of the largest of seven roots of a matrix in multivariate analysis. *Biometrika* **51**, 270–275.

Pillai, K. C. S. (1965). On the distribution of the largest characteristic root of a matrix in multivariate analysis. *Biometrika* **52**, 405–414.

Pillai, K. C. S. and Flury, B. N. (1984). Percentage points of the largest characteristic root of the multivariate beta matrix. *Communications in Statistics—Series A, Theory and Methods* **13**(18), 2199–2237.

Pillai, K. C. S. and Jayachandran, K. (1967). Power comparisons of tests of two multivariate hypotheses based on four criteria. *Biometrika* **54**, 195–210.

Pillai, K. C. S. and Sampson, P. (1959). On Hotelling's generalization of T-squared. *Biometrika* **46**, 160–168.

Pollock, M. L., Jackson, A. S., and Pate, R. R. (1980). Discriminant analysis of physiological differences between good and elite runners. *Research Quarterly* **51**, 521–532.

Pope, P. T. and Webster, J. T. (1972). The use of an F-statistic in stepwise regression procedures. *Technometrics* **14**, 327–340.

Posse, C. (1990). An effective two-dimensional projection pursuit algorithm. *Communications in Statistics: Simulation and Computation* **19**, 1143–1164.

Prabhu, S. S. and Runger, G. C. (1997). Designing a multivariate ewma control chart. *Journal of Quality Technology* **29**(1), 8–15.

Pregibon, D. (1981). Logistic regression diagnostics. *Annals of Statistics* **9**, 705–724.

Puri, M. L. and Sen, P. K. (1971). *Nonparametric Methods in Multivariate Analysis*. New York: Wiley.

Pynnonen, S. (1987). Selection of variables in nonlinear discriminant analysis by information criteria. In Tarmo Pukkila, Simo Puntanen (eds.), *Proceedings of the Second International Tampere Conference in Statistics*, pp. 627–636, Univ. Tempere, Finland.

Ramsey, F. L. (1986). A fable of PCA. *American Statistician* **40**, 323–324.

Ramsier, S. W. (1991). A graphical method for detection of influential observations in principal component analysis. In *Proceedings of the Section on Statistical Graphics*. Annual Meeting of the American Statistical Association, Atlanta, GA, August 18–22, 1991.

Randles, R. H., Broffitt, J. D., Ramberg, J. R., and Hogg, R. V. (1978). Generalized linear and quadratic discriminant functions using robust estimates. *Journal of the American Statistical Association* **73**, 564–568.

Rao, C. R. (1951). An asymptotic expansion of the distribution of Wilks' criterion. *Bulletin of the International Statistical Institute* **33**, 177–180.

Rao, C. R. (1952). *Advanced Statistical Methods in Biometric Research.* New York: Wiley.

Rao, C. R. (1964). The use and interpretation of principal component analysis in applied research. *Sankhya (Series B)* **26**, 329–358.

Rao, C. R. (1965). *Linear Statistical Inference and Its Applications.* New York: Wiley.

Rao, C. R. (1982). Likelihood ratio tests for relationships betwen two covariance matrices. Technical Report 82-36, University of Pittsburgh, Center for Multivariate Analysis.

Rao, C. R. (1983). Multivariate analysis: Some reminiscences on its origin and development. *Sankhya (Ser. B), Pt. 2* **45**, 284–299.

Ratkowsky, D. A. and Martin, D. (1974). The use of multivariate analysis in identifying relationships among disorder and mineral element content in apples. *Australian Journal of Agricultural Research* **25**, 783–790.

Raudenbush, S. W. (1993). Hierarchical linear models and experimental design. In L. K. Edwards (ed.), *Applied Analysis of Variance in Behavioral Science.* New York: Marcel-Dekker.

Rayens, W. and Greene, T. (1991). Covariance pooling and stabilization for classification. *Computational Statistics and Data Analysis* **11**, 17–42.

Reader, M. W. (1973). The analysis of covariance with a single linear covariate having heterogeneous slopes. Master's thesis, Brigham Young University, Department of Statistics.

Reaven, G. M. and Miller, R. G. (1979). An attempt to define the nature of chemical diabetes using a multidimensional analysis. *Diabetologia* **16**, 17–24.

Relles, D. A. and Rogers, W. H. (1977). Statisticians are fairly robust estimators of location. *Journal of the American Statistical Association* **72**, 107–111.

Remme, J., Habbema, J. D. F., and Hermans, J. (1980). A simulative comparison of linear quadratic and kernel discrimination. *Journal of Statistical Computation and Simulation* **11**, 87–106.

Rencher, A. C. (1988). On the use of correlations to interpret canonical functions. *Biometrika* **75**, 363–365.

Rencher, A. C. (1992a). Bias in apparent classification rates in stepwise discriminant analysis. *Communications in Statistics—Series B, Simulation and Computation* **21**, 373–389.

Rencher, A. C. (1992b). Interpetation of canoncial discriminant functions, canonical variates and principal components. *American Statistician* **46**, 217–225.

Rencher, A. C. (1993). The contribution of individual variables to Hotelling's T^2, Wilks' Λ and R^2. *Biometrics* **49**, 217–225.

Rencher, A. C. (1995). *Methods of Multivariate Analysis.* New York: Wiley.

Rencher, A. C. and Larson, S. F. (1980). Bias in Wilks' Λ in stepwise discriminant analysis. *Technometrics* **22**, 349–356.

Rencher, A. C. and Pun, F. C. (1980). Inflation of R^2 in best subset regression. *Technometrics* **22**, 49–53.

Rencher, A. C. and Scott, D. T. (1990). Assessing the contribution of individual variables following rejection of a multivariate hypothesis. *Communications in Statistics—Series B, Simulation and Computation* **19**, 535–553.

Reynolds, T. J. and Jackosfsky, E. F. (1981). Interpreting canonical analysis: The use of orthogonal transformations. *Educational and Psychological Measurement* **41**, 661–671.

Riccia, G. D. and Shapiro, A. (1983). Fisher discriminant analysis and factor analysis. *IEEE Transactions on Pattern Analysis and Machine Intelligence* **PAMI–5**(1), 99–104.

Rigdon, S. E. (1996). Preliminary sample size requirements for multivariate control charts. Presented at the Annual Meeting of the American Statistical Association, Chicago, IL, August 1996.

Rijckevorsel, J. L. A. V. and Leeuw, J. D. (1988). *Component and Correspondence Analysis*. New York: Wiley.

Rocke, D. M. (1992). On M-, and S-estimators of multivariate location and shape. Working Paper, Graduate School of Management, University of California at Davis.

Rocke, D. M. and Woodruff, D. L. (1993). Computation of robust estimates of multivariate location and shape. *Statistica Neerlandica*, **47**, 27–42.

Rocke, D. M. and Woodruff, D. L. (1996). Identification of outliers in multivariate data. *Journal of the American Statistical Association* **91**(435), 1047–1060.

Roebruck, P. (1982). Canonical forms and tests of hypotheses; Part II: Multivariate mixed linear models. *Statistica Neerlandica* **36**, 75–80.

Roes, K. C. B. and Does, R. J. M. M. (1995). Shewhart-type charts in nonstandard situations. *Technometrics* **37**, 15–24.

Rom, D. M. (1990). A sequentially rejective test procedure based on a modified Bonferroni inequality. *Biometrika* **77**, 663–665.

Romanazzi, M. (1991). Influence in canonical variates analysis. *Computational Statistics and Data Analysis* **11**, 143–164.

Romanazzi, M. (1992). Influence in canonical correlation analysis. *Pyschometrika* **57**, 237–259.

Rousseeuw, P. J. (1984). Least median of squares regression. *Journal of the American Statistical Association* **79**, 871–880.

Rousseeuw, P. J. and Yohai, V. (1984). Robust regression by means of S-estimators. In *Robust and Nonlinear Time Series Analysis, Lecture Notes in Statistics*, Vol. 26, pp. 256–272. New York: Springer-Verlag.

Roy, S. N. (1939). p-Statistics, or some generalizations in analysis of variance appropriate to multivariate problems. *Sankhya* **4**, 381–396.

Roy, S. N. (1945). The individual sampling distribution of the maximum, the minimum and any intermediate of the p-statistics on the null-hypothesis. *Sankhya* **7**, 133–158.

Roy, S. N. (1953). On a heuristic method of test construction and its use in multivariate analysis. *Annals of Mathematical Statistics* **24**, 220–238.

Roy, S. N. (1957). *Some Aspects of Multivariate Analysis*. New York: Wiley.

Roy, S. N., Gnanadesikan, R., and Srivastava, J. N. (1971). *Analysis and Design of Certain Quantitative Multiresponse Experiments*. Oxford: Pergamon Press.

Rubin, D. B. (1976). Inference and missing data. *Biometrika* **63**, 581–592.

Rubin, D. B. (1978). Multiple imputations in sample surveys—a phenomenological Bayesian approach. In *Proceedings of the Section on Survey Research Methods*. Annual Meeting of the American Statistical Association, San Diego, CA, August 14–17, 1978.

Rubin, D. B. (1987). *Multiple Imputation for Nonresponse in Surveys*. New York: Wiley.

Rubin, D. B. (1991). EM and beyond. *Psychometrika* **56**(2), 241–254.

Rubin, D. B. and Schenker, N. (1987). Interval estimation from multiply-inputed data: A case study using census agriculture industry codes. *Journal of Official Statistics* **3**(4), 375–387.

Rubin, D. B. and Thayer, D. T. (1982). EM algorithms for ML factor analysis. *Psychometrika* **47**(1), 69–76.

Rubin, D. B. and Thayer, D. T. (1983). More on EM for ML factor analysis. *Psychometrika* **48**(2), 253–257.

Rudolph, P. M. and Karson, M. (1988). The effect of unequal priors and unequal misclassification costs on MDA. *Journal of Applied Statistics* **15**, 69–83.

Ruiz-Velasco, S. (1991). Asymptotic efficiency of logistic regression relative to linear discriminant analysis. *Biometrika* **78**, 235–243.

Runger, G. C. (1996). Projections and the U^2 multivariate control chart. *Journal of Quality Technology* **28**(3), 313–319.

Runger, G. C. and Prabhu, S. S. (1996). A Markov chain model for the multivariate exponentially weighted moving averages control chart. *Journal of the American Statistical Association* **91**(436), 1701–1706.

Rutter, C., Flack, V., and Lachenbruch, P. (1991). Bias in error rate estimates in discriminant analysis when stepwise variable selection is employed. *Communications in Statistics—Series B, Simulation and Computation* **20**, 1–22.

Saito, T., Kariya, T., and Otsu, T. (1988). A generalization of principal component analysis. *Journal of the Japanese Statistical Society* **18**, 187–193.

Sarkar, S. K. (1984). A note on the power of the likelihood ratio test for MANOVA. *Sankhya: The Indian Journal of Statistics* **46**, 303–308.

Sato, M. (1990). Some remarks on principal component analysis as a substitute for factor analysis in monofactor cases. *Journal of the Japanese Statistical Society* **20**, 23–31.

Saunders, D. R. (1953). An analytic method for rotation to orthogonal simple structure. Research Bulletin 53-10, Education Testing Services, Princeton, NJ: Educational Testing Service.

Saunders, D. R. (1962). Trans-varimax. *American Psychologist* **17**, 395.

Schaafsman, W. (1982). Selecting variables in discriminant analysis for inproving upon classical procedures. In P. R. Krishnaiah and L. M. Kanal (eds.), *Handbook of Statistics*, Vol. 2, pp. 857–881. Amsterdam: North-Holland Publishing Company.

Schall, R. and Dunne, T. T. (1987). On outliers and influence in the general multivariate normal linear model. In Tarmo Pukkila, Simo Putanen (eds.), *Proceedings of the Second International Tampere Conference in Statistics*, pp. 665–678, Univ. of Tampere, Finland.

Schatzoff, M. (1966). Sensitivity comparisons among tests of the general linear hypothesis. *Journal of the American Statistical Association* **61**, 415–435.

Scheffe, H. (1953). A method of judging all contrasts in the analysis of variance. *Biometrika* **40**, 87–104.

Scheffe, H. (1959). *The Analysis of Variance*. New York: Wiley.

Schervish, M. J. (1986). A predictive derivation of principal components. Technical Report 378, Carnegie Mellon University, Department of Statistics.

Schervish, M. J. (1987). A review of multivariate analysis. *Statistical Science* **2**, 396–433.

Schmidhammer, J. L. (1982). On the selection of variables under regression models using Krishnaiah's finite intersection tests. In *Handbook of Statistics*, pp. 821–833. Amsterdam: North-Holland Publishing Company.

Schott, J. R. (1987). An improved chi-squared test for a principal component. *Statistics and Probability Letters* **5**, 361–365.

Schott, J. R. (1988). Testing the equality of the smallest latent roots of a correlation matrix. *Biometrika* **75**, 794–796.

Schott, J. R. (1990). Canonical mean projections and confidence regions in canonical variate analysis. *Biometrika* **77**, 587–596.

Schott, J. R. (1991). A test for a specific principal component of a correlation matrix. *Journal of the American Statistical Association* **86**, 747–751.

Schott, J. R. (1993). Dimensionality reduction in quadratic discriminant analysis. *Computational Statistics and Data Analysis* **16**, 161–174.

Schott, J. R. and Saw, J. G. (1984). A multivariate one-way classification model with random effects. *Journal of Multivariate Analysis* **15**, 1–12.

Schuenemeyer, J. H. and Bargmann, R. E. (1978). Maximum eccentricity as a union-intersection test statistic in multivariate analysis. *Journal of Multivariate Analysis* **8**, 268–273.

Schuurmann, F. J., Krishnaiah, P. R., and Chattopadhyay, A. K. (1975). Exact percentage points of the distribution of the trace of a multivariate beta matrix. *Journal of Statistical Computation and Simulation* **3**, 331–343.

Schwertman, N. C. and Allen, D. M. (1973). The smoothing of an indefinite matrix with applications to growth curve analysis with missing observations. Technical Report No. 56, University of Kentucky, Department of Statistics.

Schwertman, N. C. and Allen, D. M. (1979). Smoothing an indefinite variance-covariance matrix. *Journal of Statistical Computation and Simulation* **9**, 183–194.

Seaman, S. L. and Young, D. M. (1990). A non-parametric variable selection algorithm for allocatory linear discriminant analysis. *Educational and Psychological Measurement* **50**, 837–841.

Searle, S. R. (1971). *Linear Models*. New York: Wiley.

Searle, S. R. (1977). Analysis of variance of unbalanced data from 3-way and higher-order classifications. Technical Report BU-606-M, Cornell University, Biometrics Units.

Searle, S. R. (1979). Alternative covariance models for the 2-way crossed classification. *Communications in Statistics—Series A, Theory and Methods* **8**, 799–818.

Searle, S. R. (1982). *Matrix Algebra Useful for Statistics*. New York: Wiley.

Searle, S. R., Speed, F. M., and Henderson, H. V. (1981). Some computational and model equivalencies in analysis of variance of unequal-subclass-numbers data. *American Statistician* **35**, 16–33.

Seber, G. A. F. (1977). *Linear Regression Analysis*. New York: Wiley.

Seber, G. A. F. (1984). *Multivariate Observations*. New York: Wiley.

Sen, P. K. (1968). Robustness of some nonparametric procedures in linear models. *Annals of Mathematical Statistics* **39**(6), 1913–1922.

Shaffer, J. P. (1986). Modified sequentially rejective multiple test procedures. *Journal of the American Statistical Association* **81**, 826–831.

Share, D. L. (1984). Interpreting the output of multivariate analyses: A discussion of current approaches. *British Journal of Psychology* **75**, 349–362.

Shi, S. G. and Taam, W. (1992). Non-linear canonical correlation analysis with a simulated annealing solution. *Journal of Applied Statistics* **19**, 155–165.

Shillington, E. R. (1979). Testing for lack of fit in regression without replication. *Canadian Journal of Statistics* **7**, 137–146.

Shiraishi, T. (1990). *M*-tests in multivariate models. *Metrika* **37**, 198–197.

Sibson, R. (1984). Present position and potential developments: Some personal views. Multivariate analysis (with discussion). *Journal of the Royal Statistical Society, Series A* **147**, 198–207.

Silvey, S. D. (1969). Multicollinearity and imprecise estimation. *Journal of the Royal Statistical Society, Series B* **31**, 539–552.

Simes, R. J. (1986). An improved Bonferroni procedure for multiple tests of significance. *Biometrika* **73**, 751–754.

Siotani, M., Hayakawa, T., and Fujikoshi, Y. (1985). *Modern Multivariate Statistical Analysis*. Columbus, OH: American Sciences Press.

Skinner, H. A. (1977). Exploring relationships among multiple data sets. *Multivariate Behavioral Research* **12**, 199–222.

Skinner, H. A. (1978). The art of exploring predictor-criterion relationships. *Psychological Bulletin* **85**, 327–337.

Smidt, R. K. and McDonald, L. L. (1976). Ridge discriminant analysis. Research Paper No. 108, University of Wyoming, Department of Statistics.

Smith, C. A. B. (1947). Some examples of discrimination. *Annals of Eugenics* **18**, 272–283.

Smith, H., Gnanadesikan, R., and Hughes, J. B. (1962). Multivariate analysis of variance (MANOVA). *Biometrics* **18**, 22–41.

Smith, P. J. and Choi, S. C. (1982). Simple tests to compare two dependent regression lines. *Technometrics* **24**, 123–126.

Snapinn, S. M. and Knoke, J. D. (1984). Classification error rate estimators evaluated by unconditional mean squared error. *Technometrics* **26**, 371–378.

Snapinn, S. M. and Knoke, J. D. (1985). An evaluation of smoothed classification error-rate estimators. *Technometrics* **27**, 199–206.

Snapinn, S. M. and Knoke, J. D. (1988). Bootstrapped and smoothed classification error rate estimators. *Communications in Statistics—Series B, Simulation and Computation* **17**, 1135–1153.

Snapinn, S. M. and Knoke, J. D. (1989). Estimation of error rates in discriminant analysis with selection of variables. *Biometrics* **45**, 289–299.

Snedecor, G. W. and Cochran, W. G. (1967). *Statistical Methods* (6th edition). Ames, IA: The Iowa State University Press.

Snee, R. D. (1977). Validation of regression models: Methods and examples. *Technometrics* **19**, 415–428.

Somers, K. M. (1986). Multivariate allometry and removal of size with principal components analysis. *Systematic Zoology* **35**, 359–368.

Somers, K. M. (1989). Allometry, isometry and shape in principal components analysis. *Systematic Zoology* **38**, 169–173.

Sorum, M. J. (1971). Estimating the conditional probability of misclassification. *Technometrics* **13**, 333–343.

Sparks, R. S. (1992). Quality control with multivariate data. *Australian Journal of Statistics* **34**(3), 375–390.

Spearman, C. (1904). The proof and measurement of association between two things. *American Journal of Psychology* **15**, 72, 202.

Speed, F. M. (1969). A new approach to the analysis of linear models. Technical Report, National Aeronautics and Space Administration, Houston, TX. NASA Technical memo, NASA TM X-58030.

Speed, F. M. and Hocking, R. R. (1976). The use of the *R*( )-notation with unbalanced data. *American Statistician* **30**, 30–33.

Speed, F. M., Hocking, R. R., and Hackney, O. P. (1978). Methods of analysis of linear models with unbalanced data. *Journal of the American Statistical Association* **73**, 105–112.

Srivastava, M. S. (1985). Multivariate data with missing observations. *Communications in Statistics—Series A, Theory and Methods* **14**, 775–792.

Srivastava, M. S. and Khatri, C. G. (1979). *An Introduction to Multivariate Statistics*. New York: North Holland Publishing Company.

Stauffer, D. F., Garton, E. O., and Steinhorst, R. K. (1985). A comparison of principal components from real and random data. *Ecology* **66**(6), 1693–1698.

Steiger, J. H. (1980). Tests for comparing elements of a correlation matrix. *Psychological Bulletin* **87**, 245–251.

Stevens, J. P. (1979). Comment on Olson: Choosing a test statistic in multivariate analysis of variance. *Psychological Bulletin* **86**, 355–360.

Stevens, J. P. (1980). Power of the multivariate analysis of variance tests. *Psychological Bulletin* **88**, 728–737.

Stewart, D. K. and Love, W. A. (1968). A general canonical correlation index. *Psychological Bulletin* **70**, 160–163.

Stipak, B. and McDavid, J. C. (1981). Canonical correlation analysis of crimes and sentences. In *Proceedings of the Section on Social Statistics*, pp. 294–298, Annual Meeting of the American Statistical Association, Detroit, MI, August 10–13.

Styan, G. P. H. (1989). Three useful expressions for expectations involving a Wishart matrix and its inverse. *Statistical Data Analysis and Inference* 283–296.

Subrahmaniam, K. and Chinganda, E. F. (1978). Robustness of the linear discriminant function to nonnormality: Edgeworth series distribution. *Journal of Statistical Planning and Inference* **1**, 39–57.

Subrahmaniam, K. and Subrahmaniam, K. (1973). On the multivariate Behrens-Fisher problem. *Biometrika* **60**, 107–111.

Sullivan, J. H. and Woodall, W. H. (1996). A comparison of multivariate control charts for individual observations. *Journal of Quality Technology* **28**(4), 398–408.

Sutter, J. M., Kalivas, J. H., and Lang, P. M. (1992). Which principal components to utilize for principal component regression. *Journal of Chemometrics* **6**, 217–255.

Swalberg, K. D. (1995). A comparison of univariate approximations to the Pillai and Lawley-Hotelling trace test statistics in multivariate analysis of variance. Master's thesis, Brigham Young University, Department of Statistics.

Szatrowski, T. H. (1983). Missing data in the one-population multivariate normal patterned mean and covariance matrix testing and estimation problem. *Annals of Statistics* **11**, 947–958.

Tan, W. Y. and Gupta, R. P. (1983). On approximating a linear combination of central Wishart matrices with positive coefficients. *Communications in Statistics—Series A, Theory and Methods* **12**, 2589–2600.

Tanaka, Y. (1983). Some criteria for variable selection in factor analysis. *Behaviormetrika* **13**, 31–45.

Tanaka, Y. (1988). Sensitivity analysis in principal component analysis: Influence on the subspace spanned by principal components. *Communications in Statistics: Theory and Methods* **17**, 3157–3175.

Tanaka, Y. and Odaka, Y. (1989). Influential observations in principal factor analysis. *Psychometrika* **54**, 475–485.

Tang, K. L. and Algina, J. (1993). Performing of four multivariate tests under variance-covariance heteroscedasticity. *Multivariate Behavioral Research* **28**, 391–405.

Theil, H. and Chung, C. (1988). Information-theoretic measures of fit for univariate and multivariate linear regressions. *American Statistician* **42**, 249–252.

Thomas, D. R. (1992). Interpreting discriminant functions: A data analytic approach. *Multivariate Behavioral Research* **27**(3), 335–362.

Thomas, W. and Cook, R. D. (1989). Assessing influence on regression coefficients in generalized linear models. *Biometrika* **76**, 741–749.

Thomson, G. H. (1934). Hotelling's method modified to give Spearman's *g*. *Journal of Educational Psychology* **25**, 366–374.

Thomson, G. H. (1951). *The Factorial Analysis of Human Ability*. London: London University Press.

Thorndike, R. M. (1976). Studying canonical analysis: Comments on Barcikowski and Stevens. *Multivariate Behavioral Research* **11**, 249–253.

Thorndike, R. M. (1977). Canonical analysis and predictor selection. *Multivariate Behavioral Research* **12**, 75–87.

Thorndike, R. M. (1978). *Correlational Procedures for Research*. New York: Gardner Press.

Thorndike, R. M. and Weiss, D. J. (1973). A study of the stability of canonical correlations and canonical components. *Educational and Psychological Measurement* **33**, 123–134.

Thorndike, R. M. and Weiss, D. J. (1983). An empirical investigation of step-down canonical correlation with cross-validation. *Multivariate Behavioral Research* **18**, 183–196.

Thurstone, L. L. (1931). Multiple factor analysis. *Psychological Review* **38**, 406–427.

Tibshirani, R. (1992). Principal curves revisited. *Statistics and Computing* **2**, 183–190.

Tibshirani, R. (1996). Regression shrinkage and selection via the lasso. *Journal of the Royal Statistical Society, Series B* **58**(1), 267–288.

Tiku, M. L. (1967). Tables of the power of the F-test. *Journal of the American Statistical Association* **62**, 525–539.

Tiku, M. L. and Balakrishnan, N. (1985). Testing the equality of variance-covariance matrices the robust way. *Communications in Statistics: Theory and Methods* **14**(12), 3033–3051.

Timm, N. H. (1975). *Multivariate Analysis: With Applications in Education and Psychology*. Monterey, CA: Brooks/Cole.

Timm, N. H. (1996). Multivariate quality control using finite intersection tests. *Journal of Quality Technology* **28**(2), 233–243.

Titterington, D. M., Murray, G. D., Murray, L. S., Speigelhalter, D. J., and Skene, A. M. (1981). Comparison of discrimination of techniques applied to a complex data set of head injured patients. *Journal of the Royal Statistical Society, Ser. A* **144**, 145–175.

Todeschini, R. (1990). Weighted k-nearest neighbor method for the calculation of missing values. *Chemometrics and Intelligent Laboratory Systems* **9**, 201–205.

Todorov, V. K., Neykov, N. M., and Neytchev, P. N. (1990). Robust selection of variables in the discriminant analysis based on MVE and MCD estimators. *Compstat* **9**, 193–198.

Tong, Y. L. (1990). *The Multivariate Normal Distribution*. New York: Springer-Verlag.

Troendle, J. F. (1993). A stepwise resampling method of multiple hypothesis testing. Annual Meeting of the American Statistical Association, San Francisco, CA.

Troendle, J. F. (1996). A most powerful similar test for a simple normal alternative and its use in the multivariate two sample location shift problem. Presented at the Annual Meeting of the American Statistical Association, Chicago, IL, August 1996.

Tucker, L. R. (1966). Some mathematical notes on three-mode factor analysis. *Psychometrika* **31**, 279–311.

Tucker, L. R., Kooperman, R. F., and Linn, R. L. (1969). Evaluation of factor analytic research procedures by means of simulated correlation matrices. *Psychometrika* **34**, 421–459.

Turner, D. L. (1990). An easy way to tell what you are testing in analysis of variance. *Communications in Statistics—Series A, Theory and Methods* **19**, 4807–4832.

Twedt, D. J. and Gill, D. S. (1992). Comparison of algorithms for replacing missing data in discriminant analysis. *Communications in Statistics—Series A, Theory and Methods* **21**, 1567–1578.

Tyler, D. E. (1991). Some issues in the robust estimation of multivariate location and scatter. In W. Stahel and S. Weisberg (eds.), *Directions in Robust Statistics and Diagnostics, Part II*. New York: Springer-Verlag.

Urquhart, N. S. (1982). Adjustment in covariance when one factor affects the covariate. *Biometrics* **38**, 651–660.

Urquhart, N. S., Weeks, D. L., and Henderson, C. R. (1973). Estimation associated with linear models: A revisitation. *Communications in Statistics* **1**, 303–330.

Van De Geer, J. P. (1984). Linear relations among *k* sets of variables. *Psychometrika* **49**, 79–94.

Van Den Wollenberg, A. L. (1977). Redundancy analysis: An alternative for canonical analysis. *Psychometrika* **42**, 207–219.

Van Houwelingen, J. C. and LeCessie, S. (1988). Logistic regression, a review. *Statistica Neerlandica* **42**, 215–232.

Van Ness, J. (1979). On the effects of dimension in discriminant analysis for unequal covariance populations. *Technometrics* **21**, 119–127.

Veldhuisen, P. C. V. and Temkin, N. (1994). Cart versus logistic regression in predicting a binary outcome. Annual Meeting of the American Statistical Association, Toronto, Canada.

Velicer, W. F. and Jackson, D. N. (1990). Component analysis versus common factor analysis: Some issues in selecting an appropriate procedure. *Multivariate Behavioral Research* **25**, 1–28.

Velilla, S. (1993). A note on the multivariate Box-Cox transformation to normality. *Statistics and Probability Letters* **90**, 945–951.

Velilla, S. (1994). A goodness-of-fit test for autoregressive moving-average models based on the standardized sample spectral distribution of the residuals. *Journal of Time Series Analysis* **15**, 637–647.

Velilla, S. (1995). Diagnostics and robust estimation in multivariate data transformations. *Journal of the American Statistical Association* **90**, 945.

Velilla, S. and Barrio, J. A. (1994). A discriminant rule under transformation. *Technometrics* **36**(4), 348–353.

Villarroya, A., Rios, M., and Oller, J. M. (1995). Discriminant analysis algorithm based on a distance function and on a Bayesian decision. *Biometrics* **51**, 908–919.

Vinod, H. D. (1976). Canonical ridge and econometrics of joint production. *Journal of Econometrics* **4**, 147–166.

Wahl, P. W. and Kronmal, R. A. (1977). Discriminant functions when covariances are unequal and sample sizes are moderate. *Biometrics* **33**, 479–484.

Wakaki, H. (1990). Comparison of linear and quadratic discriminant functions. *Biometrika* **77**, 227–229.

Wald, A. (1943). Tests of statistical hypotheses concerning several parameters when the number of observations is large. *Transactions of the American Mathematical Society* **54**, 426–483.

Wall, F. J. (1967). *The Generalized Variance Ratio or U-Statistic*. Albuquerque, NM.

Wang, S. and Nyquist, H. (1991). Effects on the eigenstructure of a data matrix when deleting an observation. *Computational Statistics and Data Analysis* **11**, 179–188.

Wang, S. G. and Chow, S. C. (1994). *Advanced Linear Models: Theory and Applications*. New York: Marcel Dekker.

Watanabe, S. (1967). Karhunen-Loeve expansion and factor analysis. In J. Kozesnik (ed.), *Transactions of the Fourth Prague Conference*. Prague: Czechoslovakia Academy of Science.

Webster, J. T., Gunst, R. F., and Mason, R. L. (1974). Latent root regression analysis. *Technometrics* **16**, 513–522.

Weis, D. J. (1972). Canonical correlation analysis in counseling psychology research. *Journal of Counseling Psychology* **19**, 241–252.

Weisberg, S. (1985). *Applied Linear Regression*. New York: Wiley.

Welch, B. L. (1937). The significance of the difference between two means when the population variances are unequal. *Biometrika* **29**, 350–360.

Welch, B. L. (1939). Note on discriminant functions. *Biometrika* **31**, 218–220.

Welch, B. L. (1947). The generalization of "Student's" problem when several different population variances are involved. *Biometrika* **34**, 28–35.

Wernecke, K. D. (1992). A coupling procedure for the discriminant of mixed data. *Biometrics* **48**, 497–506.

Wernecke, K. D., Haerting, J., Kalb, G., and Stuerzebecher, E. (1988). On model-choice in discrimination with categorical variables. *Biometrical Journal* **31**, 289–296.

Whitcomb, K. M. and Lahiff, M. (1993). An investigation of sample influence functions for the DS estimate of the optimum error rate in discriminant analysis. *Communications in Statistics—Series A, Theory and Methods* **22**(1), 41–56.

Widaman, K. F. (1993). Common factor analysis versus principal component analysis: Differential bias in representing model parameters? *Multivariate Behavioral Research* **28**(3), 263–311.

Wilks, S. S. (1932a). Certain generalizations in the analysis of variance. *Biometrika* **24**, 471–494.

Wilks, S. S. (1932b). Moments and distributions of estimates of population parameters from fragmentary samples. *Annals of Mathematical Statistics* **3**, 163–195.

Wilks, S. S. (1946). Sample criteria for testing equality of means, equality of variances and equality of covariances in a normal multivariate distribution. *Annals of Mathematical Statistics* **17**, 257–281.

Williams, B. K. (1982). A simple demonstration of the relationship between classification and canonical variates analysis. *American Statistician* **36**, 363–365.

Williams, B. K. (1983). Some observations on the use of discriminant analysis in ecology. *Ecology* **64**, 1283–1291.

Williams, E. J. (1970). Comparing means of correlated variates. *Biometrika* **57**, 459–461.

Williams, J. S. (1979). A synthetic basis for a comprehensive factor-analysis theory. *Biometrics* **35**, 719–733.

Wilson, G. A. and Martin, S. A. (1983). An empirical comparison of two methods for testing the significance of a correlation matrix. *Educational and Psychological Measurement* **43**, 11–14.

Winer, B. J. (1971). *Statistical Principles in Experimental Design* (2nd edition). New York: McGraw-Hill.

Wiseman, M. (1993). An efficient method for computing factor similarity coefficients: A computational note and an algorithm. *Computational Statistics and Data Analysis* **15**, 111–114.

Wold, S., Esbensen, K., and Geladi, P. (1987). Principal component analysis. *Chemometrics and Intelligent Laboratory Systems* **2**, 37–52.

Wood, D. A. and Erskine, J. A. (1976). Strategies in canonical correlation with application to behavioral data. *Educational and Psychological Measurement* **36**, 861–878.

Woodard, D. E. (1931). Healing time of fractures of the jaw in relation to delay before reduction, infection, syphilis and blood calcium and phosphorus content. *Journal of the American Dental Association* **18**, 419–442.

Woodruff, D. L. and Rocke, D. M. (1994). Computable robust estimation of multivariate location and shape in high dimension using compound estimators. *Journal of the American Statistical Association* **89**, 888–896.

Wright, S. P. (1992). Adjusted P-values for simultaneous inference. *Biometrics* **48**, 1005–1013.

Wythoff, B. J. (1993). Backpropagation neural networks, a tutorial. *Chemometrics and Intelligent Laboratory Systems* **18**, 115–155.

Xie, Y., Wang, J., Liang, Y., Sun, L., Song, X., and Yu, R. (1993). Robust principal component analysis by projection pursuit. *Journal of Chemometrics* **7**, 527–541.

Yao, Y. (1965). An approximate degrees of freedom solution to the multivariate Behrens-Fisher problem. *Biometrika* **52**, 139–147.

Yates, F. (1934). The analysis of multiple classifications with unequal numbers in the different classes. *Journal of the American Statistical Association* **29**, 52–66.

Yates, F. (1939). Tests of significance of the differences between regression coefficients derived from two sets of correlated variates. *Proceedings of the Royal Society of Edinburgh* **59**, 184–194.

Yau, H. C. and Manry, M. T. (1992). Automatic determination of reject threshholds in classifiers employing discriminant functions. *IEEE Transactions on Acoustics, Speech, and Signal Processing* **40**, 711–713.

Yenyukov, I. S. (1988). Detecting structures by means of projection pursuit. *Compstat* **88**, 47–58.

Yeomans, K. A. and Golder, P. A. (1982). The Guttman-Kaiser criterion as a predictor of the number of common factors. *Statistician* **31**(3), 221–229.

Young, D. M. and Odell, P. L. (1986). Feature subset selection for statistical classification problems involving unequal covariance matrices. *Communications in Statistics—Series A, Theory and Methods* **15**, 137–157.

Zellner, A. (1962). An efficient method of estimating
for aggregation bias. *Journal of the American St*

Zeng, Y. and Hopke, P. K. (1990). Methodological stud
to three-way chemical data sets. *Chemometrics an*
237–250.

Zhou, D. (1989). ROPCA: A Fortran program for robust pri
puters and Geosciences **15**, 59–78.

Zhou, L. and Mathew, T. (1993). Hypotheses tests for variance c
mixed models. *Journal of Statistical Planning and Inferen*

Zinkgraf, S. A. (1983). Performing factorial multivariate analysis
correlation analysis. *Educational and Psychological Measure*

Zwick, R. (1986). Rank and normal scores alternatives to Hotelling's
ioral Research **21**, 169–186.

Index

WILEY SERIES IN PROBABILITY AND STATISTICS

Probability and Statistics

ANDERSON · An Introduction to Multivariate Statistical Analysis, *Second Edition*
*ANDERSON · The Statistical Analysis of Time Series
ARNOLD, BALAKRISHNAN, and NAGARAJA · A First Course in Order Statistics
BACCELLI, COHEN, OLSDER, and QUADRAT · Synchronization and Linearity: An Algebra for Discrete Event Systems
BARTOSZYNSKI and NIEWIADOMSKA-BUGAJ · Probability and Statistical Inference
BERNARDO and SMITH · Bayesian Statistical Concepts and Theory
BHATTACHARYYA and JOHNSON · Statistical Concepts and Methods
BILLINGSLEY · Convergence of Probability Measures
BILLINGSLEY · Probability and Measure, *Second Edition*
BOROVKOV · Asymptotic Methods in Queuing Theory
BRANDT, FRANKEN, and LISEK · Stationary Stochastic Models
CAINES · Linear Stochastic Systems
CAIROLI and DALANG · Sequential Stochastic Optimization
CHEN · Recursive Estimation and Control for Stochastic Systems
CONSTANTINE · Combinatorial Theory and Statistical Design
COOK and WEISBERG · An Introduction to Regression Graphics
COVER and THOMAS · Elements of Information Theory
CSÖRGŐ and HORVÁTH · Weighted Approximations in Probability Statistics
*DOOB · Stochastic Processes
DUDEWICZ and MISHRA · Modern Mathematical Statistics
DUPUIS and ELLIS · A Weak Convergence Approach to the Theory of Large Deviations
ETHIER and KURTZ · Markov Processes: Characterization and Convergence
FELLER · An Introduction to Probability Theory and Its Applications, Volume 1, *Third Edition,* Revised; Volume II, *Second Edition*
FREEMAN and SMITH · Aspects of Uncertainty: A Tribute to D. V. Lindley
FULLER · Introduction to Statistical Time Series, *Second Edition*
FULLER · Measurement Error Models
GHOSH, MUKHOPADHYAY, and SEN · Sequential Estimation
GIFI · Nonlinear Multivariate Analysis
GUTTORP · Statistical Inference for Branching Processes
HALD · A History of Probability and Statistics and Their Applications before 1750
HALL · Introduction to the Theory of Coverage Processes
HANNAN and DEISTLER · The Statistical Theory of Linear Systems
HEDAYAT and SINHA · Design and Inference in Finite Population Sampling
HOEL · Introduction to Mathematical Statistics, *Fifth Edition*
HUBER · Robust Statistics
IMAN and CONOVER · A Modern Approach to Statistics
JOHNSON and KOTZ · Leading Personalities in Statistical Sciences: From the Seventeenth Century to the Present

*Now available in a lower priced paperback edition in the Wiley Classics Library.

Probability and Statistics (Continued)

JUREK and MASON · Operator-Limit Distributions in Probability Theory
KASS and VOS · Geometrical Foundations of Asymptotic Inference
KAUFMAN and ROUSSEEUW · Finding Groups in Data: An Introduction to Cluster Analysis
LAMPERTI · Probability: A Survey of the Mathematical Theory, *Second Edition*
LARSON · Introduction to Probability Theory and Statistical Inference, *Third Edition*
LESSLER and KALSBEEK · Nonsampling Error in Surveys
LINDVALL · Lectures on the Coupling Method
McFADDEN · Management of Data in Clinical Trials
MANTON, WOODBURY, and TOLLEY · Statistical Applications Using Fuzzy Sets
MARDIA · The Art of Statistical Science: A Tribute to G. S. Watson
MORGENTHALER and TUKEY · Configural Polysampling: A Route to Practical Robustness
MUIRHEAD · Aspects of Multivariate Statistical Theory
OLIVER and SMITH · Influence Diagrams, Belief Nets and Decision Analysis
*PARZEN · Modern Probability Theory and Its Applications
PRESS · Bayesian Statistics: Principles, Models, and Applications
PUKELSHEIM · Optimal Experimental Design
PURI and SEN · Nonparametric Methods in General Linear Models
PURI, VILAPLANA, and WERTZ · New Perspectives in Theoretical and Applied Statistics
RAO · Asymptotic Theory of Statistical Inference
RAO · Linear Statistical Inference and Its Applications, *Second Edition*
*RAO and SHANBHAG · Choquet-Deny Type Functional Equations with Applications to Stochastic Models
RENCHER · Methods of Multivariate Analysis
ROBERTSON, WRIGHT, and DYKSTRA · Order Restricted Statistical Inference
ROGERS and WILLIAMS · Diffusions, Markov Processes, and Martingales, Volume I: Foundations, *Second Edition;* Volume II: Îto Calculus
ROHATGI · An Introduction to Probability Theory and Mathematical Statistics
ROSS · Stochastic Processes
RUBINSTEIN · Simulation and the Monte Carlo Method
RUBINSTEIN and SHAPIRO · Discrete Event Systems: Sensitivity Analysis and Stochastic Optimization by the Score Function Method
RUZSA and SZEKELY · Algebraic Probability Theory
SCHEFFE · The Analysis of Variance
SEBER · Linear Regression Analysis
SEBER · Multivariate Observations
SEBER and WILD · Nonlinear Regression
SERFLING · Approximation Theorems of Mathematical Statistics
SHORACK and WELLNER · Empirical Processes with Applications to Statistics
SMALL and McLEISH · Hilbert Space Methods in Probability and Statistical Inference
STAPLETON · Linear Statistical Models
STAUDTE and SHEATHER · Robust Estimation and Testing
STOYANOV · Counterexamples in Probability
STYAN · The Collected Papers of T. W. Anderson: 1943–1985
TANAKA · Time Series Analysis: Nonstationary and Noninvertible Distribution Theory
THOMPSON and SEBER · Adaptive Sampling
WELSH · Aspects of Statistical Inference
WHITTAKER · Graphical Models in Applied Multivariate Statistics
YANG · The Construction Theory of Denumerable Markov Processes

*Now available in a lower priced paperback edition in the Wiley Classics Library.

Applied Probability and Statistics

ABRAHAM and LEDOLTER · Statistical Methods for Forecasting
AGRESTI · Analysis of Ordinal Categorical Data
AGRESTI · Categorical Data Analysis
AGRESTI · An Introduction to Categorical Data Analysis
ANDERSON and LOYNES · The Teaching of Practical Statistics
ANDERSON, AUQUIER, HAUCK, OAKES, VANDAELE, and WEISBERG · Statistical Methods for Comparative Studies
ARMITAGE and DAVID (editors) · Advances in Biometry
*ARTHANARI and DODGE · Mathematical Programming in Statistics
ASMUSSEN · Applied Probability and Queues
*BAILEY · The Elements of Stochastic Processes with Applications to the Natural Sciences
BARNETT and LEWIS · Outliers in Statistical Data, *Second Edition*
BARTHOLOMEW, FORBES, and McLEAN · Statistical Techniques for Manpower Planning, *Second Edition*
BATES and WATTS · Nonlinear Regression Analysis and Its Applications
BECHHOFER, SANTNER, and GOLDSMAN · Design and Analysis of Experiments for Statistical Selection, Screening, and Multiple Comparisons
BELSLEY · Conditioning Diagnostics: Collinearity and Weak Data in Regression
BELSLEY, KUH, and WELSCH · Regression Diagnostics: Identifying Influential Data and Sources of Collinearity
BERRY · Bayesian Analysis in Statistics and Econometrics: Essays in Honor of Arnold Zellner
BERRY, CHALONER, and GEWEKE · Bayesian Analysis in Statistics and Econometrics: Essays in Honor of Arnold Zellner
BHAT · Elements of Applied Stochastic Processes, *Second Edition*
BHATTACHARYA and WAYMIRE · Stochastic Processes with Applications
BIEMER, GROVES, LYBERG, MATHIOWETZ, and SUDMAN · Measurement Errors in Surveys
BIRKES and DODGE · Alternative Methods of Regression
BLOOMFIELD · Fourier Analysis of Time Series: An Introduction
BOLLEN · Structural Equations with Latent Variables
BOULEAU · Numerical Methods for Stochastic Processes
BOX · R. A. Fisher, the Life of a Scientist
BOX and DRAPER · Empirical Model-Building and Response Surfaces
BOX and DRAPER · Evolutionary Operation: A Statistical Method for Process Improvement
BOX, HUNTER, and HUNTER · Statistics for Experimenters: An Introduction to Design, Data Analysis, and Model Building
BOX and LUCEÑO · Statistical Control by Monitoring and Feedback Adjustment
BROWN and HOLLANDER · Statistics: A Biomedical Introduction
BUCKLEW · Large Deviation Techniques in Decision, Simulation, and Estimation
BUNKE and BUNKE · Nonlinear Regression, Functional Relations and Robust Methods: Statistical Methods of Model Building
CHATTERJEE and HADI · Sensitivity Analysis in Linear Regression
CHATTERJEE and PRICE · Regression Analysis by Example, *Second Edition*
CLARKE and DISNEY · Probability and Random Processes: A First Course with Applications, *Second Edition*
COCHRAN · Sampling Techniques, *Third Edition*
*COCHRAN and COX · Experimental Designs, *Second Edition*
CONOVER · Practical Nonparametric Statistics, *Second Edition*
CONOVER and IMAN · Introduction to Modern Business Statistics

*Now available in a lower priced paperback edition in the Wiley Classics Library.

Applied Probability and Statistics (Continued)

CORNELL · Experiments with Mixtures, Designs, Models, and the Analysis of Mixture Data, *Second Edition*
COX · A Handbook of Introductory Statistical Methods
*COX · Planning of Experiments
COX, BINDER, CHINNAPPA, CHRISTIANSON, COLLEDGE, and KOTT · Business Survey Methods
CRESSIE · Statistics for Spatial Data, *Revised Edition*
DANIEL · Applications of Statistics to Industrial Experimentation
DANIEL · Biostatistics: A Foundation for Analysis in the Health Sciences, *Sixth Edition*
DAVID · Order Statistics, *Second Edition*
*DEGROOT, FIENBERG, and KADANE · Statistics and the Law
*DEMING · Sample Design in Business Research
DETTE and STUDDEN · The Theory of Canonical Moments with Applications in Statistics, Probability, and Analysis
DILLON and GOLDSTEIN · Multivariate Analysis: Methods and Applications
DODGE and ROMIG · Sampling Inspection Tables, *Second Edition*
DOWDY and WEARDEN · Statistics for Research, *Second Edition*
DRAPER and SMITH · Applied Regression Analysis, *Second Edition*
DUNN · Basic Statistics: A Primer for the Biomedical Sciences, *Second Edition*
DUNN and CLARK · Applied Statistics: Analysis of Variance and Regression, *Second Edition*
ELANDT-JOHNSON and JOHNSON · Survival Models and Data Analysis
EVANS, PEACOCK, and HASTINGS · Statistical Distributions, *Second Edition*
FISHER and VAN BELLE · Biostatistics: A Methodology for the Health Sciences
FLEISS · The Design and Analysis of Clinical Experiments
FLEISS · Statistical Methods for Rates and Proportions, *Second Edition*
FLEMING and HARRINGTON · Counting Processes and Survival Analysis
FLURY · Common Principal Components and Related Multivariate Models
GALLANT · Nonlinear Statistical Models
GLASSERMAN and YAO · Monotone Structure in Discrete-Event Systems
GNANADESIKAN · Methods for Statistical Data Analysis of Multivariate Observations, *Second Edition*
GREENWOOD and NIKULIN · A Guide to Chi-Squared Testing
GROSS and HARRIS · Fundamentals of Queueing Theory, *Second Edition*
GROVES · Survey Errors and Survey Costs
GROVES, BIEMER, LYBERG, MASSEY, NICHOLLS, and WAKSBERG · Telephone Survey Methodology
HAHN and MEEKER · Statistical Intervals: A Guide for Practitioners
HAND · Discrimination and Classification
*HANSEN, HURWITZ, and MADOW · Sample Survey Methods and Theory, Volume I: Methods and Applications
*HANSEN, HURWITZ, and MADOW · Sample Survey Methods and Theory, Volume II: Theory
HEIBERGER · Computation for the Analysis of Designed Experiments
HELLER · MACSYMA for Statisticians
HINKELMAN and KEMPTHORNE: · Design and Analysis of Experiments, Volume 1: Introduction to Experimental Design
HOAGLIN, MOSTELLER, and TUKEY · Exploratory Approach to Analysis of Variance
HOAGLIN, MOSTELLER, and TUKEY · Exploring Data Tables, Trends and Shapes
HOAGLIN, MOSTELLER, and TUKEY · Understanding Robust and Exploratory Data Analysis
HOCHBERG and TAMHANE · Multiple Comparison Procedures
HOCKING · Methods and Applications of Linear Models: Regression and the Analysis of Variables

*Now available in a lower priced paperback edition in the Wiley Classics Library.

Applied Probability and Statistics (Continued)

HOEL · Elementary Statistics, *Fifth Edition*
HOGG and KLUGMAN · Loss Distributions
HOLLANDER and WOLFE · Nonparametric Statistical Methods
HOSMER and LEMESHOW · Applied Logistic Regression
HØYLAND and RAUSAND · System Reliability Theory: Models and Statistical Methods
HUBERTY · Applied Discriminant Analysis
IMAN and CONOVER · Modern Business Statistics
JACKSON · A User's Guide to Principle Components
JOHN · Statistical Methods in Engineering and Quality Assurance
JOHNSON · Multivariate Statistical Simulation
JOHNSON and BALAKRISHNAN · Advances in the Theory and Practice of Statistics: A Volume in Honor of Samuel Kotz
JOHNSON and KOTZ · Distributions in Statistics
Continuous Multivariate Distributions
JOHNSON, KOTZ, and BALAKRISHNAN · Continuous Univariate Distributions, Volume 1, *Second Edition*
JOHNSON, KOTZ, and BALAKRISHNAN · Continuous Univariate Distributions, Volume 2, *Second Edition*
JOHNSON, KOTZ, and BALAKRISHNAN · Discrete Multivariate Distributions
JOHNSON, KOTZ, and KEMP · Univariate Discrete Distributions, *Second Edition*
JUDGE, GRIFFITHS, HILL, LÜTKEPOHL, and LEE · The Theory and Practice of Econometrics, *Second Edition*
JUDGE, HILL, GRIFFITHS, LÜTKEPOHL, and LEE · Introduction to the Theory and Practice of Econometrics, *Second Edition*
JUREČKOVÁ and SEN · Robust Statistical Procedures: Aymptotics and Interrelations
KADANE · Bayesian Methods and Ethics in a Clinical Trial Design
KADANE AND SCHUM · A Probabilistic Analysis of the Sacco and Vanzetti Evidence
KALBFLEISCH and PRENTICE · The Statistical Analysis of Failure Time Data
KASPRZYK, DUNCAN, KALTON, and SINGH · Panel Surveys
KISH · Statistical Design for Research
*KISH · Survey Sampling
LAD · Operational Subjective Statistical Methods: A Mathematical, Philosophical, and Historical Introduction
LANGE, RYAN, BILLARD, BRILLINGER, CONQUEST, and GREENHOUSE · Case Studies in Biometry
LAWLESS · Statistical Models and Methods for Lifetime Data
LE · Applied Survival Analysis
LEBART, MORINEAU., and WARWICK · Multivariate Descriptive Statistical Analysis: Correspondence Analysis and Related Techniques for Large Matrices
LEE · Statistical Methods for Survival Data Analysis, *Second Edition*
LEPAGE and BILLARD · Exploring the Limits of Bootstrap
LEVY and LEMESHOW · Sampling of Populations: Methods and Applications
LINHART and ZUCCHINI · Model Selection
LITTLE and RUBIN · Statistical Analysis with Missing Data
LYBERG · Survey Measurement
MAGNUS and NEUDECKER · Matrix Differential Calculus with Applications in Statistics and Econometrics
MAINDONALD · Statistical Computation
MALLOWS · Design, Data, and Analysis by Some Friends of Cuthbert Daniel
MANN, SCHAFER, and SINGPURWALLA · Methods for Statistical Analysis of Reliability and Life Data
MASON, GUNST, and HESS · Statistical Design and Analysis of Experiments with Applications to Engineering and Science

*Now available in a lower priced paperback edition in the Wiley Classics Library.

Applied Probability and Statistics (Continued)

McLACHLAN and KRISHNAN · The EM Algorithm and Extensions
McLACHLAN · Discriminant Analysis and Statistical Pattern Recognition
McNEIL · Epidemiological Research Methods
MILLER · Survival Analysis
MONTGOMERY and MYERS · Response Surface Methodology: Process and Product in Optimization Using Designed Experiments
MONTGOMERY and PECK · Introduction to Linear Regression Analysis, *Second Edition*
NELSON · Accelerated Testing, Statistical Models, Test Plans, and Data Analyses
NELSON · Applied Life Data Analysis
OCHI · Applied Probability and Stochastic Processes in Engineering and Physical Sciences
OKABE, BOOTS, and SUGIHARA · Spatial Tesselations: Concepts and Applications of Voronoi Diagrams
OSBORNE · Finite Algorithms in Optimization and Data Analysis
PANKRATZ · Forecasting with Dynamic Regression Models
PANKRATZ · Forecasting with Univariate Box-Jenkins Models: Concepts and Cases
PIANTADOSI · Clinical Trials: A Methodologic Perspective
PORT · Theoretical Probability for Applications
PUTERMAN · Markov Decision Processes: Discrete Stochastic Dynamic Programming
RACHEV · Probability Metrics and the Stability of Stochastic Models
RENCHER · Multivariate Statistical Inference and Applications
RÉNYI · A Diary on Information Theory
RIPLEY · Spatial Statistics
RIPLEY · Stochastic Simulation
ROSS · Introduction to Probability and Statistics for Engineers and Scientists
ROUSSEEUW and LEROY · Robust Regression and Outlier Detection
RUBIN · Multiple Imputation for Nonresponse in Surveys
RYAN · Modern Regression Methods
RYAN · Statistical Methods for Quality Improvement
SCHOTT · Matrix Analysis for Statistics
SCHUSS · Theory and Applications of Stochastic Differential Equations
SCOTT · Multivariate Density Estimation: Theory, Practice, and Visualization
*SEARLE · Linear Models
SEARLE · Linear Models for Unbalanced Data
SEARLE · Matrix Algebra Useful for Statistics
SEARLE, CASELLA, and McCULLOCH · Variance Components
SKINNER, HOLT, and SMITH · Analysis of Complex Surveys
STOYAN, KENDALL, and MECKE · Stochastic Geometry and Its Applications, *Second Edition*
STOYAN and STOYAN · Fractals, Random Shapes and Point Fields: Methods of Geometrical Statistics
THOMPSON · Empirical Model Building
THOMPSON · Sampling
TIERNEY · LISP-STAT: An Object-Oriented Environment for Statistical Computing and Dynamic Graphics
TIJMS · Stochastic Modeling and Analysis: A Computational Approach
TITTERINGTON, SMITH, and MAKOV · Statistical Analysis of Finite Mixture Distributions
UPTON and FINGLETON · Spatial Data Analysis by Example, Volume 1: Point Pattern and Quantitative Data
UPTON and FINGLETON · Spatial Data Analysis by Example, Volume II: Categorical and Directional Data
VAN RIJCKEVORSEL and DE LEEUW · Component and Correspondence Analysis

*Now available in a lower priced paperback edition in the Wiley Classics Library.

Applied Probability and Statistics (Continued)

WEISBERG · Applied Linear Regression, *Second Edition*

WESTFALL and YOUNG · Resampling-Based Multiple Testing: Examples and Methods for *p*-Value Adjustment

WHITTLE · Optimization Over Time: Dynamic Programming and Stochastic Control, Volume I and Volume II

WHITTLE · Systems in Stochastic Equilibrium

WONNACOTT and WONNACOTT · Econometrics, *Second Edition*

WONNACOTT and WONNACOTT · Introductory Statistics, *Fifth Edition*

WONNACOTT and WONNACOTT · Introductory Statistics for Business and Economics, *Fourth Edition*

WOODING · Planning Pharmaceutical Clinical Trials: Basic Statistical Principles

WOOLSON · Statistical Methods for the Analysis of Biomedical Data

*ZELLNER · An Introduction to Bayesian Inference in Econometrics

Tracts on Probability and Statistics

BILLINGSLEY · Convergence of Probability Measures

KELLY · Reversability and Stochastic Networks

TOUTENBURG · Prior Information in Linear Models

*Now available in a lower priced paperback edition in the Wiley Classics Library.